Linear Integrated Circuits

*The Saunders College Publishing
Series in Electronics Technology*

Bennett
Advanced Circuit Analysis
ISBN: 0-15-510843-4

Brey
*The Motorola Microprocessor
Family: 68000, 68010, 68020,
68030, and 68040;
Programming and Interfacing
with Applications*
ISBN: 0-03-026403-5

Carr
*Elements of Microwave
Technology*
ISBN: 0-15-522102-7

Carr
*Integrated Electronics:
Operational Amplifiers and
Linear IC's with Applications*
ISBN: 0-15-541360-0

Driscoll
Data Communications
ISBN: 0-03-026637-8

Filer/Leinonen
*Programmable Controllers and
Designing Sequential Logic*
ISBN: 0-03-032322-3

Garrod/Borns
*Digital Logic: Analysis,
Application and Design*
ISBN: 0-03-023099-3

Greenfield
The 68HC11 Microcontroller
ISBN: 0-03-051588-2

Grant
*Understanding Lightwave
Transmission: Applications
of Fiber Optics*
ISBN: 0-15-592874-0

Harrison
*Transform Methods
in Circuit Analysis*
ISBN: 0-03-020724-X

Hazen
Exploring Electronic Devices
ISBN: 0-03-028533-X

Hazen
*Fundamentals of DC
and AC Circuits*
ISBN: 0-03-028538-0

Hazen
*Experiencing Electricity
and Electronics*
Conventional Current Version
ISBN: 0-03-007747-8
Electron Flow Version
ISBN: 0-03-003427-2

Ismail/Rooney
*Digital Concepts
and Applications*
ISBN: 0-03-026628-9

Laverghetta
*Analog Communications
for Technology*
ISBN: 0-03-029403-7

Leach
*Discrete and Integrated
Circuit Electronics*
ISBN: 0-03-020844-0

Ludeman
*Introduction to Electronic
Devices and Circuits*
ISBN: 0-03-009538-7

McBride
*Computer Troubleshooting
and Maintenance*
ISBN: 0-15-512663-6

Oppenheimer
Survey of Electronics
ISBN: 0-03-020842-4

Prestopnik
*Digital Electronics:
Concepts and Applications
for Digital Design*
ISBN: 0-03-026757-9

Seeger
*Introduction to
Microprocessors
with the Intel 8085*
ISBN: 0-15-543527-2

Spiteri
Robotics Technology
ISBN: 0-03-020858-0

Wagner
Digital Electronics
ISBN: 0-15-517636-6

Winzer
Linear Integrated Circuits
ISBN: 0-03-032468-8

Yatsko/Hata
*Circuits: Principles, Analysis,
and Simulation*
ISBN: 0-03-000933-2

Linear Integrated Circuits

Jack Winzer
*Georgian College
Barrie, Ontario*

SAUNDERS COLLEGE PUBLISHING
A Harcourt Brace Jovanovich College Publisher

Fort Worth   Philadelphia   San Diego   New York   Orlando
Austin   San Antonio   Toronto   Montreal   London   Sydney   Tokyo

Copyright © 1992 by Saunders College Publishing.

All rights reserved. No part of this publication may be reproduced or transmitted in any form or by any means, electronic or mechanical, including photocopy, recording, or any information storage and retrieval system, without permission in writing from the publisher.

Requests for permission to make copies of any part of the work should be mailed to: Permissions Department, Harcourt Brace Jovanovich, Publishers, 8th Floor, Orlando, FL 32887

Text Typeface: Times Roman
Compositor: Maryland Composition
Acquisitions Editor: Barbara Gingery
Assistant Editor: Laura Shur
Managing Editor: Carol Field
Senior Project Manager: Marc Sherman
Manager of Art and Design: Carol Bleistine
Associate Art Director: Doris Bruey
Text Designer: Rebecca Lemna
Cover Designer and Cover Credit: Lawrence R. Didona
Text Artwork: Grafacon
Director of EDP: Tim Frelick
Production Manager: Bob Butler
Product Manager: Monica Wilson

Printed in the United States of America

LINEAR INTEGRATED CIRCUITS

0-03-032468-8

Library of Congress Catalog Card Number 91-053094

1234 016 987654321

THIS BOOK IS PRINTED ON **ACID-FREE, RECYCLED** PAPER

Dedication

To Ernest and Violet Winzer, my grandparents
They will never read this text, but
they contributed immensely to it.

Preface

Not too many years ago, electronics was relatively simple. It was based on essentially one type of active device—the vacuum tube. True, there were many different types of vacuum tube, but the functions that they performed were relatively few, and the basic design procedures were relatively straightforward. Some tubes were designed to perform two or three functions—in other words, several different functional configurations were placed in the same glass envelope—but, because of the bulky nature of the tube elements, this process was limited. The invention of the transistor in 1947 heralded a new era. Although the transistor was used initially simply as a replacement for the vacuum tube, being much smaller and consuming far less power, it was not long before its true potential became apparent. With the transistor, size was no problem and many transistors could be placed on a single chip. Diodes and resistors could also be placed on the same chip—the integrated circuit was born.

Now, toward the end of the twentieth century, the vacuum tube is gone. In its place are a tremendously wide variety of devices that can perform almost any function conceivable. This profusion of devices is the power of modern electronics, but the potential nemesis of the engineer, technologist, or technician who must routinely design, analyze, and service equipment built around these integrated devices. This book is intended to introduce integrated circuits, and to present practical circuit applications for a representative sample of these devices.

Text Philosophy

This text is intended as a basic introduction to linear integrated circuits. My approach to this topic, however, is quite different from most books on op-amps and integrated circuits—I use a practical systems approach that minimizes theoretical design of specific components, and stresses the practical design of complete *working* electronic systems. There are many books on the market that develop the theory of op-amps and related circuits to a far greater extent than is done in this text. Theory is easy to teach and to demonstrate with simple examples and experiments. Industry—the world beyond the protective walls of universities and colleges—however, requires people who know real devices and can design complex systems without many hours of theoretical calculation.

Instructors accustomed to the theoretical approach used in other texts may find this book somewhat less than rigorous in the theory. In many derivations and examples, various simplifying approximations are frequently made. Sometimes the approximations have been justified; occasionally they have not even been mentioned. Often, the equations are not theoretically precise, but they provide a fast and efficient method of designing practical circuits—*circuits that work*.

The emphasis in this book is on designing complete working circuits, and choosing real component values. The examples are often long and complex because a complete circuit is being developed. The design and analysis problems at the end of each chapter are also quite long. Most texts use many short problems that can be answered with only a few lines of work. In the real world, designs that require pages of work are the rule, not the exception.

The advantage to the student should be obvious. Instead of designing one resistor in a hypothetical circuit with theoretical precision, the student must design a complete working circuit within the typical 10% precision dictated with real components. The student builds real solutions to realistic problems—not theoretical solutions to idealized problems. Student interest is stimulated because these are real practical circuits—circuits that can be used for various projects and applications.

Text Contents

The word "integrated circuit" covers an extremely wide variety of devices. There are, in fact, two quite distinct families of integrated circuits—the linear or analog integrated circuits and the digital integrated circuits—and the processes for designing circuits using each family are quite different. This text is devoted entirely to linear (or analog) integrated circuits.

The emphasis of most texts on linear integrated circuits is on op-amps, with linear integrated circuits developed as a side line. Unfortunately, many instructors tend to overuse the op-amp. While the op-amp can be used as an audio amplifier, a power supply regulator, an active filter, or an oscillator, there are special-purpose integrated circuits to perform all of these functions, both with much better performance, and with far simpler circuits than op-amps. In this text, I have kept the coverage of op-amps to a minimum, and have introduced as many other types of integrated circuits as possible.

Nevertheless, the op-amp is a very important integrated circuit, and is the building block for many more complicated integrated circuits. The op-amp is introduced in Chapter 2, with basic amplifier circuits using op-amps introduced in Chapter 3, and advanced amplifier applications, including summing, integrating, and logarithmic amplifiers, introduced in Chapter 4. The application of op-amps as comparators is introduced in Chapter 5, their use in active filters in Chapter 6, their use as regulators in Chapter 7, and their use as oscillators in Chapter 10. Most textbooks still present the μA741 as the standard op-amp. The μA741 is actually quite obsolete, and most designers use the more modern BiFET op-amps. This text briefly introduces the μA741, but uses the TL081 BiFET op-amp as the main device in all op-amp applications.

Other linear integrated circuits are introduced for most applications. Special purpose op-amps including the LM321 single-supply op-amp, the LM3900 Norton op-amp, and the LH0038 instrumentation amplifier are introduced in Chapter 3. The TL810 true voltage comparator integrated circuit is introduced in Chapter 5. Switched-capacitor filter integrated circuits are introduced in Chapter 6. A variety of regulator integrated circuits are introduced in Chapters 8 and 9. Oscillator integrated circuits including the well-known 555 timer, the LM566 voltage-controlled oscillator, the LM565 phase-locked loop, and the ICL8038 function generator are introduced in Chapter 11.

This text has three chapters dedicated to topics either omitted entirely in other texts, or at best discussed superficially. The first is Chapter 1, which introduces the reader to data sheets, databooks, cross-reference indices, and other sources of information on linear integrated circuits. Data sheets are also presented for each device as it is introduced. The second is Chapter 9, which introduces switch-mode voltage regulators. There are *no* good discussions of switch-mode regulators that I have read, yet these devices are increasingly being used. This chapter is at least as important as the preceding two chapters on linear regulators. Finally, Chapter 12 discusses audio integrated circuits. A very large fraction of all integrated circuits is used in audio (and video) applications, yet this topic is usually neglected in most textbooks on integrated circuits, or if included, is included as an extension of op-amp applications. Standard op-amps are not really suitable for audio applications—the special audio integrated circuits introduced in Chapter 12 are far superior.

One omission that some instructors may note is A to D and D to A converters. Most texts on linear integrated circuits include a discussion of these devices. However, A to D and D to A devices *are not linear devices*—they are only found in applications relating to digital systems, and are properly covered in a course on digital systems. To include them in this text would require the cutting of more relevant linear integrated circuit applications.

Application

This book is intended for a one- or two-semester course for students at a sophomore or junior level in an Electronics Engineering Technology program. University engineering programs in electronics may also find this book useful as a practical course in integrated circuits paralleling a theoretical course in the same subject, typically in the third year. Practicing engineers or technicians who do not have a good working knowledge of linear integrated circuits would find this a very useful addition to their bookshelves, as it provides a thorough summary of current linear integrated circuits with a wide variety of working circuits. The electronics hobbyist may also find this book useful for the various circuit applications developed throughout the text.

Suggested prerequisites include basic AC and DC theory, a course in electronic devices and/or discrete transistor circuits, and basic college algebra. Calculus, although helpful, is not necessary.

The material in this text can be covered in a number of ways. Each chapter is generally complete by itself, and does not depend on the contents of other chapters. Hence, chapters (and sections of chapters) can be omitted at the discretion of the instructor.

A basic one-semester course stressing operational amplifiers would include Chapters 2, 3 (excluding the last section), 4, 5 (excluding the last section), 6 (excluding the last section), 7 and 10. An advanced course, or a course stressing integrated circuit applications, would include Chapters 6, 8, 9, 11, and 12. Individual chapters may be relevant for certain applications. For example, for technicians being trained for audio and video servicing, Chapter 12 would be essential.

Teaching Aids

This book is intended basically as a practical teaching text. The presentation tries not to confuse the student with a lot of secondary detail, but to simply describe the main points of operation and design using each device, then to work through one or more examples. Each example presents the design or analysis of a complete working circuit that could be built and used for a laboratory experiment, or as a student construction project.

At the end of most chapters are a series of review questions and a number of design and analysis problems. The review questions are simply one-line questions which expect one-line answers. They cover some of the highlights of the chapter and represent an excellent way for the student to determine how much he has learned from studying the chapter. The design and analysis problems are long. Many represent complete designs or analyses of circuits and some may take several hours to complete. They are based on examples in the book and the student is encouraged to locate the relevant example and follow it as he performs the design. These design problems make excellent laboratory assignments.

A separate laboratory manual containing over 30 practical experiments is available. Most of these experiments are design oriented, and are based directly on the design procedures introduced in the text. Basic equipment required for prototyping and testing any of the designs in the text, and for performing any of the experiments in the laboratory manual include a ±15 volt DC power supply, a variable power supply, a digital voltmeter, a function generator, and an oscilloscope. Designs for a ±15 volt DC power supply, a variable power supply, and a function generator are given in the text. A solutions manual containing solutions to all problems is available upon adoption.

Computer Applications

One topic that is totally missing in this text is the now almost mandatory "computer applications," usually accompanied by an applications diskette generating all sorts of tones and fancy graphics. Students using this text should be well past the stage of writing a BASIC program to calculate the gain of an amplifier. Including the simple programming exercises found in many textbooks is a waste of space. Most students will have only a single course in linear integrated circuits, and there is too much important data to learn without wasting time and space doing the elementary computer programming that they should have had in earlier programming courses.

There is, however, one very relevant computer application that should at least be mentioned—computer simulation. Simulation programs such as PSPICE are an important tool for circuit design and analysis. Unfortunately, effective use of PSPICE (or an equivalent simulation program) for the circuits developed in this text requires a sound knowledge of modelling and a thorough operating knowledge of the simulation program. This is neither a text on computer simulation, nor a text on circuit modelling, and most students do not have sufficient background to effectively use computer simulation for the integrated circuit applications developed in this text. To properly develop computer simulation procedures for integrated circuits would detract from the main purpose of this text, namely the introduction of linear integrated circuits. Therefore, computer simulations are omitted. Computer simulation of linear integrated circuits is best covered in a separate text devoted to that one topic. I hope someday to write such a text.

Acknowledgments

Several reviewers provided valuable comments and suggestions for this text. These include: David McDonald, Lake Superior State University; Ellis Nuckolls, Oklahoma State University; Jerry Sorenson, South Dakota State University; and Richard Sturtevant, Springfield Technical Community College. Their suggestions greatly improved the text and helped to shape the book into its present form.

Most importantly, it was the help and advice of the Saunders College Publishing staff that made this whole project a reality. In particular, I wish to thank: Barbara Gingery, Senior Acquisitions Editor; Laura Shur, Assistant Editor; and Marc Sherman, Senior Project Manager. Their patient guidance and encouragement shaped the many pages of rough manuscript and preliminary figures into a well-organized, complete text.

Jack Winzer
Barrie, Ontario

Contents

Preface   vii

Chapter 1
Introduction to Integrated Circuits   1

    1-1   Historical Development of Integrated Circuits   2
    1-2   Digital versus Linear Integrated Circuits   6
    1-3   Packaging of Integrated Circuits   7
    1-4   Manufacturers of Linear Integrated Circuits   10
    1-5   Identification of Linear Integrated Circuits   10
    1-6   Sources of Data on Integrated Circuits   16

Chapter 2
Operational Amplifiers   34

    2-1   Amplifier Characteristics   35
    2-2   Differential Amplifiers   48
    2-3   Characteristics of Operational Amplifiers   53
    2-4   Feedback in Operational Amplifiers   56
    2-5   Bipolar Integrated Circuit Operational Amplifiers   62
    2-6   BiFET Integrated Circuit Operational Amplifiers   67
    2-7   Uncompensated Operational Amplifiers   72

Chapter 3
Amplifier Circuits   79

    3-1   Inverting Amplifiers   80
    3-2   Noninverting Amplifiers   87
    3-3   Buffer Amplifiers   93
    3-4   Single-Supply Amplifiers   94
    3-5   The TL321C Single-Supply Op-Amp   98
    3-6   The Norton or Current-Mode Op-Amps   106
    3-7   Instrumentation Amplifiers   114
    3-8   The LH0038 Instrumentation Amplifier   120

Chapter 4
Applications of Operational Amplifiers 138

- 4-1 Summing Amplifiers 139
- 4-2 Subtracting Amplifiers 144
- 4-3 Integrating Amplifiers 149
- 4-4 Differentiating Amplifiers 153
- 4-5 Logarithmic Amplifiers 158
- 4-6 Exponential Amplifiers 163
- 4-7 Precision Rectifiers 165

Chapter 5
Comparators 172

- 5-1 Op-Amps as Comparators 173
- 5-2 Zero-Crossing Comparator Circuits 177
- 5-3 Level-Sensing Comparator Circuits 179
- 5-4 Comparators with Hysteresis 184
- 5-5 Window Comparator Circuits 190
- 5-6 True Comparator Integrated Circuits 194

Chapter 6
Active Filters 210

- 6-1 Basic Filter Types 211
- 6-2 Low-Pass and High-Pass Active Filters 227
- 6-3 Bandpass and Band-Rejection Active Filters 242
- 6-4 Switched-Capacitor Filters 259

Chapter 7
Linear Power Supplies 280

- 7-1 Rectifier Circuits 281
- 7-2 Regulation 289
- 7-3 Voltage References 294
- 7-4 Op-Amp Regulators 301

Chapter 8
Integrated Circuit Regulators 324

- 8-1 Multipin Regulators 325
- 8-2 Three-Terminal Fixed Regulators 346
- 8-3 Three-Terminal Variable Regulators 364
- 8-4 Voltage Converters 372

Chapter 9
Switch-Mode Power Supplies 385

    9-1   Principles of Switch-Mode Power Supplies 386
    9-2   Pulse-Width Modulators 402
    9-3   Single-Package Switching Regulators 432

Chapter 10
Oscillator Circuits 445

    10-1   Theory of Oscillation 446
    10-2   Relaxation Oscillators 449
    10-3   Bootstrap Oscillators 459
    10-4   Sine-Wave Oscillators 467
    10-5   Other Oscillator Circuits 476

Chapter 11
Integrated Circuit Oscillators 489

    11-1   The LM566C Voltage Controlled Oscillator 490
    11-2   The LM555C Timer/Oscillator 503
    11-3   The ICL8038 Single-Chip Function Generator 517
    11-4   The LM565C Phase-Locked Loop 522

Chapter 12
Audio Integrated Circuits 541

    12-1   The LM387 Audio Preamplifier 542
    12-2   The LM1877 Power Amplifier 563
    12-3   Volume and Tone Control Circuits 576
    12-4   AM and FM Radio Circuits 586

Appendix A
Reference Data A-1

    A-1   Standard Resistor Values A-1
    A-2   Typical Capacitor Values A-3
    A-3   Power Transformer Data A-5
    A-4   Zener Diode Data A-6
    A-5   Transistor Data A-7

Appendix B
Design of Toroidal Inductors and Transformers A-22

Appendix C
Design of Wave-Shaping Circuits   A-33

Appendix D
Glossary   A-40

Appendix E
Answers to Design and Analysis Problems   A-49

Appendix F
Device Index   A-54

Index   I-1

Introduction to Integrated Circuits   1

OUTLINE

1-1 Historical Development of Integrated Circuits

1-2 Digital versus Linear Integrated Circuits

1-3 Packaging of Integrated Circuits

1-4 Manufacturers of Linear Integrated Circuits

1-5 Identification of Linear Integrated Circuits

1-6 Sources of Data on Integrated Circuits

KEY TERMS

The following new terms will be introduced in this chapter. Definitions of these terms appear in the Glossary at the end of the book.

Integrated circuit
Nuvistor tube
Micromodule
Mesa process
Planar process
Monolithic process
Gallium-arsenide
Metal-oxide-semiconductor field-effect transistor (MOSFET)
Linear integrated circuit
Analog integrated circuit
Digital integrated circuit
Dual in-line package (DIP)
Surface-mount package
Single in-line package (SIP)
Data sheet
Data book
Selection guide
Cross-reference guide
Replacement guide

OBJECTIVES

On completion of this chapter, the reader should be able to:

- discuss the historical development of integrated circuits.
- name the materials used for integrated circuits, describe the processes used in their production, and identify the various types of packages.
- differentiate between linear (or analog) integrated circuits and digital integrated circuits.

- list a number of functions performed by the various linear integrated circuits on the market.
- name some of the major manufacturers of linear integrated circuits and identify their logos.
- use manufacturers' data books, selection guides, cross-reference guides, and replacement guides to identify, select, and replace linear integrated circuits.

Introduction

Before we actually begin to design circuits using integrated devices, it is important to understand the nature of the devices that we will be working with—how historically they developed as a natural extension of the art of making discrete transistors, and physically how they are manufactured by the monolithic diffusion process. This is not a book on the history of electronics, nor a book on the design of the actual integrated circuit chips, but a book on the design of circuits using these devices. It is important, nevertheless, to have a general idea how these devices developed and how they are manufactured both for an understanding of their limitations and for a glimpse at their future potential.

Integrated circuits are traditionally grouped into two classes: digital integrated circuits, and linear or analog integrated circuits. Although many engineers and technologists tend to look at electronics entirely from one point of view—digital—or from the other point of view—analog, this is not a wise process, and the modern engineer or technologist should be aware of the capabilities and applications of both digital and analog devices. The emphasis of this book is on linear circuits, although it is difficult to separate entirely the two classes of integrated circuits.

It is impossible for a single textbook such as this to discuss more than a few of the more representative types of integrated circuits, yet the circuit designer must have at his or her disposal at least minimal data for many of the thousands of devices on the market, to be able to select the best device for a specific application. This information is listed in various manufacturers' data books, as well as in selection guides and replacement guides. One of the first things we must do is to learn what data sources are available and how to use these data sources.

1-1 Historical Development of Integrated Circuits

Although integrated circuits have only been around for about 25 years, the concept of the **integrated circuit** is much older. The idea of having a complete circuit on a single substrate has existed for a long time. The first true electronic device was the vacuum tube. As the vacuum tube developed, it soon became apparent that several tube functions could be built into the same container. Unfortunately, this first step was limited by the nature of the electronic components then available and by the physical construction of the tube. As tube technology advanced, it became possible to reduce the size of the tube itself. The RCA **Nuvistor tubes**, for example, were of a size comparable to the first transistors which were then starting to appear. These devices could be mounted on a printed circuit board along with a few additional components and form a compact modular circuit.

As the transistor developed, the same ideas were applied. Because a transistor required such a small amount of power and had a very long lifetime compared with a vacuum tube, the circuits could be made true modules by encapsulating the components in epoxy.

Most of the volume of a discrete transistor or resistor is the package of the device. The United States Army developed a system in the mid-1950s of stacking the discrete functional devices, connecting them with fine wires, and encapsulating the entire circuit. These devices were called **micromodules**. They had a relatively small physical size, were mechanically very durable, and could be mass-produced at moderate price.

Micromodular devices were a first step, but they were still somewhat bulky and required considerable time (and expense) to produce. The first true integrated devices started to appear in the late 1950s. Jack S. Kilby, an engineer working for Texas Instruments, first experimented with fabricating all of the electronic components of a complete circuit onto a single crystal of germanium. The components were then wired together on the crystal, and packaged to produce the integrated circuit. A picture of Kilby's first device, a two-transistor phase-shift oscillator, is shown in Figure 1-1. Kilby's process, although viable, had some problems. It proved difficult to obtain reproducible component values, and the wiring was tedious and time-consuming. Fortunately, a better method of producing integrated circuits was soon forthcoming.

To understand the development of the modern integrated circuit, we must look at the way transistors are produced. The first transistor produced at Bell Telephone labs in 1947 by William Shockley, Walter Brattain, and John Bardeen, was a point contact device where the different junctions were made by physical contact. This process was not satisfactory for mass production, and the junction transistor consisting of sandwiched layers of doped germanium soon appeared. Further developments through the 1950s led to the mesa transistor. By this time, silicon had replaced germanium as the fundamental material for producing semiconductors.

In the **mesa process**, a single crystal of silicon was successively doped and etched to produce a mesalike structure to which wires were then attached. In actual manufacturing, a mask was used to allow many transistors to be produced on the same crystal simultaneously. If many transistors could be produced on the same chip, then why not interconnect them to form a complete circuit? Complete circuits, however, require other components, mainly resistors, and, unfortunately, although the mesa process was fine for transistors, it was not suitable for producing devices such as resistors.

In 1959, Jean Hoerni and Robert Noyce, founders of Fairchild Semiconductors, developed a method of producing a flat, or **planar**, transistor. Their process involved diffusing first the base into the substrate of doped silicon that was to become the collector, then diffusing the emitter into the base, covering the entire device, except for contact areas, with silicon dioxide, and vacuum-coating alu-

Figure 1-1 The first true integrated circuit chip, a phase-shift oscillator, produced by Jack S. Kilby at Texas Instruments in 1958. Actual size 1.0 mm by 1.5 mm. *(Photo courtesy of Texas Instruments Incorporated)*

minum as wires. With proper masking, it was possible to produce many transistors on a single crystal wafer, and it soon became apparent that this diffusion process was far more practical for mass production than the mesa process. Again, it was obvious that the masking process could be used to produce complete circuits. With the planar process, however, other components were easy to produce. Diodes were simply one less diffusion step than transistors. Resistors were simply a channel of lightly doped silicon (lightly doped silicon is a high-resistance conductor). Insulators, where required, were produced by diffusing oxygen into the crystal (silicon plus oxygen yields silicon dioxide, the major constituent of glass, a very good insulator). The only things that proved difficult to produce in integrated form were capacitors and inductors. In fact, with modern processes, it is possible to produce small-value capacitors (a few picofarads) and small-value inductors (less than a microhenry), although ultimately the solution was to develop circuits that did not require large inductors or capacitors, or to add these devices externally if required.

One final and important step in the development of integrated circuits was the introduction of the **metal-oxide-semiconductor field-effect transistor (MOSFET)**, first produced by Steven R. Hofstein and Frederic P. Heiman at RCA in 1962. The field-effect transistor uses an insulated gate for control, replacing the two junctions of the bipolar transistor. For use in integrated circuits, the field-effect transistor has several advantages over the bipolar transistor: it is considerably simpler to produce and, hence, is cheaper; it lends itself to large-scale integration more readily; and it consumes far less power than bipolar devices. Its main disadvantage is that it is slower than bipolar devices. MOSFET technology has led to a profusion of inexpensive battery-operated electronic devices such as pocket calculators and digital watches, microprocessors, and megabyte memories.

Modern integrated circuits are almost universally made by the planar process introduced by Hoerni and Noyce. The first integrated circuit produced by the planar process appeared in 1961, a simple digital bistable multivibrator (flip-flop), developed by Fairchild. The first linear integrated circuit was an operational amplifier, the μA702 produced in 1964, also developed by Fairchild. Improved operational amplifier circuits followed rapidly, with the very popular μA741 being developed, again by Fairchild, in 1968.

Since the development of these early devices, there has been a tremendous increase in the density of integration and the complexity of the circuits, and a dramatic reduction in the net cost. The most dynamic increases have been in the area of computer technology. The first microprocessor chip, the 4004 produced by Intel, appeared in 1971 and had over 2000 MOS transistors fabricated in an area approximately 2 mm by 2 mm. The well-known 8080 single-chip microprocessor, with 4800 transistors, came three years later in 1974. The 1980s saw first the introduction of 16 bit microprocessor chips, then 32 bit chips, and the introduction of memory chips capable of storing up to a million bits of data. As well, the 1980s produced single-chip voltmeters, frequency counters, and function generators. Programmable logic chips and programmable memory chips were produced. The increase in the complexity and diversity of integrated circuits has been, and still is, phenomenal!

With the exception of extremely high speed devices, the base or substrate material used in integrated circuits is always silicon—in fact, a flat wafer of a single crystal of extremely pure silicon. All of the parts or components of the integrated circuit are fabricated on this single piece of silicon, and consequently this type of device is called **monolithic** (from the Greek *mono* meaning *one* and *lithos* meaning *stone*). Whether the device is linear or digital, whether the device uses bipolar components or MOSFET components, the same basic method of production is used. Different manufacturers have their own variations of the planar process, but the overall results are the same.

The one main shortcoming of silicon technology is the maximum frequency at which integrated circuits can be used. If higher speed (or higher frequency response) is required than silicon devices are capable of providing, other types of substrates can be used. In particular, **gallium-arsenide** can support much higher electron velocities and hence is used for high-speed devices. Gallium-arsenide is more expensive, is somewhat more complicated to work with, and the technology is not as advanced as for silicon, yet gallium-arsenide devices are on the market, and more are in the development stage. Silicon technology devices are unlikely to be replaced by gallium-arsenide devices because of the cost and complexity of the process, but certainly gallium-arsenide devices are an important compliment to silicon devices where very high frequency performance is necessary.

Currently, research is being performed in such areas as hybrid silicon/gallium-arsenide devices and silicon-germanium alloy devices. Sometime in the future, one of these may emerge as a replacement for the silicon integrated circuit. One thing for certain—the integrated circuit has not reached its ultimate form. The future holds promise of new and improved devices!

1-2 Digital versus Linear Integrated Circuits

Integrated circuits can be broadly grouped into two distinct classes: **linear integrated circuits** (sometimes called **analog integrated circuits**), and **digital integrated circuits**. It is very hard to give a simple definition that will always distinguish one class from the other. If we use the standard definition of a linear integrated circuit as a device that accepts a continuous range of input levels, then we must consider analog-to-digital converters and frequency counters as analog devices—but these are basically digital devices. If we use the standard definition of a digital integrated circuit as a circuit operating between two signal levels, high and low, then we must consider the comparator as a digital integrated circuit—but it is definitely a linear integrated circuit. The rule that we will use in this book is simply that of application: if the integrated circuit is used primarily for digital applications and for handling digital signals, then it is a digital integrated circuit, otherwise it is a linear integrated circuit.

Digital systems operate by the manipulation of signals by logic gates: AND, NAND, OR, NOR, inverters, and bistable devices. Devices containing these gates are digital. Interface devices such as line drivers and buffers, designed primarily for digital applications, are also digital. The various microprocessor circuits and related devices, calculator chips, and clock chips are digital devices. We will also consider analog-to-digital converters and digital-to-analog converters, devices for interfacing between the analog and digital worlds, as digital integrated circuits.

Most other devices will be considered as linear integrated circuits. These include amplifiers, comparators, voltage regulators, oscillators, and a wide variety of special audio, video, and communications integrated circuits.

The most common linear circuit employed in electronics is the amplifier. Amplifiers are used for increasing signal levels and are found in audio systems, radio and television receivers, and electronic instrumentation. The most common form of integrated circuit amplifier is the operational amplifier, and we will be introducing many applications for the operational amplifier in the following chapters.

The comparator is a special form of amplifier which functions as a voltage driven switch—the output changes from low to high in response to the input voltage level. We will introduce a number of interesting applications for comparators in Chapter 5.

Every piece of electronic equipment requires a power supply. Most power supplies use a linear integrated circuit voltage regulator to hold the output voltage constant despite changes in the input voltage or output load. Many modern power supplies switch power as required by the load for greater efficiency, and use a different type regulator, the switch-mode regulator, to control the switching. Switch-mode regulators contain another form of linear integrated circuit—the pulse-width modulator.

Often it is necessary to generate some form of waveform or clocking signal. Signal generating circuits, generally called oscillators, are found in radio and television transmitters and receivers, in many types of test equipment, and as clocking circuits in most digital systems. Although oscillators can be produced using simple operational amplifiers circuits, many special oscillator integrated circuits are available. We will demonstrate in Chapter 11 how a complete laboratory function generator can be constructed from a single integrated circuit.

Finally, there are a wide variety of audio and communications integrated circuits. In addition to audio preamplifiers and power amplifiers, there are single-chip AM radio receivers, FM radio receivers, FM detectors, stereo demodulators, tone controls, equalizers, television video amplifiers, and sync separators. The communications/audio/video field has probably the widest variety of integrated circuit applications, and consequently the largest number of linear integrated circuits of any field of electronics.

1-3 Packaging of Integrated Circuits

The actual integrated circuits are small rectangular chips of silicon with dimensions on the order of a millimeter (not much larger than the head of a pin). External leads must be provided so that connections can easily be made to external circuit boards and external components. Extremely fine wires are soldered directly to the chip and connected to the much larger external leads. Integrated circuit and connecting wires are mounted in a suitable package which protects the internal chip and fine wires. There are a number of different packages used. Most of the common packages are illustrated in Figure 1-2.

The earliest integrated circuits were packaged in transistor containers. Currently popular at the time were the TO-5 metal packages, and typically integrated circuits were produced in 8, 10, or 12 pin configurations instead of the 3 pin transistor configuration. The TO-5 package is still found today for some of the older types of integrated circuits. It is illustrated in Figure 1-2(a). The limitations of this type of packaging became apparent with the development of more complex circuits. Increased circuit complexity meant that more leads were necessary, too many for the simple TO-5 can, and consequently the **dual in-line package (DIP)** was developed. Most integrated circuits today are produced in some form of DIP. The more common DIP packages are illustrated in Figure 1-2(b)-(i), where 8, 14,

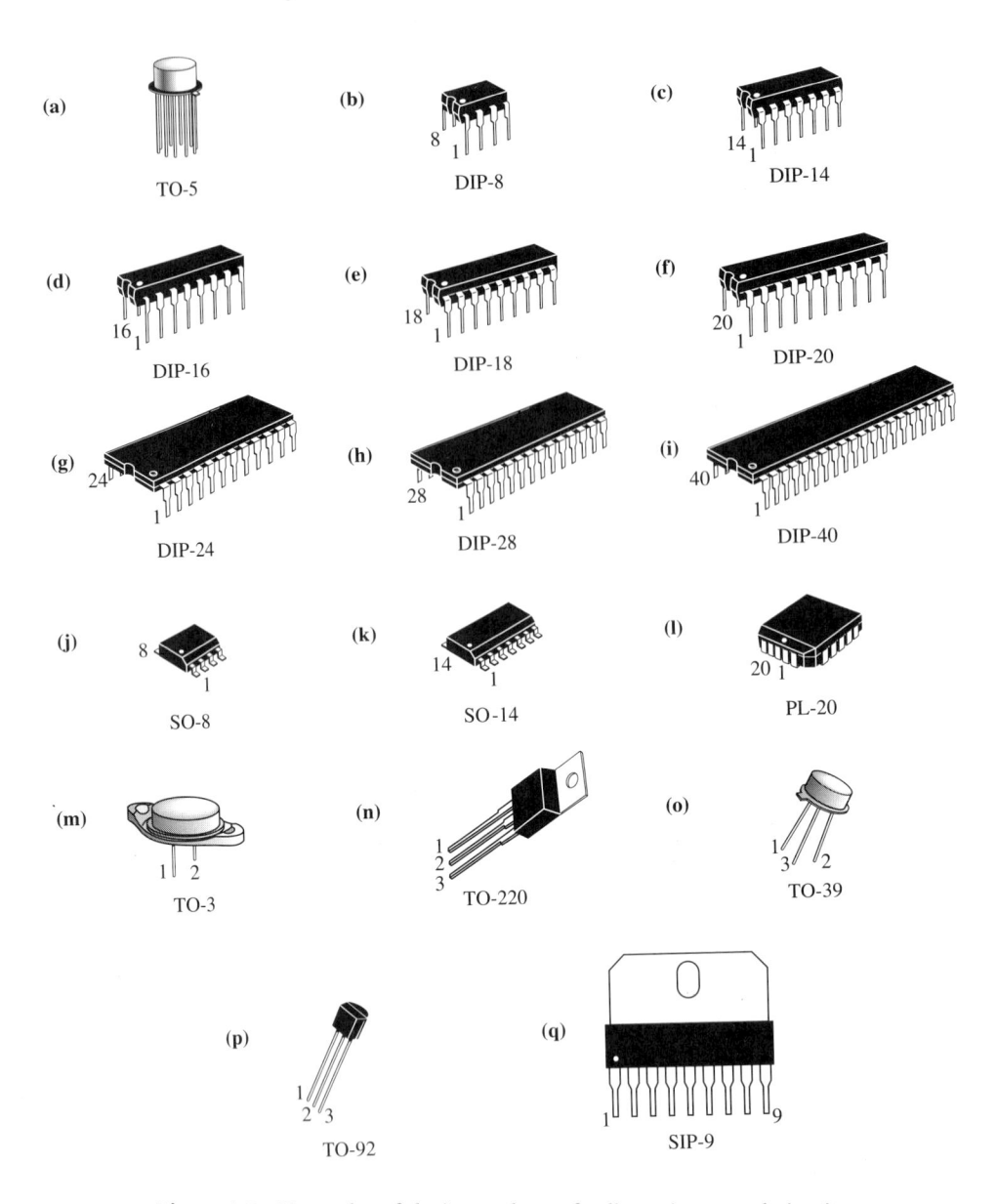

Figure 1-2 Examples of device packages for linear integrated circuits.

16, 18, 20, 24, 28, and 40 pin packages are shown. In the 8, 14, 16, 18, and 20 pin DIP packages, the rows of pins on each side of the package are 0.3 in. apart, whereas in the 24, 28, and 40 pin DIP packages the pin rows are twice as far apart (0.6 in.). Adjacent pins are always 0.1 in. apart in both sizes of package.

In the past few years an increasing number of devices have been produced in **surface-mount packages**. As the name implies, surface-mount packaged integrated circuits are designed to be mounted on the surface of a printed circuit board without the usual pins which must go through holes in the board. This mounting procedure eliminates the need for drilling holes in the circuit board, allows more compact packages to be used, and permits automated assembly methods. Figure 1-2(j) and Figure 1-2(k) show 8 and 16 pin SOIC (*s*mall *o*utline *i*ntegrated *c*ircuit) surface-mount packages, which are half the size of the equivalent DIP packages. Figure 1-2(l) shows a 20 contact PLCC (*p*lastic *l*eaded *c*hip *c*arrier) surface mount package. With contacts on all four sides of a square package, this package is approximately a quarter the size of a standard 20 pin DIP. Surface-mount packaging is becoming increasingly important for mass production and for increased packing density on circuit boards. Although the surface-mount devices shown in Figure 1-2 appear to be of similar size as the DIP packages, they are several times smaller.

Power-handling circuits including regulators and power amplifiers are usually packaged differently. For power dissipations of 20 W or more, the TO-3 case style is preferred. This is shown as Figure 1-2(m). A simple regulator has only 2 pins, with the third terminal (usually ground) being the case. Amplifiers in this case style have more than 2 pins. For power dissipations up to 15 W, the TO-220 style shown as Figure 1-2(n) is most popular. The illustration shows a 3 pin regulator. Amplifiers in this package have more pins. The TO-202 style (not shown), similar in appearance to the TO-220 but slightly smaller, is used for power dissipations up to 5 W. Some regulators with ratings up to 2 W come in the TO-39 case pictured as Figure 1-2(o). The TO-92 case style shown as Figure 1-2(p) is used for regulators up to 0.5 W and for voltage references. Power amplifiers for consumer products are sometimes found in the **single in-line package (SIP)** shown in Figure 1-2(q).

Many devices are available in a variety of packages. For example, a simple operational amplifier may be available in a 10 pin TO-5 can, in an 8 pin DIP, in a 14 pin DIP, or in a 8 pin SOIC. The DIP form of packaging has a number of variations, depending on the type of case (plastic or ceramic) and the manufacturer. These variations are slight. A few package styles are not given here, including very high frequency devices, but Figure 1-2 shows all the most common forms in which linear integrated circuits can be found.

Pin identification for the multipin packages is quite simple. For the DIP package, the package should be viewed from the top. There is always a notch or other identifying mark on one end. With the notch end pointing up, the pins are numbered starting from the leftmost pin closest to the notch, counterclockwise around the package with the highest-numbered pin on the right of the notch. The highest and lowest pin numbers are illustrated in Figure 1-2. Sometimes pin 1 is indicated with a dot on the case. The same system applies to surface-mount devices. For the SIP package, there will always be a notch or dot on one end. The pins are numbered from this end. For TO-5 packages, there is always a tab on the edge of the lip of the can. This is always opposite the highest-numbered pin. Viewed from the top, the next pin counterclockwise is pin 1, and the numbering increases as you progress counterclockwise around the case. For the TO-220 and TO-202 packages, the pins are always numbered from the left when the case is placed with the heatsink down as shown in Figure 1-2. Pin numbering is seldom a concern in 2 or 3 pin devices—they are usually simply identified by function such as input, output, or ground.

The function of each pin must be determined from data sheets, and will be discussed in later sections. For devices that perform a similar operation, the functional pin configurations are usually identical. For example, operational amplifiers in the 8 pin DIP package, regardless of manufacturer, always have pin 2 as the inverting input, pin 3 as the noninverting input, pin 6 as the output, pin 7 as the positive supply, and pin 4 as the negative supply.

1-4 Manufacturers of Linear Integrated Circuits

Many companies manufacture linear integrated circuits. Table 1-1 lists 144 companies. This list is not necessarily current or complete. Some of the companies listed have gone bankrupt or have merged with other companies, and new companies have started up. Some important mergers since Table 1-1 was produced include Fairchild, taken over by National Semiconductor, and RCA and Intersil acquired by Harris Semiconductor.

Some of the listed companies produce only a few highly specialized integrated circuits, whereas others, such as National Semiconductor, Motorola, and Texas Instruments, produce a full line of linear integrated circuits, digital integrated circuits, and discrete components. Some of the companies develop their own devices, others simply manufacture integrated circuits under license from the companies that developed them. Most of the larger companies manufacture popular devices developed by other companies as well as their own line of devices.

Is it important to know the manufacturer or developer of a particular integrated circuit? Often not. For example, if you want a simple operational amplifier, it is not important to know the manufacturer of the device. Any manufacturer's product for a particular device will always meet standard published specifications. On the other hand, if you want an enhanced performance device, one manufacturer's product may not provide as good performance as another's, and it is important to distinguish between them. Also, if you require a specialized integrated circuit, it may be available from only one manufacturer, and in this case you will need to know the manufacturer to obtain data sheets and sources of the device.

1-5 Identification of Linear Integrated Circuits

Each separate type of integrated circuit is identified by its own specific device number, and in the next section we will see how device numbers can be used to locate integrated circuit specifications in data books and cross-reference catalogs. First, however, we must learn the significance of device numbers.

In later chapters we will be making extensive use of the TL081C operational amplifier. The symbol TL081C is the device number or name. Notice that it consists of some letters (English or Greek), a number, and more letters. Most (but not all) linear integrated circuits are identified in this way.

1-5 Identification of Linear Integrated Circuits

Table 1-1 Manufacturers of linear integrated circuits

Telefunken Electronic GmbH	International Rectifier	Silicon General, Inc.
Alpha Industries	In-Phase Electronics	SGS-ATES Componenti Elet. S.p.A.
Advanced Micro Devices, Inc.	Int'l Power Semiconductor	Signetics Corp.
Gould AMI Semiconductors	Intech, Inc.	Siemens Aktiengesellschaft
Apex Microtechnology Corp.	Intel Corporation	Silicon Systems, Inc.
Analog Devices, Inc.	ITT Semiconductors Intermetall	Siemens Corp.
Analog Systems	Keltron Power Devices Ltd.	Siliconix, Inc.
Amperex Electronic Corp.	Lambda Semiconductors	Solid State Industries, Inc.
Advanced Semiconductors, Inc.	LSI Computer Systems, Inc.	Semelab
Avantek, Inc.	Linear Technology Corp.	Solitron Devices, Inc.
Bharat Electronics Ltd.	Linear Technology, Inc.	SONY Corp
Burr-Brown Corp.	Matsushita Electronics Corp.	Solid Power Corporation
Continental Device India Ltd.	Micro Electronics Ltd.	Space Power Electronics, Inc.
Calvert Electronics, Inc.	Mistral S.p.A. Via Dell'irto 32	Sprague Electric Company
Curtis Electromusic Spec, Inc.	Mitel Semiconductor	Solid State Electronics Corp.
Central Semiconductor Corp.	Mitsubishi Electric Corp.	Solid State Devices, Inc.
Cermetek, Inc.	Thomson Components-Mostek Corp.	SSM Microcomputer Products
Cherry Semiconductor Corp.	Motorola Semiconductor Products, Inc.	Solid State Microtechnology For Music
CTS Microelectronics, Inc.	Micropac Industries, Inc.	Sprague Solid State
Crimson Semiconductor, Inc.	Micro Power Systems	Solid State, Inc.
Conditioning Semiconductor Device Corp.	Micra Corp.	Solid State Systems
CTS Electronics Corp.	Mullard Ltd.	Silicon Transistor Corporation
ILC Data Devices Corp.	North American Semiconductor Co., Ltd.	Semiconductor Technology, Inc.
Dionics, Inc.	NEC Electronics U.S.A., Electron Div.	Stow Laboratories, Inc.
Diode Transistor Co., Inc.	Nippon Electric Co., Ltd.	Syntar Industries, Inc.
DYMEC	New Japan Radio Co., Ltd.	Supertex Inc.
G.E. Datel	New Jersey Semiconductor Prod. Inc.	Swampscott Electronics Co., Inc.
Eudoria GesmbH Components	National Semiconductor, Microcircuits Div.	Toko America, Inc.
Elantec	OKI Semiconductor	Teledyne Crystalonics, Inc.
Elm State Electronics, Inc.	Opamp Labs, Inc.	Thomson-CSF/EFCIS
Electronic Transistors Corp.	Origin Electric Co., Ltd.	TIC Semiconductor
Exar Corporation	Philco Radio Televisao Ltda.	Texas Instruments, Inc.
Fujitsu Ltd.	Philips Electronic Components & Materials Div.	Texas Instruments Ltd. (England)
Ferranti Electronics Ltd.	Plessey Semiconductors	Texas Instruments Ltd. (Canada)
Fairchild Linear Div.	Precision Monolithics, Inc	Teletone
GigaBit Logic Inc.	Punjab Wireless Systems Ltd.	Toshiba Corp.
General Diode Corp.	RCA Corporation, Solid State Div.	Teledyne Philbrick
General Instrument Corp.	EG & G Reticon	TRW LSI Products
Germanium Power Devices Corp.	Rohm Co., Ltd.	TRW Optoelectronics Div.
General Transistor Corporation	RICOH Co., Ltd.	Tokyo Sanyo Electric Co., Ltd.
GTE Microcircuits Div.	R.S. Components Ltd.	Teledyne Semiconductor Corp.
Harris Semiconductor	RTC-Compelec	Transistor Specialtys, Inc.
Hybrid Systems Corp.	Raytheon Company	Unitrode Corp.
Hitachi Ltd., Semicon & IC Div.	Sanken Electric Co., Ltd.	UPI Semiconductor
Hybrid Semiconductors & Elect., Inc.	Semicoa	V/O Electronzagranpostavka
Hi-Tron Semiconductor Corp.	Semiconductors, Inc./Walbern Devices	Valvo GmbH
International Devices Inc.		Walbern Devices/Semiconductors, Inc.
Interfet Corp.		Western Digital Corp.
Industria Mexicana Toshiba S.A.		Powerex, Inc.
Intersil, Inc.		

First consider the leading letter portion, or prefix, of the device number. The letter prefix generally refers to the company that developed the particular device. For example, the TL081C can be identified as an integrated circuit developed by Texas Instruments from the prefix TL. Sometimes, the device may be manufactured by a company different from that which developed it. The manufacturing company usually retains the original prefix, and lists the device in their data books under the original name. Any such integrated circuit will meet the specifications laid down by the company that developed it, regardless of who manufactured it. Most companies have distinctive letter prefixes. For example, the prefix μA for the μA741C operational amplifier denotes Fairchild as the developer, while the prefix LF for the LF357C operational amplifier indicates that this device was developed by National Semiconductor. There are a wide variety of prefixes—some manufacturers will have the same prefixes as others, some do not use them. Table 1-2 lists common prefixes used by some of the major manufacturers.

The next portion of the device number is usually a number. What is the significance of this number? Why is a μA741C numbered 741, and not 742, or 1022, or some other number? The answer to this question is far from easy. First consider the TL081C discussed above. This is one of a series of Texas Instruments operational amplifiers, and the series includes the TL080C, TL081C, TL082C, TL083C, and TL084C. These are all operational amplifiers representing different configurations of the same basic device, and hence form a family of similar devices. Similarly, Texas Instruments TL070C, TL071C, TL072C, TL073C, and TL074C form another family, this time of low-noise operational amplifiers. Now consider the following Fairchild devices: the μA702C is an operational amplifier, the μA709C is an operational amplifier, the μA714C is an operational amplifier, the μA741C is an operational amplifier, the μA747C is an operational amplifier. Are these members of a family of operational amplifiers? No, the μA723C is a

Table 1-2 Common prefixes for linear integrated circuits used by major manufacturers

Manufacturer	Prefixes
Analog Devices	AD
Exar Corporation	XR
Fairchild Linear Division	μA
Intersil, Inc.	ICL
Linear Technology Corp.	LT
Motorola Semiconductor	MC
National Semiconductor Corp.	LF, LH, LM, TBA
Precision Monolithics, Inc.	OP
RCA Corporation	CA
Signetics Corporation	NE, SE
Silicon General, Inc.	SG
Texas Instruments	TL

1-5 Identification of Linear Integrated Circuits

voltage regulator, the µA733C is a wide-band video amplifier. The numbers in this series appear to be chronological—they are numbered in the order in which Fairchild developed each device.

The number of a specific device is determined by the original manufacturer. It may be a number in a sequence of similar devices, it may be a number in a chronological sequence, or it may just be a random number. As a general rule, it is best to consider the numerical portion of the device number as an arbitrary number. For other than well-established families, do not look for any logical numbering scheme.

Lastly, there are often one or more letters attached to the end of the device number. Notice that all of the above device numbers ended in C. The C designates the devices as commercial grade, to differentiate the devices from military M or industrial I grade. The industrial and military models are the same basic device, but generally have tighter specifications and can operate over a wider temperature range. Often there is a second letter that refers to the type of package. For some integrated circuits, there may be four or more letters forming a suffix.

There is absolutely no standardization for these suffixes, and it is necessary to determine the manufacturer of the device, then go to its data book to determine the meaning. For example, the suffix P indicates a plastic DIP for Motorola, yet the same package is designated N by National Semiconductor.

Most device numbers adhere to the above rules, but there are exceptions. When a popular device is manufactured by several different companies, it is not necessarily marketed under the same name by all the companies. Large companies often market their own version of the more popular devices *and use their own prefix*. Consider the 741 operational amplifier. This device was originally developed by Fairchild as the µA741C. It is produced by Texas Instruments with the same Fairchild designation, µA741C. Motorola, however, markets this device as the MC1741C, and National Semiconductor markets it as the LM741C. How do we determine the manufacturer? The manufacturer of a device can usually be determined by the logo stamped on the case. The logos of a few of the larger manufacturers are shown in Figure 1-4. For the device shown in Figure 1-3, the manufacturer is Motorola, as indicated by the logo.

When we look at an actual integrated circuit package, such as illustrated in Figure 1-3, we find that there are two (or more) letter/number combinations stamped on the device. One of these numbers is the device number. How do we determine which number is the device number? Generally, the device number

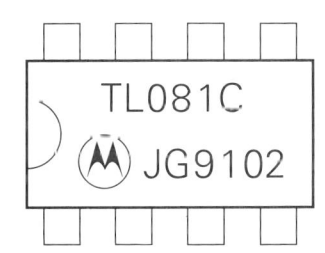

Figure 1-3 Typical linear integrated circuit showing package labeling.

14   1 Introduction to Integrated Circuits

Analog Devices	▶ ANALOG DEVICES
Exar Corporation	XR
Fairchild Linear Division	FAIRCHILD — A Schlumberger Company
Intersil, Inc.	INTERSIL
Linear Technology Corporation	LT LINEAR TECHNOLOGY
Motorola Semiconductor Products	(M)
National Semiconductor Corporation	N
Precision Monolithics, Inc.	PMI
RCA Corporation	RCA
Signetics Corporation	signetics
Silicon General, Inc.	SILICON GENERAL
Texas Instruments	TI

Figure 1-4 Logos of some of the major linear integrated circuit manufacturers.

1-5 Identification of Linear Integrated Circuits

will be obvious, being a known device number such as the TL081C shown in Figure 1-3, or easily identified by a standard letter prefix.

Difficulties may arise if both numbers have apparently standard prefixes. In this case, check if one number is four digits and begins with a 7, 8, or 9. Such a number will probably be the date of production, given in the form of YYWW where YY corresponds to the last two digits of the year and WW corresponds to the week. For example, 9102 on the package shown in Figure 1-3 indicates the second week of the year 1991 as the date of production. If the date number can be established, the other number is the device number.

On some devices, there may be a third number. The third number may be an alternative device number or manufacturer's internal part number, the registration date of the device, or some production number. Usually, there is no problem distinguishing the various numbers and identifying the part number, but you will probably, some day, come across a device on which you cannot determine which number is the device number. In this case, you must simply look up both (or all) numbers in some index, as described in the next section, and select the number that most closely represents the device function in the circuit.

To make life even more complicated, manufacturers sometimes will simply stamp a part number on the device. The part number refers to a piece of electronic equipment and has no indication of device number. This problem is encountered frequently in consumer electronics such as radio and television receivers and stereo equipment, where it is often impossible to service the device without the manual for that piece of equipment.

The device numbering system used for integrated circuits may seem totally irrational and confusing. It is! However, in most cases it is possible to identify the device without too much difficulty.

EXAMPLE 1-1

An integrated circuit is shown in Figure 1-5. Identify the integrated circuit by determining the following:

(a) the device number
(b) the company that developed the integrated circuit
(c) the manufacturer
(d) the date of production
(e) the type of package

```
F8144B
UA733C
```

Figure 1-5

Solution

All of the required information can be determined by inspecting Figure 1-5.

(a) The device number is µA733C. For this device, the part number was easy to identify because it had a recognizable prefix. Note that the device number always includes prefix and suffix letters. You will occasionally encounter different devices with the same numerical portion but different prefixes.
(b) The developer of this device is determined from the prefix as Fairchild Electronics. Note that this device is part of the Fairchild µA700 series of devices. A note of caution here—some manufacturers may use their own prefix. For example, National Semiconductor markets this device as the LM733C.
(c) The manufacturer is Texas Instruments from the logo displayed on the package.
(d) The second number on the package gives the date of production. The number is 8144, which means that the device was produced in the 44th week of 1981.
(e) The package is a 14 pin DIP.

1-6 Sources of Data on Integrated Circuits

As we saw in the last section, there are a tremendous number of manufacturers, an even larger number of different devices, and an extremely complex and confusing system for numbering devices. This section will hopefully put some order to this chaos. To do so, it is necessary to become familiar with some of the sources of data on integrated circuits.

The first, and most direct, source of information is the manufacturer's **data sheet**. For any specific device, data sheets can be obtained directly from the manufacturer by phoning or writing. Such data sheets are presented in the text for a number of devices. All the major manufacturers also produce **data books** listing the data sheets for all of their devices, usually grouped according to function and indexed in numerical order, often cross-referenced to other manufacturers' devices. In fact, most major manufacturers produce a series of between 10 and 20 data books to cover their entire line of products. Figure 1-6 shows the Texas Instruments *Linear Circuits Data Book* along with the first page of the data sheet for the µA741 operational amplifier. This device is found on page 223 in Chapter 3 of the data book. Chapter 3 in this data book is devoted entirely to operational amplifiers, and the data book itself is devoted entirely to Texas Instruments' line of linear circuits. There are separate chapters for regulators, comparators, oscillators, etc. If you have some specific device and want all of the published data, including electronic characteristics, response curves, maximum operating conditions, packaging, and typical applications, then this is the source. We will be making extensive use of data sheets in later sections to determine the characteristics of the various integrated circuits introduced in this text.

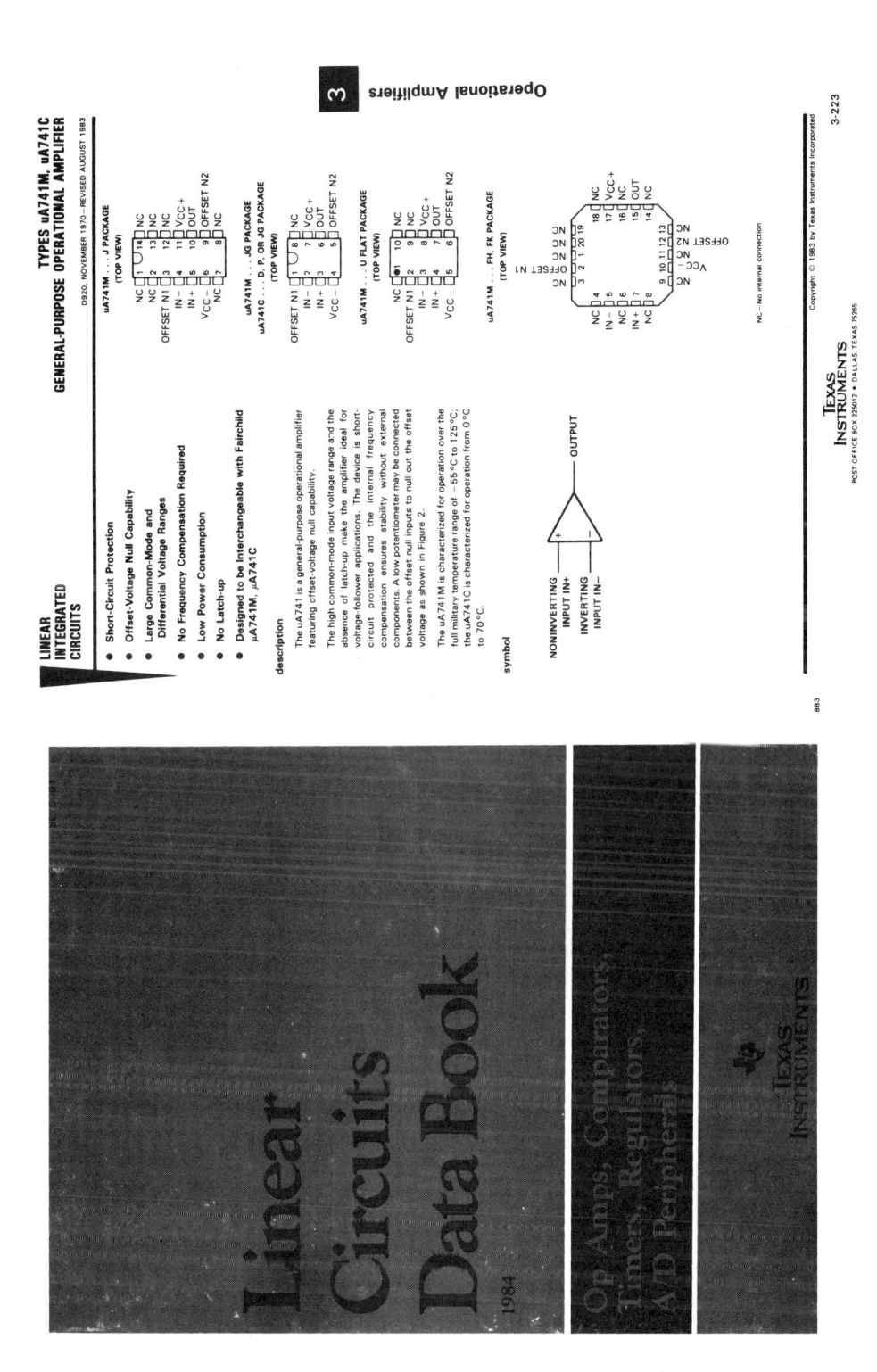

Figure 1-6 Texas Instruments *Linear Circuits Data Book* with sample data sheet. *(Reprinted by permission of Texas Instruments)*

The second source of information on integrated circuits is the manufacturer's **selection guide**. In conjunction with their data books, most manufacturers also publish selection guides. These guides list the entire line of products for the manufacturer and provide tables of devices grouped under application. The Motorola *Master Selection Guide* is shown in Figure 1-7. Page 88 from this guide, which contains tables of single, noncompensated operational amplifiers has also been reproduced in this figure. The selection guide serves as a very useful reference book for the circuit designer trying to determine which device to use for an application. Since the main electrical characteristics of all of the devices used for one specific application are listed together, they can be easily compared by the designer, and the optimum device selected. Once a device has been selected, then the designer will refer to the more extensive data sheets of that device for complete specifications.

The third source of data is the general **cross-reference guide**. There are a variety of these. One widely used series is the *D.A.T.A.BOOK*s published semiannually by D.A.T.A., Inc. in California. The *Linear Integrated Circuits D.A.T.A.BOOK* is shown as Figure 1-8 along with samples of some of the tables. These books list all of the devices (in this case, linear integrated circuits) currently in production. The devices are first listed by part number, and referenced to the page number for a second table where the device is listed under function and generally in order of power rating. The second table lists all similar devices in the same part of the table; hence, it is easy to compare, cross-reference, and substitute devices. This second table provides further reference to a drawing of the IC giving its pinout and, often, its equivalent circuit. The manufacturer is also identified, so that the manufacturer's data book can be referred to if more data are required. This type of book is most useful for identifying a device and obtaining general characteristics and pinout when only a device number is known. It is also useful for selecting replacement devices.

The fourth source of data is the **replacement guide**. Several companies produce these. The Philips *ECG Master Replacement Guide* is shown in Figure 1-9. Replacement guides list in numerical order all the devices for which replacements are available. A sample page of the *ECG Replacement Guide* (page 469) has been included in Figure 1-9. This page lists both the 741 op-amp and the 723 voltage regulator, devices to be discussed in later chapters. The ECG replacement for the 741 is the ECG941D, and the ECG replacement for the 723 is the ECG923D. This type of data source is intended for finding replacement devices, but it is also very useful for comparing device characteristics. If two devices have the same replacement, then they obviously must have approximately the same characteristics. Replacement catalogs also list the characteristics of the replacement device and give pinouts and, often, equivalent circuits.

Anyone working with integrated circuits, either the designer who must specify which device to use or the technician who needs the data to troubleshoot a circuit, must be familiar with these sources of data, and must have at his or her disposal a variety of data books, selection guides, general cross-references, and replacement guides. A well-equipped electronics shop or educational institution would normally have as many as 50 to 100 such books!

Amplifiers

Operational Amplifiers	Page
Single	88
Dual	90
Quad	91
High Frequency Amplifiers	
AGC	93
Non-AGC	93
Miscellaneous Amplifiers	
CMOS	94
Quad Programmable Op Amp	94
Quad Programmable Comparator	94
Dual/Dual Prog Op Amp-Comp	94
Bipolar	
Dual/Dual Op Amp-Comp	94
Power Amplifiers Variable Gain	94

Operational Amplifiers

Motorola offers a broad line of bipolar operational amplifiers to meet a wide range of applications. From low-cost industry-standard types to high precision circuits, the span encompasses a large range of performance capabilities. These linear integrated circuits are available as single, dual, and quad monolithic devices in a variety of temperature ranges and package styles. Most devices may be obtained in unencapsulated "chip" form as well. For price and delivery information on chips, please contact your Motorola Sales Representative or Distributor.

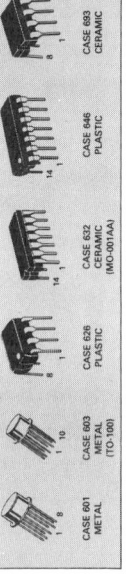

CASE 601 METAL
CASE 603 METAL (TO-100)
CASE 626 PLASTIC
CASE 632 CERAMIC (MO-001AA)
CASE 646 PLASTIC
CASE 693 CERAMIC

Single Operational Amplifiers

Noncompensated

Commercial Temperature Range (0°C to +70°C)

Device	I_B μA Max	V_{IO} mV Max	TC_{VIO} μV/°C Typ	I_{IO} nA Max	A_{VOL} V/mV Min	BW (A_V=1) MHz Typ	SR (A_V=1) V/μs Typ	Supply Voltage V Min	Supply Voltage V Max	Description	Packages
LM301A	0.25	7.5	10	50	25	1.0	0.5	±3.0	±18	General Purpose	601, 626, 693
LM308	7.0	7.5	15	1.0	25	1.0	0.3	±3.0	±18	Precision	601, 626, 693
LM308A	7.0	0.5	5.0	1.0	80	1.0	0.3	±3.0	±18	Precision	601, 626, 693
MC1439	1.0	7.5	15	100	15	2.0	4.2	±6.0	±18	High Slew Rate	601, 626
MC1709C	1.5	7.5	15	500	15	1.0	0.5	±3.0	±18	General Purpose	601, 626, 693
MC1748C	0.5	6.0	15	200	20	1.0	0.5	±3.0	±18	General Purpose	601, 626, 693

Industrial Temperature Range (−25°C to +85°C)

Device	I_B	V_{IO}	TC_{VIO}	I_{IO}	A_{VOL}	BW	SR	Min	Max	Description	Packages
LM201A	0.075	2.0	10	10	50	1.0	0.5	±3.0	±22	General Purpose	601, 626, 693
LM208	0.002	2.0	3.0	0.2	50	1.0	0.3	±3.0	±20	Precision	601, 632, 693
LM208A	0.002	0.5	1.0	0.2	80	1.0	0.3	±3.0	±20	Precision	601, 632, 693

Military Temperature Range (−55°C to +125°C)

Device	I_B	V_{IO}	TC_{VIO}	I_{IO}	A_{VOL}	BW	SR	Min	Max	Description	Packages
LM101A	0.075	2.0	10	10	50	1.0	0.5	±3.0	±22	General Purpose	601, 693
LM108	0.002	2.0	3.0	0.2	50	1.0	0.3	±3.0	±20	Precision	601, 626, 693
LM108A	0.002	0.5	1.0	0.2	80	1.0	0.3	±3.0	±20	Precision	601, 693
MC1539	0.5	3.0	15	60	50	2.0	4.0	±4.0	±18	High Slew Rate	601
MC1709	0.5	3.0	15	200	18	1.0	0.5	±3.0	±18	General Purpose	601, 693
MC1709A	0.6	3.0	5.0	100	25	1.0	0.5	±3.0	±18	High Performance MC1709	601
MC1748	0.5	5.0	15	200	50	1.0	0.5	±3.0	±22	General Purpose	601, 693

Figure 1-7 Motorola Semiconductor *Master Selection Guide* with a sample table listing op-amps. *Reprinted with permission of National Semiconductor Corporation)*

Figure 1-8 D.A.T.A. Inc. *Linear Integrated Circuits D.A.T.A.BOOK* with sample tables. *(Reprinted with permission of D.A.T.A. Business Publishing)*

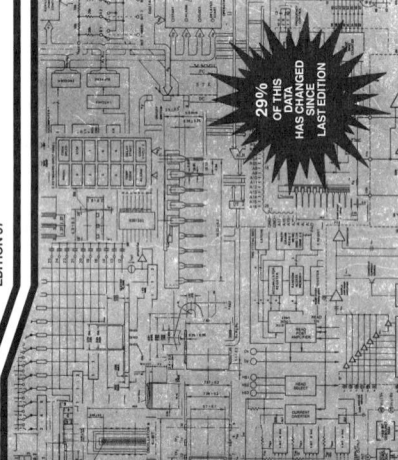

Figure 1-9 Philips *ECG Master Replacement Guide* with a sample page. *(Reprinted with permission of Philips ECG, Inc.)*

EXAMPLE 1-2

Select a positive voltage regulator capable of regulating a DC supply of 25 V and giving a fixed output of 15 V at 1 A. Use the *Motorola Master Selection Guide and Catalog*.

Solution

Although we have not studied regulators yet, we can still select a suitable device from the selection guide.

First we look in the index of the guide for the term "regulator." The selection guide index then directs us to "power supplies," found starting on page 113 (for the 1987 edition of the *Motorola Master Selection Guide and Catalog*). The information on this page tells us that regulators are listed from pages 114 to 119. We simply leaf through the selection guide until we come to "fixed output voltage regulators—15 volts." The correct page is shown in Figure 1-10. Since we must select a device capable of at least 1 A output current, we must use one of the devices in the blocked portion of Figure 1-10.

At this point, our selection is completed. Note that there are 10 suitable devices. From the requirements given in the problem, any of these are suitable. If further requirements were stipulated, such as device tolerance, then we could narrow the selection. In practice, cost and availability would also affect the selection.

EXAMPLE 1-3

Determine the maximum operating frequency, the maximum supply voltage, the typical input impedance, the typical output impedance, and the output square-wave amplitude for the LM565C.

Solution

At this point, we know nothing about the LM565C. We can find out virtually anything we need to know from the manufacturer's data book.

First, we must determine the manufacturer or developer. The prefix LM tells us that the device is produced by National Semiconductor, so refer to the *National Semiconductor Linear Databook*.

The data book lists the parts in two ways: under function and alphanumerically. Since we do not know the function, we must use the more general alphanumerical index. This tells us that the LM565C data sheet can be found on page 9-42 (in the 1982 issue of the *National Semiconductor Linear Databook*).

Figure 1-11 shows the data sheet for the LM565/LM565C phase-locked loop. Although we do not yet know anything about phase-locked loops, we can read the specifications given on the right-hand page shown in Figure 1-11 to determine the required quantities.

1-6 Sources of Data on Integrated Circuits   23

Fixed Output Voltage Regulators (continued)

Vout Volts	Tol.† Volts	Io mA Max	Device Positive Output	Device Negative Output	Vin Min/Max	Regline mV	Regload mV	ΔVo/ΔT mV/°C Typ	Case
15	±1.5	100	MC78L15C	MC79L15C	16.7/35	300	150	—	29, 79
	±0.75		MC78L15AC	MC79L15A					
		500	MC78M15C	MC79M15C	17/35	100	300	1.0	79, 221A
		1500	MC7815*	—	18.5/35	150	150	1.8	1
			MC7815B#	—		300	300		1, 221A
			MC7815C	MC7915C	17.5/35				
	±0.6		MC7815A*	—	17.9/35	22	50		1
			MC7815AC	—			100		1, 221A
	±0.75		LM140-15*	—	17.5/35	150	150		1
	±0.6		LM140A-15*	—		22	35		
	±0.75		LM340-15	—		150	150		1, 221A
	±0.6		LM340A-15	—		22	35		
	±0.3		TL780-15C	—		15	60	0.18	221A
	±0.75	3000	MC78T15*	—	17.5/40	55	30	0.3	1
			MC78T15C	—					1, 221A
	±0.6		MC78T15A*	—		22	25		1
			MC78T15AC	—					1, 221A
18	±1.8	100	MC78L18C	MC79L18C	19.7/35	325	170	—	29, 79
	±0.9		MC78L18AC	MC79L18AC					
		500	MC78M18C	—	20/35	100	360	1.0	79, 221A
		1500	MC7818*	—	22/35	180	180	2.3	1
			MC7818B#	—		360	360		1, 221A
			MC7818C	MC7918C	21/35				
	±0.7		MC7818A*	—		31	50		1
			MC7818AC	—			100		1, 221A
	±0.9		LM140-18*	—		180	180		1
			LM340-18	—					1, 221A
		3000	MC78T18*	—	20.6/40	80	30	0.36	1
			MC78T18C	—					1, 221A
20	±1.0	500	MC78M20C	—	22/40	10	400	1.1	79, 221A
24	±2.4	100	MC78L24C	MC79L24C	25.7/40	350	200	—	29, 79
	±1.2		MC78L24AC	MC79L24AC		300			
		500	MC78M24C	—	26/40	100	480	1.2	79, 221A
		1500	MC7824*	—	28/40	240	240	3.0	1
			MC7824B#	—		480	480		1, 221A
			MC7824C	MC7924C	27/40				
	±1.0		MC7824A*	—	27.3/40	36	50		1
			MC7824AC	—			100		1, 221A
	±1.2		LM140-24*	—		240	240		1
			LM340-24	—					1, 221A
		3000	MC78T24*	—	26.7/40	90	30	0.48	1
			MC78T24C	—					1, 221A

#TJ = −40 to +125°C   *TJ = −55 to +150°C   †Output Voltage Tolerance for Worst Case

MOTOROLA MASTER SELECTION GUIDE

Figure 1-10 Voltage regulator data from the *Motorola Master Selection Guide*. (© Motorola, Inc. Used by permission)

From the *Absolute Maximum Ratings* at the top of the page, we find that the maximum supply voltage is ±12 V. From the *Electrical Characteristics* we find that the typical input impedance is 5 kΩ, the typical maximum operating frequency is 500 kHz, the typical output impedance is 5 kΩ, and the typical peak-to-peak square-wave output level is 5.4 V.

We have now found all the information required by the problem. Much more information is provided in the data sheet if we needed it. The following pages list graphs of *Typical Performance Characteristics*, provide an *AC Test Circuit*, show several circuits that demonstrate *Typical Applications*, and provide design formulae under *Applications Information*.

National Semiconductor

LM565/LM565C Phase Locked Loop

Industrial Blocks

General Description

The LM565 and LM565C are general purpose phase locked loops containing a stable, highly linear voltage controlled oscillator for low distortion FM demodulation, and a double balanced phase detector with good carrier suppression. The VCO frequency is set with an external resistor and capacitor, and a tuning range of 10:1 can be obtained with the same capacitor. The characteristics of the closed loop system—bandwidth, response speed, capture and pull in range—may be adjusted over a wide range with an external resistor and capacitor. The loop may be broken between the VCO and the phase detector for insertion of a digital frequency divider to obtain frequency multiplication.

The LM565H is specified for operation over the $-55°C$ to $+125°C$ military temperature range. The LM565CH and LM565CN are specified for operation over the $0°C$ to $+70°C$ temperature range.

Features

- 200 ppm/°C frequency stability of the VCO
- Power supply range of ±5 to ±12 volts with 100 ppm/% typical
- 0.2% linearity of demodulated output
- Linear triangle wave with in phase zero crossings available
- TTL and DTL compatible phase detector input and square wave output
- Adjustable hold in range from ±1% to > ±60%.

Applications

- Data and tape synchronization
- Modems
- FSK demodulation
- FM demodulation
- Frequency synthesizer
- Tone decoding
- Frequency multiplication and division
- SCA demodulators
- Telemetry receivers
- Signal regeneration
- Coherent demodulators.

Schematic and Connection Diagrams

Metal Can Package

Order Number LM565H or LM565CH
See NS Package H10C

Dual-In-Line Package

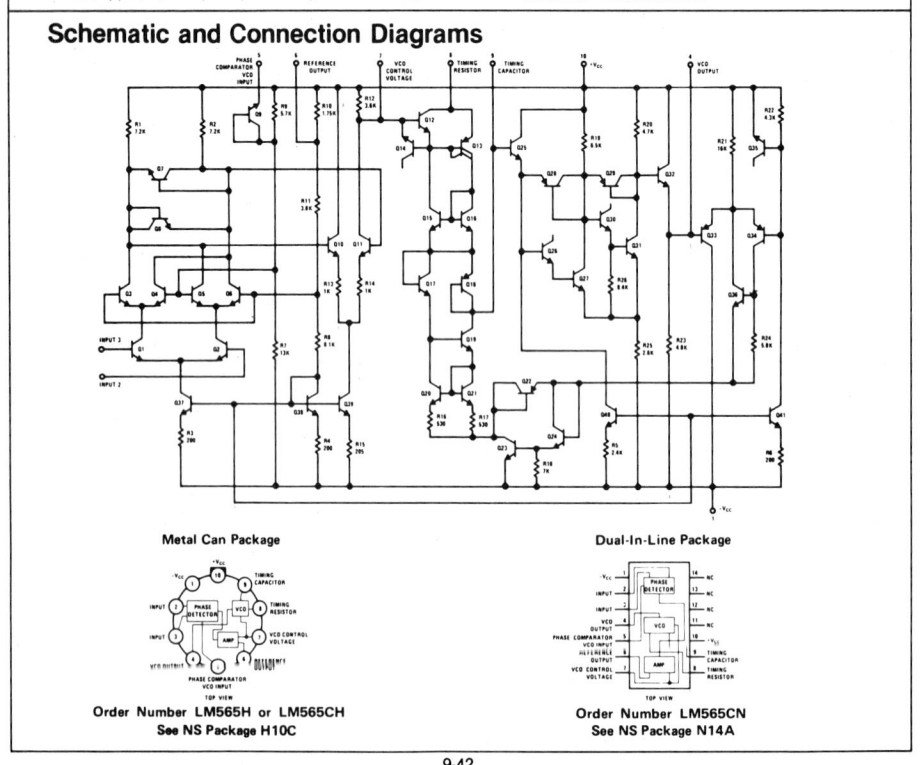

Order Number LM565CN
See NS Package N14A

Figure 1-11 Data sheets for the LM565/LM565C Phase-Locked Loop. *(Reprinted with permission of National Semiconductor Corporation)*

Absolute Maximum Ratings

Supply Voltage	±12V
Power Dissipation (Note 1)	300 mW
Differential Input Voltage	±1V
Operating Temperature Range LM565H	−55°C to +125°C
LM565CH, LM565CN	0°C to 70°C
Storage Temperature Range	−65°C to +150°C
Lead Temperature (Soldering, 10 sec)	300°C

Electrical Characteristics (AC Test Circuit, T_A = 25°C, V_C = ±6V)

PARAMETER	CONDITIONS	LM565 MIN	LM565 TYP	LM565 MAX	LM565C MIN	LM565C TYP	LM565C MAX	UNITS
Power Supply Current			8.0	12.5		8.0	12.5	mA
Input Impedance (Pins 2, 3)	$-4V < V_2, V_3 < 0V$	7	10			5		kΩ
VCO Maximum Operating Frequency	C_o = 2.7 pF	300	500		250	500		kHz
Operating Frequency Temperature Coefficient			−100	300		−200	500	ppm/°C
Frequency Drift with Supply Voltage			0.01	0.1		0.05	0.2	%/V
Triangle Wave Output Voltage		2	2.4	3	2	2.4	3	V_{p-p}
Triangle Wave Output Linearity			0.2	0.75		0.5	1	%
Square Wave Output Level		4.7	5.4		4.7	5.4		V_{p-p}
Output Impedance (Pin 4)			5			5		kΩ
Square Wave Duty Cycle		45	50	55	40	50	60	%
Square Wave Rise Time			20	100		20		ns
Square Wave Fall Time			50	200		50		ns
Output Current Sink (Pin 4)		0.6	1		0.6	1		mA
VCO Sensitivity	f_o = 10 kHz	6400	6600	6800	6000	6600	7200	Hz/V
Demodulated Output Voltage (Pin 7)	±10% Frequency Deviation	250	300	350	200	300	400	mV_{pp}
Total Harmonic Distortion	±10% Frequency Deviation		0.2	0.75		0.2	1.5	%
Output Impedance (Pin 7)			3.5			3.5		kΩ
DC Level (Pin 7)		4.25	4.5	4.75	4.0	4.5	5.0	V
Output Offset Voltage $\|V_7 - V_6\|$			30	100		50	200	mV
Temperature Drift of $\|V_7 - V_6\|$			500			500		μV/°C
AM Rejection		30	40			40		dB
Phase Detector Sensitivity K_D		0.6	.68	0.9	0.55	.68	0.95	V/radian

Note 1: The maximum junction temperature of the LM565 is 150°C, while that of the LM565C and LM565CN is 100°C. For operation at elevated temperatures, devices in the TO-5 package must be derated based on a thermal resistance of 150°C/W junction to ambient or 45°C/W junction to case. Thermal resistance of the dual-in-line package is 100°C/W.

Figure 1-11 (*Continued*)

LM565/LM565C

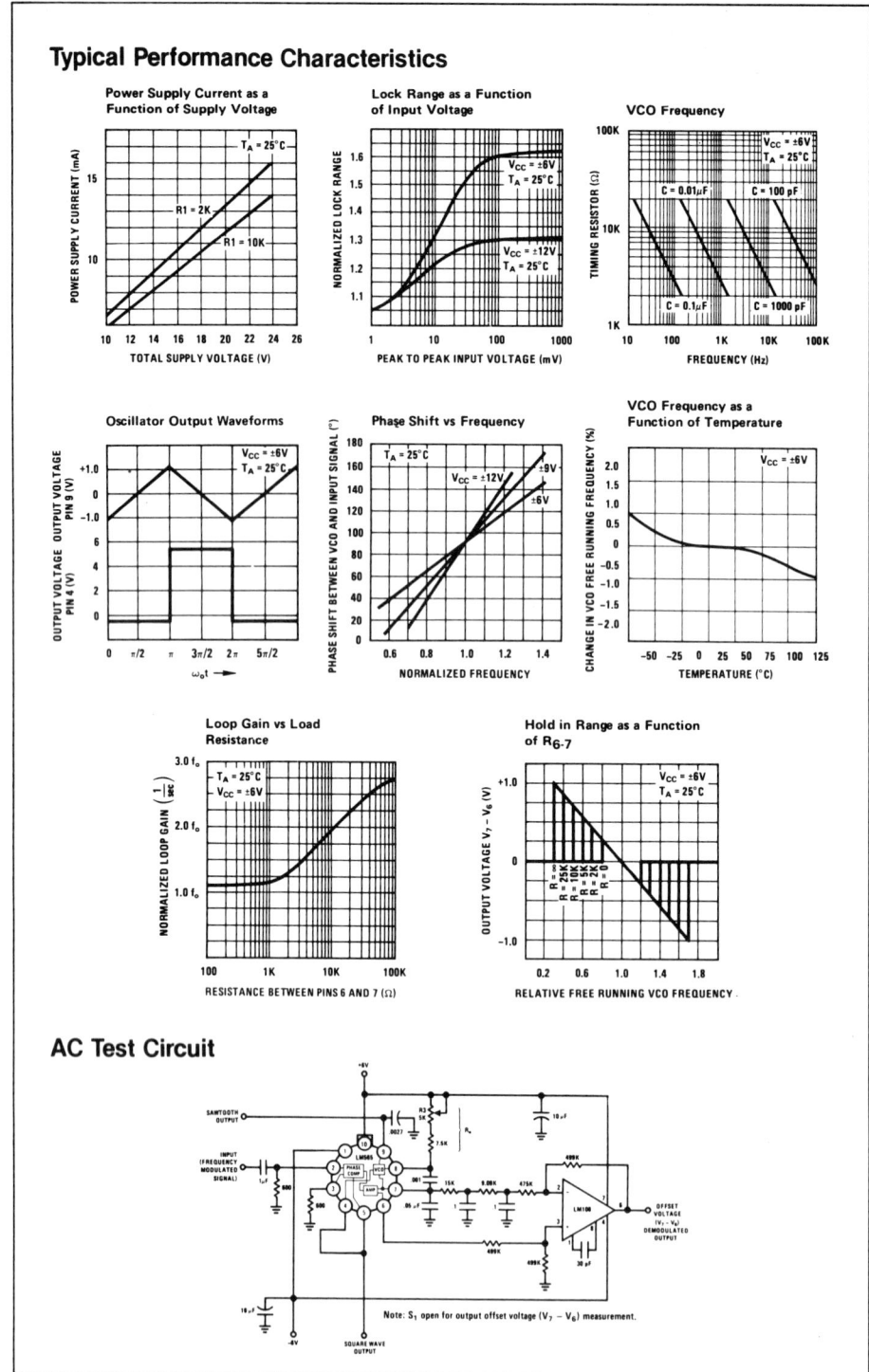

Typical Performance Characteristics

AC Test Circuit

Figure 1-11 (*Continued*)

Typical Applications

2400 Hz Synchronous AM Demodulator

FSK Demodulator (2025-2225 cps)

FSK Demodulator with DC Restoration.

Frequency Multiplier (x10)

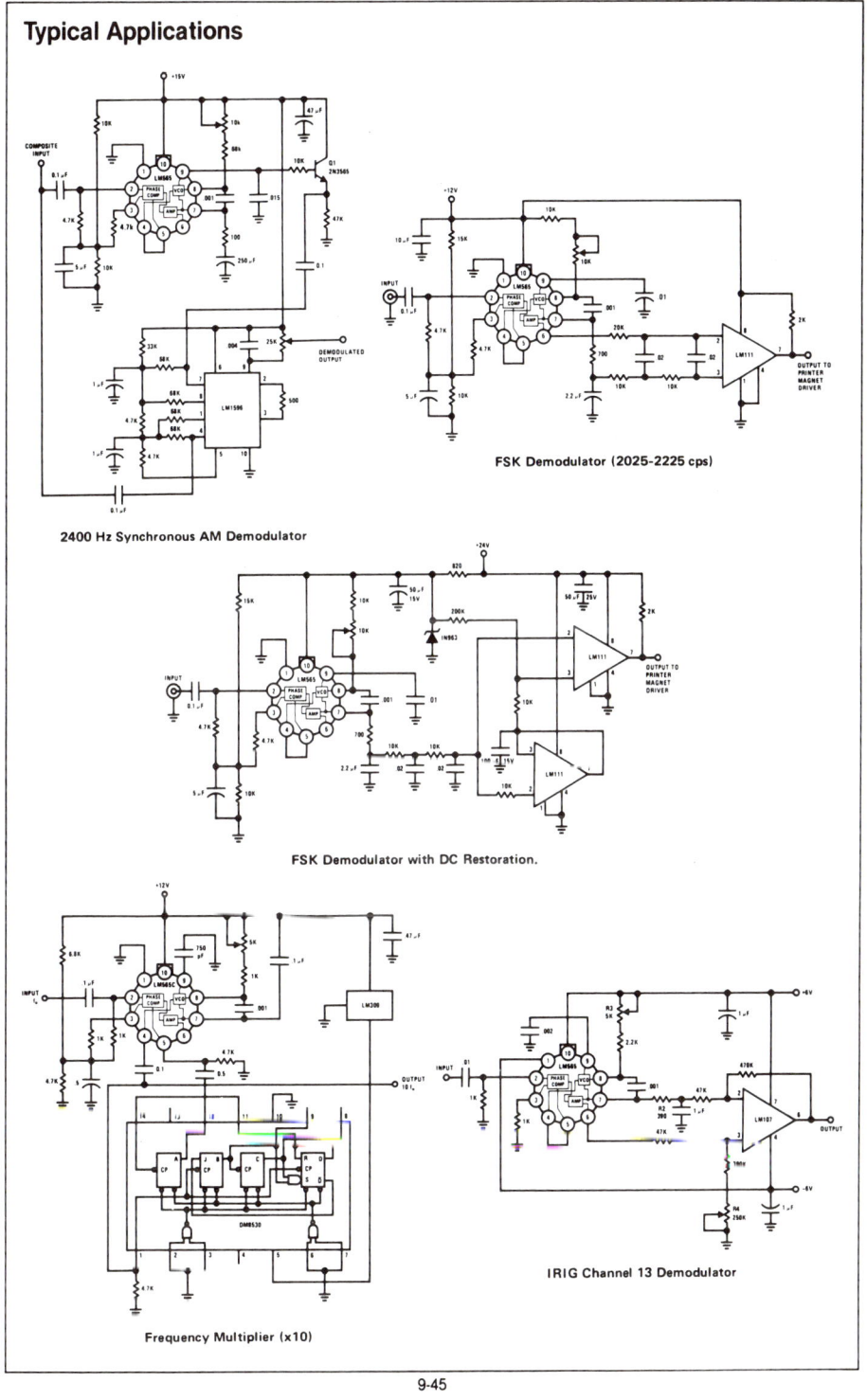

IRIG Channel 13 Demodulator

Figure 1-11 (*Continued*)

Applications Information

In designing with phase locked loops such as the LM565, the important parameters of interest are:

FREE RUNNING FREQUENCY

$$f_o \cong \frac{1}{3.7 R_0 C_0}$$

LOOP GAIN: relates the amount of phase change between the input signal and the VCO signal for a shift in input signal frequency (assuming the loop remains in lock). In servo theory, this is called the "velocity error coefficient".

$$\text{Loop gain} = K_o K_D \left(\frac{1}{\text{sec}}\right)$$

$$K_o = \text{oscillator sensitivity} \left(\frac{\text{radians/sec}}{\text{volt}}\right)$$

$$K_D = \text{phase detector sensitivity} \left(\frac{\text{volts}}{\text{radian}}\right)$$

The loop gain of the LM565 is dependent on supply voltage, and may be found from:

$$K_o K_D = \frac{33.6 \, f_o}{V_c}$$

f_o = VCO frequency in Hz
V_c = total supply voltage to circuit.

Loop gain may be reduced by connecting a resistor between pins 6 and 7; this reduces the load impedance on the output amplifier and hence the loop gain.

HOLD IN RANGE: the range of frequencies that the loop will remain in lock after initially being locked.

$$f_H = \pm \frac{8 \, f_o}{V_c}$$

f_o = free running frequency of VCO
V_c = total supply voltage to the circuit.

THE LOOP FILTER

In almost all applications, it will be desirable to filter the signal at the output of the phase detector (pin 7) this filter may take one of two forms:

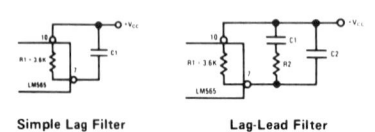

Simple Lag Filter   Lag-Lead Filter

A simple lag filter may be used for wide closed loop bandwidth applications such as modulation following where the frequency deviation of the carrier is fairly high (greater than 10%), or where wideband modulating signals must be followed.

The natural bandwidth of the closed loop response may be found from:

$$f_n = \frac{1}{2\pi} \sqrt{\frac{K_o K_D}{R_1 C_1}}$$

Associated with this is a damping factor:

$$\delta = \frac{1}{2} \sqrt{\frac{1}{R_1 C_1 K_o K_D}}$$

For narrow band applications where a narrow noise bandwidth is desired, such as applications involving tracking a slowly varying carrier, a lead lag filter should be used. In general, if $1/R_1 C_1 < K_o K_d$, the damping factor for the loop becomes quite small resulting in large overshoot and possible instability in the transient response of the loop. In this case, the natural frequency of the loop may be found from

$$f_n = \frac{1}{2\pi} \sqrt{\frac{K_o K_D}{\tau_1 + \tau_2}}$$

$$\tau_1 + \tau_2 = (R_1 + R_2) C_1$$

R_2 is selected to produce a desired damping factor δ, usually between 0.5 and 1.0. The damping factor is found from the approximation:

$$\delta \simeq \pi \tau_2 f_n$$

These two equations are plotted for convenience.

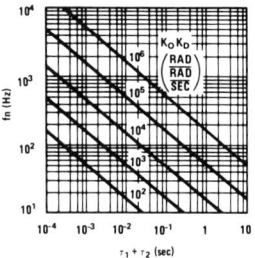

Filter Time Constant vs Natural Frequency

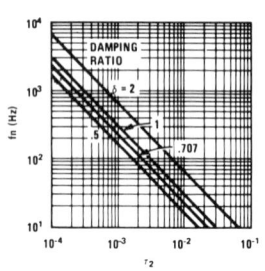

Damping Time Constant vs Natural Frequency

Capacitor C_2 should be much smaller than C_1 since its function is to provide filtering of carrier. In general $C_2 \leq 0.1 \, C_1$.

Figure 1-11 (*Continued*)

EXAMPLE 1-4

You are fixing a radio and you find a device on the circuit board listed as TDA2020DAC2. What is the device, who manufactures it, and what is its pinout? Select a suitable replacement from the Phillips *ECG Semiconductor Master Replacement Guide*.

Solution

This is a problem frequently encountered by anyone who does service work. The prefix of the device is not immediately identified with a specific manufacturer. We will have to refer to a master data book. Although the following data are taken from the *Consumer Integrated Circuits D.A.T.A.BOOK, 1983 Edition*, any of a number of general data books could have been used. The procedure is similar.

First, locate the number in the master cross-index. A portion of the relevant page is shown as Figure 1-12 with the specific part identified. This tells us the manufacturer code is SGAI and the device data is on page 10, entry number 31. A portion of page 10 is shown as Figure 1-13. The manufacturer's codes are listed at the end of the reference. The relevant portion is shown as Figure 1-14 and shows that SGAI is the code for SGS-ATES Componenti Elet. S.p.A. in Italy.

TYPE No.	MFRS	Pg&Line
TDA440Q	NSC	17 - 1
TDA4401	♦ALGG	13 - 49
TDA440TQ	♦ALGG	13 - 50
TDA0820	PHIN	15 - 4
	VALG	
TDA1001A	♦RTCF	12 - 61
	VALG	
TDA1001AT	RTCF	12 - 62
	VALG	
TDA1002	♦MULB	6 -109
	♦VALG	
TDA1002A	♦PHIN	6 -108
♦RTCF	♦VALG	
TDA1003#ai	♦MULB	29 - 2
♦PHIN	VALG	
TDA1003A#1ai	MULB	29 - 7
PHIN	RTCF	
	VALG	
TDA1003A#2ai	MULB	28 - 52
PHIN	RTCF	
	VALG	
TDA1004	♦MULB	8 - 49
♦PHIN	♦VALG	
TDA1004A	PHIN	8 - 40
♦RTCF	SIC	
	♦VALG	
TDA1005	♦MULB	14 -100
TDA1005A	MULB	13 -101
RTCF	♦VALG	
TDA1005A#ai	MULB	14 -101
♦PHIN	VALG	
TDA1005AT	PHIN	13 -102
	RTCF	
TDA1006#ai	PHIN	29 - 8
TDA1006A	♦PHIN	29 - 9
RTCF	♦VALG	
TDA1008	PHIN	25 - 60
♦RTCF	VALG	
TDA1009#ai	PHIN	9 - 50
	♦VALG	
TDA1010	♦PHIN	8 - 27
RTCF	SIC	
	VALG	
TDA1011	♦PHIN	7 - 53
RTCF	VALG	

TYPE No.	MFRS	Pg&Line
TDA1950	ITTG	17 - 50
TDA1950F	ITTG	17 - 51
TDA2002	♦ALGG	8 - 45
TDA2002#at	♦SPR	9 - 89
TDA2002A	♦ALGG	7 - 36
TDA2002A#at	SPR	9 - 90
TDA2002AH#at	♦SGAI	8 - 35
	SPR	
TDA2002AV#at	♦SGAI	8 - 36
	SPR	
TDA2002H#at	SGAI	8 - 37
	SPR	
TDA2002V#at	SGAI	8 - 38
	SPR	
TDA2003	♦ALGG	7 - 37
TDA2003H	♦SGAI	10 - 27
TDA2003V	♦SGAI	10 - 28
TDA2004	SGAI	7 - 80
TDA2005M	♦SGAI	9 - 11
TDA2005S	♦SGAI	9 - 12
TDA2006H	SGAI	9 -103
TDA2006V	SGAI	9 -104
TDA2010B82	♦SGAI	9 - 92
TDA2010B92	♦SGAI	9 - 93
TDA2010BC2	♦SGAI	9 - 94
TDA2010BD2	♦SGAI	9 - 95
TDA2020A82	♦SGAI	10 - 6
TDA2020A92	♦SGAI	10 - 7
TDA2020AC2	♦SGAI	10 - 8
TDA2020AD2	♦SGAI	10 - 9
TDA2020D	♦SGAI	10 - 43
TDA2020DA82	♦SGAI	10 - 29
TDA2020DA92	♦SGAI	10 - 30
TDA2020DAC2	♦SGAI	10 - 31
TDA2020DAD2	♦SGAI	10 - 32
TDA2030	♦ALGG	9 - 96
TDA2030H	♦SGAI	10 - 20
TDA2030V	♦SGAI	10 - 21
TDA2104(A)	VALG	14 -103
TDA2105(A)	VALG	14 -104
TDA2107(A)	VALG	14 -105
TDA2108(A)	VALG	14 -106
TDA2110(A)	VALG	14 -107
TDA2140	♦ALGG	21 - 30
	♦SGAI	

Figure 1-12

1 Introduction to Integrated Circuits

2. AUDIO AMPLIFIERS

IN ORDER OF (1) TOT VOLT (2) MAX IDLE POWER (3)MIN UPPER 3dB BW (4)MIN VOLT GAIN (5)TYPE

LINE No.	TYPE NO.	PWR SUP@25°C RATED SPECS 1 TOT VOLT ΔV	2 MAX IDLEP (W)	TRANSFER CHARACTERISTICS @ 25°C 3dB BANDWIDTH 3 MIN UPPER (Hz)	MAX LOWER (Hz)	4 MIN VOLT GAIN (dB)	MAX NOISE FIGURE (dB)	MAX THD (%)	ΔTEMP GAIN VAR. (dB)	INPUT @25°C MIN RESIST (Ω)	OUTPUT CHAR. @ 25°C MAX VOLTS P-P (ΔV)	MAX RESIST (Ω)	MIN VOLT P-P (ΔV)	MIN POWER (W)	LOAD RES. (Ω)	TC EO MD PE	DRAWINGS CIRCUIT	OUT LINE M=MO
1#	ESM432C	30 §	15 Ø	20 kt$	20 †	60 $	4.0 †*	2.0		100k	26		26	15	4.0	0E	4-32	4
2	4020#at	30	600M	35 k	20	6		120	0.1	34	100k	26		11	800m	75	07	
3	5004	30	90 M	200k	20	6		130	0.1		100k	26		11		75	07	
4#	ESM532C	32 §	15 Ø	20 kt$	20 †	60 $	4.0 †*	2.0			28		28	10 †	8.0	0E	5-32	SL11c
5#	AN272	34	6.0	30 k	30		38	60	1.0	2.0	50 k	20		6.3	5.0	8.0	16	2-72	DL10a
6#	TDA2020A82	34	25 Ø	160k	10		100†		1.0		30	5.0M†	44	4.0 †		15	4.0	4E	20-20	QL14s
7#	TDA2020A92	34	25 Ø	160k	10		100†		1.0		30	5.0M†	44	4.0 †		15	4.0	4E	20-20	TB14d
8#	TDA2020AC2	34	25 Ø	160k	10		100†		1.0		30	5.0M†	44	4.0 †		15	4.0	4E	20-20	TB14d
9#	TDA2020D2	34	25 Ø	160k	10		100†		1.0		30	5.0M†	44	4.0 †		15	4.0	4E	20-20	TB14d
10#	AN370	35		210m	100k$		90					130k	42		20			27	3-60	SL7c
11	uPC1023H	35 *		210m	100k$	10	87 Ø	1.8 †	15 m†		78 k			20			27	1-023	SL7b
12	uPC1024H	35 *		210m	100k$	10	87 Ø	1.8 †	15 m†		78 k			20			27	1-024	SL7b
13#	ML120	35 *		250mZ			70	900m†*						19		47 k	17	1-20l	SL5l
14#	M5212L	35 *		450mZ			87	1.5 *						19		47 k	17	52-12	SL7f
15#	TDA2612	35 *	2.5 Ø	16 k	40			72 †$	1.0 †					10		4.0	F	2-612	DL16ae
16#	TDA1013	35	5.5 Δ				30 †		.50 †		200k			4.5 †	6.0		17	1-013	SL9a
17#	TDA1512	35	13 †				74 †Ø		10 †		100k	16 mΔ	1.0k	.10	13	4.0	27	A370	SL9j
18#	L045T9	36	30 m				83	120 Ø	3.0		2.0M†	24	75 †	8	26 m		28	4-5T9	TO99
19	MA-80391CP	36	576m§	55 kØ			60 Ø		10 m†			72		58			57	3-91	CHØ
20#	TDA2030H	36	800m†	140	10		90 †	3.0 *†	500m		5.0M†				8.0	4.0	4E	
21#	TDA2030V	36	800m†	140	10		90 †	3.0 *†	500m		5.0M†				8.0	4.0	4E	
22#	SE540H	40		500k†			80 %	.50	.10 †	20 k†						300	5C	5-40h	CNØ
23#	TBA810ACB	40	200m†	20	40		80 *†	2.0 *†	300m†		5.0M†				5.5	4.0	28	8-10	TB12E
24#	TBA810CB	40	200m†	20	40		80 *†	2.0 *†	300m†		5.0M†				5.5	4.0	28	8-10	TB12F
25	NE540H	40		800m	100k§		70 %		1.0	100m†	20 k				10		07	5-40h	CN1g
26	NE540L	40		800m	100k§		70 %		1.0	100m†	20 k				250		07	5-40h	CN10Hb
27#	TDA2003H	40		1.2	15	40	80 †		1.0 *†		150k†				10	2.0	4E		SL5e
28#	TDA2003V	40		1.2	15	40	80 †		1.0 *†		150k†				10	2.0	4E		SL5e
29#	TDA2020DA82	40	25 Ø	160k	10		100†Ø	4.0 *†	10		5.0M†				15	4.0		20-20	QL14s
30#	TDA2020DA92	40	25 Ø	160k	10		100†Ø	4.0 *†	10		5.0M†				15	4.0		20-20	QL14s
31#	TDA2020DAC2	40	25 Ø	160k	10		100†Ø	4.0 *†	10		5.0M†				15	4.0		20-20	TB14d
32#	TDA2020DAD2	40	25 Ø	160k	10		100†Ø	4.0 *†	10		5.0M†				15	4.0		20-20	TB14d
33#	M5213L	42 §	450mZ				87	1.5 *						33			47 k	17	52-13	SL7f
34#	HA1397#at	44	456m	120k§	5.0		88 †	500mΔ	200m	38 †	55 k†	30 Δ	2.3	4.0 †	15	8.0	26	13-97	
35#	uPC1188H	44	30 Ø	20 k	20		40 †	1.0m&	1.0		56 k§				20 †	8.0	27	11-88	SL10j
36	MA332CP#ai	48	432m	30 M%			94 %	5 .$	2 m.					40			07	3-32	DL8Ø

12. MANUFACTURERS' CODES, NAMES & ADDRESSES

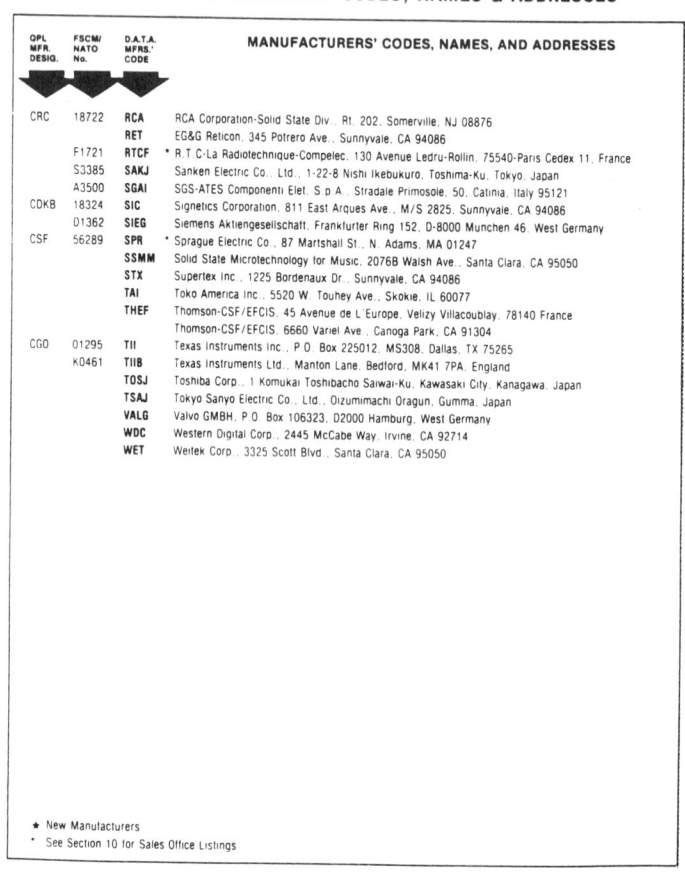

QPL MFR. DESIG.	FSCM/ NATO No.	D.A.T.A. MFRS.' CODE	MANUFACTURERS' CODES, NAMES, AND ADDRESSES
CRC	18722	RCA	RCA Corporation-Solid State Div., Rt. 202, Somerville, NJ 08876
		RET	EG&G Reticon, 345 Potrero Ave., Sunnyvale, CA 94086
	F1721	RTCF	* R.T.C-La Radiotechnique-Compelec, 130 Avenue Ledru-Rollin, 75540-Paris Cedex 11, France
	S3385	SAKJ	Sanken Electric Co., Ltd., 1-22-8 Nishi Ikebukuro, Toshima-Ku, Tokyo, Japan
	A3500	SGAI	SGS-ATES Componenti Elet. S.p.A., Stradale Primosole, 50, Catinia, Italy 95121
CDKB	18324	SIC	Signetics Corporation, 811 East Arques Ave., M/S 2825, Sunnyvale, CA 94086
	D1362	SIEG	Siemens Aktiengesellschaft, Frankfurter Ring 152, D-8000 Munchen 46, West Germany
CSF	56289	SPR	* Sprague Electric Co., 87 Martshall St., N. Adams, MA 01247
		SSMM	Solid State Microtechnology for Music, 2076B Walsh Ave., Santa Clara, CA 95050
		STX	Supertex Inc., 1225 Bordenaux Dr., Sunnyvale, CA 94086
		TAI	Toko America Inc., 5520 W. Touhey Ave., Skokie, IL 60077
		THEF	Thomson-CSF/EFCIS, 45 Avenue de L'Europe, Velizy Villacoublay, 78140 France
			Thomson-CSF/EFCIS, 6660 Variel Ave., Canoga Park, CA 91304
CGO	01295	TII	Texas Instruments Inc., P.O. Box 225012, MS308, Dallas, TX 75265
	K0461	TIIB	Texas Instruments Ltd., Manton Lane, Bedford, MK41 7PA, England
		TOSJ	Toshiba Corp., 1 Komukai Toshibacho Saiwai-Ku, Kawasaki City, Kanagawa, Japan
		TSAJ	Tokyo Sanyo Electric Co., Ltd., Oizumimachi Oragun, Gumma, Japan
		VALG	Valvo GMBH, P.O. Box 106323, D2000 Hamburg, West Germany
		WDC	Western Digital Corp., 2445 McCabe Way, Irvine, CA 92714
		WET	Weitek Corp., 3325 Scott Blvd., Santa Clara, CA 95050

* New Manufacturers
* See Section 10 for Sales Office Listings

Figure 1-14

1-6 Sources of Data on Integrated Circuits

Now go to page 10 (Figure 1-13) and find entry number 31. This tells us we have a 40 V, 25 W power amplifier with a frequency response from 10 Hz to 160 kHz. The pinout is referred to drawings 20-20 and TB14d. Figure 1-15 shows both of these drawings. This device comes in a 14 pin DIP specially designed for mounting on a heatsink.

There are two procedures for locating a replacement. The first is to return to page 10 (Figure 1-13) and check all the entries under 40 V for a device with similar characteristics. There are several: the TDA2010B82, the TDA2010B92, and the TDA2010BD2. Only the TDA2010BD2 has an identical package, although the others are similar and could be fit into the circuit. All four are made by the same company, and if it is not possible to find a TDA2020DAC2 replacement, then likely the alternatives would also be hard to find.

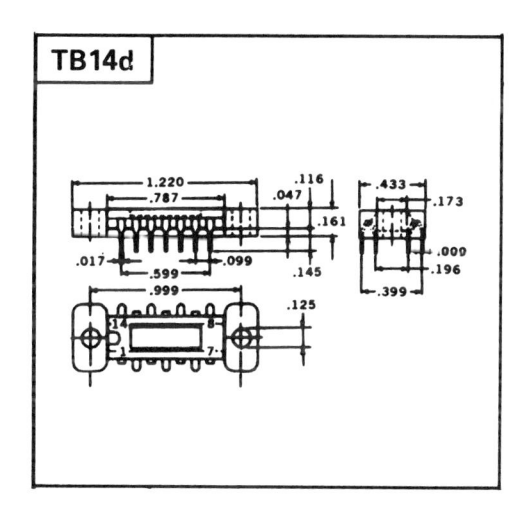

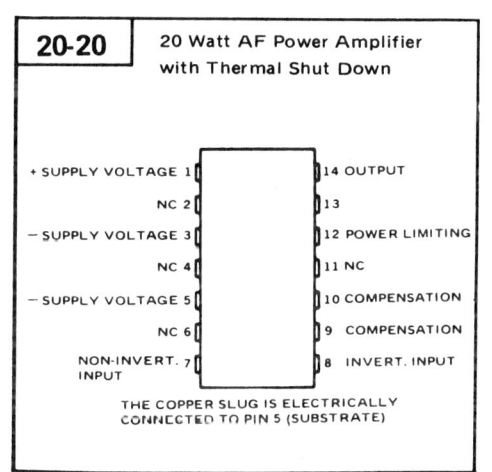

Figure 1-15

The second approach is to refer to the Phillips *ECG Semiconductor Master Replacement Guide*. The ECG catalog lists a wide variety of replacements for many integrated circuits. Search through the cross-reference in the back half of the replacement guide to find the TDA2020DAC2. A portion of the proper page is shown as Figure 1-16 with the TDA2020DAC2 identified. The replacement is ECG1287. The pinout and outline of the replacement can be found in the first part of the catalog if necessary. If you check, you will find that the ECG1287 has a different package, and its ratings are a little different from the TDA2020DAC2, but it should be a suitable replacement in most cases.

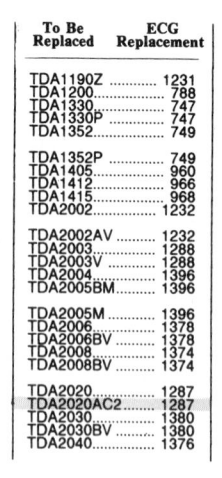

To Be Replaced	ECG Replacement
TDA1190Z	1231
TDA1200	788
TDA1330	747
TDA1330P	747
TDA1352	749
TDA1352P	749
TDA1405	960
TDA1412	966
TDA1415	968
TDA2002	1232
TDA2002AV	1232
TDA2003	1288
TDA2003V	1288
TDA2004	1396
TDA2005BM	1396
TDA2005M	1396
TDA2006	1378
TDA2006BV	1378
TDA2008	1374
TDA2008BV	1374
TDA2020	1287
TDA2020AC2	1287
TDA2030	1380
TDA2030BV	1380
TDA2040	1376

Figure 1-16

Summary

In this introductory chapter, we saw how the integrated circuit developed from the crude two-transistor phase-shift oscillator produced by Jack Kilby at Texas Instruments in 1958 to the wide spectrum of modern devices, some containing many thousands of transistors, capable of performing almost any operation. Some of these modern devices are digital, operating between two discrete states. Others are linear or analog, operating over a continuous range of signal levels. This text will be concerned only with the linear integrated circuits.

Linear integrated circuits come in a variety of packages, with a number of identifying symbols and logos. The first major step in our study of linear integrated circuits was to learn how to use manufacturers' data books, selection guides, cross-reference indices, and replacement guides for identifying individual integrated circuits and locating information about them, for selecting suitable devices to satisfy specific applications, and for finding suitable equivalent or replacement devices.

Now that linear integrated circuits have been thoroughly introduced, we will proceed in the next chapter to study the first, and most fundamental, of the many integrated circuits to be encountered in this book—the operational amplifier.

Review Questions

1-1 Historical development of integrated circuits

1. Who invented the transistor?
2. Who made the first true integrated circuit?
3. Who developed the planar process?
4. What process is most widely used in producing integrated circuits?
5. What types of devices are hard to fabricate as part of an integrated circuit?

6. What are the advantages of MOSFET technology?
7. What was the first successful linear integrated circuit, and who produced it?
8. What type of material is used for the construction of most linear integrated circuits? What material is used for very high frequency integrated circuits?

1-2 Digital versus linear integrated circuits

9. How is it best to define a digital integrated circuit?
10. What functions are considered to be the domain of linear integrated circuits?

1-3 Packaging of integrated circuits

11. Name the most common types of packages for linear integrated circuits.
12. In what type of case would a 5 W power amplifier most likely be packaged?
13. In what number of pins are DIP integrated circuits found? TO-5 integrated circuits?
14. How are the pins identified and numbered in a DIP? TO-5?
15. What is the standard spacing between pins in a DIP?
16. What type of devices are found in SIPs?
17. What is the functional pin configuration of any type of operational amplifier packaged in an 8 pin DIP?

1-4 Manufacturers of linear integrated circuits

18. Approximately how many companies manufacture linear integrated circuits?
19. What former major linear integrated circuit manufacturing companies are no longer independent companies?
20. Why might it be important to know the manufacturer of a specific linear integrated circuit?

1-5 Identification of linear integrated circuits

21. What is the most direct way of determining the manufacturer of some unknown integrated circuit found on a circuit board?
22. The LF351 is an operational amplifier. Which company developed this device? From the supplied information, is it possible to tell which company manufactured it?
23. What order is used for numbering integrated circuits?
24. What two numbers usually appear on the case of an integrated circuit? How can the correct device number be determined?
25. The date code on a linear integrated circuit package reads 9123. When was the device manufactured?
26. What standard is used for adding suffix letters to a device number?

1-6 Sources of data on integrated circuits

27. What four sources are used to locate data on linear integrated circuits?
28. What sort of data is presented in a manufacturer's data book?
29. For what would a manufacturer's selection guide be used?
30. How does a replacement guide differ from a D.A.T.A.BOOK?
31. What information can be obtained from a D.A.T.A.BOOK?
32. For what purpose other than simply finding replacements can a replacement guide be used?

2 Operational Amplifiers

OUTLINE

2-1 Amplifier Characteristics
2-2 Differential Amplifiers
2-3 Characteristics of Operational Amplifiers
2-4 Feedback in Operational Amplifiers
2-5 Bipolar Integrated Circuit Operational Amplifiers
2-6 BiFET Integrated Circuit Operational Amplifiers
2-7 Uncompensated Operational Amplifiers

KEY TERMS

Amplifier
AC coupling
DC or direct coupling
Amplification
Gain
Decibel
Input impedance
Output impedance
Frequency response
Bandwidth
Differential amplifier
Common-mode gain
Common-mode rejection ratio
Input bias current
Input offset current
Output offset voltage
Operational amplifier
Analog computer
Feedback
Slew rate
Uncompensated operational amplifier
Feedforward compensation

OBJECTIVES

On completion of this chapter, the reader should be able to:

- define an amplifier and determine the classification of an amplifier.
- determine various amplifier characteristics, including voltage gain, current gain, power gain, input impedance, output impedance, the −3 dB bandwidth, and the unity-gain bandwidth.
- describe the operation of a differential amplifier and determine various characteristics of a differential amplifier, including input bias current, input offset current, offset voltage, differential gain, common-mode gain, and common-mode rejection ratio.
- understand the concept of feedback and explain how feedback is used to modify and improve the characteristics of operational amplifiers.

- explain the difference between bipolar operational amplifiers and BiFET operational amplifiers.
- understand the need for frequency compensation, and where external frequency compensation is preferred over internal compensation.

Introduction

One of the most fundamental devices in analog electronic circuits is the amplifier. Amplifiers are used in most areas of electronics. They are used in radio transmitters to amplify microphone inputs of a few milliwatts to output powers of many kilowatts. Radio or television receivers use amplifiers to increase input signal levels of a few microwatts to levels of several watts sufficient for driving speakers and the picture tube. Test instruments such as oscilloscopes and recorders must amplify very small input signals to levels sufficient for driving their displays. Function generators have amplifiers to produce and shape the various waveforms and to control the level of the output signal. Power supplies use amplifiers in their regulator circuits. Even digital systems use amplifiers for buffers and drivers.

It should not be surprising, then, to learn that many of the first integrated circuit chips on the market were amplifiers, nor to learn that even now the most common integrated circuits are amplifiers. Amplifiers are a logical place to begin the study of linear integrated circuits, not only because they are the most common form of integrated circuit, but also because they form a building block for the development of more complicated circuits.

Simple transistor amplifiers generally include capacitors to block DC bias currents. Capacitors are difficult to incorporate into integrated circuits; hence, a special type of amplifier—the differential amplifier—is used in integrated circuits to eliminate the need for capacitors and to avoid problems of drift. Because differential amplifiers were once used to simulate mathematical operations, they are often called operational amplifiers.

The operational amplifier, or more simply, op-amp, is a very flexible device. It can be used as a simple amplifier, as a differential amplifier, as an instrumentation amplifier, or as a current amplifier. It can be used to sum, multiply, square, or take the logarithm of analog signals. It forms the heart of many filtering circuits, oscillators, and regulators.

In this chapter, the characteristics of operational amplifiers will be developed, and several common integrated circuit operational amplifiers will be introduced. The many applications of operational amplifiers will be left for later chapters.

2-1 Amplifier Characteristics

An **amplifier** is a circuit or device that takes some input signal and outputs the same signal, unchanged in shape, but increased in voltage, and/or current, and power. Amplifiers are widely used in audio systems, radio and television receivers, communication equipment, electronic test equipment, and process instrumentation applications.

Amplifier Classification

Amplifiers are usually classified by their specific application. There are two main classes of amplifier: the *radio frequency* (or *RF*) *amplifier*, operating at frequencies typically greater than 100 kHz, and the *audio frequency* (or *AF*) *amplifier* operating at frequencies typically less than 100 kHz. Radio frequency amplifiers are further subdivided into *broadband amplifiers* and *tuned* or *narrow band amplifiers*. Broadband RF amplifiers are sometimes called *video amplifiers*. Audio frequency amplifiers are subdivided into *audio amplifiers, operational amplifiers*, and *instrumentation amplifiers*. You may encounter other names, but all will fit into these two main classes. A block diagram illustrating the general classification of amplifier types is shown in Figure 2-1.

Amplifiers are also described by their construction. Discrete component amplifiers are usually **AC-coupled**; that is, they are designed to pass AC signals only. Audio frequency amplifiers are usually capacitively coupled—the input signal is coupled into the amplifier through a capacitor, it is coupled from stage to stage through capacitors, and it is coupled to the load through a capacitor. AC signals can also be coupled from one stage to the next by the use of transformers. Audio frequency amplifiers usually do not use transformers because, for low frequencies, transformers are large, heavy, and expensive. Radio frequency amplifiers, however, generally do use transformer coupling, since at high frequencies transformers are small with only a few windings and are inexpensive. More importantly, transformers can be tuned to pass only a narrow band of frequencies, making them ideal for channel or station selection.

Capacitor or transformer coupling is necessary to prevent DC biasing signals of one stage from interfering with adjacent stages. There are two disadvantages to AC coupling: AC-coupled amplifiers cannot be used to amplify DC or low-frequency AC signals, and AC-coupled amplifiers are difficult to make as integrated circuits since coupling capacitors and transformers are difficult to fabricate

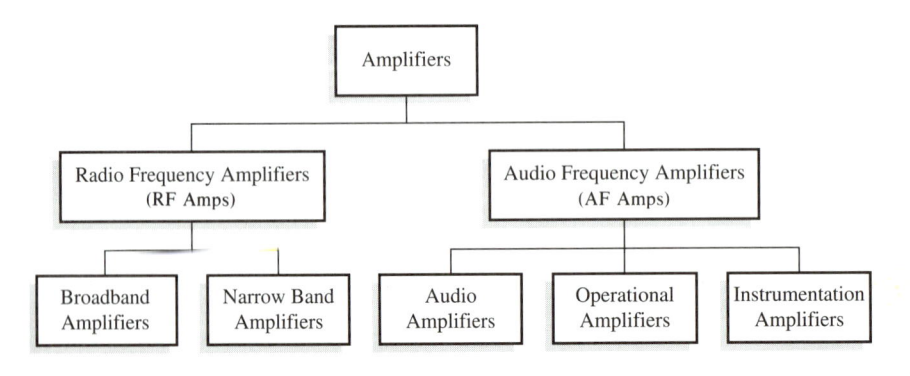

Figure 2-1 Classification of amplifier types.

onto integrated circuit chips. They are usually included as external circuit elements, making the overall circuit bulky and complex.

Amplifiers which are produced on integrated circuit chips are **direct-coupled**. They use a more complex circuit consisting entirely of resistors, diodes, and transistors to overcome the bias problems that make the use of capacitors or transformers necessary in discrete component amplifiers. This more complex circuit is usually too expensive for use in discrete component circuits, but is easy and inexpensive to fabricate onto an integrated circuit chip. Because there are no capacitors or transformers to block DC signal levels, direct-coupled amplifiers can pass all low-frequency signals and DC.

The actual circuits used for direct-coupled amplifiers and integrated circuit amplifiers will be introduced in the next section, but before we look at integrated circuit amplifiers it is necessary to describe basic amplifier characteristics and introduce some important terminology.

Schematic Symbol

First we will introduce the symbol that we will be using for an amplifier. The symbol for a simple amplifier, shown in Figure 2-2(a), is an isosceles triangle on

(a) Simple amplifier with power supply connections shown

(b) Usual method of drawing a simple amplifier; power leads are omitted

(c) Differential amplifier with power supply connections shown

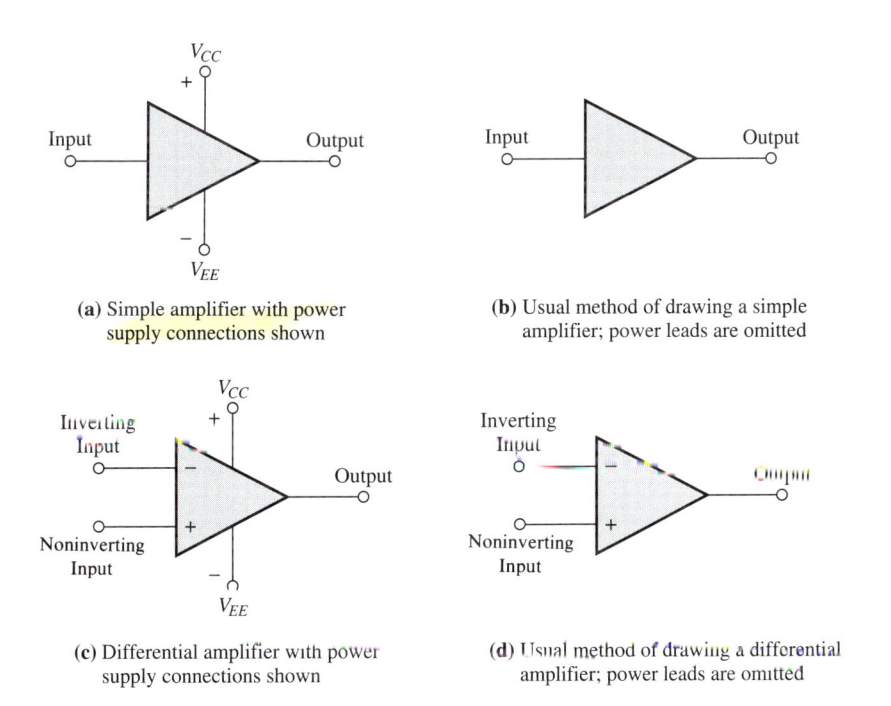

(d) Usual method of drawing a differential amplifier; power leads are omitted

Figure 2-2 Schematic symbols for simple amplifiers and differential amplifiers.

its side. The input is to the base of the triangle, and the output is from the apex. Power supply connections are shown going to the sides of the triangular symbol. Frequently, power supply connections are omitted as in Figure 2-2(b).

As we shall see in the next section, most integrated circuit amplifiers are differential amplifiers with two inputs. The symbol for a differential amplifier is shown in Figure 2-2(c). Note that there are two inputs and a dual power supply. The power supply connections are shown going to the sides of the amplifier symbol, but, as in the case of the simple amplifier, they are frequently omitted in actual schematic diagrams, and the symbol shown in Figure 2-2(d) is normally used. *Do not forget to connect the power supply when assembling the circuit even if the leads are not shown in the schematic.*

Amplifier Gain

The basic function of an amplifier is to produce a larger output for a given input. This is called **amplification**. The amount that the input is amplified is called the **gain** of the amplifier. Since an amplifier can amplify voltage, current, or power, there are three ways of defining gain, namely as *voltage gain, current gain*, and *power gain*. Each is defined in a similar manner. The voltage gain A_V is defined as the ratio of the output voltage signal amplitude to the input voltage signal amplitude.

$$A_V = \text{voltage gain} = \frac{\text{output voltage amplitude}}{\text{input voltage amplitude}} = \frac{V_{out}}{V_{in}} \qquad (2\text{-}1)$$

The current gain A_I is defined as the ratio of the output current signal amplitude to the input current signal amplitude.

$$A_I = \text{current gain} = \frac{\text{output current amplitude}}{\text{input current amplitude}} = \frac{I_{out}}{I_{in}} \qquad (2\text{-}2)$$

The power gain A_P is defined as the ratio of the output power to the input power.

$$A_P = \text{power gain} = \frac{\text{output power amplitude}}{\text{input power amplitude}} = \frac{P_{out}}{P_{in}} \qquad (2\text{-}3)$$

Since the power is simply the voltage times the current, the power gain can be written

$$A_P = \text{power gain} = \frac{V_{out} \times I_{out}}{V_{in} \times I_{in}} = A_V \times A_I \qquad (2\text{-}4)$$

The power gain is simply the product of the voltage gain and the current gain. We will see later that the current gain is also related to the voltage gain.

The voltage, current, and power gains are sometimes expressed as **decibels** (dB). A decibel is simply 20 times the logarithm (base 10) of the voltage or current gain, or 10 times the logarithm of the power gain. The definitions of gain in terms of decibels are

$$\text{dB voltage gain} = 20 \log \frac{V_{out}}{V_{in}} \qquad (2\text{-}5)$$

$$\text{dB current gain} = 20 \log \frac{I_{out}}{I_{in}} \qquad (2\text{-}6)$$

$$\text{dB power gain} = 10 \log \frac{P_{out}}{P_{in}} \qquad (2\text{-}7)$$

Note the factor of 10 in the power gain as opposed to the factor of 20 in the current and voltage gains.

It is not always possible to define a specific gain for an amplifier. For example, a **MOSFET** transistor amplifier or a MOSFET integrated circuit amplifier draws virtually no current from the source, yet provides current to the load. The current gain and, hence, the power gain would be infinite in this case. Such an amplifier is described as a voltage amplifier, with current and power gain left undefined. Sometimes, the voltage gain is less than 1. For example, a common-collector transistor amplifier always has a voltage gain less than 1. It does have an appreciable current gain and a power gain. Such an amplifier is described as a power amplifier. Even though gain may be less than 1, it is still called gain.

Input and Output Impedance

The next quantities that we will define are the **input impedance** Z_{in} and the **output impedance** Z_{out}. If an amplifier draws any power from the source, then it has finite input impedance. The source supplies current to this impedance as if it were a resistor between the input and ground, as shown in Figure 2-3(a). This impedance should either be very much larger than the source impedance, so that it does not load the source, or matched to the source impedance for maximum power transfer and proper source termination. The type of source dictates which case applies.

If the input impedance is finite, it can be measured by placing an external source resistor R_S in series with the source and the amplifier input as shown in Figure 2-3(b). Measure the voltage V_S at the input to resistor R_S, and V_{in} at the input to the amplifier. The voltage drop across the resistor is $V_S - V_{in}$, and hence the current through the resistor is this voltage divided by the resistance; that is,

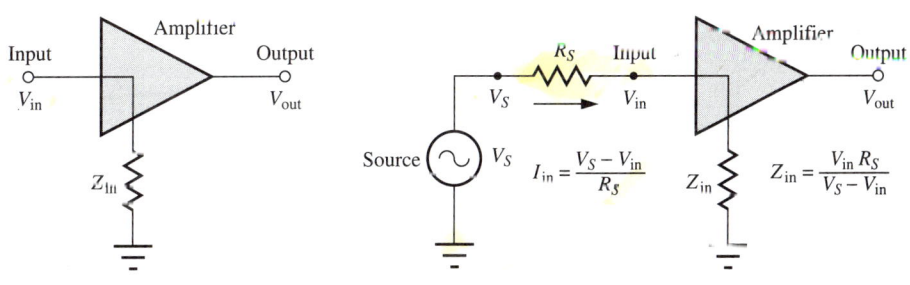

(a) Equivalent input circuit of an amplifier

(b) Determination of input impedance

Figure 2-3 Input impedance of an amplifier.

40 2 Operational Amplifiers

$I_{in} = (V_S - V_{in})/R_S$. The input impedance Z_{in} is the input voltage divided by the input current and is given by

$$Z_{in} = R_S \times \frac{V_{in}}{V_S - V_{in}} \tag{2-8}$$

The output of an ideal amplifier acts as a voltage source with $V_{out} = A_V \times V_{in}$. In practice, no amplifier is ideal, and there is always some resistance in series with the output. This resistance is called the output impedance Z_{out}. The output circuit, consisting of source and output impedance, is shown in Figure 2-4(a). Normally, the amplifier would be driving a load R_L as shown in Figure 2-4(b).

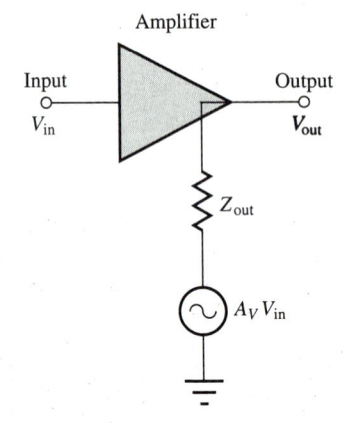

(a) Equivalent output circuit of an amplifier

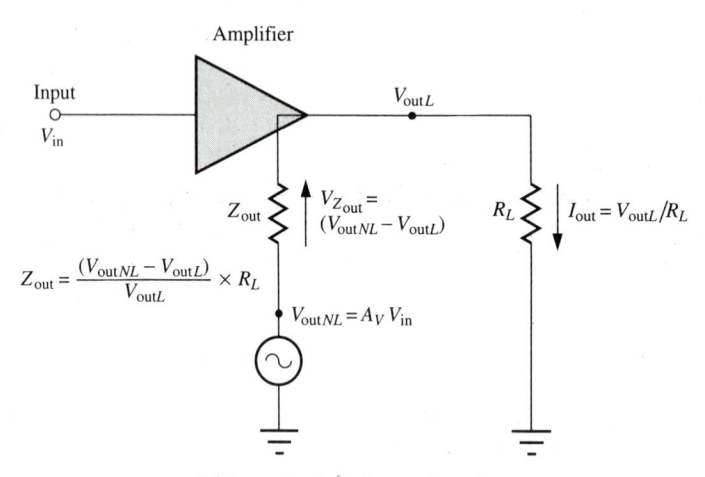

(b) Determination of output impedance

Figure 2-4 Output impedance of an amplifier.

The output impedance should either be very much smaller than the load impedance, so that power dissipated in the output impedance does not significantly reduce the output voltage of the amplifier to the load, or matched to the load impedance for maximum power transfer or proper amplifier termination. The type of load dictates which is the case.

If the output impedance is nonzero, it can be measured by the following procedure. Measure V_{outL}, the output voltage of the amplifier with the load attached. Remove the load and measure V_{outNL}, the no-load output voltage. The difference in output voltage between loaded and unloaded conditions, $V_{outNL} - V_{outL}$, is due to the voltage drop across the internal impedance Z_{out} when current is being drawn by the load. The current through the internal impedance is the same as the current through the load and hence is equal to V_{outL}/R_L. The output impedance is simply the voltage drop across the impedance divided by the current through it and is given by the formula

$$Z_{out} = R_L \times \frac{V_{outNL} - V_{outL}}{V_{outL}} \tag{2-9}$$

We can combine the information in Figures 2-3 and 2-4 to draw the equivalent circuit of an amplifier. This equivalent circuit is shown as Figure 2-5. If we are doing any calculations or circuit analysis, we simply replace the amplifier by this equivalent circuit. Figure 2-5 shows the equivalent circuit of an amplifier consisting of an input resistance Z_{in} connected between input and ground and a voltage source $A_V V_{in}$ in series with an output resistance Z_{out} connected between output and ground.

With the definition of input impedance and output impedance, it is now possible to determine the relationship between the voltage gain and the current gain. The input current is equal to the input voltage divided by Z_{in}. The output current

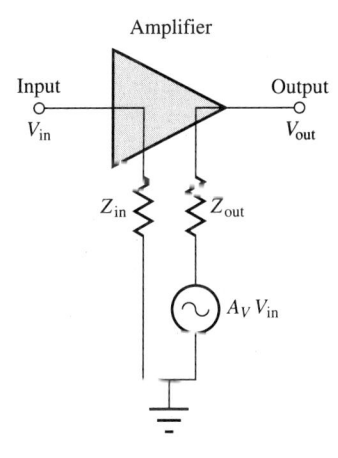

Figure 2-5 Equivalent circuit of an amplifier.

42   2   Operational Amplifiers

is equal to the output voltage divided by the load resistance R_L. Hence, the current gain can be written as

$$A_I = \frac{I_{out}}{I_{in}} = \frac{V_{out}/R_L}{V_{in}/Z_{in}} = \frac{Z_{in}}{R_L} \times \frac{V_{out}}{V_{in}} = \frac{Z_{in}}{R_L} \times A_V \qquad (2\text{-}10)$$

The current gain is equal to the voltage gain times the ratio of the input impedance to the load impedance. Since the power gain is equal to the voltage gain times the current gain, the power gain can be written as

$$A_P = A_I \times A_V = \frac{Z_{in}}{R_L} \times A_V^2 \qquad (2\text{-}11)$$

We need only know the voltage gain, the load, and the input impedance. Current gain and power gain are derivable from these. Throughout the remainder of this book, unless stated otherwise, gain will always refer to voltage gain.

EXAMPLE 2-1

The operating characteristics of an audio preamplifier are to be determined. The test circuit is shown in Figure 2-6. A function generator set at a frequency of 1000 Hz is connected to the input of the amplifier through a 10 kΩ resistor. The voltage measured at the input to the resistor is 95.2 mV RMS. The voltage measured at the input to the amplifier is 58.3 mV RMS. The output is loaded with a 3.3 kΩ resistor. The output voltage is measured to be 4.27 V RMS. When the load resistor is removed, the output voltage is measured to be 7.35 V RMS. Calculate the following quantities:

(a) the voltage gain
(b) the input impedance
(c) the output impedance
(d) the current gain
(e) the power gain.

Solution

The test circuit shown in Figure 2-6 and the measurements obtained demonstrate a typical procedure for determining the characteristics of an audio amplifier.

(a) Calculate the voltage gain, using Equation 2-1:

$$A_V = \frac{V_{out}}{V_{in}} = \frac{4.27 \text{ V}}{0.0583 \text{ V}} = 73.24$$

Note that we used the voltage at the input to the amplifier as V_{in}, and not the voltage at the input to the series input resistor. We used the loaded output voltage for V_{out}.

2-1 Amplifier Characteristics

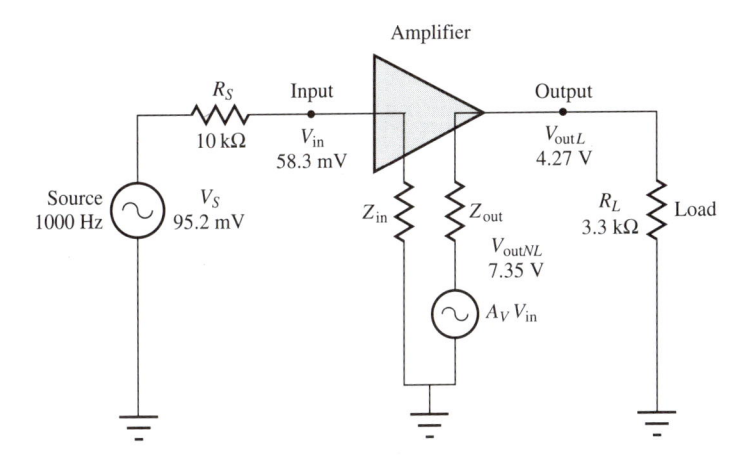

Figure 2-6

(b) Calculate the input impedance Z_{in}, using Equation 2-8:

$$Z_{in} = R_S \times \frac{V_{in}}{V_S - V_{in}} = 10 \text{ k}\Omega \times \frac{58.3 \text{ mV}}{95.2 \text{ mV} - 58.3 \text{ mV}} = 15.8 \text{ k}\Omega$$

(c) Calculate the output impedance Z_{out}, using Equation 2-9:

$$Z_{out} = R_L \times \frac{V_{outNL} - V_{outL}}{V_{outL}}$$

$$= 3.3 \text{ k}\Omega \times \frac{7.35 \text{ V} - 4.27 \text{ V}}{4.27 \text{ V}} = 2.38 \text{ k}\Omega$$

(d) Calculate the current gain, using Equation 2-10:

$$A_I = \frac{Z_{in}}{R_L} \times A_V = \frac{15.8 \text{ k}\Omega}{3.3 \text{ k}\Omega} \times 73.24 = 350.7$$

(e) Calculate the power gain, using either Equation 2-11 or Equation 2-4. We will use Equation 2-4.

$$A_p = A_V \times A_I = 73.24 \times 350.7 = 25{,}680$$

Our analysis of the amplifier is completed. Normally, we do not need to find the current gain or the power gain. We could also convert the voltage, current, and power gains into decibels by Equations 2-5, 2-6, and 2-7, respectively, if required. This is left as an exercise for the reader.

Amplifier Frequency Response

The next characteristic of importance is the amplifier **frequency response**. The frequency response of an amplifier is a description of how the gain changes with the input signal frequency. Ideally, an amplifier should have the same gain for all frequencies. In any real amplifier, however, the gain always drops off for high frequencies because of signal loss through stray capacitance. For an AC-coupled amplifier, the gain will also drop off for very low frequencies because of capacitive reactance. The typical frequency response curves for both AC-coupled and direct-coupled amplifiers are shown in Figure 2-7.

(a) Response curve for an AC coupled amplifier

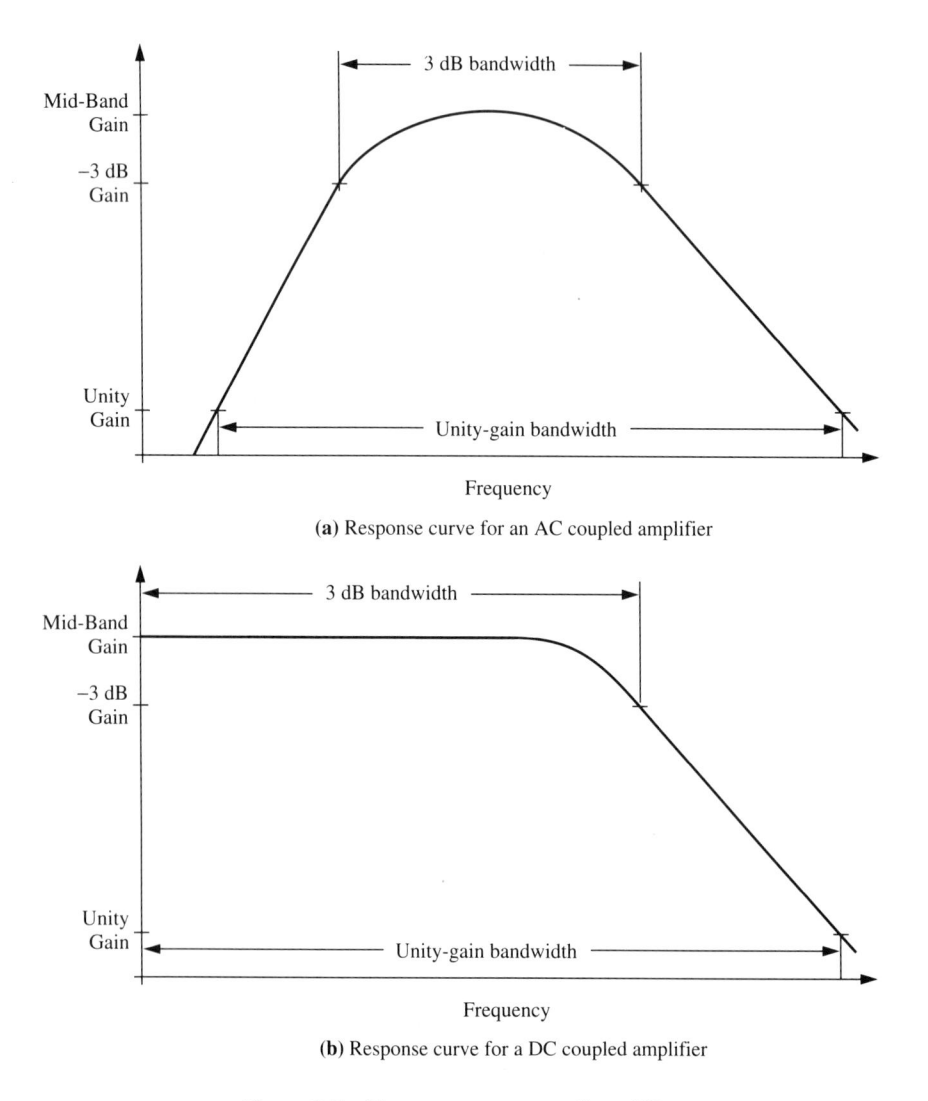

(b) Response curve for a DC coupled amplifier

Figure 2-7 **Frequency response of amplifiers.**

The frequency response of an amplifier is generally described by the **bandwidth**. The bandwidth is a measure of the range of frequencies over which the amplifier will have a gain greater than some specified value. Generally, amplifier frequency response is described by the −3 *dB bandwidth*, meaning that the voltage gain is greater than 0.707 of the midband gain. If a voltage ratio of 0.707 is converted to decibels, the result is close to −3 dB, hence the point where the gain is 0.707 of the midband gain is known as the −3 *dB point*. It is also known as the *half-power point*, since a power ratio of 0.5 also converts to −3 dB. Sometimes the amplifier frequency response is described by the *unity-gain bandwidth* instead of the −3 dB bandwidth. The unity-gain bandwidth is the range of frequencies for which the gain is greater than 1.

If the amplifier is direct-coupled, it can amplify all frequencies down to zero (that is, DC), and hence the bandwidth is simply the upper −3 dB frequency or the upper unity-gain frequency, depending on which definition is being used. This will be the case for most integrated circuit amplifiers. If the amplifier is AC-coupled, then the bandwidth is the upper frequency minus the lower frequency. The definitions of −3 dB bandwidth and unity-gain bandwidth are illustrated in Figure 2-7 for AC-coupled and direct-coupled amplifiers.

For frequencies greater than the upper −3 dB frequency (and lower than the lower −3 dB frequency for AC-coupled amplifiers), the gain generally drops uniformly. The term *roll-off* is used to describe this drop in gain. For a single-transistor AC amplifier, the high-frequency gain usually drops, or rolls off, by a factor of 100 for an increase in frequency of 10. In decibels this is described as a roll-off of 40 dB per decade, a decade being a frequency ratio of 10. For op-amps, as we will see shortly, the roll-off is 20 dB per decade, and the gain drops by a factor of 10 as the frequency increases by a factor of 10. We will discuss this concept of gain roll-off in more detail in Chapter 6 when we introduce filter circuits.

EXAMPLE 2-2

The frequency response of the amplifier tested in Example 2-1 was determined by measuring the input voltage V_{in} and the output voltage V_{out} for a range of frequencies. The circuit is the same one as shown in Figure 2-6. The measured values are shown in Table 2-1.

From these measurements:

(a) Calculate the voltage gain at each frequency.
(b) Calculate the voltage gain in dB for each frequency.
(c) Plot the logarithmic frequency response, that is, dB gain, against frequency on semilog paper.
(d) Determine the −3 dB bandwidth from the plot.
(e) Determine the unity-gain bandwidth from the plot.

2 Operational Amplifiers

Table 2-1 Experimental data for Example 2-2

Frequency (Hz)	V_{in} (volts)	V_{out} (volts)
20	0.100	0.031
30	0.100	0.074
50	0.100	0.195
100	0.100	0.806
200	0.100	3.12
300	0.100	5.18
500	0.100	6.68
1,000	0.100	7.33
2,000	0.100	7.33
3,000	0.100	7.33
5,000	0.100	6.72
10,000	0.100	3.51
20,000	0.100	0.880
30,000	0.100	0.389
50,000	0.100	0.141
100,000	0.100	0.034

Solution

(a) The voltage gain is calculated by Equation 2-1 for each frequency. For example, at a frequency of 1000 Hz the gain is calculated as follows:

$$A_V = \frac{V_{out}}{V_{in}} = \frac{7.330 \text{ V}}{0.100 \text{ V}} = 73.3$$

This value is in close agreement with the value for the gain found in Example 2-1. The gain values for the different frequencies are summarized in Table 2-2.

(b) The decibel voltage gain is defined by Equation 2-5. The dB gain must be calculated for each frequency. As an example, we will calculate the dB gain for a frequency of 100 Hz.

$$A_{VdB} = 20 \log \frac{V_{out}}{V_{in}} = 20 \log \frac{0.806 \text{ V}}{0.100 \text{ V}} = 18.1 \text{ dB}$$

The decibel voltage gains are summarized in Table 2-2.

(c) If we plot the dB gain against the logarithm of the frequency, we get the results shown in Figure 2-8. This is the conventional way of presenting the frequency response of an amplifier.

(d) To find the −3 dB bandwidth, we simply read from the graph the frequencies where the gain is −3 dB below the maximum or midband gain. Since the midband gain is 37.3 dB, the −3 dB frequencies occur when the gain is 37.3 − 3 = 34.3 dB. The lower frequency is approximately 300 Hz, and the upper

Table 2-2 Summary of the calculations performed in Example 2-2

Frequency (Hz)	V_{in} (volts)	V_{out} (volts)	Voltage Gain	dB Gain
20	0.100	0.031	0.310	−10.2
30	0.100	0.074	0.740	−2.60
50	0.100	0.195	1.95	5.80
100	0.100	0.806	8.06	18.1
200	0.100	3.12	31.2	29.9
300	0.100	5.18	51.8	34.3
500	0.100	6.68	66.8	36.5
1,000	0.100	7.33	73.3	37.3
2,000	0.100	7.33	73.3	37.3
3,000	0.100	7.33	73.3	37.3
5,000	0.100	6.72	67.2	36.5
10,000	0.100	3.51	35.1	30.9
20,000	0.100	0.880	8.80	18.9
30,000	0.100	0.389	3.89	11.8
50,000	0.100	0.141	1.41	2.98
100,000	0.100	0.034	0.340	−9.37

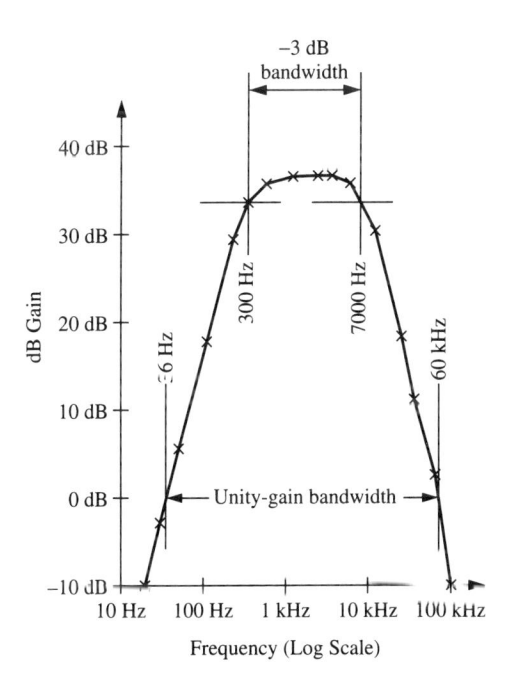

Figure 2-8

frequency is approximately 7000 Hz. The frequency range is 300 Hz to 7000 Hz, and the −3 dB bandwidth is 6300 Hz. In this example, the bandwidth is typical of a telephone amplifier. A good-quality consumer stereo amplifier would have a −3 dB bandwidth from 50 Hz to at least 20 kHz.

(e) The unity-gain bandwidth is found in the same manner as the −3 dB bandwidth. A gain of 1 has a dB gain of 0, so we need to find the frequencies corresponding to 0 dB. From the graph, unity gain occurs at frequencies of approximately 36 Hz and approximately 60 kHz. The unity-gain bandwidth of the amplifier is approximately 60 kHz.

Other Amplifier Characteristics

The amplifier characteristics that we have introduced thus far are the ones that will be of most concern in this book. There are other less important characteristics, such as the *maximum peak-to-peak output voltage*, the *input sensitivity*, the *noise figure*, the *slew rate*, and the *power supply rejection ratio*. We will encounter these later and discuss them at that time.

There are also several characteristics that are unique to differential amplifiers such as the *input bias current*, the *input offset current*, the *offset voltage*, the *common-mode gain*, and the *common-mode rejection ratio*. These will be introduced after we discuss differential amplifiers.

2-2 Differential Amplifiers

Having introduced the general characteristics of amplifiers, we will now look at some specific properties of the amplifiers that are used for integrated circuits. These amplifiers are direct-coupled differential amplifiers, and are quite different from the discrete component capacitor-coupled transistor amplifiers generally encountered. To demonstrate the differences, and to appreciate the advantages of differential amplifiers, we will first review AC-coupled amplifiers and then look in detail at differential amplifiers.

AC-Coupled Amplifiers

A typical discrete component common-emitter transistor amplifier is shown in Figure 2-9. All bipolar transistors require a bias current into the base to establish the operating point. Resistors R_{B1} and R_{B2} in Figure 2-9 form a voltage divider for biasing the base of the transistor, resistor R_E is used for stabilizing the transistor against variations in β, and resistor R_C is the impedance across which the output voltage is developed. Three capacitors are necessary. Capacitor C_{in} isolates the amplifier input from the source so that any DC voltages at the source (or output of a previous stage) will not interfere with the biasing. Capacitor C_{out} isolates the amplifier output so that the collector voltage from the amplifier will

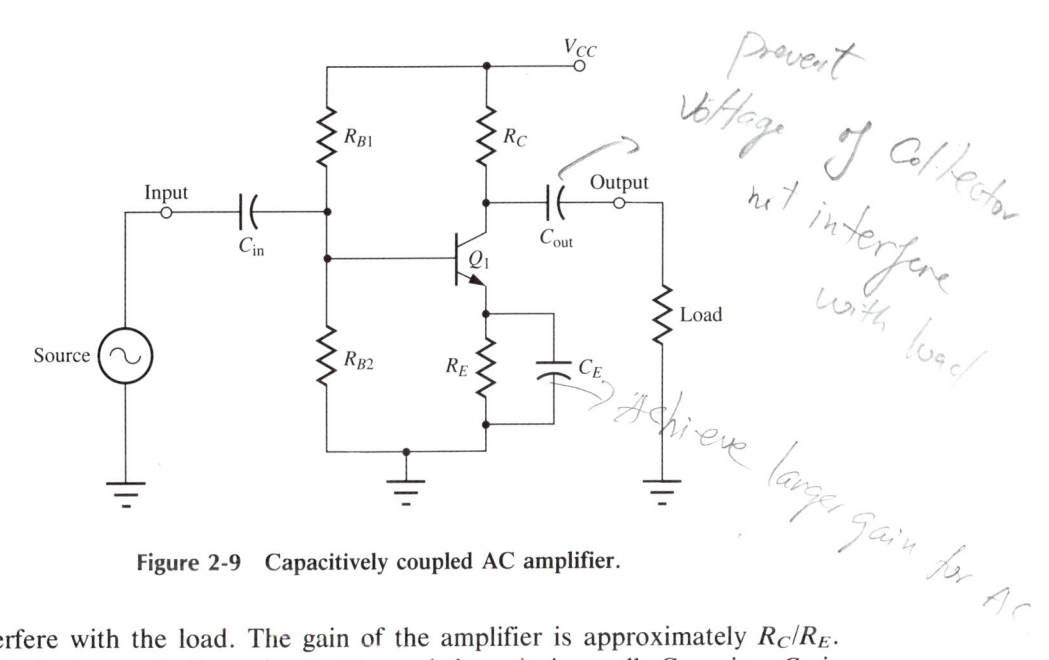

Figure 2-9 Capacitively coupled AC amplifier.

not interfere with the load. The gain of the amplifier is approximately R_C/R_E. Normally R_E has a similar value to R_C, and the gain is small. Capacitor C_E is used to bypass all or part of R_E to AC so that larger gains may be achieved.

This type of amplifier is difficult to fabricate as an integrated circuit because of the capacitors. C_{in} and C_{out} have typical values of 0.1 µF, while C_E may be as large as 100 µF. These are relatively large and bulky capacitors and would have to be added separately from the integrated circuit. There would be little advantage in integrating such a circuit.

It is possible to redesign this type of amplifier to eliminate the capacitors. When C_E is removed, the gain becomes quite small. This can be offset by using several stages of amplification. It is possible to design the stages so that the voltage levels from one stage to the next would not interfere with biasing levels, eliminating the need of coupling capacitors between stages. Input and output capacitors could be eliminated by using a separate negative power supply for biasing. Unfortunately, such an amplifier would be quite critical with regard to operating point. More importantly, it would be susceptible to DC drift. A change in any DC voltage, such as a change in transistor V_{BE} due to temperature, would be amplified as a signal by each subsequent stage and would appear as a large and variable DC offset at the output.

Differential Amplifiers

To avoid the problem of drift, it is necessary to use an entirely different concept in amplifier design—the **differential amplifier**. The circuit for a simple differential amplifier is shown in Figure 2-10. The differential amplifier consists of two identical transistors Q_1 and Q_2 connected in parallel. Bias current flows through the two base resistors R_B. The operating point is established by the single emitter resistor R_E and the negative power supply $-V_{EE}$. The output signal is developed

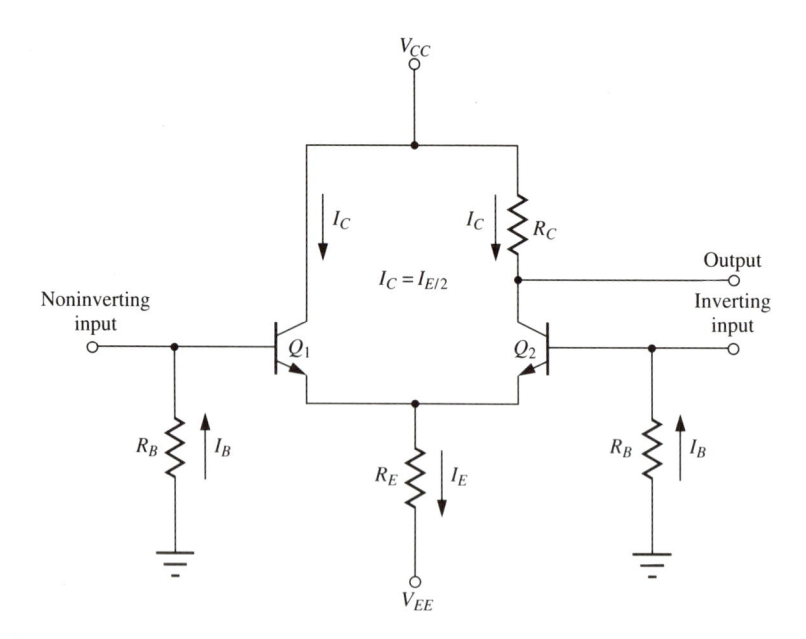

Figure 2-10 Simplified circuit of a differential amplifier.

across R_C. As we will see shortly, any DC drift in the circuit affects both transistors, yet is effectively cancelled at the output.

Notice that the differential amplifier requires a dual-polarity power supply consisting of V_{CC} and $-V_{EE}$. This is necessary to allow bias voltages at both inputs and at the output to be set to approximately zero volts. If the amplifier input and output are not at zero volts potential, capacitors must be placed at the input and output to prevent DC currents from flowing from the amplifier into the source and load. Including capacitors would take us right back to the limitations of the AC-coupled amplifier.

Operation of a Differential Amplifier

The simple differential amplifier shown in Figure 2-10 has many characteristics in common with the operational amplifiers that we will be using a little later, and a study of this differential amplifier will make understanding operational amplifiers easier.

The differential amplifier in Figure 2-10 has two almost identical amplifiers (note that the amplifier on the right has a collector resistor, whereas the one on the left does not). The amplifier on the right acts as a standard common-emitter transistor amplifier and produces an output signal 180° out of phase with the input signal. The amplifier on the left acts as a common-collector amplifier driving a common-base amplifier, and produces an output signal that is in phase with the input signal. Hence, in a differential amplifier there are two inputs, one of which is always *inverting*, and the other is always *noninverting*.

The gains of the two amplifiers are approximately equal. The output signal across R_C is the sum of the two amplified inputs, but since one is inverted with respect to the other, the effective output is the amplified difference of the two inputs. This is why the name "differential amplifier" is applied to this circuit. Any DC drift, which appears equally in both sides of the amplifier, is cancelled at the output, as is any DC bias voltage, or any noise signal present equally at both inputs.

We must qualify the statements made in the preceding paragraph—the gains of the two amplifiers are very close, but *not* exactly equal. A signal common to both inputs will produce a *small* output signal. The difference in gain between the two amplifiers is called **common-mode gain**. The measure of the ability of a differential amplifier to amplify a desired signal present at one input while rejecting a signal present at both inputs is called the **common-mode rejection ratio**, and is defined as

$$\text{common-mode rejection ratio} = 20 \log \frac{\text{single-input gain}}{\text{common-mode gain}} \quad (2\text{-}12)$$

The common-mode gain is the difference in gain between the two amplifier inputs. The single-input gain is the average gain for a signal applied to either input and is usually called the *differential gain*.

Because the differential amplifier will amplify any difference between the two inputs, we must be careful not to introduce any differences as we set DC bias levels. In a differential amplifier the collector currents in each transistor are identical. For collector current to flow, base current must flow into each transistor through resistor R_B connected between the base and ground. Because the emitter is biased negatively with respect to ground by $-V_{EE}$, grounding the base through R_B biases the base positively with respect to the emitter and establishes proper transistor biasing.

Recall that the collector current of a transistor is β times the base current. If the two transistors have identical values of β, then the bias currents for each transistor are equal since the collector currents are the same. The bias current for each transistor produces a small *negative* voltage at the base because of the voltage drop across the base resistor due to the flow of base current. If the base resistors are equal in value, the voltage at the base of each transistor is the same. The base voltages are amplified differentially and effectively cancel. If the two base resistors are different, the voltage at the base of each transistor will be different. This difference voltage will be amplified by the circuit, giving an **output offset voltage**. It is important to use equal base resistors to minimize offset.

Even if the base resistors are the same, however, β will generally not be the same for the two transistors; hence, the bias currents will not be the same. We will define the average base current by a quantity called the **input bias current**.

$$\text{input bias current} = \frac{I_{B\text{inv}} + I_{B\text{noninv}}}{2} \quad (2\text{-}13)$$

The difference in base currents is called the **input offset current** and is defined as

$$\text{input offset current} = I_{Binv} - I_{Bnoninv} \tag{2-14}$$

The input offset current multiplied by the base resistance (assumed to be the same for both inputs) gives an input offset voltage, the voltage difference between the two inputs that is amplified and produces the output offset voltage. To eliminate any output offset voltage, differential amplifiers often contain an offset compensating circuit.

Integrated Circuit Differential Amplifiers

One disadvantage of a differential amplifier over an AC-coupled amplifier is that the circuit is more complex. This complexity is not evident in Figure 2-10; however, the circuit in this figure does not represent a complete differential amplifier. Careful examination of Figure 2-10 will show that the amplifier output has a positive DC bias voltage present. In a true differential amplifier, the input and output bias levels must be at ground potential so that there is no current flow from the amplifier to the source or load. This requires the more complex circuit shown in Figure 2-11. This circuit includes a three-transistor driver to ensure zero bias voltage at the output. In many cases, differential amplifiers use even more transistors to enhance their performance. The complexity of a complete differential amplifier over the AC-coupled amplifier of Figure 2-9 should now be obvious.

Fortunately, the complexity of a differential amplifier is a disadvantage only as far as discrete circuits are concerned. For integrated circuits, the complexity is within the integrated circuit chip and not in the external circuit.

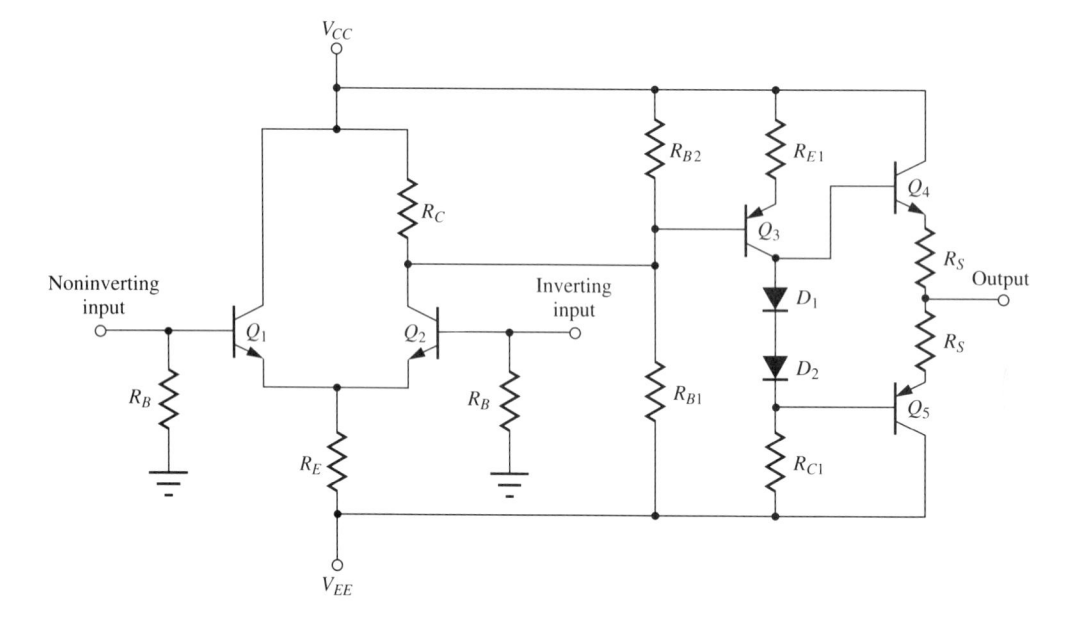

Figure 2-11 Circuit of a complete differential amplifier.

2-3 Characteristics of Operational Amplifiers

The most common form of integrated circuit amplifier is the **operational amplifier**, or more simply, **op-amp**. The integrated op-amp circuit is basically that of a differential amplifier, generally enhanced to give the op-amp very high input impedance, very low output impedance, very high differential gain, and very high common-mode rejection. Like a true differential amplifier, op-amps have two inputs—one inverting, the other noninverting—and a single output. Op-amps can normally be operated over a very wide range of supply voltages. The typical unity-gain bandwidth for a modern op-amp is 1 MHz or more.

Origin of the Name "Op-Amp"

Why the name "operational amplifier"? These near-perfect amplifying devices were first developed for use in systems called **analog computers**. An analog computer did not use switching circuits and very rapid numerical calculations, but was constructed in the form of an electronic analog of some physical system that was to be investigated. The operational amplifiers could be configured to perform various mathematical operations such as summing, squaring, differentiating, or integrating. Because the amplifier performed mathematical *operations*, it was called an "operational amplifier." The analog computer is seldom used today because modern digital computers are now fast enough and complex enough to do any of the modelling for which analog computers were designed, and they can do it more accurately. Although the operational amplifier is no longer used for "operations" in a computer, many of the mathematical operations that it was once designed to perform are still useful in a variety of analog circuits.

Understanding Op-Amp Operation

Most circuit designs using op-amps can be understood by recognizing three main characteristics. *These are important!*

1. Both the inverting input and the noninverting input have very high impedances and hence will act like open circuits for any input signals.

2. The inverting input will always try to be at the same voltage as the noninverting input. With negative feedback, the output will always be such as to ensure this condition.

3. If the noninverting input is grounded, the inverting input will act as a virtual ground.

We will use these characteristics later in developing formulae for the design and analysis of op-amp circuits.

Types of Op-Amps

There are two main types of op-amp, and we will be discussing practical examples of each in later sections of this chapter. The first is the bipolar op-amp, built using bipolar technology, and typified by the μA741C. The internal circuit is quite similar to the differential amplifier shown in Figure 2-11. The other type is the FET input or BiFET op-amp. The circuitry of this op-amp is basically that of a differential amplifier, but the input uses field-effect transistors to give an extremely high input impedance. Most modern op-amps are BiFET, and we will use the TL081C as a typical example.

Schematic Symbol for Op-Amps

Figure 2-12 shows the pinout of the 8 pin DIP used for most single op-amps and the schematic symbol used for most standard bipolar and BiFET op-amps. The pin numbering shown on the schematic symbol is standard for single op-amps in the 8 pin DIP package. It will differ for dual and quad op-amp packaging. Pins 1, 5, and 8 are usually not shown on the schematic. They are used for offset adjustment or for external compensation, concepts that will be introduced in later sections.

(a) Pinout of single op-amp in 8 pin DIP

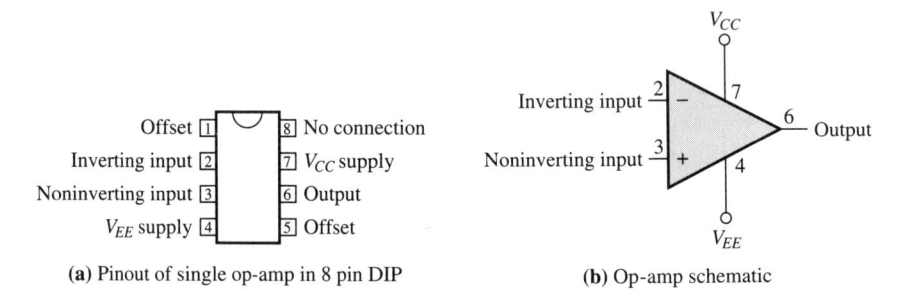

(b) Op-amp schematic

Figure 2-12 Standard op-amp package and schematic.

Power Supply Limitations

One very important characteristic of op-amps is their ability to operate over a wide range of power supply voltages. Like differential amplifiers, most op-amps require dual power supplies. Op-amps are normally operated from either ±15 V supplies or ±12 V supplies; however, they typically can operate from any dual supply in the range from ±3 V to ±20 V. Although balanced supplies normally are used, the voltages do not need to be the same for the two supplies.

Op-amps are designed to be insensitive to changes in supply voltage. This is described on data sheets by a characteristic called the *supply voltage rejection ratio*, equal to the base 10 logarithm of the output change divided by the supply change.

Although op-amps are insensitive to changes in supply voltage, the supply voltage does affect the amplitude of the output voltage swing. The maximum voltage that an op-amp can output is typically 1 V less than the supply voltage. For example, if the op-amp were operated from ±15 V, the maximum output voltage swing would be from approximately +14 V to −14 V for a total swing

2-3 Characteristics of Operational Amplifiers

of 28 V peak-to-peak. Output voltages larger than this would be clipped at ±14 V. The maximum output voltages are called the saturation voltages of the op-amp and are designated V_{sat+} and V_{sat-}.

Bandwidth Limitations

Another important characteristic of op-amps is their frequency response. Since an op-amp is direct-coupled, we would expect a frequency response similar to that shown in Figure 2-7(b). The actual frequency response of a typical op-amp is shown in Figure 2-13. Notice that the -3 dB frequency is only about 5 Hz. At frequencies higher than 5 Hz, the gain drops at the rate of 20 dB per decade, reaching unity at approximately 1 MHz. The -3 dB bandwidth of 5 Hz is very unimpressive, but we will see shortly that it can be improved considerably with feedback.

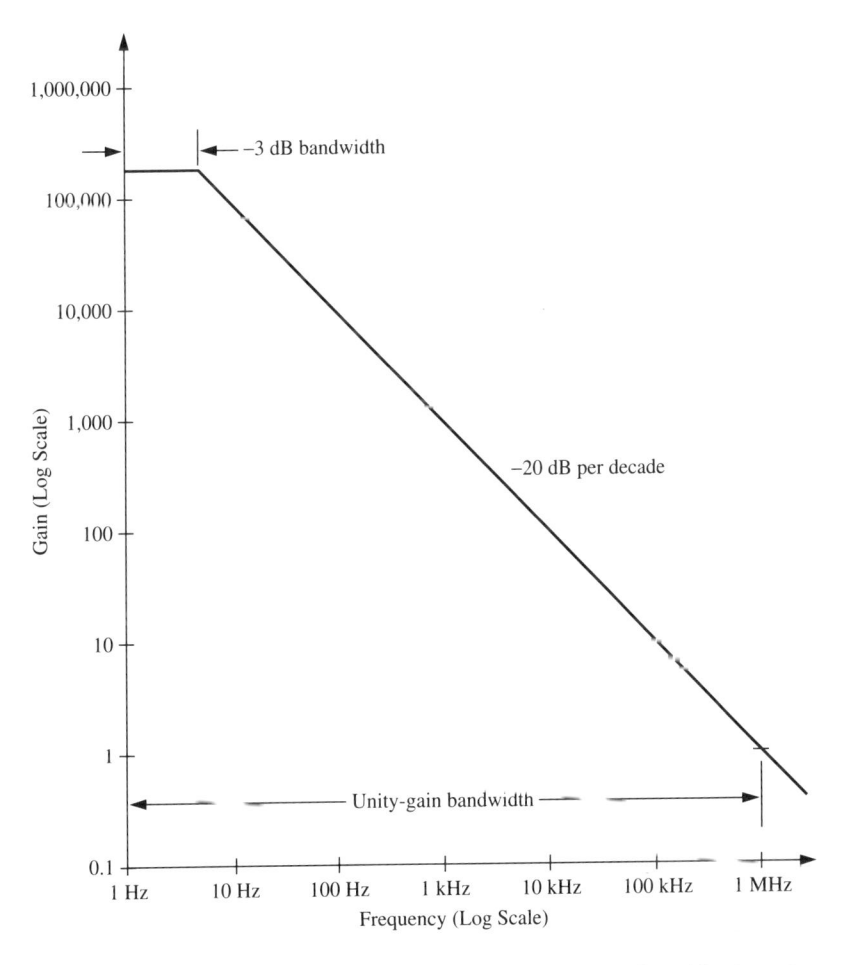

Figure 2-13 Typical op-amp frequency response plotted on logarithmic scales.

2 Operational Amplifiers

The 20 dB drop-off in gain is built into most op-amps by means of a small capacitor (typically between 15 and 30 pF forming part of the integrated circuit. This capacitor is called a compensating capacitor and is included to stabilize the op-amp against a tendency to oscillate. If the capacitor were omitted, the amplifier would have a wider -3 dB bandwidth, but would be unstable against oscillations. Some op-amps are manufactured without this internal capacitor, but with provision for external compensation. By using external compensation, as we will see later, it is sometimes possible to achieve a wider bandwidth.

2-4 Feedback in Operational Amplifiers

We will find in the next few sections that a typical op-amp may have a gain of 200,000. Very seldom would we want an amplifier with a gain of 200,000. Most typical amplifiers have gains between say 10 and 100. The high gain built into an op-amp is neither a fault of the designer nor a problem with op-amps. In fact, the very high gain is an important feature of op-amps—it allows the working gain to be set to almost any desired practical level by the use of **feedback**.

Feedback is the most important concept to grasp in the understanding of operational amplifiers. Feedback occurs when part of the output signal is routed back to the input.

Positive Feedback

If the feedback signal is in phase with the input signal, we have *positive feedback*. The feedback signal adds to the input signal, giving a larger input which, amplified by the op-amp, gives a larger output. This results in a larger feedback signal added to the input signal, producing an even larger input. This cycle repeats until a stable output level is reached or until the circuit starts to oscillate. The effect of positive feedback at producing oscillations will be studied in Chapter 10. Except for producing oscillations, positive feedback is generally undesirable. Positive feedback tends to magnify distortion present in the amplifier, reduces the effective bandwidth, and makes the circuit susceptible to oscillation.

Negative Feedback

If the feedback signal is 180° out of phase with the input, then we have *negative feedback*. This signal subtracts from the input, giving a smaller input which, amplified by the op-amp, gives a smaller output. Again, eventually a stable output level is attained. The amplifier gain is reduced, but distortion is reduced, bandwidth increased, and circuit stability increased.

Negative feedback is accomplished in op-amp circuits by connecting the output back to the inverting input, usually through a resistive voltage divider consisting of a feedback resistor R_F connected from the output to the inverting input, and an input resistor R_I connected either between the inverting input and the source (for an inverting amplifier) or between the inverting input and ground (for a noninverting amplifier). Figure 2-14 shows the feedback voltage dividers for an

2-4 Feedback in Operational Amplifiers

(a) Inverting amplifier

(b) Noninverting amplifier

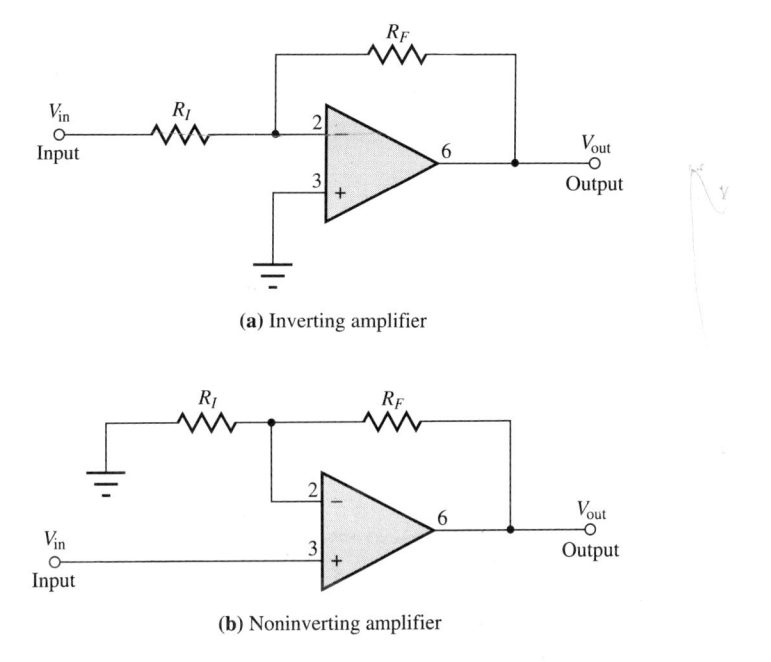

Figure 2-14 Feedback circuits for op-amps.

inverting amplifier (Figure 2-14(a)) and for a noninverting amplifier (Figure 2-14(b)).

The voltage divider feeds back a fraction of the output signal to the input. We call this the feedback fraction B. The fraction B is different for an inverting amplifier and for a noninverting amplifier, and we will see the effect of this difference in the next chapter. An amplifier has two gains: the *open-loop gain*, that is, the gain without any feedback, A_{VOL}; and the *closed-loop gain*, or gain with feedback applied, A_{VF}. Figure 2-15 shows a block diagram illustrating feedback. There are four signals in this figure: the input signal V_{in}, the output signal V_{out}, the feedback signal V_F, and the effective input signal V_E. The output signal V_{out} is $A_{VOL} \times V_E$, where the effective signal V_E is $V_{in} - V_F = V_{in} - BV_{out}$. The gain with feedback is given by

$$A_{VF} = \frac{V_{out}}{V_E} = \frac{A_{VOL}(V_{in} - BV_{out})}{V_{in}} \quad (2\text{-}15)$$

$$= \frac{A_{VOL}V_{in}}{V_{in}} - \frac{A_{VOL}BV_{out}}{V_{in}}$$

Dividing out V_{in} in the first term and substituting A_{VOL} for the factor V_{out}/V_{in}, we obtain the expression

$$A_{VF} = A_{VOL} - BA_{VOL}A_{VF}$$

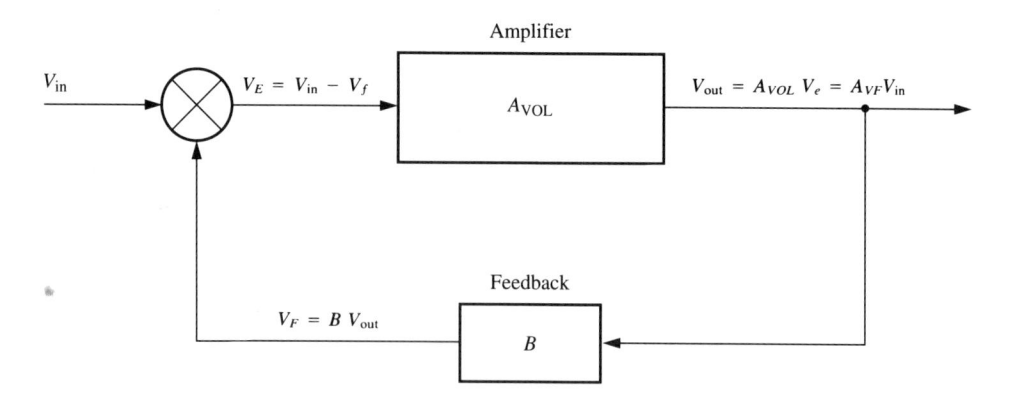

Figure 2-15 Block diagram illustrating feedback.

Collecting terms in A_{VF} from this equation gives

$$A_{VF} + BA_{VOL}A_{VF} = A_{VOL}$$

Finally, rearranging and solving for A_{VF} gives the working equation

$$A_{VF} = \frac{A_{VOL}}{1 + BA_{VOL}} \qquad (2\text{-}16)$$

If the open-loop gain A_{VOL} is sufficiently large, the gain with feedback A_{VF} is approximately $1/B$ and depends only on the characteristics of the feedback loop. This illustrates why op-amps are designed with such high gains—their gain can be precisely set to any low stable value by feedback, and the feedback is determined purely by an external voltage divider.

EXAMPLE 2-3

An op-amp has a feedback factor of 0.1.

(a) Assume the open-loop gain is 200,000. Calculate the gain.
(b) Assume the open-loop gain is 100,000. Recalculate the gain.

Solution

(a) We will use Equation 2-16 to calculate gain with feedback.

$$A_{VF} = \frac{A_{VOL}}{1 + BA_{VOL}}$$

$$= \frac{200{,}000}{1 + 0.1 \times 200{,}000} = \frac{200{,}000}{20{,}001} = 9.9995$$

(b) Repeating the calculations with an open-loop gain of 100,000 and the same value of B; we obtain

$$A_{VF} = \frac{100{,}000}{1 + 0.1 \times 100{,}000} = \frac{100{,}000}{10{,}001} = 9.9990$$

Again, this gain is almost exactly 10 and differs almost insignificantly from the previous value. This illustrates one advantage of negative feedback—the stabilization of gain.

Effect of Negative Feedback on Distortion

Negative feedback also reduces distortion. The output signal with any distortion introduced by the amplifier is fed back to the input 180° out of phase, introducing a negative distortion at the input which, when amplified, tends to cancel any distortion introduced by the amplifier. The reduction factor for the distortion is approximately

$$\text{distortion reduction factor} = \frac{\text{closed-loop gain}}{\text{open-loop gain}} \qquad (2\text{-}17)$$

Hence, if the amplifier had a 5% distortion for an open-loop gain of 200,000, it would have a distortion of 0.0005% for a gain of 20.

Effect of Negative Feedback on Bandwidth

Negative feedback increases the -3 dB bandwidth of an amplifier. We saw in Figure 2-13 that a typical op-amp has an open-loop -3 dB bandwidth of only about 5 Hz. If we reduce the gain with feedback, we will increase the -3 dB bandwidth. In any amplifier the product of gain and -3 dB bandwidth is a constant approximately equal to the unity-gain bandwidth:

$$\text{unity-gain bandwidth} = \text{gain} \times -3 \text{ dB bandwidth}$$

This equation can be solved for the -3 dB bandwidth.

$$-3 \text{ dB bandwidth} = \frac{\text{unity-gain bandwidth}}{\text{gain}} \qquad (2\text{-}18)$$

EXAMPLE 2-4

An amplifier has a unity gain bandwidth of 1.0 MHz. Calculate the -3 dB bandwidth:

(a) for an open-loop gain of 200,000
(b) for a closed-loop gain of 20.

2 Operational Amplifiers

Solution

The −3 dB bandwidth for an amplifier is given by Equation 2-18:

$$-3 \text{ dB bandwidth} = \frac{\text{unity-gain bandwidth}}{\text{gain}}$$

(a) For the open-loop gain of 200,000, the −3 dB bandwidth is

$$-3 \text{ dB bandwidth} = \frac{1 \text{ MHz}}{200,000} = 5 \text{ Hz}$$

(b) If the gain with feedback is 20, the −3 dB bandwidth is

$$-3 \text{ dB bandwidth} = \frac{1 \text{ MHz}}{20} = 50 \text{ kHz}$$

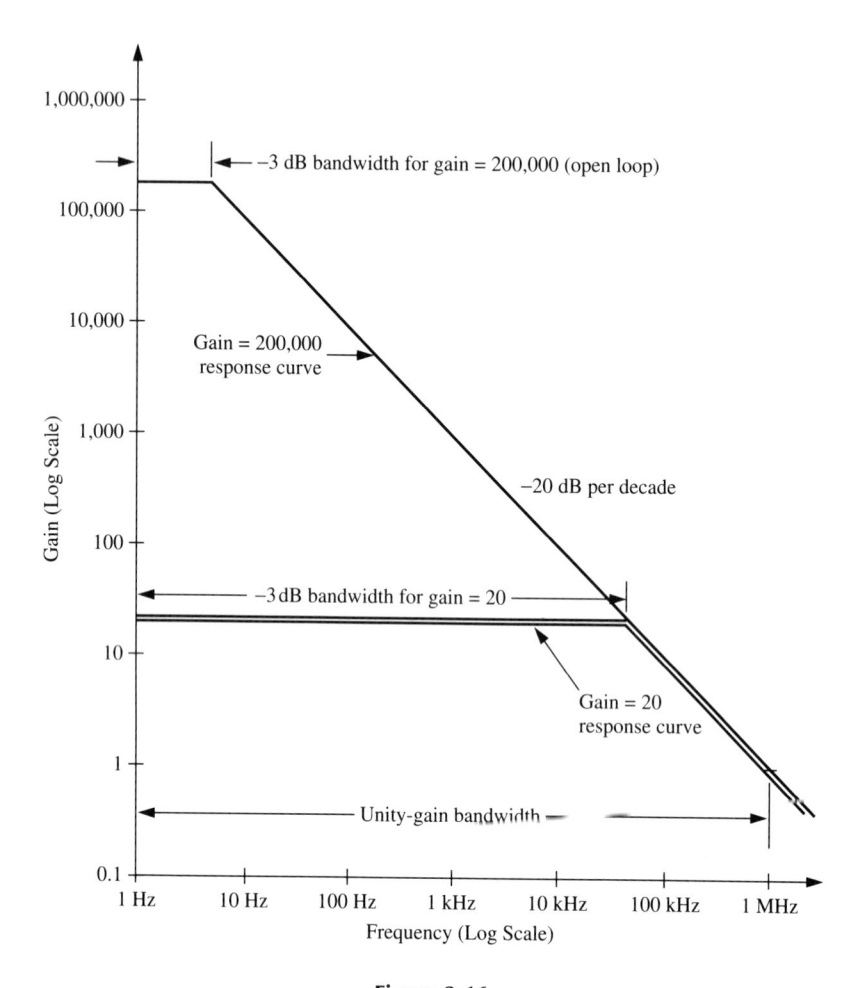

Figure 2-16

Notice the tremendous increase in bandwidth when feedback is added to the circuit. The frequency response for each case is shown in Figure 2-16.

The unity-gain bandwidth is not affected by feedback.

Effect of Negative Feedback on Input and Output Impedance

Finally, feedback affects both the input impedance and the output impedance of the op-amp. First we will consider the input impedance. From Equation 2-16, negative feedback reduces the effective input voltage by the factor $1 + BA_{VOL}$. Consequently, it effectively increases the input impedance by the same factor. The effective input impedance is given by

$$Z_{in}^* = Z_{in} \times (1 + BA_{VOL}) \qquad (2\text{-}19)$$

It is important to qualify this statement. Equation 2-19 refers to the input impedance *of the op-amp itself*, not to the input impedance of any amplifier circuit using the op-amp. In the next chapter, we will be designing amplifier circuits using op-amps, and we will see that the input impedance of the *amplifier* is determined by external resistors, not by feedback.

The effect of negative feedback on the output impedance is similarly determined. The output impedance is reduced by the factor $1 + BA_{VOL}$, and the effective output impedance is

$$Z_{out}^* = Z_{out} \times \frac{1}{1 + BA_{VOL}} \qquad (2\text{-}20)$$

Equation 2-20 also refers to the output impedance *of the op-amp itself*. For most amplifier circuits, the amplifier output impedance *will be equal* to the op-amp output impedance.

In most practical circuit designs, we will not be concerned with the effect of negative feedback on either input or output impedance. The input impedance of an op-amp is already very large. Negative feedback makes it much larger, and it can be considered as infinite in most design calculations. The output impedance of an op-amp is very low, and with negative feedback it can usually be considered as zero.

EXAMPLE 2-5

An op-amp has an open-loop gain of 200,000 and is used in a circuit with a feedback factor of 0.1 as in the previous example. If the op-amp input impedance is 4 MΩ and the output impedance is 50 Ω, calculate

(a) the effective input impedance
(b) the effective output impedance.

Solution

(a) Use Equation 2-19 to calculate the effective input impedance:

$$Z_{in}^* = Z_{in} \times (1 + BA_{VOL}) = 4 \text{ M}\Omega \times (1 + 0.1 \times 200{,}000) = 80 \text{ G}\Omega$$

We see that, at 80,000 million ohms, the input impedance is effectively infinite.

(b) Use Equation 2-20 to calculate the effective output impedance:

$$Z_{out}^* = Z_{out} \times \frac{1}{1 + BA_{VOL}} = 50 \text{ }\Omega \times \frac{1}{1 + 0.1 \times 200{,}000} = 0.0025 \text{ }\Omega$$

For most practical applications, this output impedance can be considered negligible.

2-5 Bipolar Integrated Circuit Operational Amplifiers

The earliest op-amps were constructed using bipolar technology, with circuitry similar to the differential amplifier shown in Figure 2-11. We will begin our study of op-amp properties with these bipolar op-amps. Modern BiFET op-amps are superior in many respects to the bipolar op-amps; however, some designers still adhere to the older bipolar devices, and many other types of integrated circuits still use bipolar technology.

By far the most popular bipolar op-amp was, and still is, the μA741, for many years the industry standard. Portions of the data sheets for the μA741 op-amp are shown in Figure 2-17. Since the data sheets are from the National Semiconductor data book, the device is identified as the LM741 with the National LM prefix instead of the Fairchild μA prefix. The LM741 is available in a number of different packages, with the LM741 in the 8 pin DIP being the most common form. We will use the 8 pin DIP version exclusively throughout this book. The LM741 op-amp is available in a number of different ratings; namely, the M or military rating, the I or industrial rating, and the C or commercial rating. The performance of each is almost the same except for temperature range. We will use C suffix op-amps exclusively in this text.

The various electrical characteristics of the LM741C operational amplifier are listed on the manufacturer's data sheets shown in Figure 2-17. We will investigate a number of these characteristics in detail to learn the operation of this op-amp.

Power Supply Requirements

Most operational amplifiers, including the LM741C, use dual positive/negative power supplies. The LM741C can operate from any voltage from approximately ± 3 V to ± 18 V, although typically it is operated from either ± 15 V supplies or ± 12 V supplies. The LM741C is insensitive to changes in supply voltage. The data sheets shown list a power supply rejection of 96 dB for the LM741C. A 1 V change in power supply voltage would change the output by less than -96 dB, or a factor of 0.000016 (16 μV for a 1 V power supply change).

National Semiconductor

Operational Amplifiers/Buffers

LM741/LM741A/LM741C/LM741E Operational Amplifier

General Description

The LM741 series are general purpose operational amplifiers which feature improved performance over industry standards like the LM709. They are direct, plug-in replacements for the 709C, LM201, MC1439 and 748 in most applications.

The amplifiers offer many features which make their application nearly foolproof: overload protection on the input and output, no latch-up when the common mode range is exceeded, as well as freedom from oscillations.

The LM741C/LM741E are identical to the LM741/LM741A except that the LM741C/LM741E have their performance guaranteed over a 0°C to +70°C temperature range, instead of −55°C to +125°C.

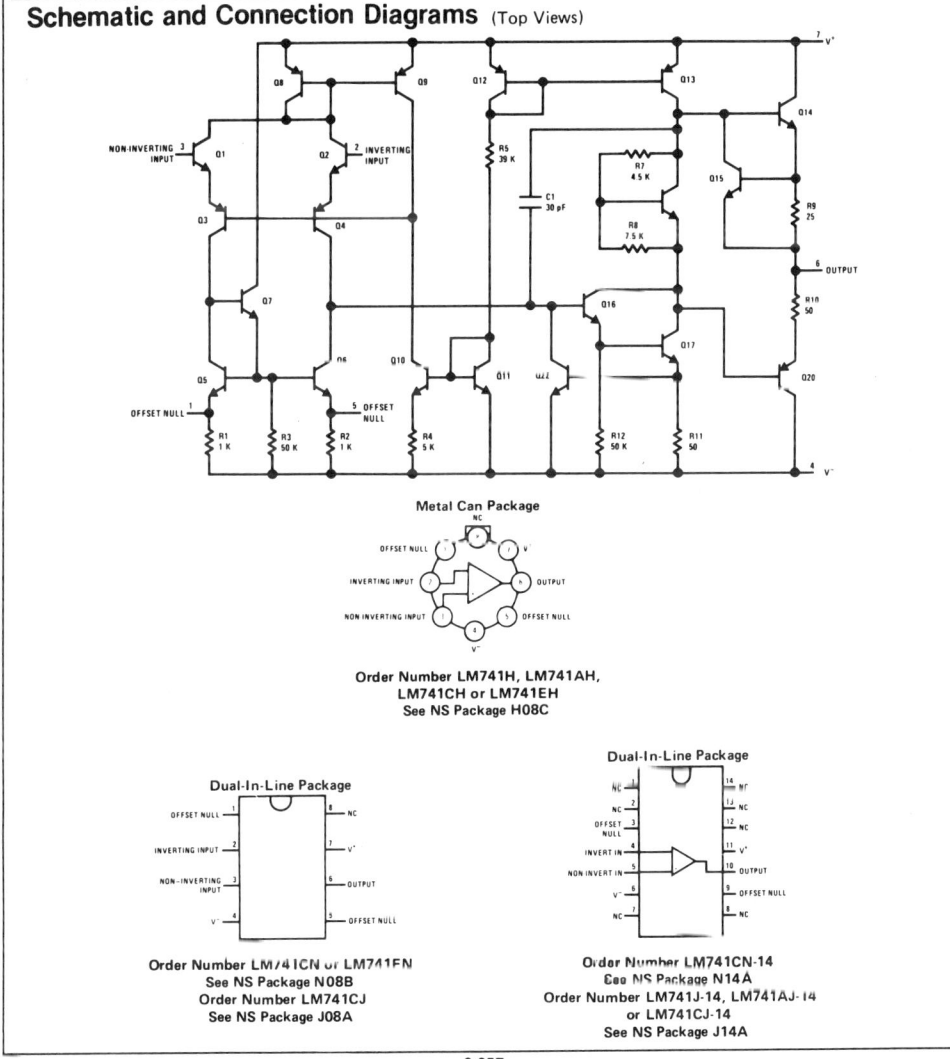

Schematic and Connection Diagrams (Top Views)

Figure 2-17 Data sheets for the LM741C operational amplifier. *(Reprinted with permission of National Semiconductor Corporation)*

Absolute Maximum Ratings

	LM741A	LM741E	LM741	LM741C
Supply Voltage	±22V	±22V	±22V	±18V
Power Dissipation (Note 1)	500 mW	500 mW	500 mW	500 mW
Differential Input Voltage	±30V	±30V	±30V	±30V
Input Voltage (Note 2)	±15V	±15V	±15V	±15V
Output Short Circuit Duration	Indefinite	Indefinite	Indefinite	Indefinite
Operating Temperature Range	−55°C to +125°C	0°C to +70°C	−55°C to +125°C	0°C to +70°C
Storage Temperature Range	−65°C to +150°C	−65°C to +150°C	−65°C to +150°C	−65°C to +150°C
Lead Temperature (Soldering, 10 seconds)	300°C	300°C	300°C	300°C

Electrical Characteristics (Note 3)

PARAMETER	CONDITIONS	LM741A/LM741E MIN	TYP	MAX	LM741 MIN	TYP	MAX	LM741C MIN	TYP	MAX	UNITS
Input Offset Voltage	$T_A = 25°C$					1.0	5.0		2.0	6.0	mV
	$R_S \leq 10\ k\Omega$										
	$R_S \leq 50\Omega$		0.8	3.0							mV
	$T_{AMIN} \leq T_A \leq T_{AMAX}$										
	$R_S \leq 50\Omega$			4.0							mV
	$R_S \leq 10\ k\Omega$						6.0			7.5	mV
Average Input Offset Voltage Drift				15							µV/°C
Input Offset Voltage Adjustment Range	$T_A = 25°C, V_S = ±20V$		±10			±15			±15		mV
Input Offset Current	$T_A = 25°C$		3.0	30		20	200		20	200	nA
	$T_{AMIN} \leq T_A \leq T_{AMAX}$			70		85	500			300	nA
Average Input Offset Current Drift				0.5							nA/°C
Input Bias Current	$T_A = 25°C$		30	80		80	500		80	500	nA
	$T_{AMIN} \leq T_A \leq T_{AMAX}$			0.210			1.5			0.8	µA
Input Resistance	$T_A = 25°C, V_S = ±20V$	1.0	6.0		0.3	2.0		0.3	2.0		MΩ
	$T_{AMIN} \leq T_A \leq T_{AMAX}$, $V_S = ±20V$	0.5									MΩ
Input Voltage Range	$T_A = 25°C$							±12	±13		V
	$T_{AMIN} \leq T_A \leq T_{AMAX}$				±12	±13					V
Large Signal Voltage Gain	$T_A = 25°C, R_L \geq 2\ k\Omega$										
	$V_S = ±20V, V_O = ±15V$	50									V/mV
	$V_S = ±15V, V_O = ±10V$				50	200		20	200		V/mV
	$T_{AMIN} \leq T_A \leq T_{AMAX}$, $R_L \geq 2\ k\Omega$,										
	$V_S = ±20V, V_O = ±15V$	32									V/mV
	$V_S = ±15V, V_O = ±10V$				25			15			V/mV
	$V_S = ±5V, V_O = ±2V$	10									V/mV
Output Voltage Swing	$V_S = ±20V$										
	$R_L \geq 10\ k\Omega$	±16									V
	$R_L \geq 2\ k\Omega$	±15									V
	$V_S = ±15V$										
	$R_L \geq 10\ k\Omega$				±12	±14		±12	±14		V
	$R_L \geq 2\ k\Omega$				±10	±13		±10	±13		V
Output Short Circuit Current	$T_A = 25°C$	10	25	35		25			25		mA
	$T_{AMIN} \leq T_A \leq T_{AMAX}$	10		40							mA
Common-Mode Rejection Ratio	$T_{AMIN} \leq T_A \leq T_{AMAX}$										
	$R_S \leq 10\ k\Omega, V_{CM} = ±12V$				70	90		70	90		dB
	$R_S \leq 50\ k\Omega, V_{CM} = ±12V$	80	95								dB

Figure 2-17 (*Continued*)

Input and Output Impedance

The input impedance for both inputs of the LM741C is typically 2 MΩ. The output impedance is not listed, although typically is between 50 Ω and 100 Ω. Recall from the discussion of feedback in the previous section that both of these values are altered by feedback, and that both are normally ignored in the design equations.

Open-Loop Gain

The open-loop differential voltage gain is listed on the data sheet as the *large-signal voltage gain* and has a typical value of 200 V/mV. This definition of gain is somewhat different than defined in Equation 2-1. If we express gain in terms of Equation 2-1, which lists the gain without units, or, more precisely, as volts/volt, we get a value of 200,000.

Common-Mode Rejection Ratio

Recall from the differential amplifier discussion that the gains of the two inputs are only approximately equal. The difference in gain is described by the common-mode rejection ratio, defined in exactly same way as for differential amplifiers. The LM741C is listed on the data sheets as having a typical common-mode rejection ratio of 90 dB. Since it has a typical differential gain of 200,000, we can calculate the common-mode gain from Equation 2-12:

$$90 = 20 \log \frac{200,000}{\text{common-mode gain}}$$

$$\text{common-mode gain} = 200,000 \times 10^{-90/20}$$

$$= 200,000 \times 0.0000316 = 6.325$$

A signal applied to both inputs simultaneously would be amplified approximately by a factor of 6, and a signal applied to one input only would be amplified approximately by 200,000. Normally we will not be concerned with the actual common-mode gain, and will simply use the listed common-mode rejection ratio as a means of comparing the performance of different op-amps.

Input Bias Current/Input Offset Current

Refer to the schematic of the internal circuit of the LM741C shown in Figure 2-17. Both inputs are connected to the base of a Darlington transistor. In normal operation, bias current must flow into the base of each input transistor. The average bias current for the two inputs is called the **input bias current**, and is defined in exactly the same way as for a differential amplifier, namely by Equation 2-13:

$$\text{input bias current} = \frac{I_{B\text{inv}} + I_{B\text{noninv}}}{2} \quad (2\text{-}13)$$

The input bias current of the LM741C has a typical value of 80 nA. Although this is a very tiny current, if it flows through different resistances to each input, it may cause a small voltage difference, which if amplified by the full gain of 200,000, can produce a large output offset voltage. We will see in the next chapter that it is necessary to consider this in the design of the input resistors.

Recall also from the discussion of differential amplifiers that the two input transistors may have different characteristics due to the manufacturing process and, hence, different bias currents. As in the case of the differential amplifier, this difference is called the **input offset current** and is defined in exactly the same way, by Equation 2-14:

$$\text{input offset current} = I_{B\text{inv}} - I_{B\text{noninv}} \tag{2-14}$$

From the data sheets, the typical input offset current for the LM741C is 20 nA. Again, this is a tiny current, but because of the high gain of the op-amp, it can cause an appreciable output offset voltage. To eliminate the effect of this offset current, we can connect an external potentiometer between pins 1 and 5 of the 8 pin DIP package. The third lead (wiper arm) of this potentiometer goes to the negative supply. By adjusting the potentiometer, we can null the output offset voltage. Figure 2-18 shows a simple amplifier circuit including the offset adjustment potentiometer. Typically, a 10 kΩ trim pot would be used. Many applications can tolerate a small DC offset, so external offset adjustment is rarely used.

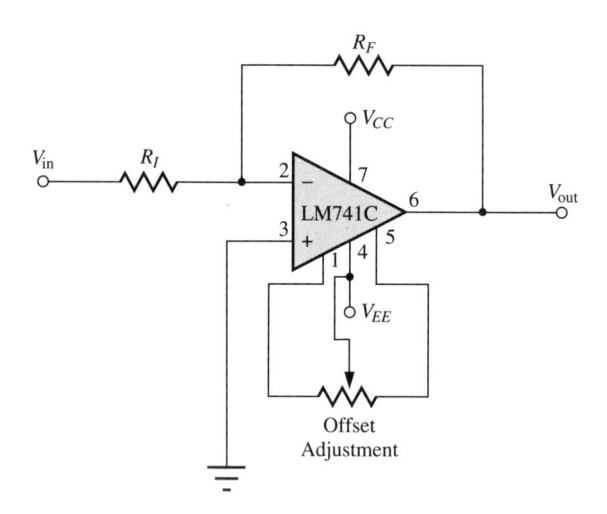

Figure 2-18 Circuit showing external offset adjustment with the LM741C op-amp.

Frequency Response

Another important characteristic of the LM741C is the frequency response. We saw earlier that the gain of the LM741C was typically 200,000. This is the gain from DC to only about 5 Hz, at which point it starts to roll off at 20 dB/decade. This drop-off in gain is due to the 30 pF compensating capacitor seen near the center of the device schematic shown in Figure 2-17. The open-loop frequency response of the LM741C is identical to the frequency response shown in Figure 2-13, with an open-loop gain of 200,000, a −3 dB bandwidth of 5 Hz, and a unity-gain bandwidth of 1 MHz.

Slew Rate

There is one last characteristic that we will include in this discussion of the LM741C—the **slew rate**. The slew rate of an amplifier is the time the amplifier takes to respond to a step input. In the case of the LM741C, the slew rate is listed as 0.5 V/µs. If the amplifier input voltage were to suddenly increase, the output voltage would only change at a rate of 0.5 V/µs. The importance of this quantity will become apparent in Chapter 5 when we discuss comparators. It is an important quantity for comparing the performance of different op-amps.

2-6 BiFET Integrated Circuit Operational Amplifiers

Although the LM741C and other bipolar op-amps are still widely used, they are basically obsolete. Bipolar technology has been replaced by BiFET technology. The term "BiFET" stands for bipolar–field-effect transistor. It is a combination of two technologies, bipolar and junction field-effect, making use of the advantages of each. Bipolar devices are good for power handling and speed—field-effect devices have very high input impedances and low power consumptions. Most modern general-purpose operational amplifiers are now produced with BiFET technology.

BiFET op-amps use field-effect transistors (FETs) for the inputs to provide extremely high input impedances, and use bipolar transistors for amplification and output. Since field-effect transistors are voltage-operated devices, bias currents and offset currents are basically negligible.

BiFET op-amps, being more modern, generally have enhanced performance characteristics over bipolar op-amps, including wider bandwidth, higher slew rate, and larger power output. Most of the applications discussed in this book, other than simple amplifiers, will use BiFET op-amps because of their superior performance.

There are quite a variety of BiFET op-amps available: the TL060 low-power, TL070 low-noise, and TL080 general-purpose series from Texas Instruments, the LF350 and LF440 series from National Semiconductor, the MC34000 and MC35000 series from Motorola, to name but a few of the major manufacturers. Some of these are summarized in Table 2-3, taken from various sources. The performance of most of these op-amps is similar, and they are all pinout-compatible with the LM741C.

In this book, we will use the TL080 series of BiFET op-amps. The decision to use these over the other types is purely arbitrary. They are typical of BiFET amplifiers, and any of the other amplifiers listed in Table 2-3 could be used in their place.

Data sheets for the TL080 series of BiFET op-amps are shown in Figure 2-19. There are five different numbers in this series: TL080, TL081, TL082, TL083,

Table 2-3 Summary of FET input op-amps available from a number of manufacturers

Device	Manufacturer	Bias Current (pA)	Gain	Bandwidth (MHz)	Slew Rate (V/μs)	Supply Voltage (V)	Features
LF351	National	50	100,000	4.0	13	±5–±18	
LF355	National	30	200,000	2.5	5	±5–±22	
LF356	National	30	200,000	4.5	12	±5–±18	
LF411	National	50	200,000	3.0	10	±6–±18	
LF441	National	50	100,000	1.0	1	±6–±18	Low power
LF451	National	50	200,000	4.0	13	±6–±18	
MC34001	Motorola	50	150,000	4.0	13	±5–±18	
TL061	Texas Inst	30	6,000	1.0	4	±5–±18	Low power
TL071	Texas Inst	30	200,000	3.0	13	±5–±18	Low noise
TL081	Texas Inst	30	200,000	3.0	13	±5–±18	

and TL084. The TL081 is representative of the family, and we will look at its characteristics shortly. The TL081 is supplied in an 8 pin DIP with a pinout identical to the LM741C. The TL080 is the same op-amp with the internal compensating capacitor omitted. We shall see later how to add an external compensating capacitor to improve the bandwidth of this op-amp. The TL082 and TL083 are both dual TL081 op-amps—the TL082 in an 8 pin DIP without provision for offset nulling, the TL083 in a 14 pin DIP with provision for offset nulling. The pinout of the TL082 is identical to that of the LM1458 bipolar dual op-amp, and the pinout of the TL083 is identical to that of the LM747 bipolar dual op-amp. Finally, the TL084 is a quad TL081 op-amp in a 14 pin DIP.

Like the LM741C, the TL080 series op-amps are available in a number of different ratings, namely the M or military rating, the I or industrial rating, and the C or commercial rating. The performance of each is almost the same except for temperature range. We will use the C suffix op-amps exclusively in this text.

Power Supply Requirements

Like most standard op-amps, the TL081C requires a dual positive/negative power supply and can operate from any supply between approximately ±3 V and ±18 V. As with bipolar op-amps such as the LM741C, typically it is operated from either a ±15 V supply or a ±12 V supply. The power supply rejection of the TL081C is 86 dB, slightly less than the LM741C.

Input and Output Impedance

Because of the field-effect transistors on the inputs, the input impedance for both inputs of the TL081C is typically 10^{12} Ω, six orders of magnitude higher than a LM741C. The output impedance is not listed, although like the LM741C, is typically between 50 and 100 Ω. Feedback in a practical amplifier circuit would make

LINEAR INTEGRATED CIRCUITS

TYPES TL080 THRU TL085, TL080A THRU TL084A, TL081B, TL082B, TL084B
JFET-INPUT OPERATIONAL AMPLIFIERS

BULLETIN NO. DL-S 12484, FEBRUARY 1977–REVISED JUNE 1978

- **27 DEVICES COVER COMMERCIAL, INDUSTRIAL, AND MILITARY TEMPERATURE RANGES**
- Low Power Consumption
- Wide Common-Mode and Differential Voltage Ranges
- Low Input Bias and Offset Currents
- Output Short-Circuit Protection
- High Input Impedance . . . JFET-Input Stage
- Internal Frequency Compensation (Except TL080, TL080A)
- Latch-Up-Free Operation
- High Slew Rate . . . 13 V/μs Typ

description

The TL081 JFET-input operational amplifier family is designed to offer a wider selection than any previously developed operational amplifier family. Each of these JFET-input operational amplifiers incorporates well-matched, high-voltage JFET and bipolar transistors in a monolithic integrated circuit. The devices feature high slew rates, low input bias and offset currents, and low offset voltage temperature coefficient. Offset adjustment and external compensation options are available within the TL081 Family.

Device types with an "M" suffix are characterized for operation over the full military temperature range of $-55°C$ to $125°C$, those with an "I" suffix are characterized for operation from $-25°C$ to $85°C$, and those with a "C" suffix are characterized for operation from $0°C$ to $70°C$.

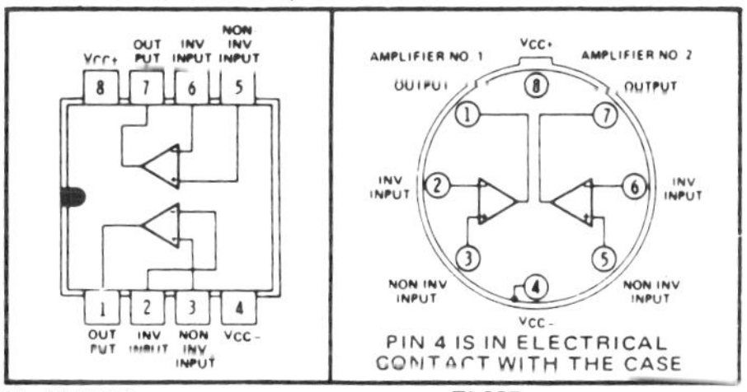

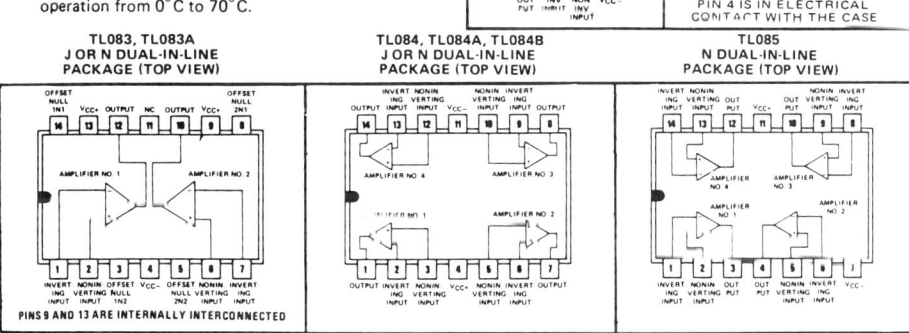

NC – No internal connection

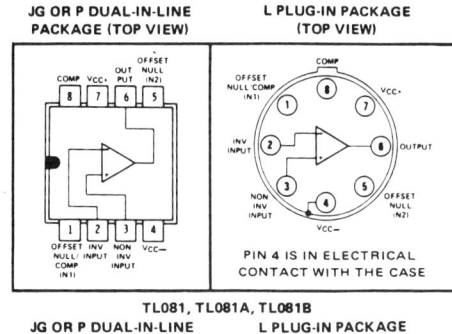

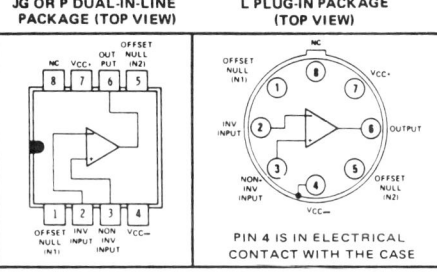

Copyright © 1978 by Texas Instruments Incorporated

TEXAS INSTRUMENTS
INCORPORATED
POST OFFICE BOX 225012 • DALLAS, TEXAS 75265

Figure 2-19 Data sheets for the TL080 series of operational amplifiers. *(Reprinted by permission of Texas Instruments)*

TYPES TL080 THRU TL085, TL080A THRU TL084A, TL081B, TL082B, TL084B JFET-INPUT OPERATIONAL AMPLIFIERS

electrical characteristics, $V_{CC\pm} = \pm 15$ V

PARAMETER		TEST CONDITIONS[†]		TL08_M TL08_AM MIN TYP MAX	TL08_I MIN TYP MAX	TL08_C TL08_AC TL08_BC MIN TYP MAX	UNIT
V_{IO}	Input offset voltage	$R_S = 50\ \Omega$, $T_A = 25°C$	TL08_ [‡]	3 6	3 6	5 15	mV
			TL08_A[‡]	2 3		3 6	
			'81B,'82B,'84B			2 3	
		$R_S = 50\ \Omega$, T_A = full range	TL08_ [‡]	9	9	20	
			TL08_A[‡]	5		7.5	
			'81B,'82B,'84B			5	
αV_{IO}	Temperature coefficient of input offset voltage	$R_S = 50\ \Omega$, T_A = full range		10	10	10	µV/°C
I_{IO}	Input offset current[§]	$T_A = 25°C$	TL08_ [‡]	5 100	5 100	5 200	pA
			TL08_A[‡]	5 100		5 100	
			'81B,'82B,'84B			5 100	
		T_A = full range	TL08_ [‡]	20	10	5	nA
			TL08_A[‡]	20		3	
			'81B,'82B,'84B			3	
I_{IB}	Input bias current[§]	$T_A = 25°C$	TL08_ [‡]	30 200	30 200	30 400	pA
			TL08_A[‡]	30 200		30 200	
			'81B,'82B,'84B			30 200	
		T_A = full range	TL08_ [‡]	50	20	10	nA
			TL08_A[‡]	50		7	
			'81B,'82B,'84B			7	
V_{ICR}	Common-mode input voltage range	$T_A = 25°C$	TL08_ [‡]	±12	±12	±10	V
			TL08_A[‡]	±12		±12	
			'81B,'82B,'84B			±12	
V_{OPP}	Maximum peak-to-peak output voltage swing	$T_A = 25°C$	$R_L = 10\ k\Omega$	24 27	24 27	24 27	V
		T_A = full range	$R_L \geq 10\ k\Omega$	24	24	24	
			$R_L \geq 2\ k\Omega$	20 24	20 24	20 24	
A_{VD}	Large-signal differential voltage amplification	$R_L \geq 2\ k\Omega$, $V_O = \pm 10$ V, $T_A = 25°C$	TL08_ [‡]	50 200	50 200	25 200	V/mV
			TL08_A[‡]	50 200		50 200	
			'81B,'82B,'84B			50 200	
		$R_L \geq 2\ k\Omega$, $V_O = \pm 10$ V, T_A = full range	TL08_ [‡]	25	25	15	
			TL08_A[‡]	25		25	
			'81B,'82B,'84B			25	
B_1	Unity-gain bandwidth	$T_A = 25°C$		3	3	3	MHz
r_i	Input resistance	$T_A = 25°C$		10^{12}	10^{12}	10^{12}	Ω
CMRR	Common-mode rejection ratio	$R_S \leq 10\ k\Omega$, $T_A = 25°C$	TL08_ [‡]	80 86	80 86	70 76	dB
			TL08_A[‡]	80 86		80 86	
			'81B,'82B,'84B			80 86	
k_{SVR}	Supply voltage rejection ratio ($\Delta V_{CC\pm}/\Delta V_{IO}$)	$R_S \leq 10\ k\Omega$, $T_A = 25°C$	TL08_ [‡]	80 86	80 86	70 76	dB
			TL08_A[‡]	80 86		80 86	
			'81B,'82B,'84B			80 86	
I_{CC}	Supply current (per amplifier)	No load, No signal, $T_A = 25°C$		1.4 2.8	1.4 2.8	1.4 2.8	mA
V_{o1}/V_{o2}	Channel separation	$A_{VD} = 100$, $T_A = 25°C$		120	120	120	dB

[†] All characteristics are specified under open-loop conditions unless otherwise noted. Full range for T_A is −55°C to 125°C for TL08_M and TL08_AM; −25°C to 85°C for TL08_I; and 0°C to 70°C for TL08_C, TL08_AC, and TL08_BC.
[‡] Types TL080AM, TL083AM, TL085I, and TL085M are not defined by this data sheet.
[§] Input bias currents of a FET-input operational amplifier are normal junction reverse currents, which are temperature sensitive as shown in Figure 18. Pulse techniques must be used that will maintain the junction temperature as close to the ambient temperature as is possible.

TEXAS INSTRUMENTS
INCORPORATED
POST OFFICE BOX 225012 • DALLAS, TEXAS 75265

2-6 BiFET Integrated Circuit Operational Amplifiers

the input impedance even higher, and the output impedance much lower. Both of these quantities are normally ignored in the design equations.

Open-Loop Gain

The large-signal voltage gain, or open-loop differential voltage gain is the same as the LM741C, with a typical value of 200 V/mV, or 200,000 as defined in Equation 2-1. As with any op-amp in a practical amplifier circuit, the actual amplifier gain will be determined by feedback.

Common-Mode Rejection Ratio

The common-mode rejection ratio of the TL081C is 86 dB. This is slightly less than the LM741C, which has a value of 90 dB, but for all practical purposes is identical.

Input Bias Current/Input Offset Current

A bipolar transistor requires a bias current into the base in order to operate. Consequently, a bipolar op-amp such as the LM741C must have an input bias current into each input. Although this is a very tiny current, with a typical value of 80 nA, it is still significant. Any differences in the bias currents to the two inputs produce an input offset current which may be as large as 20 nA.

The two inputs to a BiFET op-amp such as the TL081C are the gates of JFETs. Field-effect transistors are voltage-operated devices. They use the voltage applied to the transistor gate to control the drain current. In an ideal FET device there is no current flow into the gate. Bias currents theoretically do not exist in a BiFET op-amp. In practice, however, the JFETs used in the BiFET op-amps are not perfect but have a small leakage current. This leakage current is called the input bias current, but has a typical value of 30 pA, more than 1000 times smaller than the bias current of a bipolar op-amp. The difference between the leakage currents of the two inputs is called the input offset current, and has a value of 5 pA, also more than 1000 times smaller than in a bipolar op-amp. These currents are leakage currents only and are not necessary for biasing. We will see later that this makes the design of amplifier circuits using BiFET op-amps simpler.

There should not be a need for external offset adjustment in a BiFET op-amp, but if there is, exactly the same procedure as used for the LM741C can be employed. Pins 1 and 5 in the TL081C are reserved for offset correction.

Frequency Response

One of the improvements of the TL081C over the LM741C is the frequency response. The TL081C has a unity-gain bandwidth of 3 MHz, compared with 1 MHz for the LM741C. This higher unity gain bandwidth is typical of the newer BiFET op-amps. The TL081C has internal compensation, provided by an 18 pF internal capacitor, and the overall response curve is similar to that shown in Figure 2-13, except that the -3 dB bandwidth is approximately 15 Hz and unity gain occurs at 3 MHz.

Slew Rate

Finally, the slew rate is considerably enhanced in BiFET op-amps over that in bipolar op-amps. The TL081C has a slew rate of 13 V/μs, far better than the 0.5 V/μs of the LM741C.

2-7 Uncompensated Operational Amplifiers

In the discussion of op-amp characteristics, we saw that most op-amps contain a small internal capacitor with a value from 15 to 30 pF. This capacitor, called a compensating capacitor, causes the gain to roll off at a rate of 20 dB/decade for all frequencies above a few Hz. The capacitor essentially provides frequency-dependent negative feedback. As frequency increases, capacitive reactance decreases and feedback increases, causing the gain to be reduced.

If this capacitor were not included, the gain at high frequencies would be relatively high. At high frequencies, stray capacitance allows some positive feedback. If the gain is high enough, the positive feedback would be amplified to the point of instability, and the op-amp would oscillate. The compensating capacitor ensures that the high-frequency gain is never large enough to cause oscillations.

It is the compensating capacitor that determines the shape of the frequency response curve shown in Figure 2-13, and that ultimately determines the unity-gain bandwidth of the op-amp. If a smaller capacitor were used, the response curve would still have the same relative shape, but the unity-gain bandwidth, and hence the maximum operating frequency of the op-amp, would be increased. Decreasing the capacitor, however, may leave enough high-frequency gain to cause the op-amp to oscillate. Op-amp designers always provide enough compensating capacitance to ensure op-amp stability under all normal operating conditions.

There are applications, however, where maximum bandwidth is required. Often a standard op-amp with a smaller compensating capacitor would be stable *for a specific application*. For such cases, manufacturers produce some op-amps without the internal compensating capacitor, but with provision for adding an external compensating capacitor. These op-amps are called **uncompensated op-amps**. The TL080C op-amp is an example. It is a standard TL081C op-amp without the internal 18 pF capacitor. Pins 1 and 8 are used for connecting an external compensating capacitor.

Conventional Compensation

Because the TL080 does not have an internal compensating capacitor, an external capacitor must be added for this op-amp to operate without oscillating. Conventional compensation is achieved by connecting a capacitor with a capacitance between 10 and 20 pF between pins 1 and 8 on the op-amp as shown in Figure 2-20. The smaller this capacitor, the wider the bandwidth, but stability against oscillations is reduced. As a general rule, the smaller the compensating capacitor, the more gain that the amplifier should have for stability. For example, if a 10

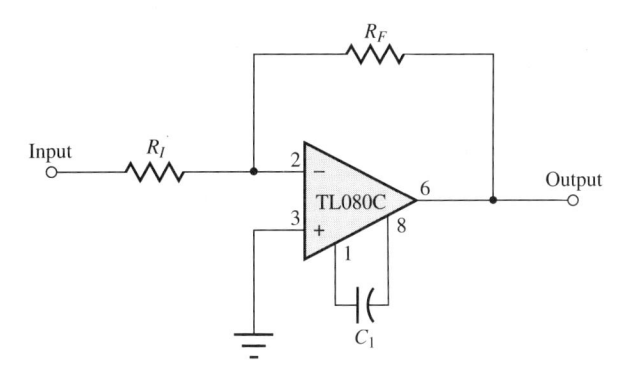

Figure 2-20 TL080C uncompensated op-amp with conventional external compensation.

pF capacitor is used for external compensation on a TL081C, the minimum gain that must be designed into the amplifier would be at least 10. Generally, the minimum gain would have to be determined by experimentation.

Feedforward Compensation

An alternative means of frequency compensation is shown in Figure 2-21. This form of compensation is called **feedforward compensation**. A feedforward capacitor C_1 with a typical value between 100 to 150 pF is connected from pin 1 (one of the external compensation inputs) to pin 2 (the inverting input). A second capacitor C_2 with a value of approximately 3 pF is connected in parallel with the

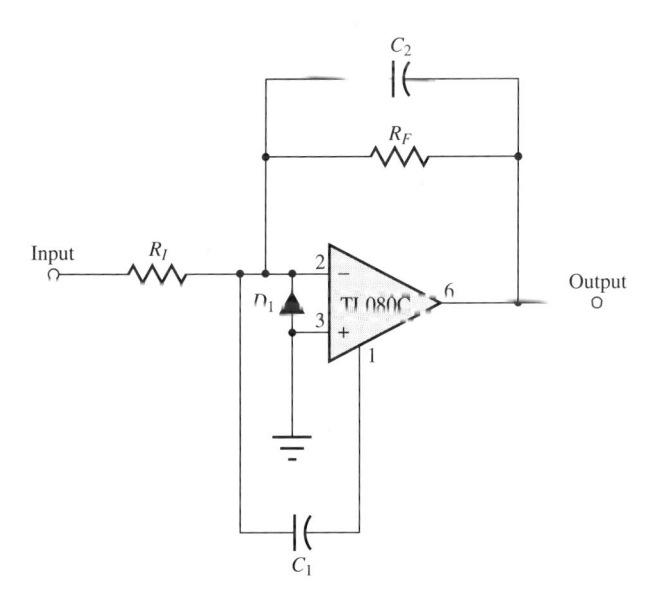

Figure 2-21 TL080C uncompensated op-amp with feedforward compensation.

feedback resistor R_F. Sometimes a diode is connected between the inverting input and ground to further improve the slew rate. This last connection is optional.

How does feedforward compensation work? Some of the input signal is fed to a later stage in the op-amp through capacitor C_1. At low frequencies the capacitive reactance is quite large, and a very small signal is fed forward. At higher frequencies, the capacitive reactance is smaller, and a larger signal is fed forward. This feedforward signal is amplified by the output stages of the op-amp, boosting the normal op-amp output at high frequencies. This is illustrated in Figure 2-22 where an amplifier response with standard compensation and a gain of 20 is compared with the same amplifier response with feedforward compensation. Unfortunately, the feedforward capacitor C_1 also introduces a phase shift. Phase-shift compensation is provided by capacitor C_2 in parallel with the feedback resistor R_F.

Feedforward compensation can provide a substantial increase in bandwidth, as much as a factor of 5 to 10. The exact value of feedforward capacitor C_1 to obtain optimum compensation depends on the internal resistance of the op-amp seen by the capacitor. Both capacitors are best selected according to recommendations in the manufacturer's data sheets, or by trial and error.

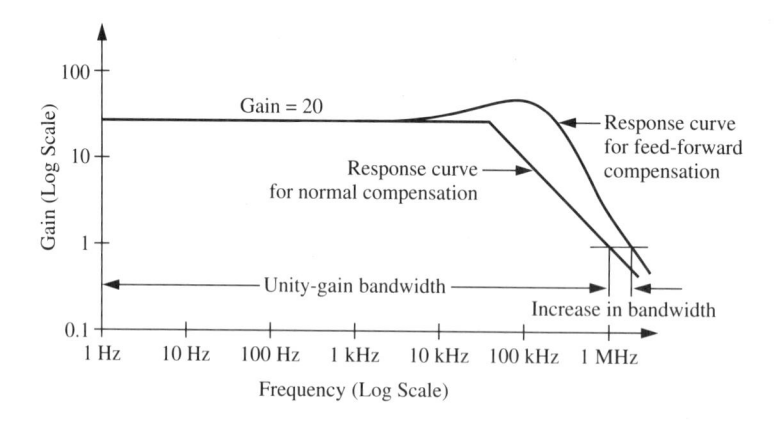

Figure 2-22 Effect of feedforward compensation on frequency response.

Summary

This chapter presented the concepts of amplification and introduced the different types of amplifiers. The differential amplifier, not requiring any capacitors, was shown to be the basis of the integrated circuit operational amplifier (op-amp). We saw that the op-amp is a two-input device with very high gain. By using negative feedback—that is, returning a portion of the amplified signal from the output to the inverting input—we can control the gain and bandwidth of the actual amplifier circuit.

Two types of operational amplifiers were introduced. The first, the μA741C bipolar op-amp, was the industry standard for many years. The other, the TL081C BiFET op-amp, is representative of most modern op-amps. It has virtually infinite input impedance and has many enhanced characteristics over the bipolar op-amps.

Most op-amps have an open-loop gain that falls off with increasing frequency at the rate of 20 dB per decade. This is due to an internal compensating capacitor, included to prevent oscillations and to generally stabilize the circuit. Some op-amps, such as the TL080C, do not have this internal capacitor and can be compensated with an external capacitor to provide a wider frequency response.

Review Questions

2-1 Amplifier characteristics

1. Define an amplifier.
2. Name the two classes of amplifiers.
3. What are the two methods of coupling various amplifier stages together?
4. Give two disadvantages of AC coupling.
5. What type of coupling is always used in integrated circuits? Why?
6. Draw the circuit symbol for a simple amplifier and for a differential amplifier.
7. Define voltage amplification in words.
8. What other types of amplification are possible besides voltage amplification?
9. What is a decibel? What is it used for?
10. Define the different types of amplification or gain in terms of decibels.
11. Define input impedance of an amplifier.
12. Describe a procedure for measuring the input impedance of an amplifier.
13. Define output impedance of an amplifier.
14. Describe a procedure for measuring the output impedance of an amplifier.
15. Describe how the current gain and the power gain of an amplifier can be determined from the voltage gain if the input and load impedances are known.
16. Define bandwidth as applied to an amplifier.
17. What are the −3 dB points of an amplifier?
18. What is the difference between an AC-coupled amplifier and a DC-coupled amplifier with respect to bandwidth?
19. What is unity-gain bandwidth?

2-2 Differential amplifiers

20. What are the functions of the capacitors in an AC-coupled amplifier?
21. What would occur if you converted an AC-coupled amplifier to DC simply by removing the capacitors and making the connections direct?
22. What is the disadvantage of discrete component direct-coupled amplifiers over discrete component AC-coupled amplifiers? Why is this not a problem if the direct-coupled amplifier is made as an integrated circuit?
23. What special type of power supply is required by a differential amplifier?
24. What are the open-circuit input and output voltages in a differential amplifier?
25. In operating a differential amplifier, why is it not possible to leave one of the inputs open?

26. Define input bias current.
27. Define input offset current.
28. Define output offset voltage.
29. What is the difference between the two inputs to a differential amplifier?
30. What is the difference in gain between the two input amplifiers called?
31. Define common-mode rejection ratio.
32. Why is a very large common-mode rejection ratio desirable?

2-3 Characteristics of operational amplifiers

33. What is the basic circuit of most operational amplifiers?
34. For what purpose were operational amplifiers originally developed?
35. Why was the name "operational" applied to integrated circuit differential amplifiers?
36. What are the two main types of op-amps?
37. Name the most popular of the early integrated circuit operational amplifiers.
38. How does it differ from modern operational amplifiers?
39. Describe the three important characteristics of the op-amp circuit model?
40. What are the most commonly used power supply voltages for op-amps?
41. What is the name of the characteristic that describes the insensitivity of op-amps to power supply variations?
42. What is the effect of the power supply voltage on the output amplitude?
43. What is the typical -3 dB bandwidth of an open-loop op-amp?

2-4 Feedback in operational amplifiers

44. Why is the gain of an operational amplifier purposely made extremely large?
45. What is feedback?
46. What are the two different types of feedback used in amplifiers?
47. What are the disadvantages of positive feedback?
48. Where is positive feedback useful?
49. What are the advantages of negative feedback?
50. How is negative feedback usually achieved in an amplifier circuit using an op-amp?
51. How does negative feedback affect the bandwidth of an amplifier circuit?
52. How does negative feedback affect the input impedance of an op-amp? Does it affect the input impedance of an amplifier constructed from the op-amp?
53. How does negative feedback affect the output impedance of an op-amp? Does it affect the output impedance of an amplifier constructed from the op-amp?
54. How does negative feedback affect the distortion in an amplifier?

2-5 Bipolar integrated circuit operational amplifiers

55. Sketch the pinout of the 741 op-amp.
56. What is the difference between the μA741 and the LM741 op-amps?
57. What is the difference between the LM741 and the LM741C op-amps?
58. Sketch the schematic symbol of the LM741C. Label and identify with pin numbers the inputs, outputs, and supply connections.
59. Over what range of power supply voltages can the LM741C operate?
60. What is the typical operating voltage of the LM741C?
61. What is the supply voltage rejection ratio?
62. What is the typical open-loop input impedance for the LM741C? What is the typical open-loop output impedance?

63. What is the open-loop voltage gain of the LM741C?
64. What is the typical common-mode rejection ratio for the LM741C?
65. How can the effects of the input offset current be compensated for in the LM741C?
66. What causes the gain of the LM741C to fall off at high frequencies. Why is this necessary?
67. What is the unity-gain bandwidth for the LM741C?
68. What is slew rate? What is the value for the LM741C?

2-6 BiFET integrated circuit operational amplifiers

69. What does the name "BiFET" represent?
70. How does the design of a BiFET op-amp differ from a bipolar op-amp?
71. What are the members of the TL081C family of op-amps? How do they differ?
72. How do the offset and bias currents of the TL081C differ from the LM741C? Why do they differ?
73. Give three ways in which the TL081C op-amp is superior to the LM741C op-amp.
74. How does the slew rate of the TL081C op-amp compare with that of the LM741C op-amp?

2-7 Uncompensated operational amplifiers

75. How does the TL080 op-amp differ from the TL081 op-amp?
76. What is the purpose of the compensating capacitor? How does it improve op-amp performance?
77. How does the introduction of a compensating capacitor limit the performance of an op-amp?
78. What is the advantage of using an external compensating capacitor?
79. Between which pins on the 8 pin DIP is the external compensating capacitor connected for conventional compensation.
80. What is feedforward compensation?
81. Describe how feedforward compensation works.

Design and Analysis Problems

2-1 Amplifier characteristics

1. A 7.5 kΩ resistor is connected in series with the input of an amplifier, and a load of 2000 Ω is connected across the output. The following measurements are made when a function generator set at a frequency of 1000 Hz is connected to the input of the amplifier through the input resistor: the voltage measured at the input to the resistor is 45.5 mV RMS; the voltage measured at the input to the amplifier is 21.7 mV RMS; when the output is loaded with the 2000 Ω resistor, the output voltage is measured to be 3.25 V RMS; when the load resistor is removed, the output voltage is measured to be 10.72 V RMS. Calculate the following quantities:
   (a) the voltage gain
   (b) the input impedance
   (c) the output impedance
   (d) the current gain
   (e) the power gain.

2. The frequency response of the amplifier is determined by measuring the input voltage V_{in} and the output voltage V_{out} for a range of frequencies. The measured values are listed in Table 2-4.
   (a) Calculate the voltage gain at each frequency.
   (b) Calculate the voltage gain in dB for each frequency.
   (c) Plot the logarithmic frequency response, that is, dB gain, against frequency on semilog paper.
   (d) Determine the −3 dB bandwidth from the plot.
   (e) Determine the unity-gain bandwidth.

Table 2-4 Experimental data for Problem 2

Frequency (Hz)	V_{in} (volts)	V_{out} (volts)
100	1.205	0.0054
200	1.205	0.0852
400	1.205	0.760
700	0.355	0.796
1,000	0.355	1.680
2,000	0.355	3.09
4,000	0.355	3.55
7,000	0.355	3.55
10,000	0.355	3.55
20,000	0.355	3.26
40,000	0.355	2.11
70,000	0.355	1.035
100,000	0.355	0.631
200,000	1.125	0.729
400,000	1.125	0.224
700,000	1.125	0.0710
1,000,000	3.45	0.1090

2-4 Feedback in operational amplifiers

3. Assume an op-amp has an open-loop gain of 150,000 and a feedback factor of 0.03. Calculate the gain with feedback.
4. An amplifier has a unity-gain bandwidth of 5.0 MHz and an open-loop gain of 250,000. Calculate the −3 dB bandwidth:
   (a) for the open-loop gain of 250,000
   (b) if the gain with feedback is 20
   (c) if the gain with feedback is 500
5. Assume an op-amp has an open-loop gain of 80,000 and is used in an amplifier circuit with a feedback factor of 0.05. If the op-amp input impedance is 3 MΩ and the output impedance is 80 Ω, calculate:
   (a) the effective input impedance
   (b) the effective output impedance.

Amplifier Circuits   3

OUTLINE

3-1 Inverting Amplifiers
3-2 Noninverting Amplifiers
3-3 Buffer Amplifiers
3-4 Single-Supply Amplifiers
3-5 The TL321C Single-Supply Op-Amp
3-6 The Norton or Current-Mode Op-Amp
3-7 Instrumentation Amplifiers
3-8 The LH0038 Instrumentation Amplifier

KEY TERMS

Inverting amplifier
Noninverting amplifier
Buffer amplifier
Norton op-amp
Current-mode op-amp
Instrumentation amplifier
Process variable
Transducer

OBJECTIVES

On completion of this chapter, the reader should be able to:

- design and analyze a variety of basic op-amp circuits, including inverting amplifiers, noninverting amplifiers, and buffer amplifiers.
- employ standard bipolar op-amps, such as the LM741C, and BiFET op-amps, such as the TL081C, in various amplifier circuits.
- design amplifier circuits using standard op-amps operating from a single supply.
- use the TL321C special single-supply op-amp in different amplifier circuits.
- design various amplifier circuits using Norton or current-mode op-amps.
- design instrumentation amplifiers from op-amps, and use instrumentation amplifiers to amplify typical transducer outputs in noisy industrial environments.
- use the LH0038 true instrumentation amplifier in a number of applications.

Introduction

This chapter begins the design of practical circuits. In the preceding chapter, we saw the characteristics of the various types of op-amps and how negative feedback could be used to set the operating characteristics. Now we will proceed to develop amplifier circuits using op-amps. The design, as we will discover shortly, is usually very simple.

Amplifiers using operational amplifiers can be constructed in two different configurations, inverting and noninverting. Although amplification can be provided by either configuration, the operating characteristics are somewhat different, and for some applications one configuration may be superior. Often, an amplifier circuit is used for buffering or power amplification. The noninverting buffer amplifier configuration, also known as a follower amplifier, is a particularly simple circuit. Although we will be designing circuits with both bipolar and BiFET op-amps, the superiority of the BiFET op-amp will soon become apparent.

The standard differential amplifier circuit used in most op-amps requires a dual power supply. In many amplifier applications, the requirement of a dual supply is a nuisance. One widely used solution is to use a conventional op-amp such as a TL081C or a LM741C with a voltage divider across the power supply to establish a floating ground. Another, and often better solution, is to use a special single-supply op-amp such as the TL321C. The third option is to use an LM3900 Norton or current-mode op-amp.

The op-amp has some very impressive specifications, yet in some applications the simple op-amp is not sufficient for the job. One area where the op-amp is not sufficient is process instrumentation. In many industrial applications, the noise levels are high, the gain requirements are severe, and precision is vital. Special-purpose amplifiers, called instrumentation amplifiers, have been developed to meet these conditions.

3-1 Inverting Amplifiers

Amplifier circuits using op-amps can be configured either as inverting amplifiers or as noninverting amplifiers. In an **inverting amplifier**, the output signal is 180° out of phase with the input signal. A positive input signal produces a negative output signal.

Any amplifier circuit using an op-amp makes use of negative feedback to establish the amplifier gain and bandwidth. Negative feedback was introduced in the previous chapter, and the circuit configuration of an inverting amplifier was shown in Figure 2-14(a). We will use this circuit, shown as Figure 3-1, to develop the equations that we will use for designing an inverting amplifier. The design of an amplifier requires the specification of the resistors in the feedback circuit.

In an inverting amplifier, there are two resistors, the feedback resistor R_F, connected from the output to the op-amp inverting input, and the input resistor R_I, connected from the source to the op-amp inverting input. The op-amp noninverting input is connected to ground. Because this input is connected to ground,

3-1 Inverting Amplifiers

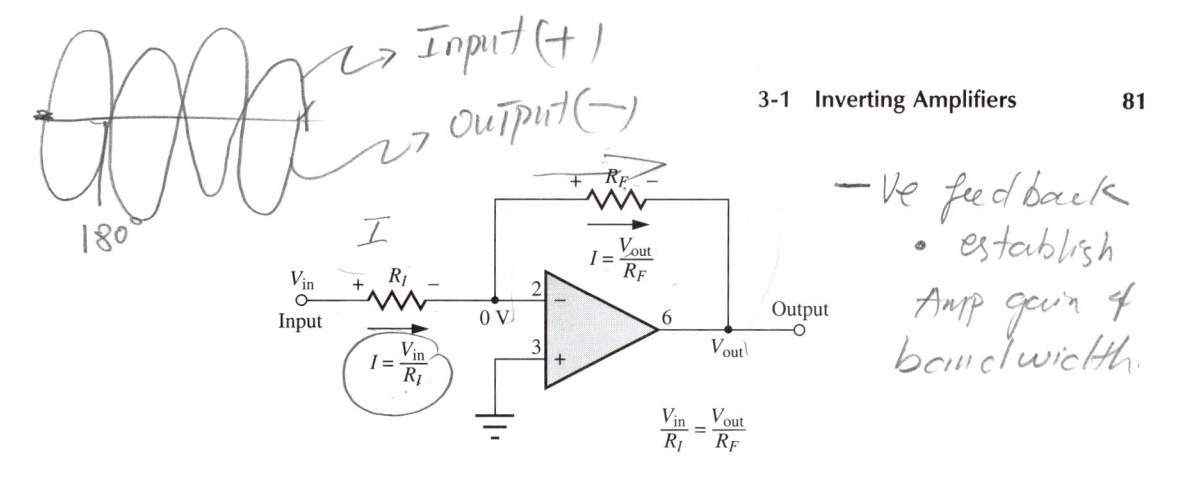

Figure 3-1 Circuit of a basic inverting amplifier.

and because the difference in voltage between the two op-amp inputs is zero, the inverting input acts as if it were a "virtual ground." The actual input current to the op-amp is extremely small because of the very high op-amp input impedance, and it can be ignored. Hence, whatever current flows through R_I must also flow through R_F. The voltage drop across R_I is $V_{in} - 0$. Hence, the current I through R_I is V_{in}/R_I. Because no current flows into the op-amp, the current through R_F is also I, and the voltage drop across R_F is

$$V_{RF} = I \times R_F = \frac{V_{in}}{R_I} \times R_F$$

Since the end of R_F connected to the inverting input is at effective ground, the other end, at the output, must have a voltage of $-V_{RF}$. Hence, the output voltage $V_{out} = -V_{RF}$. Substituting this in the above equation, we obtain

$$-V_{out} = V_{in} \times \frac{R_F}{R_I}$$

Dividing by V_{in}, we obtain

$$-\frac{V_{out}}{V_{in}} = \frac{R_F}{R_I}$$

But V_{out}/V_{in} is simply the amplifier gain with feedback (A_{VF}). The minus sign tells us that we have an inverting amplifier. To avoid later confusion, we will always treat the gain as positive. We now have the equation for establishing the gain of an inverting amplifier:

$$A_{VF} = \frac{R_F}{R_I} \qquad (3\text{-}1)$$

The gain is determined entirely by the external resistors R_I and R_F. One cautionary note should be injected here. The use of R_I in Equation 3-1 assumes that the *source has zero impedance*. If the source has significant impedance, that impedance must be added to R_I for determining gain.

Because the op-amp inverting input is a virtual ground, the only resistance seen by the input signal is R_I. This means that the amplifier input impedance is simply R_I.

$$Z_{in} = R_I \qquad (3\text{-}2)$$

The output impedance of the amplifier depends on the output impedance of the op-amp and, as discussed in the previous chapter, is generally taken as zero. If it needs to be calculated, use Equation 2-20.

Finally, the bandwidth depends on the unity-gain bandwidth of the op-amp and the gain of the inverting amplifier as discussed in the previous chapter. The equation for the amplifier bandwidth is

$$\text{bandwidth} = \frac{\text{unity-gain bandwidth}}{A_{VF} + 1} \qquad (3\text{-}3)$$

With these equations, we can now start to design and analyze amplifiers.

Inverting Amplifier Using a Bipolar Op-Amp

The circuit for an inverting amplifier using the LM741C bipolar op-amp is shown in Figure 3-2. There is one difference between this circuit and the general inverting amplifier circuit shown in Figure 3-1—the inclusion of resistor R_K between the noninverting input and ground.

The LM741C bipolar op-amp requires a bias current for each input. As we saw in the previous chapter, the typical bias current for the LM741C is 80 nA. This current flows into the inverting input through resistors R_I and R_F, producing a negative DC voltage at the inverting input equal to the bias current times the resistance of the parallel combination of R_I and R_F. Without resistance R_K, the input voltage at the noninverting input would be zero. The op-amp would amplify the differential input voltage to give a DC output offset voltage. Although the differential input voltage is very small, because the amplifier has such a large gain, it can lead to an appreciable output offset voltage. To minimize this problem, it is important to ensure that the bias current to each input goes through the same

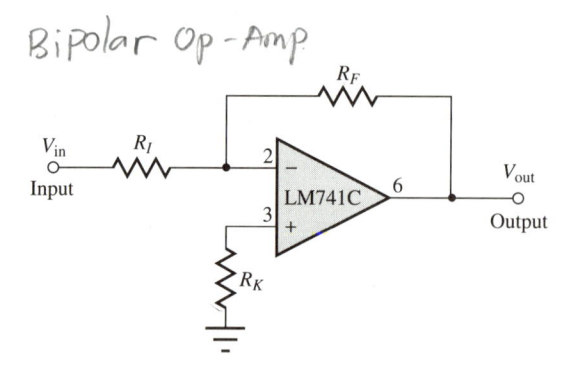

Figure 3-2 Inverting amplifier using a LM741C bipolar op-amp.

amount of resistance. Consequently, we insert resistor R_K between the noninverting input and ground. Resistor R_K must have a value equal to the parallel combination of R_I and R_F:

$$R_K = \frac{R_I R_F}{R_I + R_F} \qquad (3\text{-}4)$$

The design of an amplifier circuit using the LM741C is quite simple. The required gain and input impedance are usually stipulated, and R_F is calculated from Equation 3-1. If R_I is not stipulated, choose a value of R_F between 10 and 100 kΩ and calculate R_I from Equation 3-1. Finally, calculate R_K by using Equation 3-4 with the values determined for R_I and R_F. The bandwidth can be calculated from Equation 3-3. If bandwidth of the amplifier is important, then limitations on the gain may be imposed by Equation 3-3. Normally, a potentiometer to correct for offset, as described in Chapter 2, is not used, and we will not include one in any example. Should offset correction be desired, refer to Figure 2-18 for the circuit.

The analysis of an inverting amplifier circuit using the LM741C op-amp is also quite simple. The gain of the circuit is given by Equation 3-1, including source resistance with R_I if necessary, the input impedance is given by Equation 3-2, the output impedance is zero, and the amplifier bandwidth is given by Equation 3-3.

EXAMPLE 3-1

Design an inverting amplifier using a LM741C op-amp for a gain of 25 and an input impedance of 3000 Ω. Use a ±15 V power supply.

(a) Draw the circuit for an inverting amplifier and prepare a parts list.
(b) Calculate values for R_I, R_F, and R_K.
(c) Calculate the bandwidth.

Solution

(a) We will use the basic inverting amplifier circuit shown in Figure 3-2. This has been redrawn as Figure 3-3, and a parts list is included. The power supply connections are shown in this figure.
(b) We require an input impedance of 3000 Ω. For an inverting amplifier, the input impedance is equal to R_I from Equation 3-2. Hence, our value for R_I is 3.0 kΩ. This is a standard value from the table in Appendix A-1. Since a gain of 25 is stipulated, we can calculate R_F from Equation 3-1:

$$A_{VF} = \frac{R_F}{R_I} \quad \text{or} \quad R_F = A_{VF} \times R_I$$
$$= 25 \times 3.0 \text{ k}\Omega = 75 \text{ k}\Omega$$

3 Amplifier Circuits

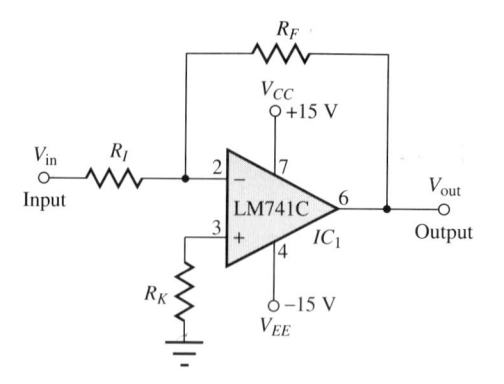

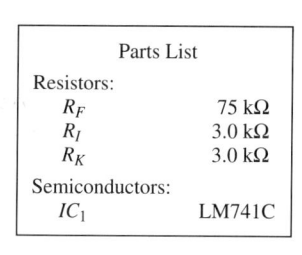

Figure 3-3

From the table of resistance values given in Appendix A-1, 75 kΩ is a standard value.

The resistor R_K is calculated from Equation 3-4:

$$R_K = \frac{R_I R_F}{R_I + R_F} = \frac{3.0 \text{ k}\Omega \times 75 \text{ k}\Omega}{3.0 \text{ k}\Omega + 75 \text{ k}\Omega} = 2885 \text{ }\Omega$$

The closest standard value in the table in Appendix A-1 is 3.0 kΩ. We will use this value.

(c) The unity-gain bandwidth of the LM741C is 1 MHz. This value can be used in Equation 3-3 to calculate the bandwidth of the inverting amplifier:

$$\text{bandwidth} = \frac{\text{unity-gain bandwidth}}{A_{VF} + 1} = \frac{1 \text{ MHz}}{25 + 1} = 38.5 \text{ kHz}$$

The design of the amplifier is complete.

EXAMPLE 3-2

Analyze the amplifier shown in Figure 3-4 to determine the gain, input impedance, and bandwidth.

Solution

By comparison with Figure 3-2, we identify the circuit shown in Figure 3-4 as a basic inverting amplifier circuit using an LM741C op-amp with $R_I = 3.0$ kΩ and $R_F = 150$ kΩ. The gain of an inverting amplifier is given by Equation 3-1:

$$A_{VF} = \frac{R_F}{R_I} = \frac{150 \text{ k}\Omega}{3.0 \text{ k}\Omega} = 50$$

The input impedance of the amplifier is simply $R_I = 3.0$ kΩ.

3-1 Inverting Amplifiers

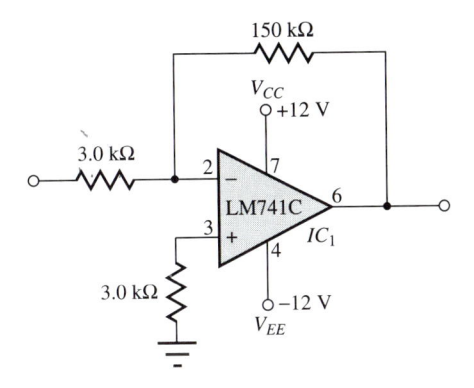

Figure 3-4

The LM741C has a unity-gain bandwidth of 1 MHz. Hence, we can determine the bandwidth of the amplifier by Equation 3-3:

$$\text{bandwidth} = \frac{\text{unity-gain bandwidth}}{A_{VF} + 1} = \frac{1 \text{ MHz}}{50 + 1} = 19.6 \text{ kHz}$$

The analysis of the amplifier is complete.

Inverting Amplifier Using a BiFET Op-Amp

The circuit for an inverting amplifier using the TL081C BiFET op-amp is shown in Figure 3-5. This circuit is identical to the general inverting amplifier circuit in Figure 3-1. Because BiFET op-amps have field-effect transistors on the inputs which are controlled by voltage rather than current, the only input current is a very tiny leakage current, typically 30 pA. A resistor R_K between the noninverting input and ground is not required, and the noninverting input is connected directly to ground.

The design of an amplifier circuit using the TL081C BiFET op-amp is even simpler than the design using a LM741C. If gain and input impedance are stip-

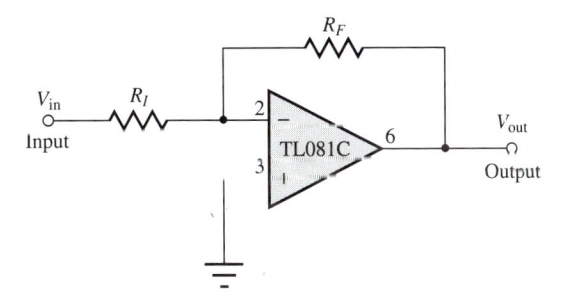

Figure 3-5 Inverting amplifier using a TL081C BiFET op-amp.

ulated, calculate R_F from Equation 3-1. Otherwise, choose a value of R_F between 10 and 100 kΩ and calculate R_I from Equation 3-1. Calculate the bandwidth from Equation 3-3 as with the LM741C.

The analysis of an inverting amplifier circuit using the TL081C BiFET op-amp is identical to that for the LM741C.

EXAMPLE 3-3

Design an inverting amplifier using a TL081C op-amp for a gain of 25 and an input impedance of 3000 Ω. Use a ±15 V power supply.

(a) Draw the circuit for an inverting amplifier and prepare a parts list.
(b) Calculate values for R_I and R_F.
(c) Calculate the bandwidth.

Solution

This is the same problem as Example 3-1, except that we are using a TL081C instead of a LM741C. This problem is intended to demonstrate the similarities and differences in the designs.

(a) We will use the BiFET inverting amplifier circuit shown in Figure 3-5. This has been redrawn as Figure 3-6, and a parts list is included. The power supply connections are shown.

(b) We require an input impedance of 3000 Ω. Since the input impedance is equal to R_I for any inverting amplifier, choose R_I to be 3.0 kΩ, as in Example 3-1. Since a gain of 25 is stipulated, we can calculate R_F from Equation 3-1, exactly as in Example 3-1.

$$A_{VF} = \frac{R_F}{R_I} \quad \text{or} \quad R_F = A_{VF} \times R_I$$

$$= 25 \times 3.0 \text{ k}\Omega = 75 \text{ k}\Omega$$

Parts List

Resistors:
- R_F 75 kΩ
- R_I 3.0 kΩ

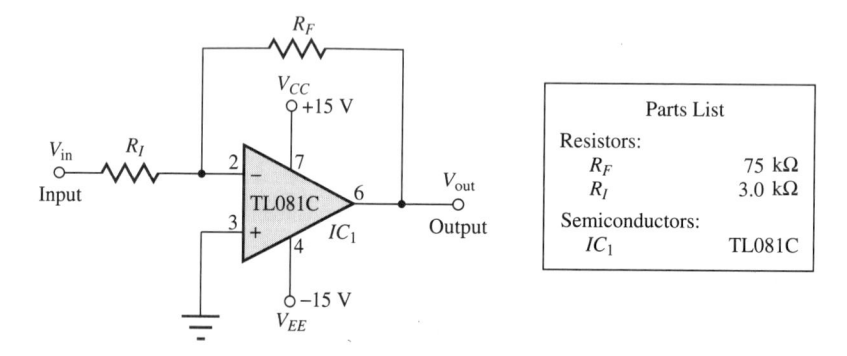

Semiconductors:
- IC_1 TL081C

Figure 3-6

We will choose $R_F = 75$ kΩ for this amplifier, the same as for the LM741C amplifier.

(c) The unity-gain bandwidth of the TL081C is 3 MHz. The bandwidth of the inverting amplifier can be calculated by Equation 3-3:

$$\text{bandwidth} = \frac{\text{unity-gain bandwidth}}{A_{VF} + 1} = \frac{3 \text{ MHz}}{25 + 1} = 115 \text{ kHz}$$

The design of the amplifier is complete. Notice that in this example there is no R_K to calculate. The bandwidth of this amplifier is three times larger than that in Example 3-1 because the TL081C has a unity-gain bandwidth of 3 MHz as compared to 1 MHz for the LM741C.

3-2 Noninverting Amplifiers

The **noninverting amplifier** is somewhat different from the inverting amplifier. Although the feedback loop for a noninverting amplifier still uses resistors R_I and R_F, the circuit is different and we require another set of equations for determining R_I and R_F.

The circuit configuration of a noninverting amplifier was shown in Figure 2-14(b). We will use this circuit, shown here as Figure 3-7, to develop the equations that we will use for designing a noninverting amplifier. Compare this circuit to Figure 3-1 for the inverting amplifier. The feedback resistor R_F is connected from the output to the op-amp inverting input as in the inverting amplifier, but now the input resistor R_I is connected from the op-amp inverting input to ground. The source is connected to the op-amp noninverting input. Because the noninverting input is not connected to ground, the inverting input is no longer a "virtual ground." Resistors R_F and R_I act as a voltage divider supplying $V_{out} \times R_I/(R_F + R_I)$ to the inverting input. Because the op-amp requires both inputs to be at the same voltage, we can write

$$V_{in} = V_{out} \times \frac{R_I}{R_F + R_I}$$

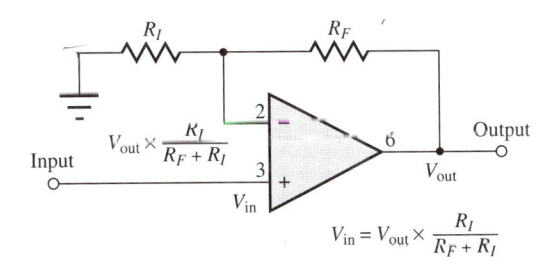

Figure 3-7 Circuit of a basic noninverting amplifier

Dividing by V_{out} and inverting the equation to obtain V_{out}/V_{in}, we get

$$\frac{V_{out}}{V_{in}} = \frac{R_F + R_I}{R_I} = \frac{R_F}{R_I} + 1$$

But V_{out}/V_{in} is simply the amplifier gain with feedback A_{VF}. We now have the equation for establishing the gain of a noninverting amplifier:

$$A_{VF} = \frac{R_F}{R_I} + 1 \qquad (3\text{-}5)$$

As for an inverting amplifier, the gain is determined entirely by the external resistors R_I and R_F, but the equation is different.

Because the op-amp noninverting input has an extremely large input impedance, the input impedance can be considered infinite. Resistors in series between the source and noninverting input in Figure 3-7 will not affect the input impedance. However, if a resistor is placed from the noninverting input to ground, then the input impedance will be this resistance plus any series resistance. We will see the implications of this a little later.

As in the case of the inverting amplifier, the output impedance of the noninverting amplifier depends on the output impedance of the op-amp and is generally taken as zero.

Also, as in the case of the inverting amplifier, the bandwidth depends on the unity-gain bandwidth of the op-amp and the gain of the noninverting amplifier. The equation for the noninverting amplifier bandwidth, however, is different:

$$\text{bandwidth} = \frac{\text{unity-gain bandwidth}}{A_{VF}} \qquad (3\text{-}6)$$

With these equations, we can now design and analyze noninverting amplifiers.

Noninverting Amplifier Using a Bipolar Op-Amp

The circuit for a noninverting amplifier using the LM741C bipolar op-amp is shown in Figure 3-8. There is one difference between this circuit and the general inverting amplifier circuit in Figure 3-7—the inclusion of resistor R_K between the source and noninverting input.

Like the inverting amplifier, the noninverting amplifier using the LM741C requires the same DC resistance into each input to minimize the differential input

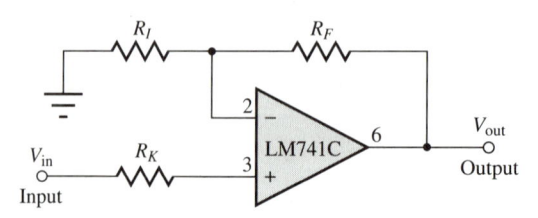

Figure 3-8 Noninverting amplifier using a LM741C bipolar op-amp.

voltage due to bias current. To achieve this, resistor R_K is placed in series between the source and noninverting input. Because R_K must equal the resistance at the inverting input, R_I in parallel with R_F, it has a value of

$$R_K = \frac{R_I R_F}{R_I + R_F} \tag{3-4}$$

This is the same equation as for the inverting amplifier.

There is a potential problem in using R_K. This resistor can only be calculated by Equation 3-4 and placed as shown in Figure 3-8, if the source impedance is much less than the calculated value of R_K. In the configuration shown, the bias current flows from the source through R_K into the noninverting input. Obviously, a DC voltage drop will be produced in any source resistance and added to that produced in R_K. If the source impedance is large, but less than R_K, its value should be subtracted from the calculated R_K value for determining the actual R_K to be used in the circuit. If the source impedance is larger than R_K, or if the source is capacitively coupled, then R_K *must be* placed between the noninverting input and ground. Resistor R_K then acts as the input impedance of the amplifier. In the examples and applications discussed in this book, we will always assume that the source has a low impedance and is DC-coupled.

The same design procedure is used for a noninverting amplifier circuit using the LM741C as was used for the inverting amplifier. Choose a value of R_F between 10 and 100 kΩ and calculate R_I from Equation 3-5, using the required gain. Alternatively, choose a value of R_I between 1 and 10 kΩ and calculate R_F. Calculate R_K by using Equation 3-4 with the values determined for R_I and R_F. The bandwidth can be calculated from Equation 3-6. Should offset correction be desired, refer to Figure 2-18 for the circuit.

The analysis of a noninverting amplifier circuit using the LM741C op-amp is also similar. The gain of the circuit is given by Equation 3-5, the input impedance is infinite (or at least very large), the output impedance is zero, and the amplifier bandwidth is given by Equation 3-6.

EXAMPLE 3-4

Design a noninverting amplifier using a LM741C op-amp with a gain of 25. Choose $R_I = 3000\ \Omega$. Use a ± 15 V power supply.

(a) Draw the circuit for a noninverting amplifier and prepare a parts list.
(b) Calculate values for R_F and R_K.
(c) Calculate the bandwidth of the amplifier.

Solution

This example is essentially a repeat of Example 3-1, except now we are designing a noninverting amplifier instead of an inverting amplifier. This example is intended to demonstrate the differences between the two types of amplifiers.

(a) We will use the basic LM741C noninverting amplifier circuit shown in Figure 3-8. This has been redrawn as Figure 3-9, and a parts list is included. The power supply connections are shown in this figure.

(b) Since R_I has been stipulated, and we want a gain of 25, we can calculate R_F from Equation 3-5.

$$A_{VF} = \frac{R_F}{R_I} + 1$$

$$A_{VF} - 1 = \frac{R_F}{R_I}$$

$$\frac{R_F}{R_I} = 24$$

Solving for R_F, we obtain

$$R_F = 24\, R_I = 24 \times 3.0 \text{ k}\Omega = 72 \text{ k}\Omega$$

We will select the closest standard value from the table in Appendix A-1, which is 75 kΩ.

The biasing resistor R_K is calculated using Equation 3-4. Substitute values for R_I and R_F:

$$R_K = \frac{R_I R_F}{R_I + R_F} = \frac{3.0 \text{ k}\Omega \times 75 \text{ k}\Omega}{3.0 \text{ k}\Omega + 75 \text{ k}\Omega} = 2.88 \text{ k}\Omega$$

Choose the closest standard value, which is 3.0 kΩ.

(c) The unity-gain bandwidth of the LM741C is 1 MHz. This value can be used in Equation 3-6 to calculate the bandwidth of the inverting amplifier:

$$\text{bandwidth} = \frac{\text{unity-gain bandwidth}}{A_{VF}} = \frac{1 \text{ MHz}}{25} = 40 \text{ kHz}$$

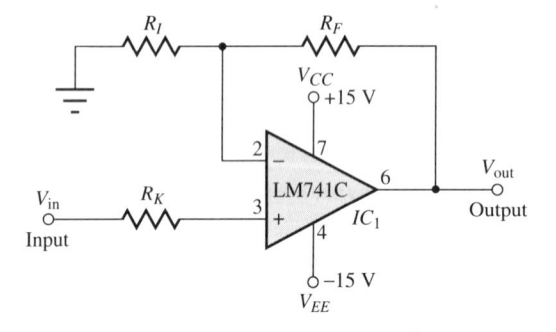

Figure 3-9

Our design is now complete. Notice that although the formulae were different, the design procedure was almost the same. The values for R_I, R_F, and R_K also turned out to be the same. This was only because R_I was stipulated as 3.0 kΩ.

Noninverting Amplifier Using a BiFET Op-Amp

The circuit for a noninverting amplifier using the TL081C BiFET op-amp is shown in Figure 3-10. Because of the very small leakage currents into the inputs of a BiFET op-amp, the resistor R_K used in the bipolar noninverting amplifier circuit between the source and noninverting input is not required, and the noninverting input is connected directly to the source.

If the source has a very high impedance, however, an additional resistor should be included in this circuit between the noninverting input and ground. Because of the extremely large impedance at the gate of a field-effect device, there is a tendency for a static charge to accumulate. Normally, this would discharge through a low-impedance source. For a very high impedance source (or capacitively coupled input), however, discharge cannot occur, and a high-value resistor should be placed between the noninverting input and ground to allow any static charge to drain to ground. This resistor, if used, determines the input impedance. Some designers automatically place a 1 MΩ resistor between noninverting input and ground.

The design of an amplifier circuit using the TL081C BiFET op-amp is identical to that of the LM741C bipolar op-amp, with the exception that R_K does not need to be calculated. Simply choose R_I or R_F and calculate the other, using the required gain and Equation 3-5. Calculate the bandwidth from Equation 3-6 as for the LM741C.

The analysis of a noninverting amplifier circuit using the TL081C BiFET op-amp is identical to that for the LM741C.

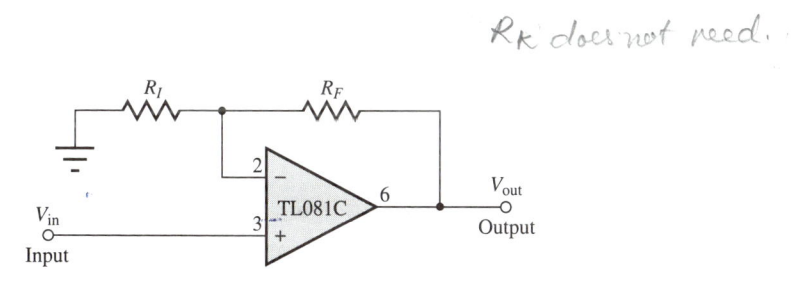

Figure 3-10 Noninverting amplifier using a TL081C BiFET op-amp.

EXAMPLE 3-5

Design a noninverting amplifier using a TL081C op amp for a gain of 40. Choose R_F = 100 kΩ. Use a ±15 V power supply.

(a) Draw the circuit for a noninverting amplifier and prepare a parts lists.
(b) Calculate the value of R_I.
(c) Calculate the bandwidth of the amplifier.

Solution

This example is similar to Example 3-4, except now we are designing for a gain of 40 and have stipulated R_F instead of R_I.

(a) We will use the basic TL081C noninverting amplifier circuit shown in Figure 3-10. This has been redrawn as Figure 3-11, and a parts list is included. The power supply connections are shown in this figure.

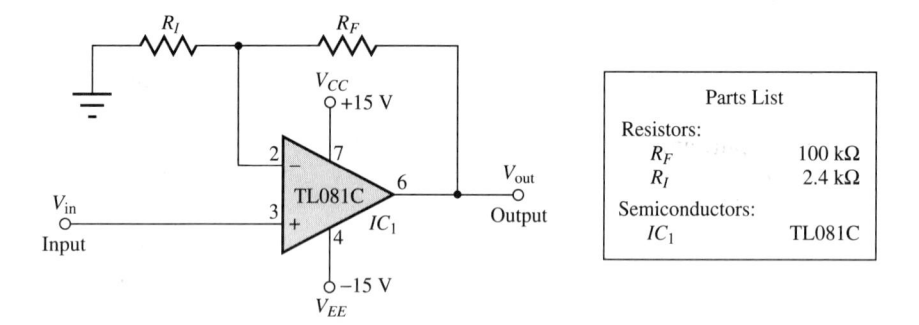

Figure 3-11

(b) Since R_F has been stipulated, and we want a gain of 40, we can calculate R_I from Equation 3-5.

$$A_{VF} = \frac{R_F}{R_I} + 1$$

$$A_{VF} - 1 = \frac{R_F}{R_I}$$

$$\frac{R_F}{R_I} = 39$$

Solving for R_I, we obtain

$$R_I = \frac{R_F}{39} = \frac{100 \text{ k}\Omega}{39} = 2.56 \text{ k}\Omega$$

Selecting the closest standard value from the table in Appendix A-1 gives $R_I = 2.4 \text{ k}\Omega$.

(c) The unity-gain bandwidth of the TL081C is 3 MHz. This value can be used in Equation 3-6 to calculate the bandwidth of the noninverting amplifier:

$$\text{bandwidth} = \frac{\text{unity-gain bandwidth}}{A_{VF}} = \frac{3 \text{ MHz}}{40} = 75 \text{ kHz}$$

Our design is now complete.

3-3 Buffer Amplifiers

Buffer amplifiers are simply unity-gain amplifiers used for protecting the load or for boosting the power to the load. They are usually noninverting, partly because of circuit simplicity, but more importantly because of the high input impedance and because there is no phase change from input to output. The noninverting buffer amplifier is sometimes called a follower amplifier.

The Noninverting Buffer Amplifier

The circuit for a noninverting buffer or follower amplifier, both for bipolar op-amps and for BiFET op-amps, is shown in Figure 3-12. Notice the simplicity of this circuit. There are no resistors to design, the output is connected directly to the inverting input, and input signal is applied directly to the noninverting input. Since V_{in} is applied directly to the noninverting input, and V_{out} is applied directly to the inverting input, and since the two inputs of an op-amp must be at the same voltage, V_{out} must equal V_{in}. Hence, $V_{out}/V_{in} = 1$, giving the amplifier unity gain.

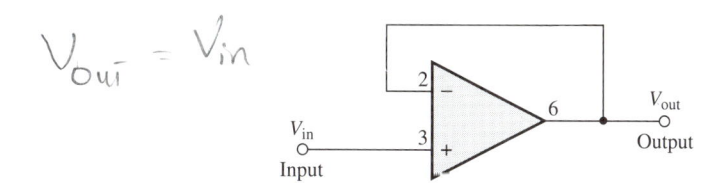

Figure 3-12 Following or noninverting buffer amplifier.

This configuration always has unity gain. The input impedance is maximum, and the output impedance is minimum and extremely close to zero. The bandwidth is the full unity-gain bandwidth (1 MHz for an LM741C, 3 MHz for a TL081C). This is an extremely simple and widely used amplifier.

Because the gain is unity and the amplifier is noninverting, the output voltage signal is identical to the input voltage signal. The amplifier acts as a current amplifier. Because of the extremely high input impedance, it draws virtually no current from the source, but it can deliver considerable output current. It effectively buffers the source from the load.

Inverting Buffer Amplifiers

Occasionally an inverting buffer amplifier is required. Such an amplifier is simply an inverting amplifier configured for unity gain. Figure 3-13 shows the circuits for a bipolar inverting buffer amplifier and a BiFET inverting buffer amplifier. These are identical to the inverting amplifiers in Figures 3-2 and 3-5, respectively. Resistors R_I and R_F are designed using Equation 3-1.

$$A_{VF} = \frac{R_F}{R_I} \tag{3-1}$$

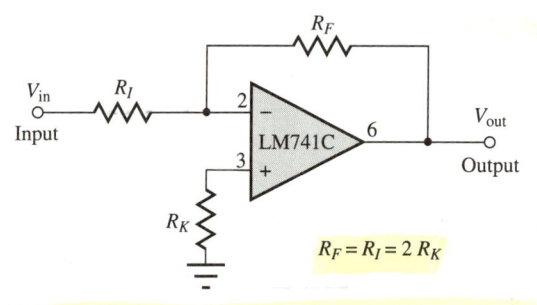

(a) Inverting buffer amplifier using a LM741C bipolar op-amp

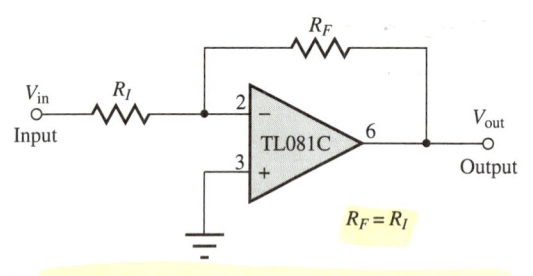

(b) Inverting buffer amplifier using a TL081C BiFET op-amp

Figure 3-13 Inverting buffer amplifiers.

If the voltage gain A_{VF} is to be 1, then R_F must equal R_I. These resistors can be any values. Normally a large input impedance is required for buffer amplifiers. Since the input impedance is equal to R_I, choose a suitably large value for R_I, typically 1 MΩ, and make R_F the same value. It is left to the reader to show that R_K for a bipolar buffer amplifier is always $\frac{1}{2}R_F$ or $\frac{1}{2}R_I$. Because this is an inverting amplifier, the bandwidth is given by Equation 3-3, and will always be one-half of the unity-gain bandwidth.

3-4 Single-Supply Amplifiers

Up to this point, we have used amplifier circuits that required dual-polarity supplies. This is the usual case. The differential amplifier circuits used in operational amplifiers are biased so that the input and output no-signal levels are at 0 V. An input AC signal can swing positive and negative with respect to zero.

Sometimes, however, it is necessary to operate an amplifier from a single power supply. For example, in portable instruments often only a single battery is available. There are several methods of making a single-supply amplifier. The most common approach is to use a conventional op-amp such as a TL081C or

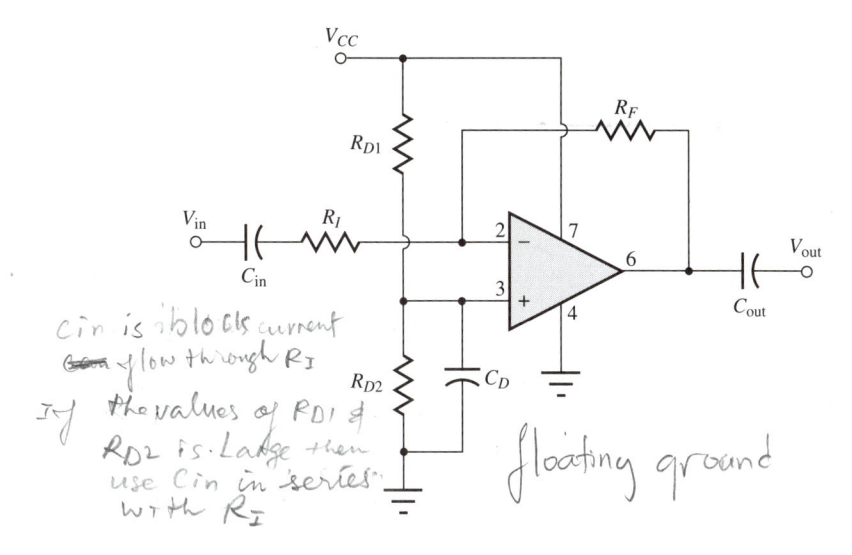

Figure 3-14 Inverting amplifier using a single power supply.

LM741C and put a voltage divider across the supply to create a floating ground. The circuit is shown in Figure 3-14. This is an inverting amplifier. The noninverting input is connected to the center of the voltage divider, which is treated as the floating ground. Pin 4 on the op-amp (the negative supply pin) is connected to the true power supply ground. Because the effective op-amp ground is different from true ground, both the input and the output of the op-amp will be offset from true ground. Hence, both the source and the load must be capacitively coupled to the amplifier. This is the price that must be paid for using a single supply.

The input resistor R_I and feedback resistor R_F are designed using Equation 3-1 in exactly the same manner as for a standard inverting amplifier with a dual supply. The voltage divider consisting of R_{D1} and R_{D2} sets the floating ground. Usually the floating-ground voltage is set equal to half of the supply voltage to simulate a standard dual supply and to provide for maximum output voltage swing. In this case, $R_{D1} = R_{D2}$. For a bipolar op-amp such as the LM741C, the voltage divider must also serve as R_K for bias offset control. For the single-supply circuit, however, $R_K = R_F$ and is not equal to $R_I \parallel R_F$ as in a standard bipolar op-amp circuit. The reason is that the input is capacitively coupled through R_I; hence, no bias current can flow through R_I. Each of the divider resistors should equal $2R_F$ so that $R_{D1} \parallel R_{D2} = R_K$ for proper offset control. For a BiFET op-amp, which does not need a bias offset control resistor R_K, any relatively large value of resistance could be chosen for R_{D1} and R_{D2}.

A capacitor C_{in} must be placed in series with R_I. For any amplifier using a conventional op-amp with a single supply, the inverting input is no longer at ground potential, so the input must be AC-coupled. To calculate a value for C_{in}, a lower cutoff frequency f_L must be determined. This frequency will set the lower

−3 dB point in the amplifier frequency response, and is, by definition, the frequency at which the capacitive reactance is equal to the source impedance R_S plus the input impedance R_I. Normally the source resistance will be much less than R_I, but we will include it in the equation.

$$X_C = \frac{1}{2\pi f_L C_{in}} = R_S + R_I$$

Solving for C_{in}, we find

$$C_{in} = \frac{1}{2\pi f_L (R_S + R_I)} \tag{3-7}$$

Always choose a larger capacitor value than the calculated value of C_{in}.

The output of the amplifier should also be capacitively coupled since there will always be a DC offset voltage at the output. The output capacitor C_{out} is determined in the same manner as the input capacitor, namely by setting the capacitive reactance at the lower cutoff frequency equal to the output impedance of the amplifier, Z_{out}, plus the load resistance R_L. Normally the output impedance of the amplifier is zero, but we will include it in the equation.

$$X_{C_{out}} = \frac{1}{2\pi f_L C_{out}} = Z_{out} + R_L$$

Solving for C_{out}, we find

$$C_{out} = \frac{1}{2\pi f_L (Z_{out} + R_L)} \tag{3-8}$$

The actual capacitor chosen should be larger than this value.

The floating ground *must be* effectively an AC ground. This is achieved by connecting capacitor C_D between the floating-ground point of the voltage divider and the signal ground. The capacitive reactance of C_D at the lower cutoff frequency should be much less than the resistance of R_{D1} in parallel with R_{D2}.

$$X_{C_D} = \frac{1}{2\pi f_L C_D} \ll \frac{R_{D1} R_{D2}}{R_{D1} + R_{D2}}$$

Solving for C_D, we get

$$C_D \gg \frac{1}{2\pi f_L R_{D1} R_{D2}/(R_{D1} + R_{D2})} \tag{3-9}$$

Typically, we would make C_D at least 100 times larger than the value that we calculate.

This amplifier will operate like a normal dual-supply amplifier with the one exception that it can only be used for AC signals. It cannot be used as a DC amplifier. The frequency response curve will be similar to that in Figure 2-7(a) rather than that in Figure 2-7(b).

3-4 Single-Supply Amplifiers

EXAMPLE 3-6

Design an AC-coupled amplifier circuit using a TL081C op-amp to run from a single 12 V supply. The amplifier should have a gain of 20, an input impedance of 2000 Ω, and a lower cutoff frequency of 75 Hz. Assume that the input is driven by a 50 Ω source and that the output is driving a 50 Ω transmission line (i.e., R_L = Z_{out} = 50 Ω). Bias the amplifier for maximum AC swing.

Solution

This is a typical single-supply inverting amplifier. Since we are using a standard BiFET op-amp designed for operation from a dual supply, we must use the circuit shown in Figure 3-14. This has been redrawn as Figure 3-15, with a parts list included. Since the input impedance is to be 2000 Ω, and since $Z_{in} = R_I$ for an inverting amplifier, we know the value for R_I. We will choose a standard 2.0 kΩ resistor for R_I. The source resistance R_S is much less than this, and its effect can be ignored.

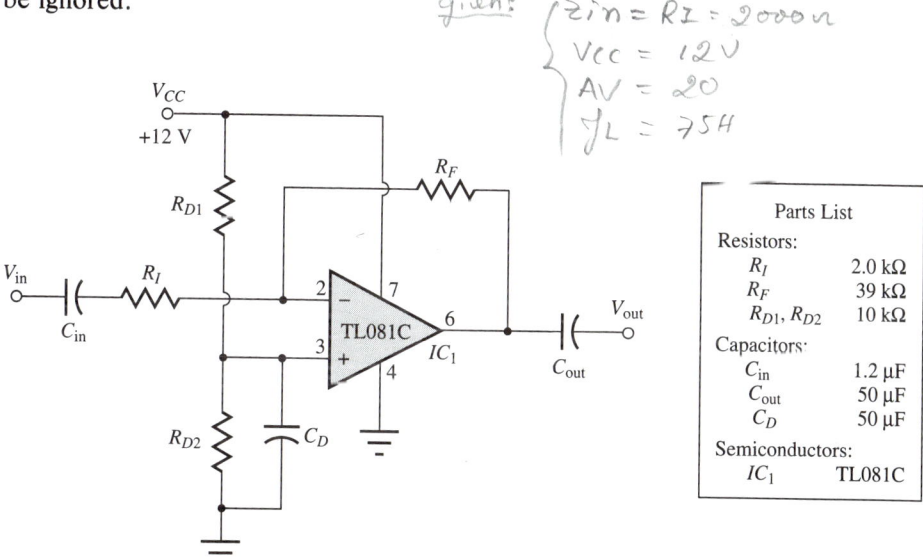

given:
$Z_{in} = R_I = 2000\,\Omega$
$V_{CC} = 12\,V$
$AV = 20$
$f_L = 75\,H$

Figure 3-15

Parts List
Resistors:
R_I — 2.0 kΩ
R_F — 39 kΩ
R_{D1}, R_{D2} — 10 kΩ
Capacitors:
C_{in} — 1.2 μF
C_{out} — 50 μF
C_D — 50 μF
Semiconductors:
IC_1 — TL081C

To calculate R_F, we use the gain equation for an inverting amplifier, Equation 3-1:

$$A_V = \frac{R_F}{R_I} \quad \therefore \quad R_F = R_I A_V$$

Substituting values for R_I and A_V, we obtain

$$R_F = 2000\,\Omega \times 20 = 40{,}000\,\Omega$$

Select a 39 kΩ resistor for R_F as the closest standard value from the table in Appendix A-1.

Since we want the maximum output voltage swing, we must set the floating ground to $\frac{1}{2}V_{CC}$, that is, to 6 V. The voltage divider is easy to design since R_{D1} and R_{D2} must have the same value. We will choose 10 kΩ for both of them.

Now we must design the capacitors. Select C_{in} such that $X_{C_{in}} \leq R_S + Z_{in}$ at 75 Hz. We know that $R_S = 50\ \Omega$ and $R_I = Z_{in} = 2000\ \Omega$. Use Equation 3-7:

$$C_{in} = \frac{1}{2\pi f_L(R_S + R_I)}$$

[↳ Resistor from source]

$$= \frac{1}{2 \times \pi \times 75\ \text{Hz} \times (50\ \Omega + 2000\ \Omega)} = 1.035\ \mu F$$

We will choose the next larger standard value, which from the table in Appendix A-2 is 1.2 μF.

Next, calculate C_{out} such that $X_{C_{out}} \leq R_L$ at 75 Hz. We will assume that $Z_{out} = 0$ for an inverting amplifier with feedback. Use Equation 3-8:

$$C_{out} = \frac{1}{2\pi f_L(Z_{out} + R_L)}$$

$$= \frac{1}{2 \times \pi \times 75\ \text{Hz} \times 50\ \Omega} = 42.44\ \mu F$$

We will choose an electrolytic capacitor from Appendix A-2 with a value of 50 μF for C_{out}.

Finally, calculate C_D such that $X_{C_D} \ll R_{D1} \parallel R_{D2}$. Since $R_{D1} = R_{D2} = 10\ \text{k}\Omega$, then $R_{D1} \parallel R_{D2} = 5000\ \Omega$. Use Equation 3-9:

$$C_D \gg \frac{1}{2\pi f_L R_{D1} R_{D2}/(R_{D1} + R_{D2})} = \frac{1}{2\pi f_L R_{D1} \parallel R_{D2}}$$

$$= \frac{1}{2 \times \pi \times 75\ \text{Hz} \times 5000\ \Omega} = 0.42\ \mu F$$

We will choose C_D to be 100 times this value, and select a 50 μF electrolytic capacitor from Appendix A-2.

Our design is now complete.

3-5 The TL321C Single-Supply Op-Amp

Some op-amps are specifically designed to operate from a single supply. These have the advantage over conventional op-amps operated from a single supply in that they can be direct-coupled and, hence, can be used to amplify DC signal levels. The TL321, for which the data sheets are given in Figure 3-16, is a typical example of a single-supply op-amp.

The TL321 has the operational characteristics of a standard bipolar op-amp such as the LM741C, but is designed to operate from a single supply with a voltage

LINEAR INTEGRATED CIRCUITS

TYPES TL321M, TL321I, TL321C
OPERATIONAL AMPLIFIERS
BULLETIN NO. DL-S 7712515, APRIL 1977

- Wide Range of Supply Voltages
  Single Supply . . . 3 V to 30 V
  or Dual Supplies
- Low Supply Current Drain
  Independent of Supply Voltage
  . . . 0.8 mA Typ
- Common-Mode Input Voltage
  Range Includes Ground Allowing
  Direct Sensing near Ground

- Low Input Bias and Offset Parameters
  Input Offset Voltage . . . 2 mV Typ
  Input Offset Current . . . 3 nA Typ (TL321M)
  Input Bias Current . . . 45 nA Typ
- Differential Input Voltage Range
  Equal to Maximum-Rated
  Supply Voltage . . . ±32 V
- Open-Loop Differential Voltage
  Amplification . . . 100 V/mV Typ
- Internal Frequency Compensation

schematic

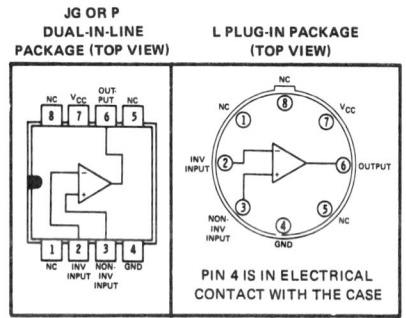

JG OR P DUAL-IN-LINE PACKAGE (TOP VIEW)

L PLUG-IN PACKAGE (TOP VIEW)

PIN 4 IS IN ELECTRICAL CONTACT WITH THE CASE

NC—No internal connection

description

The TL321 is a high-gain, frequency-compensated operational amplifier that was designed specifically to operate from a single supply over a wide range of voltages. Operation from split supplies is also possible so long as the difference between the two supplies is 3 volts to 30 volts and Pin 7 is at least 1.5 volts more positive than the input common-mode voltage. The low supply current drain is independent of the magnitude of the supply voltage.

Applications include transducer amplifiers, d-c amplification blocks, and all the conventional operational amplifier circuits that now can be more easily implemented in single-supply-voltage systems. For example, the TL321 can be operated directly off of the standard five-volt supply that is used in digital systems and will easily provide the required interface electronics without requiring additional ±15-volt supplies.

absolute maximum ratings over operating free-air temperature range (unless otherwise noted)

Supply voltage, V_{CC} (see Note 1) . . . . 32 V
Differential input voltage (see Note 2) . . . . ±32 V
Input voltage range (either input) . . . . −0.3 V to 32 V
Duration of output short-circuit to ground at (or below 25°C
  free-air temperature ($V_{CC} \leq 15$ V) (see Note 3) . . . . unlimited
Continuous total dissipation at (or below) 25°C free-air temperature (see Note 4):
  JG or P package . . . . 680 mW
  L package . . . . 600 mW
Operating free-air temperature range: TL321M . . . . −55°C to 125°C
                                       TL321I . . . . −25°C to 85°C
                                       TL321C . . . . 0°C to 70°C
Storage temperature range . . . . −65°C to 150°C
Lead temperature 1/16 inch from case for 60 seconds: JG or L package . . . . 300°C
Lead temperature 1/16 inch from case for 10 seconds: P package . . . . 260°C

NOTES: 1. All voltage values, except differential voltages, are with respect to the network ground terminal.
       2. Differential voltages are at the noninverting input terminal with respect to the inverting input terminal.
       3. Short circuits from the output to V_{CC} can cause excessive heating and eventual destruction.
       4. For operation above 25°C free-air temperature, refer to Dissipation Derating Table.

TEXAS INSTRUMENTS
INCORPORATED
POST OFFICE BOX 5012 • DALLAS, TEXAS 75222

Figure 3-16 Data sheets for the TL321C single-supply op-amp. *(Reprinted by permission of Texas Instruments)*

TYPES TL321M, TL321I, TL321C
OPERATIONAL AMPLIFIERS

electrical characteristics at specified free-air temperature, V_{CC} = 5 V (unless otherwise noted)

PARAMETER		TEST CONDITIONS[†]		TL321M, TL321I MIN TYP MAX			TL321C MIN TYP MAX			UNIT
V_{IO}	Input offset voltage	V_O = 1.4 V, V_{CC} = 5 V to 30 V	25°C		2	5		2	7	mV
			Full range			7			9	
I_{IO}	Input offset current	V_O = 1.4 V	25°C		3	30		5	50	nA
			Full range			100			150	
I_{IB}	Input bias current	V_O = 1.4 V, See Note 5	25°C		−45	−150		−45	−250	nA
			Full range			−300			−500	
V_{ICR}	Common-mode input voltage range	V_{CC} = 30 V	25°C	0 to V_{CC}−1.5			0 to V_{CC}−1.5			V
			Full range	0 to V_{CC}−2			0 to V_{CC}−2			
V_{OH}	High-level output voltage	V_{CC} = 30 V, R_L = 2 kΩ	Full range	26			26			V
		V_{CC} = 30 V, R_L ⩾ 10 kΩ	Full range	27	28		27	28		
V_{OL}	Low-level output voltage	R_L ⩽ 10 kΩ	Full range		5	20		5	20	mV
A_{VD}	Large-signal differential voltage amplification	V_{CC} = 15 V, V_O = 1 V to 11 V, R_L ⩾ 2 kΩ	25°C	50	100		25	100		V/mV
			Full range	25			15			
CMRR	Common-mode rejection ratio	R_S ⩽ 10 kΩ	25°C	70	85		65	85		dB
$\Delta V_{CC}/\Delta V_{IO}$	Supply voltage rejection ratio	R_S ⩽ 10 kΩ	25°C	65	100		65	100		dB
I_O	Output current	V_{CC} = 15 V, V_{ID} = 1 V, V_O = 0 V	25°C	−20	−40		−20	−40		mA
			Full range	−10	−20		−10	−20		
		V_{CC} = 15 V, V_{ID} = −1 V, V_O = 2.5 V	25°C	10	20		10	20		
			Full range	5	8		5	8		
		V_{ID} = −1 V, V_O = 200 mV	25°C	12	50		12	50		μA
I_{CC}	Supply current	No load, No signal	25°C		0.4			0.4		mA
			Full range			1			1	

[†]All characteristics are specified under open-loop conditions. Full range is −55°C to 125°C for TL321M, −25°C to 85°C for TL321I, and 0°C to 70°C for TL321C.

NOTE 5: The direction of the bias current is out of the device due to the P-N-P input stage. This current is essentially constant, regardless of the state of the output, so no loading change is presented to the input lines.

DISSIPATION DERATING TABLE

PACKAGE	POWER RATING	DERATING FACTOR	ABOVE T_A
JG	680 mW	7.4 mW/°C	58°C
L	600 mW	5.0 mW/°C	30°C
P	680 mW	8.0 mW/°C	65°C

TEXAS INSTRUMENTS
INCORPORATED
POST OFFICE BOX 5012 • DALLAS, TEXAS 75222

Figure 3-16 (*Continued*)

between 3 and 30 V. (It can also operate from a dual supply as long as the voltage difference is at least 3 V and less than 30 V.) The internal circuit uses a differential amplifier but uses a current regulator in the emitter circuit to bias the amplifier. The schematic of the internal circuitry of the op-amp is shown on the data sheets. The pinout of the TL321 is identical to the standard op-amp pinout. Like standard op-amps, the TL321 is available in a number of ratings including the M (military), I (industrial), and C (commercial). We will use the commercial device, namely the TL321C, with an operating temperature range of 0 to 70°C.

Like standard op-amps, the TL321C can be used to produce inverting amplifiers and noninverting amplifiers. The two types of amplifiers are quite different, and only the noninverting amplifier is capable of being direct-coupled. Since the noninverting amplifier is the simpler, we will look at it first.

Noninverting Amplifier Using a TL321C Single-Supply Op-Amp

The circuit for a noninverting amplifier using the TL321C single-supply op-amp is shown in Figure 3-17. This circuit is exactly the same as a noninverting amplifier using a standard bipolar op-amp. The feedback resistor R_F and the input resistor R_I are determined from the gain by using Equation 3-5 with the same procedure for a normal noninverting amplifier. Resistor R_K, used for offset correction, is calculated by Equation 3-4, and the same restrictions apply to this single-supply amplifier circuit as for the standard bipolar noninverting amplifier circuit.

The only difference is in operation of the amplifier—only positive signals will produce an output. For zero input, the output is zero. For a positive input signal, the output signal is a positive amplification of the input signal. For a negative input signal, since the output cannot go negative, the output is zero. For many applications, this operation is entirely satisfactory. Most signals from industrial transducers have only one polarity. For example, a standard temperature transducer will output a positive voltage proportional to the temperature.

To amplify a pure AC input, such as an audio signal, it is necessary to add a positive offset so that the input signal never goes negative. This offset will appear amplified at the output. The noninverting amplifier is not well suited for such applications. The inverting amplifier, however, is.

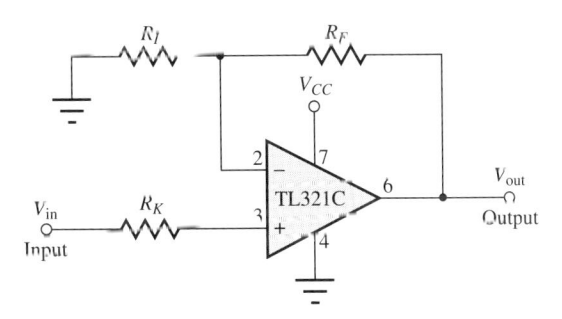

Figure 3-17 Noninverting amplifier circuit using the TL321C single-supply op-amp.

EXAMPLE 3-7

Design a DC-coupled noninverting amplifier with a gain of 50 and a high-impedance input. Use a TL321C single-supply op-amp operating from a single 24 V supply. Choose $R_I = 2.0 \text{ k}\Omega$.

Solution

This is a typical example of a noninverting amplifier design using a single-supply op-amp. We must use the circuit in Figure 3-17. This has been redrawn as Figure 3-18, and a parts list is included. The design procedure will be the same as for a standard noninverting amplifier.

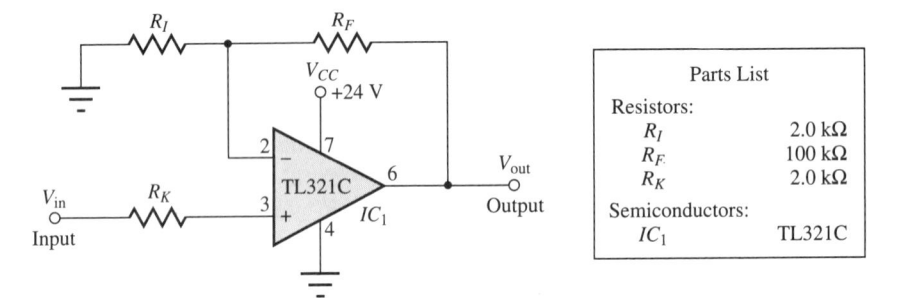

Figure 3-18

Since R_I has been stipulated, and we want a gain of 50, we can calculate R_F from Equation 3-5.

$$A_{VF} = \frac{R_F}{R_I} + 1$$

$$A_{VF} - 1 = \frac{R_F}{R_I}$$

$$\frac{R_F}{R_I} = 49$$

Solving for R_F, we obtain

$$R_F = 49\, R_I = 49 \times 2.0 \text{ k}\Omega = 98 \text{ k}\Omega$$

We will select the closest standard value from the table in Appendix A-1, which is 100 kΩ.

The biasing resistor R_K is given by Equation 3-4. Substitute values for R_I and R_F:

$$R_K = \frac{R_I R_F}{R_I + R_F} = \frac{2.0 \text{ k}\Omega \times 100 \text{ k}\Omega}{2.0 \text{ k}\Omega + 100 \text{ k}\Omega} = 1.96 \text{ k}\Omega$$

Choose the closest standard value, which is 2.0 kΩ.

Our amplifier design is now complete. Note that we do not require any capacitors in this circuit.

Inverting Amplifier Using a TL321C Single-Supply Op-Amp

The circuit for an inverting amplifier using the TL321C single-supply op-amp is shown in Figure 3-19. While this circuit appears similar to an inverting amplifier using a standard bipolar op-amp powered by a single supply, the design is somewhat different.

The feedback resistor R_F and the input resistor R_I are determined from Equation 3-1, using the same procedure as for a normal inverting amplifier. The voltage divider consisting of R_{D1} and R_{D2}, however, functions differently than for a standard op-amp operating from a single supply. In this case, the voltage divider provides a DC offset voltage that is added to the signal applied to the inverting input. Both the offset voltage at the noninverting input and the signal at the inverting input are amplified. The optimum value of DC offset at the output is $\frac{1}{2}V_{CC}$, to allow for maximum swing in the amplified signal voltage. To give this output offset, the voltage at the noninverting input should be $\frac{1}{2}V_{CC}/A_V$.

The DC offset voltage at the noninverting input is found by the standard voltage divider formula. Equate this voltage to $\frac{1}{2}V_{CC}/A_V$ to obtain

$$\frac{R_{D2}}{R_{D1} + R_{D2}} \times V_{CC} = \frac{1}{2} \frac{V_{CC}}{A_V}$$

Thus,

$$\frac{R_{D2}}{R_{D1} + R_{D2}} = \frac{1}{2A_V} \qquad (3\text{-}10)$$

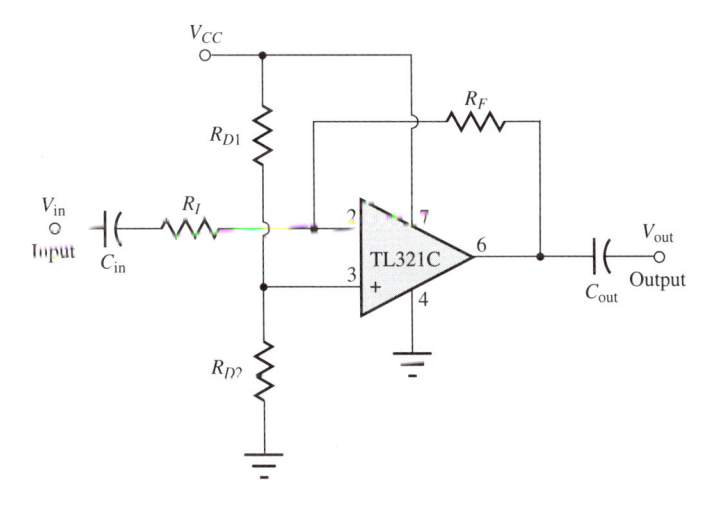

Figure 3-19 Inverting amplifier circuit using the TL321C single-supply op-amp.

For bias offset as in any bipolar op-amp, a resistor R_K is required. In this circuit, R_K consists of $R_{D1} \| R_{D2}$ and must equal the resistance through which bias current flows to the inverting input. The resistance to the inverting input is simply R_F because the input is capacitively coupled through R_I; hence, no bias current can flow through R_I. Equating $R_{D1} \| R_{D2}$ to R_F gives us a second equation in R_{D1} and R_{D2}.

$$\frac{R_{D1}R_{D2}}{R_{D1} + R_{D2}} = R_F \qquad (3\text{-}11)$$

Solving Equations 3-10 and 3-11 simultaneously, we arrive at the following equations for R_{D1} and R_{D2} after some algebra:

$$R_{D1} = 2A_V R_F \qquad (3\text{-}12)$$

$$R_{D2} = \frac{R_F R_{D1}}{R_{D1} - R_F} \qquad (3\text{-}13)$$

We will have a DC offset at the input and at the output, so we must use AC coupling. The coupling capacitors are designed by exactly the same equations as for a standard op-amp using a single supply, namely Equation 3-7 and Equation 3-8. These equations are repeated below for convenience. The equation for C_{in} is

$$C_{in} = \frac{1}{2\pi f_L (R_S + R_I)} \qquad (3\text{-}7)$$

The equation for C_{out} is

$$C_{out} = \frac{1}{2\pi f_L (Z_{out} + R_L)} \qquad (3\text{-}8)$$

There is actually little advantage in using a TL321C as an AC inverting amplifier since the circuit is almost the same as for a standard op-amp, while the design of the voltage divider is somewhat more complicated. The real advantage of the TL321C is that it can be used as a single-supply noninverting DC amplifier.

EXAMPLE 3-8

Design an inverting amplifier with a gain of 8 using a TL321C single-supply op-amp operating from a 12 V supply. The amplifier should have an input impedance of 3000 Ω. Bias the amplifier for maximum output swing. Assume a source impedance of 0 Ω, a load impedance of 10 kΩ, and a cutoff frequency of 60 Hz.

Solution

We must use the circuit shown in Figure 3-19. This has been redrawn as Figure 3-20, with a parts list included. Since the input impedance is stipulated as 3000 Ω, we know that the value for R_I must be 3000 Ω, and from Appendix A-1 we see that 3.0 kΩ is a standard value.

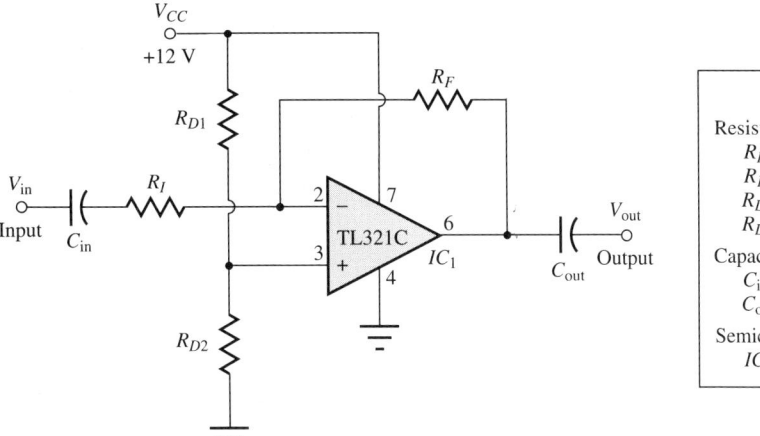

3-5 The TL321C Single-Supply Op-Amp

Figure 3-20

Parts List
Resistors:
R_I 3.0 kΩ
R_F 24 kΩ
R_{D1} 390 kΩ
R_{D2} 27 kΩ
Capacitors:
C_{in} 1.0 μF
C_{out} 0.33 μF
Semiconductors:
IC_1 TL321C

To calculate R_F, we use the gain equation for an inverting amplifier, Equation 3-1:

$$A_V = \frac{R_F}{R_I} \quad \therefore \quad R_F = R_I A_V$$

Substituting values for R_I and A_V gives

$$R_F = 3.0 \text{ k}\Omega \times 8 = 24 \text{ k}\Omega$$

The table in Appendix A-1 shows this to be a standard value, so we will make $R_F = 24$ kΩ.

Since we want the maximum output voltage swing, we must bias the noninverting input to $V_{CC}/2A_V$ using R_{D1} and R_{D2}. Use Equations 3-12 and 3-13 to calculate these resistors:

$$R_{D1} = 2A_V R_F = 2 \times 8 \times 24 \text{ k}\Omega = 384 \text{ k}\Omega$$

$$R_{D2} = \frac{R_F R_{D1}}{R_{D1} - R_F} = \frac{24 \text{ k}\Omega \times 384 \text{ k}\Omega}{384 \text{ k}\Omega - 24 \text{ k}\Omega} = 25.6 \text{ k}\Omega$$

We will select a 390 kΩ resistor for R_{D1} and a 27 kΩ resistor for R_{D2} from the table in Appendix A-1.

Next, we must calculate the coupling capacitors C_{in} and C_{out}. Use Equations 3-7 and 3-8.

$$C_{in} = \frac{1}{2\pi f_L (R_S + R_I)} = \frac{1}{2\pi \times 60 \text{ Hz} \times 3.0 \text{ k}\Omega} = 0.884 \text{ μF}$$

$$C_{out} = \frac{1}{2\pi f_L (Z_{out} + R_L)} = \frac{1}{2\pi \times 60 \text{ Hz} \times 10 \text{ k}\Omega} = 0.265 \text{ μF}$$

Choose $C_{in} = 1.0$ μF and $C_{out} = 0.33$ μF from the values listed in Appendix A-2. Note that we chose larger values of capacitance.

Our amplifier design is now complete.

3-6 Norton or Current-Mode Op-Amps

There is a another form of op-amp that is sometimes encountered. This is the **Norton op-amp,** sometimes called a **current-mode op-amp.** The LM741C and TL081C op-amps are voltage amplifiers, amplifying the differential voltage applied between the inputs while drawing negligible current. The Norton op-amp, however, is designed as a current amplifier. It amplifies the difference in current between the inputs. The Norton op-amp was originally introduced as an alternative to the bipolar op-amp, having a larger unity-gain bandwidth and being capable of operating from a single supply.

Current amplifiers are not as widely used as they could be because they are *different*—many designers simply do not know how to use them. Yet current amplifiers are just as easy, if not easier, to use than conventional op-amps, and have several quite useful features.

(a) Simple inverting amplifier

(b) True current amplifier

(c) Differential current amplifier

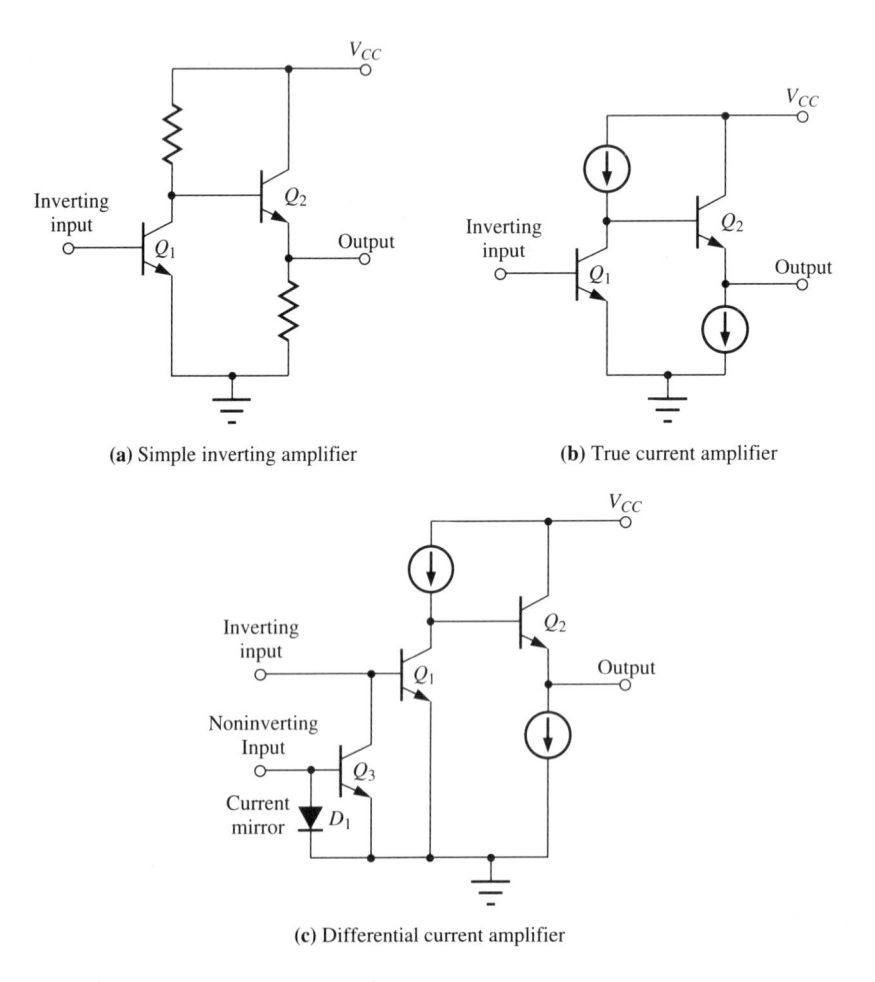

Figure 3-21 Basic circuit of a discrete component current amplifier.

3-6 Norton or Current-Mode Op-Amps

The theory of current amplifiers is significantly different from that of the operational amplifiers we have been studying. The basic circuit of a current amplifier is shown in Figure 3-21. Figure 3-21(a) shows a simple inverting amplifier (single input). The current input to the base of Q_1 is amplified by Q_1, and part of this current flows into the base of Q_2. This current is in turn amplified by Q_2, and the output current produces a voltage drop across R_2. This amplifier circuit is not a true current amplifier. The true current amplifier circuit shown in Figure 3-21(b) drives both Q_1 and Q_2 from constant-current sources. To obtain a noninverting input, we add a current mirror as shown in Figure 3-21(c). The current mirror consists of a transistor biased with a diode connected between base and emitter. (In practice, to give better stability, two matched transistors are used; the first transistor is connected as a diode with collector tied to the base.) The current mirror draws a collector current through Q_3 that is almost identical to the input current through the diode. The base current to Q_1 is equal to the difference between the two input currents.

Data sheets for the LM3900 Norton op-amp are shown in Figure 3-22. There are very few Norton-type op-amps on the market, and the data sheets in Figure 3-22 show the four most common: LM2900, LM3900, LM3301, and LM3401. The LM3900 is the most readily available of these, and we will use it exclusively. Note that the LM3900 is only available in a quad (four-device) package. The simplified schematic of the actual amplifier circuit used in this integrated circuit is shown on the data sheets and can be compared to that developed in Figure 3-21(c). Other than a more complicated output stage, the circuits are the same. The Norton op-amp has its own special symbol, shown in Figure 3-23.

The Norton op-amp can be used for any application for which a standard op-amp is used. Amplifier circuits using the Norton op-amp will be introduced shortly, but first we will look at some of the various characteristics of this op-amp.

Characteristics of the Norton or Current-Mode Op-Amp

One of the main features of the Norton op-amp is that it can operate from a single supply. All four op-amps in the 14 pin DIP of the LM3900 are powered from a common input, pin 14 for V_{CC} and pin 7 for the ground. The power supply range of the LM3900, as listed on the data sheets in Figure 3-22, is from 4 to 32 V. It can also operate from dual supplies from ± 2 to ± 16 V.

The input impedance of a Norton amplifier is 1 MΩ, somewhat less than that of a standard op-amp such as the LM741C; however, comparison of the input impedance of a current amplifier with the impedance of a voltage amplifier is not particularly relevant. The current amplifier is designed for current flow into both inputs. The voltage amplifier ideally has very little current flow. The output impedance is 8 kΩ, much larger than standard op-amps. Although this is reduced considerably by feedback in a practical amplifier circuit, it will still be a few ohms.

One weakness of the Norton op-amp over a conventional op-amp such as the LM741C is that the open-loop gain is only 2800, much less than conventional op-amps with open-loop gains of typically 200,000. In actual practice, this poses little

LM2900/LM3900, LM3301, LM3401

National Semiconductor

Operational Amplifiers/Buffers

LM2900/LM3900, LM3301, LM3401 Quad Amplifiers

General Description

The LM2900 series consists of four independent, dual input, internally compensated amplifiers which were designed specifically to operate off of a single power supply voltage and to provide a large output voltage swing. These amplifiers make use of a current mirror to achieve the non-inverting input function. Application areas include: ac amplifiers, RC active filters, low frequency triangle, squarewave and pulse waveform generation circuits, tachometers and low speed, high voltage digital logic gates.

Features

- Wide single supply voltage range or dual supplies — 4 V_{DC} to 36 V_{DC} / ±2 V_{DC} to ±18 V_{DC}
- Supply current drain independent of supply voltage
- Low input biasing current — 30 nA
- High open-loop gain — 70 dB
- Wide bandwidth — 2.5 MHz (Unity Gain)
- Large output voltage swing — (V^+ −1) Vp-p
- Internally frequency compensated for unity gain
- Output short-circuit protection

Schematic and Connection Diagrams

Typical Applications (V^+ = 15 V_{DC})

Inverting Amplifier

Triangle/Square Generator

Frequency-Doubling Tachometer

Low V_{IN} − V_{OUT} Voltage Regulator

Non-Inverting Amplifier

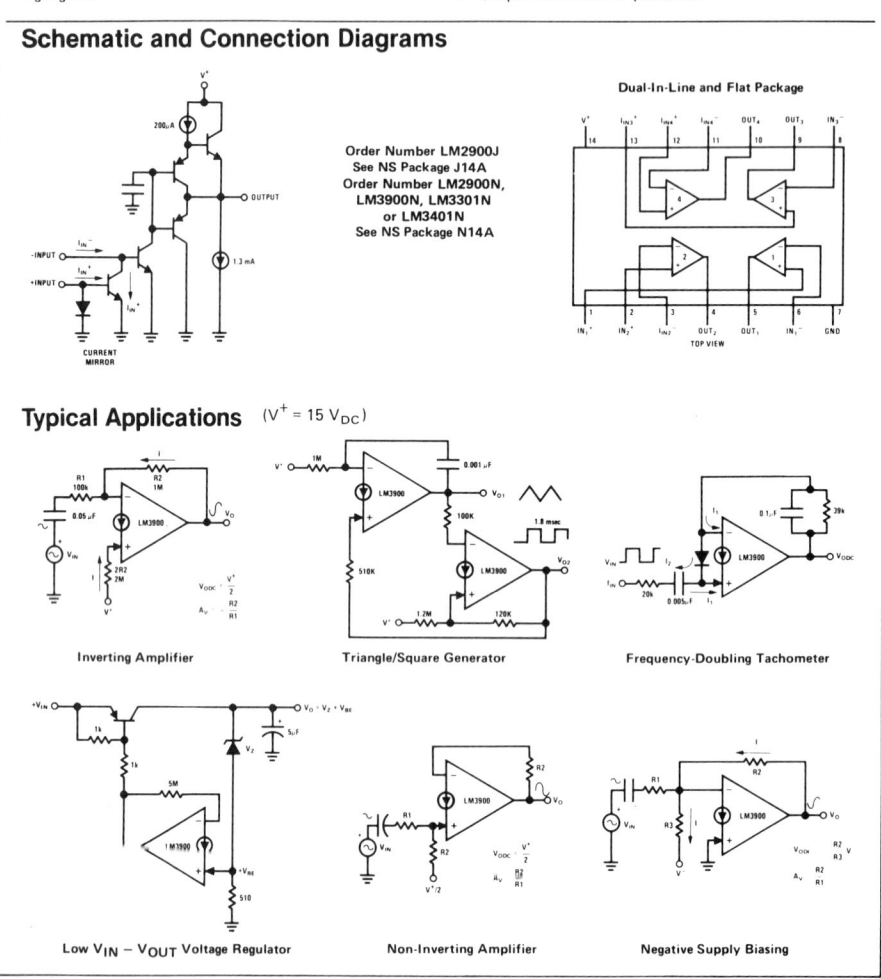

Negative Supply Biasing

Figure 3-22 Data sheets for the LM3900 Norton op-amp. *(Reprinted with permission of National Semiconductor Corporation)*

Absolute Maximum Ratings

	LM2900/LM3900	LM3301	LM3401
Supply Voltage	32 V_{DC}	28 V_{DC}	18 V_{DC}
	±16 V_{DC}	±14 V_{DC}	±9 V_{DC}
Power Dissipation (T_A = 25°C) (Note 1)			
Cavity DIP	900 mW		
Flat Pack	800 mW		
Molded DIP	570 mW	570 mW	570 mW
Input Currents, I_{IN}^+ or I_{IN}^-	20 mA_{DC}	20 mA_{DC}	20 mA_{DC}
Output Short Circuit Duration — One Amplifier	Continuous	Continuous	Continuous
T_A = 25°C (See Application Hints)			
Operating Temperature Range		−40°C to +85°C	0°C to +75°C
LM2900	−40°C to +85°C		
LM3900	0°C to +70°C		
Storage Temperature Range	−65°C to +150°C	−65°C to +150°C	−65°C to +150°C
Lead Temperature (Soldering, 10 seconds)	300°C	300°C	300°C

Electrical Characteristics (Note 6)

PARAMETER	CONDITIONS	LM2900 MIN	LM2900 TYP	LM2900 MAX	LM3900 MIN	LM3900 TYP	LM3900 MAX	LM3301 MIN	LM3301 TYP	LM3301 MAX	LM3401 MIN	LM3401 TYP	LM3401 MAX	UNITS
Open Loop														
Voltage Gain												800		V/mV
Voltage Gain	T_A = 25°C, f = 100 Hz		1.2	2.8		1.2	2.8		1.2	2.8		1.2	2.8	V/mV
Input Resistance	T_A = 25°C, Inverting Input		1			1			1			0.1	1	MΩ
Output Resistance			8			8			8			8		kΩ
Unity Gain Bandwidth	T_A = 25°C, Inverting Input		2.5			2.5			2.5			2.5		MHz
Input Bias Current	T_A = 25°C, Inverting Input		30	200		30	200		30	300		30	300	nA
	Inverting Input												500	nA
Slew Rate	T_A = 25°C, Positive Output Swing		0.5			0.5			0.5			0.5		V/μs
	T_A = 25°C, Negative Output Swing		20			20			20			20		V/μs
Supply Current	T_A = 25°C, R_L = ∞ On All Amplifiers		6.2	10		6.2	10		6.2	10		6.2	10	mA_{DC}
Output Voltage Swing	T_A = 25°C, R_L = 2k, V_{CC} = 15.0 V_{DC}													
V_{OUT} High	$I_{IN}^- = 0$, $I_{IN}^+ = 0$		13.5			13.5			13.5			13.5		V_{DC}
V_{OUT} Low	$I_{IN}^- = 10μA$, $I_{IN}^+ = 0$		0.09	0.2		0.09	0.2		0.09	0.2		0.09	0.2	V_{DC}
V_{OUT} High	$I_{IN}^- = 0$, $I_{IN}^+ = 0$ $R_L = ∞$, V_{CC} = Absolute Maximum Ratings		29.5			29.5			25.5			15.5		V_{DC}
Output Current Capability	T_A = 25°C													
Source		6	18		6	10		5	18		5	10		mA_{DC}
Sink	(Note 2)	0.5	1.3		0.5	1.3		0.5	1.3		0.5	1.3		mA_{DC}
I_{SINK}	V_{OL} = 1V, I_{IN} = 5μA		5			5			5			5		mA_{DC}

Electrical Characteristics (Continued) (Note 6)

PARAMETER	CONDITIONS	LM2900 MIN	LM2900 TYP	LM2900 MAX	LM3900 MIN	LM3900 TYP	LM3900 MAX	LM3301 MIN	LM3301 TYP	LM3301 MAX	LM3401 MIN	LM3401 TYP	LM3401 MAX	UNITS
Power Supply Rejection	T_A = 25°C, f = 100 Hz		70			70			70			70		dB
Mirror Gain	@ 20μA (Note 3)	0.90	1.0	1.1	0.90	1.0	1.1	0.90	1	1.10	0.90	1	1.10	μA/μA
	@ 200μA (Note 3)	0.90	1.0	1.1	0.90	1.0	1.1	0.90		1.10	0.90	1	1.10	μA/μA
ΔMirror Gain	@ 20μA To 200μA (Note 3)		2	5		2	5		2	5		2	5	%
Mirror Current	(Note 4)		10	500		10	500		10	500		10	500	$μA_{DC}$
Negative Input Current	T_A = 25°C (Note 5)		1.0			1.0			1.0			1.0		mA_{DC}
Input Bias Current	Inverting Input			300			300							nA

Note 1: For operating at high temperatures, the device must be derated based on a 125°C maximum junction temperature and a thermal resistance of 175°C/W which applies for the device soldered in a printed circuit board, operating in a still air ambient.
Note 2: The output current sink capability can be increased for large signal conditions by overdriving the inverting input. This is shown in the section on Typical Characteristics.
Note 3: This spec indicates the current gain of the current mirror which is used as the non-inverting input.
Note 4: Input V_{BE} match between the non-inverting and the inverting inputs occurs for a mirror current (non-inverting input current) of approximately 10μA. This is therefore a typical design center for many of the application circuits.
Note 5: Clamp transistors are included on the IC to prevent the input voltages from swinging below ground more than approximately −0.3 V_{DS}. The negative input currents which may result from large signal overdrive with capacitance input coupling need to be externally limited to values of approximately 1 mA. Negative input currents in excess of 4 mA will cause the output voltage to drop to a low voltage. This maximum current applies to any one of the input terminals. If more than one of the input terminals are simultaneously driven negative smaller maximum currents are allowed. Common-mode current biasing can be used to prevent negative input voltages; see for example, the "Differentiator Circuit" in the applications section.
Note 6: These specs apply for −55°C ≤ T_A ≤ +125°C, unless otherwise stated.

Figure 3-22 (*Continued*)

110   3 Amplifier Circuits

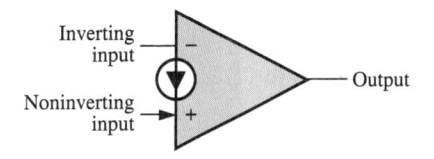

Figure 3-23 Symbol for a Norton op-amp.

restriction on the performance of an amplifier circuit using the LM3900, because, due to feedback, the operating gain is always less than the open-loop gain. The unity-gain bandwidth of LM3900 is 2.5 MHz, significantly better than the LM741C and almost as good as the TL081C.

Inverting Amplifier Using a Norton Op-Amp

The circuit of an inverting amplifier using a Norton op-amp is shown in Figure 3-24. As would be expected, this circuit is somewhat different from the inverting amplifier circuit for a standard op-amp; however, some parts can be immediately recognized. The feedback resistor R_F and the input resistor R_I are the same as for inverting amplifiers using the LM741C and the TL081C op-amps. The input impedance Z_{in} is equal to R_I, as defined in Equation 3-2, and the voltage gain A_V is equal to R_F/R_I, as defined in Equation 3-1.

What is different about this circuit is the resistor labelled R_B connected between the power supply and the noninverting input. This resistor is used for biasing the op-amp and for setting the DC output offset voltage V_{OFF}. Under operating conditions, the Norton amplifier always has the same DC bias current flowing into both inputs. The bias current through R_B to the noninverting input is $(V_{CC} - 0.7)/R_B$. The bias current to the inverting input flows through R_F from the output and is $(V_{OFF} - 0.7)/R_F$. Note the 0.7 V in each of these terms. Refer to the schematic of a current amplifier shown in Figure 3-21 or on the data sheets (Figure 3-22). Each input is to the base of a transistor where the emitter is

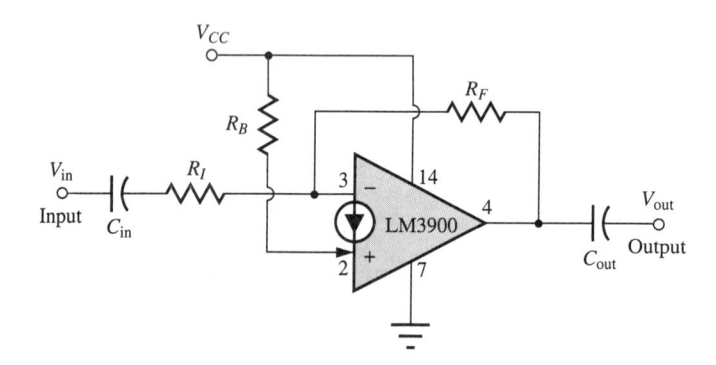

Figure 3-24 Inverting amplifier using a LM3900 Norton op-amp.

grounded. The corresponding base voltage is the standard transistor base-emitter voltage of 0.7 V. This voltage appears at each input.

If we equate the currents into the two inputs, we have

$$\frac{V_{CC} - 0.7}{R_B} = \frac{V_{OFF} - 0.7}{R_F}$$

This equation can be used in two ways. First, we can solve the equation for R_B and use it for a design equation. Normally we want to design for maximum output swing in the amplified signal. For this, the offset voltage V_{OFF} should be $\frac{1}{2} V_{CC}$. Substituting $\frac{1}{2} V_{CC}$ for V_{OFF} and solving for R_B gives the design equation

$$R_B = \frac{V_{CC} - 0.7}{\frac{V_{CC}}{2} - 0.7} \times R_F \tag{3-14}$$

If V_{CC} is large compared with 0.7 V, we get the approximation

$$R_B = 2R_F \tag{3-15}$$

Alternatively, if we solve for V_{OFF}, we can calculate the output offset voltage as determined by resistors R_B and R_F. The equation for V_{OFF} is

$$V_{OFF} = \frac{R_F}{R_B}(V_{CC} - 0.7) + 0.7 \tag{3-16}$$

There are limits on the resistance values that can be used for R_F and R_B. In most applications, there should be no problem, but it is important to be aware that the limits exist. From the data sheets in Figure 3-22, the maximum input current is 20 mA DC. Assuming an output offset voltage of $\frac{1}{2}V_{CC}$, the minimum value of R_F is

$$R_F(\text{min}) = \frac{\frac{V_{CC}}{2} - 0.7}{I_{max}} = \frac{V_{CC} - 1.4}{0.04 \text{ A}} \tag{3-17}$$

If $V_{CC} = 15$ V, then $R_F(\text{min}) = 13.6$ V/0.04 A = 340 Ω. Obviously we cannot replace R_F with a short as we did in an op-amp follower circuit.

There is also a maximum value for R_F since the LM3900 may require an input bias current as large as 200 nA (maximum value of input bias current from the data sheet). The maximum value of R_F is given by

$$R_F(\text{max}) = \frac{\frac{V_{CC}}{2} - 0.7}{I_{bias}} = \frac{V_{CC} - 1.4}{400 \text{ μA}} \tag{3-18}$$

If $V_{CC} = 15$ V, then $R_F(\text{max}) = 13.6$ V/400 nA = 34 MΩ. For a buffer amplifier where the input impedance should be very large, then choose R_F large, say typically 1 to 10 MΩ.

Unfortunately, amplifiers using the Norton op-amp must always be capacitively coupled. The input must be coupled through capacitor C_{in} to prevent bias

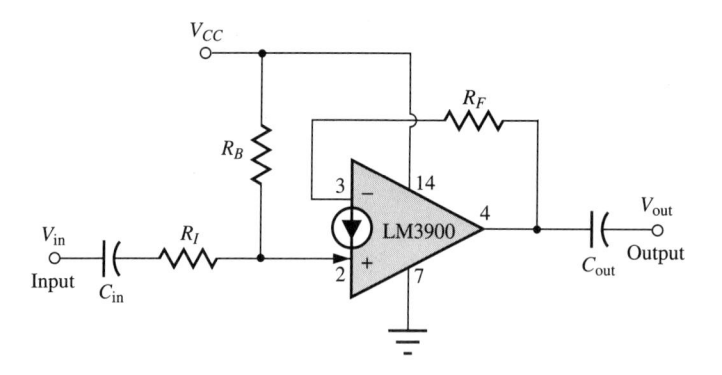

Figure 3-25 Noninverting amplifier using a LM3900 Norton op-amp.

current from flowing out through R_I. The output must be coupled through capacitor C_{out} because there is a DC output offset voltage. Operating the op-amp from a dual supply can make the output offset zero and eliminate the need for C_{out}, but the amplifier will still need C_{in}. Capacitors C_{in} and C_{out} are calculated using the same equations as developed for standard op-amps using a single supply, namely Equation 3-7 and Equation 3-8. These equations are repeated below for convenience. The equation for C_{in} is

$$C_{in} = \frac{1}{2\pi f_L (R_S + R_I)} \qquad (3\text{-}7)$$

The equation for C_{out} is

$$C_{out} = \frac{1}{2\pi f_L (Z_{out} + R_L)} \qquad (3\text{-}8)$$

To design an inverting amplifier using a Norton op-amp, a similar procedure is used as for a normal inverting amplifier. The required gain and input impedance are usually stipulated, and R_F is calculated from Equation 3-1. If R_I is not stipulated, choose a value of R_F between 10 and 100 kΩ and calculate R_I from Equation 3-1. Calculate R_B from Equation 3-15. Finally calculate the input and output capacitors using Equations 3-7 and 3-8. The bandwidth can be calculated from Equation 3-3.

To analyze an inverting amplifier design, calculate the gain from Equation 3-1, the input impedance from Equation 3-2, the DC offset of the output from Equation 3-16, and the bandwidth from Equation 3-3.

Noninverting Amplifier Using a Norton Op-Amp

The circuit of a noninverting amplifier using a Norton op-amp is shown in Figure 3-25. Here there is a surprise. The circuit is the same as the inverting amplifier circuit shown in Figure 3-24 except that the input resistor R_I is connected to the noninverting input. There is absolutely no difference in the design of a noninverting amplifier using a Norton op-amp and the design of an inverting amplifier. The same equations and procedures are used!

To switch from an inverting amplifier to a noninverting amplifier involves simply switching the input resistor R_I from the inverting input to the noninverting input. The input impedance is unchanged, the gain is unchanged, only the phase of the output signal is changed by 180°. Sometimes such a polarity switch is useful. For example, many oscilloscopes have a polarity switch for switching the polarity of the display by 180°.

Buffer Amplifier Using a Norton Op-Amp

The circuit for a unity-gain buffer amplifier, either inverting or noninverting, should now be no surprise. The buffer amplifier circuit is exactly the same as an inverting amplifier or noninverting amplifier with $R_F = R_I$ to give unity gain. An inverting buffer amplifier is completely switchable to a noninverting amplifier by simply moving R_I. Buffer amplifiers using Norton op-amps are not widely used since the circuits are more complex and they have to be capacitively coupled.

EXAMPLE 3-9

Design an inverting amplifier with a gain of 12 using an LM3900 Norton op-amp. The amplifier is to operate from a single 9 V supply and should have an input impedance of 100 kΩ. Make the lower cutoff frequency 100 Hz. Assume a source impedance of zero and a load impedance of 20 kΩ.

Solution

The circuit for an inverting amplifier using a Norton op-amp was shown in Figure 3-24. This has been redrawn as Figure 3-26, and a parts list is included.

First design R_I. Since the input impedance is 100 kΩ, we choose this value for R_I.

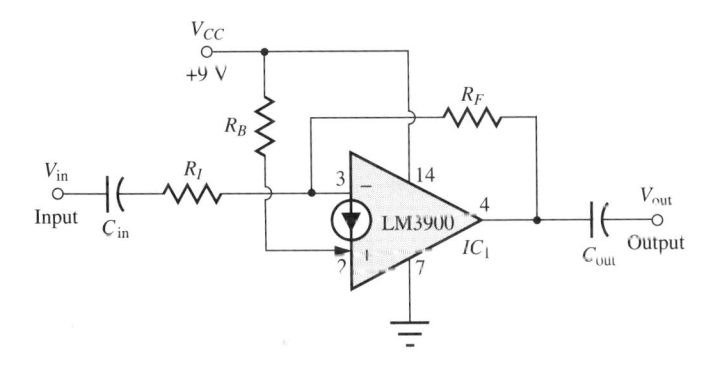

Parts List

Resistors:
R_I 100 kΩ
R_F 1.2 MΩ
R_B 2.4 MΩ

Capacitors:
C_{in} 0.022 μF
C_{out} 0.01 μF

Semiconductors:
IC_1 LM3900

Figure 3-26

Next design R_F. Since the gain is 12, we have, using Equation 3-1, $R_F = 12 \times R_I = 1.2$ MΩ.

Now design R_B, using the approximation formula $R_B = 2R_F$. This gives $R_B = 2.4$ MΩ. All of the above resistors are standard values.

Finally, determine C_{in} and C_{out} from Equations 3-7 and 3-8.

$$C_{in} = \frac{1}{2\pi f_L (R_S + R_I)} = \frac{1}{2\pi \times 100 \text{ Hz} \times 100 \text{ k}\Omega} = 0.0159 \text{ }\mu\text{F}$$

$$C_{out} = \frac{1}{2\pi f_L (Z_{out} + R_L)} = \frac{1}{2\pi \times 100 \text{ Hz} \times 20 \text{ k}\Omega} = 0.0796 \text{ }\mu\text{F}$$

Choose $C_{in} = 0.022$ μF and $C_{out} = 0.10$ μF from the values listed in Appendix A-2. Note that we chose larger values of capacitance.

Our design is complete. The same procedure would be used for a noninverting amplifier or a buffer amplifier. To convert the inverting amplifier to a noninverting amplifier, simply connect R_I to the noninverting input.

3-7 Instrumentation Amplifiers

The modern operational amplifier is almost a perfect amplifier. It has a very high gain, virtually infinite input impedance at both inputs, and very good rejection of common-mode noise. Yet in spite of all these excellent features, for some applications a simple operational amplifier is not adequate, and an amplifier with enhanced characteristics must be used. Since the most common area requiring enhanced performance amplifiers is industrial process instrumentation, these amplifiers are known as **instrumentation amplifiers.**

In industrial instrumentation applications, various **process variables** must be measured. These include such quantities as flow rate, pressure, temperature, density, weight, mass, viscosity, and moisture content. The process variable is converted to an electrical signal by a device called a **transducer.** For example, temperature may be sensed with a thermocouple, pressure may be sensed by the deformation of a piezoelectric crystal, flow rate may be sensed by the motion of the core in a tuning coil, weight may be sensed by the change in resistance of a strain gauge.

We will use a thermocouple for our examples because it is commonly used and is a very good example of a typical transducer. A thermocouple is simply the junction of two different metals, iron and constantan being most widely used in industry. In practice, there must always be two junctions. The thermocouple produces a voltage proportional to the temperature difference between two junctions, typically a few millivolts. If current flows through a thermocouple junction, the junction temperature changes and an incorrect temperature is obtained. Hence, care must be taken to ensure that the thermocouple is never loaded. The thermocouple may be installed many meters away from the actual metering circuit, frequently in an electrically noisy environment and is often subject to a large noise pickup.

3-7 Instrumentation Amplifiers

Instrumentation Amplifier Requirements

These are the three typical problems that we must contend with and must design our instrumentation amplifier to handle:

1. Very small signals
2. A source that cannot be loaded
3. Long input leads in a noisy environment.

These source characteristics demand that the amplifier circuit have

1. Large gain, typically 100 to 1000
2. Extremely large input impedance
3. High common-mode rejection.

While a standard op-amp is quite good in these respects, it often is not adequate for many critical instrumentation applications, and special solutions must be developed.

A Simple Instrumentation Amplifier

Figure 3-27 shows two amplifier circuits. Figure 3-27(a) shows a transducer connected to a standard noninverting amplifier. We will assume that the transducer is a thermocouple and that it is a considerable distance away from the amplifier in an electrically noisy environment. There will be noise induced onto the transmission lines. This noise will act as an independent source appearing in series with the desired signal from the thermocouple.

Assume that the amplifier has a gain of 100, a common-mode rejection of 76 dB (that is, a factor of 6310), an input signal of 2 mV, and a noise level of 0.1 V. These are typical numbers that might be encountered in practice. The output signal from the amplifier is

signal output = gain × input signal = 100 × 2 mV = 0.2 V

The output noise from the amplifier is

noise output = gain × noise = 100 × 0.1 V = 10 V

The common-mode rejection has no effect on this noise because the noise is present at only one input. The signal is totally lost in the noise.

In the circuit shown in Figure 3-27(b), both inputs are designed to appear identical to the transducer. Since the inverting input has an input resistor R_I and a feedback resistor R_F, the noninverting input is designed with an input resistor equal to R_I and a resistor to ground equal to R_F. The gain is R_F/R_I as in a normal inverting amplifier. This circuit is actually a differencing or subtracting amplifier, which will be introduced in the next chapter, and analysis of the circuit will be deferred until then. Because lines are now connected from the source to each of the inputs, there will be noise present at each input. Normally, in installation, the lines are twisted together to ensure that the noise pickup is identical for each line.

Assume that the amplifier has the same characteristics as the simple noninverting amplifier, namely a gain of 100, a common-mode rejection of 6310, an

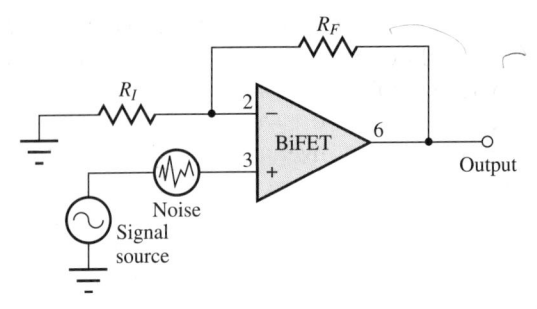

(a) Standard single-ended noninverting amplifier

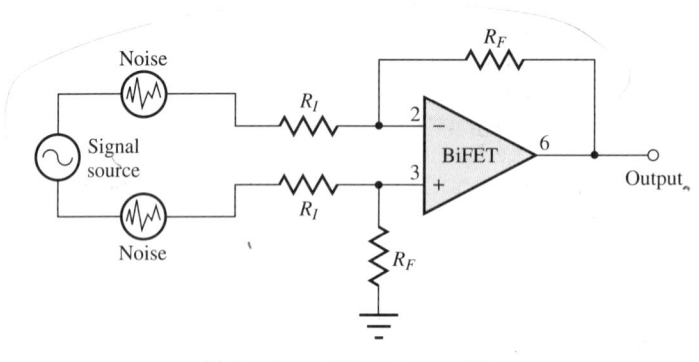

(b) Two input differencing amplifier

Figure 3-27 Amplifiers measuring transducer signals in the presence of noise.

input signal of 2 mV, and a noise level of 0.1 V on each line. The output signal from the amplifier is

$$\text{signal output} = \text{gain} \times \text{input signal} = 100 \times 2 \text{ mV} = 0.2 \text{ V}$$

The output noise from the amplifier is

$$\text{noise output} = \frac{\text{gain} \times \text{noise}}{\text{common mode rejection}}$$

$$= \frac{100 \times 0.1 \text{ V}}{6310} = 0.00158 \text{ V}$$

This demonstrates quite dramatically the improvement of signal to noise in the two-input differencing amplifier over the single-input noninverting amplifier. In this case, the output signal is well above the noise level.

An Instrumentation Amplifier Using Standard Op-amps

The simple differencing or subtracting op-amp configuration introduced above has many of the characteristics required for an instrumentation amplifier. It does, however, have limitations. The input impedance is equal to R_I, which, if the gain is to be large, as is often required in instrumentation amplifiers, cannot be very high. The value of R_I would typically be 10 kΩ or less. This circuit may load the transducer—for a thermocouple, it definitely would load the transducer.

To eliminate the loading problem, we use the circuit in Figure 3-28. Two noninverting buffer amplifiers (follower amplifiers) are added to isolate the inputs from the differencing amplifier. The circuit in Figure 3-28 uses BiFET op-amps. Because we may be dealing with high-impedance sources, resistors R_G are inserted between the noninverting input and ground to prevent any static charge buildup. The input resistance of each buffer is equal to R_G, which can be as large as desired, typically several megohms. The R_G values should be identical to ensure equal input impedances. The amplification is provided by the third op-amp, which is configured as a normal differencing amplifier. The gain is simply R_F/R_I. The operation of this circuit is identical to that of Figure 3-27(b), except that the inputs are buffered and can have very high input impedances.

There are two important points to consider in selecting components for this circuit. All op-amps must be low noise. In this respect, the TL070 series of low-noise op-amps are superior to the TL080 series that we have been using. Second, both inputs to the differencing amplifier must have the same impedance for optimum noise rejection. This means that for the noninverting input we must use an input resistor with a value of R_I and a resistor to ground with a value of R_F to ensure equal impedances to each input. These should be precisely matched to the corresponding resistors connected to the inverting input.

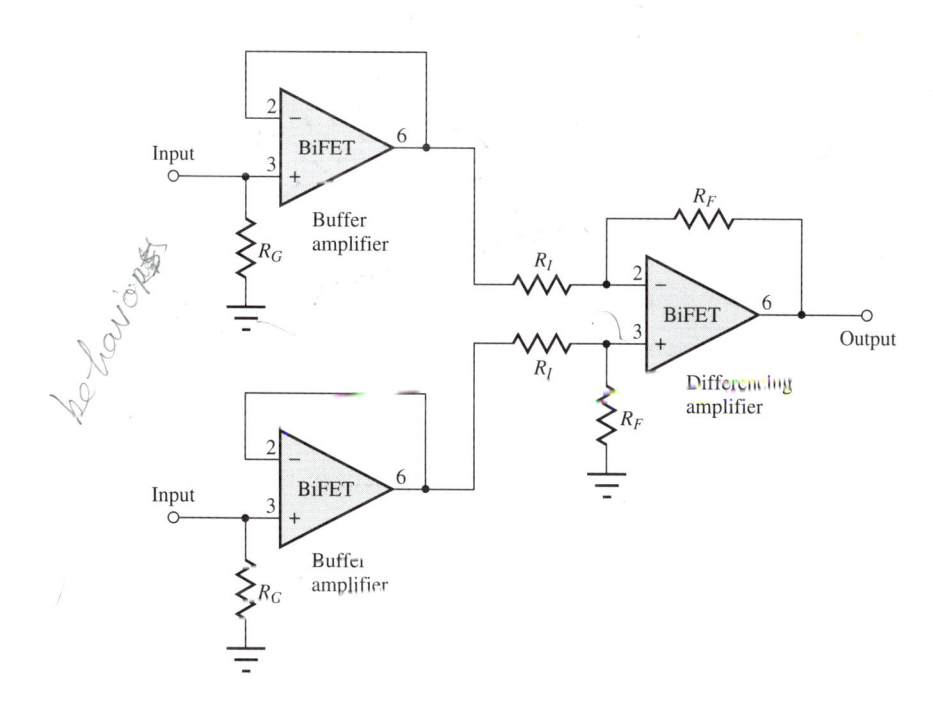

Figure 3-28 Buffered instrumentation amplifier using op-amps.

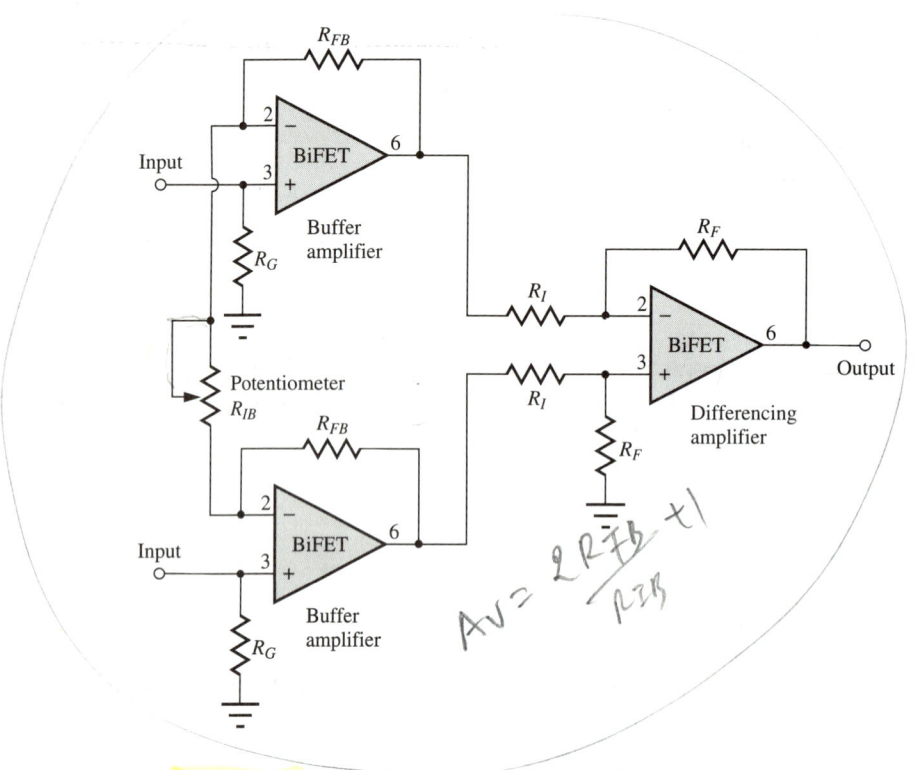

Figure 3-29 Buffered instrumentation amplifier with variable gain.

There is still one limitation with the circuit shown in Figure 3-28. If the gain is to be adjustable, as it must be in a practical instrumentation amplifier, either both R_F or both R_I resistors must be varied simultaneously and identically. It is very hard to get two potentiometers to track identically. Figure 3-29 shows an ingenious procedure for adding variable gain by using a single potentiometer.

Our variable gain amplifier consists of a fixed-gain differencing amplifier providing most of the circuit gain and two variable-gain noninverting amplifiers sharing a common R_I potentiometer. The differencing amplifier is identical to that used in the fixed-gain instrumentation amplifier, with the two resistors for R_F and the two resistors for R_I precisely matched. The noninverting amplifiers replacing the simple follower amplifiers are a little different from a single noninverting amplifier in that they share a common potentiometer R_{IB}. Because of this, the gain cannot be determined by Equation 3-4.

The easiest way to determine the gain of either noninverting amplifier is to assume some voltage V_{in} applied across the differential input, with $\frac{1}{2}V_{in}$ applied to the top amplifier and $-\frac{1}{2}V_{in}$ applied to the bottom amplifier. Both inputs of the top op-amp will be at a voltage of $+\frac{1}{2}V_{in}$, and both inputs of the bottom op-amp will be at $-\frac{1}{2}V_{in}$. The total difference V_{in} appears across R_{IB}, and the current through R_{IB} is V_{in}/R_{IB}. The pair of noninverting amplifiers produces a combined output signal V_{out}. The output of the top amplifier alone will be half of the combined output signal, $\frac{1}{2}V_\text{out}$. For the top amplifier, the voltage across R_{FB} is $\frac{1}{2}V_\text{out} - \frac{1}{2}V_\text{in}$. Hence, the current through it is $(\frac{1}{2}V_\text{out} - \frac{1}{2}V_\text{in})/R_{FB}$. Equating the currents through R_{IB} and R_{FB}, we get

$$\frac{V_\text{in}}{R_{IB}} = \frac{V_\text{out} - V_\text{in}}{2R_{FB}}$$

Rearranging the equation and solving for V_out/V_in, we get the gain equation

$$A_V = \frac{2R_{FB}}{R_{IB}} + 1 \tag{3-19}$$

Although this equation was derived for the top amplifier, the same equation applies to the bottom amplifier. Notice the factor of 2 in this equation, as compared to Equation 3-4. As the potentiometer R_{IB} is varied, the gain of each of the noninverting amplifiers changes by exactly the same amount if the two R_{FB} resistors are precisely matched.

EXAMPLE 3-10

Design an instrumentation amplifier using BiFET op-amps. The amplifier should have a gain of 500 with an input impedance of 2 MΩ.

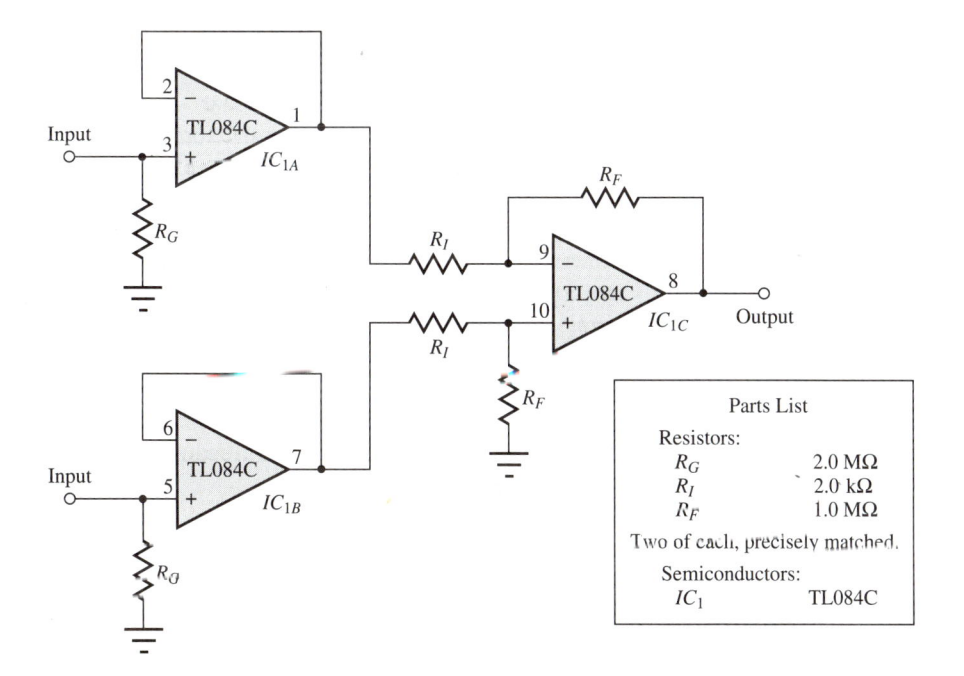

Figure 3-30

Solution

We will use the basic op-amp instrumentation amplifier in Figure 3-28. The circuit has been redrawn as Figure 3-30, and a parts list is included.

We must select a BiFET op-amp. All of our designs to this point have used the TL081C. Because we require three op-amps, it would be best to use a quad op-amp such as the TL084C. The pinout of this op-amp can be found on the TL080 series data sheets shown in Figure 2-19, and the pin identification in Figure 3-30 refers to this pinout. Because we have four independent op-amps, we can select any one for any operation. In practice, we would select a configuration that leads to the best circuit board layout. Because we are not concerned with circuit board layout in this text, the choice of op-amps is arbitrary.

We will use a TL084C because we are familiar with the TL080 series devices and the data sheets are shown in the text. In practice, it would be better to use a low-noise op-amp such as the TL074C. The pinout and most of the operating characteristics of the TL074C are identical to the TL084, so if you are building this circuit and have a TL074C, use it.

The design of our instrumentation amplifier is quite simple. Since we want an input impedance of 2 MΩ, we select R_G to be 2.0 MΩ. Since we want a gain of 500, we select R_F and R_I such that $R_F = 500\, R_I$. We will arbitrarily select $R_I = 2$ kΩ, so $R_F = 1.0$ MΩ.

Our design is now complete.

3-8 The LH0038 Instrumentation Amplifier

Because the instrumentation amplifier is a widely used circuit, special integrated circuit instrumentation amplifiers have been developed. These are very high performance amplifier circuits that have been optimized for amplifying low-level transducer signals in noisy environments. The LH0038 instrumentation amplifier produced by National Semiconductor is a typical example. The data sheets for the LH0038 are shown in Figure 3-31.

There are actually two versions of the LH0038, namely the LH0038 and the LH0038C. The LH0038 is the industrial/military model with wider operating temperature range and slightly better common-mode rejection and power supply rejection. The LH0038C is the commercial model. Because these instrumentation amplifiers are largely used for industrial applications, we will discuss the LH0038 industrial model.

The pinout of the LH0038 is shown on the data sheets and has been reproduced as Figure 3-32(a) for clarity. Figure 3-32(b) shows the schematic symbol for the LH0038. The internal circuit for the LH0038 is shown on the data sheets. It bears some similarity to the circuit that we designed using conventional op-amps. There are, however, several differences.

Note that the input signals (pins 12 and 13) go to a matched differential transistor pair rather than straight to an operational amplifier circuit. This arrangement

LH0038/LH0038C

National Semiconductor

Instrumentation Amplifiers

LH0038/LH0038C True Instrumentation Amplifier

General Description

The LH0038/LH0038C is a precision true instrumentation amplifier (TIA) capable of amplifying very low level signals, such as thermocouple and low impedance strain guage outputs. Precision thin film gain setting resistors are included in the package to allow the user to set the closed-loop gain from 100 to 2000. Since the resistors are of a homogeneous single chip construction, they track almost perfectly so that temperature variations of closed loop gain are virtually eliminated.

LH0038 exhibits excellent CMRR, PSRR, gain linearity, as well as extremely low input offset voltage, offset voltage drift and input noise voltage.

The devices are provided in a hermetically sealed 16-lead DIP. The LH0038 is guaranteed from $-55°C$ to $+125°C$; whereas the LH0038C is guaranteed from $-25°C$ to $+85°C$.

Features

- Ultra-low offset voltage   25 µV typ., 100 µV max
- Ultra-low offset drift   0.25 µV/C max
- Ultra-low input noise   0.2 µVp-p
- Pin strap gain options 100, 200, 400, 500, 1k, 2k
- Excellent PSRR and CMRR   120 dB

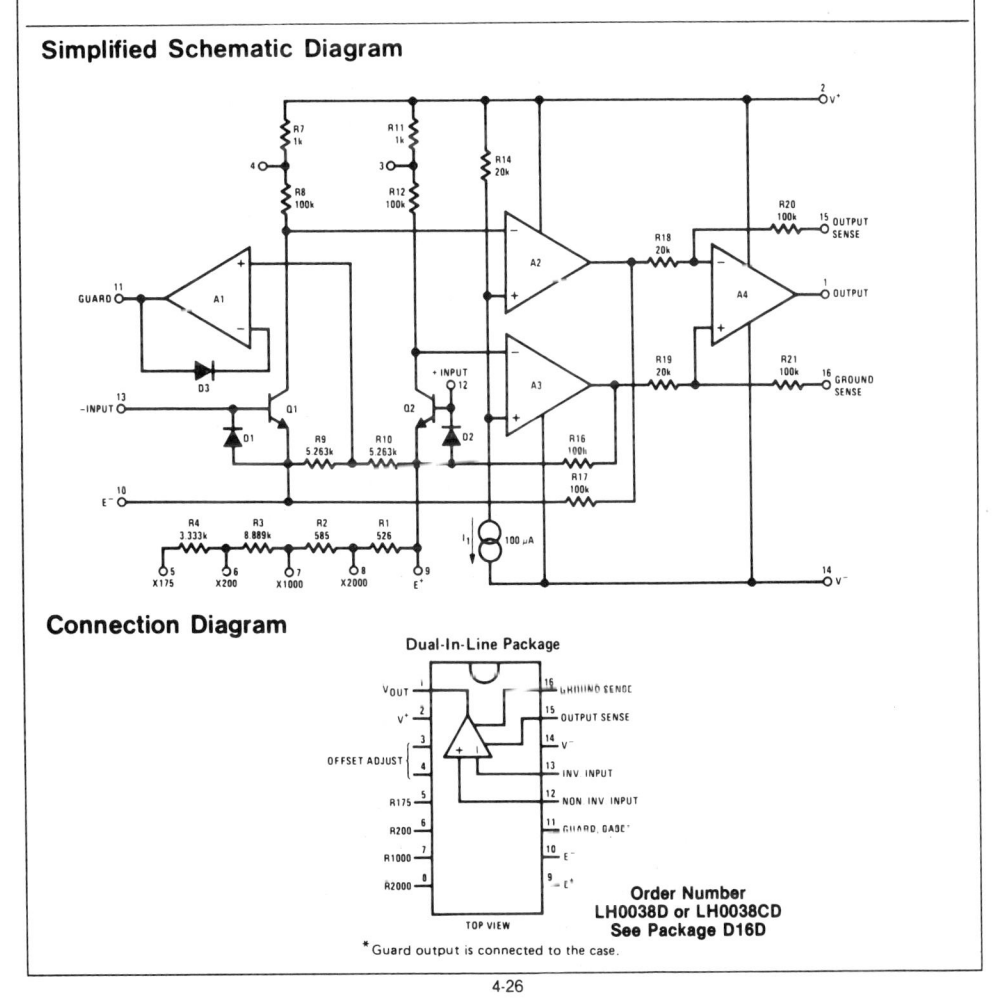

Simplified Schematic Diagram

Connection Diagram

Dual-In-Line Package

Order Number
LH0038D or LH0038CD
See Package D16D

*Guard output is connected to the case.

4-26

Figure 3-31 Data sheets for the LH0038 instrumentation amplifier. *(Reprinted with permission of National Semiconductor Corporation)*

121

Absolute Maximum Ratings

Supply Voltage	±18V
Differential Input Voltage (Note 1)	±1V
Input Voltage	±V_S
Power Dissipation (See Curve)	500 mW
Short Circuit Duration	Continuous
Operating Temperature Range	
LH0038	−55°C to +125°C
LH0038C	−25°C to +85°C
Storage Temperature	−65°C to +150°C
Lead Temperature (Soldering, 20 seconds)	300°C

DC Electrical Characteristics (Note 2)

PARAMETER		CONDITIONS	LH0038 MIN	LH0038 TYP	LH0038 MAX	LH0038C MIN	LH0038C TYP	LH0038C MAX	UNITS
V_{IOS}	Input Offset Voltage	T_A = 25°C		25	100		30	150	μV
					125			220	
$\Delta V_{IOS}/\Delta T$	Input Offset Voltage Tempco			0.1	0.25		0.2	1.0	μV/°C
V_{OOS}	Output Offset Voltage	R_S = 50Ω, V_{CM} = 0V, T_A = 25°C		3	10		5	25	mV
					15			30	
$\Delta V_{OOS}/\Delta T$	Output Offset Voltage Tempco			25			25		μV/°C
I_B	Input Bias Current	T_A = 25°C		50	100		50	100	nA
					200			200	
I_{OS}	Input Offset Current	T_A = 25°C		2	5		7	10	nA
		V_{CM} = 0V			8			15	
$\Delta I_B/\Delta T$	Input Bias Current Tempco			500			500		pA/°C
A_{VCL}	Closed Loop Gain	Gain Pins Jumpered							
		None		100			100		
		6−10		200			200		
		6−9, 10−5		400			400		V/V
		6−10, 5−9		500			500		
		7−10		1000			1000		
		8−10		2000			2000		
	Closed Loop Gain Error	A_{VCL} = 100, 200		0.1	0.3		0.1	0.4	
		A_{VCL} = 400, 500		0.2	0.3		0.2	0.6	%
		A_{VCL} = 1000		0.3	0.5		0.5	1.0	
		A_{VCL} = 2000		1.0	2.0		1.5	3.0	
	Gain Temperature Coefficient	A_{VCL} = 1k		7			7		ppm/°C
	Gain Nonlinearity	100 ≤ A_{VCL} ≤ 2k		1			1		ppm
V_{INCM}	Common-Mode Input Voltage Range		±10	±12		±10	±12		
V_O	Output Voltage	R_L ≥ 10 kΩ	±10	±12		±10	±12		V
V_S	Supply Voltage Range		±5		±18	±5		±18	
	Guard Voltage Error	−10V < V_{CM} < +10V		±10	±100		±10	±100	mV

4-27

Figure 3-31 (*Continued*)

DC Electrical Characteristics (Note 2) (Continued)

PARAMETER		CONDITIONS		LH0038 MIN	LH0038 TYP	LH0038 MAX	LH0038C MIN	LH0038C TYP	LH0038C MAX	UNITS
CMRR	Common-Mode Rejection Ratio	$V_{IN} = \pm 10V$	$A_{VCL} = 100$	94	110		86	110		dB
			$A_{VCL} = 1000$	114	120		106	110		
PSRR	Power Supply Rejection Ratio	$\pm 5V \leq \Delta V_S \leq \pm 15V$	$A_{VCL} = 100$	94	110		94	110		
			$A_{VCL} = 1000$	110	120		100	110		
I_{OSC}	Output Short Circuit Current	$T_A = 25°C$		± 2	± 5	± 10	± 2	± 5	± 10	mA
I_S	Supply Current	$T_A = 25°C$			1.6	2.0		1.6	3.0	
R_{IN} DIFF	Input Resistance	$A_{VCL} = 1000$, $T_A = 25°C$			5			5		$M\Omega$
R_{IN} CM	Common-Mode Input Resistance				1			1		$G\Omega$
R_{OUT}	Output Resistance				1			1		$m\Omega$

AC Electrical Characteristics $V_S = \pm 15V$, $T_A = 25°C$

PARAMETER		COMMENT	CONDITIONS		TYP	UNITS
e_n	Equivalent Input Noise Voltage	Figure 1	$R_S = 0$, f = 0.1 to 10 Hz		0.2	$\mu Vp\text{-}p$
$\overline{e_n}$	Equivalent Input Spot Noise Voltage	Figure 1	$R_S = 100\Omega$	f = 10 Hz	6.5	$nV/\sqrt{Hz}$
				f = 100 Hz	6.0	
				f = 1 kHz	6.0	
				f = 10 kHz	6.0	
BW	Large Signal Bandwidth		$V_{OUT} = \pm 10V$		1.6	kHz
S_r	Slew Rate		$V_{OUT} = \pm 10V$		0.3	$V/\mu s$
t_s	Settling Time to 0.01%	Figure 13		20V Step	120	μs
				−10V Step	80	
				+10V Step	60	
t_r	Rise Time		$\Delta V_{OUT} = 100$ mV	$A_{VCL} = 100$	6	μs
				$A_{VCL} = 1000$	13	
$\overline{i_n}$	Equivalent Input Spot Noise Current		$R_S = 100\ M\Omega$	f = 10 Hz	0.1	$pA/\sqrt{Hz}$

Note 1: The inputs are protected by diodes for overvoltage protection. Excessive currents will flow for differential voltages in excess of ±1V. Input current should be limited to less than 10 mA.

Note 2: Unless otherwise noted these specifications apply for $V_S = \pm 15.0V$, pin 16 connected to pin 1, pin 16 connected to ground, over the temperature range −55°C to +125°C for the LH0038 and −25°C to +85°C for LH0038C.

Figure 3-31 (*Continued*)

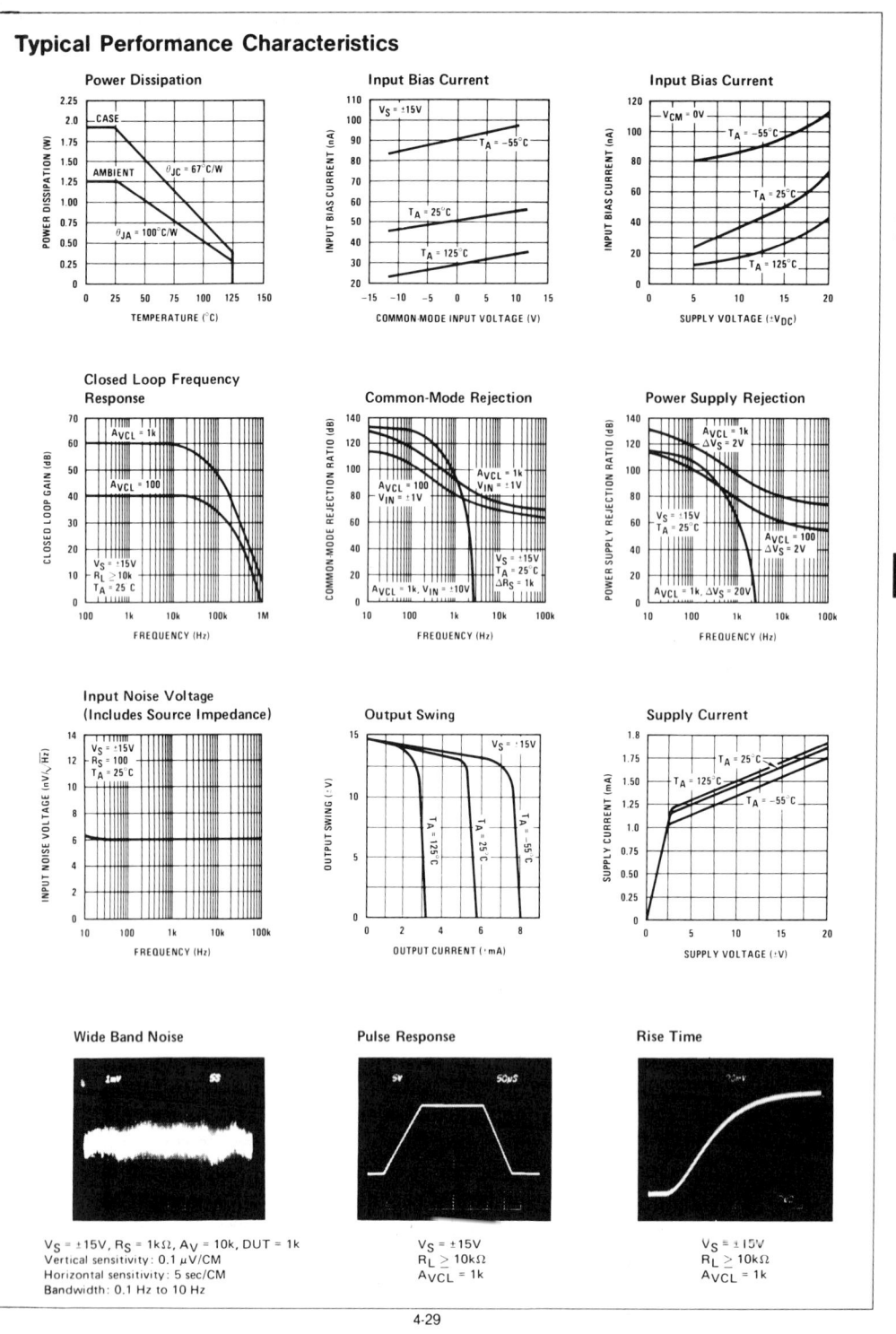

Figure 3-31 (*Continued*)

3-8 The LH0038 Instrumentation Amplifier

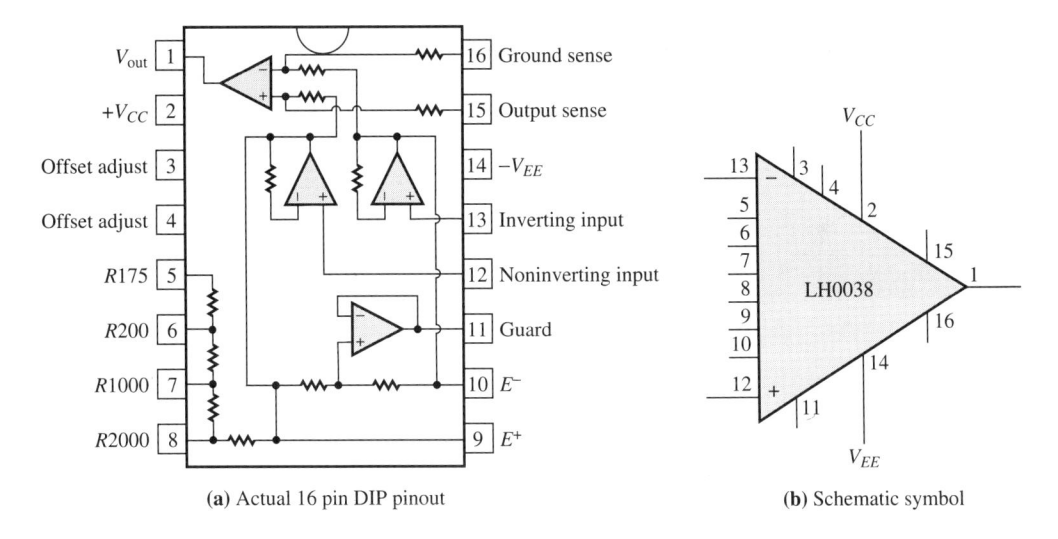

Figure 3-32 Pinout of the LH0038 instrumentation amplifier.

reduces amplifier noise by providing low-noise preamplification. Op-amps always introduce more noise than simple transistor amplifiers. This configuration, however, requires a larger input bias current than does a standard op-amp. Bias currents are carefully matched so that the offset current is kept very small. The bias current in a bipolar transistor is essentially constant with temperature, unlike a FET amplifier where the leakage current increases with temperature. The consequence of this modified input circuitry is a very low offset voltage (typically 25 μV), very low offset temperature drift (less than 0.25 μV/°C), and very low input amplifier noise (0.2 μV_{p-p}). The offset can be adjusted by placing a 10 kΩ potentiometer between pins 3 and 4 and connecting the pot adjustment terminal to the positive power supply (similar to the method of compensation used for the LM741 op-amps).

There is a fourth op-amp in the circuit. This op-amp acts as a guard drive amplifier. The guard drive output is always equal to the common-mode voltage. It is connected to the metal case of the IC to provide electrostatic shielding for the system, and should be tied to the shield of the shielded input cable to maintain the shield at the common-mode voltage, thus reducing noise pickup and improving AC common-mode rejection. The LH0038 can achieve a common-mode rejection ratio of 120 dB, which is typically 40 dB—a factor of 100—better than a standard op-amp.

The amplifier has a high gain which can be set to 100, 200, 400, 500, 1000, or 2000 by various connections of pins 5, 6, 7, 8, 9, and 10, as summarized in Table 3-1. Precision thin-film resistors fabricated onto a single chip to track perfectly with temperature are used for setting the gain.

The variable gain is obtained by a method similar to that used in Figure 3-29. Figure 3-33 shows a simplified schematic of the gain adjustment circuitry

126   3 Amplifier Circuits

(a) Setting for minimum gain of 100

(b) Setting for gain of 200, pin 6 connected to pin 10

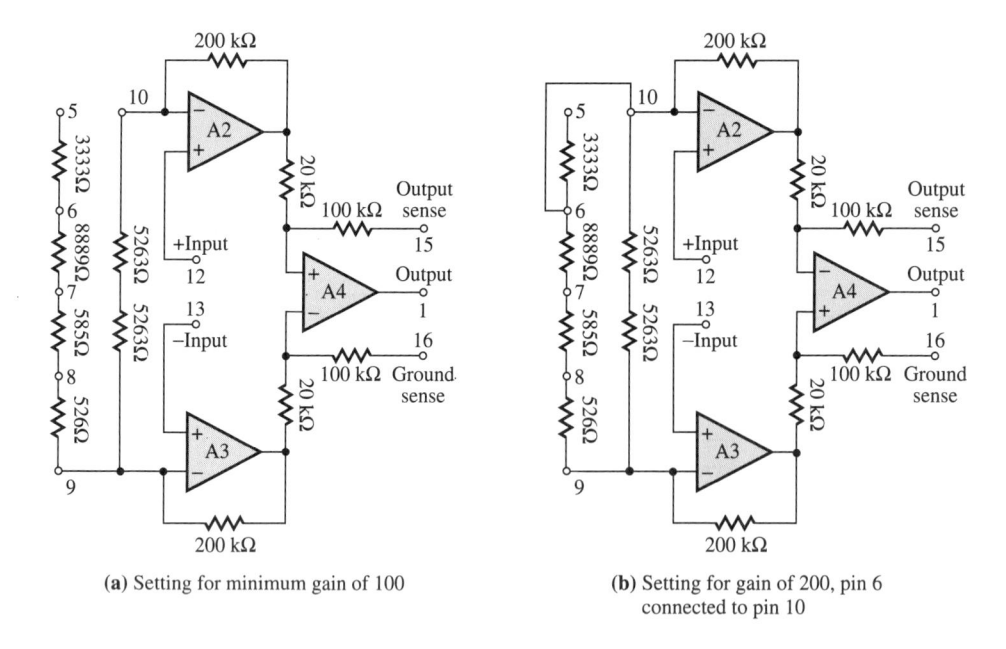

Figure 3-33 Simplified circuit showing gain settings for the LH0038 instrumentation amplifier.

(the separate differencing amplifier has been omitted for clarity). There is a fixed resistance of 10,526 Ω between pins 9 and 10, which is the maximum R_{IB} resistance of the two input amplifiers. The feedback resistor R_F is 100 kΩ, to give a gain of exactly 20 for a noninverting amplifier (use Equation 3-19). This gain is with no external connections between pins 5, 6, 7, 8, 9, or 10, as in Figure 3-33(a). If pin 6 is connected to pin 10 as in Figure 3-33(b), an additional resistance of 10,000 Ω is placed in parallel with the 10,526 Ω resistor, to give an effective resistance of 5128 Ω between pins 9 and 10, thus increasing the gain by a factor of 2 to exactly 40. If pin 6 is shorted to pin 9, and pin 5 is connected to pin 10 as in

Table 3-1 Gain setting connections for the LH0038 instrumentation amplifier

Overall Gain	Pin Connections
100	All pins open
200	pin 6 to 10
400	pin 6 to 9, pin 5 to 10
500	pin 6 to 10, pin 5 to 9
1000	pin 7 to 10
2000	pin 8 to 10

3-8 The LH0038 Instrumentation Amplifier 127

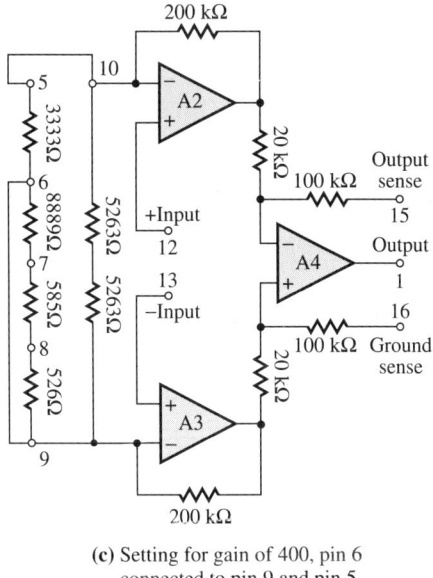

(c) Setting for gain of 400, pin 6 connected to pin 9 and pin 5 connected to pin 10

Figure 3-33 (*Continued*)

Figure 3-33(c), the parallel resistance becomes 3390 Ω, the total effective resistance between pins 9 and 10 becomes 2532 Ω, and the gain becomes 80. Similarly for the other configurations: a gain of 100 is obtained when pin 6 is connected to pin 10 and pin 5 to pin 9; a gain of 200 is obtained when pin 7 is connected to pin 10; and a gain of 400 is obtained when pin 8 is connected to pin 10. The output amplifier has a gain of exactly 5 so that the effective output gains are 100, 200, 400, 500, 1000, and 2000, respectively.

The output amplifier has two external connections, pins 15 and 16, labelled *output sense* and *ground sense*, respectively. These pins allow the feedback loop of the output amplifier to be connected directly to the load, to eliminate errors in gain due to lead resistance.

The LH0038 requires a dual-polarity power supply with values from ±5 to ±18 V. Power supply connections are pins 2 (positive) and 14 (negative). The device has an excellent power supply rejection, with a typical power supply rejection ratio (PSRR) of 120 dB for DC. Because this deteriorates for AC, the manufacturer recommends that both power supply connections be bypassed with 10 μF electrolytic capacitors in parallel with 0.01 μF ceramic disk capacitors placed no more than 1 in. from the device. One can use 1 μF tantalum capacitors instead.

Because both inputs to the LH0038 are bipolar transistors, provision must be provided to allow bias current to flow into the inputs. Since the bias current may be as large as 200 nA, the input impedance is limited to a few megohms, and a

128 3 Amplifier Circuits

path to the ground or power supply with a resistance less than this must be provided.

Because of the high gain of this amplifier, the bandwidth is not particularly wide and is typically around 20 kHz. As shown in the performance characteristics on the data sheets, the common-mode rejection and the power supply rejection drop at even lower frequencies, reducing the performance at frequencies above

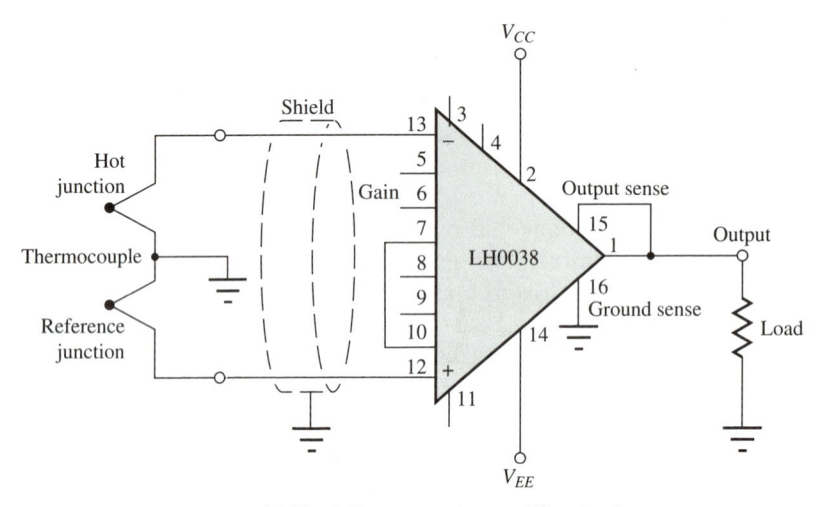

(a) Simple instrumentation amplifier circuit

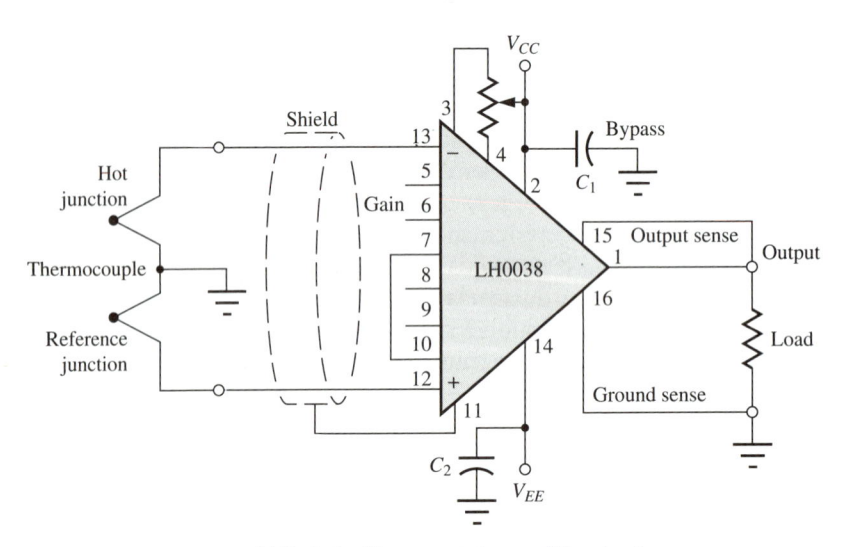

(b) Optimized instrumentation amplifier circuit

Figure 3-34 Thermocouple amplifier using the LH0038 instrumentation amplifier.

a few hundred hertz. The slew rate is also quite poor compared to a modern op-amp, being only 0.3 V/μs. None of these characteristics have much effect on the performance of the amplifier, since instrumentation signals are generally quite low frequency.

Circuits using the LH0038 are shown in Figure 3-34. Figure 3-34(a) shows a minimum circuit using the LH0038 as a thermocouple amplifier with a gain of 1000. Note the jumper between pins 7 and 10 to set the gain, the connection of pin 15 to the output and pin 16 to ground to complete the feedback and ground return path for the output amplifier, and the grounding of the center of the thermocouple to provide a path to ground for the input bias currents. The circuit is actually quite simple.

Figure 3-34(b) shows an optimized circuit using the LH0038. Note the various additions to enhance the performance of the amplifier. The output sense and ground sense pins (pins 15 and 16) are now connected to the load to minimize effects of resistance in the output leads on the circuit gain. A 10 kΩ compensating potentiometer has been connected between pins 2, 3, and the +15 V supply to provide for offset adjustment. The guard terminal, pin 11, has been connected to a shield around the input wires to maximize common-mode rejection. The power supply connections have been bypassed to ground to maximize the power supply rejection ratio. This optimized circuit is still relatively simple.

EXAMPLE 3-11

Design a thermocouple amplifier using a LH0038 instrumentation amplifier. The amplifier should have a gain of 500 with an input impedance of approximately 2 MΩ. Optimize the operation of the circuit as much as possible.

Solution

The optimized circuit of an instrumentation amplifier using the LH0038 was shown in Figure 3-34(b). That circuit was designed for a gain of 1000. We will use a similar circuit configured for a gain of 500. The circuit is drawn as Figure 3-35.

There is very little to design in this circuit. The ±15 V power supply is connected to pins 2 and 14, respectively, and is bypassed to ground with 1.0 μF tantalum capacitors to maximize the power supply rejection ratio. The output sense and ground sense pins (pins 15 and 16) are connected to the load to minimize effects of resistance in the output leads on the circuit gain. A 10 kΩ compensating potentiometer has been connected between pins 2, 3, and the +15 V supply to provide for offset adjustment. The guard terminal, pin 11, has been connected to a shield around the input wires to maximize common-mode rejection. Inputs to pins 12 and 13 are connected to the thermocouple, and the center of the thermocouple is connected to ground to provide a path for the input bias currents. The amplifier gain is set to 500 by connecting pin 6 to pin 10 and pin 5 to pin 9.

Our design is complete. We will now compare the design in Example 3-11 with the design in Example 3-10.

130   3 Amplifier Circuits

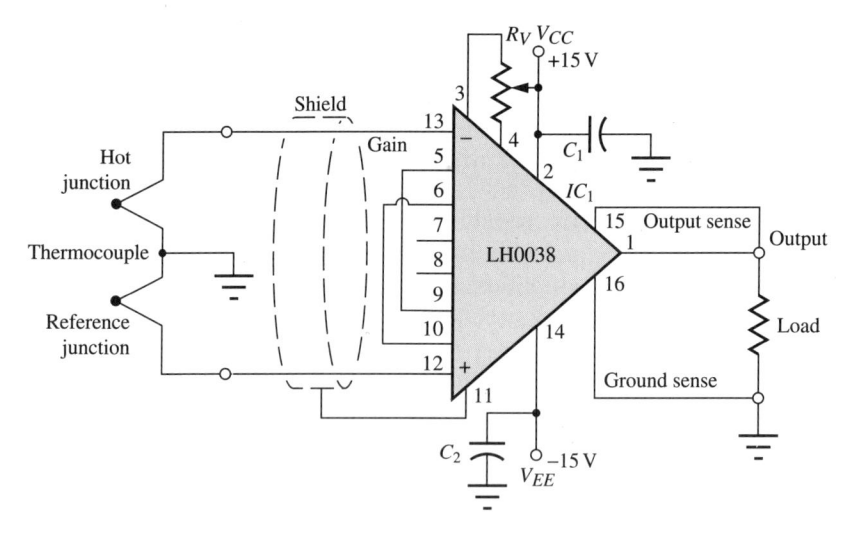

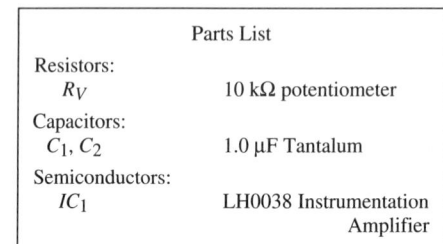

Figure 3-35

1. The cost of a TL084C or TL074C quad op-amp is typically $1 and the cost of a LH0038 is typically $30. If we used military or industrial versions of the TL084, the price would be higher. To obtain precise gain control for our op-amp circuit, we would have to use precision resistors which would add several dollars to the cost. The LH0038 has the necessary resistors built in.
2. Both common-mode rejection ratio and the power supply rejection ratio of the TL084 are 86 dB as compared to 120 dB for the LH0038. The LH0038 is better by a factor of 50.
3. The equivalent noise voltage of the TL084C is 47 $nV/Hz^{1/2}$ compared to 6 $nV/Hz^{1/2}$ for the LH0038. The LH0038 is better by a factor of 8.
4. The offset drift of the TL084C is typically 10 $\mu V/°C$ compared with 0.2 $\mu V/°C$ for the LH0038. The LH0038 is more stable by a factor of 50.

Although more expensive, the LH0038 clearly has the superior performance of the two instrumentation amplifiers.

Summary

This chapter introduced our first actual circuits. Procedures for designing inverting, noninverting, and buffer amplifiers using bipolar op-amps and BiFET op-amps were developed. Design procedures for these amplifiers were quite easy, with at most three resistors to select. This dramatically illustrates the power of integrated circuits—design is usually minimal since most of the design work has already been done and is incorporated into the integrated circuit.

One problem tackled in this chapter was to operate a standard bipolar or BiFET op-amp from a single supply. The procedure was simply to create a floating ground by placing a voltage divider across the power supply. A special op-amp designed for operation from a single supply, the TL321C, was introduced. For DC-coupled noninverting applications this is the best solution. Another alternative is to use the LM3900 Norton, or current-mode, op-amp. Amplifiers using the Norton op-amp have several unique features. In particular, it is possible to switch from inverting to noninverting amplification simply by switching the input resistor.

Finally, we looked at the most demanding of amplifier applications, process instrumentation, and saw how to construct an instrumentation amplifier using standard op-amps. As well, a special-purpose instrumentation amplifier integrated circuit, the LH0038, was introduced as the ultimate solution for tough instrumentation problems.

Review Questions

3-1 Basic inverting amplifier

1. In an inverting amplifier, what does the inverting input of the op-amp act as?
2. What is the equation for gain in an inverting amplifier?
3. How is the input impedance determined for an inverting amplifier?
4. What is the effect of source impedance on the gain of an inverting amplifier?
5. What is the difference in amplifier circuits using bipolar op-amps and BiFET op-amps?
6. Why is the compensating resistor R_K not required by BiFET op-amps?

3-2 Basic noninverting amplifier

7. What is the equation for gain in a noninverting amplifier?
8. How is the input impedance determined for a noninverting amplifier?
9. How does the input impedance of an inverting amplifier differ from a noninverting amplifier?
10. How should R_K be placed in the circuit for noninverting amplifiers constructed from bipolar op-amps if the source impedance is very high?

3-3 Buffer amplifiers

11. What is a follower amplifier? What are its characteristics?
12. What components are required for a follower amplifier?

13. What is the difference between an inverting buffer amplifier and a noninverting buffer amplifier. Which is the simpler circuit?
14. What is the input impedance of a follower amplifier? Of an inverting buffer amplifier?

3-4 Single-supply amplifier circuits

15. Why do most amplifier circuits constructed from op-amps use dual power supplies?
16. Why is it sometimes desirable to use an op-amp with a single supply?
17. How is the circuit of an inverting amplifier using a single supply different from the circuit of an inverting amplifier operating from a dual supply?
18. What is the purpose of the voltage divider in a single-supply amplifier?
19. Why must the center point of the voltage divider be connected to ground by a capacitor? How is this capacitor chosen?
20. What is the disadvantage of operating a standard op-amp from a single supply?

3-5 The TL321C single-supply op-amp

21. What is the advantage of using a special single-supply op-amp?
22. Compare the circuit of a noninverting amplifier constructed using a TL321C with that of a noninverting amplifier constructed using a LM741C.
23. What are the limitations of a noninverting amplifier constructed using a TL321C?
24. For what type of applications is a noninverting amplifier constructed from a TL321C suitable?
25. Compare an inverting amplifier using a TL321C with an inverting amplifier using a LM741C.

3-6 The Norton or current-mode op-amp

26. What is the difference between a Norton op-amp and a conventional op-amp?
27. What is the most common Norton op-amp integrated circuit?
28. Sketch the symbol for a Norton op-amp and explain how it is different from that of a standard op-amp such as the LM741C.
29. Give one limitation of a Norton op-amp.
30. What is the difference between a Norton inverting amplifier and a Norton noninverting amplifier?
31. What are the limitations on R_F for a Norton amplifier used as a buffer amplifier?

3-7 Instrumentation amplifiers

32. What is a process variable?
33. What device is used to sense a process variable?
34. What are the special problems that an instrumentation amplifier must handle?
35. What are the characteristics required for an instrumentation amplifier?
36. Why must an instrumentation amplifier always be used as a differencing amplifier?
37. How must a standard op-amp amplifier circuit be modified to convert it to an instrumentation amplifier?
38. Why must precision resistors be used in an instrumentation amplifier?
39. How can the gain of an instrumentation amplifier constructed from op-amps be made variable?

3-8 The LH0038 instrumentation amplifier

40. What differences are found between a true instrumentation amplifier and one constructed from op-amps.
41. How is noise reduced in a true instrumentation amplifier?
42. What is a guard drive amplifier?
43. How is the gain of a true instrumentation amplifier set?
44. Why does an instrumentation amplifier have ground sense and output sense connections?
45. Compare the bandwidth of an instrumentation amplifier with that of a standard op-amp. Compare the slew rate of an instrumentation amplifier with that of a standard op-amp. Why are these not critical for an instrumentation amplifier?

Design and Analysis Problems

3-1 Basic inverting amplifier

1. Design an inverting amplifier using a LM741C op-amp with a gain of 20 and an input impedance of 2000 Ω. Use a ± 15 V power supply. Do not use a compensating potentiometer.
   (a) Draw the circuit for an inverting amplifier and prepare a parts list.
   (b) Calculate values for R_I, R_F, and R_K.
   (c) Calculate the bandwidth.
2. Design an inverting amplifier using a TL081C op-amp with a gain of 100 and an input impedance of 1000 Ω. Use a ± 15 V power supply.
   (a) Draw the circuit and prepare a parts list.
   (b) Calculate values for R_I and R_F.
3. For the amplifier in Figure 3-36, determine the
   (a) type of amplifier
   (b) gain
   (c) input impedance
   (d) bandwidth

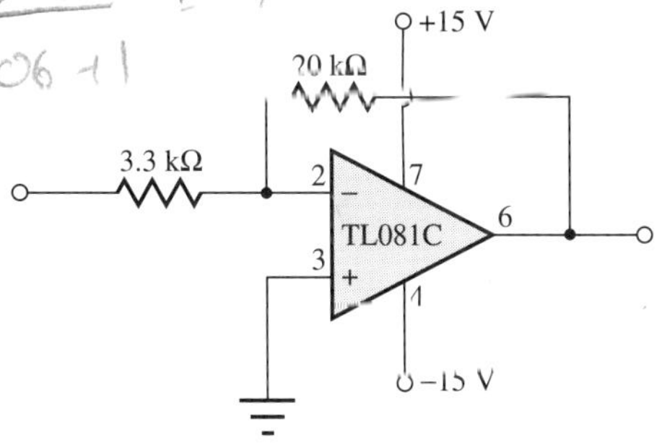

Figure 3-36

3-2 Basic noninverting amplifier

4. Design a noninverting amplifier using a LM741C op-amp with a gain of 10. Use a ±15 V power supply. Do not use a compensating potentiometer.
   (a) Draw the circuit for a noninverting amplifier and prepare a parts list.
   (b) Calculate values for R_I, R_F, and R_K.
5. Design a noninverting amplifier using a TL081C op-amp with a gain of 50. Use a ±15 V power supply.
   (a) Draw the circuit and prepare a parts list.
   (b) Calculate values for R_I and R_F.
6. For the amplifier in Figure 3-37, determine the
   (a) type of amplifier
   (b) gain
   (c) input impedance
   (d) bandwidth.

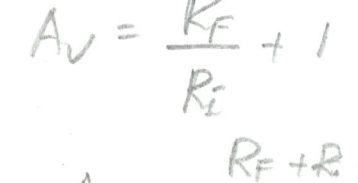

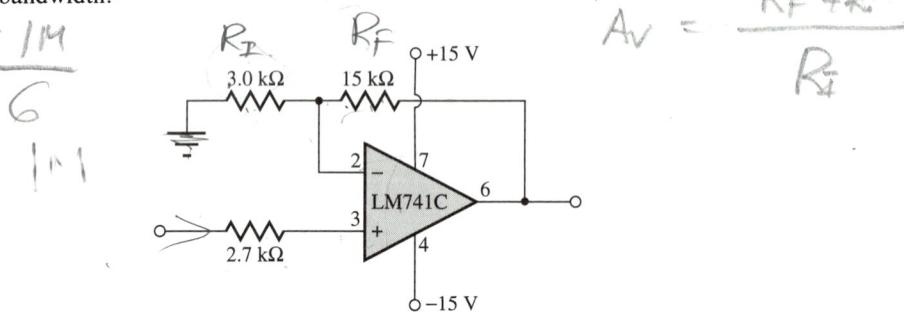

Figure 3-37

3-3 Buffer amplifiers

7. For the amplifier in Figure 3-38, determine the
   (a) type of amplifier
   (b) gain
   (c) input impedance
   (d) bandwidth.

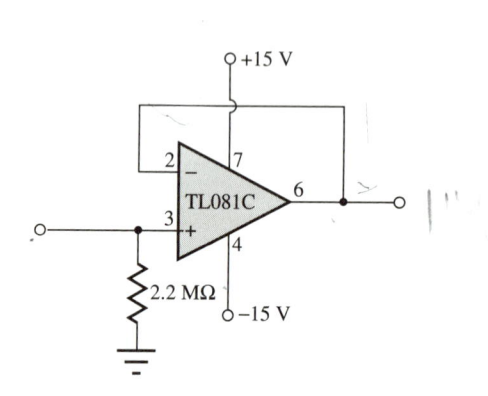

Figure 3-38

3-4 Single-supply amplifier circuits

8. Design an inverting amplifier circuit using a TL081C op-amp to run from a single 18 V supply. The amplifier should have a gain of 25, an input impedance of 3000 Ω, and a lower cutoff frequency of 50 Hz. Assume that the input impedance is zero and the load is 25 kΩ. Bias the amplifier for maximum AC swing.
9. Analyze the amplifier circuit in Figure 3-39 to determine
   (a) gain
   (b) input impedance
   (c) DC output offset voltage.

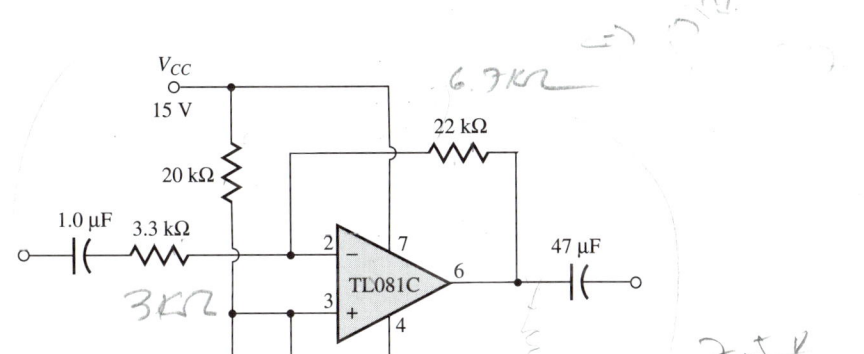

Figure 3-39

3-5 The TL321 single-supply op-amp

10. Design a noninverting amplifier using a TL321C op-amp and a single 20 V supply. The gain should be 45, and choose $R_I = 3300$ Ω.
11. Design an inverting amplifier with a gain of 25 using a TL321C single-supply op-amp to operate from an 18 V supply. The amplifier should have an input impedance of 3000 Ω. Assume the source has zero impedance and the load is 7000 Ω. Make the lower cutoff frequency 200 Hz. Bias the amplifier for maximum output swing.
12. Analyze the amplifier circuit in Figure 3-40 to determine
    (a) gain
    (b) input impedance
    (c) DC output offset voltage.

3-6 The Norton or current-mode op-amp

13. Design a noninverting amplifier with a gain of 20 using a LM3900 Norton op-amp. The amplifier is to operate from a single 12 V supply and have an input impedance of 200 kΩ.

3 Amplifier Circuits

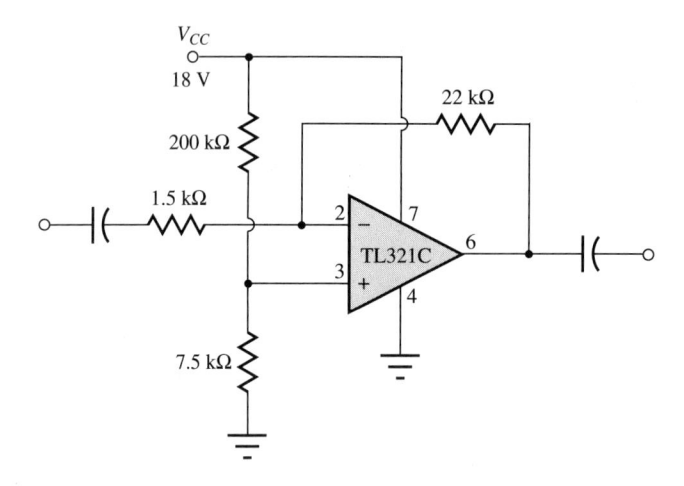

Figure 3-40

14. Analyze the amplifier circuit in Figure 3-41 to determine
    (a) gain
    (b) input impedance
    (c) output offset voltage.

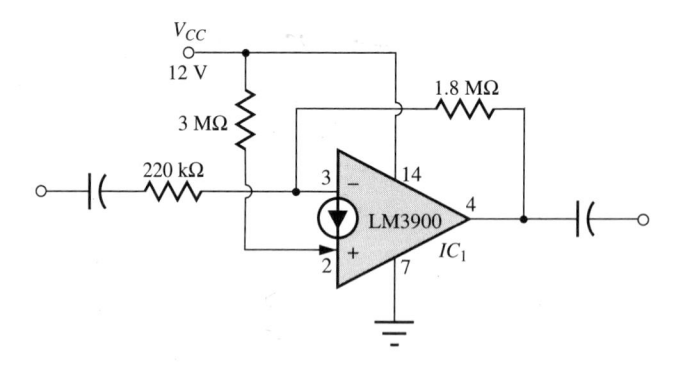

Figure 3-41

3-7 Instrumentation amplifiers

15. Design an instrumentation amplifier using BiFET op-amps. The amplifier should have a gain of 1000 with an input impedance of approximately 5 MΩ.

3-8 The LH0038 instrumentation amplifier

16. Design an instrumentation amplifier using an LH0038 instrumentation amplifier. The amplifier should have a gain of 1000 with an input impedance of approximately 5 MΩ. Optimize the operation of the circuit as much as possible.

17. Analyze the instrumentation amplifier in Figure 3-42 to determine the gain and the input impedance.

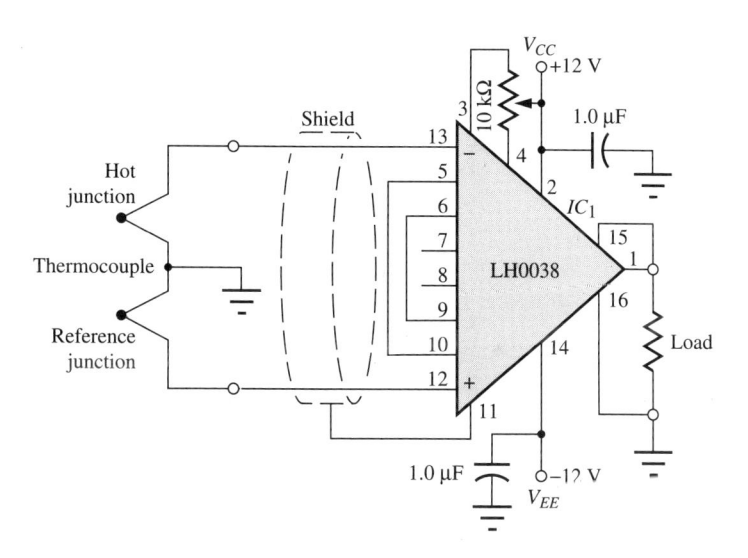

Figure 3-42

4 | Applications of Operational Amplifiers

OUTLINE

4-1 Summing Amplifiers
4-2 Subtracting Amplifiers
4-3 Integrating Amplifiers
4-4 Differentiating Amplifiers
4-5 Logarithmic Amplifiers
4-6 Exponential Amplifiers
4-7 Precision Rectifiers

KEY TERMS

Summing amplifier
Subtracting amplifier
Integration
Integrating amplifier
Differentiation
Differentiating amplifier
Logarithmic amplifier
Exponential amplifier
Precision rectifier

OBJECTIVES

On completion of this chapter, the reader should be able to:

- design summing amplifier circuits and use them as audio mixers.
- design subtracting and differencing amplifiers.
- understand the concept of integration, design an integrating amplifier, and use an integrating amplifier to convert a square wave to a triangle wave.
- understand the concept of differentiation, design a differentiating amplifier, and use a differentiating amplifier to convert a triangle wave to a square wave.
- design logarithmic and exponential amplifiers, and understand how they can be used for square root extraction, squaring, multiplying, and dividing.
- design a precision rectifier.

Introduction

We saw in the previous chapter how operational amplifiers could be used to produce inverting, noninverting, and buffer amplifiers. In future chapters, we shall see more complex applications of op-amps. In this chapter, we will introduce a number of different amplifier circuits to illustrate the flexibility of operational amplifiers in a variety of interesting and useful applications.

As described earlier, operational amplifiers got their name from the fact that they were originally used to perform mathematical "operations." Most of the amplifier circuits described in this chapter perform mathematical operations on the input signals. We will see how to make a summing amplifier that can be used as a mixer to combine two or more signals, a subtracting amplifier that can subtract one signal from another, an integrating amplifier that can generate the integral of an input waveform, and a differentiating amplifier that can generate the derivative of a waveform. Two more amplifier types, the logarithmic amplifier and the exponential amplifier, form the basis of circuits that extract square roots, square, multiply, and divide. Lastly, although not a mathematical operation, precision rectifier circuits are a useful application of op-amps.

Although any of the types of op-amps introduced in the previous chapters—bipolar, BiFET, single supply, or Norton—can be used for the amplifiers discussed in this chapter, we will use only the BiFET op-amps, since these are clearly the superior type of device, and they represent current technology. The reader is encouraged, however, to experiment with these circuits using bipolar op-amps, single-supply op amps, or Norton op-amps.

4-1 Summing Amplifiers

The first application that we will introduce is the **summing amplifier**. Summing amplifiers are amplifiers that add or combine two or more signals and find a variety of uses, including audio mixers and circuits for adding a DC offset voltage. It is possible to construct both an inverting summing amplifier and a noninverting summing amplifier. As we shall see shortly, the inverting summing amplifier is the more practical of the two.

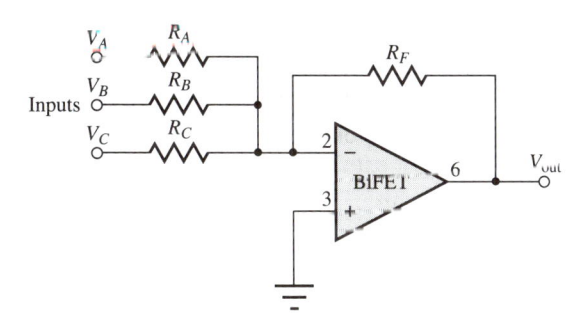

Figure 4-1 Circuit for an inverting summing amplifier using a BiFET op-amp.

Inverting Summing Amplifier

Figure 4-1 shows an inverting summing amplifier with three inputs. In theory, any number of inputs can be summed. For a straight summing circuit, we make $R_A = R_B = R_C = R_F = R$. Consider source A alone. Since the noninverting input is grounded, the inverting op-amp input is a virtual ground, and the input resistance seen by source A is R_A. The feedback resistance is R_F. Since both R_A and R_F have a value R, from Equation 3-1 for an inverting amplifier the gain is simply $R_F/R_A = 1$. The output voltage is $-V_A$, negative because this is an inverting amplifier. Each of the other input sources can be analyzed in the same way to give outputs of $-V_B$ and $-V_C$, respectively. The output is simply the sum of the individual input voltages:

$$-V_{out} = V_A + V_B + V_C \tag{4-1}$$

Because of the virtual ground at the inverting op-amp input, each source input sees the same input impedance of $R\ \Omega$. Also, because of the virtual ground, there is no crosstalk between inputs.

This circuit is not restricted to simple summing. If we have the three inputs as pictured and make $R_A = R_B = R_C = 3R_F$, we produce an averaging circuit. In this case, each input sees a gain of $\frac{1}{3}$; hence,

$$-V_{out} = \frac{1}{3}V_A + \frac{1}{3}V_B + \frac{1}{3}V_C = \frac{V_A + V_B + V_C}{3}$$

Actually, we can use whatever gain values we like. The general equation describing this is

$$-V_{out} = A_{VA}V_A + A_{VB}V_B + A_{VC}V_C \tag{4-2}$$

$$= \frac{R_F}{R_A}V_A + \frac{R_F}{R_B}V_B + \frac{R_F}{R_C}V_C$$

For example, if we make $R_A = R_F$, $R_B = \frac{1}{2}R_F$, and $R_C = \frac{1}{4}R_F$, we get an output voltage of

$$-V_{out} = V_A + 2V_B + 4V_C$$

This gives us the ability to weight the different inputs. If we make the input resistors variable, we obtain variable weighting.

EXAMPLE 4-1

Design a three-input inverting summing amplifier using a TL081C op-amp. Use a feedback resistance of 10 kΩ. Calculate all resistance values, draw the circuit, and prepare a parts list.

Solution

We will use the basic summing amplifier circuit shown in Figure 4-1. This has been redrawn as Figure 4-2, and a parts list is included. The power supply connections are shown.

4-1 Summing Amplifiers

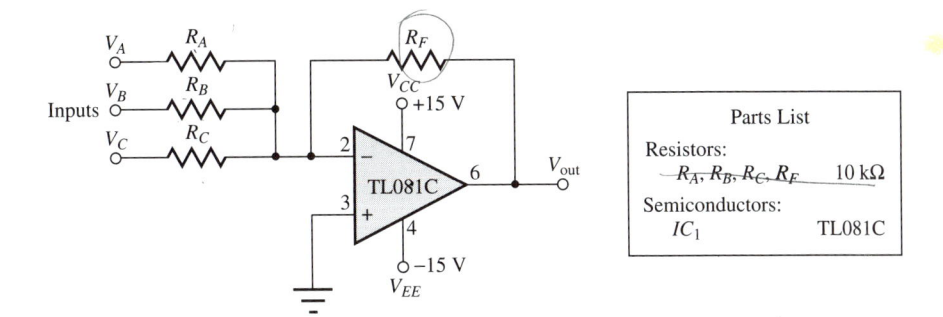

Figure 4-2

Because we want a simple summing amplifier, the gain for each input must be unity. This means that $R_F/R_I = 1$. Since $R_F = 10$ kΩ, then $R_A = R_B = R_C = R_F = 10$ kΩ. This is a standard value, so we shall select all the resistors with this value.

EXAMPLE 4-2

Design a two-input mixing circuit using an inverting summing amplifier with a TL081C op-amp. Input signal A should have a gain of 0.5, and input signal B should have a gain of 3. Use a feedback resistance of 10 kΩ. Calculate all resistance values, draw the circuit, and prepare a parts list.

Solution

A mixing circuit is simply a summing amplifier with different gains for the various inputs. We will use the inverting summing amplifier circuit shown in Figure 4-1, but with two inputs instead of three. The circuit has been redrawn as Figure 4-3, and a parts list is included.

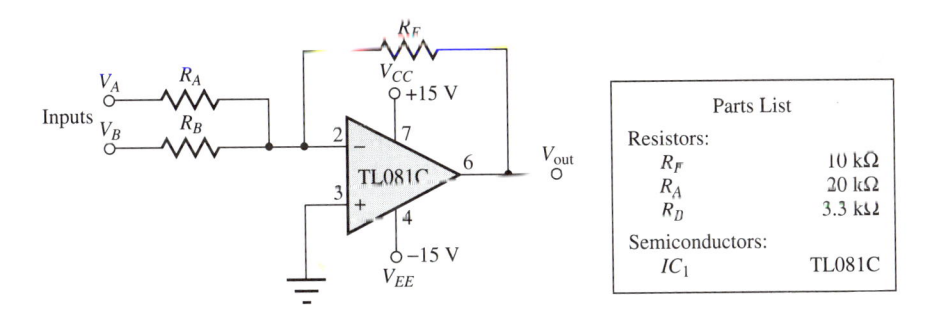

Figure 4-3

The gain for each input is given by Equation 4-2. Rewriting this equation, we get

$$A_{V_A} = \frac{R_F}{R_A} \quad \text{and} \quad A_{V_B} = \frac{R_F}{R_B}$$

Since the gain of input A is 0.5, then $R_F/R_A = 0.5$; since $R_F = 10\ \text{k}\Omega$, then $R_A = 20\ \text{k}\Omega$. This is a standard value, so we will use a 20 kΩ resistor. The gain of input B is R_F/R_B, where R_F is still 10 kΩ. Thus, $R_B = 10\ \text{k}\Omega/3 = 3333\ \Omega$. The closest standard value is 3.3 kΩ.

The design of the mixer circuit is now complete.

Noninverting Summing Amplifier

Figure 4-4 shows a noninverting summing amplifier with three inputs. We will only develop the design of a three-input noninverting summing amplifier with equal input resistances, $R_A = R_B = R_C = R$. Using unequal resistances makes the design much more complicated. First, determine the voltage at the noninverting input for each source. Assume source A is the only source and that sources B and C are grounded. Because no current flows into the noninverting input of the op-amp, current from source A flows in through R_A and out through $R_B \parallel R_C$. Since we assumed $R_B = R_C = R$, $R_B \parallel R_C = \frac{1}{2}R$. The input voltage V_A sees a voltage divider consisting of R in series with $\frac{1}{2}R$. Hence, the voltage at the noninverting input of the op-amp due to source A is

$$V_{NIA} = V_A \times \frac{\frac{1}{2}R}{R + \frac{1}{2}R} = \frac{1}{3}V_A$$

Repeating this analysis for sources B and C, we find voltages at the noninverting input of the op-amp of $\frac{1}{3}V_B$ and $\frac{1}{3}V_C$, respectively. By the superposition theorem of elementary circuit analysis, the total voltage at the noninverting input due to all three sources is

$$V_{NI} = \frac{1}{3}V_A + \frac{1}{3}V_B + \frac{1}{3}V_C = \frac{V_A + V_B + V_C}{3}$$

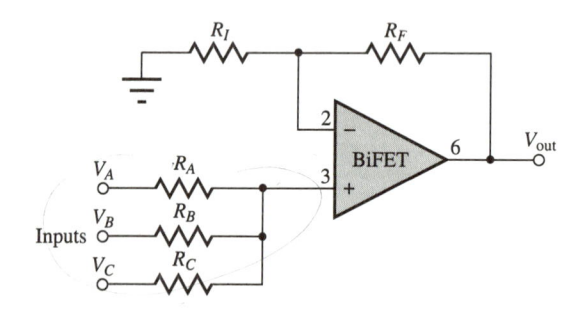

Figure 4-4 Circuit for a noninverting summing amplifier using a BiFET op-amp.

If the gain of the amplifier is 1 (feedback resistor shorted, giving a follower amplifier), we will get the average of the three inputs. To get the sum of the three inputs, we must design the amplifier with a gain of 3. Using Equation 3-4, which defines the gain for a noninverting amplifier as $A_V = R_F/R_I + 1$, we find $R_F/R_I = 2$ or $R_I = \frac{1}{2}R_F$.

In general, when designing a noninverting summing amplifier with equal input resistances, we must select the gain to be equal to the number of inputs. If the gain is 1, then the output is the average of the inputs. Remember, this only applies if the input resistors are equal. If they are not equal, the calculations are much more involved.

This circuit is actually far less useful than the inverting summing circuit. First, although it is possible to have different gains for the different inputs, the procedure for determining them is more complicated than for the inverting summing amplifier. Second, because the noninverting input is not a virtual ground and has a very high input impedance, crosstalk and loading between the different inputs can be a problem.

EXAMPLE 4-3

Design a three-input noninverting summing amplifier using a TL081C op-amp. Use 10 kΩ resistors for the feedback resistor R_F and for each of the input resistors. Calculate the gain and the value of R_I, draw the circuit, and prepare a parts list.

Solution

For the circuit, we will use the noninverting summing amplifier in Figure 4-4. This has been redrawn as Figure 4-5 with a parts list.

The problem stipulated that we use equal 10 kΩ resistors for the inputs to make a simple three-input summing amplifier. For a simple noninverting amplifier with equal input resistors, the gain must equal the number of inputs. For three

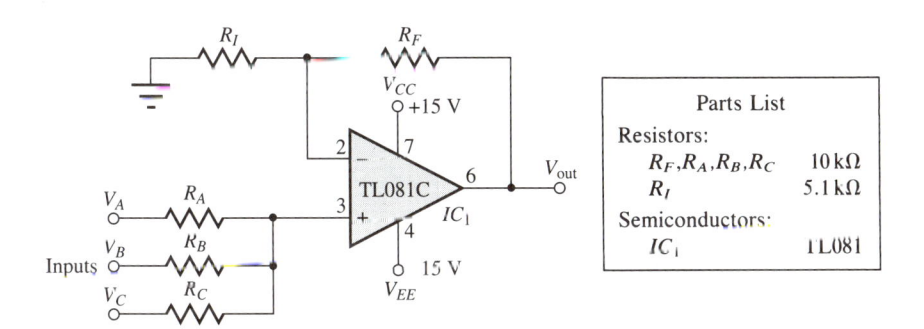

Figure 4-5

inputs we require a gain of 3. The gain of a noninverting amplifier is given by Equation 3-4:

$$A_V = \frac{R_F}{R_I} + 1 \quad \text{or} \quad \frac{R_F}{R_I} = A_V - 1 = 3 - 1 = 2$$

Since we are given that $R_F = 10 \text{ k}\Omega$, we solve the above equation for R_I:

$$R_I = \frac{R_F}{2} = \frac{10 \text{ k}\Omega}{2} = 5.0 \text{ k}\Omega$$

We will choose $R_I = 5.1 \text{ k}\Omega$, the closest standard value.

As described earlier, this circuit is susceptible to crosstalk from the various inputs and is not particularly recommended. The inverting summing amplifier designed in Example 4-1 is superior.

4-2 Subtracting Amplifiers

Although an op-amp is a differential amplifier and amplifies the difference between the signals applied to the two inputs, we have only used it as a single-ended amplifier amplifying a signal applied to the inverting input or to the noninverting input. A special amplifier, known as a **subtracting or differencing amplifier**, is used for amplifying the difference between two input signals.

The circuit of a differencing amplifier is shown in Figure 4-6. Notice that this circuit is a little different from an ordinary inverting or noninverting amplifier. The circuit should, however, be familiar: it is the same circuit that was introduced for a simple instrumentation amplifier in Chapter 3.

Before we can design this amplifier, we must derive equations for gain. Consider the inverting input first. Apply a signal V_A to the inverting input and a zero signal to the noninverting input. This places the noninverting op-amp input at ground potential and gives a standard inverting amplifier with a gain given by

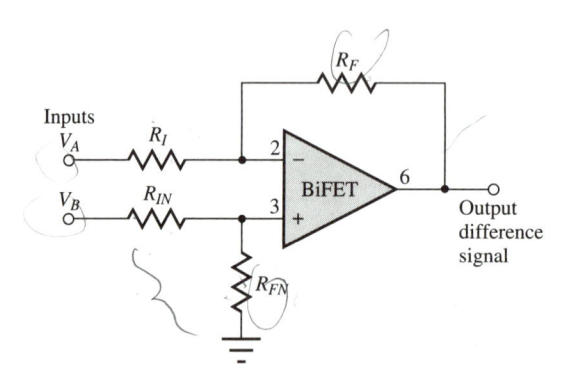

Figure 4-6 Differencing amplifier using a BiFET op-amp.

Equation 3-1:

$$A_V = \frac{R_F}{R_I} \quad \therefore V_{out} = \frac{R_F}{R_I} \times V_A$$

Next consider a signal V_B applied to the noninverting input of the amplifier and a zero signal applied to the inverting input. This grounds R_I and gives the noninverting amplifier a gain given by Equation 3-4:

$$A_V = \frac{R_F + R_I}{R_I} = \frac{R_F}{R_I} + 1$$

Hence,

$$V_{out} = \frac{R_F + R_I}{R_I} \times V_{NI}$$

The signal V_B, however, goes through a voltage divider consisting of R_{IN} and R_{FN}, and the voltage actually applied to the noninverting input of the op-amp is

$$V_{NI} = \frac{R_{FN}}{R_{FN} + R_{IN}} \times V_B \quad (4\text{-}3)$$

Multiply this equation by the noninverting gain equation:

$$V_{out} = \frac{R_{FN}}{R_{FN} + R_{IN}} \times \frac{R_F + R_I}{R_I} \times V_B$$

If $R_{FN} = R_F$ and $R_{IN} = R_I$, then this equation reduces to

$$V_{out} = \frac{R_F}{R_I} \times V_B$$

The gain for the noninverting input, V_{out}/V_{in}, is simply R_F/R_I, the same as for the inverting input. In other words, the gain of a subtracting or differencing amplifier is the same as for an inverting amplifier *provided that* $R_{FN} = R_F$ and $R_{IN} = R_I$. (Actually, if $R_{FN}/R_F = R_{IN}/R_I$, the gain conditions are also satisfied.)

$$A_{V\text{diff}} = \frac{R_F}{R_I} \quad (4\text{-}4)$$

Next, we must determine the impedance at each input. Assume that $R_{FN} = R_F$ and $R_{IN} = R_I$. The input impedance for the noninverting input $Z_{in\,NI}$ is simply

$$Z_{in\,NI} = R_I + R_F \quad (4\text{-}5)$$

The impedance at the inverting input is somewhat more complex. Because the voltage will be the same at each op-amp input, the voltage at the inverting input of the op-amp will be the same as the voltage produced by V_B at the noninverting amplifier input, namely

$$V_I = V_{NI} = \frac{R_F}{R_F + R_I} \times V_B \quad (4\text{-}6)$$

The current into the inverting input of the amplifier will be the current through the input resistor R_I. The voltage across R_I is $V_{R_I} = V_A - V_I$, and is

$$V_{R_I} = V_A - \frac{R_F}{R_F + R_I} \times V_B = \frac{V_A R_F + V_A R_I - V_B R_F}{R_F + R_I}$$

The input current I_{in} is the current through R_I and is equal to V_{R_I}/R_I. The input impedance of the inverting input $Z_{in_{INV}}$ is the input voltage divided by the input current, $V_A/I_{in} = V_A \times R_I/V_{R_I}$ and is

$$Z_{in_{INV}} = \frac{V_A R_I}{\frac{V_A R_F + V_A R_I - V_B R_F}{R_F + R_I}} = \frac{V_A R_I (R_F + R_I)}{V_A R_F + V_A R_I - V_B R_F} \qquad (4\text{-}7)$$

We see that Z_{in} for the inverting input is a complicated function of R_I, R_F, V_A, and V_B. There are two simple cases. If $V_B = 0$, the above equation reduces to $Z_{in} = R_I$, exactly what we would expect for a standard inverting amplifier. If $V_A = V_B$, then $Z_{in} = R_F + R_I$, the same as for the noninverting input. This latter condition is what we used for the instrumentation amplifier discussed in Chapter 3. In general, however, the impedances for the two inputs for a differencing amplifier will be different.

Simple Differencing Amplifier

In designing a subtracting or differencing amplifier, we have three options. The first option is to make a simple unity-gain differencing amplifier where $R_F = R_I = R_{FN} = R_{IN}$. The output of this amplifier is the difference of the two inputs $V_B - V_A$.

Multiplying Differencing Amplifier

The second option is to make a multiplying differencing amplifier by setting $R_F = R_{FN}$ and $R_I = R_{IN}$ and choosing the gain or multiplying factor as R_F/R_I. The output of this amplifier is the difference of the two inputs $V_B - V_A$ multiplied by the gain.

Differencing and Multiplying Amplifier

The third option is to make a differencing and multiplying amplifier where each input is multiplied by a different constant. Signal A to the inverting input of the op-amp will always be multiplied by the inverting gain.

$$V_{outA} = \frac{R_F}{R_I} \times V_A \qquad (4\text{-}8)$$

Signal B to the noninverting input will be multiplied by the noninverting gain and the voltage divider factor to give the equation

$$V_{outB} = \left(\frac{R_F}{R_I} + 1\right)\left(\frac{R_{FN}}{R_{IN} + R_{FN}}\right) V_B \qquad (4\text{-}9)$$

Values of R_{IN} and R_{FN} must be chosen to give the required multiplying factor. For this type of amplifier to work, the inverting gain must always be chosen larger than the noninverting gain.

EXAMPLE 4-4

Design a differencing amplifier to produce the simple difference between the two inputs. Use a TL081C op-amp operated from a ±15 V supply. Make the input impedance to the noninverting input equal to 200 kΩ.

Solution

The circuit of the differencing amplifier that we are designing is shown in Figure 4-7.

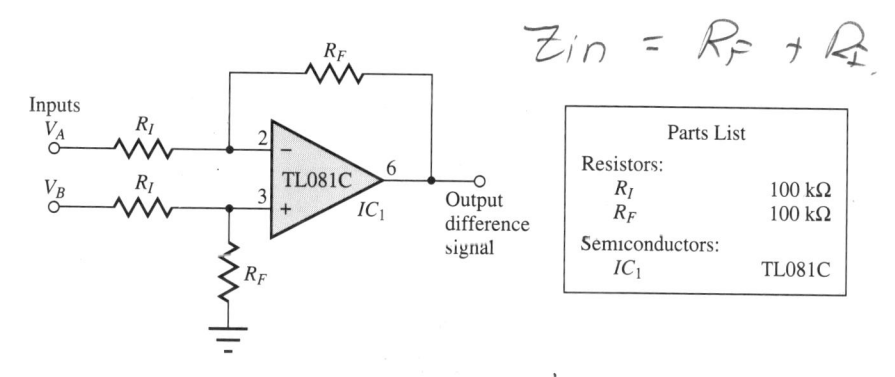

$Z_{in} = R_F + R_I$

Figure 4-7

The design of this differencing amplifier is very simple. Because we want the simple difference between the two inputs, we must use a gain of 1. The gain is defined by Equation 4-4 as $R_F/R_I = 1$; hence, $R_I = R_F$. We are also given that the input impedance to the noninverting input should be 200 kΩ. The input impedance to the noninverting input is given in Equation 4-5 as $R_I + R_F$, hence, we find that $R_I = R_F = 100$ kΩ. From the table in Appendix A-1, this is a standard value of resistance. Note that we cannot determine the input impedance to the inverting input.

The design of the differencing amplifier is complete.

EXAMPLE 4-5

Design a differencing amplifier to subtract ten times the voltage at input *A* from five times the voltage at input *B*. Use a TL081C op-amp operated from a ±15 V supply. Make R_F and R_{FN} equal to 100 kΩ.

4 Applications of Operational Amplifiers

Solution

Superficially, this is the same problem that we did in Example 4-4. However, in this example, the two inputs have different gain factors, making the problem somewhat more difficult. The circuit is exactly the same as in the previous example and is shown as Figure 4-8, including a parts list.

First set the gain of the inverting input to 10, using Equation 4-8 with $R_F = 100 \text{ k}\Omega$.

$$V_{outA} = \frac{R_F}{R_I} \times V_A$$

Since $V_{out}/V_A = 10$, then $R_F/R_I = 10$, and we calculate $R_I = 10 \text{ k}\Omega$.

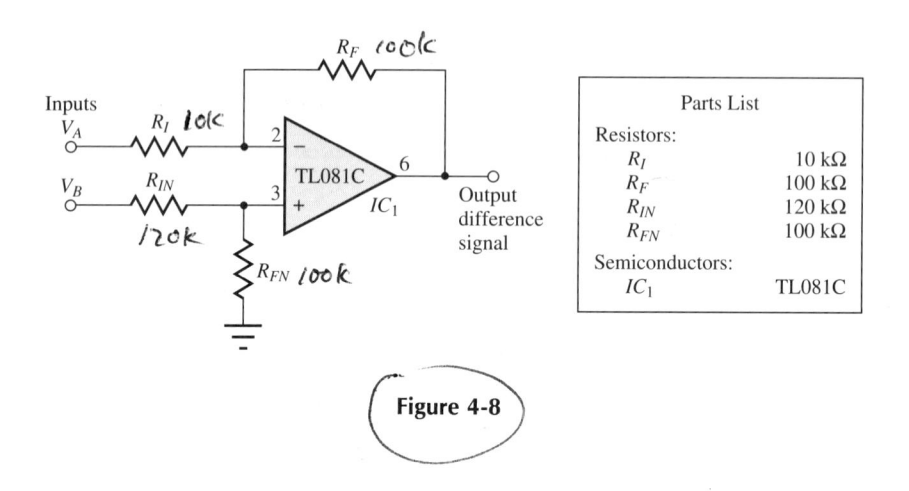

Figure 4-8

Parts List

Resistors:
- R_I 10 kΩ
- R_F 100 kΩ
- R_{IN} 120 kΩ
- R_{FN} 100 kΩ

Semiconductors:
- IC_1 TL081C

To set the gain of the noninverting input, we must use Equation 4-9. The value of R_{FN} was stipulated as 100 kΩ, we know R_I and R_F, and we require $V_{outB}/V_B = 5$. Substituting these data into Equation 4-9 gives

$$V_{outB} = \left(\frac{R_F}{R_I} + 1\right)\left(\frac{R_{FN}}{R_{IN} + R_{FN}}\right) V_B$$

$$5 = \left(\frac{100 \text{ k}\Omega}{10 \text{ k}\Omega} + 1\right)\left(\frac{100 \text{ k}\Omega}{R_{IN} + 100 \text{ k}\Omega}\right)$$

Solving for R_{IN} gives

$$R_{IN} = \frac{11}{5} \times 100 \text{ k}\Omega - 100 \text{ k}\Omega = 120 \text{ k}\Omega$$

Choose R_{IN} as 120 kΩ. This is a standard value from the table in Appendix A-1. The design of this more complicated differencing amplifier is complete.

4-3 Integrating Amplifiers

Integration is an operation from calculus by which some variable quantity is summed over a period of time (or some other variable). For example, the voltage across a capacitor at any time is proportional to the accumulated charge due to the current flow, or in calculus terms, the integral of the current flow. This is described in symbols as

$$V(t) = \frac{1}{C} \int_0^t I(t)\, dt \qquad (4\text{-}10)$$

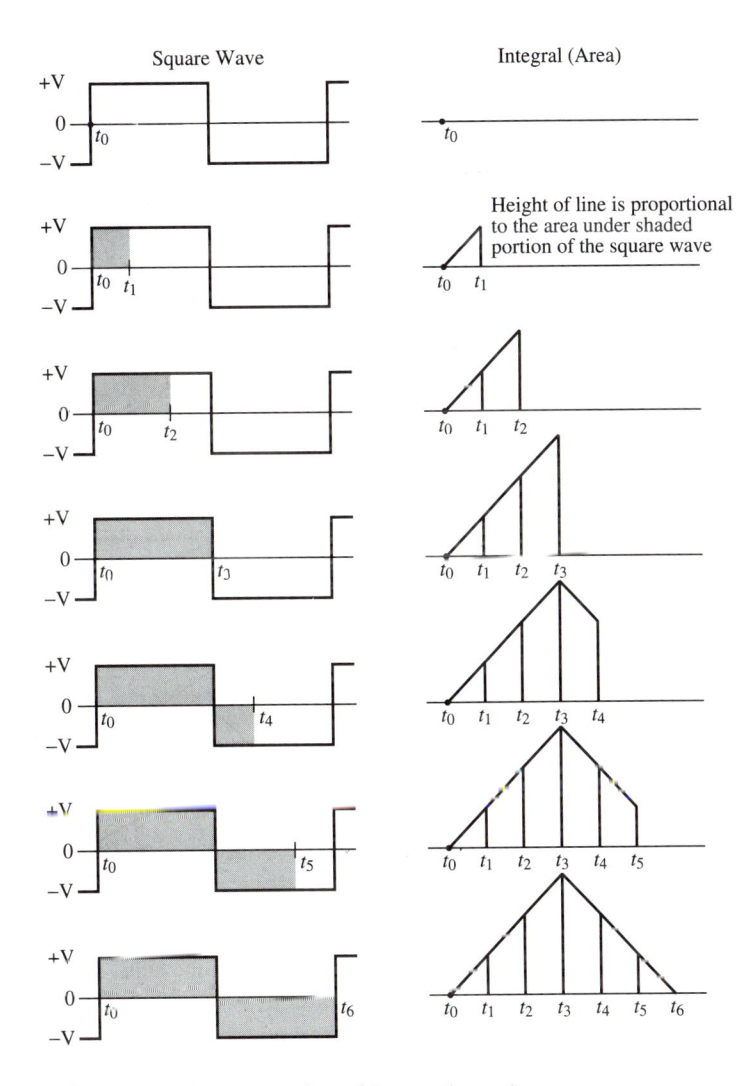

Figure 4-9 Demonstration of integration using a square wave.

The symbols $V(t)$ and $I(t)$ simply represent voltage and current as variables with time. It is possible to perform this operation on an electronic signal using an **integrating amplifier** constructed from an op-amp. A capacitor is used to store the charge resulting from the current flow through an input resistor. The resulting capacitor voltage is the voltage $V(t)$ as described by the above equation.

Integration is a useful concept, and it is important to fully understand what is happening. Figure 4-9 shows a square wave. The integral of this square wave is built up by summing very small slices of the wave with respect to time. If we take the area under the curve at t_0, we get zero area since this is the starting point of our integration, and we have not yet produced any area. If we take the area at time $t = t_1$, then we get a small positive area. This area is plotted as the height of a line on the graph to the right in Figure 4-9. If we take the area at $t = t_2$, we get a larger area and, hence, plot a proportionally longer line. The area will continue to increase with time up to the end of the positive cycle of the square wave at t_3. The variation of area with time is a line that slopes up to the right.

Now continue taking areas when the square wave has switched to negative polarity. In this case we have to subtract the areas from t_3 to t_4, from t_3 to t_5, and from t_3 to t_6, respectively. When we get to the end of the cycle, the area has been reduced to zero. For the next cycle, the process repeats. The resulting waveform of area is a triangle wave, showing that the integral of a square wave is a triangle wave. The same process can be performed for any waveform.

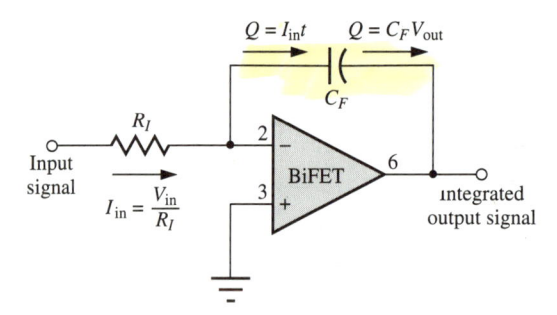

Figure 4-10 Simple integrating amplifier using a BiFET op-amp.

Figure 4-10 shows a simple integrating circuit based on a single BiFET op-amp. The feedback resistor of a conventional amplifier circuit is replaced with a capacitor. To determine an equation describing the operation of the integrator circuit, assume that a constant DC voltage V_{in} is applied to the input of R_I. Current $I_{in} = V_{in}/R_I$ will flow through R_I since the inverting input is at virtual ground potential. Since the input impedance of the op-amp is virtually infinite, all of this input current flows to charge the capacitor. Because the input current is constant, the charge on the capacitor at any time t is $Q = I_{in}t$. The voltage across the capacitor equals the output voltage, which is negative since the capacitor is connected to the virtual ground at the inverting input. The output voltage is $V_{out} =$

Q/C_F. If we substitute for Q and I_{in} in this equation, we find

$$V_{out} = \frac{Q}{C_F} = \frac{I_{in}t}{C_F} = \frac{V_{in}t}{R_I C_F} \quad (4\text{-}11)$$

This equation can be written with calculus notation to describe the integration of any type of input signal $V_i(t)$:

$$V_{out}(t) = \frac{1}{C_F R_I} \int_0^t V_i(t) \, dt \quad (4\text{-}12)$$

The circuit shown in Figure 4-10 can be used to integrate a DC or fluctuating input. For example, light detectors used in astronomy and radiation detectors used in nuclear physics produce signals that are normally integrated over a period of time by a circuit similar to that in Figure 4-10. Resistor R_I would be calculated by Equation 4-11.

More frequently, we want an integrating circuit to be used for waveform shaping—converting one waveform into its integral. We will see this application in Chapters 10 and 11, when we use an integrating circuit to convert a square wave into a triangle wave. The circuit in Figure 4-10 does not work for this type of application, and we must use the circuit in Figure 4-11.

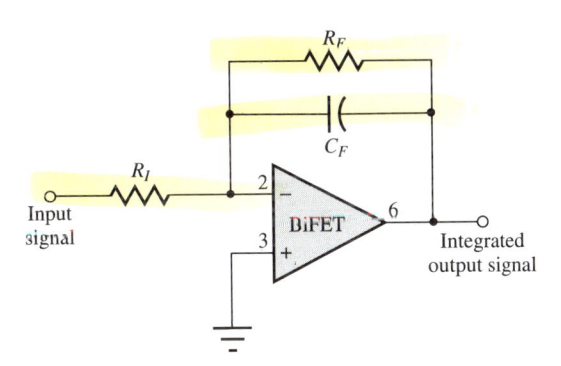

Figure 4-11 Practical integrating amplifier for periodic waveforms.

We also require a general working equation that can be used for designing R_I (or C_F) in this circuit. We will use the equation

$$V_{out} = \frac{V_{in}}{4fR_I C_F} \quad (4\text{-}13)$$

where V_{in} is the amplitude of the input waveform, V_{out} is the amplitude of the output waveform, and f is the frequency. This equation is similar to Equation 4-11, except that we have replaced the integration time t by the factor $1/4f$. This is simply an empirical formula that gives good design results. The usual procedure is to choose a value of C_F and calculate R_I. We can solve Equation 4-13 for R_I

to give a useful design equation:

$$R_I = \frac{V_{in}}{V_{out}} \times \frac{1}{4fC_F} \qquad (4\text{-}14)$$

If we choose to make $V_{out} = V_{in}$, then the equation simplifies to

$$R_I = \frac{1}{4fC_F}$$

The problem with the circuit in Figure 4-10 for integrating periodic waveforms is that the voltage gain of the amplifier A_V depends on frequency and is approximately (exactly for a sine wave)

$$A_V = \frac{X_{C_F}}{R_I} = \frac{1}{2\pi f C_F R_I} \qquad (4\text{-}15)$$

For very low frequencies, the gain can become very large. This introduces distortion into the output waveform. To limit the low frequency gain, we place R_F in parallel with C_F. For low frequencies, $R_F \ll X_{C_F}$. Hence, the gain is determined by R_F alone. For best operation, limit the gain to 25 for low frequencies by making $R_F = 25R_I$.

EXAMPLE 4-6

Design an integrating amplifier using a TL081C op-amp to integrate a 4000 Hz square wave. The peak-to-peak output voltage should be approximately the same as the peak-to-peak input voltage. Use a 0.01 µF feedback capacitor. Calculate all resistance values, draw the circuit, and prepare a parts list.

Solution

We will use the integrating amplifier for periodic waveforms shown in Figure 4-11. This has been redrawn as Figure 4-12, and a parts list is included.

We are given the operating frequency of 4000 Hz, require $V_{out} = V_{in}$, and know $C_F = 0.01$ µF. We must calculate R_I and R_F. From Equation 4-14,

$$R_I = \frac{V_{in}}{V_{out}} \times \frac{1}{4fC_F}$$

$$= \frac{1}{4fC_F} \quad \text{if} \quad V_{out} = V_{in}$$

$$= \frac{1}{4 \times 4 \text{ kHz} \times 0.01 \text{ µF}} = 6.25 \text{ k}\Omega$$

The closest standard resistor from Appendix A-1 is 6.2 kΩ, so we will select that value for R_I.

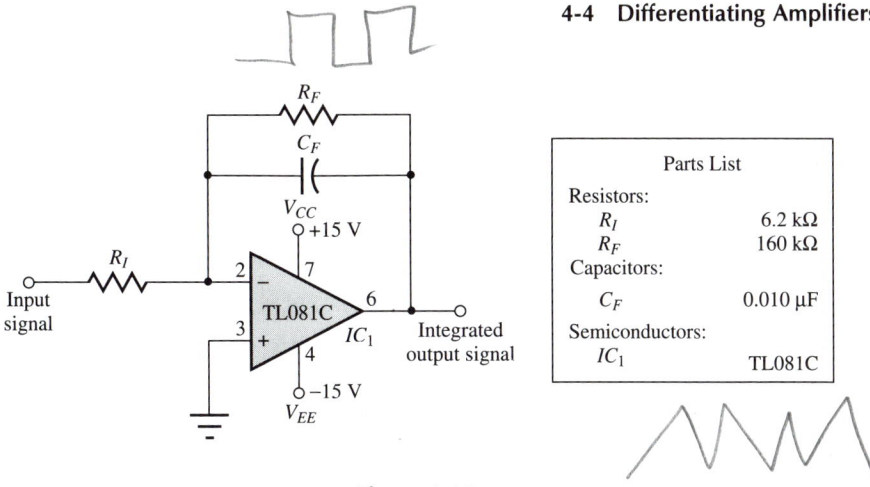

Figure 4-12

Parts List
Resistors:
 R_I 6.2 kΩ
 R_F 160 kΩ
Capacitors:
 C_F 0.010 μF
Semiconductors:
 IC_1 TL081C

For integrating a periodic waveform, we need a gain-limiting resistor R_F in parallel with C_F. Choose $R_F = 25R_I$; hence, calculate $R_F = 156$ kΩ. Select an actual value of 160 kΩ from Appendix A-1.

The design of our integrator circuit is now complete.

4-4 Differentiating Amplifiers

The other type of operation frequently encountered in calculus is **differentiation**. This operation determines the rate of change of one variable with respect to another, frequently the rate of change of voltage or current with time. For example, in an inductor the instantaneous voltage is proportional to the rate of change of current through the inductor, with time. In calculus symbols, this is

$$V(t) = L \frac{dI(t)}{dt} \tag{4-16}$$

The symbol $dI(t)/dt$ simply stands for the rate of change of current with time. As in the case of integration, $V(t)$ is the voltage expressed as a function of time and $I(t)$ is the current expressed as a function of time. It is possible to perform this operation on electronic signals using a **differentiating amplifier** (not a *differencing amplifier*) constructed from an op-amp.

Mathematically, differentiation is the inverse operation of integration. If we integrate some mathematical function, then differentiate the result, we will obtain the original function. The same applies if we perform these operations on an electronic signal. We saw that the effect of an integrating circuit on a square wave was to convert it to a triangle wave. If we differentiate a triangle wave, we should get a square wave. Figure 4-13 shows what happens when we differentiate a triangle wave. The triangle wave is shown as starting at approximately zero and

154 **4 Applications of Operational Amplifiers**

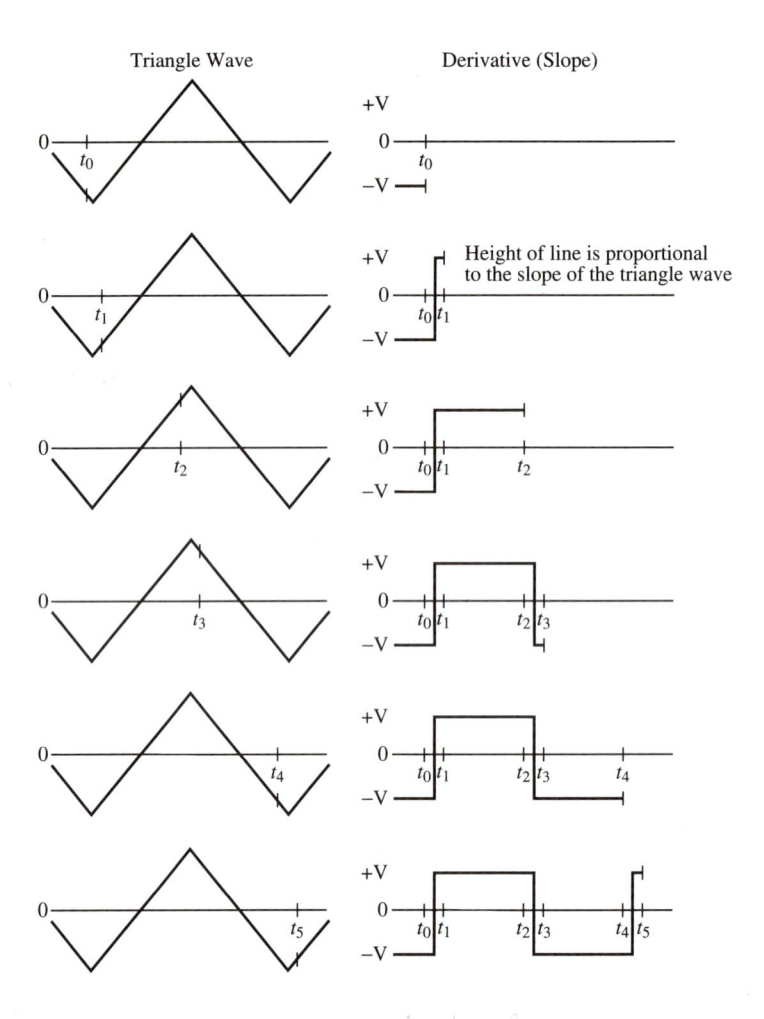

Figure 4-13 Demonstration of differentiation using a triangle wave.

decreasing. The rate of change of voltage with time (dV/dt) is constant and negative at t_0. Hence, the output of the differentiator circuit would be a negative fixed voltage. When the triangle wave reaches its negative peak shortly after t_0, it starts to increase at a constant rate. The point t_1 occurs shortly after the wave starts to increase. Now, since the rate of change is constant in a positive direction, the output of the differentiator circuit is a positive fixed voltage. From t_1 to t_2, the rate of change remains constant, so the output voltage shown on the right remains constant. Shortly after t_2, the triangle wave reaches its peak and starts to decline. At t_3, the slope is negative and the rate of change is constant. Thus, we get a negative constant voltage as the differentiated output. The output voltage remains constant until shortly after t_4, when the triangle wave reaches its negative minimum and starts to increase. Point t_5 occurs shortly after the negative minimum, when the triangle wave voltage is uniformly increasing. The differentiated

4-4 Differentiating Amplifiers   155

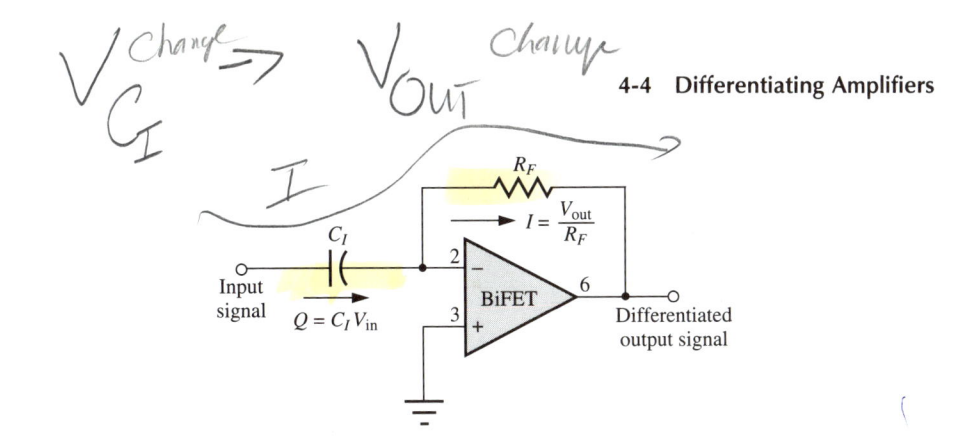

Figure 4-14 Circuit of a simple differentiating amplifier using a BiFET op-amp.

voltage is now positive and constant. If we consider a number of cycles of a differentiated triangle wave, we find that we get a square wave, exactly as predicted.

A simple differentiating amplifier is shown in Figure 4-14. The input resistor R_I of a standard inverting amplifier is replaced with an input capacitor C_I. As the input voltage to the capacitor changes, the output voltage must also change to balance the charge on the capacitor. Assume a uniformly rising input voltage to the capacitor, for example the rising portion of a triangle wave. Current flows from the source to charge the capacitor. Because the source voltage is increasing at a uniform rate, the current flow into the capacitor is constant. The other side of the capacitor is connected to the virtual ground of the inverting input. As the capacitor voltage increases, current must flow from the capacitor through the feedback resistor R_F to balance the input current. Because the input current is constant, the current through R_F must also be constant. This is only possible if the output voltage V_{out} is constant.

The voltage across the capacitor at any time t is simply the input voltage V_{in}. The charge on the capacitor for any voltage V_{in} is $Q = C_I V_{in}$. The current flowing through R_F is equal to V_{out}/R_F. The charge supplied to the capacitor by this current is equal to $(V_{out}/R_F)\,t$. If we equate the two charges, we get the following equation for the differentiator circuit:

$$C_I V_{in} = \frac{V_{out}\,t}{R_F} \quad \text{or} \quad V_{out} = \frac{V_{in} C_I R_F}{t} \tag{4-17}$$

This equation can be written in calculus notation to describe the differentiation of any type of input signal $V_i(t)$. The input current into the capacitor as a function of applied voltage is

$$I_{in}(t) = C_I \frac{dV_i(t)}{dt}$$

The output voltage is $I_{in} R_F$. Hence, the equation for V_{out} in calculus notation is

$$V_{out}(t) = C_I R_F \frac{dV_i(t)}{dt} \tag{4-18}$$

Most frequently, we use differentiating amplifiers for wave-shaping applications on periodic waveforms. The simple differentiator circuit in Figure 4-14 is not adequate for this type of application, since at high frequencies capacitor C_I has a very low impedance and the gain, given by Equation 4-19, becomes very large, causing distortion:

$$A_V = \frac{R_F}{X_C} = 2\pi f R_F C_I \tag{4-19}$$

To overcome this problem, we place a resistor R_I in series with the input capacitor C_I, as shown in Figure 4-15. This will limit the maximum gain to R_F/R_I at high frequencies and eliminate the distortion.

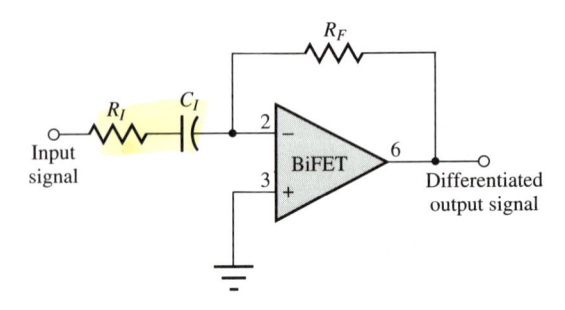

Figure 4-15 Practical differentiating amplifier for periodic waveforms.

We will require a general working equation that can be used for designing R_F (or C_I) in this circuit. Use Equation 4-17 and replace the time t with $\frac{1}{4}f$ to obtain

$$V_{out} = 4fC_I R_F V_{in} \tag{4-20}$$

where V_{in} is the amplitude of the input waveform, V_{out} is the amplitude of the output waveform, and f is the frequency. Like the similar equation that we used for the integrating amplifier, this is simply an empirical formula that gives good design results.

Normally, when using Equation 4-20, we choose a value of C_I and then calculate R_F. We can solve Equation 4-20 for R_F to get a useful design equation:

$$R_F = \frac{V_{out}}{V_{in}} \times \frac{1}{4fC_I} \tag{4-21}$$

If we stipulate that the amplitude of the differentiated waveform be the same as the input amplitude, this equation simplifies to

$$R_F = \frac{1}{2fC_I}$$

Finally, we must choose the high frequency gain to determine R_I. Typically, a gain of 25 gives good results, so we choose $R_I = R_F/25$.

EXAMPLE 4-7

Design a differentiating amplifier using a TL081C op-amp to differentiate a 7000 Hz triangle wave. The peak-to-peak output voltage should be approximately the same as the peak-to-peak input voltage. Use a 0.0022 µF input capacitor. Calculate all resistance values, draw the circuit, and prepare a parts list.

Solution

We will use the differentiating circuit shown in Figure 4-15 to differentiate the triangle wave. This circuit has been redrawn as Figure 4-16, and a parts list is included.

Parts List

Resistors:
R_F 16 kΩ
R_I 680 Ω

Capacitors:
C_I 0.0022 µF

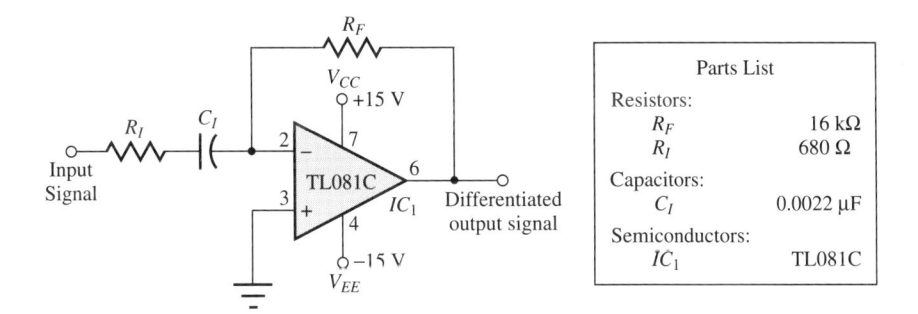

Semiconductors:
IC_1 TL081C

Figure 4-16

Since we are given C_I and the frequency, we can calculate R_F from Equation 4-21:

$$R_F = \frac{V_{out}}{V_{in}} \times \frac{1}{4fC_I}$$

$$= \frac{1}{4fC_I} \quad \text{since} \quad V_{out} = V_{in}$$

$$= \frac{1}{4 \times 7 \text{ kHz} \times 0.0022 \text{ µF}} = 16.2 \text{ kΩ}$$

The closest standard resistor value from Appendix A-1 is 16 kΩ.

Since we are differentiating a periodic waveform, we require gain-limiting resistor R_I to be placed in series with C_I. Calculate $R_I = R_F/25 = 16.2$ kΩ/25 = 650 Ω. The closest standard value from Appendix A-1 is 680 Ω.

The design of our differentiating circuit is now complete.

4 Applications of Operational Amplifiers

4-5 Logarithmic Amplifiers

Our next application is the **logarithmic amplifier**. The amplification in a standard amplifier is linear. For example, if the gain is 10, an input of 0.01 V produces an output of 0.1 V, an input of 0.1 V produces an output of 1 V, and an input of 1 V produces an output of 10 V. A logarithmic amplifier, however, has an amplification that is logarithmic. That is, the output voltage is the logarithm of the input voltage. For example, if an input of 0.01 V produces an output of 1 V, an input of 0.1 V would produce an output of 2 V, and an input of 1 V would produce an output of 3 V.

Logarithmic amplifiers have several useful applications. First, they provide a way of compressing electronic signals. For example, a range of 1 to 100 V can be compressed to a range of 0 to 2 V when processed by a logarithmic amplifier. Such signal compression is often required for scientific and industrial signals which may cover many magnitudes of intensity. Function generators use logarithmic amplifiers in their sweep circuits. Audio noise reduction circuits sometimes use logarithmic compression.

Mathematically, logarithms are used to simplify calculations. For example, the logarithm of the product of two numbers is equal to the sum of the logarithms of the two numbers, and the logarithm of a number raised to an exponent is equal to the exponent times the logarithm of the number. This characteristic allows us to design electronic circuits that can perform such functions as multiplying two input signals or producing the square root of an input. Square root extraction is important in many industrial flow measurement systems.

Simple Logarithmic Amplifier

A logarithmic amplifier can be constructed by placing a bipolar transistor in the feedback loop of an op-amp as shown in Figure 4-17. For a bipolar transistor, the relationship between the collector current I_C and the base-emitter voltage V_{BE} is given by the Shockley equation

$$I_C = I_{EO}(\epsilon^{qV_{BE}/kT} - 1) \qquad (4\text{-}22)$$

In this equation, V_{BE} is the base-emitter voltage of the transistor, k is Boltzmann's constant (1.38×10^{-23} J/K), T is the absolute temperature in kelvins, q is the charge on an electron (1.602×10^{-19} C), I_C is the collector current, and I_{EO} is the saturation current of the base-emitter diode. In practice, the exponential term is much larger than 1, so Equation 4-22 can be approximated as

$$I_C = I_{EO}\epsilon^{qV_{BE}/kT} \qquad (4\text{-}23)$$

Take logarithms of both sides and solve for V_{BE}:

$$V_{BE} = \frac{kT}{q} \log_\epsilon I_C/I_{EO} \qquad (4\text{-}24)$$

This equation shows that the variation of V_{BE} is precisely logarithmic with the

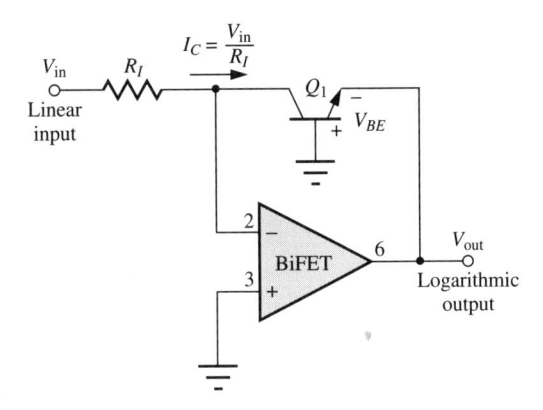

Figure 4-17 A simple logarithmic amplifier.

collector current I_C. We can convert the natural logarithm (base ϵ) to the common logarithm (base 10) by multiplying by the factor 0.4343:

$$V_{BE} = 0.4343 \frac{kT}{q} \log I_C/I_{EO} \tag{4-25}$$

In Figure 4-17, we see that the input current $I = V_{in}/R_I$. Since no current flows into the inverting input of the op-amp, this is also the collector current of the transistor. The output voltage of the amplifier $V_{out} = -V_{BE}$ since the base of the transistor is grounded. Substituting these values, we obtain the gain equation, or at least the equation relating V_{out} to V_{in}, for the logarithmic amplifier:

$$-V_{out} = 0.4343 \frac{kT}{q} \log \frac{V_{in}}{R_I I_{EO}} \tag{4-26}$$

Practical Logarithmic Amplifier

There is one major problem with the circuit in Figure 4-17—the equations contain the reverse saturation current I_{EO}, which is strongly variable with temperature. To eliminate this factor, we must use the circuit in Figure 4-18. This circuit uses a matched pair of transistors Q_1 and Q_2 to cancel the unwanted I_{EO}. Transistor Q_1 acts as the logarithmic feedback element, with the feedback applied to the emitter of Q_1 through voltage divider resistors R_1 and R_2 and the emitter-base junction of Q_2. The collector current through Q_1 is exactly equal to the input current through R_I, which is proportional to the input voltage. Transistor Q_2 acts as a reference. The collector current in Q_2 is equal to the current through the input resistor R_3 to the second op-amp, I_{C_2}. If V_{CC} is constant, then the collector current in Q_2 is constant and equal to V_{CC}/R_3. This in turn holds V_{BE} of Q_2 constant. Then V_{BE} of Q_1 will vary with the input voltage according to Equation

4 Applications of Operational Amplifiers

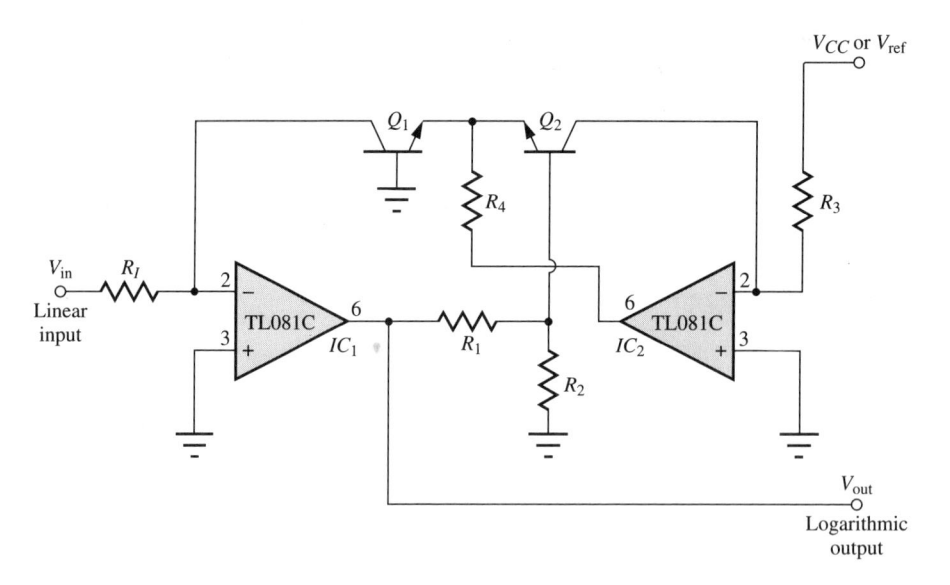

Figure 4-18 A practical logarithmic amplifier using BiFET op-amps.

4-25. The output voltage is proportional to the difference between the base-emitter voltages of the two transistors:

$$V_{out} = \frac{R_1 + R_2}{R_2}(V_{BE_2} - V_{BE_1}) \qquad (4\text{-}27)$$

If we use Equation 4-25, we find that

$$V_{BE_2} - V_{BE_1} = 0.4343\frac{kT}{q}\log\frac{I_{C_1}}{I_{C_2}} \qquad (4\text{-}28)$$

Notice that the I_{EO} term has cancelled. These two equations can be combined to give a final expression for the output voltage:

$$V_{out} = -0.4343\frac{kT}{q}\left(\frac{R_1 + R_2}{R_2}\right)\log\frac{V_{in}R_3}{V_{CC}R_I} \qquad (4\text{-}29)$$

The output voltage is proportional to the logarithm of the input voltage as it should be.

The factor multiplying the logarithm term must be equal to 1 if the output is to be the straight logarithm of the input. We will discuss later what happens when it is not equal to 1. We now have an equation relating R_1 and R_2:

$$0.4343\frac{kT}{q}\left(\frac{R_1 + R_2}{R_2}\right) = 1 \qquad (4\text{-}30)$$

Notice that this equation contains the absolute temperature T. If the amplifier is to operate over a range of temperatures, it is necessary to provide temperature compensation. This is done by using a special temperature-compensating resistor for R_2 in which the resistance also varies with the absolute temperature. The resistance of R_2 will generally be known, and hence we will need to calculate R_1 from Equation 4-30. Solving for R_1 gives

$$R_1 = \frac{q}{kT} \times \frac{R_2}{0.4343} - R_2 \qquad (4\text{-}31)$$

The zero point of the logarithmic conversion is determined by the contents of the log term in Equation 4-29. This term must be set equal to 1 for the input voltage producing zero output (recall that log 1 = 0).

$$\frac{V_{in}R_3}{V_{CC}R_I} = 1 \qquad (4\text{-}32)$$

Physically, the zero is determined by the current through R_3, so we must solve this equation for R_3.

$$R_3 = \frac{V_{CC}}{V_{in}} \times R_I \qquad (4\text{-}33)$$

The input impedance R_I of the amplifier will normally be stipulated.

Using the above equations, it is possible to design a logarithmic amplifier. The design procedure is best illustrated by the following example.

EXAMPLE 4-8

Design a logarithmic amplifier to have an output of 1 V for an input signal of 0.1 V, an output of 2 V for an input of 1.0 V, and an output of 3 V for an input of 10 V. Use two TL081C operational amplifiers and two 2N4401 transistors. Make the input impedance 1000 Ω. Use a 1.0 kΩ temperature-compensating resistor for R_2. Use the +15 V supply as the reference voltage. Assume the design temperature is 20°C.

Solution

The circuit that we will use for the logarithmic amplifier will be the same as in Figure 4-18. This circuit has been redrawn as Figure 4-19, with a parts list included.

From our design requirements, if an input of 0.1 V gives an output of 1 V, an input of 1.0 V gives an output of 2 V, and an input of 10 V gives an output of 3 V, then an input of 0.01 V should give an output of 0 V. The logarithmic term in Equation 4-29 must be 0 for an input $V_{in} = 0.01$ V. In Equation 4-33, we equated the contents of the log term to 1 and solved the resulting expression for R_3. Since we know $V_{CC} = 15$ V and $R_I = 1000$ Ω, we can now calculate R_3.

$$R_3 = \frac{V_{CC}}{V_{in}} \times R_I = \frac{15 \text{ V}}{0.01 \text{ V}} \times 1.0 \text{ k}\Omega = 1.5 \text{ M}\Omega$$

This is a standard value.

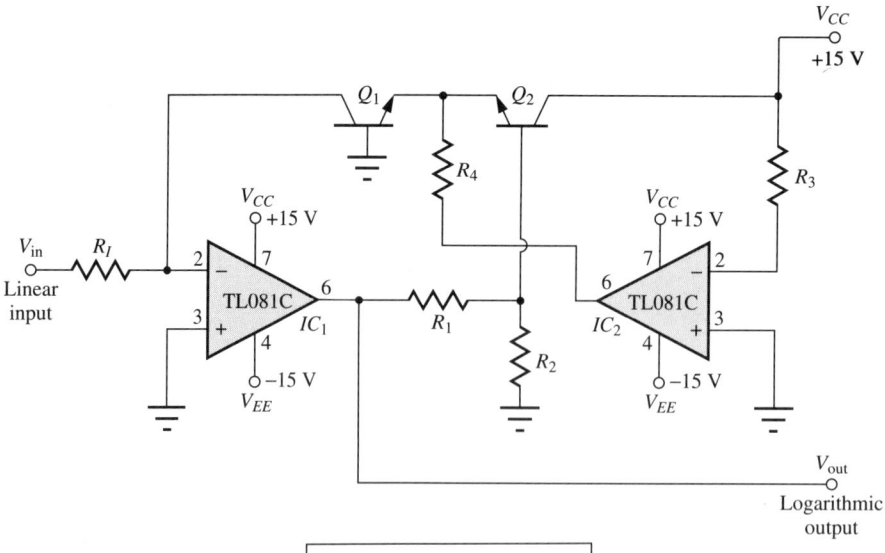

4 Applications of Operational Amplifiers

Figure 4-19

Parts List

Resistors:
- R_I 1.0 kΩ
- R_1 91 kΩ
- R_2 1.0 kΩ
- R_3 1.5 MΩ
- R_4 2.0 kΩ

Semiconductors:
- IC_1, IC_2 TL081C
- Q_1, Q_2 2N4401

We are supplied a temperature-compensating resistor R_2 with a value of 1.0 kΩ. This value is used in Equation 4-31 to solve for R_1.

$$R_1 = \frac{q}{kT} \times \frac{R_2}{0.4343} - R_2$$

$$= \frac{1.602 \times 10^{-19}}{1.381 \times 10^{-23} \times 293} \times \frac{1.0 \text{ k}\Omega}{0.4343} - 1.0 \text{ k}\Omega = 90.2 \text{ k}\Omega$$

Choose a standard 91 kΩ resistor for R_1.

The only resistor that we have not yet determined is R_4. This is simply a protective resistor to limit the base current in Q_2. We will arbitrarily assign a value of 2 kΩ.

The design of our logarithmic amplifier is complete. It is a wise practice to place reverse-biased diodes between the collector and base of each transistor to protect the transistors from large reverse voltages. This is a difficult circuit to get working, so to any ambitious reader who constructs this circuit, *be forewarned*.

4-6 Exponential Amplifiers

For many applications, the logarithmic amplifier discussed in the previous section must be combined with an **exponential** or **antilog amplifier** to regain a linear output. As an example, if we compress a signal for noise reduction, eventually we will want to expand it back to obtain the original signal.

The action of the exponential amplifier is simply the inverse of the logarithmic amplifier. For example, if a logarithmic amplifier produced an output of 1 V for an input of 0.01 V, an output of 2 V for an input of 0.1 V, and an output of 3 V for an input of 1 V, the corresponding exponential amplifier would produce an output of 0.01 V for an input of 1 V, an output of 0.1 V for an input of 2 V, and an output of 1 V for an input of 3 V.

Simple Exponential Amplifier

An exponential amplifier is very similar in construction to a logarithmic amplifier. The only difference is that the bipolar transistor replaces the input resistor instead of the feedback resistor. A simple exponential amplifier, corresponding to the simple logarithmic amplifier shown in Figure 4-17, is shown in Figure 4-20. Notice that the input voltage V_{in} must always be negative for proper transistor biasing. As for the logarithmic amplifier, the collector current I_C and the base-emitter voltage V_{BE} are related by Equation 4-23:

$$I_C = I_{EO}\epsilon^{qV_{BE}/kT} \qquad (4\text{-}23)$$

We see from Figure 4-20 that $V_{in} = V_{BE}$. The collector current I_C is simply V_{out}/R_F. Substituting these into Equation 4-23 and rearranging gives

$$V_{out} = R_F I_{EO} \epsilon^{qV_{in}/kT} \qquad (4\text{-}34)$$

This is the gain equation, or at least the equation relating V_{out} to V_{in}, for the exponential amplifier. Notice that this equation contains the factor I_{EO}, the reverse saturation current. As in the logarithmic amplifier, this is variable with temperature and limits the usefulness of the exponential amplifier shown in Figure 4-20.

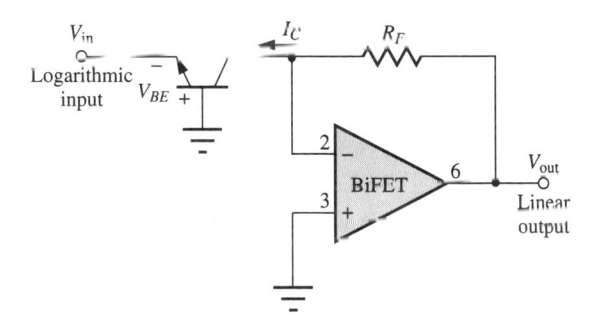

Figure 4-20 A simple exponential amplifier.

Practical Exponential Amplifier

We can eliminate the variable I_{EO} factor by using a matched transistor pair in the circuit in Figure 4-21. This circuit bears a very close resemblance to the logarithmic amplifier shown in Figure 4-18. The main difference is that transistor Q_2 is now the transistor generating the exponential response, whereas transistor Q_1 is the reference transistor biased by V_{CC} through R_3. The logarithmic input is to resistors R_1 and R_2, connected to the base of Q_1. This was the output circuit for the logarithmic amplifier. The design of the exponential amplifier is almost identical to that of the logarithmic amplifier and will not be developed here.

A few words should be said about the application of the exponential amplifier in conjunction with the logarithmic amplifier. We will not develop any circuits for these applications, as they tend to become relatively complex and are difficult to get working. Any reader interested in these applications is encouraged to develop the necessary circuits on his own. Various manufacturers' applications handbooks also provide working circuits.

The first application that we will mention is a multiplying circuit. Recall that the logarithm of a product of two numbers is equal to the sum of the logarithms of the individual numbers. To perform the electronic equivalent of this by multiplying two input voltages, two logarithmic amplifiers and one exponential amplifier are required. One input voltage is applied to the input of each logarithmic amplifier. The outputs of the logarithmic amplifiers are summed, then put through the exponential amplifier to give the linear product as an output. The same pro-

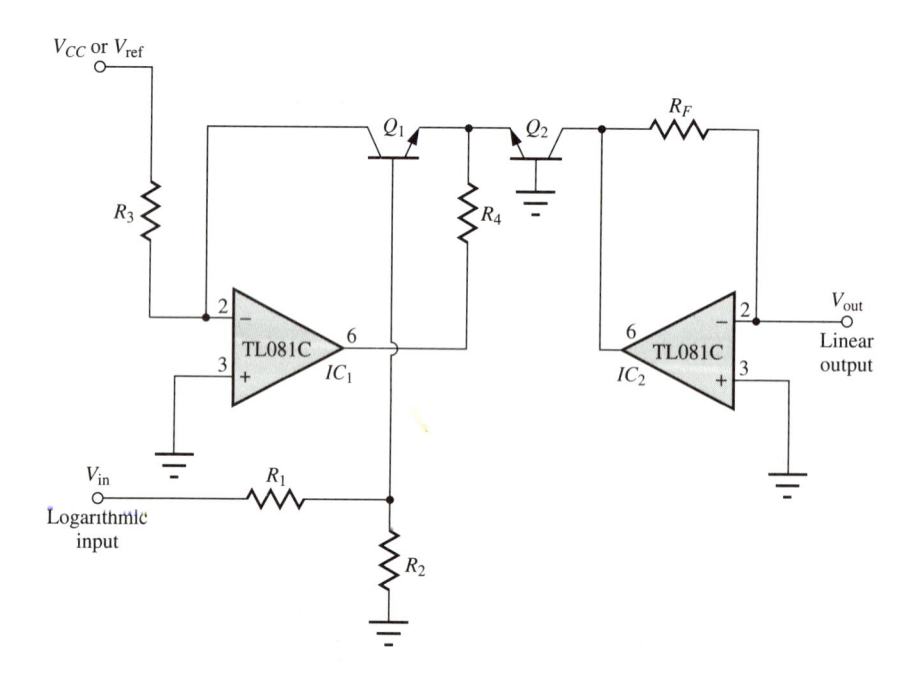

Figure 4-21 A practical exponential amplifier using BiFET op-amps.

cedure is used for division, except that the outputs of the logarithmic amplifiers are subtracted.

The other application that we will mention is a squaring circuit. The logarithm of a number squared is twice the logarithm of the number. For this, we need a single logarithmic amplifier and an exponential amplifier. The factor of 2 necessary for squaring can be built into the factor multiplying the log term in Equation 4-29 of the logarithmic amplifier design. In other words, equate Equation 4-30 to 2 instead of 1:

$$0.4343 \frac{kT}{q} \left(\frac{R_1 + R_2}{R_2} \right) = 2 \qquad (4\text{-}35)$$

Solving for R_1 gives

$$R_1 = \frac{2q}{kT} \times \frac{R_2}{0.4343} - R_2 \qquad (4\text{-}36)$$

Note that R_1 is approximately twice as large in the squaring logarithmic amplifier. Passing the output of this amplifier through a standard exponential amplifier will produce the square of the input signal. Obviously, this procedure can be expanded to cubing an input signal, finding the square root of an input signal, or any similar function.

4-7 Precision Rectifiers

The last of the special-purpose amplifier circuits using op-amps that we will introduce is the precision rectifier. For a standard discrete diode, there is always a forward voltage drop of 0.7 V (0.35 V for a germanium diode) when the diode is conducting. This means that the output voltage is always $V_{in} - 0.7$ V. Often, this small voltage drop is of little consequence. For example, the 0.7 or 1.4 V drop caused by the diodes used as rectifiers in power supplies is usually of no consequence and is compensated for in the design (see Chapter 7). In some applications, however, this voltage drop can be quite a problem. For example, a simple diode rectifier used in a voltmeter for converting AC to DC makes the bottom end of the AC scale quite nonlinear and limits the minimum AC voltage that can be read to approximately 0.7 V. Here a rectifier that has a forward voltage drop of zero is quite desirable. A rectifier with a zero-forward-voltage drop is called a **precision rectifier**.

Simple Precision Rectifier Circuit

A simple precision rectifier circuit is shown in Figure 4-22. This is a simple follower amplifier with a diode D_1 placed in the feedback loop and in series with the output. This circuit produces a half-wave rectified output with 0 V loss when forward conducting.

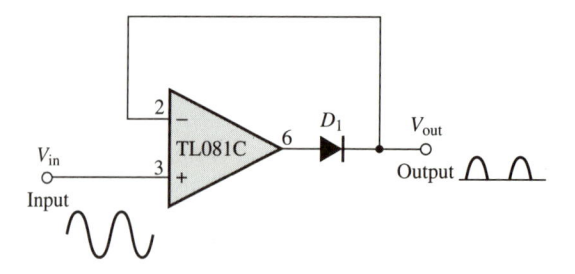

Figure 4-22 Simple precision rectifier circuit.

If a positive voltage greater than 0.7 V is applied to the input, then the output is positive and equal to the input. The condition that both op-amp inputs have the same voltage applied is satisfied. If, however, the input voltage is less than 0.7 V, the diode is initially nonconducting and the voltage at the inverting input will be zero. The two inputs will be different. The op-amp will increase its output voltage until the diode conducts and the inverting input sees a voltage equal to the noninverting input. The output of the circuit will accurately follow the input for all positive voltage levels.

When the input voltage is negative, the output of the op-amp will try to drive the diode negative to set the inverting input to the same voltage as the noninverting input. The diode, however, is reverse-biased. The op-amp output voltage goes to the negative saturation level V_{sat-} in trying to make the diode conduct, but still the diode remains nonconducting and the output stays at zero.

This circuit will provide half-wave rectification if the op-amp has sufficient gain to force the diode to conduct. The minimum V_{in} for which this occurs is

$$V_{in} = \frac{0.7}{\text{open-loop gain of the amplifier}} \qquad (4\text{-}37)$$

At low frequencies, the op-amp open-loop gain is approximately 200,000, which means that the system would respond to any input signal greater than typically 3.5 μV. At higher frequencies, the gain is less, but even at 300 kHz, where the gain is 10, this circuit should work for input voltages greater than 0.07 V. In fact, this circuit only works to a maximum frequency of a few hundred hertz. Why? Because when the input signal is negative, the op-amp is driven to the negative saturation voltage V_{sat-}. When the input signal goes positive again, the op-amp output must change from this large negative voltage to a positive voltage. This takes some time, and is determined by the op-amp slew rate. We will discuss the problem of slew rate in op-amps in the next chapter.

High Frequency Precision Rectifier Circuit

To improve the frequency response of the precision rectifier in Figure 4-22, it is necessary to prevent the op-amp output from going to the negative saturation

4-7 Precision Rectifiers 167

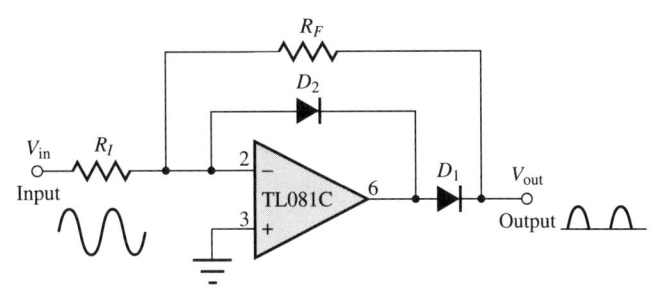

(a) Inverting precision rectifier using a TL081C op-amp

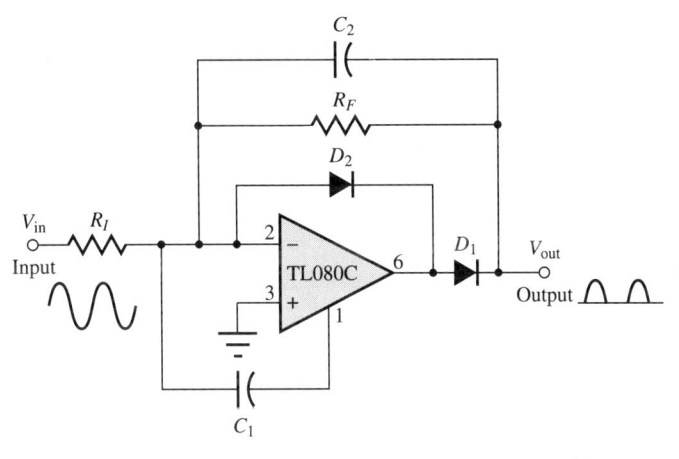

(b) Inverting precision rectifier using a TL080C op-amp with feed-forward compensation

Figure 4-23 High frequency precision rectifier circuits.

voltage V_{sat-} during the negative cycles. A circuit that achieves this is shown in Figure 4-23(a). Notice that we are using an inverting amplifier and are rectifying the negative half of the input signal.

The input resistor R_I is equal to the feedback resistor R_F, to give a gain of 1 when D_1 is forward biased. The output V_{out} will always be equal to $-V_{in}$ to ensure that the inverting input is kept at 0 V. When the input signal V_{in} goes positive, D_1 is reverse-biased and there is no output. Diode D_2, however, provides a feedback path whenever the op-amp output is more negative than -0.7 V, holding the negative output swing to -0.7 V and preventing the output from going to V_{sat-}.

By using feedforward compensation with an uncompensated op-amp such as a TL080, an even wider bandwidth can be achieved. The circuit for a fast rectifier with a bandwidth of up to 1 MHz is shown in Figure 4-23(b). This rectifier circuit operates in the same manner as the rectifier in Figure 4-23(a).

168   4 Applications of Operational Amplifiers

EXAMPLE 4-9

Design a high frequency precision rectifier circuit using a TL080 op-amp. The input impedance should be 100 kΩ. Calculate resistance values, draw the circuit, and prepare a parts list.

Solution

We will use the precision rectifier circuit shown in Figure 4-23(b). This has been redrawn as Figure 4-24, and a parts list is included.

Parts List

Resistors:
R_I, R_F 100 kΩ

Capacitors:
C_1 150 pF
C_2 3 pF

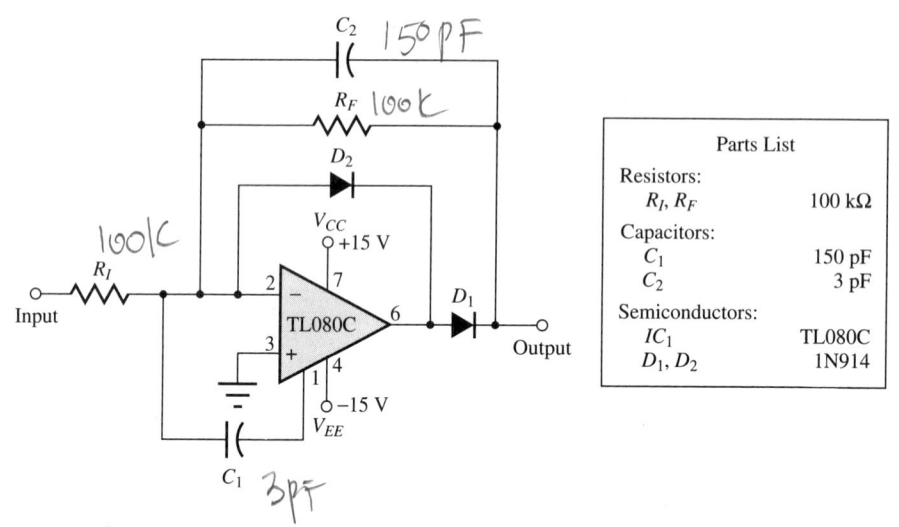

Semiconductors:
IC_1 TL080C
D_1, D_2 1N914

Figure 4-24

Note that we are using the high frequency version of the circuit to achieve the maximum bandwidth and to demonstrate the use of the TL080C uncompensated op-amp with external feedforward compensation.

There are basically no calculations required. The circuit is essentially a unity-gain inverting amplifier circuit. Since we want an input impedance of 100 kΩ, we must choose R_I to be 100 kΩ. To establish unity gain, $R_F = R_I = 100$ kΩ. We will use the standard capacitors suggested for feedforward compensation, as discussed in Chapter 2. Thus, choose $C_1 = 150$ pF and $C_2 = 3$ pF.

Summary

This chapter introduced a variety of different, but useful, amplifier applications using op-amps. We saw how to design an op-amp summing amplifier to add two or more inputs, and how to design an op-amp subtracting circuit to subtract one input from the other, both useful in signal processing systems. We designed an integrating amplifier by replacing the feedback resistor with a capacitor, and a differentiating amplifier by replacing the input resistor with a capacitor. We will see how integrating and differentiating circuits find practical use in waveform shaping in Chapters 10 and 11.

By placing a bipolar transistor in the feedback loop of an op-amp, we were able to create a logarithmic amplifier, and by moving the transistor to the input we created an exponential amplifier. This pair of amplifiers is useful in audio compression and expansion used in noise reduction systems, and for mathematical operations such as square root extraction.

Finally, by placing a diode in the feedback loop of an op-amp, we produced a precision rectifier circuit, a rectifier that starts to conduct at 0 V instead of 0.7 V.

Review Questions

4-1 Summing amplifiers

1. What is the relationship between the input resistor and the feedback resistor for an inverting summing circuit?
2. Describe how to multiply the various inputs of an inverting summing amplifier by a constant.
3. What are the disadvantages of a noninverting summing amplifier?
4. How is the gain defined for a noninverting summing amplifier if all the inputs have equal resistors?
5. Give one application of a summing amplifier.

4-2 Subtracting or differencing amplifiers

6. What is the difference between a differential amplifier and a differencing amplifier?
7. What circuit is used for a differencing circuit?
8. What three options are possible with differencing amplifiers?
9. What is the relationship among the four resistors in each of the three types of differencing amplifiers?
10. Under what conditions are the impedances the same for both inputs of a differencing amplifier?
11. Give one application for a differencing amplifier.

4-3 Integrating amplifiers

12. What is integration? Write the calculus symbol for integration.
13. What is the physical difference between an integrating amplifier and a linear inverting amplifier?
14. What addition must be added to an integrating amplifier if it is to be used for integrating periodic signals?
15. Give one application of an integrating amplifier.

4-4 Differentiating amplifiers

16. What is differentiation? Write the calculus symbol for differentiation.
17. What is the relation between integration and differentiation?
18. What is the physical difference between a differentiating amplifier and a linear inverting amplifier?
19. What modification must be made to a simple differentiating circuit to allow it to be used for periodic waveforms? Why is this modification necessary?
20. Give one application of a differentiating amplifier.

4-5 Logarithmic amplifiers

21. How does the output of a logarithmic amplifier differ from the output of a linear amplifier for the same input?
22. What applications do logarithmic amplifiers have?
23. What characteristic of a transistor makes it suitable for producing a logarithmic amplifier?
24. Where must the transistor be placed in the circuit?
25. Why must two transistors be used?
26. How does temperature affect the operation of a logarithmic amplifier, and how is it compensated?

4-6 Exponential amplifiers

27. What is the difference in operation between a logarithmic amplifier and an exponential amplifier?
28. What is the difference in the circuit between a logarithmic amplifier and an exponential amplifier?
29. Why is an exponential amplifier usually used with a logarithmic amplifier?
30. Give two applications of an exponential amplifier.

4-7 Precision rectifiers

31. What forward and reverse characteristics should an ideal diode possess?
32. What is the problem with using ordinary diodes in rectifier circuits for low-range AC voltmeters?
33. How does an op-amp in conjunction with an ordinary diode provide an improved rectifier circuit?
34. Describe the operation of a simple precision rectifier using a follower amplifier.
35. What is the limitation of the circuit in the previous question? What causes this limitation?
36. How can the performance of a simple precision amplifier be improved? What is the difference in the circuit?

Design and Analysis Problems

4-1 Summing amplifiers

1. Design a four-input inverting summing amplifier using a TL081C op-amp. Use a feedback resistance of 20 kΩ. Calculate all resistance values, draw the circuit, and prepare a parts list.

$R_A = R_B = R_C = R_D = 20 \text{k}\Omega$

Figure 4-25

Handwritten annotations on figure:
- $V_A = 1.2\text{ V}$, 30 kΩ
- $V_B = 2.1\text{ V}$, 20 kΩ
- $V_C = 0.7\text{ V}$, 10 kΩ
- Feedback: 20 kΩ
- Op-amp: TL081C

$$V_o = -\left[\left(\frac{20k}{30k} \times 1.2\right) + \left(\frac{20k}{20k} \times 2.1\text{ V}\right) + \left(\frac{20k}{10k} \times 0.7\text{ V}\right)\right]$$

$A_1 = \frac{20}{30} = 0.67$
$A_2 = 1$
$A_3 = 2$

$= -(0.8 + 2.1 + 1.4)$
$= -4.3\text{ V}$

Figure 4-25

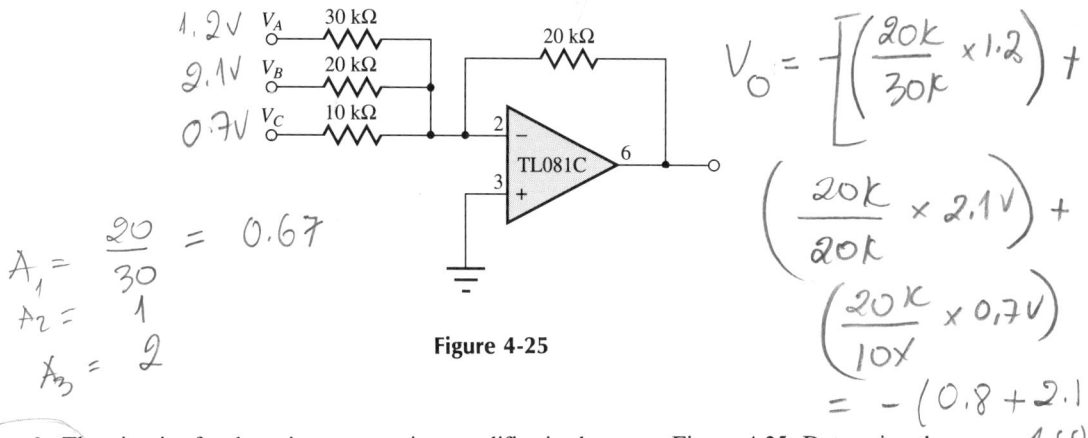

2. The circuit of a three-input summing amplifier is shown as Figure 4-25. Determine the gain seen by each of the inputs. Calculate the output voltage if 1.2 V is applied to input A, 2.1 V to input B, and 0.7 V to input C.

3. Design a two-input mixing circuit using an inverting summing amplifier with a TL081C op-amp. Input signal A should have a gain of 2.0, and input signal B should have a gain of 3.5. Use a feedback resistance of 15 kΩ. Calculate resistance values, draw the circuit, and prepare a parts list.

Handwritten: $\frac{R_{fA}}{R_A} = 2 \Rightarrow R_A = \frac{15}{2} = 7.5\text{ k}\Omega$

$R_B = \frac{15}{3.5} = 4.28\text{ k}$

4-2 Subtracting amplifiers

4. Design a subtracting amplifier to take the difference between the two inputs and multiply the result by 8. Use a TL081C op-amp operated from a ±15 V supply. Make R_F equal to 200 kΩ.

5. Design a differencing amplifier to subtract six times the voltage at input A from two times the voltage at input B. Use a TL081C op-amp operated from a ±15 V supply. Make R_F and R_{FN} equal to 30 kΩ.

Handwritten: $R_A = \frac{200}{8} = 25\text{ k}\Omega$

4-3 Integrating amplifiers

6. Design an integrating circuit using a TL081C op-amp operating at a frequency of 2000 Hz. Use a 0.022 μF feedback capacitor. Calculate all resistance values, draw the circuit, and prepare a parts list.

4-4 Differentiating amplifiers

7. Design a differentiating circuit using a TL081C op-amp operating at a frequency of 3000 Hz. Use a 0.0047 μF input capacitor. Calculate all resistance values, draw the circuit, and prepare a parts list.

4-5 Logarithmic amplifiers

8. Design a logarithmic amplifier to have an output of 1 V for an input signal of 0.1 V, an output of 2 V for an input of 1.0 V, and an output of 3 V for an input of 10 V. Use two TL081C operational amplifiers and two 2N4401 transistors. Make the input impedance 2000 Ω. Use a +10 V reference voltage. Assume the temperature is 25°C.

5 Comparators

OUTLINE

5-1 Op-Amps as Comparators
5-2 Zero-Crossing Comparator Circuits
5-3 Level-Sensing Comparator Circuits
5-4 Comparators with Hysteresis
5-5 Window Comparator Circuits
5-6 True Comparator Integrated Circuits

KEY TERMS

Comparator
Slew rate
Zero-crossing comparator
Level-sensing comparator
Hysteresis
Schmitt trigger
Upper trigger voltage
Lower trigger voltage
Window
Window comparator
Response time

OBJECTIVES

On completion of this chapter, the reader should be able to:

- understand the characteristics of comparators and use op-amps for a variety of comparator applications.
- design and use zero-crossing comparator circuits.
- design and use positive and negative voltage level-sensing comparator circuits.
- understand the concept of hysteresis and use it to produce Schmitt trigger circuits.
- design window-detecting circuits from level comparators.
- compare the operation of true comparators with op-amps configured as comparators, and design circuits using true comparators.

Introduction

A comparator is a circuit that compares an input voltage to a reference voltage and switches between a positive output and a negative (or zero) output, depending on the relative input level. Basically, a comparator is a voltage controlled switch, switching state in response to the input voltage level.

There are many different comparator circuits, ranging from simple zero-crossing comparators that switch as the input voltage increases past zero, to level-sensing comparators that switch as the input voltage increases past a preset level, to window comparators that switch when the input voltage is within a certain voltage range. A typical comparator application, developed in an example later, is a circuit to turn on a warning light if a battery voltage falls below a certain level.

Comparators can also be designed to switch at one level when the voltage is increasing and at a different level when the voltage is decreasing. This type of switching action is called hysteresis, and is important if there is noise on the input signal. The Schmitt trigger encountered in digital systems is an example of a comparator with hysteresis.

5-1 Op-Amps as Comparators

A **comparator** is a circuit (or a device as we will see later) that compares the voltage level of an input signal to a reference voltage and gives a high or low, positive or negative output response depending on the input comparison. Such circuits are widely used in electronics as alarm or warning sensors, as on-off control elements, and as systems for converting analog signals to digital.

A single op-amp can be used to make a simple comparator circuit, and the circuit is shown in Figure 5-1. The circuit consists of nothing but an op-amp! Although the circuit is simple, a full understanding of its operation is important to appreciate the design of more complicated comparator circuits and to demonstrate the limitations of the op-amp as a comparator.

Without feedback, the op-amp can be considered as a simple differential amplifier with a gain equal to the op-amp open-loop gain. Apply a reference voltage to the noninverting input and the signal voltage to the inverting input, as shown in Figure 5-1. Suppose the reference voltage is 5 V, the supply is ±15 V, and the op-amp open-loop gain is 200,000. If the input signal voltage applied to the inverting input is 5.1 V, the differential voltage seen by the op-amp is 0.1 V, positive with respect to the inverting input. The op-amp will amplify this by 200,000 and

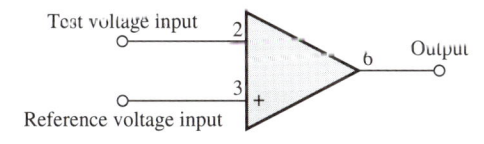

Figure 5-1 Op-amp used as a basic comparator.

try to give an output of $-20,000$ V. Obviously, the op-amp cannot give an output of $-20,000$ V since the negative supply voltage is only -15 V. The actual output voltage will be approximately -14 V, the negative saturation voltage of the op-amp V_{sat-}. If our input signal voltage was only 5.001 V, the output would still be V_{sat-} or -14 V. In other words, for virtually any voltage greater than the reference voltage, we get the saturated output level V_{sat-}.

Now lower the input signal voltage to 4.9 V. The differential voltage now seen by the op-amp is 0.1 V, *negative* with respect to the inverting input. As described above, the op-amp will try to amplify this negative voltage by 200,000 and give an output voltage of $+20,000$ V. The best it can do is approximately $+14$ V, the positive saturation voltage of the op-amp V_{sat+}.

We see that the comparator circuit will give one of two outputs. If the inverting input is positive with respect to the noninverting input, the output will be V_{sat-}. If the inverting input is negative with respect to the noninverting input, the output will be V_{sat+}. Only if the two voltages are within 70 μV of each other will the op-amp act as a linear amplifier and give some output voltage other than V_{sat-} or V_{sat+}. In most practical applications, we can ignore this region of linear amplification.

Slew Rate

The op-amp is not a perfect switching device. The output voltage cannot change instantaneously when the device switches. In fact, it changes at a rate known as the **slew rate**. The concept of slew rate as an op-amp property was introduced in Chapter 2 and was defined as the rate at which the output voltage of an op-amp

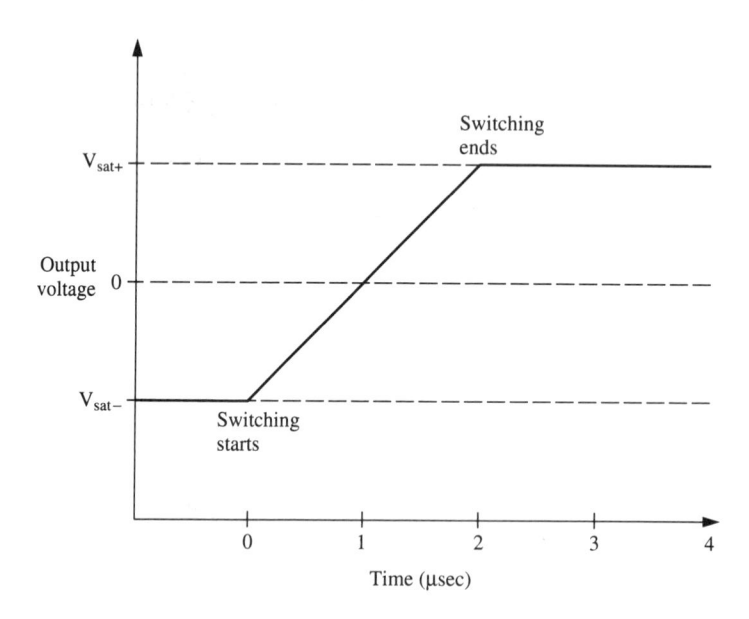

Figure 5-2 Switching curve for a TL081C op-amp showing the effects of slew rate.

changes in response to a step change in the input. The slew rate in an op-amp is illustrated in Figure 5-2. The poor frequency response of the simple precision rectifier introduced in Chapter 4 was caused by slew rate, although we did not describe it as such at that time.

The slew rate of a LM741C is 0.5 V/μs. Because the output voltage of an op-amp used in a comparator circuit must switch from V_{sat-} to V_{sat+}, a typical voltage range of approximately 28 V, the switching time is approximately 60 μs. If a comparator using a LM741C were triggered on and off with a 10 kHz square wave, the output would be a pure triangle wave because of the slow slew rate! The TL081C BiFET op-amp is much better with a slew rate of 13 V/μs, but even so it takes over 2 μs to switch. We will use op-amps as comparators in the next few sections, but remember their limitations.

Load Switching

We have seen that comparator circuits are essentially switching circuits. What do they switch? This is a complicated question, but in general they switch either indicator devices, such as LED displays, or transistors supplying power to external loads.

Two versions of a circuit using a comparator to switch a LED are shown in Figure 5-3. A typical LED draws 20 mA with a 2 V drop and, hence, requires a series resistor R_{LED} to drop the op-amp output voltage to 2 V and limit the current

(a) Positive output voltage

(b) Negative output voltage

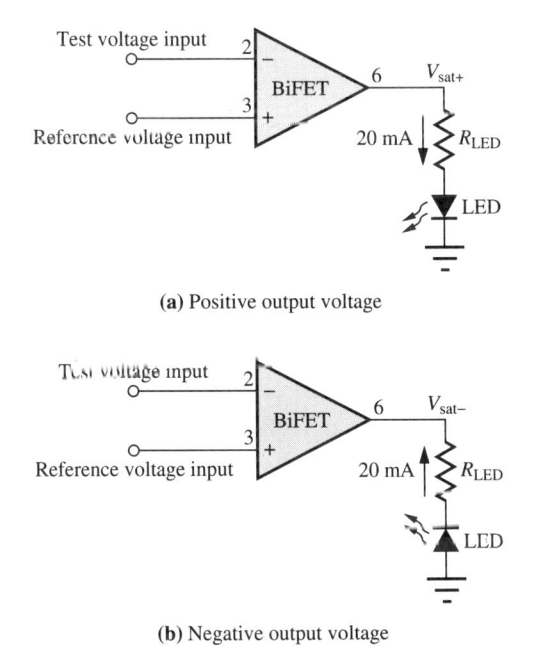

Figure 5-3 LED driven by a comparator circuit.

to 20 mA. If the comparator output is V_{sat+}, as in Figure 5-3(a), the voltage drop across the resistor will be $V_{sat+} - 2$ V, and the series resistor can be calculated from Ohm's law:

$$R_{LED} = \frac{V_{sat+} - 2.0 \text{ V}}{20 \text{ mA}} \tag{5-1}$$

If the comparator output is V_{sat-}, as in Figure 5-3(b), the voltage drop across the resistor will be $V_{sat-} + 2$ V, and the series resistor is given by

$$R_{LED} = \frac{V_{sat-} + 2.0 \text{ V}}{-20 \text{ mA}} \tag{5-2}$$

Note that both V_{sat-} and the current are negative.

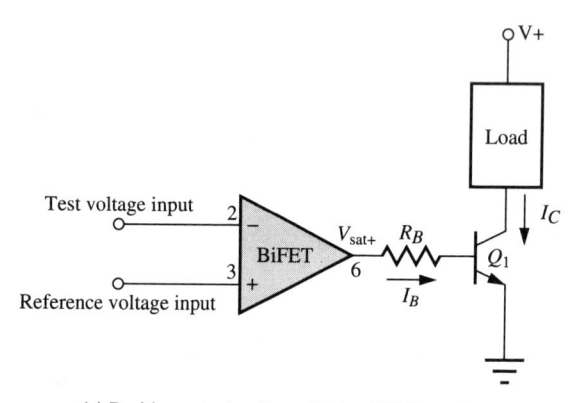

(a) Positive output voltage driving NPN transistor

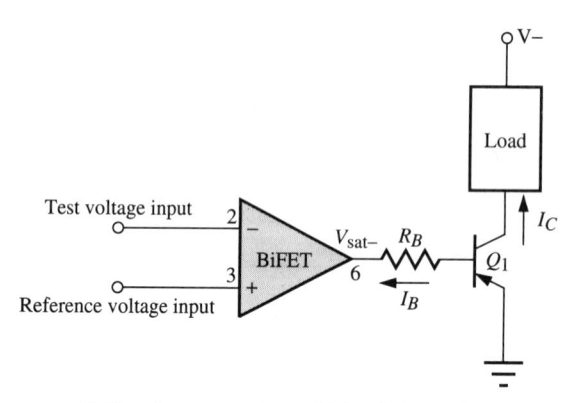

(b) Negative output voltage driving PNP transistor

Figure 5-4 Comparator driving transistor and external load.

If the op-amp is operated from a dual supply, then either circuit shown in Figure 5-3 may be used. For single-supply operation, only the circuit in Figure 5-3(a) is applicable.

If the comparator is to switch a load that requires a substantial amount of current, such as an alarm or a motor, it probably will not be able to drive the load directly. In this case, a separate switching transistor is used to supply current to the load, the transistor being switched by the comparator. Figure 5-4 shows the circuits used when a comparator is used to drive an external switching transistor.

The comparator output is connected directly to the base of the transistor through resistor R_B. This base resistor is necessary to limit the transistor base current. If power is to be switched to the load when the comparator output is positive, as in Figure 5-4(a), an NPN transistor is used. The voltage drop across R_B will be $V_{\text{sat}+} - V_{BE}$; hence, R_B can be calculated from the equation

$$R_B = \frac{V_{\text{sat}+} - V_{BE}}{I_B} \tag{5-3}$$

This equation requires I_B, the transistor base current, which should be set equal to the *maximum* load or collector current divided by the *minimum* transistor β.

If power is to be switched to the load when the comparator output is negative, as in Figure 5-4(b), a PNP transistor is used, the voltage drop across R_B is $V_{\text{sat}-} + V_{BE}$, and R_B is calculated from the equation

$$R_B = \frac{V_{\text{sat}-} + V_{BE}}{I_B} \tag{5-4}$$

Note that both $V_{\text{sat}-}$ and the current are negative.

As for the LED display, if the op-amp is operated from a dual supply, then either circuit may be used, depending on the switching requirements. For single-supply operation, only the circuit in Figure 5-4(a) is applicable.

5-2 Zero-Crossing Comparator Circuits

The simplest of all comparator circuits is the **zero-crossing comparator**. For a zero-crossing comparator, the reference voltage is zero. We have two options: the output can be positive for a positive input (noninverting comparator), or the output can be negative for a positive input (inverting comparator). The output polarity is determined by whether the inverting or noninverting input is grounded. Figure 5-5 shows both possibilities.

Zero-crossing comparators are used for applications such as alarm detectors (indicating the presence of an input signal by switching on an indicator such as an LED) and polarity indicators (turning on one LED for positive polarity and another LED for negative polarity). This latter application is demonstrated in Example 5-1.

5 Comparators

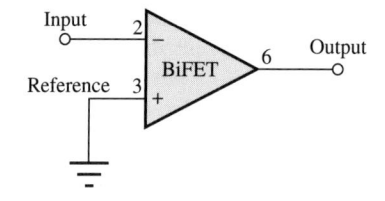

(a) Negative output for positive input

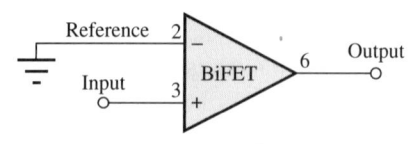

(b) Positive output for positive input

Figure 5-5 Zero-crossing comparators.

EXAMPLE 5-1

Design a polarity detector circuit using a TL081C op-amp operating from a ± 15 V supply. The detector should turn on one LED when the voltage is positive and another LED when the voltage is negative.

Solution

The polarity detector consists of a simple zero-crossing inverting comparator with two LEDs connected to the output as shown in Figure 5-6.

There is nothing to design in the comparator circuit itself, since it consists of a single TL081C op-amp.

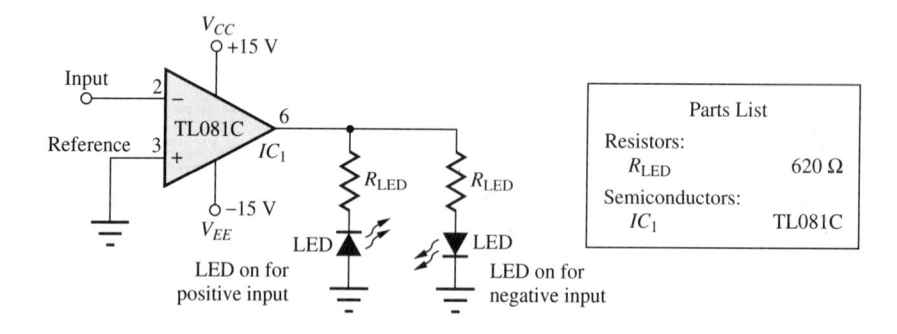

Parts List
Resistors:
R_{LED} 620 Ω
Semiconductors:
IC_1 TL081C

Figure 5-6

Two LEDs are required, one to indicate a positive polarity, the other to indicate a negative polarity. The LEDs are connected directly to the output with opposite polarities. Each LED requires a series voltage-dropping resistor to limit the LED current to 20 mA with a 2 V drop. Since the comparator is operated from a dual ±15 V supply, $V_{sat-} = -14$ V and $V_{sat+} = +14$ V. The resistors can be calculated from either Equation 5-1 or 5-2. Since the supply is balanced, both equations will yield the same result. We will use Equation 5-1:

$$R_{LED} = \frac{V_{sat+} - 2.0 \text{ V}}{20 \text{ mA}} = \frac{14 \text{ V} - 2 \text{ V}}{20 \text{ mA}} = 600 \text{ } \Omega$$

Choose the closest standard value from Appendix A-1, 620 Ω.

The design of our polarity-indicating comparator is complete.

5-3 Level-Sensing Comparator Circuits

A slightly more complicated comparator is the **level-sensing comparator**. This is a considerably more useful form of comparator, with many more applications than the zero-crossing comparator. In the level-sensing comparator, the reference is not grounded but set at some DC level.

There are several ways to establish a reference voltage. Figure 5-7 shows one method. In the circuits shown the reference voltage is determined by a voltage divider, placed either between the positive supply and ground, to produce a positive reference (Figure 5-7(a)), or between the negative supply and ground, to produce a negative reference (Figure 5-7(b)). The circuits produce a negative output if the input signal is greater (more positive) than the reference voltage, so

(a) Positive voltage detector

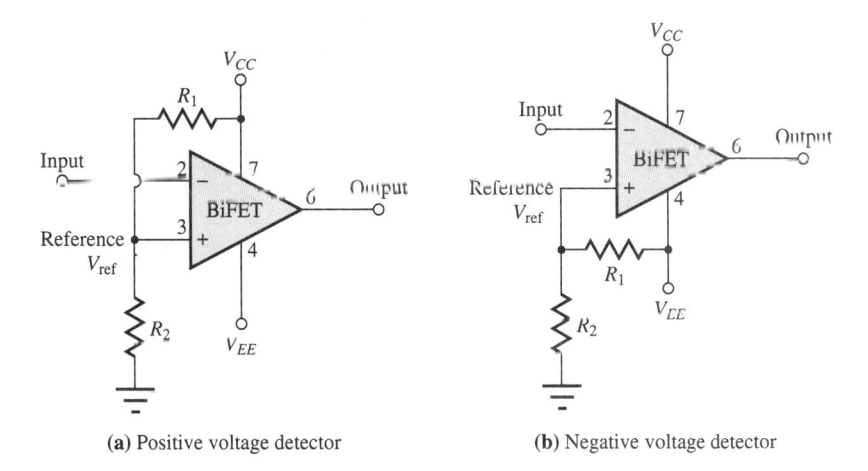

(b) Negative voltage detector

Figure 5-7 Level-sensing comparators with voltage divider reference.

they are inverting comparators. If we want a noninverting comparator, that is, one that produces a positive output when the input signal is greater (more positive) than the reference, then we simply reverse the input connections.

To design the voltage divider, knowing V_{CC} or V_{EE} and the desired reference voltage V_{ref}, use the simple voltage divider formula

$$V_{ref} = \frac{R_2}{R_1 + R_2} \times V_{CC} \tag{5-5}$$

To design values for R_1 and R_2 from this formula, assume a value for R_2, say 10 kΩ, and calculate R_1 from the formula

$$R_1 = R_2 \left(\frac{V_{CC}}{V_{ref}} - 1 \right) \tag{5-6}$$

This equation uses V_{CC} to give a positive reference. For a negative reference, simply use V_{EE} in place of V_{CC}.

One potential problem with the circuits in Figure 5-7 is that if the supply voltage changed the reference voltage would change. This would be a problem if we were using the comparator to signal when a battery voltage dropped below a certain level, if we were powering the comparator circuit from the battery (see Example 5-3). Circuits for positive and negative inverting level-sensing comparators, where the reference voltage is independent of the supply voltage, are shown in Figure 5-8. These circuits use a zener diode or a voltage reference integrated circuit to produce the reference voltage.

Zener diodes and voltage reference integrated circuits will be discussed in more detail in Chapter 7. Appendix A-4 lists typical zener diode characteristics and can be used for selecting zener diodes for applications. Briefly, a zener diode (or voltage reference) will maintain a constant voltage drop V_Z if a bias current flows through it. Zener diodes (and voltage references) have an optimum design

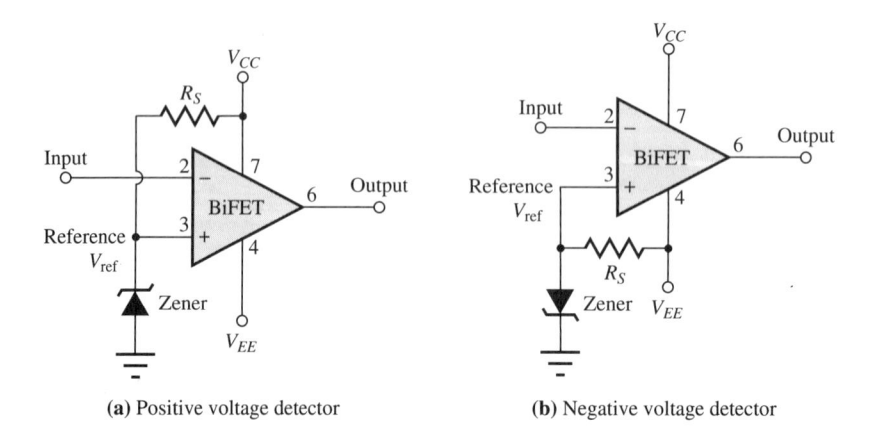

Figure 5-8 Level-sensing comparators with zener diode reference.

bias current I_{ZT}, and a series resistor R_S is always included in the circuit to set this bias current. We will see in Chapter 7 that the series resistor is designed by the following equation:

$$R_S = \frac{V_{CC} - V_Z}{I_{ZT}} \tag{5-7}$$

This is the equation for a positive reference voltage. If a negative reference voltage is required, use V_{EE} in place of V_{CC}.

The circuit in Figure 5-8(a) uses a zener diode (or voltage reference) to produce a positive reference voltage, and is equivalent to the voltage divider circuit of Figure 5-7(a). The circuit in Figure 5-8(b) uses a zener diode to produce a negative reference, and is equivalent to the voltage divider circuit of Figure 5-7(b). The circuits produce a negative output if the input signal is greater (more positive) than the reference voltage and, hence, are inverting comparators. As with the voltage divider circuits, if we want a noninverting comparator we simply reverse the input connections.

EXAMPLE 5-2

Design a simple comparator circuit to turn on an LED warning light when the voltage of a nominal 12 V battery drops below 10 V. Use a TL081C op-amp operated from a separate ±15 V supply, with a voltage divider for biasing.

Solution

This problem requires an inverting positive voltage detector with voltage divider biasing. The circuit that we shall use is similar to that shown in Figure 5-7(a). It has been redrawn as Figure 5-9, with the inclusion of a parts list.

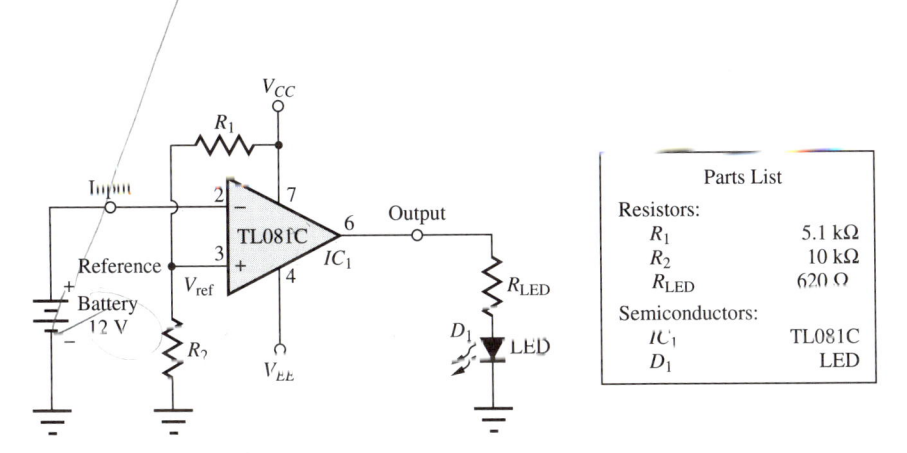

Figure 5-9

First we must calculate the resistors in the voltage divider. Assume a value for R_2 of 10 kΩ and calculate R_1 from Equation 5-6, with V_{CC} as the positive supply voltage of +15 V and V_{ref} as the desired switching point of 10 V:

$$R_1 = R_2 \times \left(\frac{V_{CC}}{V_{ref}} - 1\right) = 10 \text{ k}\Omega \times \left(\frac{15 \text{ V}}{10 \text{ V}} - 1\right) = 5.0 \text{ k}\Omega$$

We will choose the closest standard value of 5.1 kΩ for R_1.

Next, we must calculate the series LED resistor R_{LED} from Equation 5-1. Since the comparator will give an output of $V_{sat+} = +14$ V when the test voltage is less than the reference, we must calculate a suitable current-limiting resistor to protect the LED.

$$R_{LED} = \frac{V_{sat+} - 2 \text{ V}}{20 \text{ mA}} = \frac{14 \text{ V} - 2 \text{ V}}{20 \text{ mA}} = 600 \text{ }\Omega$$

Choosing the closest standard value from Appendix A-1, we will make $R_{LED} = 620 \text{ }\Omega$.

Notice the polarity of the LED in the circuit. We must insert it such that the LED lights when the input voltage is less than 10 V. Using the circuit in Figure 5-9, this produces a positive output voltage, so the LED must be inserted as shown.

Our circuit design is now complete.

In using a voltage divider to set the reference voltage, it is important to ensure that the voltage powering the voltage divider is constant. This type of circuit could not be powered from the DC supply being monitored, since the input voltage to the voltage divider would vary as the supply voltage varied; consequently the trigger point would vary. To power the comparator reference from the battery voltage, it is necessary to use a zener diode reference as demonstrated in the next example.

EXAMPLE 5-3

Design a simple comparator circuit using a TL081C op-amp to turn on an LED warning light when the voltage of a nominal 12 V battery drops below 10 V. Use a zener diode reference. The entire circuit should be powered from the battery.

Solution

This problem will require an inverting positive voltage level detector with a zener diode reference. The circuit will be similar to that in Figure 5-8(a), but because the comparator will be powered from the battery it is testing, the op-amp will be operated from a single supply with the negative supply lead V_{EE} connected to ground. This does not pose any problems. In fact, the comparator reference voltage applied to the noninverting input of the op-amp is equivalent to the voltage divider procedure of Section 3-4 for operating an op-amp from a single supply. The one restriction is that the output can never be negative.

5-3 Level-Sensing Comparator Circuits

We do have one design problem, however. If we use the circuit shown in Figure 5-8(a), we must use a 10 V zener. Unfortunately, if the battery voltage drops to 10 V, the zener will no longer give a 10 V reference. We must choose a zener with a voltage less than 10 V and use a voltage divider to step down the battery voltage to a fraction of the actual voltage.

The circuit that we will use is shown in Figure 5-10, and includes a parts list.

Parts List

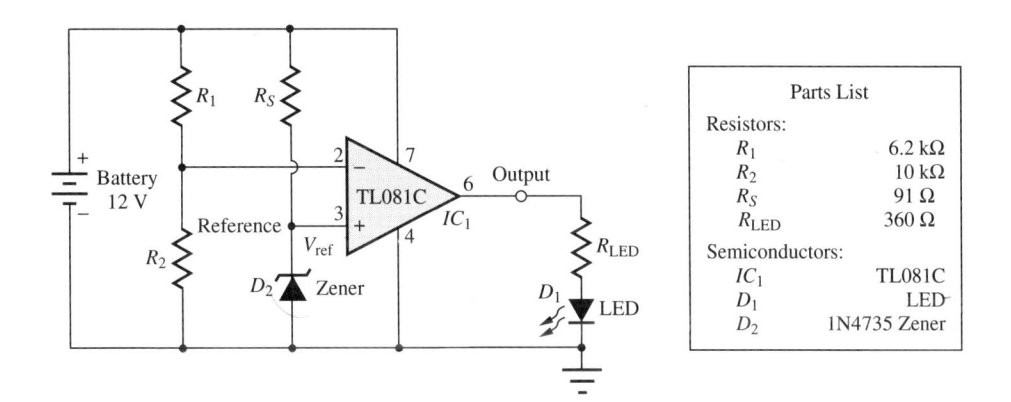

Resistors:
R_1	6.2 kΩ
R_2	10 kΩ
R_S	91 Ω
R_{LED}	360 Ω

Semiconductors:
IC_1	TL081C
D_1	LED
D_2	1N4735 Zener

Figure 5-10

We will choose a 6.2 V zener. From the data sheet in Appendix A-4, this will be a 1N4735 with a bias current of 41 mA. We can now design the series resistor R_S by using Equation 5-7. Use $V_{CC} = 10$ V because this is the nominal operating point of the circuit.

$$R_S = \frac{V_{CC} - V_Z}{I_{ZT}} = \frac{10 \text{ V} - 6.2 \text{ V}}{41 \text{ mA}} = 92.7 \text{ Ω}$$

Choose $R_S = 91$ Ω. This is the closest standard value from the table in Appendix A-1. Check that a power rating of $\frac{1}{2}$ W is adequate.

Now design a voltage divider such that when the battery voltage is 10 V, the voltage applied to the noninverting input is 6.2 V. We can use Equation 5-6 where $V_{CC} = 10$ V and $V_{ref} = 6.2$ V, and R_2 was chosen to be 10 kΩ.

$$R_1 = R_2 \times \left(\frac{V_{CC}}{V_{ref}} - 1\right) = 10 \text{ kΩ} \times \left(\frac{10 \text{ V}}{6.2 \text{ V}} - 1\right) = 6.13 \text{ kΩ}$$

The closest standard value for R_1 is 6.2 kΩ from Appendix A-1.

Finally, select a current-limiting resistor for the LED. Since the supply voltage of the op-amp is 10 V at the point where the LED is to come on, the value of V_{sat+} will be approximately 9 V. Calculate R_{LED} from Equation 5-1.

$$R_{LED} = \frac{V_{sat+} - 2 \text{ V}}{20 \text{ mA}} = \frac{9 \text{ V} - 2 \text{ V}}{20 \text{ mA}} = 350 \text{ Ω}$$

The closest standard value from Appendix A-1 is 360 Ω. We will use this value for R_{LED}.

Our circuit design is now complete. This was a relatively complicated design problem.

5-4 Comparators with Hysteresis

In our discussion of comparators thus far, we have careful avoided one serious problem. What happens if the input voltage is almost exactly equal to the reference voltage and there is noise present? With a gain of 200,000, a comparator is a very sensitive detector. A few microvolts of noise can cause jitter—the comparator switching on and off repeatedly. This can be annoying. Figure 5-11(a) shows how noise affects the switching of a zero-crossing inverting comparator, illustrating jitter. Each time the input voltage goes past zero, the comparator switches. Numerous changes in output due to the noise on the input are shown in Figure 5-11(a). It is possible to modify a comparator so that it has two different switching levels: one that causes the comparator to switch as the voltage is increasing, the other that causes the comparator to switch as the voltage is decreasing. Figure 5-11(b) shows the effectiveness of a 0.2 V upper trigger level and a −0.2 V lower trigger level at reducing jitter. In Figure 5-11(b), the comparator initially has an output of V_{sat+} and does not switch to V_{sat-} until the input voltage exceeds the *upper* trigger voltage. Once the comparator has switched to V_{sat-}, it will not switch to V_{sat+} until the input voltage drops to less than the *lower* trigger voltage. A comparator that has two different switching levels is said to have **hysteresis**.

(a) Comparator without hysteresis. Note the jitter

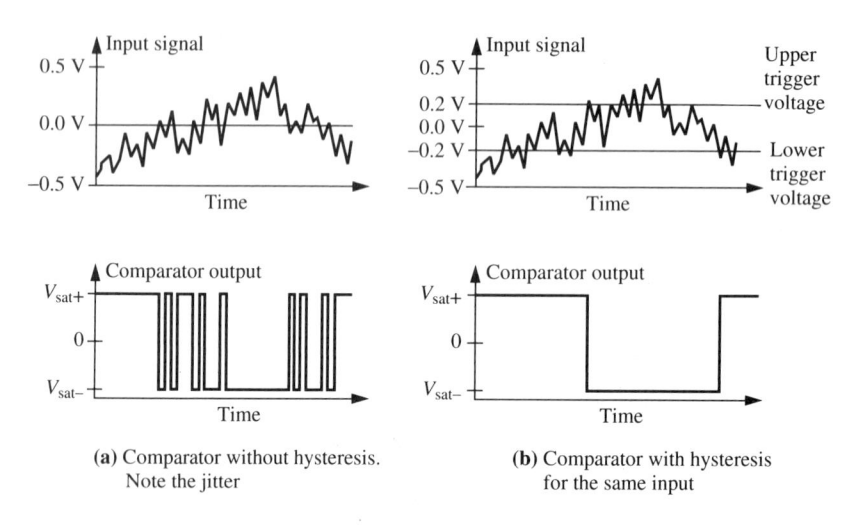

(b) Comparator with hysteresis for the same input

Figure 5-11 Effect of hysteresis on the operation of a zero-crossing inverting comparator.

5-4 Comparators with Hysteresis

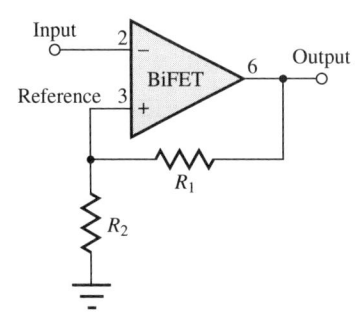

Figure 5-12 Inverting zero-crossing comparator with hysteresis.

Figure 5-12 shows a zero-detecting inverting comparator with hysteresis. This type of circuit is also known as an inverting **Schmitt trigger**. Schmitt trigger circuits are widely used in digital applications, and a number of digital integrated circuits make use of various forms of Schmitt triggers.

This comparator uses positive feedback to the noninverting input to define the trigger voltage. The trigger voltage is the fraction of the output voltage at the noninverting input produced by a voltage divider consisting of resistors R_1 and R_2. Assume that the output is high, with an output voltage of V_{sat+}. Treating R_1 and R_2 as a standard voltage divider, the voltage applied to the noninverting input is calculated as

$$V_{TU} = V_{sat+} \times \frac{R_2}{R_1 + R_2} \qquad (5\text{-}8)$$

The voltage V_{TU} is called the **upper trigger voltage**. This is the voltage at which the comparator output will switch from positive to negative as the input test voltage is increasing. Similarly, we can determine the lower trigger voltage V_{TL} by the formula

$$V_{TL} = V_{sat-} \times \frac{R_2}{R_1 + R_2} \qquad (5\text{-}9)$$

The **lower trigger voltage** is the voltage at which the comparator output will switch from negative to positive as the input test voltage is decreasing.

To illustrate the operation of an inverting zero-crossing comparator with hysteresis, again refer to Figure 5-11(b). For the comparator generating the output in this figure, the upper trigger voltage V_{TU} is 0.2 V and the lower trigger voltage V_{TL} is -0.2 V. Starting with an input voltage of approximately -0.5 V, the output will be V_{sat+} and positive. Hence, the reference voltage will be the upper trigger voltage of $+0.2$ V. Now increase the input voltage past 0.2 V. The comparator will switch as the input voltage passes 0.2 V, since at this point the inverting input becomes positive with respect to the noninverting input. As the comparator switches, the output becomes V_{sat-} and negative. Hence, the reference voltage will now be the lower trigger point of -0.2 V. Now decrease the input voltage past -0.2 V. The comparator will again switch as the input voltage decreases past -0.2 V, since the inverting input has now become negative with respect to the noninverting input. As the comparator switches, the output will become V_{sat+} and positive, and the reference voltage is again the upper trigger voltage.

To design a zero-crossing comparator with hysteresis, first select one of the trigger points, then choose some suitable value for R_2, typically around 1000 Ω. If we specify the upper trigger point, solve Equation 5-8 for resistor R_1:

$$R_1 = R_2 \times \left(\frac{V_{sat+}}{V_{TU}} - 1\right) \tag{5-10}$$

Alternatively, we could have specified the lower trigger point and solved Equation 5-9 for R_1:

$$R_1 = R_2 \times \left(\frac{V_{sat-}}{V_{TL}} - 1\right) \tag{5-11}$$

If R_1 is calculated from Equation 5-10 with the upper trigger voltage, then the lower trigger voltage can be calculated from Equation 5-9. If R_1 is found from Equation 5-11 and the lower trigger voltage, the upper trigger voltage is given by Equation 5-8.

We can also incorporate hysteresis into a level-sensing comparator as shown in Figure 5-13. Figure 5-13(a) shows the circuit of an inverting voltage-sensing comparator using a voltage divider consisting of R_{D1} and R_{D2} across the power supply to set the reference voltage. A second voltage divider, consisting of R_1 and R_2, sets the trigger points the desired amount above and below this reference voltage. This same circuit is used for single-supply operation. It is important to

(a) Comparator using voltage divider reference

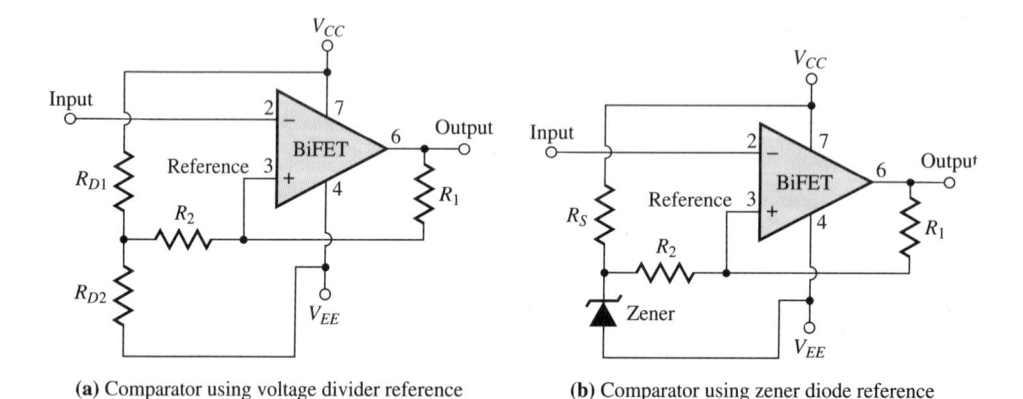

(b) Comparator using zener diode reference

Figure 5-13 Inverting level-sensing comparators with hysteresis.

make $R_{D1} + R_{D2}$ much smaller than $R_1 + R_2$ so that the reference voltage divider will not be loaded by the feedback voltage divider.

Assume that we have a power supply with voltages V_{CC} and V_{EE}, where V_{EE} is negative (or zero). The total voltage across the divider is $V_{CC} - V_{EE}$. Remember that V_{EE} is negative. Hence, if we have a ± 15 V supply, our total voltage would be 30 V. Suppose we want a reference voltage V_{ref} above ground. We must design our voltage divider such that the output voltage is $V_{ref} - V_{EE}$. For example, if we want a 5 V reference for $V_{EE} = -15$ V, our divider must be designed for 20 V. Writing the equation for the voltage divider consisting of R_{D1} and R_{D2}, we get

$$V_{ref} - V_{EE} = (V_{CC} - V_{EE}) \times \frac{R_{D2}}{R_{D1} + R_{D2}} \tag{5-12}$$

The easiest way of designing the voltage divider is to choose a convenient value for R_{D2}, say 1000 Ω, then solve Equation 5-12 for R_{D1}:

$$R_{D1} = R_{D2} \times \left(\frac{V_{CC} - V_{EE}}{V_{ref} - V_{EE}} - 1\right) \tag{5-13}$$

This formula takes into account that the value of V_{EE} will probably be negative.

Figure 5-13(b) shows the circuit of an inverting voltage-sensing comparator using a zener diode (or voltage reference integrated circuit) to set the reference voltage. The reference voltage is $V_Z + V_{EE}$. Substituting this into Equation 5-7, we obtain the equation for calculating series resistor R_S:

$$R_S = \frac{V_{CC} - V_{EE} - V_Z}{I_{ZT}} \tag{5-14}$$

Again, this formula assumes that the value of V_{EE} will probably be negative. Actually, all of these formulae will work if V_{EE} is negative, zero, or even positive. *Remember to use the sign of the actual voltage.*

To determine the equation for the upper trigger voltage V_{TU} for the circuits in Figure 5-13, assume that R_1 and R_2 are much larger than R_{D1} and R_{D2}, to avoid loading the reference level voltage divider. In this case, the voltage from the divider is V_{ref}, while the output voltage from the op-amp is V_{sat+}. The voltage across the feedback divider consisting of R_1 and R_2 is $V_{sat+} - V_{ref}$. The voltage drop across resistor R_2, using a simple voltage divider calculation, is

$$V_{R_2} = (V_{sat+} - V_{ref}) \times \frac{R_2}{R_1 + R_2}$$

The upper trigger voltage V_{TU} is V_{R_2} added to V_{ref}. Adding V_{ref} to the above equation and multiplying, we get

$$V_{TU} = (V_{sat+} - V_{ref}) \times \left(\frac{R_2}{R_1 + R_2}\right) + V_{ref} \tag{5-15}$$

Similarly, the lower trigger voltage V_{TL} is

$$V_{TL} = (V_{sat-} - V_{ref}) \times \left(\frac{R_2}{R_1 + R_2}\right) + V_{ref} \tag{5-16}$$

5 Comparators

These equations work for both the voltage divider reference and the zener reference. The total resistance of the voltage divider establishing the reference voltage must be much less than the total resistance of the feedback voltage divider establishing the hysteresis voltage, as explained above.

To design a positive or negative level-sensing comparator with hysteresis, select the upper trigger voltage, select a suitable value for R_2, typically around 10 kΩ, then solve Equation 5-15 for R_1:

$$R_1 = R_2 \times \left(\frac{V_{\text{sat}+} - V_{\text{ref}}}{V_{TU} - V_{\text{ref}}} - 1 \right) \quad (5\text{-}17)$$

Alternatively, choose the lower trigger voltage and solve Equation 5-16 for R_1:

$$R_1 = R_2 \times \left(\frac{V_{\text{sat}-} - V_{\text{ref}}}{V_{TL} - V_{\text{ref}}} - 1 \right) \quad (5\text{-}18)$$

If R_1 is calculated by using Equation 5-17 with the upper trigger voltage, then the lower trigger voltage is given by Equation 5-16. If R_1 is calculated by using Equation 5-18 with the lower trigger voltage, the upper trigger point is given by Equation 5-15.

Although we have not discussed the use of comparators with single supplies, any of the above comparator circuits will work fine with single supplies as long as we are comparing to a positive input test voltage and have a positive reference voltage. The equations that have been derived for all of the above circuits can be used for single-supply comparators simply by replacing V_{EE} by zero.

EXAMPLE 5-4

Design a comparator circuit using a TL081C op-amp to turn on an LED indicator when the input voltage exceeds 3.3 V. To avoid the possibility of spurious triggering from noise, incorporate hysteresis into the circuit so that the upper trigger point must be 0.2 V above the reference voltage. Calculate the lower trigger point. Use a zener diode for biasing. The comparator will be powered from a single +12 V supply.

Solution

We will use an inverting level-sensing comparator with hysteresis, and the circuit will be similar to Figure 5-13(b). The output is connected to an LED and a series current-limiting resistor R_{LED}. The complete circuit is shown as Figure 5-14.

The upper trigger voltage V_{TU} has been specified as 0.2 V above the reference level. Hence, $V_{TU} = 3.3 \text{ V} + 0.2 \text{ V} = 3.5 \text{ V}$. We will choose $R_2 = 10 \text{ k}\Omega$ and find resistor R_1 by using Equation 5-17. This equation requires a value for $V_{\text{sat}+}$ which we will assume to be 11 V. Substituting these values into Equation 5-17, we find

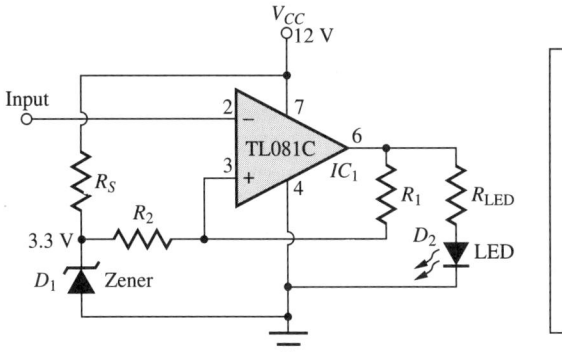

5-4 Comparators with Hysteresis

Figure 5-14

Parts List
Resistors:
- R_1 360 kΩ
- R_2 10 kΩ
- R_S 120 Ω
- R_{LED} 470 Ω

Semiconductors:
- IC_1 TL081C
- D_1 1N4728 Zener
- D_2 LED

$$R_1 = R_2 \times \left(\frac{V_{sat+} - V_{ref}}{V_{TU} - V_{ref}} - 1\right)$$

$$= 10 \text{ k}\Omega \times \left(\frac{11 \text{ V} - 3.3 \text{ V}}{3.5 \text{ V} - 3.3 \text{ V}} - 1\right) = 375 \text{ k}\Omega$$

The closest standard value from Appendix A-1 is 360 kΩ. We will use this value for R_1.

We can now calculate the lower trigger point from Equation 5-16. This equation requires V_{sat-}, which would be approximately 1 V above the lower supply value. Since we have a single supply, the lower supply value is 0 V, so V_{sat-} is approximately 1 V.

$$V_{TL} = (V_{sat-} - V_{ref}) \times \left(\frac{R_2}{R_1 + R_2}\right) + V_{ref}$$

$$= (1 \text{ V} - 3.3 \text{ V}) \times \frac{10 \text{ k}\Omega}{360 \text{ k}\Omega + 10 \text{ k}\Omega} + 3.3 \text{ V} = 3.24 \text{ V}$$

Notice that the lower trigger voltage is not 0.2 V less than V_{ref}, as we might expect. The reason is that we are operating from a single supply instead of a dual supply. With a dual supply, the trigger voltages would be symmetrically placed above and below V_{ref}.

Next, we will have to select a suitable zener and calculate the value of the series resistor R_S. The data sheets for zener diodes are shown in Appendix A-4. Since we require a 3.3 V reference, we will use a 1N4728 zener diode. This zener diode requires a bias current I_{ZT} of 76 mA, and if we ignore the small current through R_1 and R_2, this is the current through R_S. The voltage drop across R_S is the supply voltage minus the zener voltage. We can now calculate R_S from Equation 5-14:

$$R_S = \frac{V_{CC} - V_Z}{I_{ZT}} = \frac{12 \text{ V} - 3.3 \text{ V}}{76 \text{ mA}} = 114.5 \text{ }\Omega$$

Because the voltage and current for this resistor are relatively large, we should check the power. The power is simply the voltage times the current or 8.7 V × 0.076 A = 0.6612 W. Choose a 120 Ω, 1 W resistor as the closest standard value from Appendix A-1.

Finally, we must design the current-limiting resistor for the LED. Since the typical LED requires 20 mA with a 2.0 V drop, and the output voltage of the comparator is $V_{sat+} = 11$ V, we can calculate R_{LED} from Equation 5-1:

$$R_{LED} = \frac{V_{sat+} - 2\text{ V}}{20\text{ mA}} = \frac{11\text{ V} - 2\text{ V}}{20\text{ mA}} = 450\text{ }\Omega$$

We should also check the power for this resistor. The power is 9 V × 20 mA = 0.18 W. Choose a 470 Ω, $\frac{1}{2}$ W resistor as the closest standard value from Appendix A-1.

Our design is now complete.

5-5 Window Comparator Circuits

Often it is necessary to detect when a voltage is outside of, or within, a specified voltage range. The voltage range is called a **window**, and the comparator circuit that gives an output when the voltage is either within the window or not within the window is called a **window comparator**.

A variety of circuits can be used for window comparators. The simplest form of window comparator is shown in Figure 5-15. This window comparator turns on an LED whenever the input voltage is *outside* of the window. It consists of two level-sensing comparators. The upper comparator is noninverting, and its output voltage goes to positive saturation V_{sat+} when the input voltage is greater than the upper window voltage. The lower comparator is inverting, and its output voltage goes to negative saturation V_{sat-} when the input voltage is greater than the lower window voltage.

If the input voltage is less than the lower window voltage, the lower comparator has a positive output, diode D_2 conducts, and the LED is on. If the input voltage is greater than the lower window voltage but less than the upper window voltage (that is, the input voltage is within the window), both comparators have negative output voltages, both diodes are reverse-biased, and the LED is off. If the input voltage is greater than the upper window voltage, the upper comparator has a positive output, and diode D_1 conducts, turning the LED on.

The design of the individual comparators is identical to the design of an ordinary positive or negative voltage detector as described earlier. Note that while Figure 5-15 uses voltage dividers to set the reference levels, zener diodes could have been used instead.

The equation for designing the LED current-limiting resistor R_{LED} is similar to Equation 5-1, but the diode voltage drop must be included:

$$R_{LED} = \frac{V_{sat+} - V_{diode} - 2.0\text{ V}}{20\text{ mA}} \tag{5-19}$$

5-5 Window Comparator Circuits

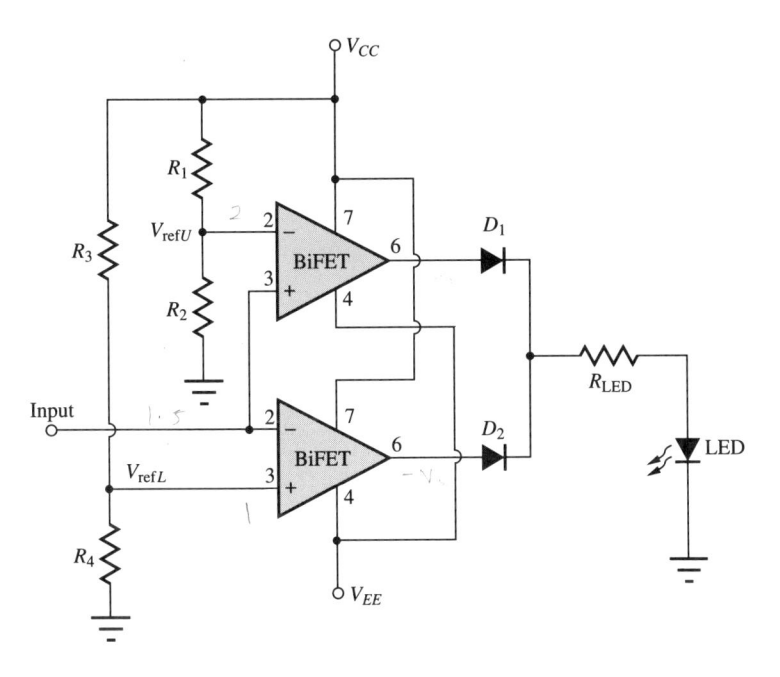

Figure 5-15 A window comparator that turns an LED off when the input voltage is within the window.

To reverse the action of the comparator so that the LED comes on when the voltage is within the window is not as easy as it might appear. Simply reversing the direction of the LED and/or the direction of the diodes does not work. A more complicated solution is required. Figure 5-16 shows a window comparator that turns on an LED when the input voltage is within the voltage window. It is the same basic circuit as in Figure 5-15, except that a transistor has been added to invert the output signal.

If the input voltage is less than the lower window voltage, the lower comparator has a positive output and diode D_2 conducts. Current flows into the base of the transistor, switching it on and shorting the collector to ground, holding the LED off. If the input voltage is greater than the lower window voltage but less than the upper window voltage (that is, the input voltage is within the window), both comparators have negative output voltages, neither diode conducts, and the transistor is biased off. Current flows through the collector resistor to the LED, causing the LED to come on. If the input voltage is greater than the upper window voltage, the upper comparator has a positive output and diode D_1 conducts. Current flows into the base of the transistor from the upper comparator, causing it to switch on and hold the LED off.

The transistor base resistor R_B must be chosen to ensure that the transistor saturates when it is turned on. The voltage across this resistor is $V_{\text{sat}+} - V_{\text{diode}} - V_{BE}$. The current through it is the maximum collector current divided by the minimum β of the transistor $I_{C\,\text{max}}/\beta_{\text{min}}$. The maximum collector current in this

192   5   Comparators

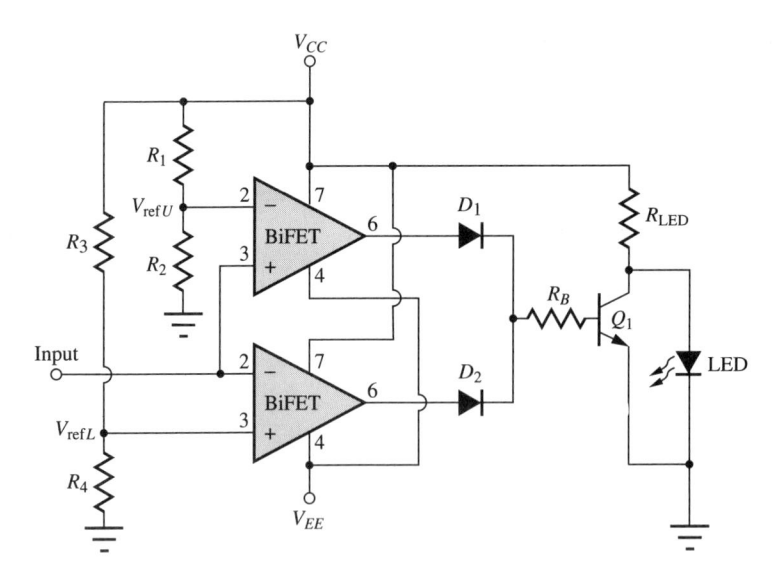

Figure 5-16 A window comparator that turns an LED on when the input voltage is within the window.

case would be the LED current. The formula for calculating the base resistor is

$$R_B = \frac{V_{sat+} - V_{diode} - V_{BE}}{I_{C\,max}/\beta_{min}} \qquad (5\text{-}20)$$

$$= \beta_{min} \times \frac{V_{sat+} - V_{diode} - V_{BE}}{I_{C\,max}}$$

The collector resistor is simply the current-limiting resistor for the LED, although the formula is slightly different from that derived before, since the LED is driven directly from the supply instead of from the op-amp. Use V_{supply} instead of V_{sat+}.

$$R_{LED} = \frac{V_{supply} - 2.0\text{ V}}{20\text{ mA}} \qquad (5\text{-}21)$$

Any small-signal NPN transistor would be suitable for this circuit. The 2N4401 transistors for which data sheets are given in Appendix A-5 would be quite adequate.

EXAMPLE 5-5

Design a window comparator using two TL081C op-amps to cause an LED to turn on whenever the input voltage is between +3 and +4 V. Use voltage dividers for biasing. Use a single +12 V supply.

5-5 Window Comparator Circuits

Solution

Since we want the window comparator to turn on an LED when the input voltage is within the window, we must use the second design that was discussed above. The circuit that we will use appears in Figure 5-17.

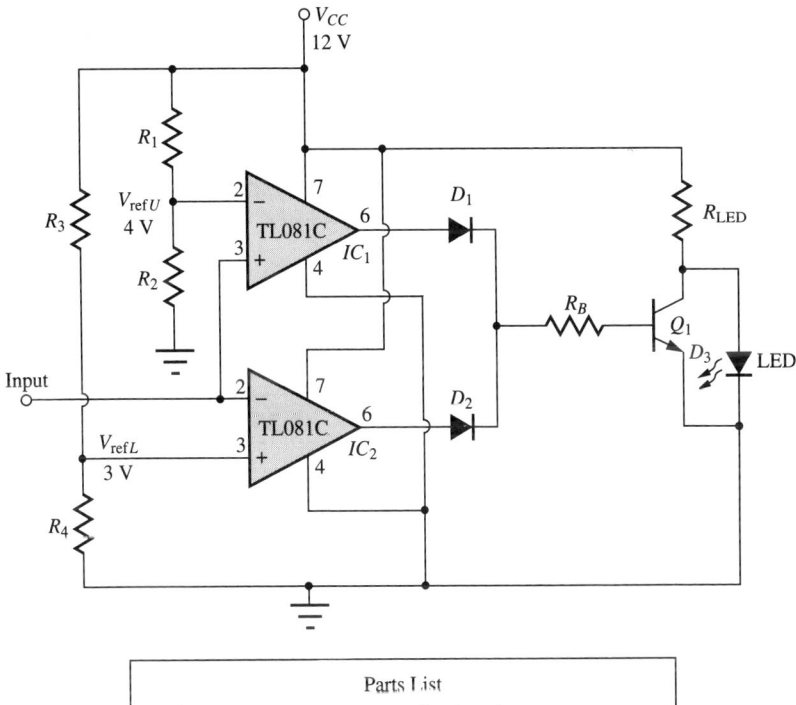

Parts List

Resistors:
- R_1 20 kΩ
- R_2 10 kΩ
- R_3 30 kΩ
- R_4 10 kΩ
- R_B 47 kΩ
- R_{LED} 510 Ω

Semiconductors:
- IC_1, IC_2 TL081C
- D_1, D_2 1N4004
- D_3 LED
- Q_1 2N4401

Figure 5-17

The upper comparator must switch on when the voltage exceeds +4 V. Note that this comparator is noninverting. Design the voltage divider for this comparator by using Equation 5-6, where we will choose R_2 to be 10 kΩ.

$$R_1 = R_2 \times \left(\frac{V_{CC}}{V_{ref}} - 1\right) = 10\ \text{k}\Omega \times \left(\frac{12\ \text{V}}{4\ \text{V}} - 1\right) = 20\ \text{k}\Omega$$

Both resistors R_1 and R_2 are standard values.

The lower comparator, which is inverting, is designed in exactly the same way, using the same equation with R_1 replaced by R_3, and R_2 replaced by R_4. Choose $R_4 = 10 \text{ k}\Omega$.

$$R_3 = R_4 \times \left(\frac{V_{CC}}{V_{\text{ref}}} - 1\right) = 10 \text{ k}\Omega \times \left(\frac{12 \text{ V}}{3 \text{ V}} - 1\right) = 30 \text{ k}\Omega$$

Again, both resistors R_3 and R_4 are standard values.

Resistor R_B must be chosen to ensure that the transistor saturates when it is turned on. We will use a 2N4401 transistor. The base resistor is calculated from Equation 5-20, where $V_{\text{sat}+} = 11$ V for a TL081C op-amp operating from a 12 V supply, $V_{\text{diode}} = V_{BE} = 0.7$ V, $I_{C \text{ max}} = 20$ mA, and $\beta_{\text{min}} = 100$.

$$R_B = \beta_{\text{min}} \times \frac{V_{\text{sat}+} - V_{\text{diode}} - V_{BE}}{I_{C \text{ max}}}$$

$$= 100 \times \frac{11 \text{ V} - 0.7 \text{ V} - 0.7 \text{ V}}{20 \text{ mA}} = 48 \text{ k}\Omega$$

The closest standard value is 47 kΩ, so we will choose this value for R_B.

The collector resistor is simply the current-limiting resistor for the LED and is found from Equation 5-21:

$$R_{\text{LED}} = \frac{V_{\text{supply}} - 2 \text{ V}}{20 \text{ mA}} = \frac{12 \text{ V} - 2 \text{ V}}{20 \text{ mA}} = 500 \text{ }\Omega$$

The closest standard value is 510 Ω. We will use this.

Our design is now complete.

5-6 True Comparator Integrated Circuits

The circuits discussed thus far have all used op-amps connected as comparators. Op-amps can work as comparators, but they are not optimized to perform the switching action required in an ideal comparator. The main weakness of an op-amp is its slew rate. Slew rate was discussed in Section 5-1, where we saw that a bipolar op-amp such as the LM741C with a slew rate of 0.5 V/μs would take about 60 μs to switch, whereas a BiFET op-amp such as the TL081C with a slew rate of 13 V/μs would switch in approximately 2 μs. Where switching speed is not important, such as in the applications discussed earlier, an op-amp makes a perfectly good comparator. Sometimes, however, particularly in digital systems, a switching time of 2 μs is not fast enough.

Special comparator integrated circuits, optimized for switching speed, are available. A typical example of a comparator integrated circuit is the TL810C, for which the data sheets are shown in Figure 5-18. Comparators are not simply a form of op-amp. Although related to op-amps, they are a distinct class of device. The input circuitry is similar to an op-amp, but the output is similar to a digital TTL logic gate. Compare the circuit shown on the data sheet for the LM741C

LINEAR INTEGRATED CIRCUITS

TYPES TL810M, TL810C
DIFFERENTIAL COMPARATORS

D993, MARCH 1971 – REVISED AUGUST 1983

- Low Offset Characteristics
- High Differential Voltage Amplification
- Fast Response Times
- Output Compatible with Most TTL Circuits

TL810M . . . JG PACKAGE
TL810C . . . JG OR P PACKAGE
(TOP VIEW)

```
GND   [ 1   8 ] V_CC+
IN+   [ 2   7 ] OUT
IN−   [ 3   6 ] NC
V_CC− [ 4   5 ] NC
```

TL810M . . . U PACKAGE
(TOP VIEW)

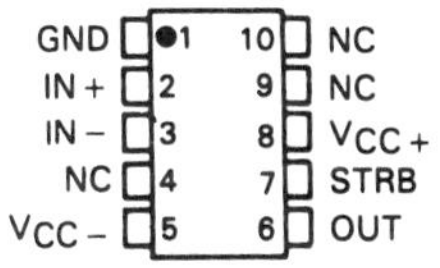

```
GND   [•1  10 ] NC
IN+   [ 2   9 ] NC
IN−   [ 3   8 ] V_CC+
NC    [ 4   7 ] STRB
V_CC− [ 5   6 ] OUT
```

NC – No internal connection

description

The TL810 is an improved version of the TL710 high-speed voltage comparator with an extra stage added to increase voltage amplification and accuracy. Typical amplification is 33,000. Component matching, inherent in monolithic integrated circuit fabrication techniques, produces a comparator with low-drift and low-offset characteristics. These circuits are particularly useful for applications requiring an amplitude discriminator, memory sense amplifier, or a high-speed limit detector.

The TL810M is characterized for operation over the full military temperature range of −55°C to 125°C; the TL810C is characterized for operation from 0°C to 70°C.

symbol

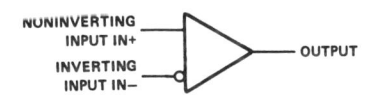

NONINVERTING INPUT IN+
INVERTING INPUT IN−
OUTPUT

Copyright © 1983 by Texas Instruments Incorporated

Texas Instruments
POST OFFICE BOX 225012 • DALLAS, TEXAS 75265

Figure 5-18 Data sheets for the TL810/TL810C comparator. *(Reprinted by permission of Texas Instruments)*

TYPES TL810M, TL810C
DIFFERENTIAL COMPARATORS

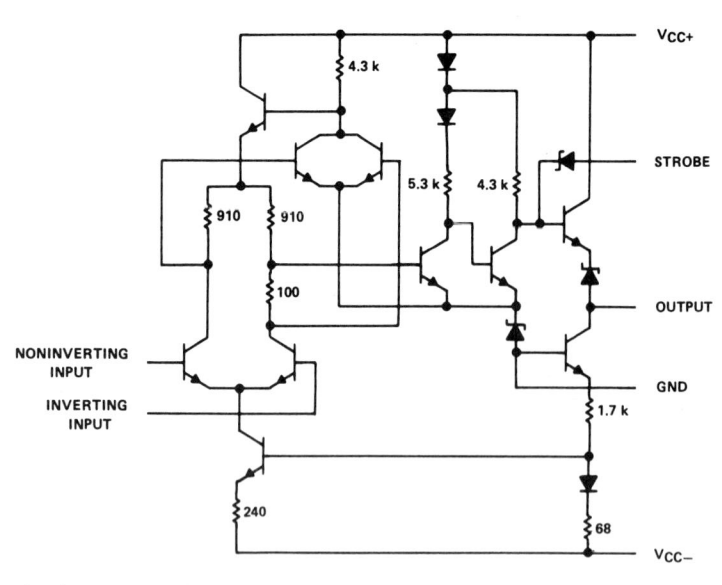

schematic

Resistor values shown are nominal in ohms.

absolute maximum ratings over operating free-air temperature range (unless otherwise noted)

Supply voltage V_{CC+} (see Note 1)	14 V
Supply voltage V_{CC-} (see Note 1)	−7 V
Differential input voltage (see Note 2)	±5 V
Input voltage (either input, see Note 1)	±7 V
Peak output current ($t_w \leq 1$ s)	10 mA
Continuous total power dissipation at (or below) 70°C free-air temperature (see Note 3)	300 mW
Operating free-air temperature range: TL810M Circuits	−55°C to 125°C
TL810C Circuits	0°C to 70°C
Storage temperature range	−65°C to 150°C
Lead temperature 1,6 mm (1/16 inch) from case for 60 seconds: JG or U package	300°C
Lead temperature 1,6 mm (1/16 inch) from case for 10 seconds: P package	260°C

NOTES: 1. All voltage values, except differential voltages, are with respect to the network ground terminal.
2. Differential voltages are at the noninverting input terminal with respect to the inverting input terminal.
3. For operation of the TL810M above 70°C free-air temperature, refer to dissipation Derating Curves, Section 2. In the JG package, TL810M chips are alloy-mounted; TL810C chips are glass-mounted.

Figure 5-18 (*Continued*)

TYPES TL810M, TL810C
DIFFERENTIAL COMPARATORS

electrical characteristics at specified free-air temperature, $V_{CC+} = 12$ V, $V_{CC-} = -6$ V (unless otherwise noted)

PARAMETER		TEST CONDITIONS†		TL810M MIN	TL810M TYP	TL810M MAX	TL810C MIN	TL810C TYP	TL810C MAX	UNIT
V_{IO}	Input offset voltage	$R_S \leq 200$ Ω, See Note 4	25°C		0.6	2		1.6	3.5	mV
			Full range			3			4.5	
αV_{IO}	Average temperature coefficient of input offset voltage	$R_S = 50$ Ω, See Note 4	MIN to 25°C		3	10		3	20	µV/°C
			25°C to MAX		3	10		3	20	
I_{IO}	Input offset current	See Note 4	25°C		0.75	3		1.8	5	µA
			MIN		1.8	7			7.5	
			MAX		0.25	3			7.5	
αI_{IO}	Average temperature coefficient of input offset current	See Note 4	MIN to 25°C		15	75		24	100	nA/°C
			25°C to MAX		5	25		15	50	
I_{IB}	Input bias current	See Note 4	25°C		7	15		7	20	µA
			MIN		12	25		9	30	
V_{ICR}	Common-mode input voltage range	$V_{CC-} = -7$ V	Full range	±5			±5			V
A_{VD}	Large-signal differential voltage amplification	No load, $V_O = 0$ to 2.5 V	25°C	12.5	33		10	33		V/mV
			Full range	10			8			
V_{OH}	High-level output voltage	$V_{ID} = 5$ mV, $I_{OH} = 0$	Full range		4§	5		4§	5	V
		$V_{ID} = 5$ mV, $I_{OH} = -5$ mA	Full range	2.5	3.6§		2.5	3.6§		
V_{OL}	Low-level output voltage	$V_{ID} = -5$ mV, $I_{OL} = 0$	Full range	-1	-0.5§	0‡	-1	-0.5§	0‡	V
I_{OL}	Low-level output current	$V_{ID} = -5$ mV, $V_O = 0$	25°C	2	2.4		1.6	2.4		mA
			MIN	1	2.3		0.5	2.4		
			MAX	0.5	2.3		0.5	2.4		
r_o	Output resistance	$V_O = 1.4$ V	25°C		200			200		Ω
CMRR	Common-mode rejection ratio	$R_S \leq 200$ Ω	Full range	80	100§		70	100§		dB
I_{CC+}	Supply current from V_{CC+}	$V_{ID} = -5$ mV, No load	Full range		5.5§	9		5.5§	9	mA
I_{CC-}	Supply current from V_{CC-}		Full range		-3.5§	-7		-3.5§	-7	mA
P_D	Total power dissipation		Full range		90§	150		90§	150	mW

†Full range (MIN to MAX) for TL810M is -55°C to 125°C and for the TL810C is 0°C to 70°C.
‡The algebraic convention, where the most-positive (least-negative) limit is designated as maximum, is used in this data sheet for logic levels only, e.g., when 0 V is the maximum, the minimum limit is a more-negative voltage.
§These typical values are at $T_A = 25$°C.
NOTE 4: These characteristics are verified by measurements at the following temperatures and output voltage levels: for TL810M, $V_O = 1.8$ V at $T_A = -55$°C, $V_O = 1.4$ V at $T_A = 25$°C, and $V_O = 1$ V at $T_A = 125$°C; for TL810C, $V_O = 1.5$ V at $T_A = 0$°C, $V_O = 1.4$ V at 25°C, and $V_O = 1.2$ V at $T_A = 70$°C. These output voltage levels were selected to approximate the logic threshold voltages of the types of digital logic circuits these comparators are intended to drive.

switching characteristics, $V_{CC+} = 12$ V, $V_{CC-} = -6$ V, $T_A = 25$°C

PARAMETER	TEST CONDITIONS	MIN	TYP	MAX	UNIT
Response time	$R_L = \infty$, $C_L = 5$ pF, See Note 5		30	80	ns

NOTE 5: The response time specified is for a 100-mV input step with 5-mV overdrive and is the interval between the input step function and the instant when the output crosses 1.4 V.

Figure 5-18 (*Continued*)

op-amp (Figure 2-17) with the circuit shown on the data sheet for the TL810C comparator (Figure 5-18). The symbol for a comparator is slightly different from an op-amp, and the pinout for the 8 pin DIP is also different. The 8 pin DIP pinout is shown in Figure 5-19(a), and the schematic symbol of a true comparator is shown in Figure 5-19(b). In the schematic symbol, the inverting input is designated with a circle. Also notice that the symbol is inverted from a standard op-amp; that is, the noninverting input is shown at the top instead of the bottom of the schematic.

A close look at the data sheets for the TL810C shown in Figure 5-18 shows that the TL810C has some quite different operating characteristics from an op-amp. The TL810C, unlike an op-amp, is not operated from a balanced dual supply but from an unbalanced dual supply. The recommended supplies are $+12$ V and -6 V. With these supplies, the output switches between approximately -0.5 V and approximately 4.0 V. These are close to nominal digital TTL voltage levels.

No slew rate is given for the TL810C on the data sheets. Instead, a quantity called the **response time** is used to describe the switching performance. The response time of the TL810C, defined as the time between the application of an input signal and the time when the output has reached 1.4 V (the low limit of a digital high signal), is typically 30 ns. The TL810 effectively switches states in 30 ns—over a thousand times faster than a LM741!

In some respects, the TL810C does resemble an op-amp. It has a typical amplification of 33,000, less than a typical op-amp but still very large. The common-mode rejection ratio is similar to that of an op-amp. The input impedance is extremely high like an op-amp, and the input bias current and input offset currents are similar to an op-amp.

The TL810C can be used in any of the op-amp comparator circuits described earlier, provided that the supply voltage, input voltage, and output voltage limits of the comparator are observed. In fact, the following examples will demonstrate the use of the TL810C in the same circuits where we used op-amps. One point to watch for, however, is that if the comparator is to have hysteresis, the reference voltage cannot exceed the output voltage of approximately 4.0 V.

The TL810C cannot be used in any circuit requiring a true op-amp, such as the circuits described in other chapters of this book. Remember this point: *an*

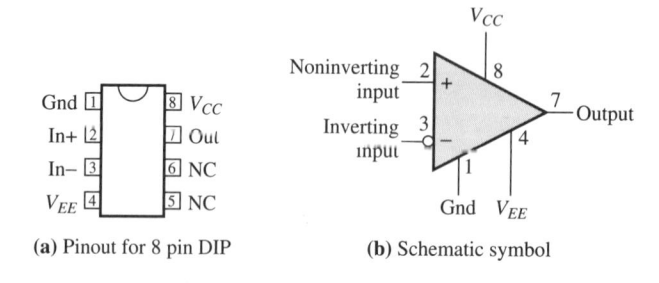

(a) Pinout for 8 pin DIP

(b) Schematic symbol

Figure 5-19 Symbol and pinout for a standard comparator.

5-6 True Comparator Integrated Circuits

op-amp can be used as a comparator, but a comparator cannot be used as an op-amp.

EXAMPLE 5-6

Design a polarity detector circuit using a TL810C comparator operating from standard +12 V and −6 V supplies. The detector should turn on one LED when the voltage is positive and another LED when the voltage is negative.

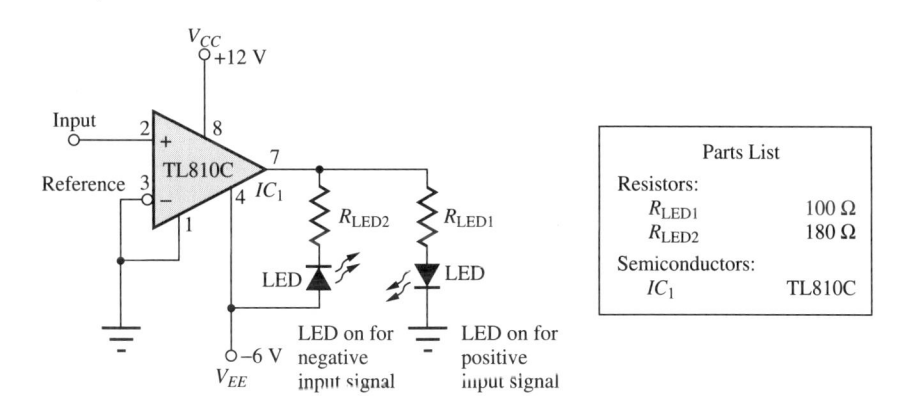

Figure 5-20

Solution

The circuit in this example is intended to produce the same results using the TL810C comparator as the circuit in Example 5-1 did with a standard op-amp. The polarity detector, shown in Figure 5-20, consists of a TL810C connected as a simple noninverting zero-crossing comparator with two LEDs connected to the output. The connection of the LEDs in this figure is quite different from that using the op-amp in Figure 5-6 of Example 5-1. Since the op-amp could assume either positive or negative output voltages, the LEDs could be connected in parallel. In the case of the comparator, however, the output is either −0.5 or 4.0 V. For a positive input signal, the output is 4.0 V, and diode D_1, connected from the output to ground, is forward-biased and hence comes on. For a negative input signal, the output is −0.5 V, and diode D_2, connected between the output and the −5 V supply, is forward-biased and hence comes on.

The series voltage-dropping resistor for each of the LEDs will be different. LED 1 is supplied from 4.0 V, which we will use as V_{sat+} in Equation 5-1.

$$R_{LED\ 1} = \frac{V_{sat+} - 2.0\ V}{20\ mA} = \frac{4.0\ V - 2\ V}{20\ mA} = 100\ \Omega$$

LED 2 is supplied from -6 V and sees an input voltage of -0.5 V. We will use the net difference in voltage, 5.5 V, as V_{sat+} in Equation 5-1.

$$R_{LED\ 2} = \frac{V_{sat+} - 2.0\ V}{20\ mA} = \frac{5.5\ V - 2\ V}{20\ mA} = 175\ \Omega$$

Using the closest standard values for both resistors from Appendix A-1, we will choose $R_{LED\ 1} = 100\ \Omega$, and $R_{LED\ 2} = 180\ \Omega$.

There is nothing to design in the comparator circuit itself, since it consists of a single TL810C comparator. Note, however, the differences in pin connections from the op-amp used in Example 5-1.

The design of our polarity indicating comparator is complete.

EXAMPLE 5-7

Design a simple circuit using a TL810C comparator to turn on an LED warning light when the voltage of a nominal 12 V battery drops below 10 V. Use a voltage divider for biasing. The comparator will be powered from a separate $+12$ V/-6 V supply.

Solution

The circuit in this example is intended to produce the same results using the TL810C comparator as did the circuit in Example 5-2 with a standard op-amp. This problem requires a simple positive voltage detector with voltage divider biasing. As in the previous example, we will connect the comparator with the signal into the noninverting input. The circuit that we will use, shown as Figure 5-21, is similar to that developed in Figure 5-9 of Example 5-2 except that we will be using a TL810C comparator instead of a TL081C op-amp. Since the comparator

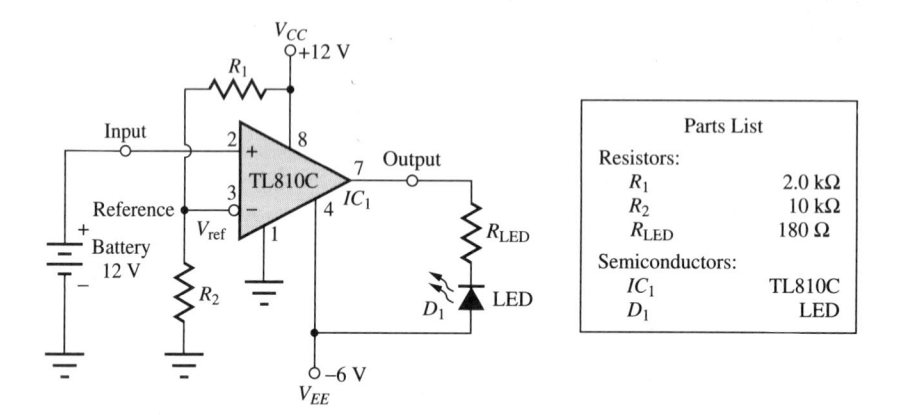

Figure 5-21

is to be powered from its own supply, the V_{CC} terminal will be at +12 V and the V_{EE} terminal will be at −6 V. For the LED to come on when the input voltage is less than the reference voltage, it will see an output voltage of −0.5 V. Consequently, we must connect the LED to the negative supply, as shown in the circuit in Figure 5-21.

First we must calculate the resistors in the voltage divider. This calculation will be the same as in Example 5-2, except that we are using a 12 V supply instead of a 15 V supply. Assume a value for R_2 of 10 kΩ and calculate R_1 from Equation 5-6, with V_{CC} as the positive supply voltage of +12 V and V_{ref} as the desired switching point of 10 V:

$$R_1 = R_2 \times \left(\frac{V_{CC}}{V_{ref}} - 1\right) = 10 \text{ k}\Omega \times \left(\frac{12 \text{ V}}{10 \text{ V}} - 1\right) = 2.0 \text{ k}\Omega$$

As this is a standard value, we will choose $R_1 = 2.0$ kΩ.

The calculation of R_{LED} is different from Example 5-2 since we will be using the −6 V supply. This resistor is calculated in the same way as $R_{LED\,2}$ was determined in Example 5-6. As in that example, we will use the difference in voltage between the −6 V supply and the comparator output voltage of −0.5 V, 5.5 V, as V_{sat+} in Equation 5-1:

$$R_{LED\,2} = \frac{V_{sat+} - 2.0 \text{ V}}{20 \text{ mA}} = \frac{5.5 \text{ V} - 2 \text{ V}}{20 \text{ mA}} = 175 \text{ }\Omega$$

Choosing the closest standard value from Appendix A-1, we will make $R_{LED} = 180$ Ω.

Our circuit design is now complete.

EXAMPLE 5-8

Design a circuit using a TL810C comparator to turn on an LED warning light when the voltage of a nominal 12 V battery drops below 10 V. Use a zener diode reference. The entire circuit should be powered from the battery.

Solution

This problem is similar to the previous example, except now the comparator is powered from the battery under test. This example is similar to Example 5-3, except that we are using a TL810C comparator instead of an op-amp. The circuit, shown as Figure 5-22, is similar to Figure 5-10 of Example 5-3 except that the comparator has been substituted for the op-amp.

As we saw in Example 5-3, we cannot use a 10 V zener diode, because if the battery voltage drops to 10 V the zener will no longer give a 10 V reference. We must choose a zener with a voltage less than 10 V and use a voltage divider to reduce the input voltage to a fraction of the actual battery voltage. As in Example 5-3, we will choose a 6.2 V zener, a 1N4735 from the data sheet in Appendix

5 Comparators

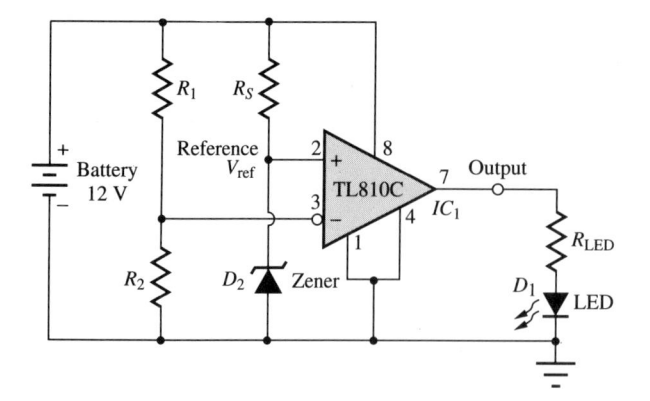

Figure 5-22

Parts List
Resistors:
R_1 6.2 kΩ
R_2 10 kΩ
R_S 91 Ω
R_{LED} 100 Ω
Semiconductors:
IC_1 TL810C
D_1 LED
D_2 1N4735 Zener

A-4. This zener requires a bias current I_{ZT} of 41 mA. Use $V_{CC} = 10$ V, because this is the point where we are concerned with an accurate zener voltage, and use Equation 5-7 to design R_S:

$$R_S = \frac{V_{CC} - V_Z}{I_{ZT}} = \frac{10 \text{ V} - 6.2 \text{ V}}{41 \text{ mA}} = 92.7 \text{ }\Omega$$

Choose R_S to be 91 Ω as the closest standard value from Appendix A-1. Verify that the power rating should be $\frac{1}{2}$ W.

Now design a voltage divider such that when the battery voltage is 10 V, the voltage applied to the noninverting input is 6.2 V. This design is the same as in Example 5-3. We can use Equation 5-6, where $V_{CC} = 10$ V and $V_{ref} = 6.2$ V. Choose $R_2 = 10$ kΩ and solve for R_1:

$$R_1 = R_2 \times \left(\frac{V_{CC}}{V_{ref}} - 1\right) = 10 \text{ k}\Omega \times \left(\frac{10 \text{ V}}{6.2 \text{ V}} - 1\right) = 6.13 \text{ k}\Omega$$

The closest standard value for R_1 is 6.2 kΩ from Appendix A-1.

Finally, select a current-limiting resistor for the LED. Notice that we are using the TL810C as an inverting comparator. Since we are applying the battery voltage through the divider circuit to the inverting input, when this voltage is less than the reference voltage the output will be 4.0 V. We will use this output voltage as V_{sat+} in Equation 5-1 to calculate R_{LED}:

$$R_{LED} = \frac{V_{sat+} - 2.0 \text{ V}}{20 \text{ mA}} = \frac{4.0 \text{ V} - 2 \text{ V}}{20 \text{ mA}} = 100 \text{ }\Omega$$

Because this is a standard value from Appendix A-1, we will use 100 Ω for R_{LED}. Our circuit design is now complete.

EXAMPLE 5-9

Design a circuit using a TL810C comparator to turn on an LED indicator when the input voltage exceeds 3.3 V. To avoid the possibility of spurious triggering from noise, incorporate hysteresis into the circuit so that the upper trigger point is 0.2 V above the reference voltage. Calculate the lower trigger point. Use a zener diode for biasing. The comparator will be powered from a single +12 V supply.

Solution

This example is similar to Example 5-4, except that we are using a TL810C comparator instead of a TL081C op-amp. The circuit that we will use is shown in Figure 5-23 and is similar to Figure 5-14 of Example 5-4, except that the reference zener is connected to the inverting input and the test voltage to the noninverting input.

The upper trigger point V_{TL} has been specified as 0.2 V above the reference level. Hence, $V_{TL} = 3.3 \text{ V} + 0.2 \text{ V} = 3.5 \text{ V}$. Choose $R_2 = 10 \text{ k}\Omega$ and calculate R_1 by using Equation 5-17. This equation requires a value for $V_{\text{sat}+}$. The output from the TL810C is typically 4.0 V, and we will use this for $V_{\text{sat}+}$. Substitute these values into Equation 5-17:

$$R_1 = R_2 \times \left(\frac{V_{\text{sat}+} - V_{\text{ref}}}{V_{TU} - V_{\text{ref}}} - 1 \right)$$

$$= 10 \text{ k}\Omega \times \left(\frac{4.0 \text{ V} - 3.3 \text{ V}}{3.5 \text{ V} - 3.3 \text{ V}} - 1 \right) = 25 \text{ k}\Omega$$

We will choose $R_1 = 24 \text{ k}\Omega$, the closest standard value.

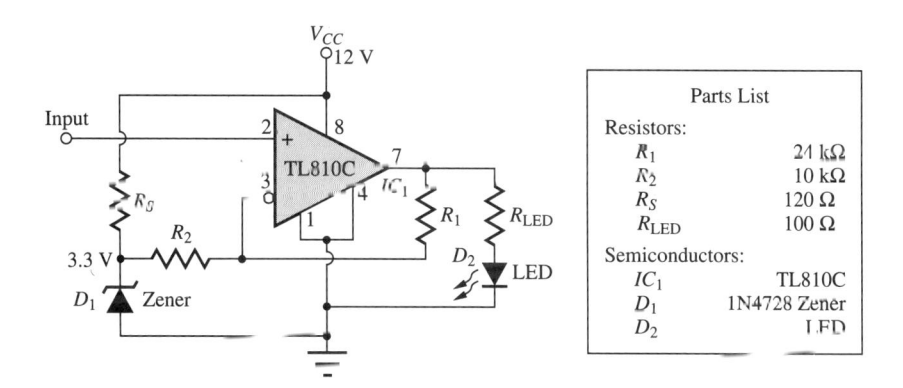

Parts List	
Resistors:	
R_1	24 kΩ
R_2	10 kΩ
R_S	120 Ω
R_{LED}	100 Ω
Semiconductors:	
IC_1	TL810C
D_1	1N4728 Zener
D_2	LED

Figure 5-23

Next, calculate the lower trigger point from Equation 5-16. This equation requires $V_{\text{sat}-}$. Since we have a single supply, we will assume $V_{\text{sat}-}$ is approximately 1 V.

$$V_{TL} = (V_{\text{sat}-} - V_{\text{ref}}) \times \left(\frac{R_2}{R_1 + R_2}\right) + V_{\text{ref}}$$

$$= (1 \text{ V} - 3.3 \text{ V}) \times \left(\frac{10 \text{ k}\Omega}{24 \text{ k}\Omega + 10 \text{ k}\Omega}\right) + 3.3 \text{ V} = 2.62 \text{ V}$$

Notice that, as in Example 5-4, the lower trigger voltage is not 0.2 V less than V_{ref} as we might expect, because we are operating from a single supply.

Next, we will have to select a suitable zener diode and calculate the value of the series resistor R_S. We will use a 1N4728 zener diode from Appendix A-4. This is a 3.3 V zener diode and requires a bias current of 76 mA. We will ignore the small current through R_1 and R_2. Calculate R_S from Equation 5-14:

$$R_S = \frac{V_{CC} - V_Z}{I_{ZT}} = \frac{12 \text{ V} - 3.3 \text{ V}}{76 \text{ mA}} = 115 \text{ }\Omega$$

Because the voltage and current for this resistor are relatively large, we should check the power. The power dissipated in the resistor is 8.7 V × 0.076 A = 0.6612 W. Choose a 120 Ω, 1 W resistor as the closest standard value from Appendix A-1.

Finally, we must design the current-limiting resistor for the LED. The output voltage of the comparator is 4.0 V, which we will use for $V_{\text{sat}+}$ in calculating R_{LED} from Equation 5-1:

$$R_{\text{LED}} = \frac{V_{\text{sat}+} - 2.0 \text{ V}}{20 \text{ mA}} = \frac{4.0 \text{ V} - 2.0 \text{ V}}{20 \text{ mA}} = 100 \text{ }\Omega$$

This is a standard resistor value. Because of the small voltage across it, the power dissipated will be quite small.

Our design is now complete.

EXAMPLE 5-10

Design a window detector to cause an LED to turn on whenever the input voltage is between +3 and +4 V. Use voltage dividers for biasing. Use two TL810C comparators operating from a single +12 V supply.

Solution

This window detector is almost identical to one using op-amps designed in Example 5-5 and illustrated in Figure 5-17. The window detector being designed in this example uses TL810C comparators instead of op-amps and is shown in Figure 5-24.

5-6 True Comparator Integrated Circuits

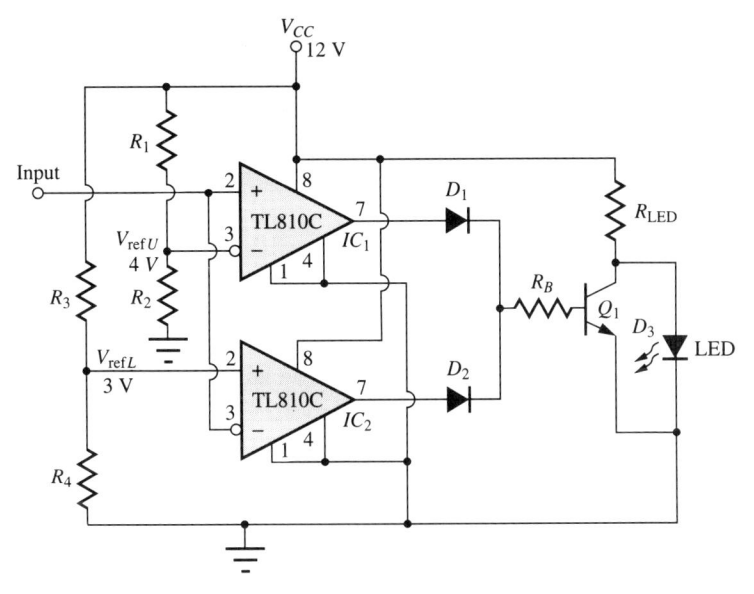

Figure 5-24

Parts List

Resistors:
- R_1 — 20 kΩ
- R_2 — 10 kΩ
- R_3 — 30 kΩ
- R_4 — 10 kΩ
- R_B — 13 kΩ
- R_{LED} — 510 Ω

Semiconductors:
- IC_1, IC_2 — TL810C
- D_1, D_2 — 1N4004
- D_3 — LED
- Q_1 — 2N4401

First we will design the upper comparator to switch on when the input exceeds the reference voltage of +4 V. To design the voltage divider for this comparator, select R_2 to be 10 kΩ, and use Equation 5-6.

$$R_1 = R_2 \times \left(\frac{V_{CC}}{V_{ref}} - 1\right) = 10 \text{ k}\Omega \times \left(\frac{12 \text{ V}}{4 \text{ V}} - 1\right) = 20 \text{ k}\Omega$$

Both R_1 and R_2 are standard values.

The lower comparator is designed in exactly the same way, to switch off when the input voltage exceeds the 3.0 V reference voltage. Use Equation 5-6 with R_1 replaced by R_3, and R_2 replaced by R_4. Choose $R_4 = 10$ kΩ.

$$R_3 = R_4 \times \left(\frac{V_{CC}}{V_{ref}} - 1\right) = 10 \text{ k}\Omega \times \left(\frac{12 \text{ V}}{3 \text{ V}} - 1\right) = 30 \text{ k}\Omega$$

Again, both resistors are standard values.

Use a 2N4401 transistor for Q_1, designing R_B to ensure that the transistor saturates when it is turned on. The base resistor is calculated from Equation 5-20, using $V_{sat+} = 4.0$ V for a TL810C comparator, $V_{diode} = V_{BE} = 0.7$ V, $I_{C\,max} = 20$ mA, and $\beta_{min} = 100$.

$$R_B = \beta_{min} \times \frac{V_{sat+} - V_{diode} - V_{BE}}{I_{C\,max}}$$

$$= 100 \times \frac{4.0 \text{ V} - 0.7 \text{ V} - 0.7 \text{ V}}{20 \text{ mA}} = 13.0 \text{ k}\Omega$$

This is a standard value of resistor, so we will use it.

Resistor R_{LED} is the current-limiting resistor for the LED and is found from Equation 5-1, using the supply voltage for V_{sat+}:

$$R_{LED} = \frac{V_{supply} - 2.0 \text{ V}}{20 \text{ mA}} = \frac{12 \text{ V} - 2.0 \text{ V}}{20 \text{ mA}} = 500 \text{ }\Omega$$

The closest standard value is 510 Ω. We will use this.

Our design is now complete.

Summary

This chapter introduced the comparator as a voltage controlled switch. We saw how an op-amp, operating without feedback, could be used as a comparator, and we developed a number of application circuits for it, including a polarity indicator, a voltage level-sensing circuit, a low-battery indicator, and a window detector. By using positive feedback, the comparator can be switched on at one level as the input voltage is increasing, and switched off at a different level as the voltage level is decreasing. This is called hysteresis, and comparators with hysteresis are sometimes called Schmitt triggers. Hysteresis is important for making the comparator insensitive to noise.

Op-amp comparator circuits are widely used, but for some applications the op-amp is not sufficiently fast in switching. When switching speed is important, true comparator integrated circuits such as the TL810 are used. These switch 100 to 1000 times faster than an op-amp and are optimized for comparator action.

Review Questions

5-1 *Op-amps as comparators*

1. What is a comparator?
2. Of what does the circuit of an op-amp configured as a comparator consist?
3. Between what levels does an op-amp switch?
4. How fast does an LM741C op-amp switch? a TL081C op-amp?

5. What term is used to describe the switching speed of an op-amp?
6. What two devices do comparators usually switch?
7. Why is a resistor always placed in series with an LED?
8. What is the purpose of the base resistor of a transistor switch?
9. What limiting values are used in the design of the base resistor of a transistor switch?

5-2 Zero-crossing comparator circuits

10. What is the reference voltage for a zero-crossing comparator?
11. What two types of zero-crossing comparator are possible? What is the difference in their responses?
12. Describe how to reverse the comparator action—that is, how to make the comparator respond to a negative-going signal instead of a positive-going signal.
13. Give two applications of a zero-crossing comparator.

5-3 Level-sensing comparator circuits

14. How can a reference voltage be derived for a comparator circuit?
15. What is the advantage of using a zener diode reference over a voltage divider reference?
16. Describe two ways of reversing the action of a level-sensing comparator.
17. Suggest two applications for a level-sensing comparator.
18. Why should you not use a voltage divider to set the reference level for a low-battery indicator if it is powered by the same battery?

5-4 Comparators with hysteresis

19. Define hysteresis.
20. Describe how hysteresis is introduced into a comparator circuit.
21. What is another name for a comparator with hysteresis?
22. What practical application does hysteresis provide?

5-5 Window comparator circuits

23. Describe the operation of a window comparator.
24. Describe the circuit for a window comparator that turns on an LED when the voltage is outside of the window.
25. How do you reverse the action of the comparator circuit described in Question 24?
26. Suggest one practical application for a window comparator.

5-6 True comparator integrated circuits

27. What are the limitations of an op-amp used as a comparator?
28. How does a true comparator overcome the limitations of an op-amp comparator?
29. Give the device number for a true comparator.
30. How does the switching time of a true comparator compare with that of an op-amp?
31. How does the symbol of a true comparator differ from that of an op amp?
32. What are the typical supply voltages for a true comparator?
33. What are the output voltage levels for a true comparator operating from the recommended supplies?
34. How can you make an LED come on for a low output of a true comparator?

Design and Analysis Problems

5-2 Zero-crossing comparator circuits

1. Design a simple zero-crossing comparator to turn on an LED when the voltage is greater than zero. Use a TL081C op-amp with a ±12 V supply.
2. Repeat Question 1, but with the LED turning on when the voltage is less than zero.

5-3 Level-sensing comparator circuits

3. Design a simple comparator circuit using a TL081C op-amp to turn on an LED warning light when the voltage of a car battery (normally 13.4 V) drops below 11.5 V. Use a zener diode for biasing. The comparator will be powered from a separate ±12 V supply.
4. Design a simple comparator circuit to turn on an LED warning light when the voltage of a nominal 9 V battery drops below 8.2 V. The entire circuit should be powered from the battery. Use a TL081C op-amp configured as a comparator.
5. Analyze the comparator circuit in Figure 5-25 to determine the minimum voltage before the LED comes on.

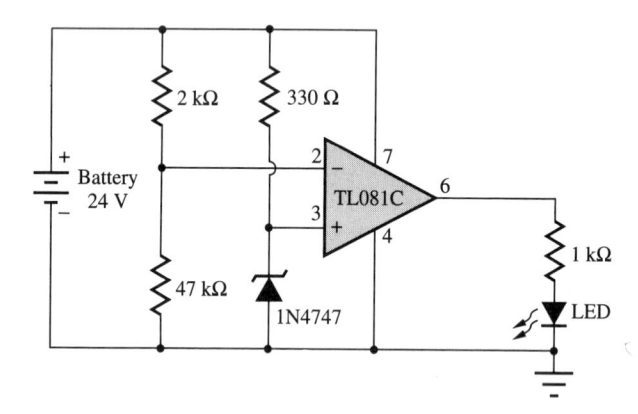

Figure 5-25

5-4 Comparators with hysteresis

6. Design a Schmitt trigger circuit using an op-amp to turn on an LED indicator when the input voltage exceeds 5 V. To avoid the possibility of spurious triggering from noise, incorporate hysteresis into the circuit so that the upper trigger point will be 0.5 V above the reference voltage. Calculate the lower trigger point. Use a voltage divider for biasing. The op-amp will be powered from a single +12 V supply.

5-5 Window comparators

7. Design a window comparator using op-amps to cause an LED to turn on whenever the input voltage is between +5 and +7 V. Use voltage dividers for biasing. Use a single +12 V supply.

8. Design a window comparator using op-amps to cause an LED to turn on whenever the input voltage is between +4.3 and +6.2 V. Use zener diodes for biasing. Use a single +12 V supply.
9. Determine the minimum and maximum window voltages for the window comparator shown in Figure 5-26.

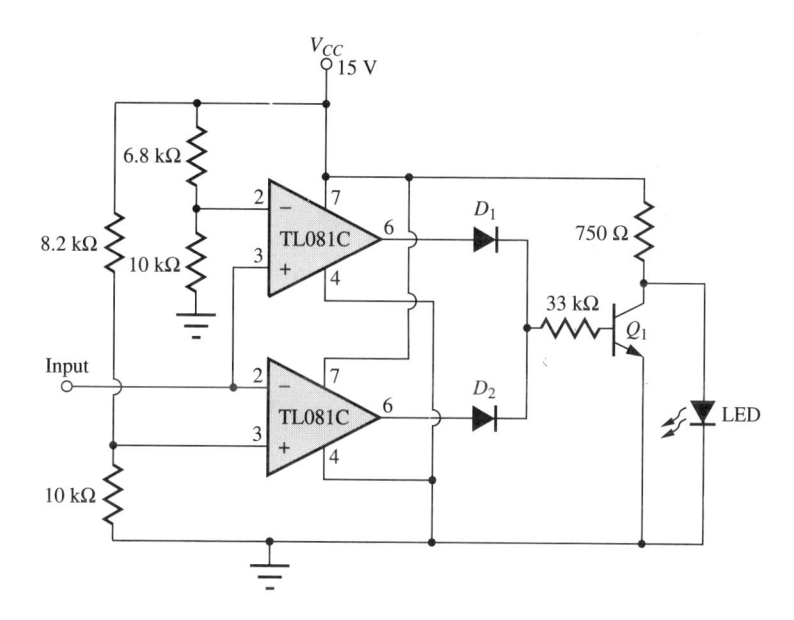

Figure 5-26

5-6 True comparator integrated circuits

10. Design a simple comparator circuit to turn on an LED warning light when the voltage of a nominal 9 V battery drops below 8.2 V. The entire circuit should be powered from the battery. Use a TL810C comparator.
11. Design a Schmitt trigger circuit using a TL810C comparator to turn on an LED indicator when the input voltage exceeds 3.3 V. Include hysteresis so that the upper trigger point will be 0.1 V above the reference voltage. Calculate the lower trigger point. Use a zener diode for biasing. Use power supplies of +12 V and −6 V.
12. Design a window detector using two TL810C comparators to turn off an LED when the input voltage is between 2.5 and 5.0 V. Use a single 12 V supply.

6 Active Filters

OUTLINE

6-1 Basic Filter Types
6-2 Low-Pass and High-Pass Active Filters
6-3 Bandpass and Band-Rejection Active Filters
6-4 Switched-Capacitor Filters

KEY TERMS

Electronic filter
Frequency response
Phase response
Critical frequency (3 dB frequency)
Passive filter
Active filter
Decade
Bode plot
Roll-off
Poles
Low-pass filter
High-pass filter
Bandpass filter
Band-rejection filter
All-pass filter
Bandwidth
Butterworth response
Bessel response
Chebyshev response
Cauer response
Broadband filter
Narrow-band filter
Center frequency
Quality factor
Switched-capacitor filter
Universal filter
State-variable filter

OBJECTIVES

On completion of this chapter, the reader should be able to:

- list the various types of filters and describe their frequency response.
- relate the frequency response of a filter to the number of poles, and prepare Bode plots for any type of filter.
- name the different filter responses commonly encountered, describe their respective frequency and phase characteristics, and suggest advantages of each type.
- design multipole Butterworth low-pass and high-pass active filters using unity-gain design and equal-component design.
- construct low-Q bandpass and band-rejection filters using combinations of Butterworth low-pass and high-pass filters.
- design high-Q bandpass and band-rejection filters.

- explain the operation of switched-capacitor filters and design low-pass or high-pass filters using the MF6 switched-capacitor filter integrated circuit.
- describe state-variable filters and explain the advantages of this type of filter as a switched-capacitor filter.

Introduction

Previous chapters have introduced a variety of simple but important op-amp applications. One application of op-amps that was not mentioned is their use in active filters. This is a major application that warrants its own chapter.

A filter is simply a circuit that passes signals within a certain frequency range and attenuates all other frequencies. Filters are widely used in electronics, particularly in the communications field. A simple transistor AM or FM radio receiver may have a dozen or more filter circuits. Telecommunications systems consist mostly of filter circuits. Even the standard op-amp integrated circuit contains a low-pass filter to cut the high-frequency gain, hence minimizing device oscillation.

Basic filters are generally simple combinations of resistors, capacitors, and inductors. Active filters are the basic filters optimized through the addition of op-amps. The addition of an op-amp to a filter circuit provides such improvements as supplying gain to overcome circuit loss, increasing the input impedance so that the filter does not load the signal source or previous filter stage, and allowing higher output power. Op-amps allow multistage filters with sharp cutoff characteristics to be produced simply and efficiently without the need of bulky inductors. They provide a flexibility impossible to achieve with passive components only.

Filter design is a whole subfield of electronics, and it is impossible, in a single chapter, to do more than introduce a few basic filter circuits. We will see how to use standard recipes to design multipole filters of various types, and we will introduce other recipes to design bandpass and band-rejection filters.

Active filter circuits using op-amps are easy to design, but, unfortunately, they require numerous capacitors and resistors, and their performance is critically dependent on these component values. Integrated circuit technology provides an alternative—the switched-capacitor filter. Switched-capacitor filters have a complete filter circuit on a single chip with few, if any, external components required. Universal filters are available, in which a single integrated circuit can be configured as any type of filter with any type of response.

6-1 Basic Filter Types

Electronic filters are circuits that pass all signals within a certain frequency range and strongly attenuate all signals with frequencies outside of that range. There are four major classes of filters: low-pass filters, high-pass filters, bandpass filters, and band-rejection filters. The idealized frequency response for each of these

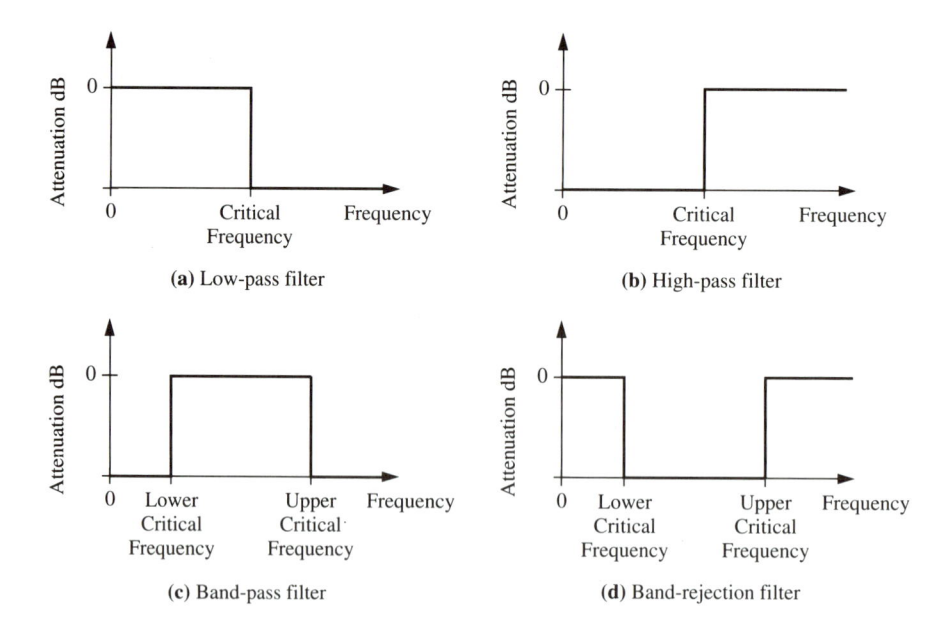

Figure 6-1 Frequency response of ideal filters.

(a) Low-pass filter
(b) High-pass filter
(c) Band-pass filter
(d) Band-rejection filter

filters is illustrated in Figure 6-1. The **frequency response** of a filter represents the attenuation of a filter at different frequencies, and is usually presented as a plot of the number of decibels of attenuation against the logarithm of the frequency. A filter also changes the phase of the output signal with respect to the input. The variation of the phase change with frequency is known as the **phase response** of the filter.

Filters may be either passive or active. **Passive filters** normally are constructed of networks of resistors and capacitors (*R-C* filters) or inductors and capacitors (*L-C* filters). **Active filters** use operational amplifiers, or other active devices, to increase the performance of simple *R-C* filters. This section will discuss the design and performance of simple passive *R-C* filters. Active filters will be introduced in the next section.

The actual frequency response of any real filter will differ considerably from the ideal response shown in Figure 6-1. There will be a gradual knee in the frequency response near the cutoff frequency, followed by a constant increase in attenuation, as shown in Figure 6-2. Notice that attenuation is plotted increasing downward. This constant increase in attenuation is called **roll-off** and is usually expressed in units of decibels per decade. A **decade** in frequency corresponds to an increase or decrease in frequency by a factor of 10. Occasionally, roll-off may be expressed in units of decibels per octave, where an octave is an increase or decrease in frequency by a factor of 2. Decibels per decade are more commonly used, and hence will be used exclusively in this book.

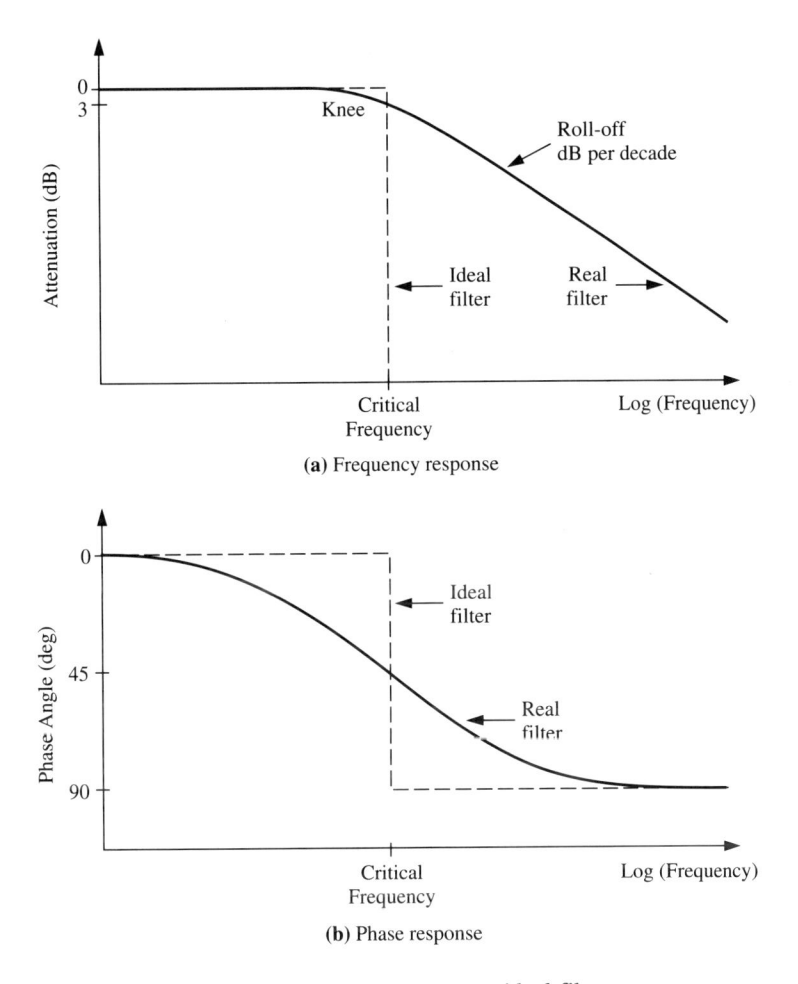

Figure 6-2 Real filter response versus ideal filter response.

Bode Plots

Frequency response curves are often represented by a straight-line approximation called a **Bode plot.** The Bode plots for two simple filters are illustrated in Figure 6-3. Figure 6-3(a) shows the Bode plot for a typical low-pass filter. The plot consists of two straight lines, one horizontal and one sloping. The horizontal straight line is drawn at 0 dB attenuation from the left side of the graph to the critical frequency of the filter and represents the frequency response for the range over which the filter does not attenuate the input signal (the filter passband). The critical frequency, plotted on the horizontal 0 dB line, marks the limit of the passband of the filter and represents the breakpoint in the frequency response. The sloping straight line starts at the breakpoint and drops uniformly with a slope

214 6 Active Filters

(a) Bode plot for a low-pass filter

(b) Bode plot for a high-pass filter

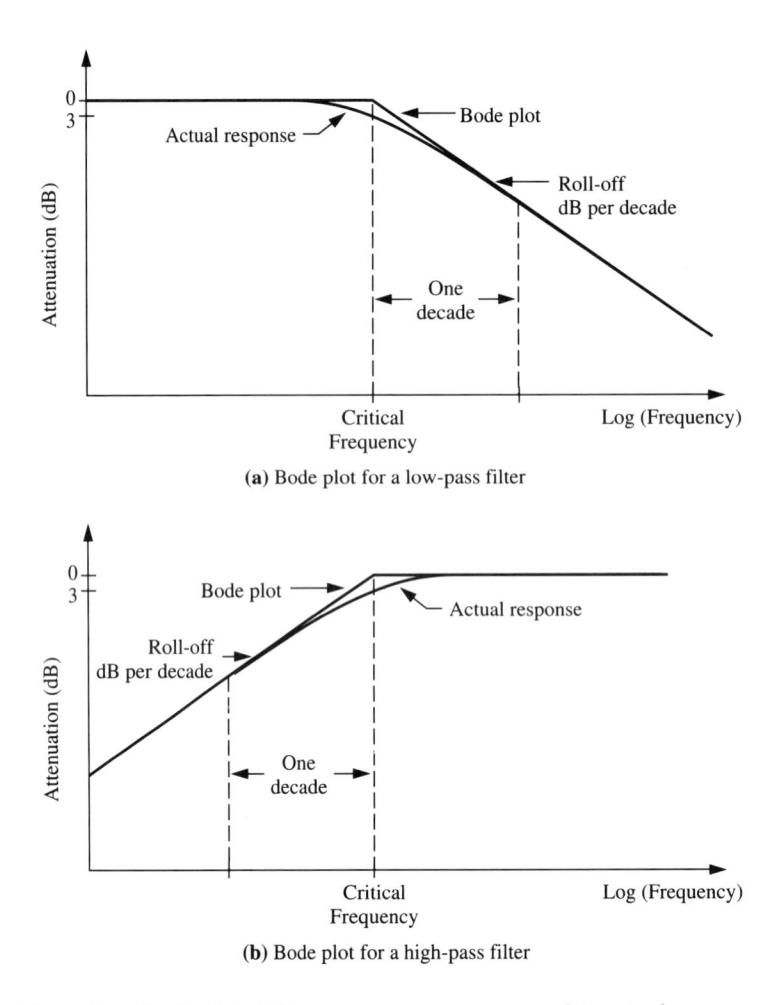

Figure 6-3 Bode plot of frequency response compared to actual response.

equal to the roll-off in decibels per decade, representing the frequency response for the range over which the filter attenuates the input signal (the filter stopband). Figure 6-3(b) shows the Bode plot for a typical high-pass filter, and is produced in a similar manner to that for the low-pass filter. The Bode plot for a bandpass filter will be discussed in Example 6-2.

For the simple *R-C* filters discussed in this section, the **critical frequency** is the frequency at which the attenuation of the real filter is 3 dB, corresponding to the filter passing 0.707 of the input signal. This frequency is sometimes called the **3 dB frequency.** We have seen the term "3 dB frequency" in Chapter 2, but there it was introduced as "−3 dB frequency." In Chapter 2 we were discussing amplifier frequency response, where the bandwidth was determined by the −3 dB points in amplifier *gain*. In this chapter, we are discussing filter frequency response, and are using as our critical frequency the point where the filter *attenuation* is +3 dB.

A Bode plot makes it easy to plot the frequency response of any filter. Although we have not yet discussed actual filter responses, if we are given the critical frequency of the filter, the roll-off rate in dB/decade, and the type of filter, we can plot the Bode approximation to the frequency response. The next two examples will illustrate this.

EXAMPLE 6-1

Sketch the frequency response of a low-pass filter with a critical frequency of 8000 Hz and an attenuation of 40 dB/decade using a Bode plot. Compare the frequency response with the response of an idea low-pass filter.

Solution

The frequency response of a filter is a plot of the attenuation in decibels against the logarithm of the frequency.

In the Bode plot of a low-pass filter, the attenuation is 0 dB up to the critical frequency. This portion of the Bode plot is easy to produce. First mark the critical frequency of 8 kHz at 0 dB. Now draw a straight line from the left-hand axis to the critical frequency plotted on the 0 dB line.

To draw the portion of the Bode plot after the breakpoint, plot a point one decade to the right of the critical frequency, at a frequency of 80 kHz and 40 dB

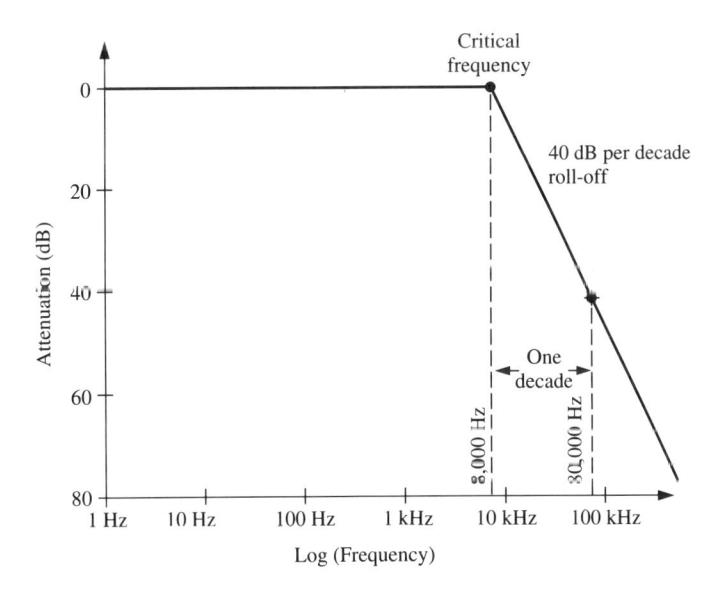

Figure 6-4

down (since the slope is 40 dB per decade). Draw a line from the critical frequency at 0 dB through the point at 80 kHz and 40 dB, and extend it off the graph.

The Bode plot is now complete and is shown in Figure 6-4. The ideal filter response drops off with a vertical slope at the critical frequency. This is also shown in the figure.

EXAMPLE 6-2

Sketch the Bode plot of the frequency response of a bandpass filter with a lower critical frequency of 200 Hz and an upper critical frequency of 5000 Hz. The roll-off is 60 dB per decade for the low frequency roll-off and 20 dB per decade for the high frequency roll-off.

Solution

In the case of a Bode plot of a bandpass filter, the horizontal 0 dB line extends from the lower critical frequency to the upper critical frequency.

First mark the two critical frequencies, 200 Hz and 5000 Hz, on the 0 dB axis of the plot. Join these two points with a line. This line represents the unattenuated portion of the filter response.

To draw the low frequency roll-off portion of the Bode plot, plot a point one decade to the left of the lower critical frequency at 20 Hz and 60 dB down (since

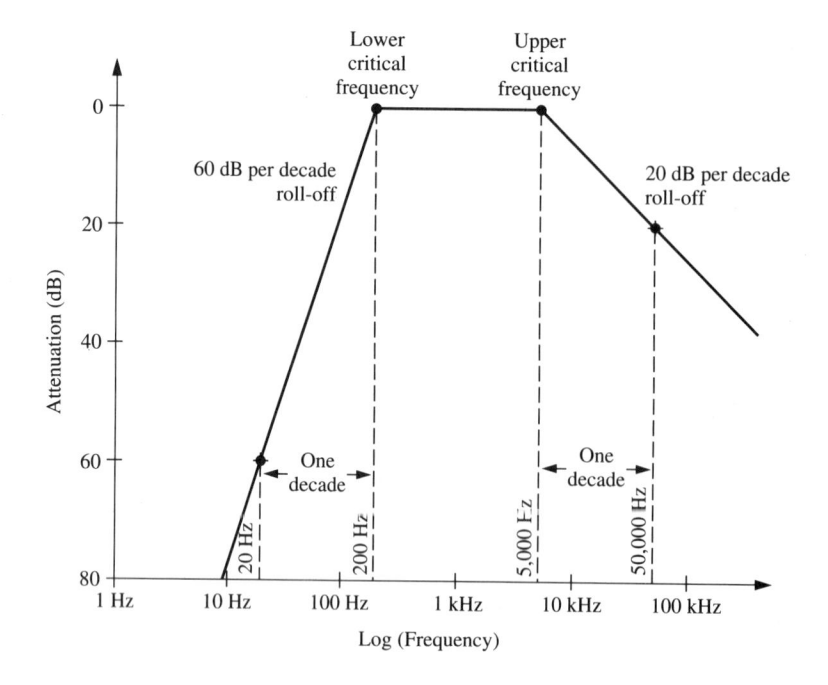

Figure 6-5

the low frequency roll-off is 60 dB per decade). Draw a straight line from the lower critical frequency plotted on the 0 dB line through the point just plotted, and continue off the left-hand side of the plot.

To draw the high frequency roll-off portion of the Bode plot, plot a point one decade to the right of the upper critical frequency at 50 kHz and 20 dB down (since the high frequency roll-off is 20 dB per decade). Draw a straight line from the upper critical frequency plotted on the 0 dB line through the point just plotted and off the right-hand side of the plot.

The Bode plot is now complete and is shown in Figure 6-5.

Although we have not yet discussed bandpass filters, it was easy to produce the Bode plot of the frequency response of such a filter.

Filter Response from Bode Plots

Given an actual filter circuit, it is often necessary to determine the attenuation of the filter for some specific frequency. The Bode plot of the filter frequency response provides an easy way to estimate the attenuation at any frequency. Figure 6-6 shows the typical Bode plot for a low-pass filter and illustrates the various quantities that we will need to determine the attenuation.

We want the filter attenuation at some frequency f and are given the critical frequency f_o and the roll-off in dB per decade. Determine the actual attenuation at f_{dec}, one decade above f_o, by converting the decibel roll-off to a straight attenuation (*not decibels*). Recall that the decibel is defined as 20 times the logarithm of the attenuation. Divide the dB attenuation by 20 and take the antilogarithm. For example, a roll-off of 20 dB/decade gives an attenuation of 10 one decade above the critical frequency. Use the ratios as illustrated in Figure 6-6 to obtain the following equation:

$$\frac{\text{attenuation at } f}{\text{attenuation at } f_{\text{dec}}} = \frac{f - f_o}{f_{\text{dec}} - f_o} \quad (6\text{-}1)$$

Now solve the equation for the attenuation at f.

$$\text{attenuation at } f = \frac{f - f_o}{f_{\text{dec}} - f_o} \times \text{attenuation at } f_{\text{dec}} \quad (6\text{-}2)$$

Equation 6-2 gives the linear attenuation. Convert it back to decibels by taking 20 times the logarithm of the attenuation. Although only a low-pass filter is illustrated in Figure 6-6, Equation 6-2 works for either low-pass or high-pass filters.

Frequently the design specifications for a filter circuit will not give the roll-off, but the desired attenuation at some frequency. To design a suitable filter, however, the roll-off will be required and must be calculated. The Bode plot provides a convenient procedure for this as well. To determine the roll off in dB per decade given the critical frequency f_o and the attenuation at any frequency f, solve Equation 6-1 for the attenuation at f_{dec}:

$$\text{attenuation at } f_{\text{dec}} = \frac{f_{\text{dec}} - f_o}{f - f_o} \times \text{attenuation at } f \quad (6\text{-}3)$$

218   6   Active Filters

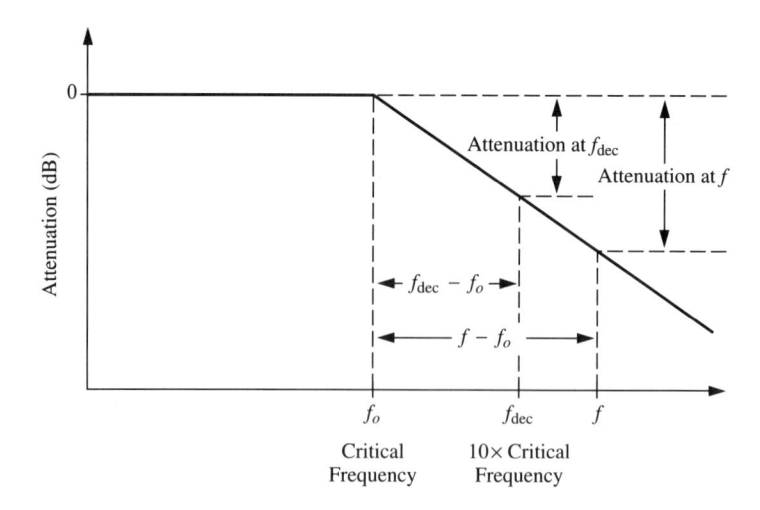

Figure 6-6   Use of Bode plot to determine attenuation at any frequency.

Note that we must use the linear attenuation at f, not the dB attenuation. This equation also works for either low-pass or high-pass filters. Convert the calculated attenuation to decibels, using the definition of decibels as 20 times the logarithm of the attenuation to obtain the roll-off in dB per decade.

EXAMPLE 6-3

A low-pass filter has a critical frequency of 150 Hz and an attenuation of 45 dB at 200 Hz.

(a) Calculate the roll-off in dB per decade.
(b) Determine the dB attenuation of a 250 Hz signal.

Solution

(a) We will need to know the linear attenuation at $f = 200$ Hz. Dividing 45 dB by 20 and taking the antilogarithm gives a value of 177.8. Since we know the critical frequency f_o is 150 Hz, the decade frequency f_{dec} is 1500 Hz, and we can solve Equation 6-3 for the roll-off.

$$\text{attenuation at } f_{dec} = \frac{f_{dec} - f_o}{f - f_o} \times \text{attenuation at } f$$

$$= \frac{1500 \text{ Hz} - 150 \text{ Hz}}{200 \text{ Hz} - 150 \text{ Hz}} \times 177.8 = 4800$$

Now convert this to decibels.

    20 log 4800 = 73.6 dB

The roll-off of the filter is 73.6 dB per decade.

(b) We can use Equation 6-2 to calculate the attenuation for a frequency $f = 250$ Hz. The attenuation at the decade frequency $f_{dec} = 1500$ Hz was determined in the previous calculation as 4800.

$$\text{attenuation at } f = \frac{f - f_o}{f_{dec} - f_o} \times \text{attenuation at } f_{dec}$$

$$= \frac{1500 \text{ Hz} - 150 \text{ Hz}}{250 \text{ Hz} - 150 \text{ Hz}} \times 4800 = 64,800$$

Now convert this to decibels.

$$20 \log 64,800 = 96.2 \text{ dB}$$

The filter will attenuate a 250 Hz signal by 96.2 dB.

Poles in Filters

The frequency response of a real *R-C* filter is usually quite a poor approximation to the ideal response shown in Figure 6-1. For a filter consisting of a single resistor and capacitor, the roll-off is only 20 dB per decade. This means that signals at a frequency ten times higher than the critical frequency in a low-pass filter, or ten times lower than the critical frequency in a high-pass filter, would only be attenuated by a factor of 10. For example, a low-pass filter with a critical frequency of 1 kHz would pass a 1 V signal at a frequency of 10 kHz with an amplitude of 0.1 V. This is not very effective rejection.

The rate of roll-off of a filter is determined by the number of reactive, or frequency-dependent, elements in the filter, that is, the number of inductors and capacitors. For a filter with a single reactive element (inductor or capacitor), the roll-off is 20 dB per decade. If there were two reactive elements in the filter, the roll-off would be 40 dB per decade. Three reactive elements would give 60 dB per decade, and so forth. Obviously, we must insert more reactive elements to obtain a sharper filter cutoff. For example, if we used three reactive elements, the roll-off would be 60 dB per decade, and only $\frac{1}{1000}$ of the signal would get through the filter at a frequency ten times higher than the critical frequency. The number of reactive elements is often referred to as the number of **poles** in the filter. A filter with four reactive elements would be described as a 4 pole filter. Figure 6-7 compares the Bode plots of filters with 1 pole, 2 poles, 3 poles, and 4 poles.

Several procedures can be used to obtain more poles in a filter. The simplest is to connect several single-pole *R-C* filters in series. This is not a particularly useful procedure. Each filter added in series places a load on the previous filter stage, causing unwanted attenuation and a shift in filter frequency.

A second procedure is to replace the resistor in a simple *R-C* filter with an inductor. The inductor is a reactive element which adds a second pole to the simple filter. Because the resistors can be eliminated, the attenuation and loading in a series of filters can be greatly reduced. This is a practical approach, and, traditionally, this was the procedure used in older communication systems. Entire

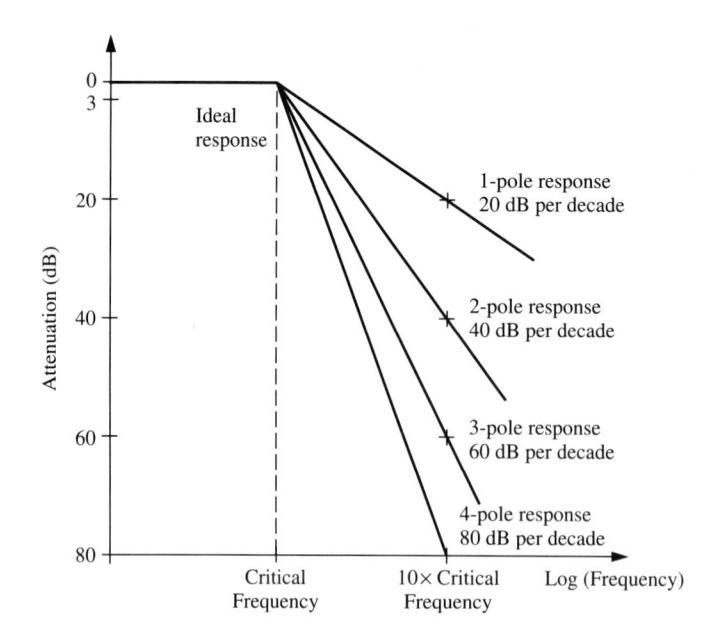

Figure 6-7 Bode plots for filters with 1 pole, 2 poles, 3 poles, and 4 poles.

books are available on the topic of the design of *L-C* filters of all types. There are, however, two limitations to their use. First, the inductors tend to be large and bulky, particularly for low frequency use, and, second, the inductors are relatively expensive to design and construct.

The introduction of integrated circuits provides a third, and the most widely used, procedure for increasing the poles in a filter. Simple inexpensive *R-C* filters are used, with op-amps between filter stages to provide buffering and preventing one stage from loading the previous. Gain can be introduced to offset any filter losses. These circuits are called *active filters* because they contain an active circuit element (the op-amp). The remainder of this section will discuss simple passive filters. The next section will describe the design of active filters.

Low-Pass Filter

Low-pass filters are filter circuits that pass all frequencies below the critical frequency with zero attenuation and reject all higher frequencies. Audio systems use low-pass filters to limit the bandwidth of an amplifier, thereby reducing the noise amplified. As discussed in Chapter 2, standard operational amplifiers have an effective low-pass filter in the form of the compensating capacitor to limit the bandwidth and prevent oscillations from high gain at high frequencies. Phase-locked loops, discussed in Chapter 11, use low-pass filters to eliminate high frequency signal components in the error amplifier output. Telephone systems and other communication systems use low-pass filters to limit the bandwidth of the input signal.

6-1 Basic Filter Types

The circuit for a simple low-pass filter consisting of a resistor and capacitor is shown in Figure 6-8(a). The frequency response for this low-pass filter, plotted in decibels of attenuation against the logarithm of the frequency, is shown in Figure 6-8(b) and is compared with the frequency response of an ideal low-pass filter (dashed lines).

The resistor and capacitor act as a voltage divider where the reactance of the capacitor is given as $1/2\pi fC$. At low frequencies, the reactance of the capacitor is much larger than the resistance of the resistor; hence, most of the input signal is dropped across the capacitor and appears at the output. At high frequencies, the reactance of the capacitor is much smaller than the resistance of the resistor; hence, most of the input signal is dropped across the resistor, and there is very little signal at the output. At intermediate frequencies, the output signal is

$$V_{out} = \frac{X_C}{\sqrt{R^2 + X_C^2}} \times V_{in}$$

$$= \frac{1/2\pi fC}{\sqrt{R^2 + 1/(2\pi fC)^2}} \times V_{in}$$

$$= \frac{1}{\sqrt{(2\pi fRC)^2 + 1}} \times V_{in} \tag{6-4}$$

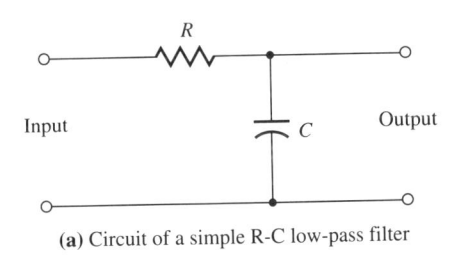

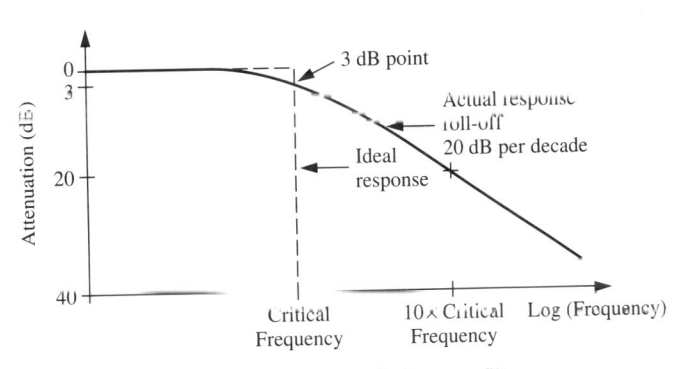

(a) Circuit of a simple R-C low-pass filter

(b) Frequency response of a low-pass filter

Figure 6-8 Low-pass filter.

There is one special case, namely when the resistance is equal to the capacitive reactance:

$$R = \frac{1}{2\pi f C}$$

This defines the critical frequency f_o.

$$f_o = \frac{1}{2\pi RC} \qquad (6\text{-}5)$$

If we substitute this value into the general equation for the filter output voltage (Equation 6-4), we find that $V_{out} = 0.7071 V_{in}$. If we now convert the voltage ratio V_{out}/V_{in} into decibels, we find that the ratio is $20 \log 0.7071 = -3.010$ dB. The attenuation of the filter at this frequency is almost exactly 3 dB.

The frequency response plotted in Figure 6-8(b) is determined by Equation 6-4. The decibel attenuation is

$$\text{attenuation (dB)} = -20 \log \frac{1}{\sqrt{(2\pi fRC)^2 + 1}}$$

$$= 10 \log((2\pi fRC)^2 + 1) \qquad (6\text{-}6)$$

At frequencies much lower than the critical frequency, the first term in parentheses is much smaller than 1, and the attenuation is simply the logarithm of 1, that is, zero. At frequencies much higher than the critical frequency, the first term is much larger than the second term, and the attenuation in decibels is equal to $20 \log(2\pi fRC)$. If the frequency increases by a factor of 10 (that is, by one decade), the attenuation increases by a factor of 20 dB. This type of filter has a roll-off of 20 dB per decade.

A simple low-pass filter with a roll-off of 20 dB per decade has a frequency response that is a very poor approximation to the ideal low-pass filter. At a frequency ten times higher than the critical frequency, the input signal is only attenuated by $\frac{1}{10}$. For most applications, this is clearly not satisfactory. The solution to this problem is obvious—we must add more poles. The procedure for doing this will be developed in the next section when we introduce active filters.

High-Pass Filter

High-pass filters are filter circuits that pass all frequencies above the critical frequency with zero attenuation and reject all lower frequencies. High-pass filters are used in single-sideband communication systems to eliminate the lower sideband and carrier, in stereo systems to separate the stereo information from the subcarrier, and in television receivers to separate the sound from the composite video signal.

The circuit for a simple high-pass filter consisting of a capacitor and a resistor is shown in Figure 6-9(a). The frequency response for this high-pass filter, plotted

6-1 Basic Filter Types

(a) Circuit of a simple R-C high-pass filter

(b) Frequency response of a high-pass filter

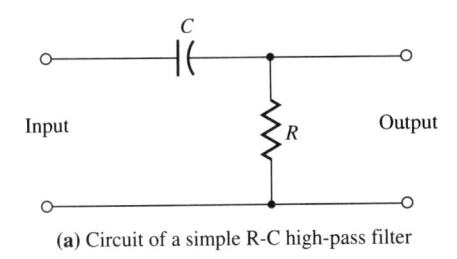

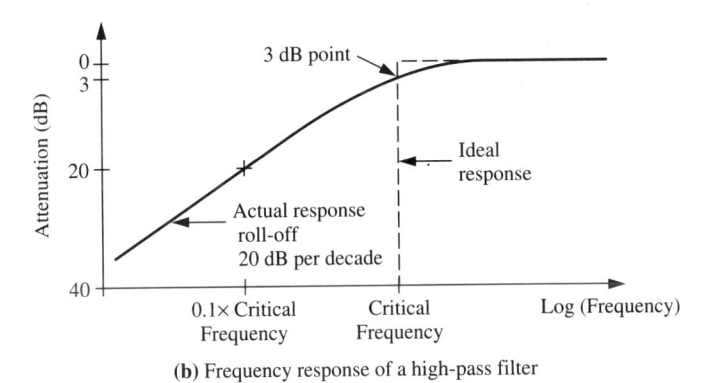

Figure 6-9 High-pass filter.

in decibels of attenuation against the logarithm of the frequency, is shown in Figure 6-9(b) and is compared with the frequency response of an ideal high-pass filter (dashed lines).

As in the case of the low-pass filter, the resistor and capacitor act as a voltage divider. At low frequencies, the reactance of the capacitor is much larger than the resistance of the resistor. Most of the input signal is dropped across the capacitor and does not reach the output. At high frequencies, the reactance of the capacitor is much smaller than the resistance of the resistor. Most of the input signal will be dropped across the resistor, and the output signal will be approximately equal to the input signal. At intermediate frequencies the output signal is given by

$$V_{out} = \frac{R}{\sqrt{R^2 + X_C^2}} \times V_{in}$$

$$= \frac{R}{\sqrt{R^2 + 1/(2\pi fC)^2}} \times V_{in}$$

$$= \frac{2\pi fRC}{\sqrt{(2\pi fRC)^2 + 1}} \times V_{in} \qquad (6\text{-}7)$$

As for the low-pass filter, the critical frequency of the high-pass filter occurs when the capacitive reactance equals the resistance. This is simply Equation 6-5 as derived earlier:

$$f_o = \frac{1}{2\pi RC} \tag{6-5}$$

If we substitute this value into the general equation for the filter output voltage (Equation 6-7), we find that $V_{out} = 0.7071\, V_{in}$, exactly the same as for the low-pass filter.

The frequency response plotted in Figure 6-9(b) is determined by Equation 6-7. The decibel attenuation is

$$\text{attenuation (dB)} = 10\log((2\pi fRC)^2 + 1) - 20\log(2\pi fRC) \tag{6-8}$$

At frequencies much lower than the critical frequency, the first term is approximately zero, and the attenuation is determined by the second term. The attenuation at low frequencies is proportional to $-20\log(2\pi fRC)$, increasing as the frequency decreases. At frequencies much larger than the critical frequency, the first term is approximately equal to the second term and gives approximately zero attenuation. This type of filter, like the low-pass filter, has a roll-off of 20 dB per decade, except that, in the case of the high-pass filter, the roll-off is toward lower frequencies.

The simple high-pass filter has the same problem as the low-pass filter, namely that the roll-off is very slow since it is only a single-pole filter. The solution is also the same as for the low-pass filter—introduce more poles. Active high-pass filters will be discussed in the next section.

Bandpass Filter

Bandpass filters are filter circuits that pass only signals with frequencies greater than a lower critical frequency and less than an upper critical frequency. Signals with frequencies outside of this range, that is, above the upper critical frequency or below the lower critical frequency, are attenuated or rejected. AM and FM radio receivers and television receivers all use bandpass filters to select a specific channel or station. In fact, bandpass filters make radio communication possible.

The most common example of a bandpass filter is the tuned *L-C* tank circuit used in all types of radio receivers. Although this type of circuit is used as a bandpass filter, it is strictly tuned for a single frequency, and all other frequencies are attenuated by varying degrees. This type of filter is suitable for communication applications where the bandwidth is typically less than 1% of the center frequency. For example, an AM radio at a frequency of 1 MHz has a bandwidth of only 10 kHz. Tuned circuits are generally not suitable for use at audio frequencies. Because most linear integrated circuits work primarily in the audio range, we will restrict our discussion of bandpass filters to filters that work in this range, and hence will not further discuss *L-C* tuned circuits.

6-1 Basic Filter Types

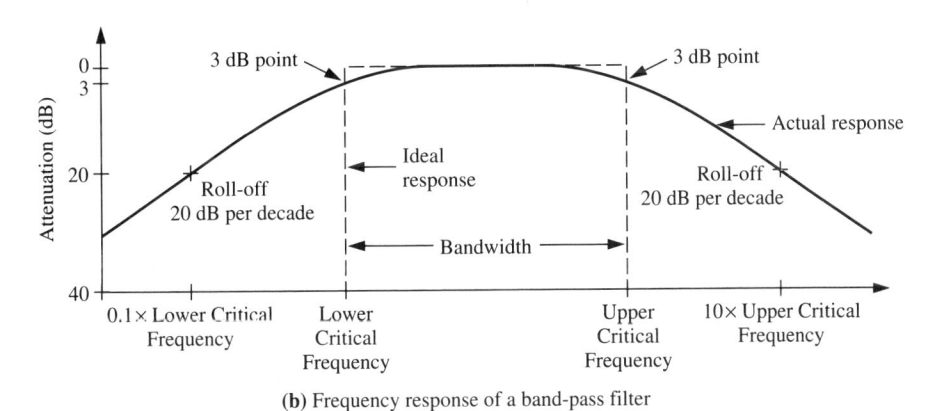

Figure 6-10 Bandpass filter.

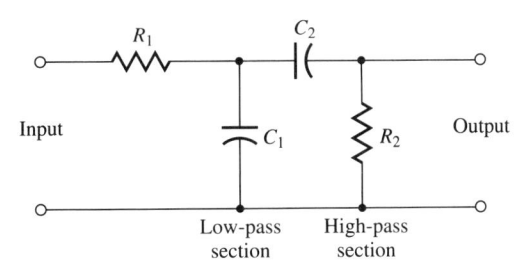

(a) Circuit of a simple R-C band-pass filter

(b) Frequency response of a band-pass filter

The circuit for a simple audio frequency bandpass filter consists of a high-pass filter in series with a low-pass filter as shown in Figure 6-10(a). The frequency response for this bandpass filter, plotted in decibels of attenuation against the logarithm of the frequency, is shown in Figure 6-10(b) and is compared with the frequency response of an ideal bandpass filter (dashed lines).

The high-pass filter rejects all frequencies less than its critical frequency. The critical frequency of the high-pass filter f_{oHP} determines the lower critical frequency f_{oL} of the bandpass filter. The low-pass filter rejects all frequencies greater than its critical frequency; hence, the critical frequency of the low-pass filter f_{oLP} determines the upper critical frequency f_{oU} of the bandpass filter. The order of the filters is not important, but the second filter in the series connection must have a high impedance so as not to load the first filter. The **bandwidth** of the bandpass filter is the range of frequencies that will be passed by the filter and is given by

$$\text{bandwidth} = f_{oU} - f_{oL} = f_{oLP} - f_{oHP} \tag{6-9}$$

The frequency response in Figure 6-10(b) is determined from the individual frequency responses of the low-pass filter and the high-pass filter. This simple bandpass filter has the usual problem of very slow roll-off (20 dB per decade), since it comprises two single-pole filters acting independently. The solution is the same as for the other filters discussed—introduce more poles.

Band-Rejection Filter

Band-rejection filters are just the opposite of bandpass filters, rejecting only those frequencies within a specified range and passing all frequencies above an upper critical frequency and below a lower critical frequency. Band-rejection filters are used to eliminate unwanted signals. They are used in FM stereo receivers to remove the 19 kHz stereo subcarrier. They are used in single-sideband transmitters to suppress the carrier. One early method of scrambling the signal for pay TV involved adding an interfering signal in the middle of the picture band. A band-rejection filter was used to remove this signal to unscramble the picture.

Like the bandpass filter, the most common example of a band-rejection filter is the tuned *L-C* tank circuit. Like the bandpass filter, this type of circuit is satisfactory for communication applications but not for the audio range. We will restrict our discussion of band-rejection filters to the audio range.

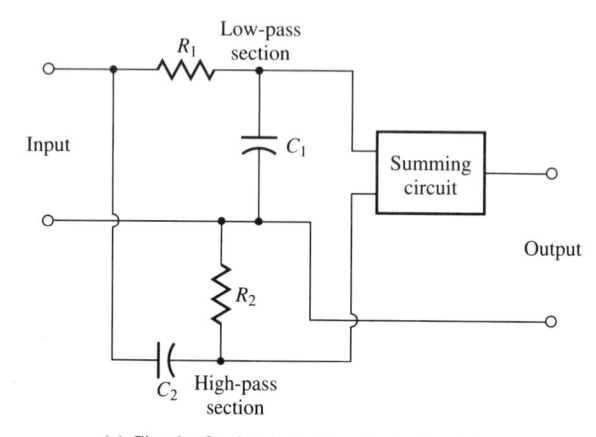

(a) Circuit of a simple R-C band-rejection filter

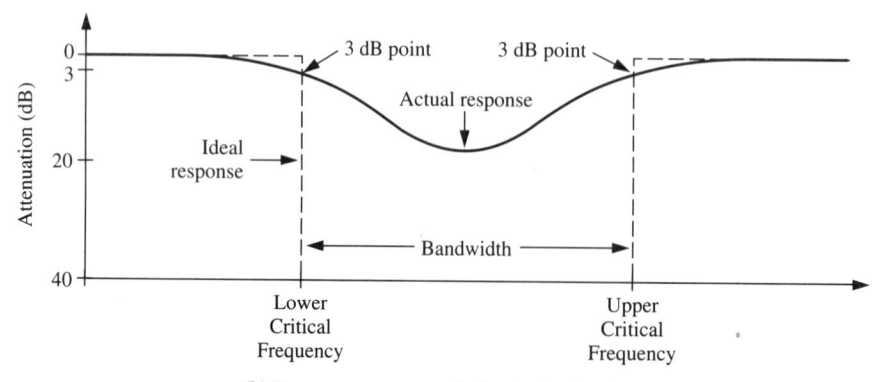

(b) Frequency response of a band-rejection filter

Figure 6-11 Band-rejection filter.

The circuit for a simple audio frequency band-rejection filter is quite different from a bandpass filter. Like a bandpass filter, it consists of a high-pass filter and a low-pass filter as shown in Figure 6-11(a), only now the two filters are connected in parallel through a summing circuit. A high-input-impedance summing circuit is necessary to isolate the two filters. The frequency response for this band-rejection filter, plotted in decibels of attenuation against the logarithm of the frequency, is shown in Figure 6-11(b) and is compared with the frequency response of an ideal band-rejection filter (dashed lines).

The high-pass filter rejects all frequencies less than its critical frequency. The critical frequency of the high-pass filter f_{oHP} determines the upper critical frequency of the band-rejection filter f_{oU}. The low-pass filter rejects all frequencies greater than its critical frequency; hence, the critical frequency of the low-pass filter f_{oLP} determines the lower critical frequency of the band-rejection filter f_{oL} (exactly the reverse of the bandpass filter). The bandwidth of the band-rejection filter is the range of frequencies that will be rejected by the filter:

$$\text{bandwidth} = f_{oU} - f_{oL} = f_{oHP} - f_{oLP} \tag{6-10}$$

The frequency response in Figure 6-8(b) is determined from the individual frequency responses of the low-pass filter and the high-pass filter. Like the bandpass filter, it has the problem of very slow roll-off (20 dB per decade) since it also comprises two single-pole filters acting independently.

All-Pass Filter

One other type of filter is occasionally encountered—the **all-pass filter.** As its name implies, this filter passes all frequencies unattenuated. Although the amplitude of the input frequencies are not affected, the all-pass filter does produce a phase shift at its critical frequency. All-pass filters find application in quadrature detectors where sine and cosine waves are both required, and in applications where a phase shift is necessary to cancel unwanted phase shifts produced by other circuitry or transmission lines. All-pass filters are beyond the scope of the present discussion and will not be considered further.

6-2 Low-Pass and High-Pass Active Filters

The previous section introduced the various types of filters. These were all single pole R-C filters, and all had the problem of very slow roll-off. The solution to slow roll-off, suggested in the previous section, is to introduce more poles into the filter, either by the use of L-C elements or by the use of active filters. The traditional approach of using inductors and capacitors leads to an expensive and bulky circuit. The modern solution is to use active filters with simple resistors and capacitors. The active filters use operational amplifiers to provide high-impedance isolation between filter stages with amplification to overcome any filter losses, and can achieve, at very low cost, compact efficient filters with any desired frequency response.

The design of multipole active filters is a complex topic. Various circuits can be used, and the solution of the circuit equations for any one of these can be quite tedious. Different combinations of components will give different filter frequency and phase response characteristics. Because of the importance of filters in electronic systems, extensive work has been done on multipole active filter design. Entire books are written on the topic. Recipes have been developed for designing filters with different response characteristics. Modern filter design involves simply choosing a filter response that gives the most suitable frequency and phase characteristics for the particular application, then following the design recipe for that specific filter response.

In this section, we will introduce several basic recipes and demonstrate how to use them to design active filters. We will not derive the basic filter design equations, however, since our object is to investigate applications of integrated circuits, not to develop filter design theory. Readers interested in the design of specific filter circuits are referred to either a book on circuit theory or a book specifically on filter design.

Four types of filter responses are commonly used: Butterworth, Bessel, Chebyshev, and Cauer. The **Butterworth filter response** is characterized by a flat frequency response up to the critical frequency, followed by a smooth roll-off of 20 dB per decade per pole. The phase shift, however, varies nonlinearly with frequency. This means that different frequency components will experience different time delays as they pass through the filter. This will cause distortion and ringing on square and pulse waves. The **Bessel filter response** has a linear variation of phase with frequency and is the best type of response for pulse waveforms. Unfortunately, Bessel filters have a somewhat slower initial roll-off than Butterworth filters and, consequently, are poorer for audio applications. The **Chebyshev filter response** has a faster initial roll-off than a Butterworth filter, allowing the design of a much sharper cutoff filter for the same number of poles. The price of this faster roll-off, however, is increased nonlinear phase shift and ripples in the amplitude response of the filter passband. The **Cauer** (or **elliptic**) **filter response** has an even sharper initial roll-off than the Chebyshev filter does, but has ripples in amplitude response in the filter passband *and* in the stopband. As well, the phase response is quite nonlinear. Figure 6-12 shows comparative frequency and phase responses of single-pole filters of each of the four types. Note that the 3 dB point does not correspond to the critical frequency of the Bessel, Chebyshev, or Cauer filters, but that the 45° phase angle does correspond to the critical frequency of all four types of filters.

Filters with Butterworth response are most widely used, and we will use only recipes for this type of response to demonstrate the design of multipole active filters. The circuits and design procedures are similar for the other types, and the reader is referred to any book on filter design for information on these.

One cautionary note must be injected at this point. The various recipes introduced in this section will produce usable filters. However, if the filters are constructed using standard 5% components, *the measured response may be quite different from that expected.* To get close to the predicted response, it may be

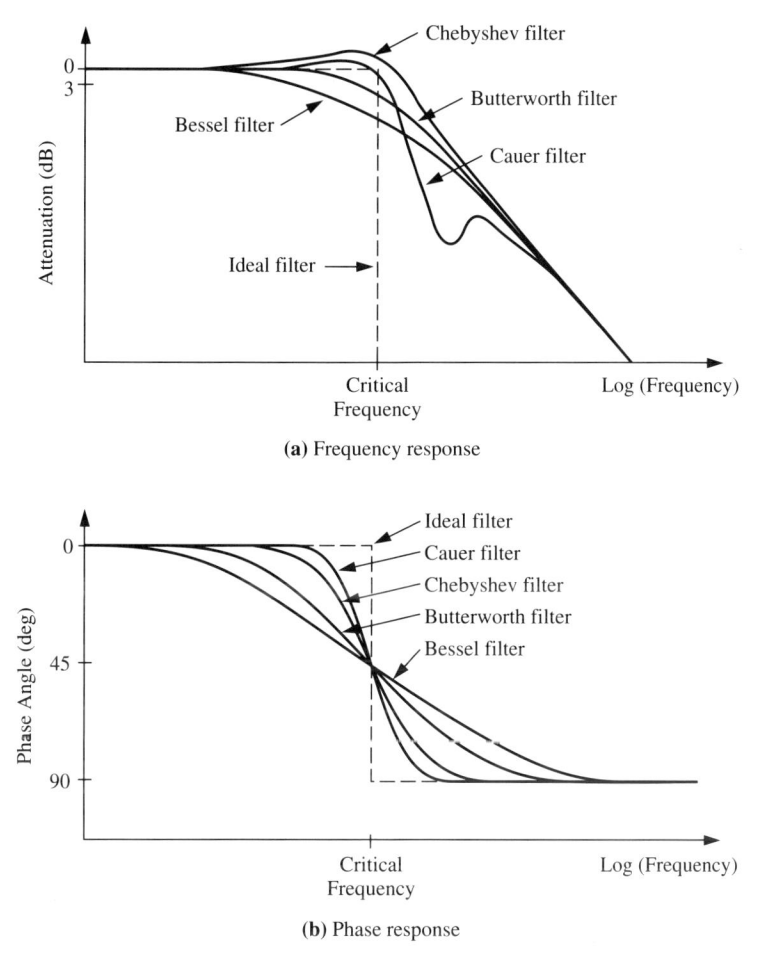

Figure 6-12 Comparison of Butterworth, Bessel, Chebyshev, and Cauer filter responses for single-pole filters.

necessary to use 1% or even 0.1% tolerance components. The production of accurate filters is an art as well as a science!

Unity-Gain Filter Design

If a low-pass filter with unity gain is required, it can be designed with the data in Table 6-1 and the appropriate circuit in Figure 6-13. All of these designs and circuits assume the use of BiFET op-amps. Table 6-1 lists normalized capacitor values (in farads) for normalized resistance values of 1 Ω and a normalized critical frequency f_o of 1 rad/s.

First select the number of poles desired for the filter. To design real component values from the normalized values in the table, we must calculate a frequency

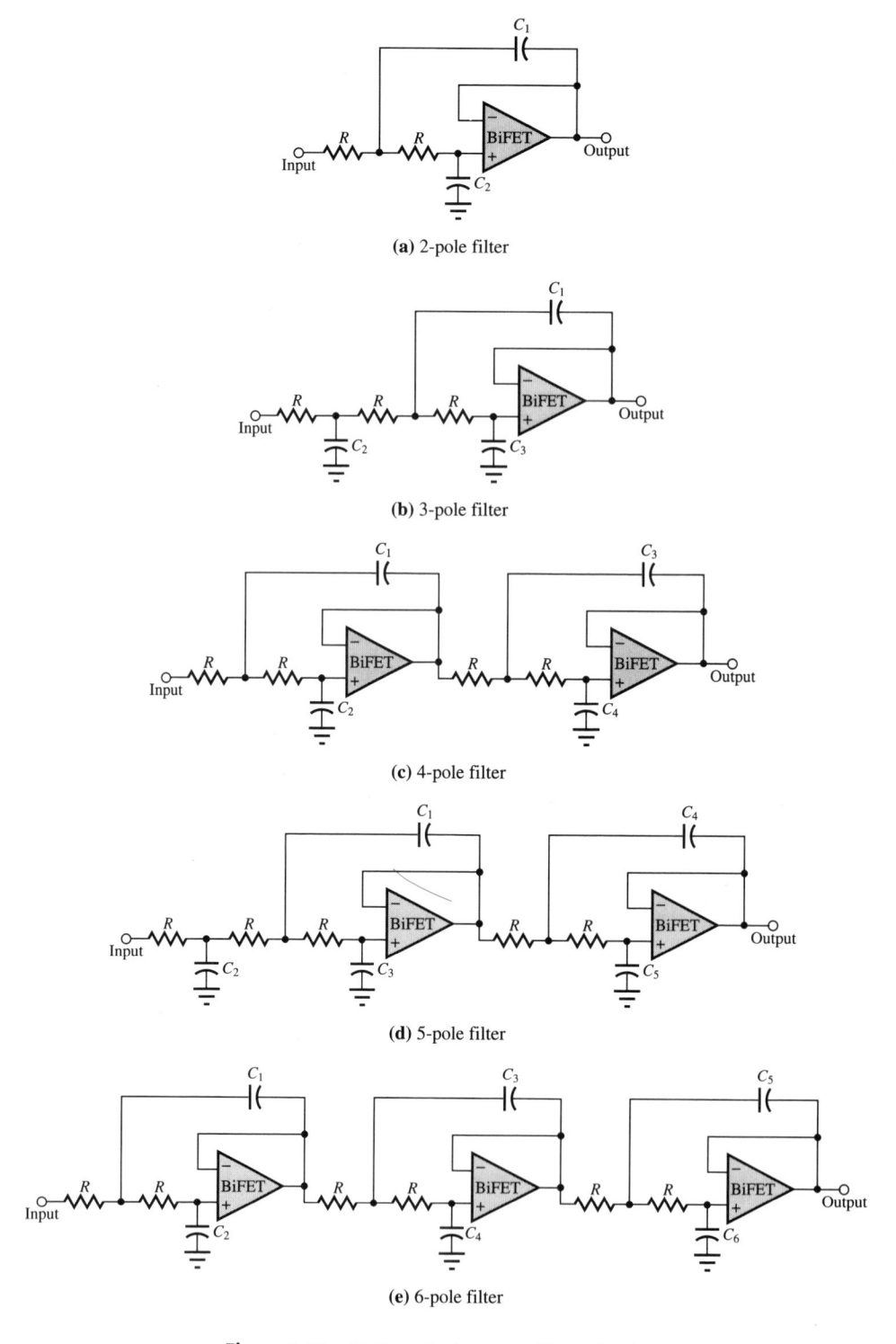

Figure 6-13 Unity-gain low-pass filter circuits.

6-2 Low-Pass and High-Pass Active Filters

Table 6-1 Normalized capacitor values for unity-gain low-pass active filters

Poles	C_1	C_2	C_3	C_4	C_5	C_6
2	1.4142	0.7071				
3	3.546	1.392	0.2024			
4	1.082	0.9241	2.613	0.3825		
5	1.753	1.354	0.4214	3.235	0.3089	
6	1.035	0.9660	1.4142	0.7071	3.863	0.2588

scaling constant K_f. This constant is simply the radian critical frequency:

$$K_f = 2\pi f_o \tag{6-11}$$

Divide the capacitor values of the selected filter by this scaling factor. These frequency-scaled capacitor values will probably still be quite unreasonable.

To convert the capacitors into more reasonable values, use an impedance scaling constant K_X. Select one of the capacitors to be scaled, and select a suitable value for this capacitor. The impedance scaling factor is simply the frequency-scaled value divided by the desired capacitance value:

$$K_X = \frac{\text{frequency-scaled capacitor value}}{\text{desired capacitor value}} \tag{6-12}$$

All of the frequency-scaled capacitor values should be divided by the impedance scaling factor to give the actual filter capacitance value.

Finally, the resistances must be scaled up to match the scaling down of the capacitors. All the resistors have an initial value of 1 Ω and are scaled up by a factor of K_X; hence, they will all have a scaled value equal to K_X Ω.

Exactly the same procedure is used to design a unity gain high-pass filter. The circuits for 2 pole, 3 pole, 4 pole, 5 pole, and 6 pole unity-gain high-pass filters are shown in Figure 6-14. Table 6-2 lists normalized resistor values (in ohms) for normalized capacitance values of 1 F and a normalized critical frequency f_o of 1 rad/s.

The scaling is done just as for the low-pass filter. First calculate the frequency scaling factor by Equation 6-11; then divide the capacitor values by this factor (all the capacitor values are identical). Next determine K_X using Equation 6-12,

Table 6-2 Normalized resistor values for unity-gain high-pass active filters

Poles	R_1	R_2	R_3	R_4	R_5	R_6
2	0.7071	1.4142				
3	0.2820	0.7184	4.941			
4	0.9242	1.082	0.3827	2.614		
5	0.5705	0.7386	2.373	0.3091	3.237	
6	0.9662	1.035	0.7071	1.4142	0.2589	3.864

(a) 2-pole filter

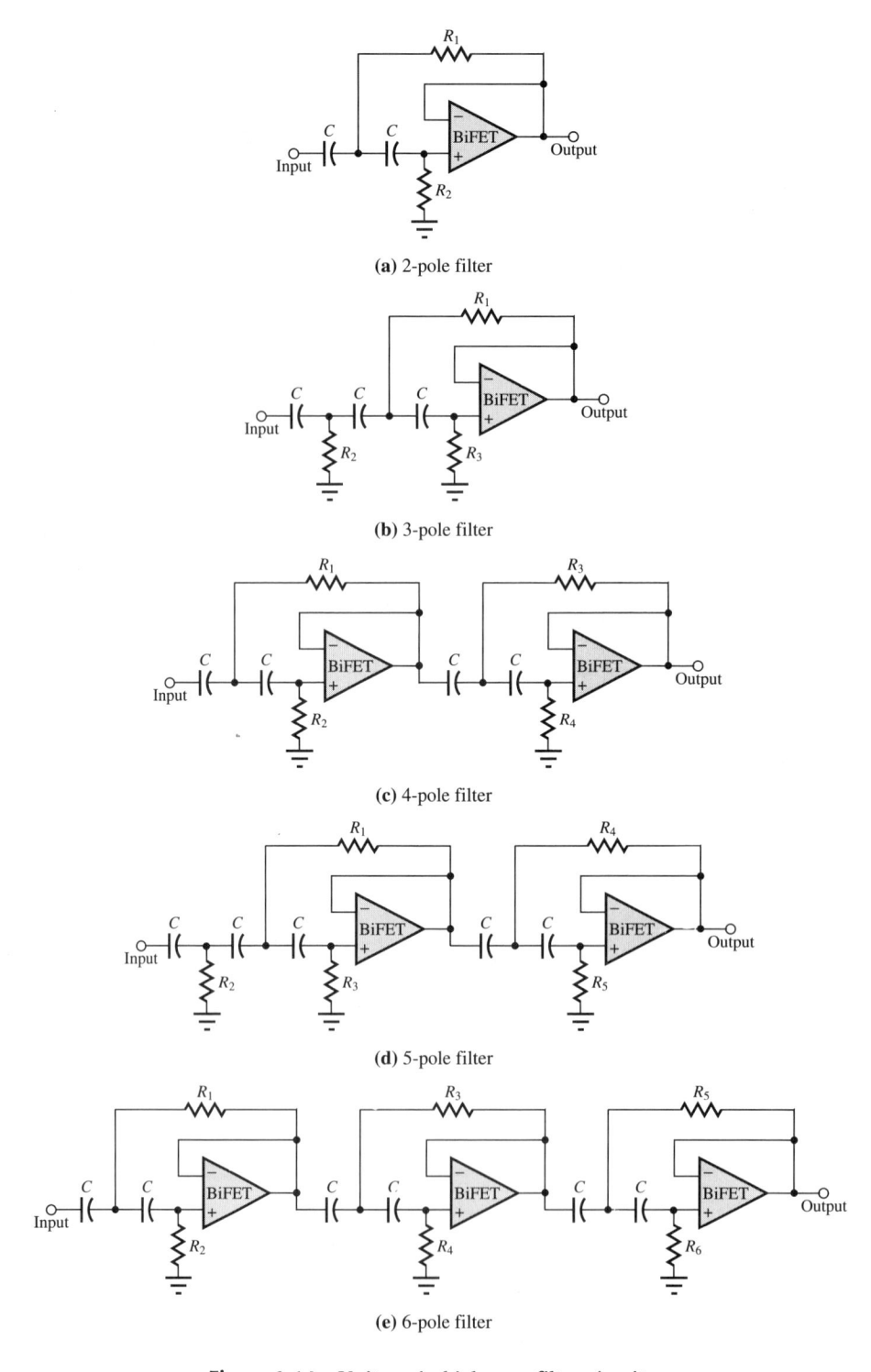

(b) 3-pole filter

(c) 4-pole filter

(d) 5-pole filter

(e) 6-pole filter

Figure 6-14 Unity-gain high-pass filter circuits.

choosing some desired capacitor value. Finally, convert all the unity-scaled resistors to actual values by multiplying each value by K_X, and select the closest standard values.

The only possible difficulty in this design is the choice of a suitable capacitor value. Try some common value, say 0.1 µF, and calculate the resistors. If the resistor values turn out to be unrealistic (for example, many megohms), simply try another capacitor value and calculate new resistor values. The following examples demonstrate the design procedure.

EXAMPLE 6-4

Design a unity-gain low-pass Butterworth filter with a critical frequency of 3000 Hz and a roll-off of 60 dB per decade.

Solution

A single-pole filter has a roll-off of 20 dB per decade, so for a roll-off of 60 dB per decade we will require a 3 pole active filter. The circuit is shown in Figure 6-13(b) and has been redrawn as Figure 6-15.

Table 6-1 lists the normalized capacitor values for a 3 pole filter as

$C_1 = 3.546$ F, $C_2 = 1.392$ F, $C_3 = 0.2024$ F

First, we must calculate the frequency scaling constant K_f from Equation 6-11:

$K_f = 2\pi f_o = 2\pi \times 3000$ Hz $= 18{,}849.6$ rad/s

Next, determine the frequency-scaled capacitor values by dividing the normalized capacitor values by K_f to obtain

$C_1 = 188.12$ µF, $C_2 = 73.848$ µF, $C_3 = 10.738$ µF

We will select a real value of 0.01 µF for C_1.

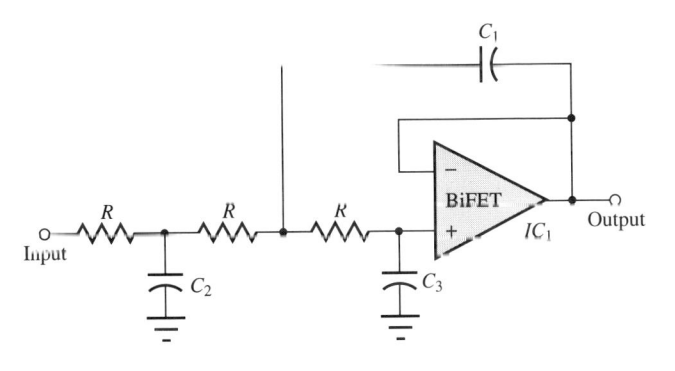

Figure 6-15

Parts List
Resistors:
  R   18 kΩ
Capacitors:
  C_1 0.01 µF
  C_2 0.0039 µF
  C_3 560 pF
Semiconductors:
  IC_1 TL081C

234   6 Active Filters

Now calculate the impedance scaling factor K_X from Equation 6-12:

$$K_X = \frac{\text{frequency-scaled capacitor value}}{\text{desired capacitor value}} = \frac{188.12 \; \mu\text{F}}{0.01 \; \mu\text{F}} = 18{,}812$$

Dividing each of the capacitor values by this gives

$$C_1 = 0.01 \; \mu\text{F}, \quad C_2 = 0.003926 \; \mu\text{F}, \quad C_3 = 0.0005708 \; \mu\text{F}$$

Choose the closest standard values of $C_1 = 0.01 \; \mu\text{F}$, $C_2 = 0.0039 \; \mu\text{F}$, and $C_3 = 560 \; \text{pF}$.

The resistors all have the same value and are equal to K_X. Since $K_X = 18{,}812$, we will make all the resistors 18 kΩ, since this is the closest standard value from the table in Appendix A-1.

Our filter design is now complete. At this point it is important to repeat the cautionary note advanced earlier in the chapter, namely that if the circuit is constructed as designed using standard 5% resistor values and 10% capacitor values, the measured response of the filter may be quite different from that expected. For a precise response, it may be necessary to use 1% or better components, or to add trimmer resistors and capacitors. This note applies to all of the following filter designs as well. This is a restriction with complex filter designs in general. There is a solution to this requirement for precise components which we will introduce in the last section of this chapter.

EXAMPLE 6-5

Design a unity-gain low-pass Butterworth filter with a critical frequency of 15 kHz. The attenuation must be at least 300 at 20,000 Hz.

Solution

We need to know the dB per decade of roll-off before we can select a filter. Calculate the attenuation one decade above the 15 kHz critical frequency, using Equation 6-3.

$$\text{attenuation at } f_{\text{dec}} = \frac{f_{\text{dec}} - f_o}{f - f_o} \times \text{attenuation at } f$$

$$= \frac{150 \; \text{kHz} - 15 \; \text{kHz}}{20 \; \text{kHz} - 15 \; \text{kHz}} \times 300 = 8100$$

Now convert this attenuation into decibels. This is simply $20 \log 8100 = 78.17$ dB. We will require a filter with a roll-off of at least 80 dB per decade, hence a 4 pole active filter. The circuit is shown in Figure 6-13(c). It has been redrawn as Figure 6-16.

The rest of the filter design follows the previous example. The normalized capacitor values are listed in Table 6-1:

$$C_1 = 1.082 \; \text{F}, \quad C_2 = 0.9241 \; \text{F}, \quad C_3 = 2.613 \; \text{F}, \quad C_4 = 0.3825 \; \text{F}$$

6-2 Low-Pass and High-Pass Active Filters

Figure 6-16

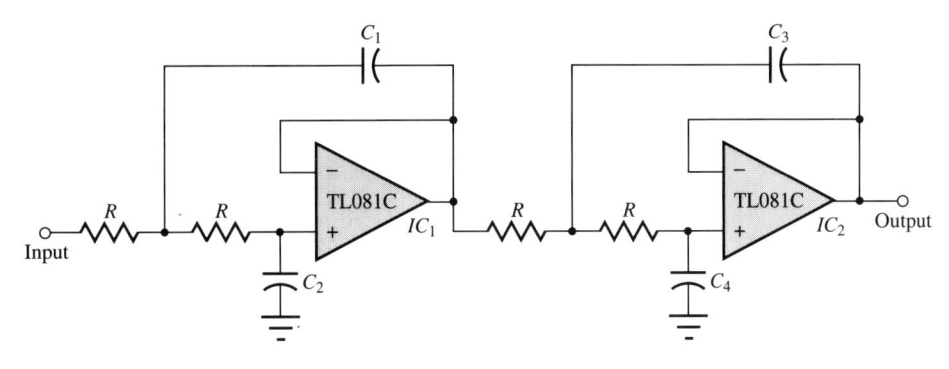

Parts List

Resistors:
R   11 kΩ

Capacitors:
C_1 0.01 µF
C_2 820 pF
C_3 0.0022 µF
C_4 330 pF

Semiconductors:
IC_1, IC_2 TL081C

Calculate the frequency scaling constant K_f from Equation 6-11:

$$K_f = 2\pi f_o = 2\pi \times 15 \text{ kHz} = 94{,}247.8 \text{ rad/s}$$

The frequency-scaled capacitor values are determined by dividing the normalized capacitor values by K_f:

$$C_1 = 11.480 \text{ µF}, \quad C_2 = 9.8050 \text{ µF}, \quad C_3 = 27.725 \text{ µF}, \quad C_4 = 4.0585 \text{ µF}$$

Select the actual value of C_1 as 0.001 µF.
Calculate the impedance scaling factor K_X from Equation 6-12:

$$K_X = \frac{\text{frequency scaled capacitor value}}{\text{desired capacitor value}} = \frac{11{,}480 \text{ µF}}{0.001 \text{ µF}} = 11{,}480$$

Dividing each of the capacitor values by this gives

$$C_1 = 0.01 \text{ µF}, \quad C_2 = 854.1 \text{ pF}, \quad C_3 = 0.002415 \text{ µF}, \quad C_4 = 353.5 \text{ pF}$$

Choose the closest standard values of $C_1 = 0.01$ µF, $C_2 = 820$ pF, $C_3 = 0.0022$ µF, and $C_4 = 330$ pF.

The resistors all have the same value and are equal to K_X. Since $K_X = 11{,}480$, we will make all the resistors 11 kΩ, the closest standard value from the table in Appendix A-1.

Our filter design is now complete.

EXAMPLE 6-6

Design a unity-gain high-pass Butterworth filter with a critical frequency of 18 kHz, and a roll-off of 100 dB per decade.

Solution

Since we require a roll-off of 100 dB per decade, we will require a 5 pole active filter. The circuit is shown in Figure 6-14(d). It has been redrawn as Figure 6-17.

Table 6-2 lists the normalized resistor values for a 5 pole unity-gain high-pass filter as

$$R_1 = 0.5705 \, \Omega, \quad R_2 = 0.7386 \, \Omega, \quad R_3 = 2.373 \, \Omega,$$
$$R_4 = 0.3091 \, \Omega, \quad R_5 = 3.237 \, \Omega$$

First, we must calculate the frequency scaling constant K_f from Equation 6-11:

$$K_f = 2\pi f_o = 2\pi \times 18 \text{ kHz} = 113{,}097 \text{ rad/s}$$

Next, determine the frequency-scaled capacitor value by dividing the normalized capacitor value by K_f to obtain $C = 8.842 \, \mu\text{F}$. Note that all the capacitor values

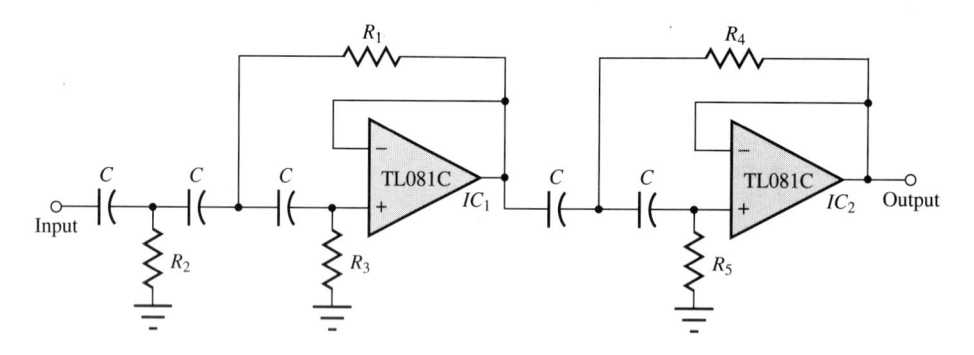

Parts List

Resistors:
R_1	5.1 kΩ
R_2	6.8 kΩ
R_3	20 kΩ
R_4	2.7 kΩ
R_5	30 kΩ

Capacitors:
C	0.001 μF

Semiconductors:
IC_1, IC_2	TL081C

Figure 6-17

are the same. Choose the actual value of C as 0.001 μF, and calculate the impedance scaling factor K_X from Equation 6-12:

$$K_X = \frac{\text{frequency-scaled capacitor value}}{\text{desired capacitor value}} = \frac{8.842 \text{ μF}}{0.001 \text{ μF}} = 8842$$

Multiplying each of the resistor values by this gives

$$R_1 = 5044 \text{ Ω}, \quad R_2 = 6531 \text{ Ω}, \quad R_3 = 20{,}982 \text{ Ω},$$
$$R_4 = 2733 \text{ Ω}, \quad R_5 = 28{,}622 \text{ Ω}$$

Choose the closest standard values from the table in Appendix A-1 to obtain $R_1 = 5.1$ kΩ, $R_2 = 6.8$ kΩ, $R_3 = 20$ kΩ, $R_4 = 2.7$ kΩ, and $R_5 = 30$ kΩ.

The capacitors all have the same value of 0.001 μF.

Our filter design is now complete.

Equal-Component Filter Design

There is a second way to design low-pass and high-pass filters—select resistors with the same value and capacitors with the same value. This method is called equal-component design. The equal-component design procedure is easier than the unity-gain design in that no scaling or calculation of scaling factors is required. All the resistors in the filter have the same value R, and all the capacitors have the same value C. Values for R and C are determined by Equation 6-5, which defines the critical frequency.

$$f_o = \frac{1}{2\pi RC} \tag{6-5}$$

Normally a suitable value for C is chosen, and Equation 6-5 is solved for R.

$$R = \frac{1}{2\pi f_o C} \tag{6-13}$$

Each stage in an equal-component filter has a different gain, and the overall gain of the filter is not unity but depends on the number of poles. The gains of the various stages are given in Table 6-3. To design the filter once R and C have been

Table 6-3 Amplifier gains for equal-component low-pass and high-pass active filters

Poles	A_{V_1}	A_{V_2}	A_{V_3}	$A_{V_{total}}$
2	1.5858			1.5858
3	1.0000	2.0000		2.0000
4	1.1523	2.2346		2.5749
5	1.0000	1.3819	2.3819	3.2917
6	1.0684	1.5858	2.4824	4.2058

(a) 2-pole filter

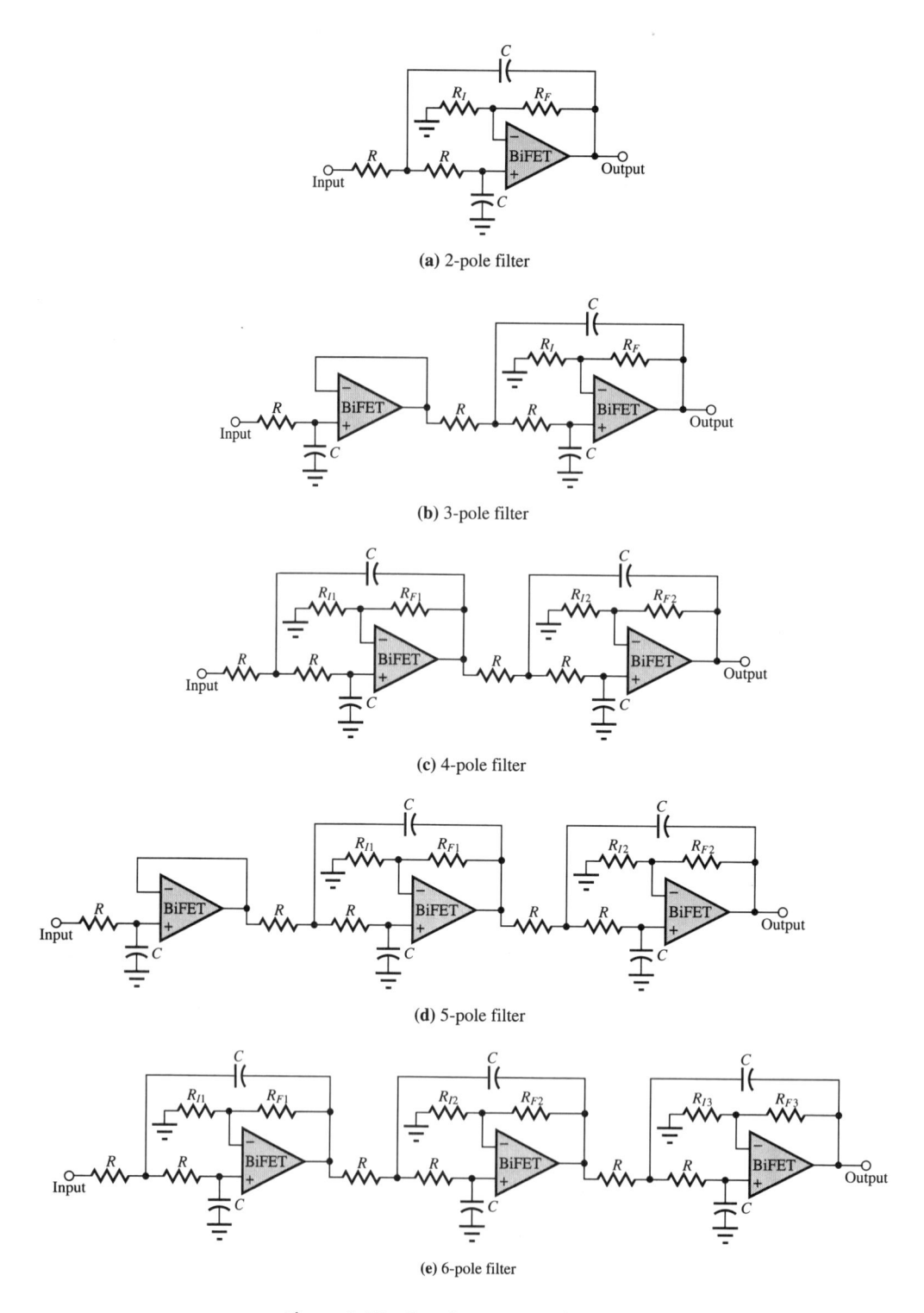

(b) 3-pole filter

(c) 4-pole filter

(d) 5-pole filter

(e) 6-pole filter

Figure 6-18 Equal-component low-pass filter.

(a) 2-pole filter

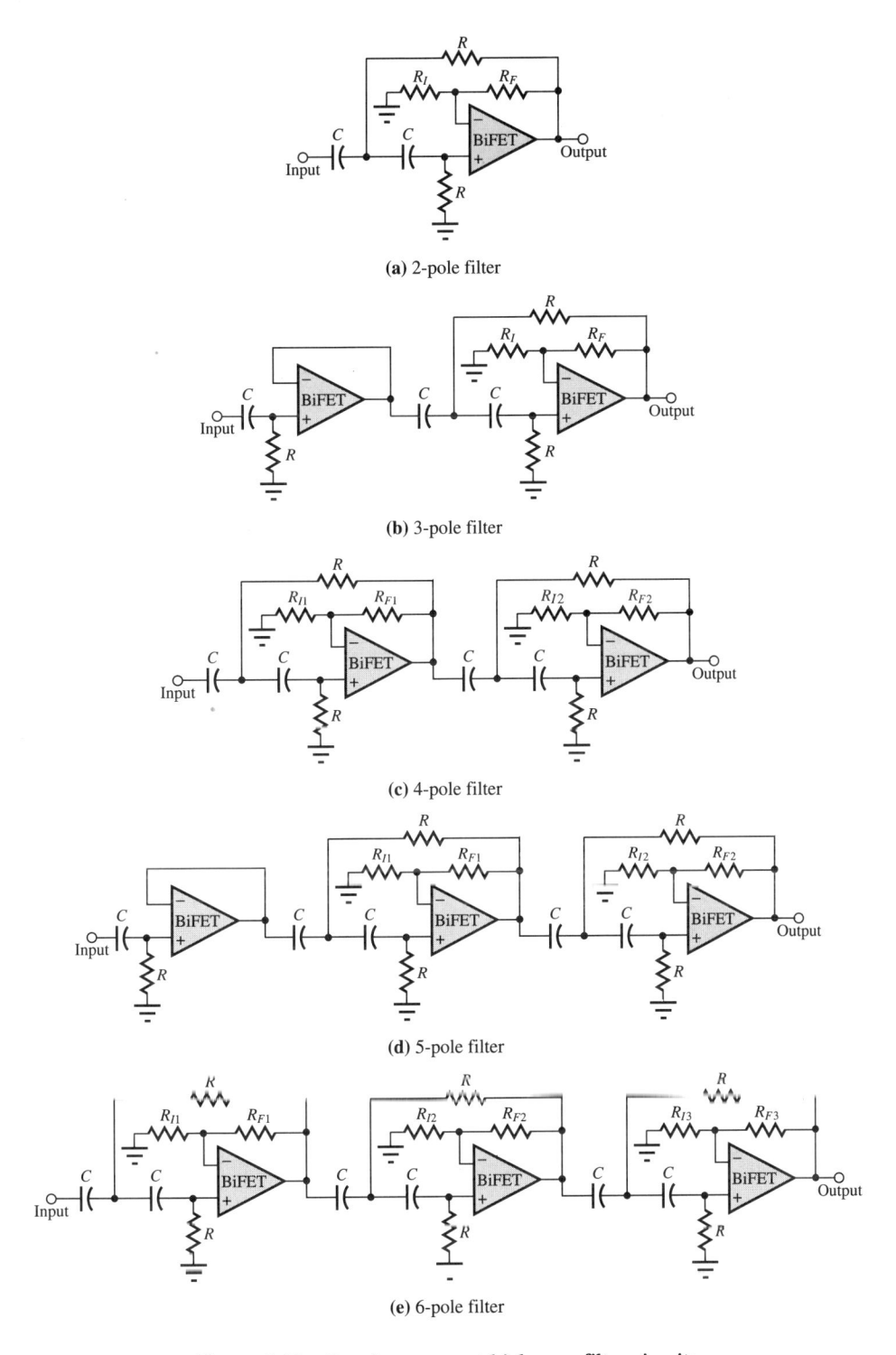

(b) 3-pole filter

(c) 4-pole filter

(d) 5-pole filter

(e) 6-pole filter

Figure 6-19 Equal-component high-pass filter circuits.

selected, you need only design the gain of each filter stage. The design of the gain is performed as for any noninverting BiFET amplifier. Any suitable values of R_I and R_F may be chosen. The amplifier gains are the same for both low-pass and high-pass filters. Figure 6-18 shows the circuits for 2 pole, 3 pole, 4 pole, 5 pole, and 6 pole equal-component low-pass active filters. Figure 6-19 shows the circuits for 2 pole, 3 pole, 4 pole, 5 pole, and 6 pole equal-component high-pass active filters. The circuits in these two figures are similar. The capacitors C and resistors R in the high-pass circuit are simply interchanged from the low-pass circuit.

EXAMPLE 6-7

Design an equal-component low-pass Butterworth filter with a critical frequency of 2000 Hz and a roll-off of 60 dB per decade.

Solution

For a roll-off of 60 dB per decade, we will require a 3 pole active filter. The circuit is shown in Figure 6-18(b). It has been redrawn as Figure 6-20.

First we must calculate values for R and C. Choose a value of 0.01 µF for C and calculate R from Equation 6-13:

$$R = \frac{1}{2\pi f_o C} = \frac{1}{2\pi \times 2000 \text{ Hz} \times 0.01 \text{ µF}} = 7958 \ \Omega$$

Parts List			
Resistors:		Capacitors:	
R	8.2 kΩ	C	0.01 µF
R_I	10 kΩ	Semiconductors:	
R_F	10 kΩ	IC_1, IC_2	TL081C

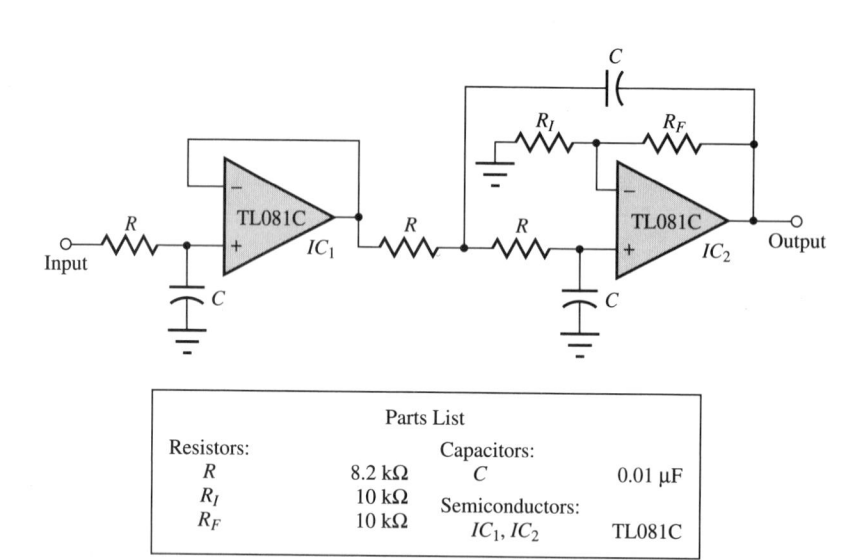

Figure 6-20

The closest standard value listed in the table in Appendix A-2 is 8.2 kΩ, which we will choose.

Table 6-3 lists the gains required for the two amplifier stages. The first stage is a unity-gain amplifier and requires no design. The second-stage amplifier must have a gain of 2. Since this is a noninverting amplifier, we must use Equation 3-5:

$$A_V = \frac{R_F}{R_I} + 1 \quad \text{or} \quad \frac{R_F}{R_I} = A_V - 1 = 1$$

This equation shows that $R_F = R_I$. Select both to be 10 kΩ.

The design of the filter is complete. Notice how much simpler this design procedure was than the design of a unity-gain low-pass filter. The only restriction is that we have an amplifier with a gain of 2 as well as a filter.

EXAMPLE 6-8

Design an equal-component high-pass Butterworth filter with a critical frequency of 3000 Hz and a roll-off of 120 dB per decade.

Solution

For a roll-off of 120 dB per decade, we will require a 6 pole active filter. The circuit is shown in Figure 6-19(e). It has been redrawn as Figure 6-21.

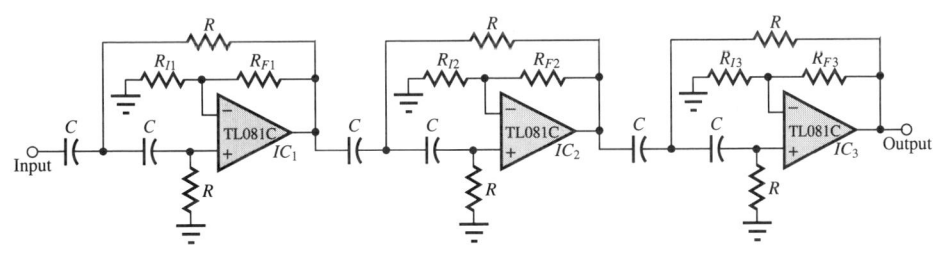

	Parts List		
Resistors:		Capacitors:	
R	5.1 kΩ	C	0.01 μF
R_{I1}	100 kΩ	Semiconductors:	
R_{F1}	6.2 kΩ	IC_1, IC_2, IC_3	TL081C
R_{I2}	10 kΩ		
R_{F2}	5.6 kΩ		
R_{I3}	10 kΩ		
R_{F3}	15 kΩ		

Figure 6-21

First we must calculate values for R and C. Choose a value of 0.01 μF for C and calculate R from Equation 6-13:

$$R = \frac{1}{2\pi f_o C} = \frac{1}{2\pi \times 3000 \text{ Hz} \times 0.01 \text{ μF}} = 5305 \text{ }\Omega$$

The closest standard value from Appendix A-1 is 5.1 kΩ, which we will choose.

The 6 pole active filter will require three amplifier stages. Table 6-3 lists the gains required for the three stages. The first-stage amplifier must have a gain of 1.0648. Since this is a noninverting amplifier, we must use Equation 3-5:

$$A_V = \frac{R_F}{R_I} + 1 \quad \text{or} \quad \frac{R_{F_1}}{R_{I_1}} = A_{V_1} - 1 = 1.0648 - 1 = 0.0648$$

From this equation $R_{F_1} = 0.0648 R_{I_1}$. Choose R_{I_1} to be 100 kΩ, then $R_{F_1} = 6480$ Ω. Select R_{F_1} as 6.2 kΩ, the closest standard value from Appendix A-1.

The second-stage amplifier must have a gain of 1.5858. By Equation 3-5,

$$\frac{R_{F_2}}{R_{I_2}} = A_{V_2} - 1 = 1.5858 - 1 = 0.5858$$

Hence, $R_{F_2} = 0.5858 R_{I_2}$. Choose R_{I_2} to be 10 kΩ. Then $R_{F_2} = 5858$ Ω. Select R_{F_2} as 5.6 kΩ, the closest standard value from Appendix A-1.

The third-stage amplifier must have a gain of 2.4824.

$$\frac{R_{F_3}}{R_{I_3}} = A_{V_3} - 1 = 2.4824 - 1 = 1.4824$$

Hence, $R_{F_3} = 1.4824 R_{I_3}$. Choose R_{I_3} to be 10 kΩ. Then $R_{F_3} = 14,824$ Ω. Select R_{F_3} as 15 kΩ, the closest standard value.

Our filter design is complete.

6-3 Bandpass and Band-Rejection Active Filters

Bandpass and band-rejection active filters pose a somewhat more difficult design problem than low-pass or high-pass active filters. If a wide band of frequencies is to be passed or rejected, the solution is simply to combine an active low-pass filter with an active high-pass filter. Such a filter is called a **broadband filter.** If a narrow band of frequencies is to be passed or rejected, however, it is necessary to design the filter to provide a suitable Q, or quality, factor. This is a **narrow-band filter,** and the design is more complicated than the designs of low-pass, high-pass, or wide-band bandpass or band-rejection filters.

The **quality factor**, or **Q**, of a filter circuit is the ratio of the center frequency of the filter to the bandwidth of the filter. The **center frequency** f_{center} is defined as the geometric mean of the upper critical frequency and the lower critical frequency, and the **bandwidth** f_{BW} is defined as the upper critical frequency minus the lower critical frequency:

$$Q = \frac{f_{\text{center}}}{f_{BW}} \tag{6-14}$$

where

$$f_{\text{center}} = \sqrt{f_{oU} \times f_{oL}} \tag{6-15}$$

and

$$f_{BW} = f_{oU} - f_{oL} \tag{6-16}$$

This Q factor determines whether we have a broadband filter or a narrow-band filter. Typically, if $Q > 1$, we will assume that the filter is narrow-band.

Broadband Bandpass Active Filters

A broadband bandpass filter is a filter that passes all frequencies over a range that is large compared with the center frequency of the filter. The Q of the circuit would be typically less than 1. For example, a telephone passes frequencies from approximately 300 to 3000 Hz. The center frequency is 950 Hz, and the bandwidth is 2700 Hz, giving a Q of 950/2700 = 0.35.

The design of a broadband bandpass active filter is quite simple. It consists of a low-pass active filter in series with a high-pass active filter, exactly the same as a bandpass filter produced with passive filters. With active filters, however, multipole filters can be used to give a much faster roll-off than is possible with simple R-C passive filters. The low-pass filter is designed such that its critical frequency is equal to the upper critical frequency of the bandpass filter. The high-pass filter is designed such that its critical frequency is equal to the lower critical frequency of the bandpass filter. The roll-off of the low-pass and high-pass filters and the filter response are selected on the basis of the characteristics required for the bandpass filter.

Figure 6-22(a) shows a typical bandpass filter designed to give a roll-off of 100 dB per decade and which uses a Butterworth unity-gain low-pass filter in series with a Butterworth unity-gain high-pass filter. Figure 6-22(b) shows the typical frequency response.

EXAMPLE 6-9

Design a bandpass filter using 3 pole Butterworth low-pass and high-pass filters to pass frequencies between 200 and 10,000 Hz. Use TL081C op-amps.

Solution

First we must determine the center frequency of the filter, using Equation 6-15:

$$f_{\text{center}} = \sqrt{f_{oU} \times f_{oL}} = \sqrt{200 \text{ Hz} \times 10 \text{ kHz}} = 1412 \text{ Hz}$$

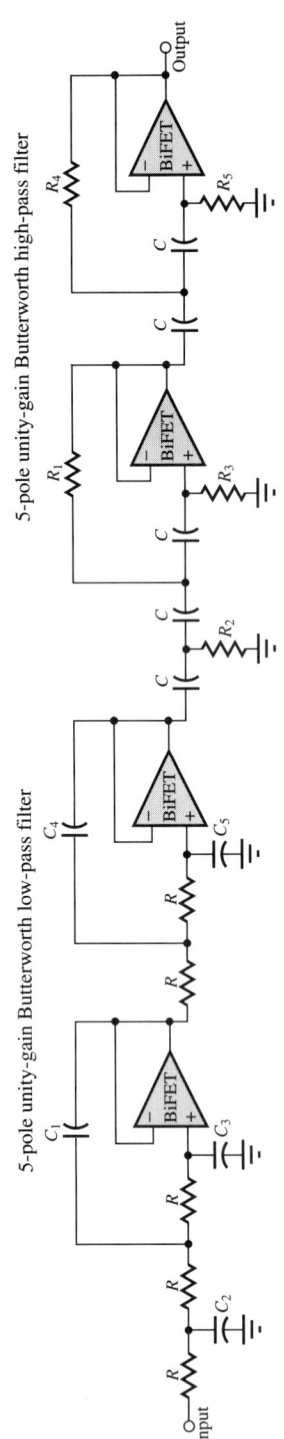

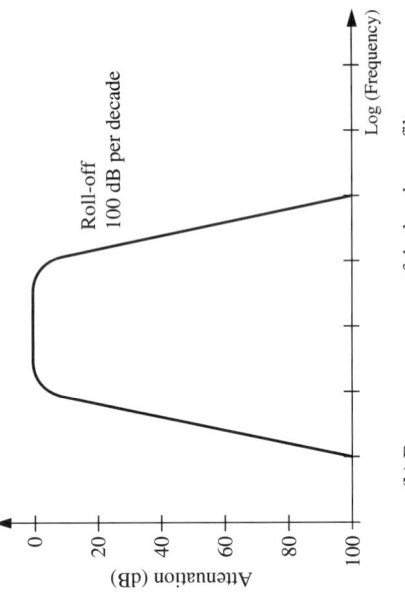

Figure 6-22 Butterworth bandpass filter with roll-off of 100 dB per decade.

6-3 Bandpass and Band-Rejection Active Filters

We find the bandwidth from Equation 6-16:

$$f_{BW} = f_{oU} - f_{oL} = 10{,}000 \text{ Hz} - 200 \text{ Hz} = 9800 \text{ Hz}$$

This is a wide-band filter since the Q of the filter, as determined by Equation 6-14, is

$$Q = \frac{f_{\text{center}}}{f_{BW}} = \frac{1414 \text{ Hz}}{9800 \text{ Hz}} = 0.144$$

The wide-band solution for a bandpass filter is to use a standard low-pass filter in series with a standard high-pass filter. We will use 3 pole unity-gain filters. The circuit of the filter is shown in Figure 6-23.

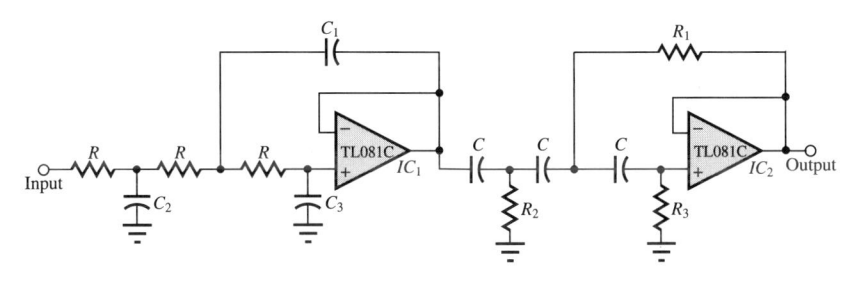

Parts List

Resistors:		Capacitors:		Semiconductors:	
R	5.6 kΩ	C	0.1 µF	IC_1, IC_2	TL081C
R_1	2.2 kΩ	C_1	0.01 µF		
R_2	5.6 kΩ	C_2	0.0039 µF		
R_3	39 kΩ	C_3	560 pF		

Figure 6-23

First we will design the high-pass filter, which determines the low frequency cutoff of the filter. The critical frequency of the filter is 200 Hz. The procedure is similar to that used in Example 6-6, except for a 3 pole filter.

Table 6-2 lists the normalized resistor values for a 3 pole unity gain high-pass filter as

$$R_1 = 0.2820 \text{ Ω}, \quad R_2 = 0.7184 \text{ Ω}, \quad R_3 = 4.941 \text{ Ω}$$

First we must calculate the frequency scaling constant K_f from Equation 6-11:

$$K_f = 2\pi f_o = 2\pi \times 200 \text{ Hz} = 1256.6 \text{ rad/s}$$

Next, determine the frequency-scaled capacitor value by dividing the normalized capacitor value of 1 F by K_f to obtain $C = 795.8$ µF. All the capacitor values are the same.

Choose an actual value for C of 0.1 μF and calculate the impedance scaling factor K_X from Equation 6-12:

$$K_X = \frac{\text{frequency-scaled capacitor value}}{\text{desired capacitor value}} = \frac{795.8\ \mu F}{0.1\ \mu F} = 7958$$

Multiplying each of the resistor values by this gives

$$R_1 = 2244\ \Omega, \quad R_2 = 5717\ \Omega, \quad R_3 = 39{,}320\ \Omega$$

Choose the closest standard values from the table in Appendix A-1 of $R_1 = 2.2$ kΩ, $R_2 = 5.6$ kΩ, and $R_3 = 39$ kΩ.

The capacitors all have the same value of 0.1 μF.

Next, we must design a 3 pole unity-gain low-pass filter to define the upper cutoff frequency of the filter as 10,000 Hz. The procedure is similar to Example 6-4.

Table 6-1 lists the normalized capacitor values for a 3 pole filter as

$$C_1 = 3.546\ F, \quad C_2 = 1.392\ F, \quad C_3 = 0.2024\ F$$

Calculate the frequency scaling constant K_f from Equation 6-11:

$$K_f = 2\pi f_o = 2\pi \times 10\ \text{kHz} = 62{,}831.9\ \text{rad/s}$$

Next, determine the frequency-scaled capacitor values by dividing the normalized capacitor values by K_f to obtain

$$C_1 = 56.436\ \mu F, \quad C_2 = 22.154\ \mu F, \quad C_3 = 3.2213\ \mu F$$

Choose an actual value of 0.01 μF for C_1.

Now calculate the impedance scaling factor K_X from Equation 6-12:

$$K_X = \frac{\text{frequency-scaled capacitor value}}{\text{desired capacitor value}} = \frac{56.436\ \mu F}{0.01\ \mu F} = 5643.6$$

Dividing each of the capacitor values by this gives

$$C_1 = 0.01\ \mu F, \quad C_2 = 0.003926\ \mu F, \quad C_3 = 0.0005708\ \mu F$$

Choose the closest standard values of $C_1 = 0.01\ \mu F$, $C_2 = 0.0039\ \mu F$, and $C_3 = 560$ pF.

The resistors all have the same value and are equal to K_X. Since $K_X = 5643.6$, we will make all the resistors 5.6 kΩ, since this is the closest standard value from the table in Appendix A-1.

These two filters are simply connected in series to give the required bandpass filter. Hence, our filter design is now complete.

Broadband Band-Rejection Active Filters

A broadband band-rejection filter is a filter that attenuates all frequencies over a range that is large compared with the center frequency of the filter; that is, the Q of the filter is less than 1.

6-3 Bandpass and Band-Rejection Active Filters 247

The design of a broadband band-rejection active filter is a little more complicated than the design of a bandpass filter. Although it consists of a low-pass active filter and a high-pass active filter, the filters must be placed in parallel and the outputs combined with a mixer or summing amplifier. The procedure is exactly the same as the procedure used with passive filters, except that the use of active

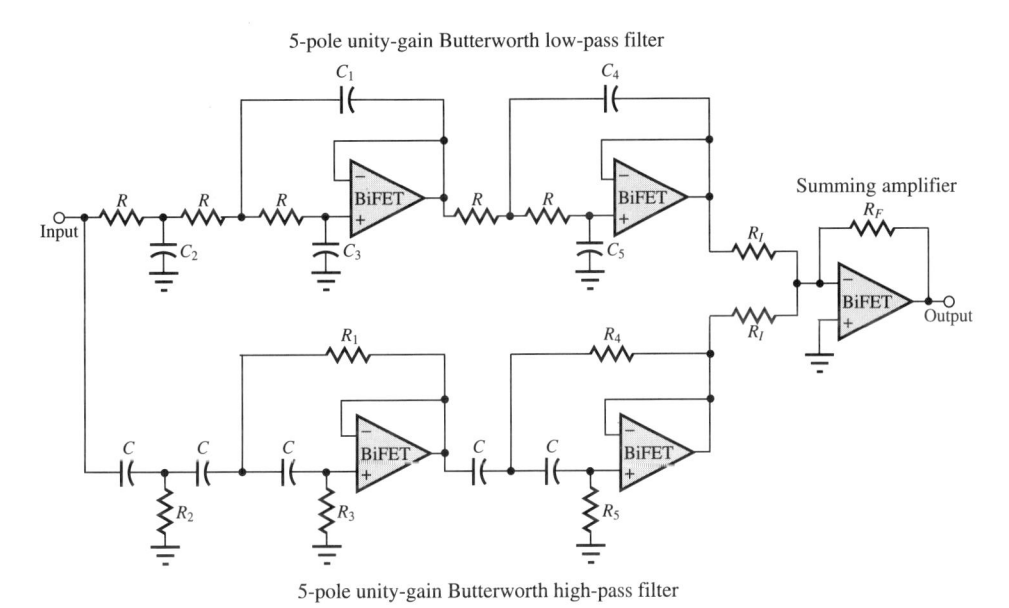

(a) Circuit consisting of low-pass filter, high-pass filter and summing amplifier

(b) Frequency response of the band-rejection filter

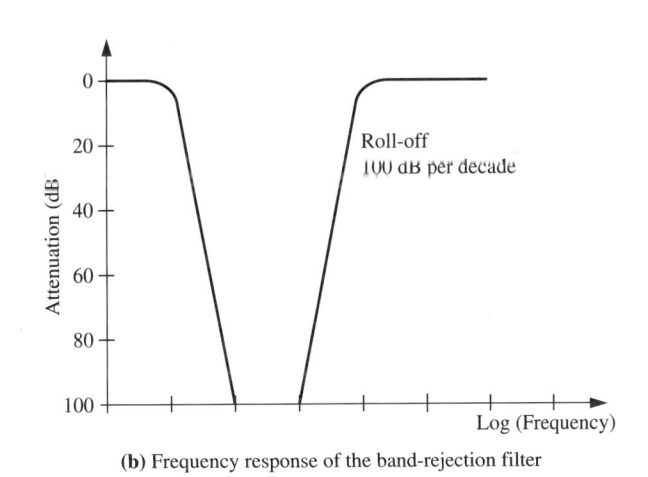

Figure 6-24 Butterworth band-rejection filter with roll-off of 100 dB per decade.

filters allows a much faster roll-off than was possible with simple R-C passive filters.

The low-pass active filter is designed such that its critical frequency is equal to the lower critical frequency of the band-rejection filter. The high-pass active filter is designed such that its critical frequency is equal to the upper critical frequency of the band-rejection filter. The roll-off of both filters is selected to satisfy the design requirements of the band-rejection filter. The mixing circuit is a standard op-amp summing circuit as discussed in Section 4-1 (see Figure 4-1). It is important to use an inverting summing circuit to avoid the crosstalk problems associated with the noninverting summing circuit. Both input resistors should be equal and have the same value as the feedback resistor to ensure unity gain for both filters. Resistor values are not important since the op-amps in both filters are capable of driving any reasonable load. Typically, 10 kΩ resistors would be used.

Figure 6-24(a) shows a typical band-rejection filter designed to give a roll-off of 100 dB per decade, using Butterworth unity-gain low-pass and Butterworth unity-gain high-pass filters connected to a two-input summing amplifier. Figure 6-24(b) shows the typical frequency response.

EXAMPLE 6-10

Design a band-rejection filter using 5 pole Butterworth low-pass and high-pass filters to reject frequencies between 10 and 50 kHz. Use TL081C op-amps.

Solution

First we must determine the Q of the filter to verify that we have a broadband filter. Calculate the center frequency of the filter by using Equation 6-15:

$$f_{center} = \sqrt{f_{oU} \times f_{oL}} = \sqrt{10 \text{ kHz} \times 50 \text{ kHz}} = 22.36 \text{ kHz}$$

Calculate the bandwidth by using Equation 6-16:

$$f_{BW} = f_{oU} - f_{oL} = 50 \text{ kHz} - 10 \text{ kHz} = 40 \text{ kHz}$$

Now calculate Q by using Equation 6-14:

$$Q = \frac{f_{center}}{f_{BW}} = \frac{22.36 \text{ kHz}}{40 \text{ kHz}} = 0.559$$

Since $Q < 1$, we have a broadband filter and can use a standard low-pass filter and a standard high-pass filter connected to a summing circuit. We will use 5 pole unity-gain filters. The circuit of the filter is shown in Figure 6-25.

Design the low-pass filter with a critical frequency of 10 kHz.
Table 6-1 lists the normalized capacitor values for a 5 pole filter as

$$C_1 = 1.753 \text{ F}, \quad C_2 = 1.354 \text{ F}, \quad C_3 = 0.4214 \text{ F},$$
$$C_4 = 3.235 \text{ F}, \quad C_5 = 0.3089 \text{ F}$$

6-3 Bandpass and Band-Rejection Active Filters

5-pole unity-gain Butterworth low-pass filter

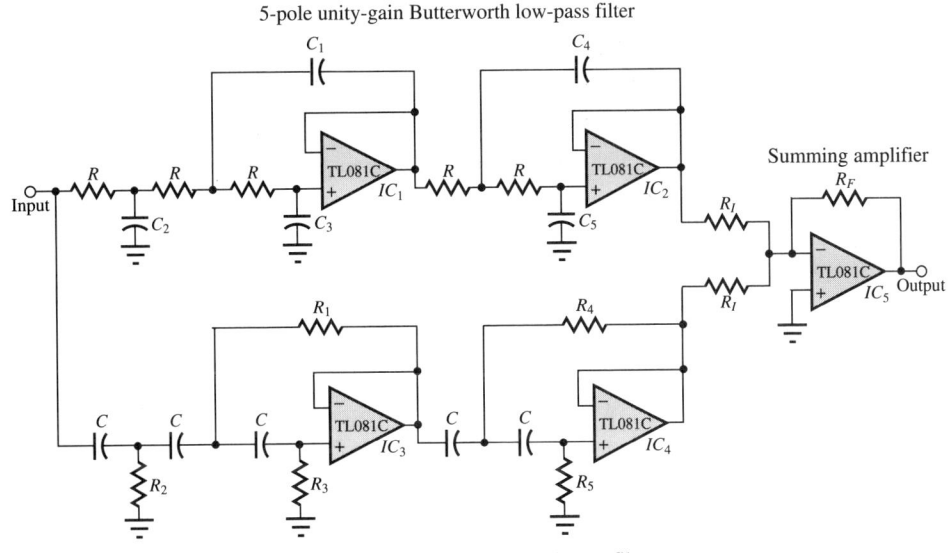

5-pole unity-gain Butterworth high-pass filter

Parts List			
Resistors:		Capacitors:	
R	2.7 kΩ	C	0.001 µF
R_1	1.8 kΩ	C_1	0.01 µF
R_2	2.4 kΩ	C_2	0.0082 µF
R_3	7.5 kΩ	C_3	0.0022 µF
R_4	1.0 kΩ	C_4	0.018 µF
R_5	10 kΩ	C_5	0.0018 µF
R_I	10 kΩ	Semiconductors:	
R_F	10 kΩ	$IC_{1,2,3,4,5}$	TL081C

Figure 6-25

Calculate the frequency scaling constant K_f from Equation 6-11:

$$K_f = 2\pi f_o = 2\pi \times 10 \text{ kHz} = 62{,}832 \text{ rad/s}$$

Next, determine the frequency-scaled capacitor values by dividing the normalized capacitor values by K_f to obtain

$$C_1 = 27.900 \text{ µF}, \quad C_2 = 21.550 \text{ µF}, \quad C_3 = 6.7068 \text{ µF},$$
$$C_4 = 51.487 \text{ µF}, \quad C_5 = 4.9163 \text{ µF}$$

Choose an actual value for C_1 of 0.01 µF and calculate the impedance scaling factor K_X from Equation 6-12:

$$K_X = \frac{\text{frequency-scaled capacitor value}}{\text{desired capacitor value}} = \frac{27.900 \text{ µF}}{0.01 \text{ µF}} = 2790$$

Dividing each of the capacitor values by this gives

$C_1 = 0.01 \ \mu F, \quad C_2 = 0.007724 \ \mu F, \quad C_3 = 0.002404 \ \mu F,$
$C_4 = 0.018454 \ \mu F, \quad C_5 = 0.001762 \ \mu F$

Choose the closest standard values of $C_1 = 0.01 \ \mu F$, $C_2 = 0.0082 \ \mu F$, $C_3 = 0.0022 \ \mu F$, $C_4 = 0.018 \ \mu F$, and $C_5 = 0.0018 \ \mu F$.

The resistors all have the same value and are equal to K_X. Since $K_X = 2790$, choose a value of 2.7 kΩ.

Now design the high-pass filter with a critical frequency of 50 kHz. We will require a 5 pole filter.

Table 6-2 lists the normalized resistor values for a 5 pole unity-gain high-pass filter as

$R_1 = 0.5705 \ \Omega, \quad R_2 = 0.7386 \ \Omega, \quad R_3 = 2.373 \ \Omega,$
$R_4 = 0.3091 \ \Omega, \quad R_5 = 3.237 \ \Omega$

First we must calculate the frequency scaling constant K_f from Equation 6-11:

$$K_f = 2\pi f_o = 2\pi \times 50 \text{ kHz} = 314{,}159 \text{ rad/s}$$

Next, determine the frequency-scaled capacitor value by dividing the normalized capacitor value of 1 F by K_f to obtain $C = 3.183 \ \mu F$. We will choose the actual value of C as 0.001 μF.

Now calculate the impedance scaling factor K_X from Equation 6-12:

$$K_X = \frac{\text{frequency-scaled capacitor value}}{\text{desired capacitor value}} = \frac{3.183 \ \mu F}{0.001 \ \mu F} = 3183$$

Multiplying each of the resistor values by this gives

$R_1 = 1816 \ \Omega, \quad R_2 = 2351 \ \Omega, \quad R_3 = 7553 \ \Omega,$
$R_4 = 983.9 \ \Omega, \quad R_5 = 10{,}303 \ \Omega$

The closest standard values from Appendix A-1 are $R_1 = 1.8$ kΩ, $R_2 = 2.4$ kΩ, $R_3 = 7.5$ kΩ, $R_4 = 1.0$ kΩ, and $R_5 = 10$ kΩ.

The capacitors all have the same value of 0.001 μF.

Finally, we must design a summing circuit to combine the response of the two filters. We will use an inverting summing amplifier to prevent any crosstalk between the filters. Since the gain for the complete filter circuit is to be unity, choose all the summing amplifier resistors to be the same value. We will choose values of 10 kΩ for R_I and R_F.

The filter design is now complete.

Narrow-Band Bandpass Active Filters

A narrow-band bandpass filter has a bandwidth that is small compared with the center frequency of the filter. Consequently, the Q of the filter will be greater than 1. For example, a filter to pass the 19 kHz subcarrier in a stereo radio receiver

6-3 Bandpass and Band-Rejection Active Filters

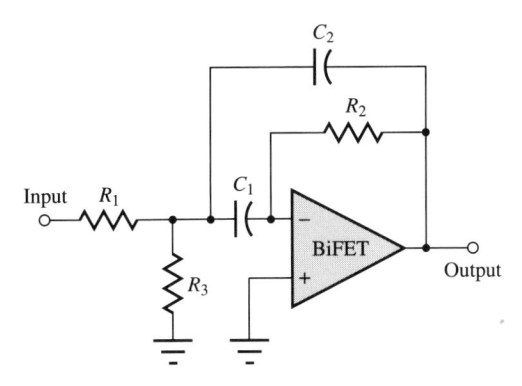

Figure 6-26 Narrow-band bandpass active filter using a single op-amp.

must have a bandwidth of 1 kHz or less. The required Q is at least 19. A series combination of a low-pass filter and a high-pass filter will not work in this case since the critical frequencies are too close—one filter would affect the response of the other.

There is a different recipe used for the design of narrow-band bandpass and band-rejection filters. The circuit in Figure 6-26 illustrates a single-op-amp single-pole bandpass filter. The bandpass filter consists of a low-pass filter formed by R_1 and C_1, and a high-pass filter formed by R_2 and C_2. Although this circuit may appear to be a single-pole low-pass filter in series with a single-pole high-pass filter, the two filters interact to give a much sharper roll-off than is possible with single-pole filters. The derivation of the equations for this type of filter is relatively complicated; hence, we will simply state the basic equations and design by recipe, as we have done for the previous types of filters.

The center frequency of the filter is determined by the values of the three resistors and two capacitors. It is given by the following equation:

$$f_{center} = \frac{1}{2\pi} \times \sqrt{\frac{R_1 + R_3}{C_1 C_2 R_1 R_2 R_3}} \qquad (6\text{-}17)$$

Normally we choose $C_1 = C_2$ to simplify the design.

The qualify factor Q of the filter depends on R_2 and C_2 and is defined as

$$Q = \pi f_{center} R_2 C_2 \qquad (6\text{-}18)$$

The maximum value of Q which can be designed into this filter is set by the bandwidth of the op-amp:

$$Q_{max} = \sqrt{\frac{f_{GB}}{20 f_{center}}} \qquad (6\text{-}19)$$

The quantity f_{GB} is simply the unity-gain bandwidth of the op-amp, which for a TL081 BiFET op-amp is approximately 3 MHz.

If the desired Q is larger than Q_{max}, then two or more filters can be connected in series to give a higher Q. The bandwidth after n identical stages is

$$f_{BWn} = f_{BW1} \sqrt{2^{1/n} - 1} \qquad (6\text{-}20)$$

Rewriting the equation to solve for the bandwidth of a single stage, we get

$$f_{BW1} = \frac{f_{BWn}}{\sqrt{2^{1/n} - 1}} \qquad (6\text{-}21)$$

Since $Q = f_{center}/f_{BW}$, we can substitute f_{center}/Q_n for f_{BWn} and f_{center}/Q_1 for f_{BW1} in Equation 6-21. Cancelling out the common factor f_{center} and rearranging gives

$$Q_n = \frac{Q_1}{\sqrt{2^{1/n} - 1}} \qquad (6\text{-}22)$$

This equation can be solved to express Q_1 in terms of Q_n.

$$Q_1 = Q_n \sqrt{2^{1/n} - 1} \qquad (6\text{-}23)$$

The Q required for the filter is Q_n. Guess some trial value for n and calculate Q_1, the Q for a single stage, using Equation 6-23. If this single-stage Q_1 is less than Q_{max}, then we have a suitable n, and the rest of the circuit can be designed. If Q_1 turns out to be greater than Q_{max}, then try a larger value of n, and repeat until a satisfactory value is found.

To design a narrow-band bandpass filter of this type for a specified frequency band, first determine f_{center} from Equation 6-15 and Q from Equation 6-14. Use Equation 6-19 to see if the calculated Q can be achieved by a single stage. If not, determine the number of stages and the Q for each stage by using Equation 6-23. Choose values for capacitors C_1 and C_2 and solve Equation 6-18 for R_2:

$$R_2 = \frac{Q}{\pi f_{center} C_2} \qquad (6\text{-}24)$$

Resistor R_1 is calculated by the following equation:

$$R_1 = \frac{R_2}{4Q^2} \qquad (6\text{-}25)$$

Finally, calculate R_3 by rearranging Equation 6-17:

$$R_3 = \frac{R_1}{4\pi^2 f_{center}^2 R_1 C_1 R_2 C_2 - 1} \qquad (6\text{-}26)$$

Use actual values for R_1 and R_2 in this equation. The first term in the denominator should be slightly greater than 1. If it should turn out to be slightly less than 1, choose a larger value for either R_1 or R_2 and recalculate. The performance of this filter is sensitive to component values, and 1% or better tolerance components should be used if the required response is to be obtained. Alternatively, trimpots could be added for precise calibration of the resistances.

EXAMPLE 6-11

Design a narrow-band bandpass filter using TL081 op-amps to pass frequencies between 1.7 and 2.0 kHz.

Solution

Before we can draw the circuit for this filter, we need to determine the number of stages. To do this, we must calculate the required Q for the filter.

First determine the center frequency f_{center} from Equation 6-15:

$$f_{center} = \sqrt{f_{oU} \times f_{oL}} = \sqrt{1700 \text{ Hz} \times 2000 \text{ Hz}} = 1844 \text{ Hz}$$

Find the bandwidth from Equation 6-16:

$$f_{BW} = f_{oU} - f_{oL} = 2000 \text{ Hz} - 1700 \text{ Hz} = 300 \text{ Hz}$$

Now we must determine the Q of the circuit and verify that we can achieve this with a single op-amp. Calculate the Q of the filter by using Equation 6-14:

$$Q = \frac{f_{center}}{f_{BW}} = \frac{1844 \text{ Hz}}{300 \text{ Hz}} = 6.15$$

The maximum Q of a single op-amp bandpass filter using a TL081C is given by Equation 6-19 as

$$Q_{max} = \sqrt{\frac{f_{GB}}{20 f_{center}}} = \sqrt{\frac{3 \text{ MHz}}{20 \times 1844 \text{ Hz}}} = 9.02$$

The required Q is less than the maximum Q; hence, the design is possible with a single op-amp. We can now draw the circuit. We will use the circuit shown in Figure 6-26 and redraw it here as Figure 6-27. Choose $C_1 = C_2 = 0.01 \ \mu\text{F}$ and calculate R_2 from Equation 6-24.

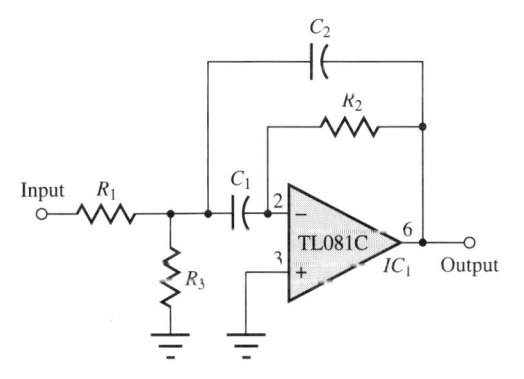

Parts List

Resistors:
- R_1 750 Ω
- R_2 110 kΩ
- R_3 6.8 kΩ

Capacitors:
- C_1 0.01 μF
- C_2 0.01 μF

Semiconductors:
- IC_1 TL081C

Figure 6-27

$$R_2 = \frac{Q}{\pi f_{center} C_2} = \frac{6.15}{\pi \times 1844 \text{ Hz} \times 0.01 \text{ μF}} = 106 \text{ k}\Omega$$

Use a value of 110 kΩ for R_2, the closest standard value from the table in Appendix A-1.

Next, use Equation 6-25 to determine R_1:

$$R_1 = \frac{R_2}{4Q^2} = \frac{106 \text{ k}\Omega}{4 \times 6.15^2} = 701 \text{ }\Omega$$

Use a value of 750 Ω for R_1.

Finally, calculate R_3 from Equation 6-26:

$$R_3 = \frac{R_1}{4\pi^2 f_{center}^2 R_1 C_1 R_2 C_2 - 1}$$

$$= \frac{750 \text{ }\Omega}{4\pi^2 \times (1844 \text{ kHz})^2 \times 110 \text{ k}\Omega \times 0.01 \text{ μF} \times 750 \text{ }\Omega \times 0.01 \text{ μF} - 1}$$

$$= 6978 \text{ }\Omega$$

With the choice of a value of 6.8 kΩ for R_3, our filter circuit design is complete.

The performance of this circuit is quite sensitive to component values. If 5% tolerance resistors and 10% tolerance capacitors are used, the response of the filter may differ considerably from the design specifications. For a response close to the original specifications, 1% or better tolerance components may be necessary. This comment applies to the next design as well.

EXAMPLE 6-12

Design a narrow-band bandpass filter using TL081C op-amps to pass the range of frequencies between 18 and 22 kHz.

Solution

As in the previous example, we must first determine the number of stages required for the filter.

First determine the center frequency f_{center} by using Equation 6-15:

$$f_{center} = \sqrt{f_{oU} \times f_{oL}} = \sqrt{18 \text{ kHz} \times 22 \text{ kHz}} = 19.9 \text{ kHz}$$

Next, find the bandwidth by using Equation 6-16:

$$f_{BW} = f_{oU} - f_{oL} = 22 \text{ kHz} - 18 \text{ kHz} = 4.0 \text{ kHz}$$

Now use Equation 6-14 to determine Q:

$$Q = \frac{f_{center}}{f_{BW}} = \frac{19.9 \text{ kHz}}{4.0 \text{ kHz}} = 4.98$$

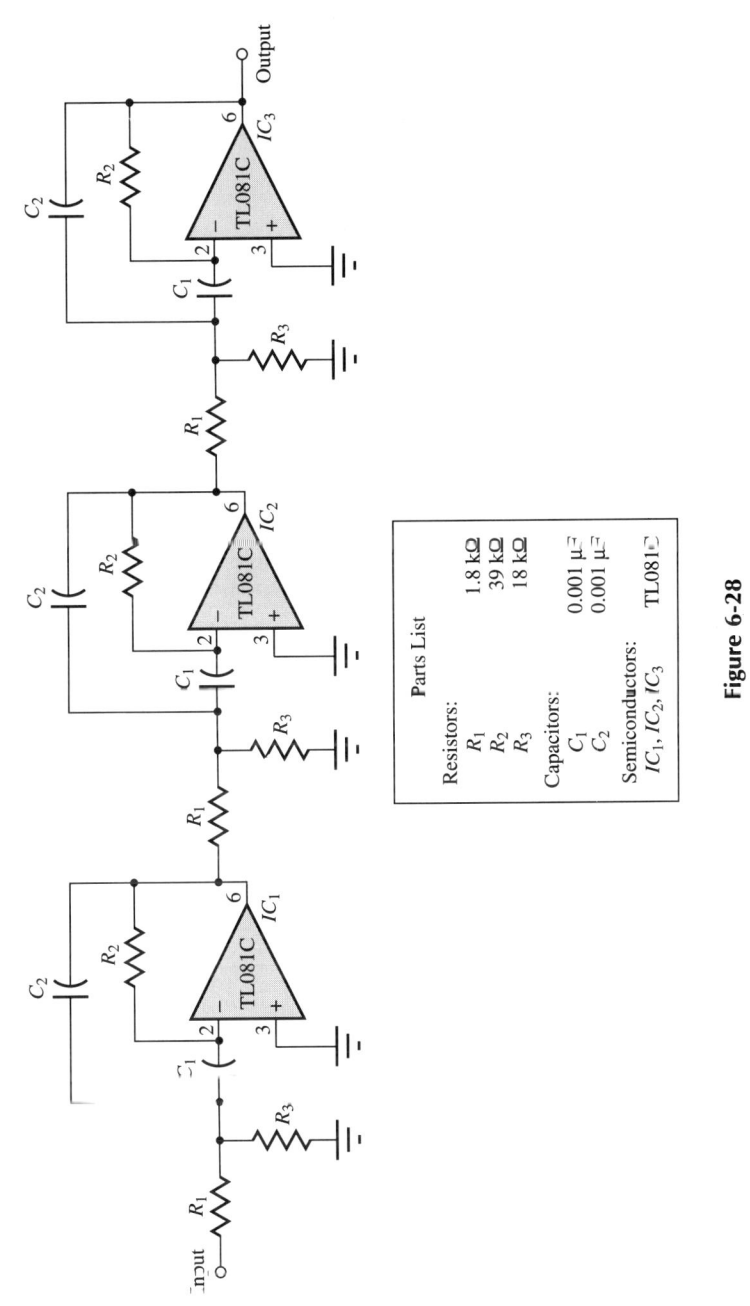

Figure 6-28

Parts List

Resistors:
R_1 1.8 kΩ
R_2 39 kΩ
R_3 18 kΩ

Capacitors:
C_1 0.001 μF
C_2 0.001 μF

Semiconductors:
IC_1, IC_2, IC_3 TL081C

The maximum Q of a single-op-amp bandpass filter using a TL081C is given by Equation 6-19 as

$$Q_{max} = \sqrt{\frac{f_{GB}}{20 f_{center}}} = \sqrt{\frac{3 \text{ MHz}}{20 \times 19.9 \text{ kHz}}} = 2.74$$

Our filter will require more than one stage. To determine the number of stages required, use Equation 6-23:

$$Q_1 = Q_n \sqrt{2^{1/n} - 1}$$

Try $n = 2$ with $Q_n = 4.98$. This gives $Q_1 = 3.21$. This is slightly too large, so try $n = 3$. This gives $Q_1 = 2.54$, which is less than the maximum value determined. We will have to design our filter with three stages, each stage having $Q = 2.54$.

We can now draw the circuit for the filter. It is shown in Figure 6-28. Choose $C_1 = C_2 = 0.001$ μF. Calculate R_2 from Equation 6-24:

$$R_2 = \frac{Q}{\pi f_{center} C_2} = \frac{2.54}{\pi \times 19.9 \text{ kHz} \times 0.001 \text{ μF}} = 40.6 \text{ k}\Omega$$

Use a value of 39 kΩ for R_2.

Next, calculate R_1 from Equation 6-25:

$$R_1 = \frac{R_2}{4Q^2} = \frac{40.6 \text{ k}\Omega}{4 \times 2.54^2} = 1573 \text{ }\Omega$$

Use a value of 1.8 kΩ for R_1.

Finally, calculate R_3 from Equation 6-26:

$$R_3 = \frac{R_1}{4\pi^2 f_{center}^2 R_1 C_1 R_2 C_2 - 1}$$

$$= \frac{1.8 \text{ k}\Omega}{4\pi^2 (19.9 \text{ kHz})^2 \times 39 \text{ k}\Omega \times 0.001 \text{ μF} \times 1.8 \text{ k}\Omega \times 0.001 \text{ μF} - 1}$$

$$= 18.5 \text{ k}\Omega$$

Choose a value of 18 kΩ for R_3.

The design of our three-stage filter circuit is now complete.

Narrow-Band Band-Rejection Active Filters

A narrow-band band-rejection filter rejects frequencies over a bandwidth that is small compared with the center frequency of the filter. Like the narrow-band bandpass filter, the Q of the filter will be greater than 1. The Q of the circuit is still defined by Equation 6-14, where the center frequency is the center of the rejection band and the bandwidth is the width of the rejection band.

It is possible to devise a circuit, equivalent to that shown in Figure 6-26, that can be used for a narrow-band band-rejection filter. Such a circuit would require an entirely new set of design equations. Instead of introducing more equations,

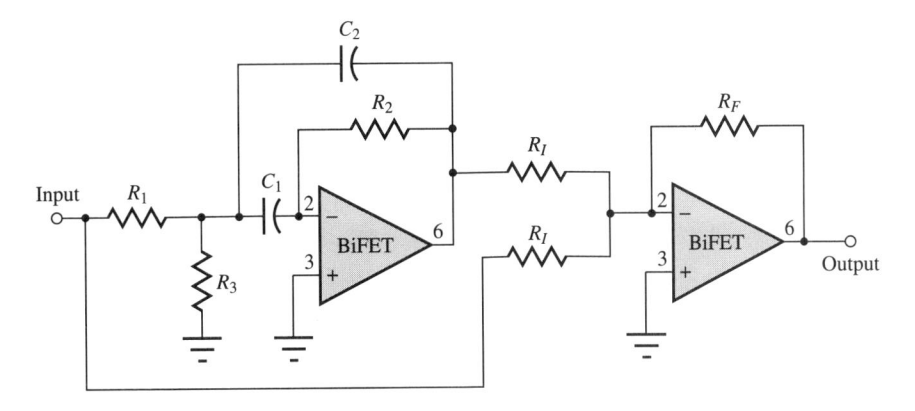

Figure 6-29 Narrow-band band-rejection active filter using a single op-amp.

we will take a different approach. It is possible to produce a band-rejection filter using a bandpass filter and a summing amplifier using the circuit shown in Figure 6-29. In this circuit, the summing amplifier combines the unattenuated input with the output of the bandpass filter. The output of the bandpass filter is inverted, since the op-amp is configured as an inverting amplifier. Any frequency in the passband of the bandpass filter cancels the corresponding frequency of the input signal in the summing stage. Hence, the output from the summing stage is the input signal attenuated by the response of the bandpass filter, in other words, a band-rejection response.

The bandpass portion of the circuit is designed exactly the same as the bandpass filter discussed previously. The same formulae and the same procedures are used. The summing circuit is the standard unity-gain inverting amplifier introduced in Section 4-1. An inverting summing amplifier is used to avoid any crosstalk between the two inputs. Because the gain is unity, the feedback resistor R_F is equal to the two input resistors R_I. Input resistor R_I must be chosen large enough so that it does not load the bandpass filter input. The input impedance of the bandpass filter is approximately R_1, so choose $R_I \gg R_1$; typically, $R_I = 100 R_1$.

As with the bandpass filter, the performance of the narrow-band band-rejection filter is quite dependent on component values.

EXAMPLE 6-13

Design a narrow-band band-rejection filter to reject all frequencies between 4.0 and 5.0 kHz. Use TL081C op-amps.

Solution

The design of this circuit is similar to the design of the narrow-band bandpass filter, except that we must sum the inverted filter output with the input signal.

First, we need to determine the number of stages in the bandpass filter. To do this, we must calculate Q for the filter.

Determine the center frequency f_{center} from Equation 6-15:

$$f_{center} = \sqrt{f_{oU} \times f_{oL}} = \sqrt{4.0 \text{ kHz} \times 5.0 \text{ kHz}} = 4.47 \text{ kHz}$$

Calculate the bandwidth from Equation 6-16:

$$f_{BW} = f_{oU} - f_{oL} = 5.0 \text{ kHz} - 4.0 \text{ kHz} = 1.0 \text{ kHz}$$

Now we can determine the Q of the circuit from Equation 6-14:

$$Q = \frac{f_{center}}{f_{BW}} = \frac{4.47 \text{ kHz}}{1.0 \text{ kHz}} = 4.47$$

The maximum Q of a single-op-amp bandpass filter using a TL081C is given by Equation 6-19:

$$Q_{max} = \sqrt{\frac{f_{GB}}{20 f_{center}}} = \sqrt{\frac{3 \text{ MHz}}{20 \times 4.47 \text{ kHz}}} = 5.79$$

The filter Q is less than Q_{max}, so it is possible to design this filter with a single op-amp with the circuit in Figure 6-29. This circuit has been redrawn as Figure 6-30.

Choose $C_1 = C_2 = 0.01 \, \mu\text{F}$. Calculate R_2 from Equation 6-24:

$$R_2 = \frac{Q}{\pi f_{center} C_2} = \frac{4.47}{\pi \times 4.47 \text{ kHz} \times 0.01 \, \mu\text{F}} = 31.8 \text{ k}\Omega$$

Use a value of 33 kΩ for R_2 as the closest standard value from Appendix A-1.

Next, calculate R_1 from Equation 6-25:

$$R_1 = \frac{R_2}{4Q^2} = \frac{31.8 \text{ k}\Omega}{4 \times 4.47^2} = 398 \, \Omega$$

Use a value of 430 Ω for R_1.

Finally, calculate R_3 from Equation 6-26:

$$R_3 = \frac{R_1}{4\pi^2 f_{center}^2 R_1 C_1 R_2 C_2 - 1}$$

$$= \frac{430}{4\pi^2 (4.47 \text{ kHz})^2 \times 33 \text{ k}\Omega \times 0.01 \, \mu\text{F} \times 430 \, \Omega \times 0.01 \, \mu\text{F} - 1} = 3604 \, \Omega$$

Use a value of 3.6 kΩ for R_3.

Lastly, we must design the summing circuit. The summing circuit must be inverting to prevent crosstalk between the inputs. Furthermore, since the input impedance of the summing circuit must not load the input to the filter, we should choose R_I to be at least 100 times the input impedance of the filter. The input impedance of the filter is approximately 3.6 kΩ, so R_I should be 360 kΩ. Since

Parts List

Resistors:
- R_1 430 Ω
- R_2 33 kΩ
- R_3 3.6 kΩ
- R_I 360 kΩ
- R_F 360 kΩ

Capacitors:
- C_1 0.01 μF
- C_2 0.01 μF

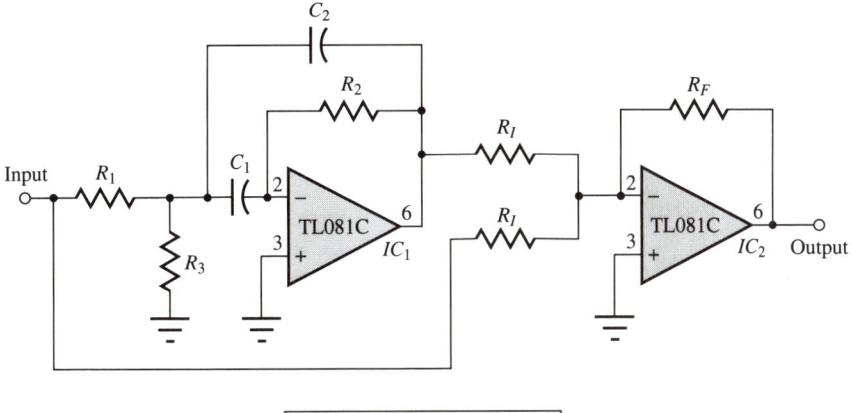

Semiconductors:
- IC_1, IC_2 TL081C

Figure 6-30

the gain for both inputs to the summing circuit is to be unity, choose resistors R_I and R_F to be the same value, namely 360 kΩ.

Our filter circuit is now designed.

6-4 Switched-Capacitor Filters

In the various types of active filters discussed thus far, there has been one serious limitation—the filters are very sensitive to component values. In general, the more complex the filter and the more poles it has, the more difficult it is to get the desired output response with standard components. For optimal performance, precise component values must be used. These can be expensive, difficult to obtain, or, if variable components are used, tedious to align.

There is, however, a solution to this problem—the **switched-capacitor filter**. Switched-capacitor filters are integrated circuits designed to simulate various types of active filters. A single integrated circuit chip can replace a complex active filter circuit. Precise component values are achieved by high frequency switching.

Some switched-capacitor filters are designed to simulate a single type of filter, but most modern switched-capacitor filters are multipurpose and can be programmed by external circuit connections to act as low-pass, high-pass, bandpass, or band-rejection filters.

Switched-Capacitor Theory

If we took a basic active filter circuit and integrated it onto a single chip, the circuit would still require precise resistance and capacitance values. Resistors fabricated as part of an integrated circuit chip are hard to produce with the required precision. Small-value capacitors, however, can be produced very precisely. Switched-capacitor filters operate as ordinary active filters but do not contain resistors. Instead, they simulate resistors by switching capacitors. Figure 6-31 illustrates how a switched capacitor can simulate a resistor. With the switch connected as in Figure 6-31(a), the capacitor charges to $Q = C \times V$. In Figure 6-31(b) the capacitor is switched to ground, and the charge that was stored in the capacitor flows to ground. Now suppose that this switching action repeats f times in a second. The amount of charge delivered through the capacitor to ground is $Q_T = Q \times f$ C/s. Current flow I is defined as coulombs per second; thus, the amount of current flowing through the switched capacitor to ground is controlled by the switching rate and is equal to $Q \times f = C \times V \times f$. Knowing the current flow I in coulombs per second and the applied voltage V, one can determine the effective resistance from Ohm's law, $R = V/I$.

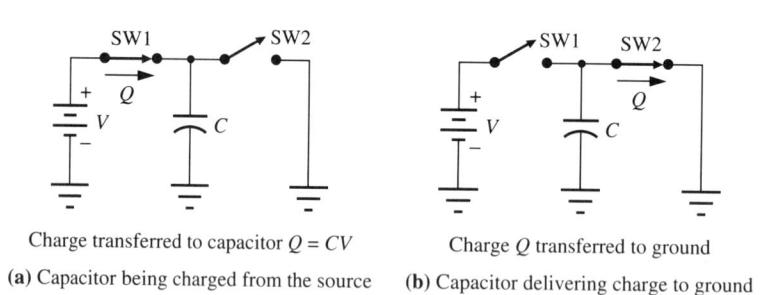

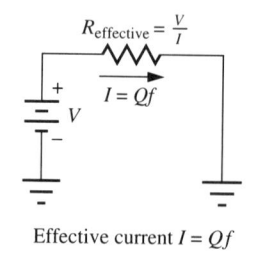

Charge transferred to capacitor $Q = CV$

(a) Capacitor being charged from the source

Charge Q transferred to ground

(b) Capacitor delivering charge to ground

$R_{\text{effective}} = \frac{V}{I}$

$I = Qf$

Effective current $I = Qf$

(c) Equivalent circuit using a resistor

Figure 6-31 Capacitor switching simulating a resistor.

6-4 Switched-Capacitor Filters 261

Suppose we had a 100 pF capacitor connected to a 5 V source. The charge on the capacitor is

$$Q = C \times V = 100 \text{ pF} \times 5 \text{ V} = 5.00 \times 10^{-10} \text{ C}$$

If the capacitor is switched at a rate of 100,000 times per second (100 kHz clock), the total charge passed to ground by the capacitor per second is

$$Q_T = Q \times f = 5.00 \times 10^{-10} \text{ C} \times 100 \text{ kHz} = 5.00 \times 10^{-5} \text{ C/s}$$

The effective current passed is 50 µA. Now use Ohm's law to determine the resistance:

$$R = \frac{V}{I} = \frac{5 \text{ V}}{50 \text{ µA}} = 100 \text{ k}\Omega$$

The switched capacitor is equivalent to a 100 kΩ resistor. If a different switching frequency is used, then a different value of resistance will be simulated. For example, if a frequency of 1 MHz is used instead of 100 kHz, the simulated resistance is 10 kΩ. The accuracy to which a resistor can be simulated depends on the accuracy to which the capacitor can be fabricated and on the precision of the clock frequency.

The MF6 Switched-Capacitor Filter

A number of manufacturers produce switched-capacitor filters. We will consider the National Semiconductor MF6 switched-capacitor filter integrated circuit as a typical example. The filter simulates a 6 pole noninverting Butterworth low-pass filter, as part of a line of filters produced by National Semiconductor, including the MF4 4 pole low-pass filter and the MF8 4 pole bandpass filter. In this series, National also produces the MF5 and MF10 universal filters, but because universal filters operate on a somewhat different principle, we will defer discussion of these until later.

The first two pages of the data sheets for the MF6 switched-capacitor filter are shown in Figure 6-32. There are several versions of the MF6 listed on the data sheets. The CN, CJ, and CWM suffixes refer to 14 pin DIP plastic package, 14 pin DIP ceramic package, and 14 pin wide-body surface-mount package, respectively. The ceramic package version is designed to industrial specifications. We will use the CN package version in our discussion. There are also two numerical suffixes, the MF6CN-50 and the MF6CN-100. These indicate that the cutoff frequency is $\frac{1}{50}$ and $\frac{1}{100}$ of the clock frequency, respectively. We will use the MF6CN-50. The filter can be operated from an internal clock or an external clock. In either case, the maximum clock frequency is 1 MHz, limiting the maximum cutoff frequency of the MF6CN-50 filter to 20 kHz. All versions normally operate from a balanced ±5 V supply, although they can operate from ±2.5 to ±7 V. They can also operate in a TTL mode from a single +5 V supply (the data for this mode are not shown and will not be discussed further). The package also contains two op-amps which are separately available and which can be used for signal conditioning or other applications. The use of these will be discussed a little later.

National Semiconductor

MF6 6th Order Switched Capacitor Butterworth Lowpass Filter

General Description

The MF6 is a versatile easy to use, precision 6th order Butterworth lowpass active filter. Switched capacitor techniques eliminate external component requirements and allow a clock tunable cutoff frequency. The ratio of the clock frequency to the lowpass cutoff frequency is internally set to 50 to 1 (MF6-50) or 100 to 1 (MF6-100). A Schmitt trigger clock input stage allows two clocking options, either self-clocking (via an external resistor and capacitor) for stand-alone applications, or an external TTL or CMOS logic compatible clock can be used for tighter cutoff frequency control. The maximally flat passband frequency response together with a DC gain of 1 V/V allows cascading MF6 sections for higher order filtering. In addition to the filter, two independent CMOS op amps are included on the die and are useful for any general signal conditioning applications.

Features

- No external components
- 14-pin DIP or 14-pin wide-body S.O. package
- Cutoff frequency accuracy of ±0.3% typical
- Cutoff frequency range of 0.1 Hz to 20 kHz
- Two uncommitted op amps available
- 5V to 14V total supply voltage
- Cutoff frequency set by external or internal clock

Block and Connection Diagrams

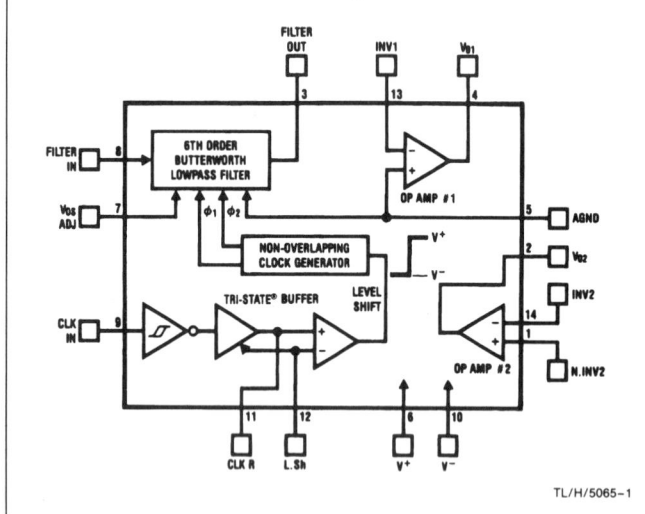

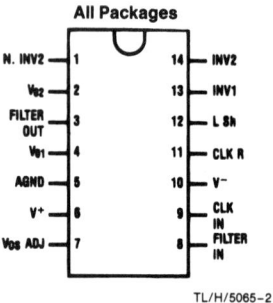

Order Number MF6CWM-50 or MF6CWM-100
See NS Package Number M14B

Order Number MF6CN-50 or MF6CN-100
See NS Package Number N14A

Order Number MF6CJ-50 or MF6CJ-100
See NS Package Number J14A

Figure 6-32 Data sheets for the MF6 switched-capacitor filter. *(Reprinted with permission of National Semiconductor Corporation)*

Absolute Maximum Ratings (Note 11)

If Military/Aerospace specified devices are required, please contact the National Semiconductor Sales Office/Distributors for availability and specifications.

Supply Voltage	14V
Voltage at Any Pin	$V^- - 0.2V$, $V^+ + 0.2V$
Input Current at Any Pin (Note 13)	5 mA
Package Input Current (Note 13)	20 mA
Power Dissipation (Note 14)	500 mW
Storage Temperature	$-65°C$ to $+150°C$
ESD Susceptibility (Note 12)	800V
Soldering Information	
N Package (10 sec.)	260°C
J Package (10 sec.)	300°C
SO Package	
Vapor Phase (60 sec.)	215°C
Infrared (15 sec.)	220°C

See AN-450 "Surface Mounting Methods and Their Effect on Product Reliability" (Appendix D) for other methods of soldering surface mount devices.

Operating Ratings (Note 11)

Temperature Range	$T_{MIN} \leq T_A \leq T_{MAX}$
MF6CN-50, MF6CN-100	$0°C \leq T_A \leq +70°C$
MF6CWM-50, MF6CWM-100	$0°C \leq T_A \leq +70°C$
MF6CJ-50, MF6CJ-100	$-40°C \leq T_A \leq +85°C$
Supply Voltage ($V_S = V^+ - V^-$)	5V to 14V

Filter Electrical Characteristics

The following specifications apply for $f_{CLK} \leq 250$ kHz (see Note 3) unless otherwise specified. **Boldface limits apply for T_{MIN} to T_{MAX}**; all other limits $T_A = T_J = 25°C$.

Parameter	Conditions	MF6CWM-50, MF6CWM-100, MF6CN-50, MF6CN-100 Typical (Note 8)	Tested Limit (Note 9)	Design Limit (Note 10)	MF6CJ-50, MF6CJ-100 Typical (Note 8)	Tested Limit (Note 9)	Design Limit (Note 10)	Units
$V^+ = +5V$, $V^- = -5V$								
f_c, Cutoff Frequency Range (Note 1) MF6-50 Min				**0.1**		**0.1**		Hz
Max				**20k**		**20k**		
MF6-100 Min				**0.1**		**0.1**		
Max				**10k**		**10k**		
Total Supply Current	$f_{CLK} = 250$ kHz	4.0	6.0	**8.5**	4.0	**8.5**		mA
Maximum Clock Feedthrough	Filter Output	30			30			mV (peak-to-peak)
	Op Amp 1 Out	25			25			
	Op Amp 2 Out	20			20			
H_o, DC Gain	$R_{source} \leq 2$ kΩ	0.0	±0.30	**±0.30**	0.0	**±0.30**		dB
f_{CLK}/f_c, Clock to Cutoff Frequency Ratio	MF6-50	49.27±0.3%	49.27±1%	**49.27±1%**	49.27±0.3%	**49.27±1%**		
	MF6-100	98.97±0.3%	98.97±1%	**98.97±1%**	98.97±0.3%	**98.97±1%**		
DC Offset Voltage	MF6-50	-200			-200			mV
	MF6-100	-400			-400			
Minimum Output Voltage Swing	$R_L = 10$ kΩ	+4.0	+3.5	**+3.5**	+4.0	**+3.5**		V
		-4.1	-3.8	**-3.5**	-4.1	**-3.5**		
Maximum Output Short Circuit Current (Note 6)	Source	50	60	**80**	50	**80**		mA
	Sink	1.5	2.0	**3.0**	1.5	**3.0**		
Dynamic Range (Note 2)	MF6-50	83			83			dB
	MF6-100	81			81			
Additional Magnitude Response Test Points (Note 4)	MF6-50 $f_{CLK}=250$ kHz							
	$f = 6000$ Hz	-9.47	-9.47±0.5	**-9.47±0.65**	-9.47	**-9.47±0.65**		dB
	$f = 4500$ Hz	-0.92	-0.92±0.2	**-0.92±0.3**	-0.92	**-0.92±0.3**		
	MF6-100 $f_{CLK}=250$ kHz							
	$f = 3000$ Hz	-9.48	-9.48±0.5	**-9.48±0.65**	-9.48	**-9.48±0.65**		dB
	$f = 2250$ Hz	-0.97	-0.97±0.2	**-0.97±0.3**	-0.97	**-0.97±0.3**		

Figure 6-32 (*Continued*)

264   6 Active Filters

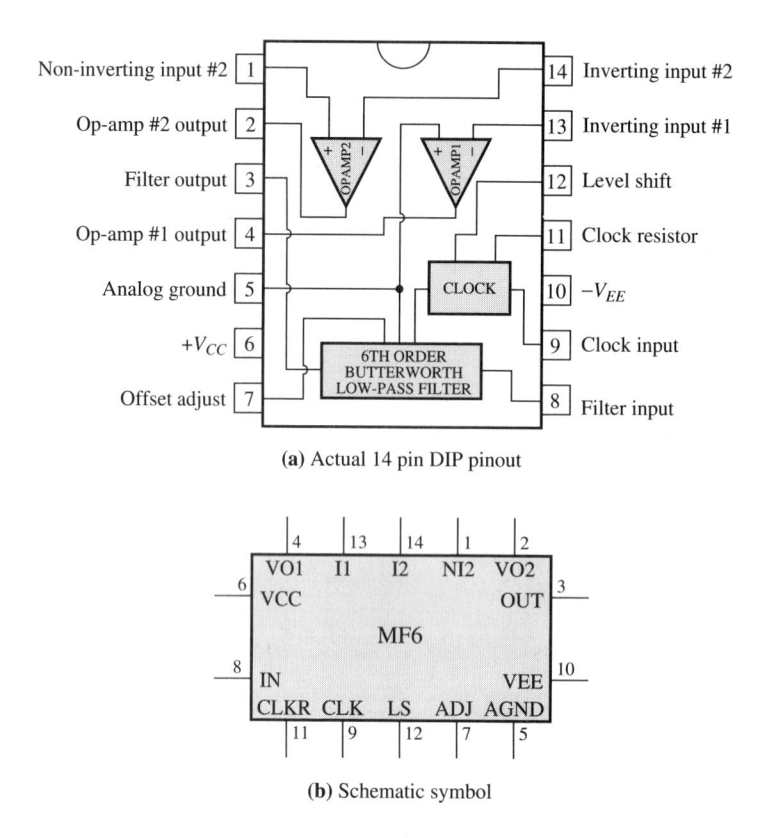

(a) Actual 14 pin DIP pinout

(b) Schematic symbol

Figure 6-33 Pinout and schematic of the MF6 switched-capacitor filter.

Although the data sheets show the package pinout, the pinout has been redrawn in Figure 6-33, along with the schematic symbol of the MF6CN-50. Circuits using the MF6CN-50 are extremely simple, as illustrated in Figure 6-34. Figure 6-34(a) shows the circuit for the filter using the internal clock. A resistor R_1 is connected between pin 11 (CLKR) and pin 9 (CLK), and a capacitor is connected from pin 9 to ground. The clock frequency is determined by the equation

$$f_{CLK} = \frac{1}{1.69 R_1 C_1} \qquad (6\text{-}27)$$

This equation is for ±5 V operation. The numerical constant will be different for other supply voltages. For the MF6CN-50, the clock frequency f_{CLK} must be 50 times higher than the desired cutoff frequency of the filter. Once this is selected, choose a value for C_1 and solve Equation 6-27 for R_1:

$$R_1 = \frac{1}{1.69 f_{CLK} C_1} \qquad (6\text{-}28)$$

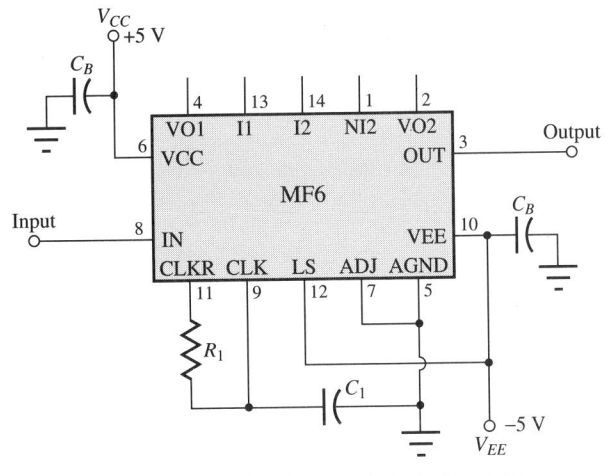

(a) Filter circuit using the internal clock of the MF6

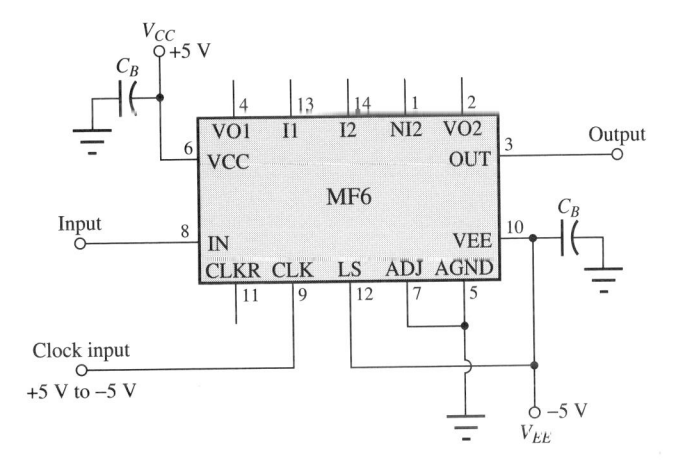

(b) Filter circuit using an external clock operating between +5 V and −5 V

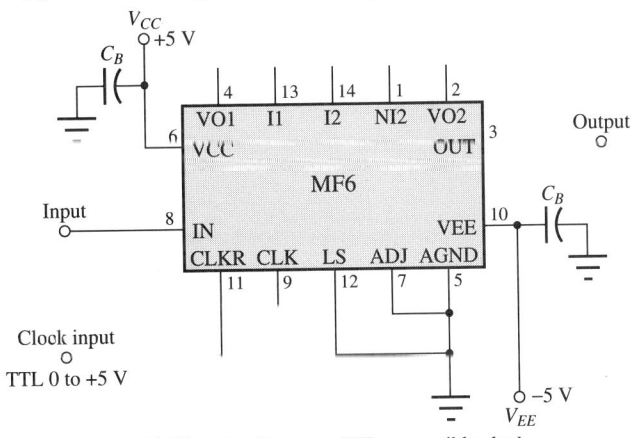

(c) Filter circuit using a TTL compatible clock

Figure 6-34 Low-pass filter circuits using the MF6 switched-capacitor filter.

When using the internal clocking circuit, the level shift LS (pin 12) must be connected to the negative supply. The only other connections required are the +5 V power supply to V_{CC} (pin 6), the −5 V power supply to V_{EE} (pin 10), the input to pin 8, the output to pin 3, the analog ground AGND (pin 5) to ground, and pin 7 (ADJ) also to ground. Pin 7 can be used to adjust the filter offset. If it is not used, it should be connected to ground.

If an external clocking circuit is used, the circuit is even simpler. Figure 6-34(b) shows the circuit for an external clock operating between ±5 V. For example, a clock similar to the relaxation oscillator or bootstrap oscillator, to be discussed in Sections 10-2 and 10-3, could be used. In this case, the clock input is applied to pin 9 and the level shift LS (pin 12) is connected to the negative supply. Pin 11 (CLKR) is not used. Figure 6-34(c) shows the circuit if a TTL-compatible clock input is used. TTL signals switch between 0 and 5 V instead of +5 and −5 V, and are generated by such circuits as that to be developed in Section 11-1 using the LM566 voltage controlled oscillator. For this type of clock signal, the clocking signal is input to pin 11 (CLKR), the level shift (pin 12) is connected to ground, and pin 9 (CLK) is left open.

Bypass capacitors, typically 0.1 μF mica disks, are normally connected between each power supply input and ground to eliminate any noise or signal on the power supply rails. This is the complete circuit of a 6 pole active low-pass filter. Notice the almost complete absence of external components. If a greater roll-off rate is required, then two (or more) MF6CN-50 switched-capacitor filters can be connected in series. If a lower roll-off is required, an MF4CN-50 with 4 poles can be used. The operation of the MF4CN-50 is identical to the MF6CN-50.

The MF8 (available as the MF8CCN commercial version or MF8CCJ industrial version) is the equivalent bandpass filter to the MF4 and MF6 low-pass filters. The one difference between a low-pass filter and a bandpass filter is that bandwidth or a Q must be specified for the bandpass filter. In the MF8, this is done by programming a 5 bit binary word into five "Q-set" pins, allowing 1 of 32 different Q values between 0.45 and 90 to be selected. Otherwise, the MF8 is similar in operation to the MF4 and the MF6, and the circuits are quite similar. There is only a single uncommitted op-amp provided with the MF8.

Thus far, we have seen only low-pass and bandpass switched-capacitor filters in the MF series. There are no high-pass or band-rejection switched-capacitor filters in this series, nor are any required. Recall the procedure used in the previous section for converting a single op-amp bandpass filter into a band-rejection filter by using a summing circuit. This same procedure can be used to convert the MF6 low-pass switched-capacitor filter into a high-pass filter, and can also be used to convert the MF8 bandpass filter into a band-rejection filter. The purpose of the op-amps included with switched-capacitor filters should now be apparent—they are used for filter conversion.

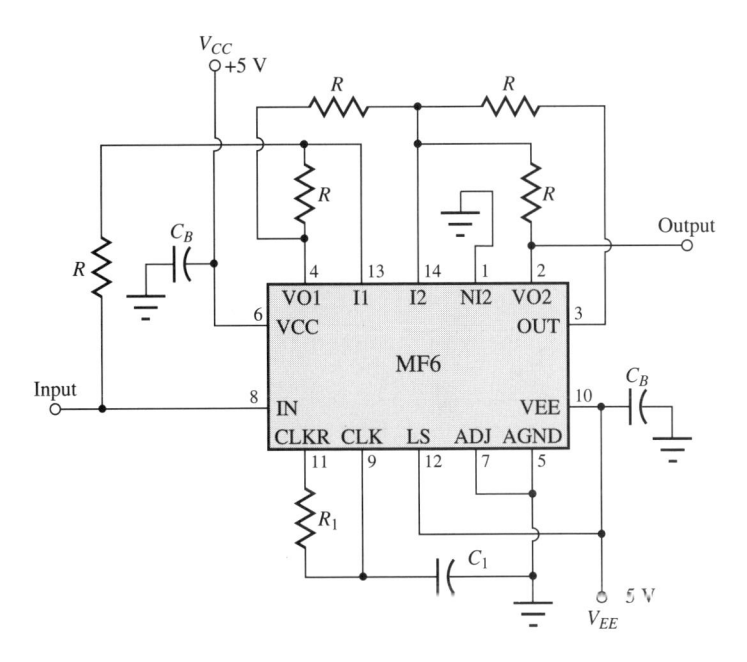

Figure 6-35 MF6 switched-capacitor filter configured as a high-pass filter using the two internal op-amps. All resistors have the same value.

The circuit shown in Figure 6-35 illustrates how these op-amps are used to convert the MF6 into a high-pass filter. The input signal is applied to the inverting input of op-amp 1, which is configured as a unity-gain amplifier, as well as to the input of the filter. The inverted signal from op-amp 1 is added to the noninverted output signal of the filter using op-amp 2. Over the frequency range where the input signal is not attenuated by the filter, the two inputs to the summing amplifier are equal but 180° out of phase, so they cancel. Over the frequency range where the filter attenuates the signal, the output of the summing amplifier is simply the inverted input signal. The net frequency response is that of a 6 pole high-pass filter.

EXAMPLE 6-14

Use a MF6CN-50 switched capacitor filter to synthesize a unity-gain low-pass Butterworth filter with a critical frequency of 3000 Hz and a roll-off of 120 dB per decade. Use the internal clocking circuit.

Solution

The design of a switched-capacitor filter circuit is much simpler than a conventional active filter circuit. Essentially, all that must be designed is the clock frequency.

Since we will be using the internal clock, the circuit that we will require is that shown in Figure 6-34(a). This circuit has been redrawn as Figure 6-36. Since the filter is to have a cutoff frequency of 3000 Hz, the clock frequency must be 50 times higher, namely 150 kHz. This clock frequency is determined by R_1 and C_1 from Equation 6-27:

$$f_{CLK} = \frac{1}{1.69 R_1 C_1}$$

Choose C_1 as 0.001 µF and solve Equation 6-28 to find R_1:

$$R_1 = \frac{1}{1.69 f_{CLK} C_1} = \frac{1}{1.69 \times 150 \text{ kHz} \times 0.001 \text{ µF}} = 3945 \text{ }\Omega$$

Choose R_1 as 3.9 kΩ, the closest standard resistor.

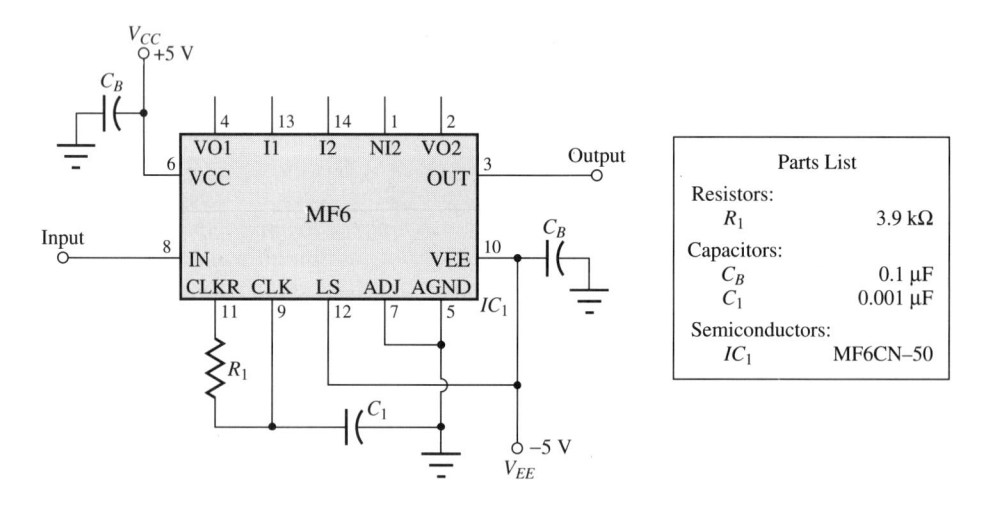

Figure 6-36

Our filter design is now complete. Note that we used a standard 5% resistor and a 10% capacitor. Our clock frequency may only be accurate to 15%. To obtain a more precise response, use 1% or better components, or use a variable resistor for R_1 and adjust to obtain the exact frequency. If the utmost precision is required, then an external, crystal controlled oscillator should be used. This will give a filter precision limited by the manufacturing tolerances of the switched-capacitor filter, typically 0.3%.

State-Variable Filters

Switched-capacitor filters such as the MF4 and the MF6 represent one method of using switched-capacitor technology to produce single-chip active filters. The limitation is that a separate chip must be designed for each application. As we have seen, the MF6CN-50 is a 6 pole low-pass filter with a Butterworth response characteristic. To generate a 6 pole low-pass filter with, say, a Chebyshev response would require a different chip. For more flexibility in filter response selection, a class of switched-capacitor filters called **universal filters** has been developed. The universal switched-capacitor filter uses a filter circuit called a **state-variable filter**. A state-variable filter uses integrating circuits as filters with provision for adding or subtracting integrator outputs to generate any desired filter response.

The simple integrating circuit, such as discussed in Chapter 4, shown in Figure 6-37(a) is a low-pass filter. The gain of the circuit is the feedback impedance, equal to the reactance of the feedback capacitor, $X_C = 1/2\pi f C$, divided by the input resistance R:

$$A_V = \frac{1}{2\pi f C R} \tag{6-29}$$

The gain decreases at 20 dB per decade exactly as in a single-pole low-pass filter. The frequency at which the gain of the integrator is unity is given by

$$f_o = \frac{1}{2\pi C R} \tag{6-30}$$

This is exactly the same as the formula for the critical frequency of a low-pass filter given as Equation 6-5.

To create a state-variable filter, two such integrator stages are connected in series with a standard unity-gain inverting amplifier, as shown in Figure 6-37(b). The two integrating amplifiers give a 2 pole 40 dB per decade roll-off. The feedback resistor R from the output of the second integrator to the inverting input of the first amplifier turns this first amplifier into a summing amplifier. For frequencies below the critical frequency, the feedback signal is identical in amplitude to the input signal but 180° out of phase (after three inversions). The summing amplifier combines these signals to produce zero output. For frequencies above cutoff, the feedback is small, and the output of the summing amplifier is equal to the input signal. The output from the summing amplifier is equivalent to a high-pass filter, and the output from the second integrator is equivalent to a low-pass filter.

To generate a bandpass filter from this circuit, we connect a voltage divider consisting of R_1 and R_2 from the output of the first integrator back to the non-inverting input of the first amplifier, as shown in Figure 6-37(c). The voltage divider effectively connects a low-pass filter in series with a high-pass filter to generate a bandpass filter. The Q of the bandpass filter is determined by the following equation:

$$Q = \frac{R_1 + R_2}{3R_1} \tag{6-31}$$

270 6 Active Filters

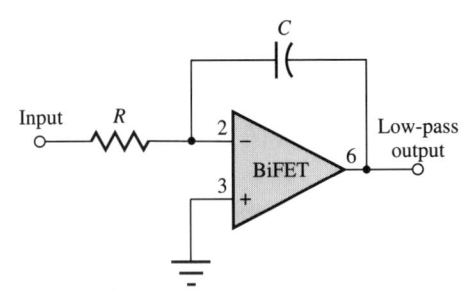

(a) Simple low-pass filter using an integrator

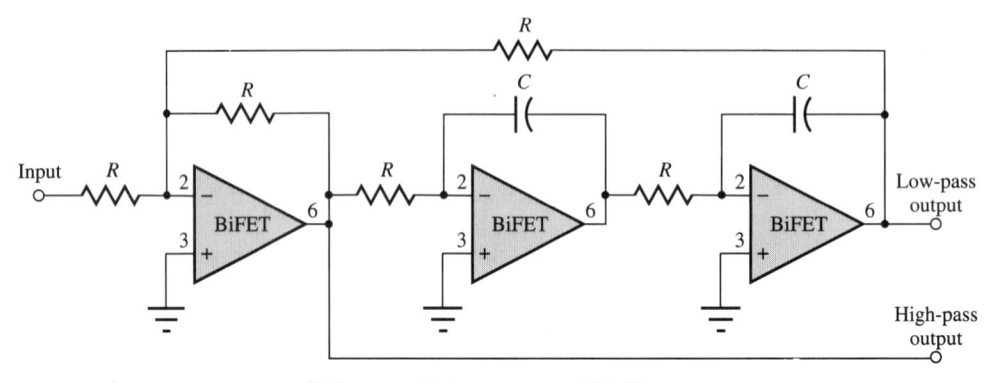

(b) Low-pass/high-pass state-variable filter

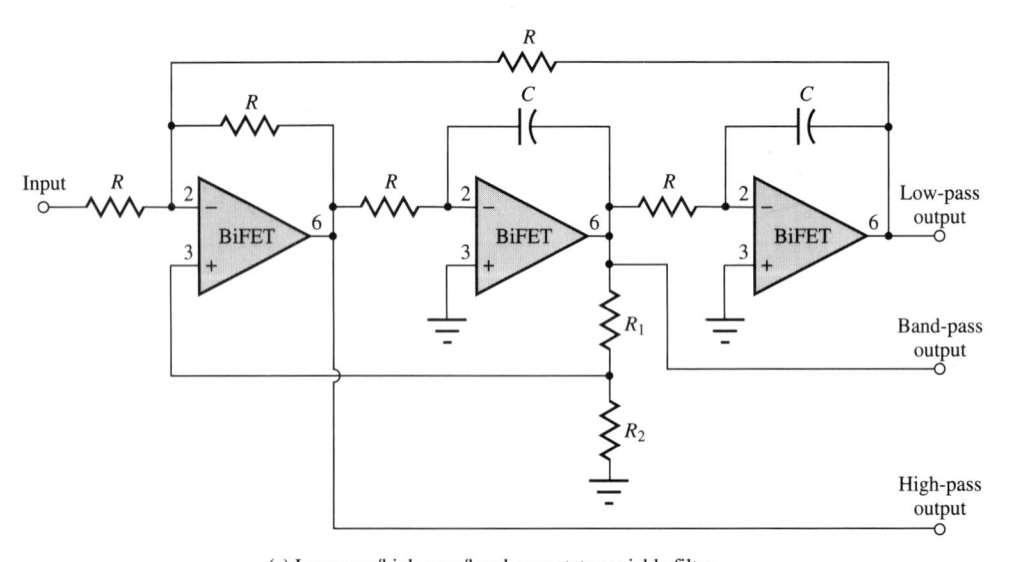

(c) Low-pass/high-pass/band-pass state-variable filter

Figure 6-37 State-variable universal filters.

The output from the first integrator stage is equivalent to the output of a bandpass filter.

To generate a notch filter, we can sum either the high-pass output and the low-pass output by a fourth op-amp, or we can sum the input and the bandpass output.

There is little advantage to using a state-variable filter constructed from discrete op-amps over the other active filter circuits discussed earlier. The state-variable filter, however, is an excellent base from which to design an integrated circuit universal switched-capacitor filter. All types of filters can be synthesized from the single-base filter. The MF5 and the MF10 universal filters are based on the state-variable concept. Rather than discuss one of these, however, we will briefly look at the LMF100 universal filter, which reflects the current state-of-the-art in universal filters.

The LMF100 Universal Filter

The first two pages of the data sheets for the LMF100 universal filter are shown in Figure 6-38. The LMF100 is pinout-compatible to the MF10, but is of much more recent design with considerably enhanced features. Compared with the MF10, the LMF100 has over three times the frequency range, with operation up to 100 kHz and overall improved performance. To operate, the LMF100 simply needs an external clock with a frequency up to a maximum of 3.5 MHz and, typically, four external resistors.

The LMF100 consists of two identical second-order state-variable filters. The internal operation of these filters is quite complex, and by the addition of external resistors in various configurations they can be set to function in one of seven different modes to generate a number of filter types. Mode 3 is the most general-purpose mode for simple filter operation, and we will discuss only this mode. Configuration of either of the second-order filters in mode 3 through the addition of four external resistors allows the LMF100 to be used as a low-pass, a high-pass, or a bandpass filter. The resistor values chosen allow selection of a Butterworth, Bessel, Chebyshev, or Cauer response. If both second-order filters are used and connected in series (eight resistors total), a 4 pole filter of any response can be produced. Any form of bandpass filter can easily be converted into a band-rejection filter by the addition of an external op-amp and three more resistors.

The circuit of a 4 pole low-pass filter designed using the LMF100 is shown in Figure 6-39. Note that there are two sets of four resistors (R_{1A}, R_{2A}, R_{3A}, R_{4A} and R_{1B}, R_{2B}, R_{3B}, R_{4B}). In proper filter design, the "A" resistors do not have the same values as the "B" resistors; in other words, the 4 pole filter is designed as an entity rather than simply as two identical 2 pole filters in series. To design a filter of this type requires an extensive knowledge of the desired filter response and access to tables of filter characteristics. The design of these filters is beyond the intended scope of this text and will not be discussed further. The reader is referred to the manufacturer's data book on the LMF100 for design details.

Without delving into the design details of the LMF100 circuit shown in Figure 6-39, we will briefly look at some of the circuit features. The configuration of the resistors shown in Figure 6-39 sets the filter operation to mode 3. The choice of

LMF100 High Performance Dual Switched Capacitor Filter

General Description

The LMF100 consists of two independent general purpose high performance switched capacitor filters. With an external clock and 2 to 4 resistors, various second-order and first-order filtering functions can be realized by each filter block. Each block has 3 outputs. One output can be configured to perform either an allpass, highpass, or notch function. The other two outputs perform bandpass and lowpass functions. The center frequency of each filter stage is tuned by using an external clock or a combination of a clock and resistor ratio. Up to a 4th-order biquadratic function can be realized with a single LMF100. Higher order filters are implemented by simply cascading additional packages, and all the classical filters (such as Butterworth, Bessel, Elliptic, and Chebyshev) can be realized.

The LMF100 is fabricated on National Semiconductor's high performance analog silicon gate CMOS process, LMCMOS™. This allows for the production of a very low offset, high frequency filter building block. The LMF100 is pin-compatible with the industry standard MF10, but provides greatly improved performance.

Features

- Wide 4V to 15V power supply range
- Operation up to 100 kHz
- Low offset voltage typically
  (50:1 or 100:1 mode) $V_{os1} = \pm 5$ mV
  $V_{os2} = \pm 15$ mV
  $V_{os3} = \pm 15$ mV
- Low crosstalk -60 dB
- Clock to center frequency ratio accuracy $\pm 0.2\%$ typical
- $f_0 \times Q$ range up to 1.8 MHz
- Pin-compatible with MF10

4th Order 100 kHz Butterworth Lowpass Filter

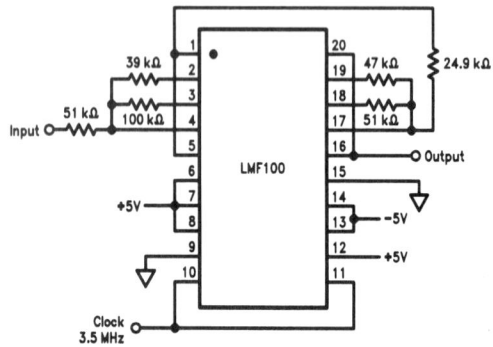

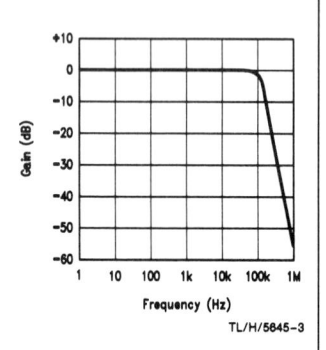

Connection Diagram

Surface Mount and Dual-In-Line Package

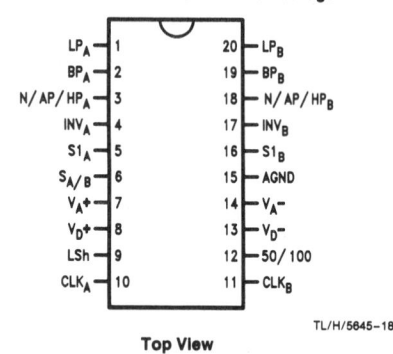

Order Number LMF100AJ, LMF100CCJ, LMF100ACN, LMF100CCN or LMF100CCWM
See NS Package Number J20A, N20A or M20B

Figure 6-38 Data sheets for the LMF100 universal filter. *(Reprinted with permission of National Semiconductor Corporation)*

Absolute Maximum Ratings (Note 1)

If Military/Aerospace specified devices are required, please contact the National Semiconductor Sales Office/Distributors for availability and specifications.

Supply Voltage (V$^+$ − V$^-$)	16V
Voltage at Any Pin	V$^+$ + 0.3V
	V$^-$ − 0.3V
Input Current at Any Pin (Note 2)	5 mA
Package Input Current (Note 2)	20 mA
Power Dissipation (Note 3)	500 mW
Storage Temperature	150°C
ESD Susceptability (Note 11)	2000V

Soldering Information
N Package: 10 sec.	260°C
J Package: 10 sec.	300°C
SO Package: Vapor Phase (60 sec.)	215°C
Infrared (15 sec.)	220°C

See AN-450 "Surface Mounting Methods and Their Effect on Product Reliability" (Appendix D) for other methods of soldering surface mount devices.

Operating Ratings (Note 1)

Temperature Range	$T_{MIN} \leq T_A \leq T_{MAX}$
LMF100ACN, LMF100CCN	0°C ≤ T_A ≤ +70°C
LMF100CCWM	0°C ≤ T_A ≤ +70°C
LMF100CCJ	−40°C ≤ T_A ≤ +85°C
LMF100AJ	−55°C ≤ T_A ≤ +125°C
Supply Voltage	4V ≤ V$^+$ − V$^-$ ≤ 15V

Electrical Characteristics

The following specifications apply for Mode 1, Q = 10 ($R_1 = R_3$ = 100k, R_2 = 10k), V$^+$ = +5V and V$^-$ = −5V unless otherwise specified. **Boldface limits apply for T_{MIN} to T_{MAX}**; all other limits $T_A = T_J$ = 25°C.

Symbol	Parameter	Conditions		LMF100ACN, LMF100CCN, LMF100CCWM Typical (Note 8)	Tested Limit (Note 9)	Design Limit (Note 10)	LMF100AJ LMF100CCJ Typical (Note 8)	Tested Limit (Note 9)	Design Limit (Note 10)	Units
I_S	Maximum Supply Current	f_{CLK} = 250 kHz No Input Signal		9	13	**13**	9	**13**		mA
f_0	Center Frequency Range	MIN		0.1			0.1			Hz
		MAX		100			100			kHz
f_{CLK}	Clock Frequency Range	MIN		5.0			5.0			Hz
		MAX		3.5			3.5			MHz
f_{CLK}/f_0	Clock to Center Frequency Ratio Deviation	V_{Pin12} = 5V or 0V f_{CLK} = 1 MHz	LMF100A	±0.2	±0.6	**±0.6**	±0.2	**±0.6**		%
			LMF100C	±0.2	±0.8	**±0.8**	±0.2	**±0.8**		%
$\Delta Q/Q$	Q Error (MAX) (Note 4)	Q = 10, Mode 1 V_{Pin12} = 5V or 0V f_{CLK} = 1 MHz	LMF100A	±0.5	±4	**±5**	±0.5	**±5**		%
			LMF100C	±0.5	±5	**±6**	±0.5	**±6**		%
H_{OBP}	Bandpass Gain at f_0	f_{CLK} = 1 MHz		0	±0.4	**±0.4**	0	**±0.4**		dB
H_{OLP}	DC Lowpass Gain	$R_1 = R_2$ = 10k f_{CLK} = 250 kHz		0	±0.2	**±0.2**	0	**±0.2**		dB
V_{OS1}	DC Offset Voltage (Note 5)	f_{CLK} = 250 kHz		±5.0	±15	**±15**	±5.0	**±15**		mV
V_{OS2}	DC Offset Voltage (Note 5)	f_{CLK} = 250 kHz	$S_{A/B}$ = V$^+$	±30	±80	**±80**	±30	**±80**		mV
			$S_{A/B}$ = V$^-$	±15	±70	**±70**	±15	**±70**		mV
V_{OS3}	DC Offset Voltage (Note 5)	f_{CLK} = 250 kHz		±15	±40	**±60**	±15	**±60**		mV
	Crosstalk (Note 6)	A Side to B Side or B Side to A Side		−60			−60			dB
	Output Noise (Note 12)	f_{CLK} = 250 kHz 20 kHz Bandwidth 100:1 Mode	N	40			40			μV
			BP	320			320			
			LP	300			300			
	Clock Feedthrough (Note 13)	f_{CLK} = 250 kHz 100:1 Mode		6			6			mV
V_{OUT}	Minimum Output Voltage Swing	R_L = 5k (All Outputs)		+4.0 / −4.7	±3.8	**±3.7**	+4.0 / −4.7	**±3.7**		V
		R_L = 3.5k (All Outputs)		+3.9 / −4.6			+3.9 / −4.6			V
GBW	Op Amp Gain BW Product			5			5			MHz
SR	Op Amp Slew Rate			20			20			V/μs

1-72

Figure 6-38 (*Continued*)

274   **6 Active Filters**

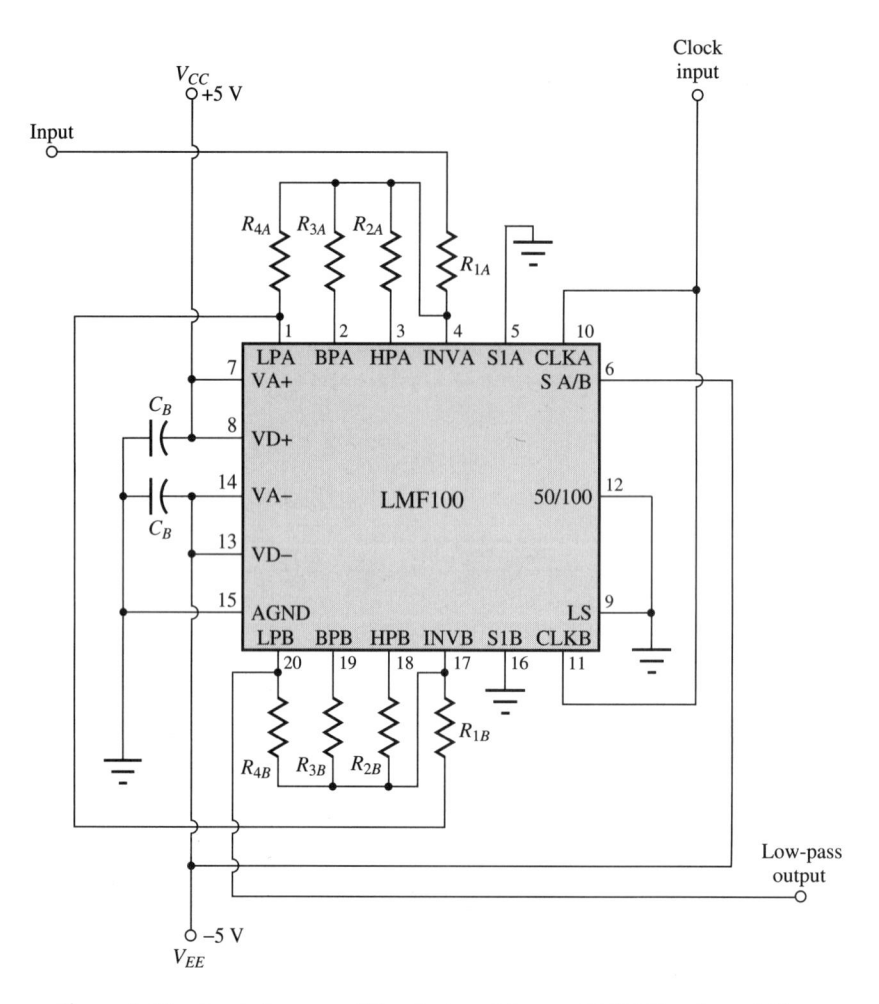

Figure 6-39 4-pole low-pass filter designed from a LMF100 universal filter.

resistor values sets the filter response to Butterworth, Bessel, Chebyshev, or Cauer. The choice of external clock frequency sets the filter critical or center frequency. The use of both filter sections gives an 80 dB per decade response. The output in Figure 6-39 is taken from pin 20 (LPB) to give a low-pass filter response. If a high-pass response is desired, the output is taken from pin 18 (HPB), and if a bandpass response is required, the output is taken from pin 19 (BPB). If only a 2 pole response is required, then only one of the filters is used. The LMF100 is unity gain, so additional circuits can be series connected to provide additional poles as required.

There are four power supply connections to the LMF100. These correspond to positive and negative supplies for the analog (amplifier) and digital (capacitor switching) functions of the chip. Normally, the analog and digital voltages are the same, and the analog and digital power supply inputs are connected as shown in Figure 6-39. Bypass capacitors are generally used at the power supply inputs.

Although this description of the LMF100 universal filter has been brief, it has introduced the concept of the state-variable universal filter and illustrates some of the potential for these devices. The LMF100 is representative of the current state-of-the-art in integrated circuit filter design.

Summary

This chapter introduced filtering circuits and described the various types of filters used in electronics, including low-pass, high-pass, bandpass, and band-rejection filters. The concept of Bode plots and the procedures for using Bode plots to represent the frequency response of filters were described.

Multipole active filters were introduced as a method of obtaining sharper cutoff characteristics, and various types of multipole filter responses were described. Butterworth response filters, with a flat passband and a uniform 20 dB/decade/pole roll-off, are most widely used, although Bessel filters with a linear phase response are used for pulse waveforms to eliminate ringing, and Chebyshev and Cauer responses provide faster initial roll-offs when a sharp cutoff filter is required. For the most part, active filter design is by recipe. Recipes for unity-gain and equal-component multipole Butterworth low-pass and high-pass filters were introduced. Broadband bandpass and band-rejection filters can be made by combining high-pass and low-pass filters. Narrow-band bandpass filters can also be produced by recipe.

Switched-capacitor integrated circuit filters were introduced as an alternative to standard active filters. These filters use switched capacitors to simulate resistors and can be programmed to any cutoff frequency by selecting the switching frequency. National Semiconductor's MF6 low-pass filter is a typical example.

State-variable filters extend the switched-capacitor filter concept by using integrating circuits as filters. It is possible to configure a single state-variable filter as a low-pass filter, as a high-pass filter, as a bandpass filter, or as a band-rejection filter, thus creating essentially a universal filter. The LMF100 state-variable filter was introduced as an example of such a universal filter.

Review Questions

6-1 Basic filter types

1. Define an electronic filter.
2. What are the four common types of filter circuits?
3. Define frequency response for a filter. Sketch the frequency response for the four common types of filters.
4. Define phase response for a filter.

5. What is the difference between a passive filter and an active filter?
6. Describe how the frequency response of a real filter differs from the frequency response of an ideal filter.
7. What is roll-off?
8. What is a decade?
9. What is a Bode plot? Describe how to produce a Bode plot for a filter.
10. How is the critical frequency of a filter defined?
11. What is a pole of a filter? What is the effect of the number of poles in a filter on the frequency response?
12. Define a low-pass filter. Sketch the circuit of a simple low-pass filter that uses passive components.
13. Give three applications of low-pass filters.
14. How is the critical frequency of a low-pass filter defined?
15. What is the main problem with simple, passive low-pass filters? How can this problem be solved?
16. Define a high-pass filter. Sketch the circuit of a simple high-pass filter using passive components.
17. Give two applications of high-pass filters.
18. Define a bandpass filter. Sketch the circuit of a simple bandpass filter that uses passive components.
19. How does the circuit of a band-rejection filter differ from the circuit of a bandpass filter?
20. Describe how the use of a summing amplifier in a band-rejection filter generates the band-rejection response.

6-2 Low-pass and high-pass active filters

21. What is the objection to using inductors and capacitors to produce a multipole filter?
22. Name the four most commonly used filter responses. Explain the main characteristics of each type of response.
23. What is the main problem with getting an active filter designed from a standard recipe to work?
24. Describe the steps in designing a unity-gain low-pass or high-pass filter.
25. What are the units of the resistors and capacitors in the unity-gain unity-frequency circuit?
26. How does the equal-component design differ from the unity-gain design? Which is the preferred design? Which is the simpler design?
27. What is the difference in the circuit configuration between a low-pass design and a high-pass design?

6-3 Bandpass and band-rejection active filters

28. How is the quality factor Q of a bandpass or band-rejection filter defined?
29. By what criterion do we determine whether a bandpass or band-rejection filter is broad-band or narrow-band?
30. Describe how a broadband bandpass filter is constructed from low-pass and high-pass filters.
31. Describe how to make a broadband band-rejection filter from a low-pass filter and a high-pass filter.
32. Describe the design procedure for making a single-pole narrow-band bandpass filter.

33. What is the restriction on Q for a narrow-band bandpass filter? How can a higher Q be obtained?
34. What additional element must be added to convert a narrow-band bandpass filter to a band-rejection filter? What is the difference in the actual filter design?
35. What is the one major limitation of the single-op-amp narrow-band bandpass or band-rejection filter?

6-4 Switched-capacitor filters

36. Describe how a capacitor can be switched to simulate a resistor. What determines the effective resistance for a fixed value of capacitor?
37. What is the advantage of a switched-capacitor filter over a filter constructed of a series of op-amps?
38. How is the critical frequency set in a switched-capacitor filter?
39. What is the difference between an MF6CN-100 and an MF6CN-50?
40. What three different clocking procedures can be used for an MF6 switched-capacitor filter?
41. If an external clock is used, what external components are required by an MF6 switched-capacitor filter?
42. What are the differences between an MF4, an MF6, and an MF8?
43. What additional devices are on the MF4, MF6, and MF8 switched-capacitor filter chips?
44. Why are only low-pass and bandpass switched-capacitor filter integrated circuits produced? How can the other types of filters be produced?
45. What is a universal filter?
46. What circuits constitute the main filtering elements of state-variable filters?
47. Explain how a state-variable filter can be used to produce a universal filter.
48. In how many modes can the LMF100 universal filter operate? What is the most common mode of operation?
49. How are the different modes in an LMF100 selected?
50. How is the LMF100 configured to give different filter types (low-pass, high-pass, bandpass)? How is it configured to give different filter response characteristics (Butterworth, Bessel, Chebyshev, Cauer)?

Design and Analysis Problems

6-1 Basic filter types

1. Prepare a Bode plot showing the frequency response of a low-pass filter with a critical frequency of 2500 Hz and a roll-off of 60 dB per decade.
2. Prepare a Bode plot showing the frequency response of a high-pass filter with a critical frequency of 1500 Hz and a roll-off of 40 dB per decade.
3. Prepare a Bode plot showing the frequency response of a bandpass filter with a lower critical frequency of 500 Hz and an upper critical frequency of 7000 Hz. Both the low-pass and high-pass roll-offs are 80 dB per decade.
4. Prepare a Bode plot showing the frequency response of a low-pass filter with a critical frequency of 2500 Hz and an attenuation of 35 at a frequency of 3000 Hz.

5. Prepare a Bode plot showing the frequency response of a bandpass filter with a lower critical frequency of 80 Hz and an upper critical frequency of 12,000 Hz if the attenuation is 30 at 70 Hz and 25 at 13,000 Hz.
6. Prepare a Bode plot showing the frequency response of a band-rejection filter with a lower critical frequency of 3000 Hz and an upper critical frequency of 14,000 Hz if the attenuation is 53 at 4000 Hz and 17 at 13,000 Hz.

6-2 Low-pass and high-pass active filters

7. Design a unity-gain low-pass Butterworth filter with a critical frequency of 1200 Hz and a roll-off of 60 dB per decade.
8. Design a unity-gain high-pass Butterworth filter with a critical frequency of 600 Hz such that the attenuation is at least 100 at a frequency of 500 Hz.
9. Design an equal-component high-pass Butterworth filter with a critical frequency of 120 Hz and a roll-off of 60 dB per decade.
10. Design an equal-component low-pass Butterworth filter with a critical frequency of 15,000 Hz such that the attenuation is at least 100 at a frequency of 16,000 Hz.
11. Design an equal-component high-pass Butterworth filter with a critical frequency of 75 Hz and a roll-off of 120 dB per decade.

6-3 Bandpass and band-rejection active filters

12. Design a unity-gain band-rejection filter with a lower critical frequency of 3000 Hz, an upper critical frequency of 9000 Hz, and a roll-off of 100 dB per decade for both the low-pass and high-pass filters.
13. Design an equal-component bandpass filter with a lower critical frequency of 250 Hz, an upper critical frequency of 7500 Hz, and a roll-off of 80 dB per decade for both the low-pass and high-pass filters.
14. Design a narrow-band bandpass filter to pass a range of frequencies between 5.5 and 6.0 kHz. Calculate the Q of the filter and determine the number of stages required.
15. Design a narrow-band bandpass filter to pass a range of frequencies between 7.5 and 9.5 kHz. Calculate the Q of the filter and determine the number of stages required.
16. Design a narrow-band band-rejection filter to reject a range of frequencies between 10 and 12 kHz. Calculate the Q of the filter and determine the number of stages required.

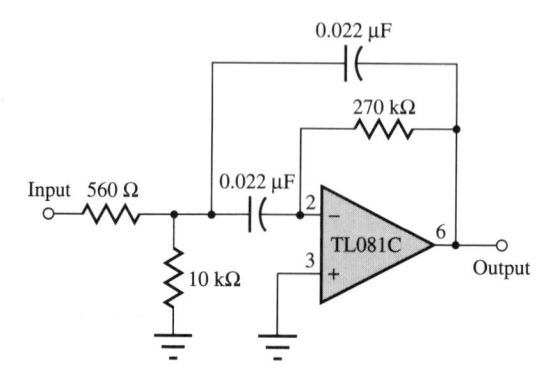

Figure 6-40

17. Analyze the circuit in Figure 6-40 to determine the
    (a) type of filter
    (b) center frequency
    (c) bandwidth.

6-4 Switched-capacitor filters

18. Design a high-pass filter with a critical frequency of 320 Hz, using the MF6 switched-capacitor filter. Use the internal oscillator circuit.

7 | Linear Power Supplies

OUTLINE

7-1 Rectifier Circuits
7-2 Regulation
7-3 Voltage References
7-4 Op-Amp Regulators

KEY TERMS

Peak reverse voltage rating
Maximum steady-state current rating
Maximum surge current
Filter circuit
Ripple voltage
Rectification
Regulation
Regulator
Line regulation
Load regulation
Ripple rejection
Power supply impedance
Voltage reference
Sensing circuit
Error detector
Zener diode
Pass transistor
Darlington configuration
Dropout voltage
Foldback current limiting

OBJECTIVES

On completion of this chapter, the reader should be able to:

- design simple transformer-operated power supplies and determine the output voltage and ripple voltage for capacitor-filtered power supplies.
- define regulation, and calculate line regulation, load regulation, and ripple rejection for a regulated power supply.
- describe the use of zener diodes for regulation and for voltage references.
- understand the advantages of voltage references over simple zener diodes.
- design and analyze simple regulator circuits constructed from op-amps.

Introduction

Virtually every piece of electronic equipment requires electric power for operation. This power is sometimes supplied by batteries, but generally it is derived from AC line power supplied by the local utility. Large generating stations distribute power in the form of alternating current because AC can be readily transformed to high voltages for economic transmission, and then reduced by the local utility to standard operating levels. Unfortunately, however, all electronic equipment requires DC power for operation. In this chapter, we will look at circuits for converting the AC line power to DC power of suitable quality for the operation of electronic equipment.

The first step is to convert the line AC to DC by a process of rectification. This process is covered in virtually any textbook on elementary electronics, but it will be repeated here, partly as a review, but mainly to emphasize the design decisions that must be made in producing a suitable rectifier circuit. Simple rectified AC is seldom adequate for electronic applications. The DC voltage produced generally contains an element of AC ripple and tends to fluctuate in level as either line voltage or load current changes. Regulator circuits are used to stabilize the DC.

This chapter will introduce the concept of regulation and the function of basic regulator circuits. Voltage regulators hold the output voltage constant despite variations in the line voltage and variations in the load.

Regulation can be performed by a simple op-amp acting as an error sensor, sensing the difference between the output and a reference level provided by a zener diode or voltage reference integrated circuit. Several regulator circuits using op-amps will be introduced. More commonly, however, special integrated circuit regulators are used. These will be introduced in Chapter 8.

7-1 Rectifier Circuits

Power supplies are circuits that convert the AC line power to DC power suitable for operating electronic devices. The first stage in this conversion process is to convert the line voltage of 115 V RMS to the level required by the electronic circuit. This operation is best performed by a transformer. Since most modern solid state circuits operate from voltages in the range of 5 to 25 V, step-down transformers are usually required. The transformer must be chosen to provide sufficient voltage for rectification, filtering, and regulation to give the desired output voltage, and must have a sufficient current rating to satisfy the requirements of the load.

Rectification

The low-voltage AC output from the transformer must be converted into DC by rectification. Semiconductor diodes are used in circuits similar to those shown in Figures 7-1(a), 7-2(a), and 7-3(a). Of these, the circuit in Figure 7-3(a) is preferred.

(a) Circuit diagram of rectifier with filter

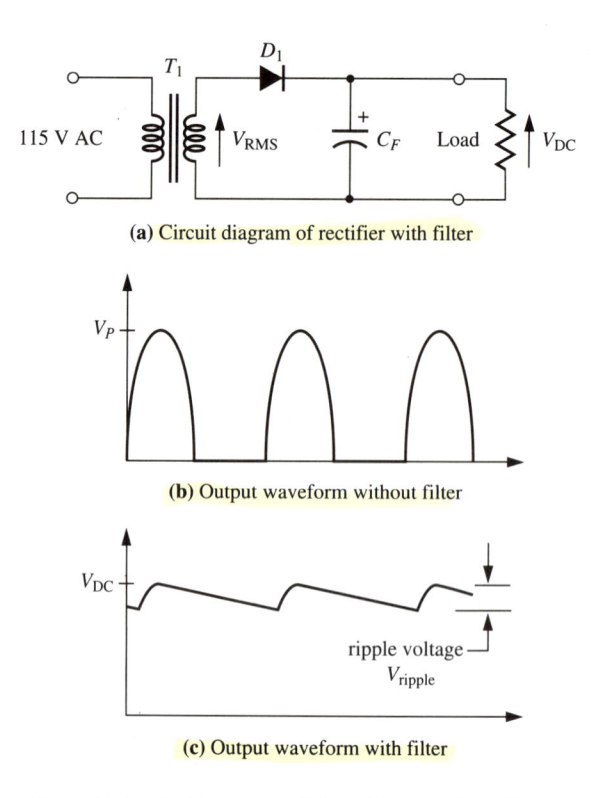

(b) Output waveform without filter

(c) Output waveform with filter

Figure 7-1 Half-wave rectifier with capacitor filter.

Although the bridge rectifier in this circuit can be constructed from discrete diodes, generally special bridge rectifiers are used. Bridge rectifiers are inexpensive, readily available in a variety of current and voltage ratings, and are easy to install.

The output voltage of a rectifier circuit is a pulsating DC waveform shown in Figures 7-1(b), 7-2(b), and 7-3(c), respectively, for the three types of rectifiers. The peak output voltages V_P for these rectifier circuits are

half-wave rectifier $\quad V_P = \sqrt{2}\, V_{RMS} - 0.7 \quad\quad (7\text{-}1)$

full-wave rectifier $\quad V_P = \dfrac{\sqrt{2}\, V_{RMS}}{2} - 0.7 \quad\quad (7\text{-}2)$

bridge-type rectifier $\quad V_P = \sqrt{2}\, V_{RMS} - 1.4 \quad\quad (7\text{-}3)$

Note that the two-diode full-wave rectifier has two disadvantages over the other forms of rectifiers—the voltage is approximately half that of the other two forms of rectifiers, and a center-tapped transformer is required.

Diodes or bridge rectifiers are selected for a particular application on the basis of three quantities: peak reverse voltage rating, maximum steady-state current, and maximum surge current. The **peak reverse voltage rating** is the maximum

7-1 Rectifier Circuits 283

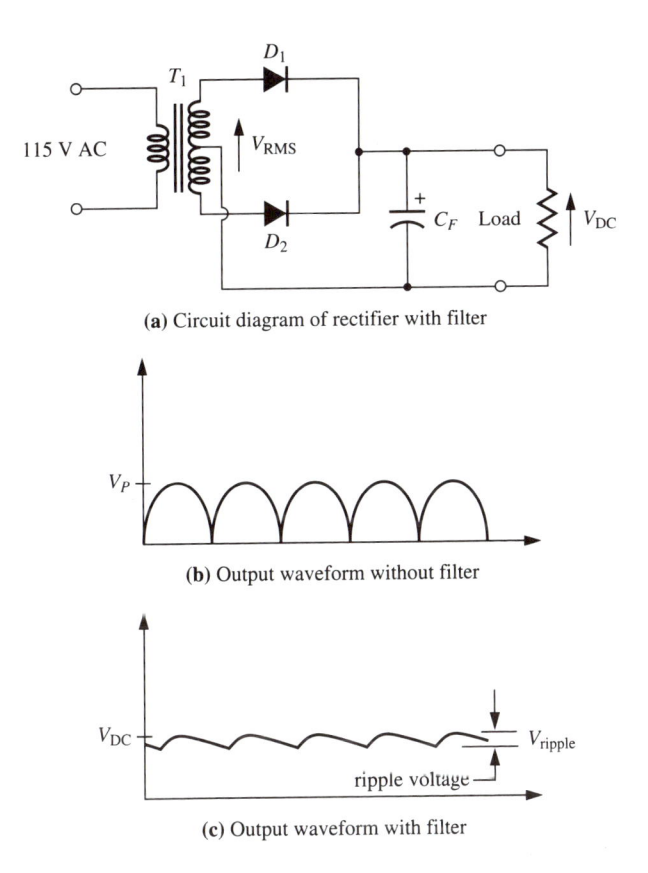

(a) Circuit diagram of rectifier with filter

(b) Output waveform without filter

(c) Output waveform with filter

Figure 7-2 Full-wave rectifier using a center-tapped transformer with a capacitor filter.

reverse voltage that a diode (or bridge rectifier) is capable of withstanding. The diode must have a rating greater than the peak output voltage of the transformer. The **maximum steady-state current rating** is the maximum continuous current the diode can pass. The diode must have a rating greater than the design current of the power supply. The **maximum surge current** is the current that would flow through the diode if the power supply were switched on at the peak of the input voltage and with the filter capacitor across the output totally discharged. The capacitor initially acts as a dead short in the circuit and the only thing limiting the current is the resistance of the diode and the resistance of the transformer windings. Typically, the diode resistance is much less than the resistance of the transformer, so the surge current is approximately

$$I_{surge} = \frac{V_P}{R_W} \tag{7-4}$$

Usually the winding resistance R_W is not known, but it can be estimated from the

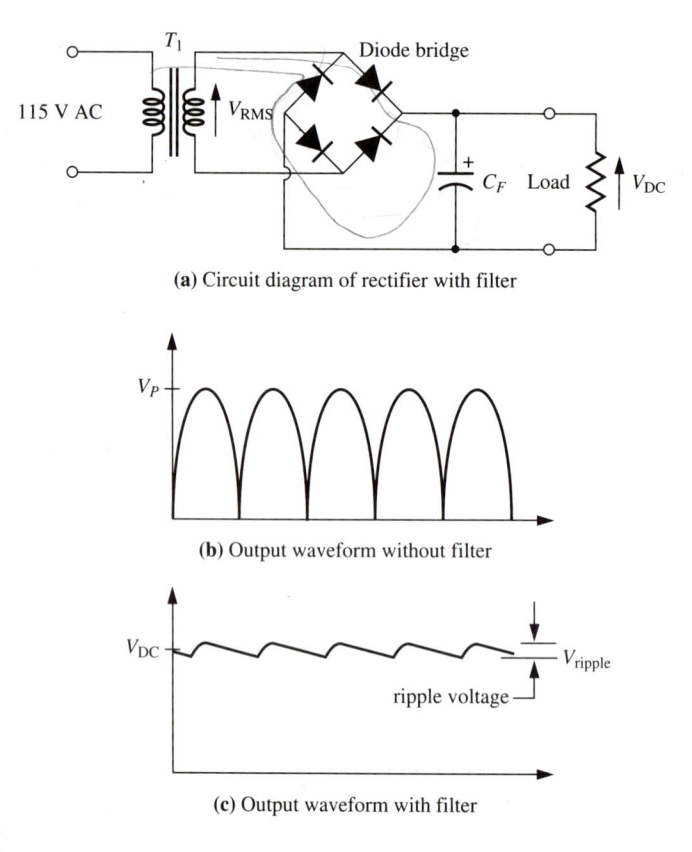

Figure 7-3 Full-wave rectifier using diode bridge with capacitor filter.

(a) Circuit diagram of rectifier with filter

(b) Output waveform without filter

(c) Output waveform with filter

transformer data. The transformer data sheets usually list the full-load voltage V_{FL} at the rated current I_{FL} and the no-load voltage V_{NL} with no current. The difference in voltage is due to the resistance drop in the windings. For full-load current I_{FL} flowing, the voltage drop across the winding resistance is simply $V_{NL} - V_{FL}$, and hence the winding resistance R_W can be approximated by

$$R_W = \frac{V_{NL} - V_{FL}}{I_{FL}} \tag{7-5}$$

Filter Circuits

The pulsating DC produced by a simple rectifier is definitely not acceptable for almost any electronic application. A **filter circuit** is used to remove the pulsations. The filter can be a capacitor, an inductor, or a combination of capacitors and inductors. Most modern power supply circuits use a simple capacitor, with the filtered output being further smoothed by a regulator. Capacitor filters are used in the circuits shown in Figures 7-1(a), 7-2(a), and 7-3(a).

7-1 Rectifier Circuits

With no load, the output voltage is pure DC. The capacitor charges up to the full peak voltage during the first pulse cycle and retains this voltage thereafter. If, however, a load is placed across the capacitor, the capacitor discharges somewhat as it supplies current to the load when the rectifier voltage is less than its peak value. The waveform that would be measured across the filter capacitor or across the load is shown for each of the three filter circuits in Figures 7-1(c), 7-2(c), and 7-3(c), respectively. How much the voltage drops depends on how much charge the capacitor is capable of holding and on how much charge the capacitor must supply to the load. Assuming the capacitor only discharges slightly between charge cycles, the DC output voltages for each of the filters are given by the following equations:

half-wave rectifier
$$V_{DC} = V_P - \frac{0.00833}{R_L C_F} \times V_P \quad (7\text{-}6)$$

full-wave rectifier
and
bridge-type rectifier
$$V_{DC} = V_P - \frac{0.00417}{R_L C_F} \times V_P \quad (7\text{-}7)$$

The derivations of these formulae are lengthy. They are given in most textbooks on electronic devices and will not be repeated here. Note that the full-wave rectifier and the bridge rectifier have the same formula, although the values of V_P will be different.

The variation of voltage across the capacitor is called the **ripple voltage** and is simply an AC component added to the DC output: the larger the AC ripple, the smaller the DC output voltage. It is important to keep the AC ripple reasonably small—just how small will depend on the regulator circuit and design requirements. The equations for calculating the AC ripple for each of the three types of rectifier circuits are

half-wave rectifier
$$V_{\text{ripple}} = \frac{0.0048}{R_L C_F} \times V_P \quad (7\text{-}8)$$

full-wave rectifier
and
bridge-type rectifier
$$V_{\text{ripple}} = \frac{0.0024}{R_L C_F} \times V_P \quad (7\text{-}9)$$

The ripple voltages are in RMS volts. Again, the derivation is quite lengthy, and the reader is referred to a textbook on electronic devices. The formulae for the full-wave rectifier and the bridge-type rectifier are the same, although they may not yield the same value of ripple because the values of V_P will be different. The advantage of full-wave rectification either by two diodes or by a bridge-type rectifier should be obvious from these formulae—the ripple is reduced to half of the half-wave ripple. The advantage of the bridge over the two-diode full-wave rectifier may not be as obvious—the bridge rectifier gives approximately twice the output DC voltage that the simple full-wave rectifier gives for the same transformer. This is apparent when the equations for V_P for these two types of rectifiers are compared.

Usually in design specifications, the output voltage V_{DC} and the ripple voltage V_{ripple} will be supplied. For selecting a suitable transformer, however, the peak voltage V_P must be known. Solve Equation 7-9 for $R_L C_F$, substitute it in Equation 7-7, and solve the resulting equation for V_P to get

$$V_P = V_{DC} + 1.736 V_{ripple} \qquad (7\text{-}10)$$

The design and analysis of a simple rectifier/capacitor filter circuit is quite straightforward, although there are a few practical aspects to consider. As mentioned, the transformer must be selected to give adequate voltage and current. The diodes or bridge rectifier must be selected to withstand the reverse voltage peaks, to pass the required maximum current, and to withstand the maximum surge current. The capacitor (almost always an aluminum electrolytic type) must be chosen to withstand the peak voltage and must have sufficient capacity to reduce the ripple to an acceptable limit. The design of a rectifier/filter circuit is best illustrated by examples.

EXAMPLE 7-1

Calculate the peak voltage, the DC voltage, the ripple voltage, and the surge current for a full-wave rectifier circuit consisting of a Hammond 166F28 transformer, two 1N4001 diodes, and a 2200 µF, 25 V electrolytic capacitor. The resistance of the load is 100 Ω. The circuit for this power supply is shown in Figure 7-4.

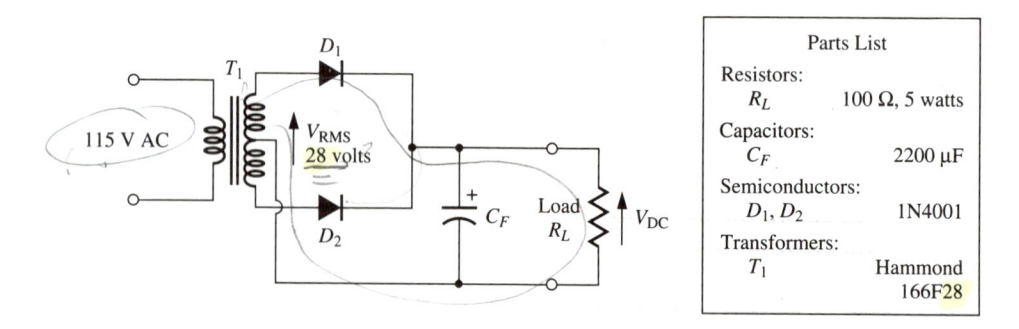

Figure 7-4

Solution

This is a typical problem to analyze the operation of an existing power supply.

First, we must determine the characteristics of the transformer. A number of transformers, including the 166F28, are listed in a table in Appendix A-3. From this table, we find that the full-load voltage of a 166F28 transformer is 28 V. We will use this information to calculate the peak voltage from Equation 7-2:

$$V_P = \frac{\sqrt{2}\, V_{RMS}}{2} - 0.7 = \frac{\sqrt{2} \times 28.0\ V}{2} - 0.7\ V = 19.1\ V$$

This will be the voltage across the capacitor, so the choice of a 25 V rating for the capacitor is adequate.

Next calculate the DC output voltage from Equation 7-5:

$$V_{DC} = V_P - \frac{0.00417}{R_L C_F} \times V_P$$

$$= 19.1\ V - \frac{0.00417}{100\ \Omega \times 2200\ \mu F} \times 19.1\ V = 18.7\ V$$

Now calculate the ripple voltage from Equation 7-9:

$$V_{ripple} = \frac{0.0024}{R_L C_F} \times V_P$$

$$= \frac{0.0024}{100\ \Omega \times 2200\ \mu F} \times 19.1\ V = 0.208\ V_{RMS}$$

Next, we must calculate the surge current. We will need the winding resistance R_W of the transformer. The table in Appendix A-3 lists the no-load voltage V_{NL} of the 166F28 transformer as 31.2 V, and the full-load voltage V_{FL} for the rated current I_{FL} of 0.25 A as 28 V. Use Equation 7-5 to determine R_W:

$$R_W = \frac{V_{NL} - V_{FL}}{I_{FL}} = \frac{31.2\ V - 28.0\ V}{0.25\ A} = 12.8\ \Omega$$

We will assume that the diode resistance is much less than the transformer winding resistance and can be ignored. Since the peak voltage is 19.1 V, the surge current, from Equation 7-4, is

$$I_{surge} = \frac{V_P}{R_W} = \frac{19.1\ V}{12.8\ \Omega} = 1.49\ A$$

Lastly, calculate the load current. This is simply the DC output voltage V_{DC} divided by the load resistance R_L:

$$I_L = \frac{V_{DC}}{R_L} = \frac{18.7\ V}{100\ \Omega} = 0.187\ A$$

The maximum surge current for a 1N4001 is 30 A. Our value of 1.49 A is well within this specification. The peak reverse voltage for the 1N4001 is 50 V. The V_P for the transformer is 19.1 V, again well within the 1N4001 limits. Finally, the maximum continuous current is 1 A, and the rectifier current of 0.187 A is much less than this value.

Our analysis of the circuit is complete. We have calculated that the rectifier circuit has an output voltage of 18.74 V with a ripple of 0.573 V peak-to-peak, and have verified that all the components have adequate ratings.

EXAMPLE 7-2

Design a rectifier circuit to supply 1.2 A at a minimum voltage of 15 V and with a maximum RMS ripple of 1 V.

Solution

This is a realistic design problem. Note the minimal information supplied and the number of decisions the designer must make.

First, we will decide to use a bridge rectifier. This is generally the preferred choice in designing a modern power supply. Bridge rectifiers are economical and easy to install. The circuit is shown in Figure 7-5.

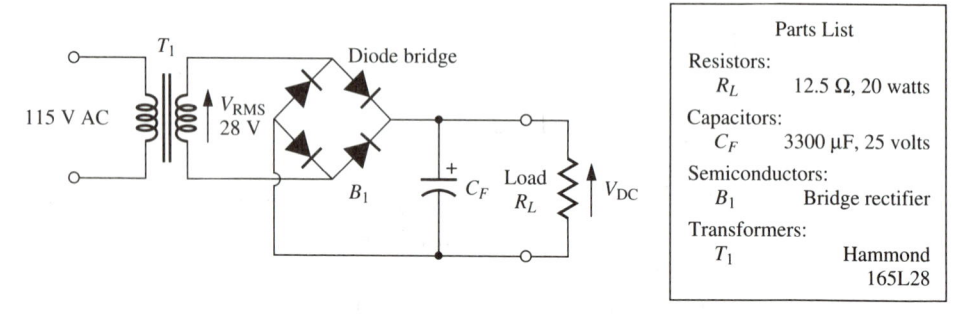

Figure 7-5

Parts List
Resistors:
  R_L 12.5 Ω, 20 watts
Capacitors:
  C_F 3300 μF, 25 volts
Semiconductors:
  B_1 Bridge rectifier
Transformers:
  T_1 Hammond 165L28

To select a bridge rectifier, assume a peak voltage rating equal to twice the output voltage of the finished supply, a current rating at least equal to the design current, and a surge current rating 25 times the design current.

We are not given a load, but we can calculate an approximate value of R_L from the specified output voltage and current by using Ohm's law. We have $I_{FL} = 1.2$ A supplied at $V_{DC} = 15$ V. Thus,

$$R_L = \frac{V_{DC}}{I_{FL}} = \frac{15 \text{ V}}{1.2 \text{ A}} = 12.5 \text{ Ω}$$

Our ripple voltage is given by Equation 7-9:

$$V_{ripple} = \frac{0.0024}{R_L C_F} \times V_P$$

We know $R_L = 12.5$ Ω and have the ripple voltage specified at 1.0 V peak-to-peak, but we do not know either C_F or V_P. We can calculate a value of V_P from Equation 7-10:

$$V_P = V_{DC} + 1.736 V_{ripple}$$
$$= 15.0 \text{ V} + 1.736 \times 1.0 \text{ V} = 16.7 \text{ V}$$

Now find C_F by rearranging Equation 7-9:

$$C_F = \frac{0.0024}{R_L V_{\text{ripple}}} \times V_P$$

$$= \frac{0.0024}{12.5\ \Omega \times 1.0\ \text{V}} \times 16.7\ \text{V} = 0.00321\ \text{F}$$

Choose the next *larger* value of C_F with a voltage rating larger than the output voltage. Appendix A-2 lists some typical capacitor values. We will choose $C_F = 3300\ \mu\text{F}$ at 25 V.

Next, we will use our estimate of V_P to calculate V_{RMS} of the transformer by rearranging Equation 7-3 to solve for V_{RMS}:

$$V_{\text{RMS}} = \frac{V_P + 1.4}{\sqrt{2}}$$

$$= \frac{(16.74\ \text{V} + 1.4\ \text{V})}{\sqrt{2}} = 12.83\ \text{V}$$

Refer to the transformer data sheet in Appendix A-3. We will choose a Hammond 166L14 rated at 14.0 V for a full-load current of 2 A.

The design is complete.

It is left as an exercise for the reader to repeat Example 7-1 to verify the satisfactory operation of the circuit. It will be necessary to use the bridge rectifier equations instead of the full-wave rectifier equations in this analysis.

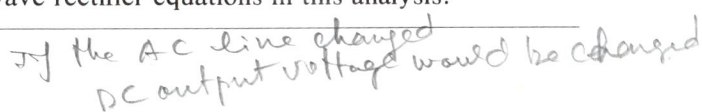

7-2 Regulation

If the AC line voltage changed for any of the simple rectifier and filter circuits shown in Figures 7-1, 7-2, or 7-3, the DC output voltage V_{DC} would also change because, by Equation 7-7, the output voltage from the filter is proportional to the peak input voltage V_P. Even if the line voltage were constant, but the load resistance R_L changed, the filter capacitor would have to deliver a different amount of current, its average charge would be different, and the output DC voltage would change. This is also shown by Equation 7-7. Generally, a DC power supply is expected to provide a constant output voltage despite variations in line voltage or load current. To achieve this, most modern power supplies use an additional circuit (or device) called a **regulator**. The measure of the effectiveness of a regulator at holding the output voltage constant is called the **regulation**.

The regulator is placed in a power supply circuit after the filter, as shown in Figure 7-6. In this figure, the regulator is pictured as a simple box, but is in fact a relatively complex circuit, as we will see later. Before we develop actual regulator circuits, we must look at the requirements of a regulator.

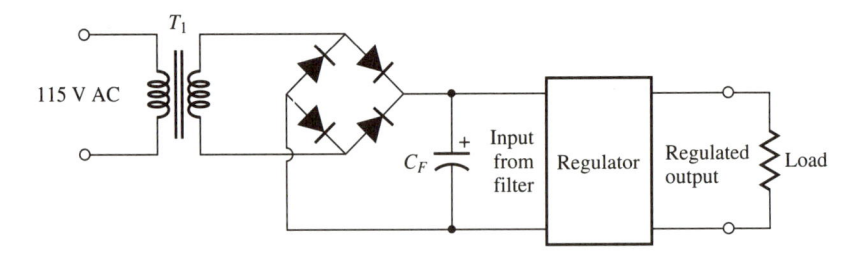

Figure 7-6 Bridge rectifier and capacitor filter with regulator circuit.

Line Regulation

The first requirement of a regulator is that it should try to maintain a constant DC output despite changes in the supply voltage. The measure of the degree to which a regulator achieves this is called the **line regulation**. Line regulation is defined as the amount that the output voltage varies for a change in input voltage:

$$\text{line regulation} = \frac{\Delta V_{out}}{\Delta V_{in}} \tag{7-11}$$

In this equation, ΔV_{in} is the change in input voltage, and ΔV_{out} is the corresponding change in output voltage. The line regulation is usually expressed as millivolts per volt, but is often (and incorrectly) given simply as millivolts.

Line regulation is sometimes given as a percent of output voltage. In this form, the equation is

$$\text{percent line regulation} = \frac{\Delta V_{out}}{\Delta V_{in}} \times \frac{1}{V_{out}} \times 100 \tag{7-12}$$

The smaller the numerical value of line regulation, the better is the regulation. Ideally, the line regulation should be zero, and with a good modern regulator it usually is very close to zero.

Load Regulation

The next requirement is that the regulator should try to maintain a constant DC output despite changes in the load current. The measure of the degree to which the regulator achieves this is called the **load regulation**. Load regulation is defined as the change in output voltage for a change in load current:

$$\text{load regulation} = \frac{\Delta V_{out}}{\Delta I_L} \tag{7-13}$$

In this equation, ΔV_{out} is the change in output voltage for the change in load current ΔI_L. Load regulation is usually expressed as millivolts per ampere but is often (and incorrectly) listed simply as millivolts.

Load regulation is also sometimes given as a percentage of the output voltage. In this case, it is defined by the equation

$$\text{percent load regulation} = \frac{\Delta V_{\text{out}}}{\Delta I_L} \times \frac{1}{V_{\text{out}}} \times 100 \qquad (7\text{-}14)$$

The smaller the load regulation, the better is the regulation.

Ripple Rejection

The ripple voltage is a result of the filtering process. If no current is drawn from the power supply filter, there will be zero ripple voltage, but if current is drawn from the circuit, the filter capacitor discharges slightly between charge cycles, producing ripple. Any regulator should reduce the output ripple voltage, and the degree to which this is achieved is called **ripple rejection**. Ripple rejection is usually expressed in decibels and is defined as

$$\text{ripple rejection} = -20 \log \frac{\text{output ripple}}{\text{input ripple}} \qquad (7\text{-}15)$$

Ripple rejection is related to the line regulation—the ripple voltage at the input to a regulator is equivalent to a variation in the line voltage. Most integrated circuit regulators achieve considerably better ripple rejection than predicted by the line regulation equation alone.

The amount of ripple depends on the choice of filter capacitor and on the ripple rejection of the regulator. In the design of a power supply, the choice of filter capacitor ultimately sets the output ripple, and it is important to properly design the filter capacitor to ensure that the output ripple from the regulator meets design specifications. We will consider this when we start to design power supplies and regulator circuits.

Power Supply Impedance

Finally, a regulator should approximate a perfect voltage source, that is, have zero source resistance. The **power supply impedance** is a measure of how much internal resistance there is in the power supply. For a simple regulator, it is directly related to the load regulation and can be derived from it:

$$\text{power supply impedance} = \frac{\Delta V_{\text{out}}}{\Delta I_L} \qquad (7\text{-}16)$$

This equation is exactly the same as Equation 7-13 given for load regulation, except that power supply impedance is given as ohms or milliohms instead of millivolts per ampere. The lower the power supply impedance, the better is the power supply.

Measurement of Regulator Performance

Line regulation, load regulation, ripple rejection, and power supply impedance are used to specify the performance of a regulator. When we study integrated circuit voltage regulators in the next chapter, we will see these quantities listed

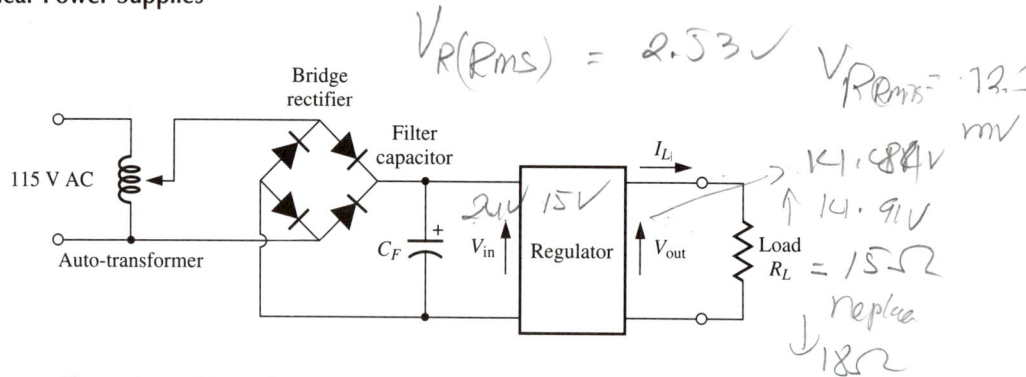

Figure 7-7 Circuit for determination of regulator characteristics.

on the device data sheets. It is possible to measure these quantities and experimentally assess the regulator performance by a relatively simple procedure.

To determine the various regulator characteristics, simply connect the regulator to a filtered rectifier circuit such as shown in Figure 7-7. To measure line regulation, vary the AC supply voltage and measure the change in output voltage from the regulator for the change in input voltage to the regulator under loaded conditions. The line regulation is given by Equation 7-11. To measure load regulation, keep the input voltage constant and measure the regulator output voltage for two different loads. Calculate the output current for each load (V_{out}/R_L) and take the difference to find ΔI. The load regulation is given by Equation 7-14. These data can also be used to determine the power supply impedance from Equation 7-16. Finally, measure the input ripple and the output ripple under full load and calculate the ripple rejection from Equation 7-15. The determination of these various quantities for an actual power supply is illustrated in Example 7-3. Integrated circuit regulator manufacturers use quite a different, and much more elaborate, procedure to measure these quantities for inclusion on their data sheets; however, the simple method described here is quite satisfactory for most purposes.

EXAMPLE 7-3

The output voltage from a 15 V regulator is measured with a digital voltmeter. First, a 15.0 Ω load resistor is connected across the output. The DC supply voltage at the input to the regulator is measured to be 24.0 V with an RMS ripple of 2.53 V, and the output voltage is measured to be 14.84 V with an RMS ripple of 12.3 mV. When the 15.0 Ω resistor is replaced with an 18.0 Ω resistor, the output voltage rises to 14.91 V. The 15 Ω resistor is reconnected and the DC supply voltage is reduced to 20 V. The voltage measured across the 15.0 Ω resistor is now measured to be 14.78 V. Determine the

(a) line regulation
(b) load regulation
(c) ripple rejection
(d) power supply impedance.

Solution

(a) The line regulation is calculated as follows

change in input voltage $\Delta V_{in} = 24\text{ V} - 20\text{ V} = 4\text{ V}$

change in output voltage $\Delta V_{out} = 14.84\text{ V} - 14.78\text{ V} = 0.06\text{ V}$

From Equation 7-11,

$$\text{line regulation} = \frac{\Delta V_{out}}{\Delta V_{in}} = \frac{0.06\text{ V}}{4\text{ V}} = 0.015\text{ V/V or 15 mV/V}$$

From Equation 7-12,

$$\text{percent line regulation} = \frac{\Delta V_{out}}{\Delta V_{in}} \times \frac{1}{V_{out}} \times 100 = \frac{0.06\text{ V}}{4\text{ V}} \times \frac{100}{14.84}$$

$$= 0.1011\%$$

(b) The load regulation is calculated as follows.
First, calculate the change in output voltage:

$$\Delta V = V_{18\,\Omega} - V_{15\,\Omega} = 14.91\text{ V} - 14.84\text{ V} = 0.07\text{ V}$$

Next, calculate the change in current:

$$\Delta I = I_{15\,\Omega} - I_{18\,\Omega} = \frac{V_{15\,\Omega}}{15\,\Omega} - \frac{V_{18\,\Omega}}{18\,\Omega} = \frac{14.85\text{ V}}{15\,\Omega} - \frac{14.91\text{ V}}{18\,\Omega}$$

$$= 0.9893\text{ A} - 0.8283\text{ A} = 0.1610\text{ A}$$

From Equation 7-13,

$$\text{load regulation} = \frac{\Delta V_{out}}{\Delta I_L} = \frac{0.07\text{ V}}{0.161\text{ A}} = 0.4358\text{ V/A or 435.8 mV/A}$$

From Equation 7-14,

$$\text{percent load regulation} = \frac{\Delta V_{out}}{\Delta I_L} \times \frac{1}{V_{out}} \times 100 = \frac{0.07\text{ V} \times 100}{0.161\text{ A} \times 14.84\text{ V}}$$

$$= 2.93\%$$

(c) To calculate the ripple rejection, use the full-load RMS output ripple of 12.3 mV and the full-load input ripple of 2.53 V in Equation 7-15:

$$\text{ripple rejection} = 20\log\frac{\text{output ripple}}{\text{input ripple}}$$

$$= 20\log\frac{12.3\text{ mV}}{2.53\text{ V}} = -46.3\text{ dB}$$

(d) The power supply impedance is numerically equal to the load regulation but is expressed in ohms or milliohms. From Equation 7-16, or from part (b) of the solution, we find that the power supply impedance is $0.4358\,\Omega$.

The performance of the regulator has now been fully determined. Although this was purely a numerical problem, the procedure described can be applied to any real regulator. A modern regulator, such as one of the regulator chips to be described in the next chapter, will probably show better performance than the above example, possibly to the point where the values are impossible to measure. The methods used in this example were approximate. Manufacturers stipulate the characteristic values of their regulators at constant temperature and measure them with pulsed signals. The method described in this example, however, is perfectly adequate for most applications.

7-3 Voltage References

The simplest voltage-regulating device is the zener diode. A **zener diode** is a device which acts as a normal diode when forward-biased, but conducts at almost constant voltage when reversed-biased. The voltage at which the zener conducts is called the zener voltage. The current versus voltage characteristic curve for a typical zener diode is illustrated in Figure 7-8. As shown on this curve, the zener voltage is not exactly constant, but varies with the amount of current flowing through the zener. In designing a circuit using a zener diode, we must design for a specific zener current I_{ZT} to give a constant zener voltage V_Z. Appendix A-4 lists the characteristics of a wide range of zener diodes, giving the optimal I_{ZT} values for the various V_Z values.

The zener voltage also depends on temperature. Zener diodes with rated voltages less than 6 V have a negative coefficient of voltage change with temperature.

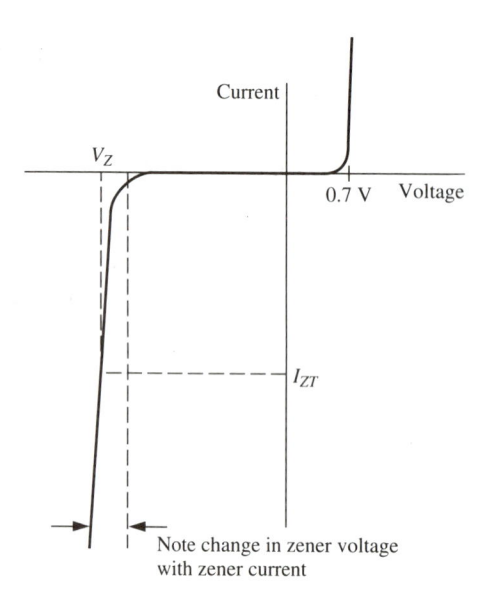

Figure 7-8 Typical zener diode current-voltage characteristic curve.

7-3 Voltage References

That is, as the temperature increases, the zener voltage decreases. Zener diodes with rated voltages more than 6 V have a positive coefficient. Zener diodes with rated voltages near 6 V have essentially zero change in voltage with temperature. Consequently, for maximum temperature stability, use 6.2 V zener diodes.

Zener Diodes as Voltage Regulators

A simple regulator circuit using a zener diode is shown in Figure 7-9. It consists of the zener diode with voltage V_Z in parallel with the load R_L and in series with resistor R_S. Resistor R_S establishes the zener bias current I_{ZT} and drops the excess voltage from the supply. Since the voltage across R_S is $V_{supply} - V_Z$ and the current through it is $I_{ZT} + I_L$, R_S can be calculated by Ohm's law:

$$R_S = \frac{V_{supply} - V_Z}{I_{ZT} + I_L} \qquad (7\text{-}17)$$

This circuit regulates as follows. Suppose the supply voltage increases. The zener voltage will remain essentially constant, and the voltage across R_S will increase, causing the current through R_S to increase. This increased current flows through the zener diode. From the zener characteristic curve in Figure 7-8, we see that as the zener current increases, there will be only a slight increase in zener voltage. In this way, the zener circuit maintains a relatively constant output voltage despite line voltage changes, and thus provides line regulation.

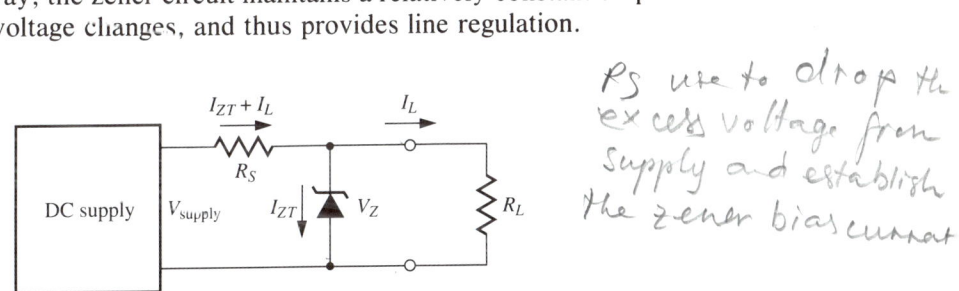

Figure 7-9 Zener diode regulator circuit.

Now suppose the load resistance decreases for constant supply voltage. The load current will increase. Since the supply voltage is unchanged, the current through R_S is constant. The current through the zener will decrease to compensate for the increased current to the load. Again referring to Figure 7-8, we see that for a current decrease, the zener voltage will decrease only slightly. In this way, the zener circuit maintains a relatively constant output voltage despite load changes, and thus provides load regulation.

For small changes in line voltage or in load current, the change in zener current is small, the zener voltage remains relatively constant, and the regulation is quite good. However, for large changes in line voltage or in load current, the change in zener current will be large, and the zener voltage will change by a significant amount, giving poor regulation. If the current changes are too large, the zener may not regulate at all.

Obviously, the zener diode is capable of acting as a regulator, but its performance is limited. The op-amp regulator circuits discussed later use zener diodes as voltage references, not as regulators.

Zener Diodes as Voltage References

We have already seen the use of zener diodes as voltage references in the level-sensing comparators discussed in Chapter 5, and we will see shortly that these comparator circuits are similar to the circuits used in op-amp regulators. The circuit for a zener diode used as a voltage reference is shown in Figure 7-10. In this circuit, we must stipulate two requirements for stable operation: the supply voltage V_{supply} should be relatively constant, and the load resistance R_L should be large so that $I_L \ll I_{ZT}$. The equation for calculating R_S is

$$R_S = \frac{V_{supply} - V_Z}{I_{ZT}} \tag{7-18}$$

This is the same equation used in Chapter 5.

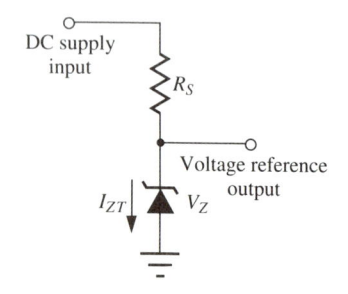

Figure 7-10 Zener diode voltage reference.

The zener diode makes a good voltage reference, and it is quite widely used in such applications. We will often use it as a voltage reference in this book. It does, however, have several limitations. The main limitation is that the zener requires a relatively large bias current I_{ZT}. For example, a 6.2 V zener requires a 41 mA bias current. This large current is necessary to bias the zener to approximately the middle of its operating curve, and it can be excessive in many applications. If a 6.2 V zener is used as the reference for a battery-level-indicating circuit, such as developed in Example 5-3, the 41 mA bias current will rapidly discharge the battery. The large current poses a second problem. The power dissipated in the zener is the voltage V_Z times the current I_{ZT}. For the 6.2 V zener, this would be approximately $\frac{1}{4}$ W, enough to cause significant heating and, consequently, potential drift in the operating characteristics.

The simple zener diode voltage reference can be improved considerably by adding transistors to pass excess current and hold the zener operating point constant. A typical circuit is shown in Figure 7-11. The best way to understand the operation of this circuit is to assume that it is operating under nominal conditions

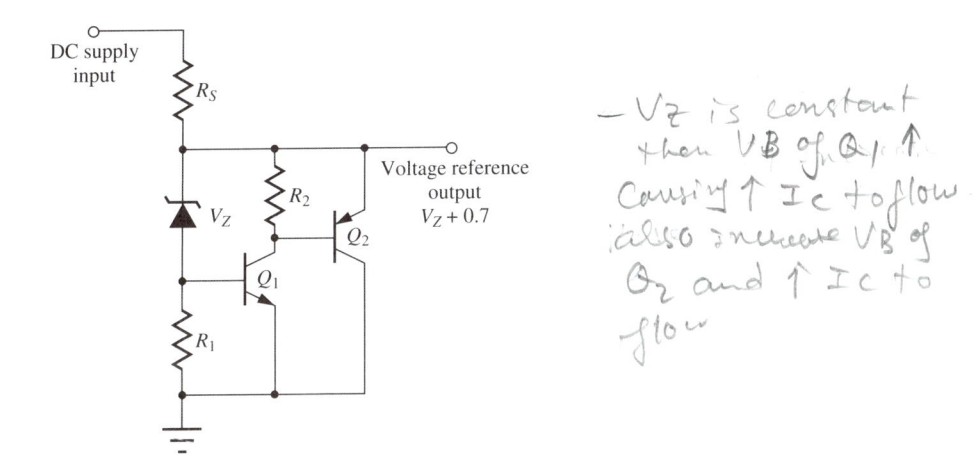

Figure 7-11 Improved zener diode voltage reference.

when the input voltage tries to increase. Since the zener voltage will remain approximately constant, the base voltage of Q_1 will increase, causing increased collector current to flow. This in turn increases the voltage at the base of Q_2, causing increased collector current to flow in this transistor as well. The excess current flowing through the circuit causes an increased voltage drop across R_S; hence, the voltage across the reference circuit remains constant. Since the zener current is held relatively constant by the transistors, the zener diode can be biased near the top of its characteristic curve. This means that the bias current can be much smaller, and the operating point is made more stable.

Integrated Circuit Voltage References

Integrated circuit voltage references simply take the modified zener diode circuit shown in Figure 7-11, add a temperature stabilizing circuit, and package the complete circuit in a 2 pin TO-100 metal case or a TO-92 plastic case. Voltage reference integrated circuits are pin-compatible replacements for zener diodes with improved temperature stability, precise operating voltages, smaller bias currents, and a very small variation of voltage with current. A voltage reference is very close to an ideal zener diode.

Several voltage reference integrated circuits are available, but, unlike zener diodes, there are relatively few choices in voltage. Most voltage references have voltages of 1.22, 2.5, 5, 6.95, or 10 V. These voltages are selected to satisfy three applications. The 2.5, 5, and 10 V references are designed for instrument calibration applications. The 6.95 V references are designed for minimum drift, taking advantage of the zero temperature variation of zener diodes at this voltage. The 1.22 V references are used for various power supply applications. We will encounter some of these applications later.

The LM313 voltage reference produced by National Semiconductor is a typical example. The data sheets for the LM113/LM313 voltage reference are shown

National Semiconductor

Voltage References

LM113/LM313 Reference Diode

General Description

The LM113/LM313 are temperature compensated, low voltage reference diodes. They feature extremely-tight regulation over a wide range of operating currents in addition to an unusually-low breakdown voltage and good temperature stability.

The diodes are synthesized using transistors and resistors in a monolithic integrated circuit. As such, they have the same low noise and long term stability as modern IC op amps. Further, output voltage of the reference depends only on highly-predictable properties of components in the IC; so they can be manufactured and supplied to tight tolerances. Outstanding features include:

- Low breakdown voltage: 1.220V
- Dynamic impedance of 0.3Ω from 500 µA to 20 mA
- Temperature stability typically 1% over $-55°C$ to $125°C$ range (LM113), $0°C$ to $70°C$ (LM313)
- Tight tolerance: ±5% standard, ±2% and ±1% on special order.

The characteristics of this reference recommend it for use in bias-regulation circuitry, in low-voltage power supplies or in battery powered equipment. The fact that the breakdown voltage is equal to a physical property of silicon—the energy-band-gap voltage—makes it useful for many temperature-compensation and temperature-measurement functions.

Schematic and Connection Diagrams

Metal Can Package

Note: Pin 2 connected to case.
TOP VIEW

Order Number LM113H or LM313H
See NS Package H02A

Typical Applications

Level Detector for Photodiode

Low Voltage Regulator

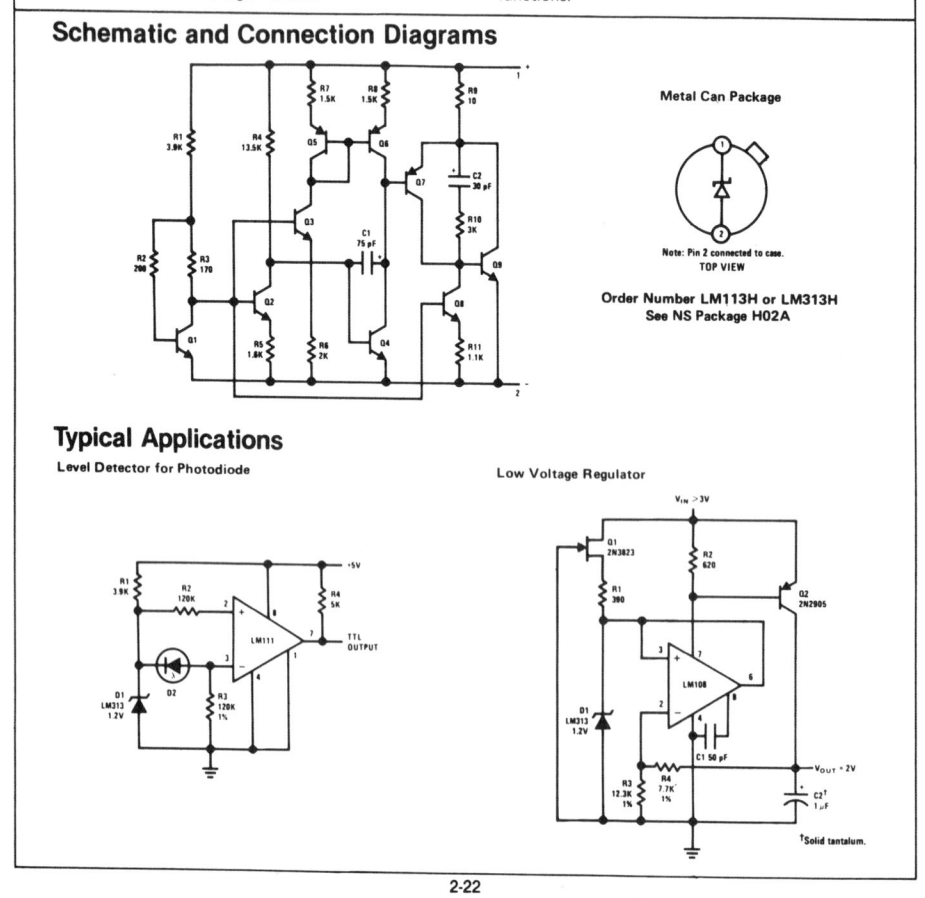

†Solid tantalum.

Figure 7-12 Data sheets for the LM313 voltage reference. *(Reprinted with permission of National Semiconductor Corporation)*

Absolute Maximum Ratings

Power Dissipation (Note 1)	100 mW
Reverse Current	50 mA
Forward Current	50 mA
Storage Temperature Range	−65°C to +150°C
Lead Temperature (Soldering, 10 seconds)	300°C

Operating Conditions

Temperature (T_A)	MIN	MAX	UNITS
LM113	−55	+125	°C
LM313	0	70	°C

Electrical Characteristics (Note 2)

PARAMETER	CONDITIONS	MIN	TYP	MAX	UNITS
Reverse Breakdown Voltage					
LM113/LM313	I_R = 1 mA	1.160	1.220	1.280	V
LM113-1		1.210	1.22	1.232	V
LM113-2		1.195	1.22	1.245	V
Reverse Breakdown Voltage Change	0.5 mA ≤ I_R ≤ 20 mA		6.0	15	mV
Reverse Dynamic Impedance	I_R = 1 mA		0.2	1.0	Ω
	I_R = 10 mA		0.25	0.8	Ω
Forward Voltage Drop	I_F = 1.0 mA		0.67	1.0	V
RMS Noise Voltage	10 Hz ≤ f ≤ 10 kHz, I_R = 1 mA		5		μV
Reverse Breakdown Voltage Change with Current	0.5 mA ≤ I_R ≤ 10 mA, T_{MIN} ≤ T_A ≤ T_{MAX}			15	mV
Breakdown Voltage Temperature Coefficient	1.0 mA ≤ I_R ≤ 10 mA, T_{MIN} ≤ T_A ≤ T_{MAX}		0.01		%/°C

Note 1: For operating at elevated temperatures, the device must be derated based on a 150°C maximum junction and a thermal resistance of 80°C/W junction to case or 440°C/W junction to ambient.

Note 2: These specifications apply for T_A = 25°C, unless stated otherwise. At high currents, breakdown voltage should be measured with lead lengths less than 1/4 inch. Kelvin contact sockets are also recommended. The diode should not be operated with shunt capacitances between 200 pF and 0.1 μF, unless isolated by at least a 100 Ω resistor, as it may oscillate at some currents.

Typical Performance Characteristics

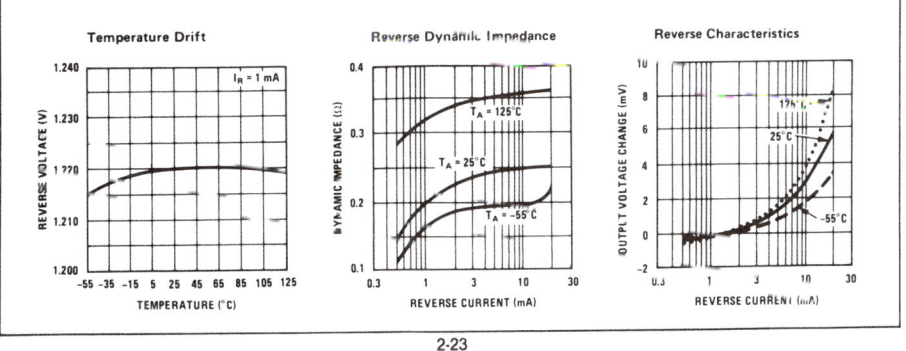

Figure 7-12 (*Continued*)

in Figure 7-12. The numbers LM113 and LM313 refer to identical devices differing only in range of operating temperatures. The LM113 has an operating range of -55 to $+125°C$, and the LM313's operating range is only from 0 to $+70°C$. This voltage reference was selected for discussion because we will be using it in various power supply circuits. It has a nominal output voltage of 1.22 V for a nominal bias current of 1 mA and can handle currents from 0.6 to 20 mA.

Circuits using integrated circuit voltage references are exactly the same as for zener diodes. Figure 7-13 shows the basic circuit. This figure is exactly the same as Figure 7-10 except for the different symbol used for a voltage reference. To design a voltage reference circuit using the LM313, we need only calculate a value for R_S. Equation 7-18 is used. Assume a bias current I_{ZT} of 1 mA and a device voltage V_Z of 1.22 V.

Finally, brief mention should be made to the LM199/LM299/LM399 voltage references. These are 4 pin devices where the two additional pins are connected to a heater/temperature stabilizer circuit to give improved temperature stability. These voltage references have a nominal output voltage of 6.95 V with a tolerance of $+1\%$, -2%, and a drift of 1 part per million in output voltage. The LM313, by comparison, has a $\pm 5\%$ tolerance and a 100 ppm drift in output voltage.

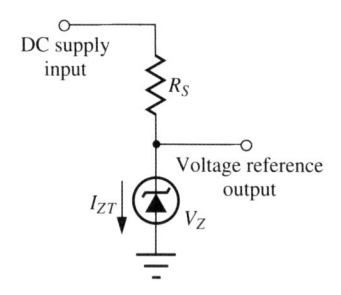

Figure 7-13 Voltage reference using an integrated circuit voltage reference.

EXAMPLE 7-4

Design a voltage reference circuit using an LM313 voltage reference integrated circuit operating from an 18 V power supply.

Solution

The circuit that we will be using is shown in Figure 7-14. This is the same circuit that was given in Figure 7-13.

The only component that we need to design in this circuit is R_S. We know $V_{supply} = 18$ V and $V_Z = 1.22$ V. The bias current of the voltage reference is not critical, but the manufacturer recommends 1.0 mA, so we will use this value.

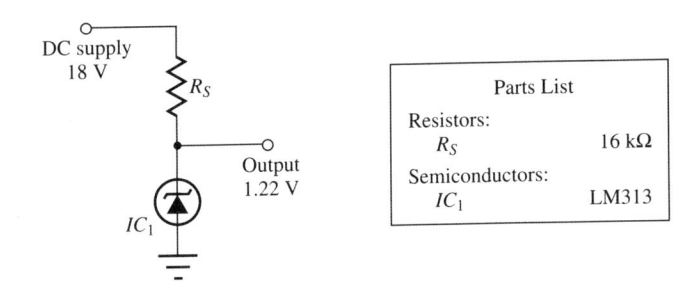

Figure 7-14

Calculate R_S from Equation 7-18:

$$R_S = \frac{V_{\text{supply}} - V_Z}{I_{ZT}} = \frac{18 \text{ V} - 1.22 \text{ V}}{1.0 \text{ mA}} = 16.8 \text{ k}\Omega$$

From the values in Appendix A-1, choose R_S as 16 kΩ.

Our design is complete. The design procedure would be identical with a zener diode, but R_S would be much smaller.

7-4 Op-Amp Regulators

A zener diode used as a voltage regulator is a poor substitute for even the simplest regulator circuit using a transistor or op-amp. In the previous section, we saw that the most useful function for a zener diode was as a voltage reference, and in this section we will use it in that capacity as we introduce the first real regulator circuits.

Although most modern power supplies use integrated circuit regulator chips, which will be discussed in the next chapter, it is quite possible to build a regulator circuit using a simple operational amplifier. Occasionally, regulator requirements are such that a commercial regulator chip is not available for a desired application or simply will not work in the proposed application. In this case, it is important to know how to design a regulator using simpler building blocks, namely, operational amplifiers. The circuits developed in this section will also serve as an introduction to the circuitry found in commercial regulator circuits.

Simple Op-Amp Regulator Circuits

A simple regulator consists of the following parts: a **voltage reference** that produces a reference voltage to which the output voltage is compared, a **sensing circuit** that detects the level of the output voltage, and an **error detector** (or **error amplifier**) that provides a corrected output if the sensed signal changes with respect to the reference signal. We can use a zener diode instead of a voltage reference to produce the reference voltage.

7 Linear Power Supplies

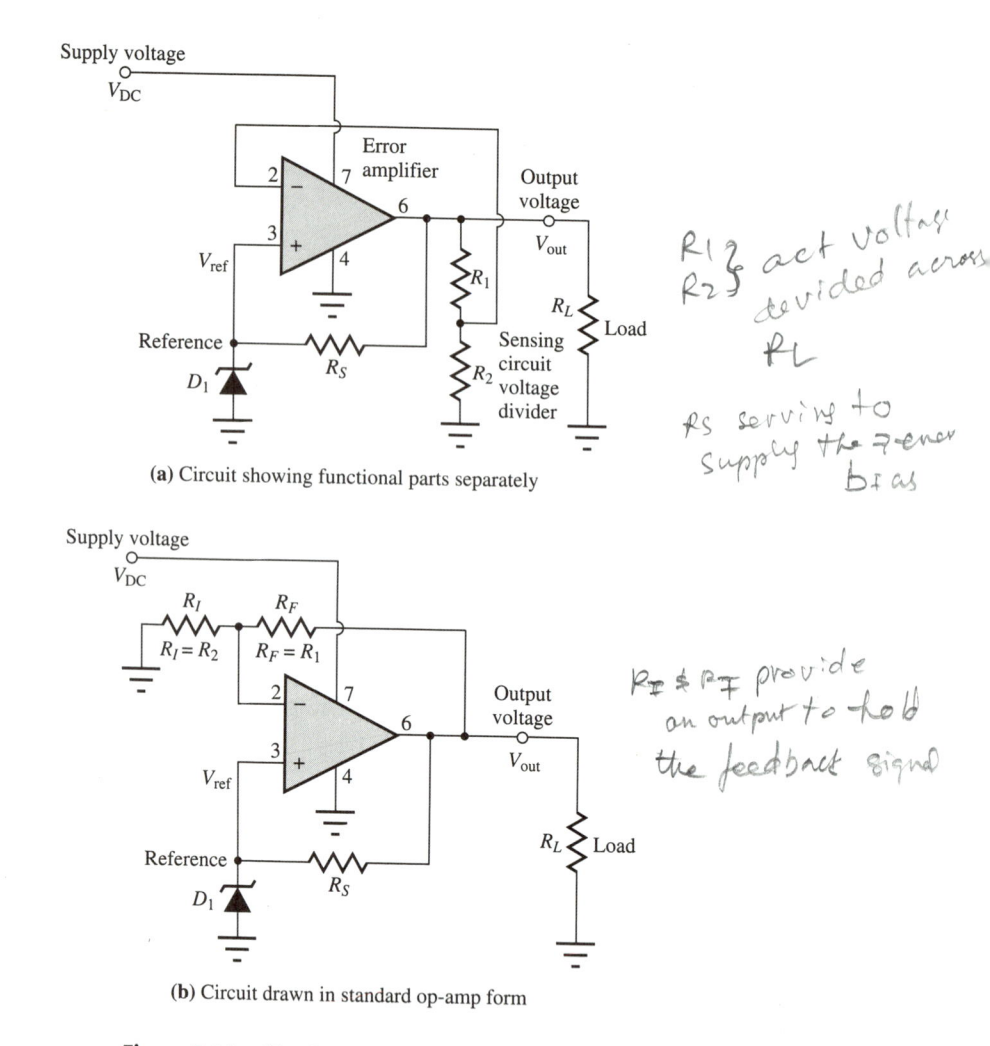

(a) Circuit showing functional parts separately

(b) Circuit drawn in standard op-amp form

Figure 7-15 Simple op-amp voltage regulator.

A simple op-amp regulator circuit is shown in Figure 7-15. The circuit is drawn in two forms. Figure 7-15(a) shows the circuit drawn with each of the three functional parts—voltage reference, sensing circuit, and error detector—shown separately. The output voltage of the op-amp is the voltage supplied to the load. Resistors R_1 and R_2 act as a voltage divider across the load resistance to sense a fraction of the output voltage. Resistor R_S and the zener (or voltage reference) form the reference voltage source, R_S serving simply to supply the zener bias current I_{ZT}.

Although the circuit shown in Figure 7-15(a) resembles a level-sensing comparator as described in Chapter 5, it is actually a noninverting amplifier. This is shown more clearly when the circuit is redrawn as in Figure 7-15(b) with $R_F = R_1$ and $R_I = R_2$. Recall that an op-amp will always provide an output to hold the feedback signal at the inverting input to the same voltage as the reference signal at the noninverting input. This is why the circuit provides a constant output voltage.

To design this regulator circuit, we will have to calculate values of R_S, R_F, and R_I to give the desired output voltage V_{out}. First, we must determine the gain of the amplifier. Since the circuit acts as a noninverting amplifier, the gain is simply the output voltage divided by the reference voltage:

$$A_V = \frac{V_{out}}{V_Z} \tag{7-19}$$

The gain of a noninverting amplifier was derived in Chapter 3 and is given by Equation 3-5:

$$A_V = \frac{R_F}{R_I} + 1 \tag{3-5}$$

Combining these two equations, we obtain an expression for R_F and R_I in terms of V_Z and V_{out}:

$$\frac{R_F}{R_I} = \frac{V_{out}}{V_Z} - 1 \tag{7-20}$$

This gives the ratio of R_F to R_I. Choose a value for R_F and solve Equation 7-20 for R_I:

$$R_I = \frac{R_F}{A_V - 1} = \frac{R_F}{V_{out}/V_Z - 1} \tag{7-21}$$

Choose a relatively large value for R_F so as not to load the output. Typically we would use 10 kΩ.

Finally, we must design the series resistor R_S for the zener or voltage reference. Notice that the zener bias resistor R_S is connected to the output of the regulator instead of to the supply. The circuit would certainly work if the zener were biased from the supply, but since the supply is unregulated the zener current may vary due to fluctuations in the supply, and hence the reference voltage may not be perfectly stable. By biasing the zener from the output, we are ensured of a stable output voltage, and hence, a stable reference voltage. Resistor R_S was calculated earlier by Equation 7-18, only now we must replace V_{supply} by V_{out}:

$$R_S = \frac{V_{out} - V_Z}{I_{ZT}} \tag{7-22}$$

There is one important point you may have noticed in the op-amp regulator design. A standard op-amp requires a dual power supply, yet the circuit in Figure 7-15 has only a single power supply, namely the voltage source that we wish to regulate. The circuit works as shown without a dual supply because both inputs are clamped at V_Z above ground by the zener diode at the noninverting input. This circuit and all of the circuits to be developed later in this section work fine from a single supply.

The circuit we have just designed can provide a regulated output for any value of voltage greater than the reference voltage if the DC supply voltage is at least 2 V greater than the desired regulated voltage. Why must the supply be at least 2 V greater than the output? Recall that the output of an op-amp ranges from V_{sat-} to V_{sat+}, where V_{sat-} is approximately 1 V more positive than the negative supply voltage, and V_{sat+} is approximately 1 V less than the positive supply voltage. The operating range is 2 V less than the supply range, exactly what we find in the regulator circuit. This 2 V difference between the supply voltage and maximum output voltage is called the **dropout voltage**. We will consider the implications of dropout voltage in the next chapter.

If we are using a zener diode as a reference, we would typically use a 6.2 V reference zener to minimize thermal drift. What if we wanted a regulated 5 V supply? We could use a zener with a voltage of 5 V or less, but this would give poorer temperature stability. Instead, it is better to use the 6.2 V zener and use a voltage divider connected to the noninverting input.

Using a zener diode as reference, however, poses a serious problem. The bias current of a zener is relatively large, yet the output current that an op-amp can supply is relatively small. Most op-amps can supply sufficient output current to provide the 41 mA required to bias a 6.2 V zener. They may not be able to provide sufficient current to bias a lower-voltage zener, and they may not be able to provide sufficient current to bias a zener while supplying current to a load. The best solution is to use a voltage reference such as the LM313. Its low-bias-current requirements (1 mA) and low output voltage (1.22 V) make it far more suitable than a zener diode as a reference.

EXAMPLE 7-5

Design a 15 V op-amp regulated power supply using a BiFET op-amp and a 6.2 V zener diode. The zener has I_{ZT} = 41 mA. Assume that the DC source which is being regulated has a voltage of 25 V and can supply any required current.

Solution

To solve this problem, we must design a simple voltage regulator similar to that shown in Figure 7-15. The circuit is redrawn as Figure 7-16, and a parts list is included.

First we must determine the gain of the regulator given that the reference is 6.2 V and the output is 15 V. The gain is given by Equation 7-20:

$$A_V = \frac{V_{out}}{V_Z} = \frac{15 \text{ V}}{6.2 \text{ V}} = 2.419$$

7-4 Op-Amp Regulators

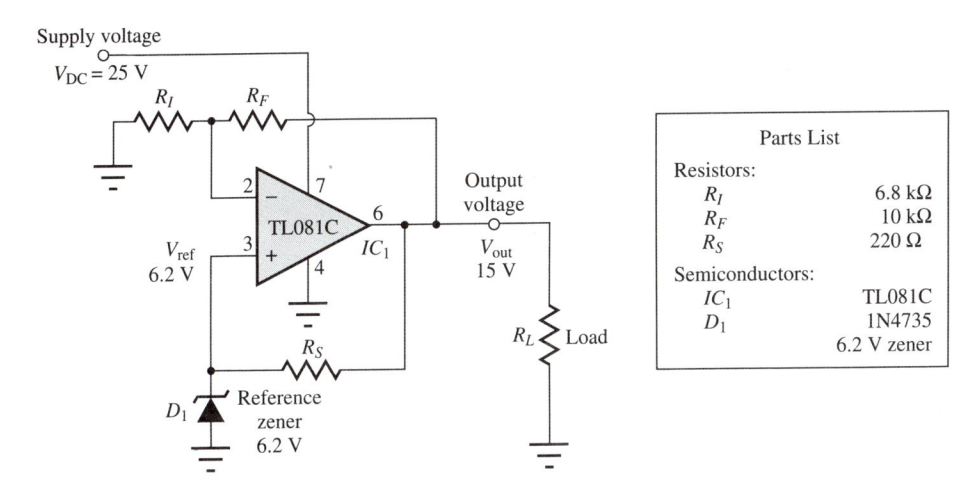

Figure 7-16

Parts List

Resistors:
- R_I 6.8 kΩ
- R_F 10 kΩ
- R_S 220 Ω

Semiconductors:
- IC_1 TL081C
- D_1 1N4735 6.2 V zener

We will choose $R_F = 10$ kΩ and use Equation 7-21 to solve for R_I:

$$R_I = \frac{R_F}{A_V - 1} = \frac{10 \text{ k}\Omega}{2.419 - 1} = 7.05 \text{ k}\Omega$$

The closest standard value from Appendix A-1 is 6.8 kΩ. We will use this value for R_I.

Finally, we must calculate the zener bias resistor R_S by using Equation 7-22:

$$R_S = \frac{V_{out} - V_Z}{I_{ZT}} = \frac{15 \text{ V} - 6.2 \text{ V}}{41 \text{ mA}} = 215 \text{ }\Omega$$

The closest standard value from Appendix A-1 is 220 Ω. Because we have a relatively large amount of current flowing through this resistor, we should check the power dissipation. The power is the voltage (15 V − 6.2 V) times the current (41 mA), and is calculated to be 360 mW. We should use at least a half-watt resistor.

Our design is complete. We shall verify the operation of the design in Example 7-6.

EXAMPLE 7-6

Analyze the circuit designed in Example 7-5 to determine the actual output voltage of the simple op-amp regulator.

Solution

The circuit that we will be analyzing was shown in Figure 7-16.

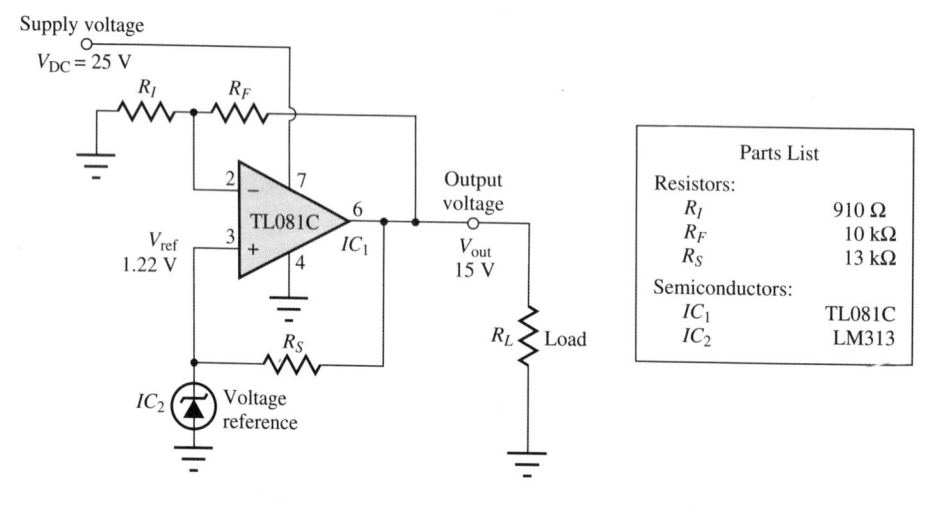

First determine the gain of the regulator op-amp, considering it to be a noninverting amplifier where $R_F = 10 \text{ k}\Omega$ and $R_I = 6.8 \text{ k}\Omega$. Use Equation 3-5:

$$A_V = \frac{R_F}{R_I} + 1 = \frac{10 \text{ k}\Omega}{6.8 \text{ k}\Omega} + 1 = 2.471$$

Next, use the definition of gain given by Equation 7-19:

$$A_V = \frac{V_{out}}{V_Z}$$

Solving for V_{out}, we have

$$V_{out} = A_V \times V_Z = 2.471 \times 6.2 \text{ V} = 15.3 \text{ V}$$

This agrees quite well with the original design value of 15 V.

EXAMPLE 7-7

Design a 15 V op-amp regulated power supply using a BiFET op-amp and an LM313 voltage reference. Use an I_{ZT} for the LM313 of 1 mA. Assume that the DC source being regulated has a voltage of 25 V and can supply any required current.

Solution

This is the same problem as Example 7-5, except that we are using an LM313 voltage reference instead of a zener diode. The circuit is the same as used in Example 7-5 (Figure 7-16) and is redrawn as Figure 7-17, with a parts list included.

Figure 7-17

7-4 Op-Amp Regulators

We will use a procedure similar to Example 7-5. First, determine the gain of the regulator by using Equation 7-20, given that the reference is 1.22 V and the output is 15 V.

$$A_V = \frac{V_{out}}{V_Z} = \frac{15 \text{ V}}{1.22 \text{ V}} = 12.30$$

Choose $R_F = 10 \text{ k}\Omega$ and use Equation 7-21 to solve for R_I:

$$R_I = \frac{R_F}{A_V - 1} = \frac{10 \text{ k}\Omega}{12.30 - 1} = 885 \text{ }\Omega$$

The closest standard value is 910 Ω.

Calculate the zener bias resistor R_S from Equation 7-22:

$$R_S = \frac{V_{out} - V_Z}{I_{ZT}} = \frac{15 \text{ V} - 1.22 \text{ V}}{1 \text{ mA}} = 13.78 \text{ k}\Omega$$

The closest standard value is 13 kΩ. In this case, we have only a small current, and we do not need to be concerned with the power rating of the resistor.

Our design is complete. Notice that the design procedure was identical to that in Example 7-5.

It is informative to construct these two regulator circuits and observe their performance under load. The regulator designed in Example 7-5 will probably cease regulation for any load resistance smaller than approximately 5000 Ω, whereas the regulator designed in Example 7-7 should regulate for a load resistance as small as 300 Ω. The reason that the regulator designed in Example 7-5 performs so poorly is that the zener uses most of the output current of the op-amp. In all future examples, we will use the LM313 as our reference.

Op-Amp Regulator with Pass Transistor

There is a restriction with the regulator circuits that we have just designed—the output current is limited by the op-amp and typically has a maximum value of about 50 mA. For most power supply applications, this would not be adequate. To improve the current-handling capability of the circuit, we simply add a **pass transistor**. Circuits of op-amp regulators with pass transistors are shown in Figure 7-18. Figure 7-18(a) shows an op-amp regulator with a simple pass transistor. The op-amp regulator itself is designed exactly as we have already discussed. Transistor Q_1 must be chosen to pass the desired load current $I_L = I_C$. It must also be able to withstand the collector-emitter voltage $V_{CE} = V_{DC} - V_{out}$ and be able to dissipate the power $P = I_C \times V_{CE}$.

If the regulator circuit is to supply a very heavy current, say 5 A or more, the base current which must be supplied by the op-amp may exceed the output of the op-amp. The solution in this case is to use a Darlington pass transistor or two transistors connected in a **Darlington configuration**. This will reduce the amount of base current by a factor equal to the β of the driver transistor, typically

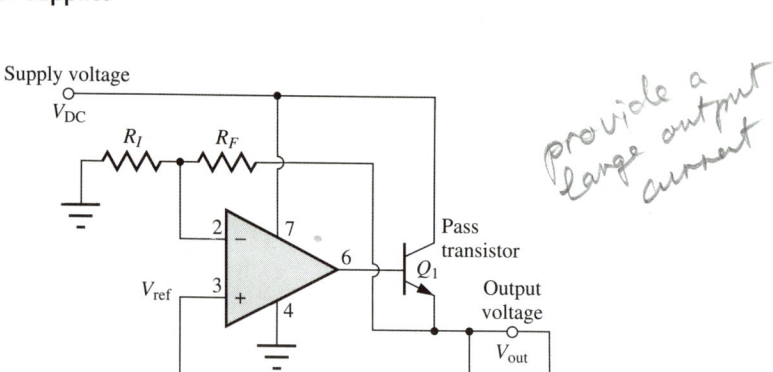

(a) Simple pass transistor

(b) Darlington pass transistor

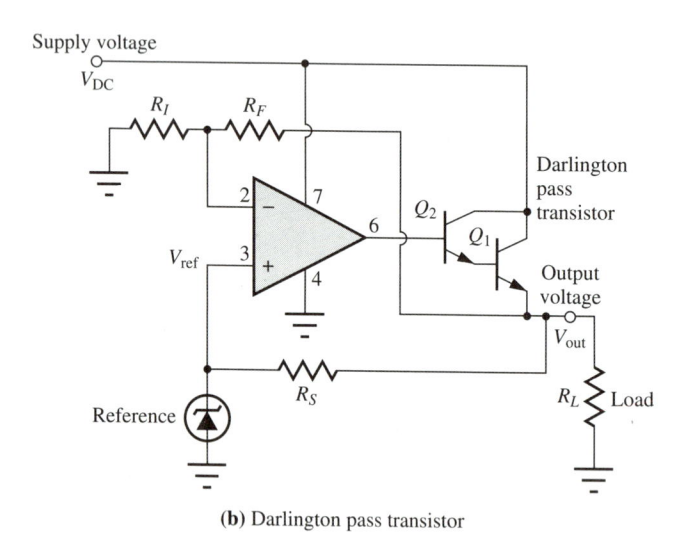

Figure 7-18 Op-amp regulator using a pass transistor to provide a larger output current.

more than 100. This circuit is shown in Figure 7-18(b). The circuit design is exactly the same as for the simple pass transistor.

EXAMPLE 7-8

Modify the circuit designed in Example 7-7 so that it can drive a 10 Ω load. Analyze the operation of the circuit to determine the actual output voltage and transistor power.

Solution

If the load resistance is 10 Ω, the load current will be 1.5 A, far beyond the rating of any standard op-amp. We must use a pass transistor. The circuit that we shall use is the one shown in Figure 7-18(a). This is redrawn as Figure 7-19, and a parts list is included. The regulator circuit is exactly the same as designed in Example 7-7. There are no additional calculations required to include the pass transistor.

We will, however, have to determine the voltage, current, and power ratings for the transistor. The collector current I_C for the transistor is 1.5 A. The collector-emitter voltage V_{CE} is 10 V, and the power dissipated in the pass transistor is $I_C \times V_{CE}$ = 15 W. These are the minimum ratings for the power transistor that we must choose. We will use a TIP31 power transistor.

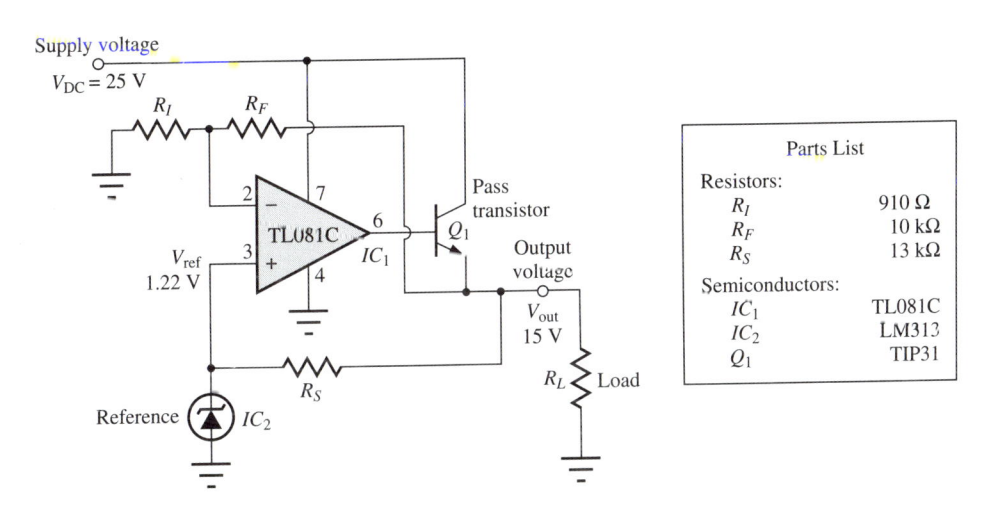

Figure 7-19

The data sheet for the TIP31 power transistor is shown in Appendix A-5. This transistor has a maximum current of 5 A, maximum V_{CE} of 40 V, and maximum power *with adequate heatsinking* of 40 W. Transistor power specifications are often confusing. The power given usually assumes an infinite heatsink. Without heatsinking, the maximum power rating is only about 5% of the listed value. For the TIP31, the maximum power without heatsinking is only 2 W, so a heatsink will definitely be necessary. A Wakefield 305 heatsink would be suitable, although the design procedure for heatsinks will not be discussed here.

The TIP31 has a minimum DC β of approximately 25. If the load current, and hence the collector current, is 1.5 A, then the base current will be 0.06 A. A modern BiFET op-amp can handle this much current, although it is approaching the limit. We will redesign the circuit to reduce the op-amp current in the next example.

To determine the actual output voltage of the circuit, we first determine the gain of the regulator op-amp by using Equation 3-5, where $R_F = 10 \text{ k}\Omega$ and $R_I = 910 \text{ }\Omega$:

$$A_V = \frac{R_F}{R_I} + 1 = \frac{10 \text{ k}\Omega}{910 \text{ }\Omega} + 1 = 11.99$$

Next, use the definition of gain given by Equation 7-19:

$$A_V = \frac{V_{out}}{V_Z}$$

Solving for V_{out} gives

$$V_{out} = A_V \times V_Z = 11.99 \times 1.22 \text{ V} = 14.63 \text{ V}$$

We can now determine the actual power dissipated in the transistor. The voltage across the transistor is simply the difference between the supply voltage and the output voltage:

$$V_{CE} = V_{supply} - V_{out} = 25 \text{ V} - 14.63 \text{ V} = 10.37 \text{ V}$$

Since the load is 10 Ω, the current is 1.463 A. Hence, the power dissipated in the transistor is

$$P = V_{CE} \times I_C = 10.37 \text{ V} \times 1.463 \text{ A} = 15.2 \text{ W}$$

This is slightly larger than our design value of 15 W.

EXAMPLE 7-9

Modify the circuit designed in Example 7-8 to use a Darlington configuration pass transistor to minimize the current that the op-amp must supply.

Solution

The current that the op-amp must supply in Example 7-8 is the maximum output current divided by the β of the pass transistor. In the case of the single TIP31 pass transistor, which has a minimum β of 25, this current could be as large as 0.06 A, close to the limit of a TL081C op-amp. If we construct a Darlington pass transistor configuration by adding a driver transistor, the current that the op-amp would have to supply is now the base current of the pass transistor divided by the β of the driver. Suppose the driver has a β of 100; then the op-amp current is reduced to 0.06/100 = 0.0006 A or 0.6 mA.

The circuit that we shall use is the Darlington configuration shown in Figure 7-20. We have the choice of simply adding a separate driver transistor to the circuit designed in Example 7-8, or replacing the simple pass transistor with a Darlington pass transistor. We will choose the latter option and use a TIP120. The TIP120 is a typical Darlington with somewhat higher voltage, current, and power ratings than the TIP31 used in the previous example.

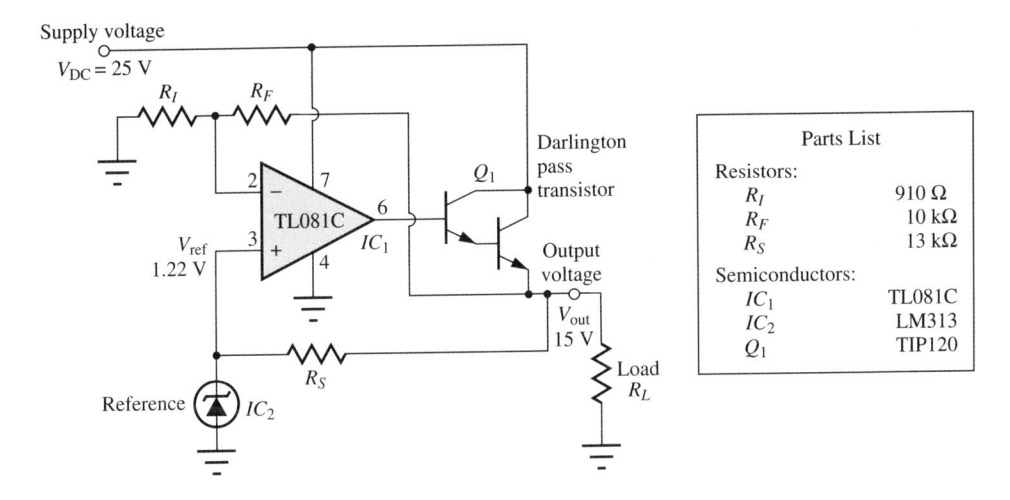

Figure 7-20

Since the output voltage and current are the same as in the previous example, the voltage, current, and power requirements for the Darlington are the same as for the TIP31 in the original design. Our TIP120 Darlington will have to pass 1.5 A with a V_{CE} of 10 V and a power of 15 W.

It is not necessary to reanalyze the circuit to determine exact values of V_{CE}, output voltage, or transistor power since these are exactly the same as for the simple TIP31 pass transistor, namely $V_{CE} = 10.37$ V, $V_{out} = 14.63$ V, and $P_{TIP120} = 15.2$ W.

Op-Amp Regulator with Current Limiting

The use of a simple pass transistor has one limitation. Suppose the output is shorted. The transistor will try to supply infinite current and will immediately blow. To eliminate this problem, we can add a current-limiting circuit. This modification is shown in Figure 7-21. In this circuit, a bypass transistor Q_2 and a sensing resistor R_{sens} are added. Transistor Q_2 is normally not conducting unless its base to emitter voltage is greater than 0.7 V. This occurs if the load current through R_{sens} is large enough to cause a 0.7 V drop. The sensing resistor can be calculated by Ohm's law:

$$R_{sens} = \frac{V_{BE}}{I_{max}} \qquad (7\text{-}23)$$

When transistor Q_2 conducts, it essentially bypasses the base current of Q_1 to the load and Q_1 will only conduct the limited maximum current.

There is, however, a potential problem with this short-circuit protection. If the output is shorted, the output voltage is reduced to zero and the entire supply voltage is dropped as V_{CE} across the pass transistor. The pass transistor would

312   **7 Linear Power Supplies**

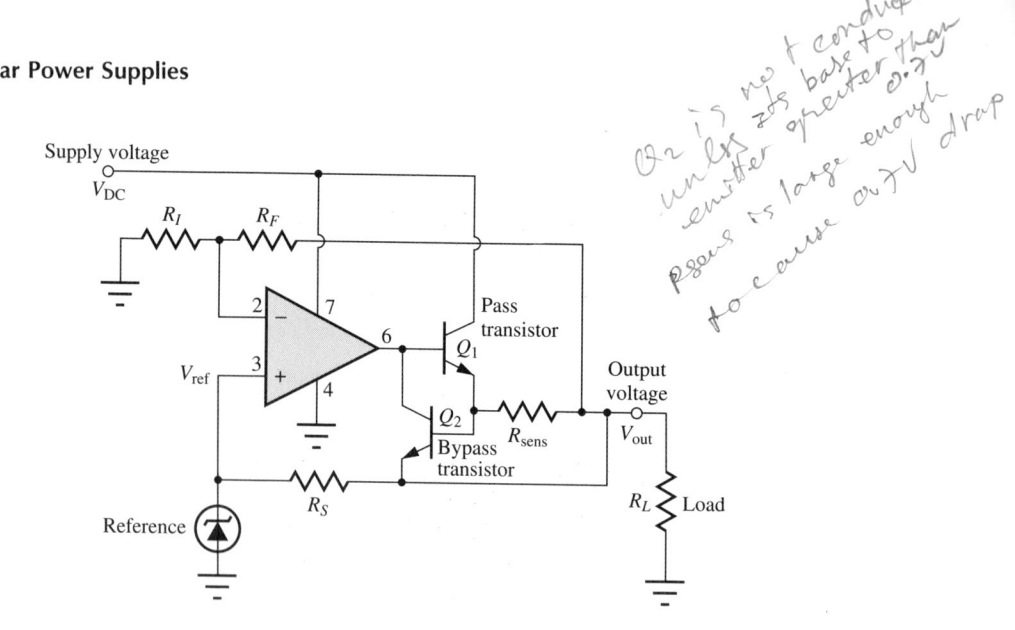

Figure 7-21   Op-amp regulator with pass transistor and simple current limiting.

be subject to a power dissipation three to four times larger than under normal operation at full load, *even though the current is limited to full-load current.* If this potential problem is not recognized, the pass transistor may burn out under short-circuit conditions. The easiest, and best, solution to this problem is simply to design the transistor and heatsink to absorb any potential overload condition.

EXAMPLE 7-10

Add current limiting to the circuit designed in Example 7-9 to limit the maximum current to 2 A. Analyze the operation of the circuit.

Solution

To incorporate current limiting to protect the transistor from accidental shorts across the load, we will use a circuit similar to that in Figure 7-21. This circuit has a single pass transistor. The circuit we are designing uses a Darlington pass transistor configuration. Our circuit, including the Darlington pass transistor, is drawn as Figure 7-22, with a parts list included.

Since we are using the pass transistor circuit that we designed in the previous example, the only thing we have to design is the value of the sensing resistor R_{sens}. We need a voltage drop of 0.7 V across the sensing resistor when the current is 2.0 A. Hence,

$$R_{sens} = \frac{V_{BE}}{I_{max}} = \frac{0.7 \text{ V}}{2.0 \text{ A}} = 0.35 \text{ }\Omega$$

Because of the extremely small value, we would probably construct R_{sens} using resistance wire wound on a suitable resistor form.

7-4 Op-Amp Regulators

Figure 7-22

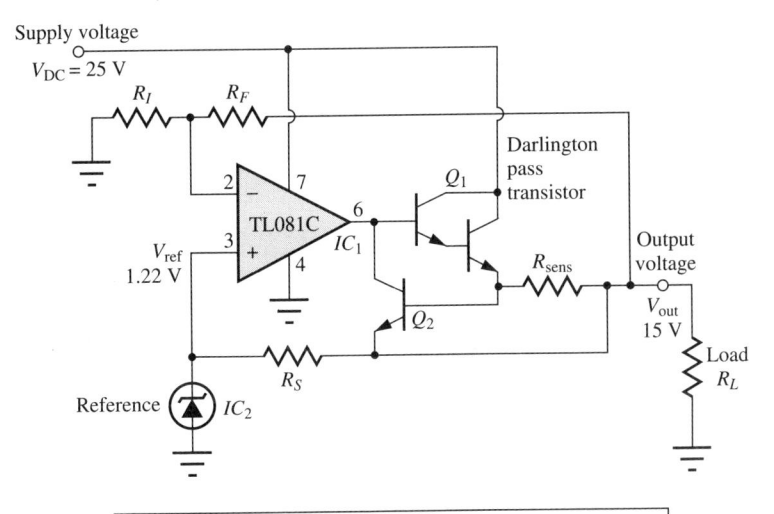

Parts List

Resistors:		Semiconductors:	
R_I	910 Ω	IC_1	TL081C
R_F	10 kΩ	IC_2	LM313
R_S	13 kΩ	Q_1	TIP120
R_{sens}	0.35 Ω	Q_2	2N4401

It is important to consider the power that would be dissipated in the pass transistor should the output be shorted. The current limiting that we have just designed would hold the maximum current to 2 A, so the current limit of the pass transistor would not be exceeded. However, if the output becomes shorted, the full voltage of the supply, 25 V, will appear as V_{CE} across the pass transistor. The short-circuit power dissipated in the pass transistor is

$$\text{transistor power} = I_C \times V_{CE} = 2 \times 25 = 50 \text{ W}$$

This is in excess of the maximum power rating of the TIP120. In normal operation, if the short only lasted for a second or so, the transistor would probably survive. If the short lasted longer than a few seconds, then the transistor would blow. We can solve this problem in two ways. The simplest solution is to choose a pass transistor with a higher power rating. The second option is foldback current limiting, which we will investigate next.

Foldback Current Limiting

There is another alternative to current limiting that protects the pass transistor in case of a shorted output. This alternative is known as **foldback current limiting**. In foldback limiting, the current will increase normally up to the maximum current as the load resistance is decreased. If the load resistance is decreased further, the current (and hence the output voltage) will begin to drop. Under short-circuit conditions (zero output voltage), the current is reduced to typically a quarter of the full-load current.

The circuit for incorporating foldback limiting in an op-amp regulated power supply is shown in Figure 7-23. The graph illustrates how the voltage and current vary under foldback operation. This circuit is a modification of the simple current-limiting circuit with the base of the bypass transistor Q_2 fed by a voltage divider connected to ground instead of directly from the sensing resistor. Under normal operation, the voltage applied to the base of Q_2 is

$$V_B = \frac{R_B}{R_A + R_B} \times (V_{out} + I_L R_{sens}) \qquad (7\text{-}24)$$

This equation assumes the current into the base of Q_2 is negligible. The voltage applied to the emitter is simply V_{out}. Hence, the base-emitter voltage is

$$V_{BE} = \frac{R_B}{R_A + R_B} \times (V_{out} + I_L R_{sens}) - V_{out} \qquad (7\text{-}25)$$

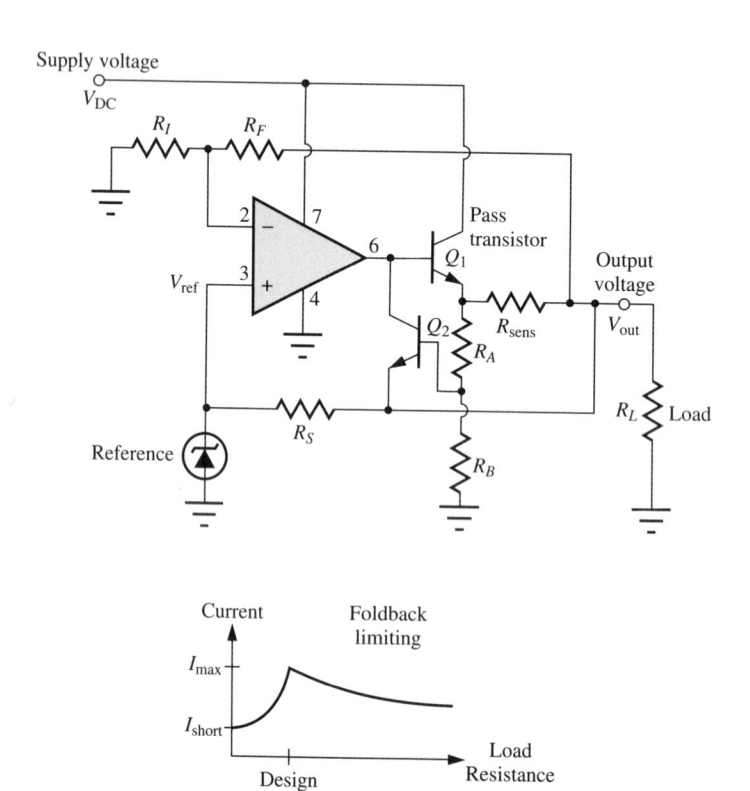

Figure 7-23 Op-amp regulator with pass transistor and foldback current limiting.

7-4 Op-Amp Regulators

Until current limiting begins, the output voltage of the regulator remains at the design level and V_{BE} will be less than 0.7 V. Thus, Q_2 will not conduct. When the current reaches the maximum current I_{Lmax}, V_{BE} will be 0.7 V, so Q_2 will start to conduct and current limiting begins. Substituting these values into Equation 7-25 gives

$$0.7 = \frac{R_B}{R_A + R_B} \times (V_{out} + I_{Lmax}R_{sens}) - V_{out} \tag{7-26}$$

If the load resistance R_L becomes smaller than the critical value of $R_{Lcrit} = V_{out}/I_{max}$, the output current does not increase but the output voltage will drop. Substitute V_{out}/R_L for I_L and 0.7 V for V_{BE} in Equation 7-25, and solve for V_{out}:

$$V_{out}^* = \frac{(R_A + R_B) \times R_L}{R_B R_{sens} - R_A R_L} \times 0.7 \text{ V} \tag{7-27}$$

V_{out}^* is the output voltage under current limiting. If the output is shorted, R_L becomes zero and the output voltage becomes zero as expected.

The output current will actually drop from the maximum value I_{Lmax} for load resistances smaller than R_{Lcrit}. Return to Equation 7-25, make $V_{BE} = 0.7$ V, and solve for I_L:

$$I_L^* = \frac{R_A + R_B}{R_B R_{sens}} \times (V_{out}^* + 0.7) - \frac{V_{out}^*}{R_{sens}} \tag{7-28}$$

I_L^* is the output current under current-limited operation. Setting the output voltage to V_{out}, the design output voltage, we get the maximum output current I_{Lmax}:

$$I_{Lmax} = \frac{R_A + R_B}{R_B R_{sens}} \times (V_{out} + 0.7) - \frac{V_{out}}{R_{sens}} \tag{7-29}$$

Setting the output voltage to zero gives the short-circuit current I_{short}:

$$I_{short} = \frac{R_A + R_B}{R_B R_{sens}} \times 0.7 \tag{7-30}$$

Solve Equation 7-30 for $R_B/(R_A + R_B)$:

$$\frac{R_B}{R_A + R_B} = \frac{0.7}{I_{short} \times R_{sens}} \tag{7-31}$$

Substitute for $R_B/(R_A + R_B)$ in Equation 7-29 and solve for R_{sens}:

$$R_{sens} = \frac{0.7 V_{out}}{V_{out} I_{short} + 0.7 I_{short} - 0.7 I_{Lmax}} \tag{7-32}$$

Calculate the required value for R_{sens}, using this equation. Substitute this value into Equation 7-31, choose a value for $R_A + R_B$ near 1 kΩ, and solve for R_B.

$$R_B = \frac{0.7}{I_{short}} \times \frac{R_A + R_B}{R_{sens}} \tag{7-33}$$

Finally, solve for R_A. The choice of value for $R_A + R_B$ is dictated by the assumption made at the start to ignore the base current of Q_2. By making $R_A + R_B$ relatively small, we make the bleeder current in the voltage divider much larger than the base current. If we include the base current in the calculations, the equations become much more complicated. The actual design of a foldback current limiter is illustrated in Example 7-11 and analyzed in Example 7-12.

EXAMPLE 7-11

Modify the current-limiting circuit designed in Example 7-10 to limit the short-circuit current to 0.5 A and hence protect the pass transistor under short-circuit conditions.

Solution

We will redesign the circuit shown in Figure 7-22 to incorporate foldback limiting. The circuit that we will use is similar to that shown in Figure 7-23, except that we are now using a Darlington pass transistor. The actual circuit is shown in Figure 7-24.

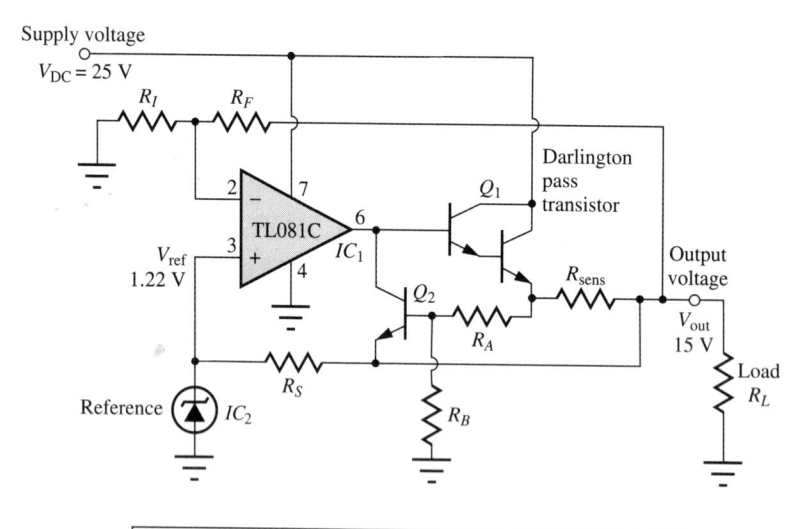

Parts List

Resistors:
- R_I 910 Ω
- R_F 10 kΩ
- R_S 13 kΩ
- R_{sens} 1.5 Ω
- R_A 130 Ω
- R_B 820 Ω

Semiconductors:
- IC_1 TL081C
- IC_2 LM313
- Q_1 TIP120
- Q_2 2N4401

Figure 7-24

Most of the regulator circuit has already been designed in previous examples. In this example, we need only design the foldback voltage divider consisting of R_A and R_B, and the sensing resistor R_{sens}.

We first must calculate R_{sens} from Equation 7-32:

$$R_{sens} = \frac{0.7 V_{out}}{V_{out} I_{short} + 0.7 I_{short} - 0.7 I_{Lmax}}$$

$$= \frac{0.7 \text{ V} \times 15 \text{ V}}{15 \text{ V} \times 0.5 \text{ A} + 0.7 \text{ V} \times 0.5 \text{ A} - 0.7 \text{ V} \times 2.0 \text{ A}} = 1.627 \text{ }\Omega$$

This will have to be a power resistor since it must carry 2 A of current at maximum load and dissipate approximately 6.5 W! The closest standard power resistor is 1.5 Ω. We will use this value for R_{sens}.

Next assume $R_A + R_B = 1.0$ kΩ and calculate R_B from Equation 7-33:

$$R_B = \frac{0.7}{I_{short}} \times \frac{R_A + R_B}{R_{sens}} = \frac{0.7 \text{ V}}{0.5 \text{ A}} \times \frac{1.0 \text{ k}\Omega}{1.63 \text{ }\Omega} = 860 \text{ }\Omega$$

This gives $R_A = 140 \text{ }\Omega$. Choosing the closest standard values, we will make $R_A = 130 \text{ }\Omega$ and $R_B = 820 \text{ }\Omega$.

Our foldback current-limiting design is completed, as well as our complete regulator. We shall verify the operation of the regulator in the next example.

EXAMPLE 7-12

Analyze the circuit designed in Example 7-11 to determine the variation of voltage and current with load for the foldback current-limiting circuit under overload conditions.

Solution

The circuit that we will be analyzing was shown in Figure 7-24.

First calculate the voltage and current for a load of 20 Ω. Because this is a larger resistance than the critical value of 10 Ω for current limiting, the voltage and current can be found simply from Ohm's law applied to the load resistance. The voltage is simply the regulated output voltage, which from Example 7-8 is 14.63 V, and the current is 14.63 V/20 Ω = 0.732 A.

Repeating these calculations for resistances of 16 Ω and 12 Ω, we find currents of 0.914 A and 1.219 A, respectively, for a constant voltage of 14.63 V.

Next, calculate the voltage and current at the point where current limiting begins. Our design value for this was 2 A, for a critical load of 7.5 Ω. Since our design, however, used real resistor values, the actual current where limiting begins may be slightly different. Use Equation 7-29 to solve for $I_{L\max}$:

$$I_{L\max} = \frac{R_A + R_B}{R_B R_{sens}} \times (V_{out} + 0.7) - \frac{V_{out}}{R_{sens}}$$

$$= \frac{130\ \Omega + 820\ \Omega}{820\ \Omega \times 1.5\ \Omega} \times (14.63\ \text{V} + 0.7\ \text{V}) - \frac{14.63\ \text{V}}{1.5\ \Omega} = 2.09\ \text{A}$$

This is very close to the design value.

Calculate the resistance to give this I_L for a voltage of 14.63 V.

$$R_L = \frac{V_L}{I_L} = \frac{14.63\ \text{V}}{2.09\ \text{A}} = 7.00\ \Omega$$

Current limiting begins when the load resistance has dropped to 7.00 Ω. The voltage is 14.63 V, and the current is 2.09 A.

Next, calculate the current and voltage for a load resistance of 4.0 Ω. Use Equation 7-27 to calculate the voltage:

$$V_{out}^* = \frac{(R_A + R_B) \times R_L}{R_B R_{sens} - R_A R_L} \times 0.7\ \text{V}$$

$$= \frac{(130\ \Omega + 820\ \Omega) \times 4.0\ \Omega}{820\ \Omega \times 1.5\ \Omega - 130\ \Omega \times 4.0\ \Omega} \times 0.7\ \text{V} = 3.75\ \text{V}$$

The corresponding current is found from Ohm's law:

$$I_L^* = \frac{V_{out}^*}{R_L} = \frac{3.75\ \text{V}}{4.0\ \Omega} = 0.938\ \text{A}$$

Finally, calculate the short-circuit current from Equation 7-30:

$$I_{short} = \frac{R_A + R_B}{R_B R_{sens}} \times 0.7$$

$$= \frac{130\ \Omega + 820\ \Omega}{820\ \Omega \times 1.5\ \Omega} \times 0.7 = 0.541\ \text{A}$$

This is quite close to our design value of 0.5 A.

Before we leave this problem, we must plot our results. The results of the foregoing calculations are plotted in Figure 7-25. This figure shows the voltage and current characteristics of the foldback current limiter.

Other Refinements to Op-Amp Regulator Circuits

Other refinements can be added to these regulator circuits. For example, we could make a variable regulator by replacing R_F with a potentiometer. This allows us to vary the voltage from approximately 1.22 V (limited by the voltage of the voltage reference integrated circuit) for $R_F = 0$ to as high as we want (limited by the

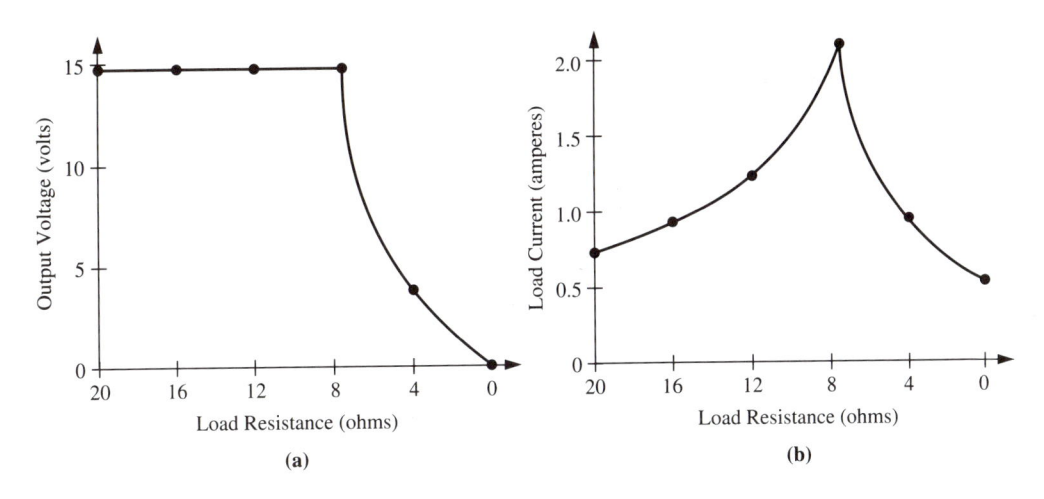

Figure 7-25

maximum available supply voltage and the gain determined by the maximum value of R_F).

The circuit could also be modified to provide variable current limiting. The best circuit for this would be the foldback current-limiting procedure, where the current limit could be incorporated by replacing the voltage divider consisting of R_A and R_B with a potentiometer. Because of the large amount of power dissipated in R_{sens}, it is not advisable to replace this with a potentiometer to make a variable current-limiting system.

These op-amp regulator circuits have considerable flexibility, but they do have the one disadvantage that the circuits are relatively complex. We will remedy this in the next chapter when we introduce integrated circuit regulators.

Summary

This chapter discussed power supplies and presented the concepts of power supply regulation. The various regulator characteristics, including line regulation, load regulation, ripple rejection, and power supply impedance, were introduced.

Power supply regulation is usually achieved by comparing the output voltage to a reference voltage. The reference voltage can be provided by a zener diode or a voltage reference integrated circuit. Voltage references were shown to be far superior to zener diodes in having much smaller bias currents, greater precision, and smaller variation of voltage with bias current or temperature change.

Different types of linear voltage regulators using op-amps were developed, all based on a noninverting amplifier and a voltage reference. High power output was achieved by adding a pass transistor, current limiting was achieved by adding a bypass transistor, and foldback limiting was introduced to protect the pass transistor from shorted outputs.

7 Linear Power Supplies

Although the regulators developed in this chapter were intended simply as an introduction to the integrated circuit regulators in the next chapter, they certainly are capable of producing acceptable regulation. Their only disadvantages are circuit complexity and the amount of design calculations required.

Review Questions

7-1 Rectifier circuits

1. Why is electric power supplied as alternating current?
2. What is the purpose of the transformer in a rectifier circuit?
3. What are the advantages of a bridge-type rectifier over a simple two-diode full-wave rectifier or a simple half-wave rectifier?
4. What data are required in order to select a diode or bridge-type rectifier?
5. What is the advantage of a bridge-type rectifier over a simple two-diode full-wave rectifier?
6. What is the advantage of full-wave rectification over half-wave rectification?
7. What is the surge current in a rectifier? How is it determined?
8. How do you estimate the winding resistance of a transformer?
9. What is the purpose of a filter capacitor in a rectifier circuit?
10. Explain how ripple voltage arises in a filtered rectifier circuit.
11. What is the relation between ripple voltage, DC output voltage, and peak rectified voltage?
12. What type of capacitor is usually used as a filter capacitor?

7-2 Regulation

13. Name four quantities that describe the quality of a regulator. What is the ideal value for each of these?
14. Define line regulation.
15. Define load regulation.
16. How do you convert line regulation to percent line regulation?
17. How do you convert load regulation to percent load regulation?
18. What are the units of load regulation and line regulation?
19. What measurements are necessary to determine the operating characteristics of a regulator?
20. How do manufacturers determine the characteristics of their regulators?

7-3 Voltage references

21. Describe the voltage versus current characteristics of a zener diode.
22. What is the difference between a zener diode and a rectifier diode?
23. What are the limitations in using a zener diode as a regulator circuit?
24. What function is a zener diode best used for in a regulator circuit?
25. Why is it best to use a zener with a voltage near 6 V?
26. How does a voltage reference differ in performance from a simple zener diode?
27. What are the advantages of using a voltage reference instead of a zener diode in a regulator circuit?
28. What is the difference between a regulator circuit using a voltage reference and a regulator circuit using a zener diode?

7-4 Op-amp regulators

29. Name the three main functional parts of a voltage regulator.
30. In a regulator circuit using a standard op-amp, why do we not need a dual supply for the op-amp?
31. How is the op-amp in a regulator circuit configured?
32. What is dropout voltage? How does it arise?
33. What limits the output current in a simple op-amp regulator? How can a larger current be achieved?
34. What quantities determine the power dissipated in a pass transistor used in a regulator circuit?
35. What additional design must be done on a regulator circuit to include a pass transistor?
36. What is the advantage of using a Darlington power transistor instead of an ordinary power transistor?
37. What happens to the pass transistor in a simple regulator without current limiting if the output is shorted?
38. Describe how simple current limiting works.
39. What is the potential problem for the pass transistor with simple current limiting?
40. Describe the variation of current with load and voltage with load for foldback current limiting. Include the case where the load is zero (shorted).

Design and Analysis Problems

7-1 Rectifier circuits

1. Design a rectifier circuit to give an output voltage of at least 20 V at 2.0 A with an RMS ripple voltage less than 3 V.
   (a) Sketch the circuit you are going to use.
   (b) Calculate the effective load resistance.
   (c) Calculate the peak voltage and the capacitance to give the required DC output and ripple. Note that it will be necessary to solve Equations 7-7 and 7-9 simultaneously for V_P and C. Select a suitable transformer and filter capacitor.
2. Analyze the rectifier circuit shown as Figure 7-26 to determine the
   (a) DC output voltage
   (b) ripple voltage
   (c) current supplied to the load.

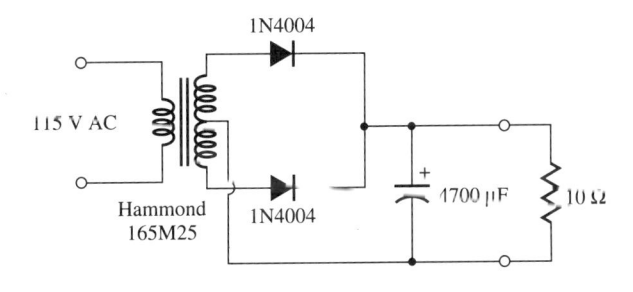

Figure 7-26

7 Linear Power Supplies

7-2 Regulation

3. An integrated circuit regulator is being tested. For a DC input voltage of 25.0 V, the output voltage for a load of 10 Ω is 18.27 V, and for a load of 12 Ω it is 18.41 V. For a DC input voltage of 20.0 V and a load of 10 Ω, the output voltage is 18.11 V. Calculate the line regulation and load regulation for the regulator.

7-3 Voltage references

4. Design a circuit to give a reference voltage of 15 V from a 24 V DC supply. Select a suitable zener diode and calculate the bias resistor R_S.

7-4 Op-amp regulators

5. Design a regulator circuit using a TL081 op-amp and an LM313 voltage reference to give a regulated output of 2.5 V at 100 mA. The DC source can supply 12 V at 100 mA. Note that a pass transistor will not be required.
   (a) Sketch the circuit you are going to use.
   (b) Calculate R_F, R_I, and R_S. Select the nearest standard values.
6. Analyze the regulator circuit in Figure 7-27 to determine the output voltage.

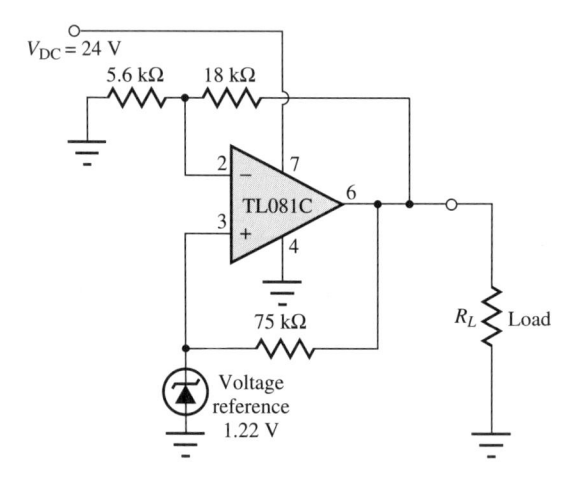

Figure 7-27

7. Analyze the regulator circuit in Figure 7-28 to determine the
   (a) output voltage
   (b) maximum current
   (c) short-circuit current
   (d) DC output voltage from the rectifier circuit
   (e) power dissipated in the pass transistor under full load.

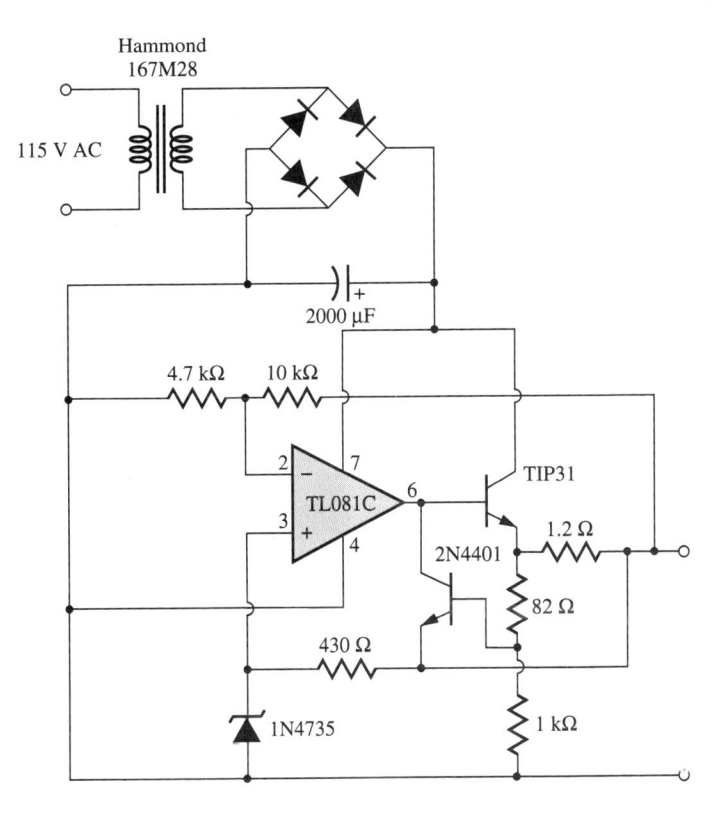

Figure 7-28

8. Design a 12 V, 1 A power supply using a TL081 op-amp and an LM313 voltage reference. The power supply should have simple current limiting. The maximum ripple voltage from the rectifier circuit should be less than 1 V RMS.
   (a) Sketch the complete circuit you are going to use, including rectifier, regulator, and pass transistor.
   (b) Determine the minimum DC voltage that must be supplied by the rectifier circuit. Hence, calculate the effective load seen by the rectifier. Design the rectifier circuit by specifying transformer, filter capacitor, and bridge rectifier.
   (c) Calculate the DC output voltage and the ripple from the rectifier circuit.
   (d) Determine values of R_I, R_F, and R_S for the regulator circuit. Choose nearest standard values.
   (e) Calculate the voltage, current, and power of the pass transistor, and select a suitable pass transistor.
   (f) Calculate the value of the sensing resistor.

8 | Integrated Circuit Regulators

OUTLINE

8-1 Multipin Voltage Regulators
8-2 Three-Terminal Fixed Regulators
8-3 Three-Terminal Variable Regulators
8-4 Voltage Converters

KEY TERMS

Multipin regulator
Three-terminal regulator
Floating regulator
Quiescent current
Voltage converter

OBJECTIVES

On completion of this chapter, the reader should be able to:

- use the LM723 linear voltage regulator in a variety of applications, including a simple low-voltage power supply, a simple high-voltage power supply, a high-current power supply, and a power supply with current limiting.
- use both positive and negative three-terminal fixed regulators for a variety of regulator applications, including simple positive and negative regulated power supplies, a dual power supply, a high-current power supply, a current-limited supply, and a floating variable-voltage regulated supply.
- explain the differences between three-terminal variable regulators and three-terminal fixed regulators, and design variable power supplies using variable regulators.
- use the ICL7660 voltage converter to produce a negative voltage from a positive source.

Introduction

In the previous chapter, we introduced the voltage regulator, a circuit or device whose primary function is to regulate or stabilize the output of a DC power supply. We saw that a simple op-amp could be used to produce a voltage regulator. More commonly, however, integrated circuits designed specifically as regulators are used.

The earliest integrated circuit voltage regulators were multipin devices such as the LM723C. These devices are still used because of the flexibility by which they can be configured to satisfy different regulator requirements, and because the regulation is superior to most other types of regulators. The most common form of integrated circuit regulator, however, is the three-terminal regulator. These are inexpensive and simple to use, yet they provide excellent regulation. Much of this chapter will be devoted to the design of practical DC power supplies using these three-terminal regulators. Although standard three-terminal regulators can be used to make variable supplies, special three-terminal regulator integrated circuits have been developed for this application.

Finally, we will look at voltage converters. Voltage converters are a relatively unknown integrated circuit, but are very useful, since they take a positive supply voltage and produce the equivalent negative voltage. We will encounter applications for these several times in later chapters.

8-1 Multipin Voltage Regulators

In the previous chapter, we saw how to design voltage regulators using operational amplifiers. Although the circuits are basically simple, recall that a complete regulator circuit consists typically of a voltage reference, a sensing circuit, an error correction op-amp, a pass transistor, and current limiting. Because integrated circuits are so easy to manufacture, the logical approach is to integrate some or all of the above parts onto a single chip and call the complete unit a voltage regulator. In fact, two approaches have been followed in producing integrated circuit regulators. The first approach is to package the basic regulator in a conventional package, such as a 14 pin DIP or 10 pin TO-100 can, with provisions for adding external sensing, external current limiting, and an external pass transistor when necessary. This approach provides the greatest flexibility and produces the highest quality regulation. The second approach is to include the sensing circuit, pass transistor, and current limiting as part of the integrated circuit and use a three-pin power transistor packaging (TO-3 or TO-220 case). This approach produces the simplest form of regulator.

Three-pin regulators are far more popular than the **multipin regulators** because they are so easy to use. Nevertheless, multipin regulators are still in use because of their greater flexibility, and we will start our study of integrated circuit regulators with a multipin regulator, the LM723. The LM723 is a typical multipin regulator and is probably the most widely used.

Data sheets for the LM723 multipin voltage regulator are shown in Figure 8-1. It is quite informative to compare the equivalent circuit of the LM723 shown

National Semiconductor

Voltage Regulators

LM723/LM723C Voltage Regulator

General Description

The LM723/LM723C is a voltage regulator designed primarily for series regulator applications. By itself, it will supply output currents up to 150 mA; but external transistors can be added to provide any desired load current. The circuit features extremely low standby current drain, and provision is made for either linear or foldback current limiting. Important characteristics are:

- 150 mA output current without external pass transistor
- Output currents in excess of 10A possible by adding external transistors

- Input voltage 40V max
- Output voltage adjustable from 2V to 37V
- Can be used as either a linear or a switching regulator.

The LM723/LM723C is also useful in a wide range of other applications such as a shunt regulator, a current regulator or a temperature controller.

The LM723C is identical to the LM723 except that the LM723C has its performance guaranteed over a 0°C to 70°C temperature range, instead of −55°C to +125°C.

Schematic and Connection Diagrams*

Dual-In-Line Package

Order Number LM723CN
See NS Package N14A

Order Number LM723J or LM723CJ
See NS Package J14A

Metal Can Package

Note: Pin 5 connected to case.

Order Number LM723H or LM723CH
See NS Package H10C

Equivalent Circuit*

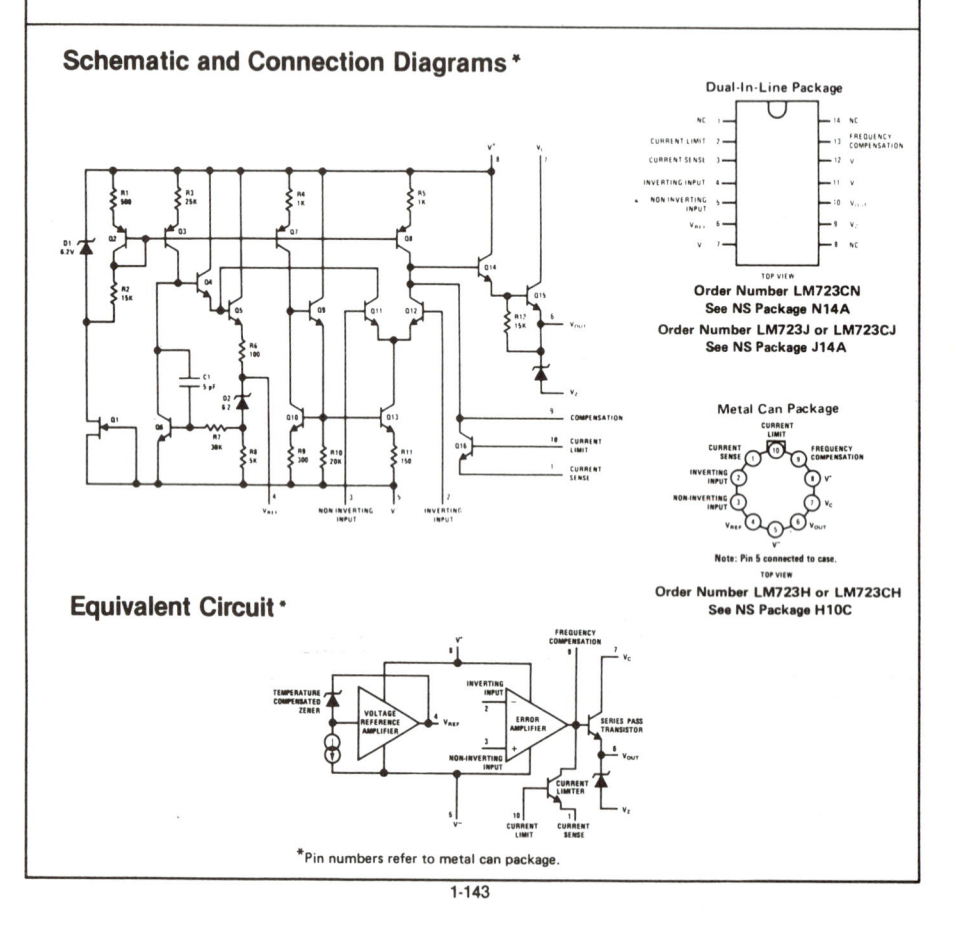

*Pin numbers refer to metal can package.

Figure 8-1 Data sheets for the LM723 voltage regulator. *(Reprinted with permission of National Semiconductor Corporation)*

LM723/LM723C

Absolute Maximum Ratings

Pulse Voltage from V⁺ to V⁻ (50 ms)	50V
Continuous Voltage from V⁺ to V⁻	40V
Input-Output Voltage Differential	40V
Maximum Amplifier Input Voltage (Either Input)	7.5V
Maximum Amplifier Input Voltage (Differential)	5V
Current from V_Z	25 mA
Current from V_{REF}	15 mA
Internal Power Dissipation Metal Can (Note 1)	800 mW
Cavity DIP (Note 1)	900 mW
Molded DIP (Note 1)	660 mW
Operating Temperature Range LM723	$-55°C$ to $+125°C$
LM723C	$0°C$ to $+70°C$
Storage Temperature Range Metal Can	$-65°C$ to $+150°C$
DIP	$-55°C$ to $+125°C$
Lead Temperature (Soldering, 10 sec)	$300°C$

Electrical Characteristics (Note 2)

PARAMETER	CONDITIONS	LM723 MIN	LM723 TYP	LM723 MAX	LM723C MIN	LM723C TYP	LM723C MAX	UNITS
Line Regulation	V_{IN} = 12V to V_{IN} = 15V		.01	0.1		.01	0.1	% V_{OUT}
	$-55°C \leq T_A \leq +125°C$			0.3				% V_{OUT}
	$0°C \leq T_A \leq +70°C$						0.3	% V_{OUT}
	V_{IN} = 12V to V_{IN} = 40V		.02	0.2		0.1	0.5	% V_{OUT}
Load Regulation	I_L = 1 mA to I_L = 50 mA		.03	0.15		.03	0.2	% V_{OUT}
	$-55°C \leq T_A \leq +125°C$			0.6				% V_{OUT}
	$0°C \leq T_A \leq +70°C$						0.6	% V_{OUT}
Ripple Rejection	f = 50 Hz to 10 kHz, C_{REF} = 0		74			74		dB
	f = 50 Hz to 10 kHz, C_{REF} = 5 μF		86			86		dB
Average Temperature Coefficient of Output Voltage	$-55°C \leq T_A \leq +125°C$		.002	.015				%/°C
	$0°C \leq T_A \leq +70°C$					.003	.015	%/°C
Short Circuit Current Limit	R_{SC} = 10Ω, V_{OUT} = 0		65			65		mA
Reference Voltage		6.95	7.15	7.35	6.80	7.15	7.50	V
Output Noise Voltage	BW = 100 Hz to 10 kHz, C_{REF} = 0		20			20		μVrms
	BW = 100 Hz to 10 kHz, C_{REF} = 5 μF		2.5			2.5		μVrms
Long Term Stability			0.1			0.1		%/1000 hrs
Standby Current Drain	I_L = 0, V_{IN} = 30V		1.3	3.5		1.3	4.0	mA
Input Voltage Range		9.5		40	9.5		40	V
Output Voltage Range		2.0		37	2.0		37	V
Input-Output Voltage Differential		3.0		38	3.0		38	V

Note 1: See derating curves for maximum power rating above 25°C.
Note 2: Unless otherwise specified, T_A = 25°C, V_{IN} = V^+ = V_C = 12V, V^- = 0, V_{OUT} = 5V, I_L = 1 mA, R_{SC} = 0, C_1 = 100 pF, C_{REF} = 0 and divider impedance as seen by error amplifier $\leq$ 10 kΩ connected as shown in Figure 1. Line and load regulation specifications are given for the condition of constant chip temperature. Temperature drifts must be taken into account separately for high dissipation conditions.
Note 3: L_1 is 40 turns of No. 20 enameled copper wire wound on Ferroxcube P36/22-3B7 pot core or equivalent with 0.009 in. air gap.
Note 4: Figures in parentheses may be used if R1/R2 divider is placed on opposite input of error amp.
Note 5: Replace R1/R2 in figures with divider shown in Figure 13.
Note 6: V^+ must be connected to a +3V or greater supply.
Note 7: For metal can applications where V_Z is required, an external 6.2 volt zener diode should be connected in series with V_{OUT}.

Figure 8-1 (*Continued*)

328 8 Integrated Circuit Regulators

on the first page of the data sheets to the simple op-amp regulators that we designed in Chapter 7. Notice that, although the configuration is slightly different, the equivalent circuit contains most of the parts of our op-amp regulator, including the voltage reference, error amplifier, pass transistor, and current limiter. Only a few external connections are necessary to produce a complete regulator circuit.

Like the op-amps discussed in earlier chapters, the LM723 is available in different ratings: the military/industrial rating listed as the LM723, and the commercial rating listed as the LM723C. We shall use the LM723C exclusively. The electrical characteristics are quite similar for the two ratings, the main difference being the temperature range for operating the device.

The LM723C is available in two types of packages: a 14 pin DIP and a 10 pin TO-100 metal can. The 14 pin DIP configuration is more common, and we will

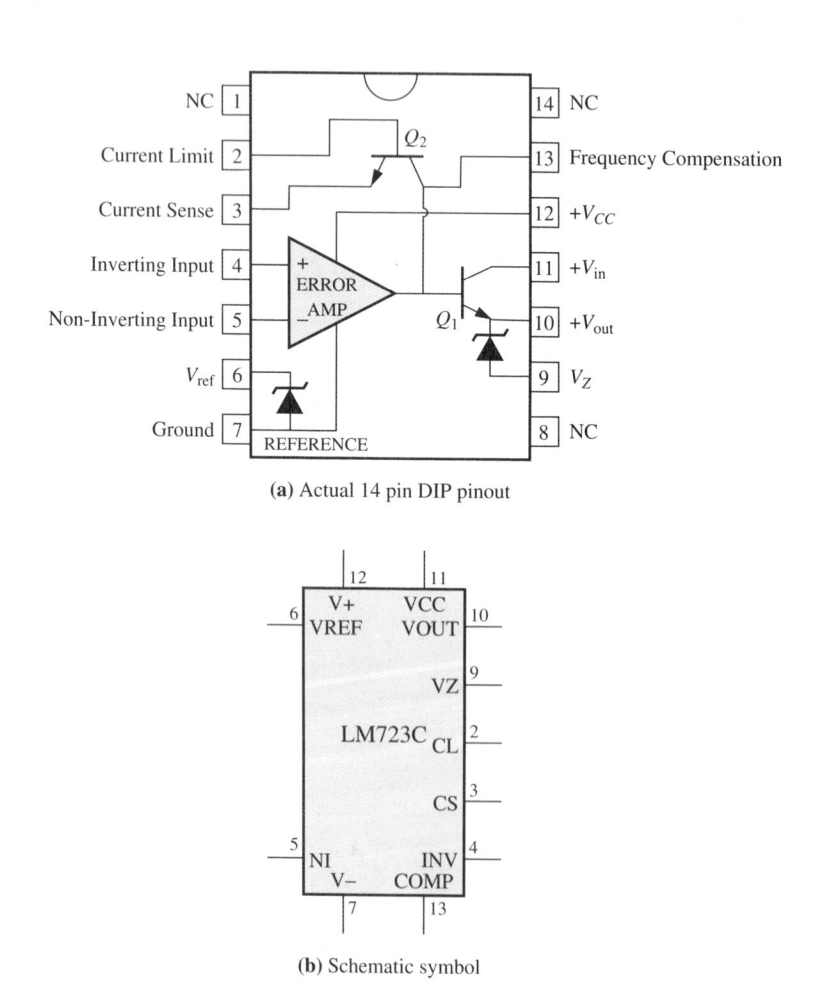

(a) Actual 14 pin DIP pinout

(b) Schematic symbol

Figure 8-2 Pinout of the LM723C voltage regulator.

use it in all of our designs. Design with the 10 pin package is identical except for the pinout. Both pinouts are shown on the data sheets in Figure 8-1.

The pinout of the 14 pin DIP configuration is repeated in Figure 8-2(a). Figure 8-2(b) shows the schematic symbol for the LM723C that we will use in all circuit diagrams. Other than the amplifiers we have encountered in previous chapters, most integrated circuits are represented by a rectangular schematic symbol, such as shown for the LM723C. This symbol is not simply a copy of the device pinout, but has the pins placed to simplify the drawing of circuit diagrams and to group related functions on the schematic.

The LM723C can be used as either a positive or negative, variable or fixed regulator for any voltage from 2 to 37 V. To make a complete regulator, we need only add an external feedback loop from the output and connect the voltage reference. If necessary, we can add either simple current limiting or foldback current limiting. The LM723C can supply an output current of 150 mA. If a larger output current is required, an additional external pass transistor can be added. With the LM723C, the designer has full control over the configuration of the regulator circuit.

As with the op-amp regulators discussed in the previous chapter, the error amplifier in the LM723C compares the output voltage to a reference voltage in order to maintain a constant output. The LM723C has an internal 7.15 V reference (pin 6), but it can be used with any external reference. If the internal reference is used, the LM723C is normally operated in one of two ways: as a high-voltage regulator for any output voltage in the range from 7.15 V to approximately 37 V, or as a low-voltage regulator for any output voltage in the range from 2 to 7.15 V.

Basic High-Voltage Regulator

Figure 8-3 shows the LM723C configured as a basic high-voltage regulator. Two connections are necessary. First, a feedback voltage divider consisting of R_1 connected from the output (pin 10) to the inverting input (pin 4), and R_2 connected from the inverting input to ground is used to configure the internal error amplifier as a noninverting amplifier like the op-amp voltage regulators developed in the previous chapter. Second, the reference voltage (pin 6) must be connected to the noninverting input (pin 5).

The output voltage is

$$V_{out} = V_{ref} \times \frac{R_1 + R_2}{R_2} \tag{8-1}$$

This equation may not be immediately familiar; however, if R_1 is renamed R_F, R_2 is renamed R_I, and V_{out}/V_{ref} is called A_V, we get the gain equation for a noninverting amplifier, exactly the same equation used for designing op-amp regulators in the previous chapter. To design for a specified output voltage, choose some value for $R_1 + R_2$, typically 10 kΩ, and solve Equation 8-1 for R_2:

$$R_2 = (R_1 + R_2) \times \frac{V_{ref}}{V_{out}} \tag{8-2}$$

8 Integrated Circuit Regulators

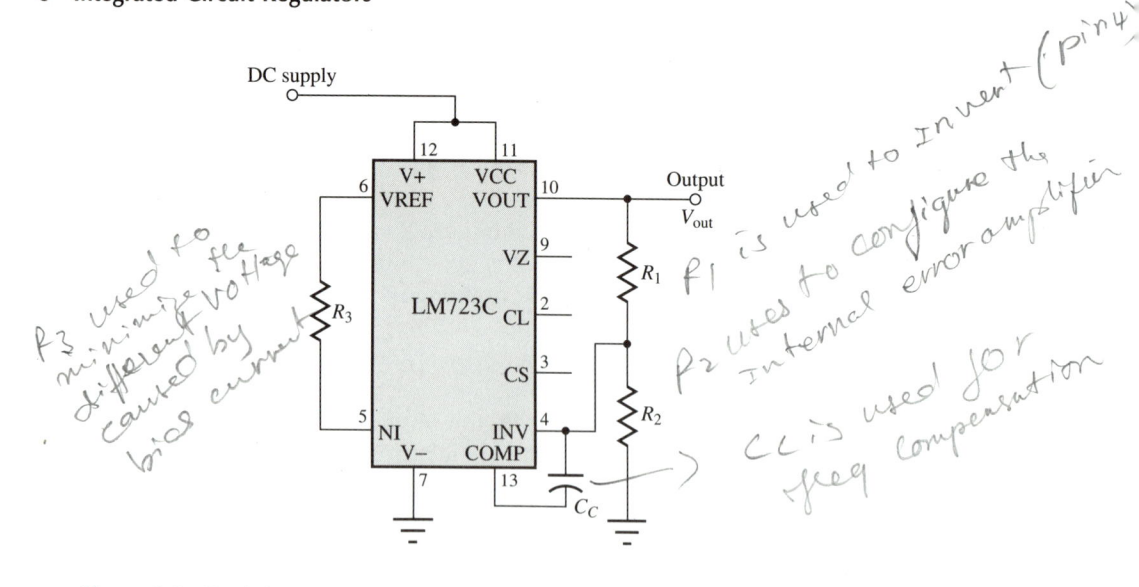

Figure 8-3 Basic high-voltage regulator circuit using an LM723C voltage regulator.

Because the error amplifier in the LM723C is bipolar, bias current must flow into each input. Resistor R_3, connecting the reference voltage (pin 6) to the non-inverting input (pin 5) must be designed to ensure that both amplifier inputs see the same resistance to minimize any differential voltage caused by the bias currents. Resistor R_3 is equivalent to resistor R_K introduced for bipolar op-amps in Chapter 3 and is designed by the same equation, namely

$$R_3 = \frac{R_1 R_2}{R_1 + R_2} \tag{8-3}$$

This value is not critical, and R_3 may be omitted if utmost precision in the output voltage is not required.

To complete the regulator circuit, a capacitor C_C is normally connected from pin 13 to the inverting input (pin 4) for frequency compensation. This capacitor has a typical value of 100 pF. Alternatively, the capacitor could be connected from pin 13 to ground (pin 7) with a typical value of 0.001 µF.

From the device specifications given in Figure 8-1 the maximum current without an external pass transistor is 150 mA. The supply voltage connected to pin 12 ($V+$) and pin 11 (V_{CC}) should be at least 3 V higher than the maximum output voltage. This 3 V is the dropout voltage of the LM723C.

To make the regulator variable, replace R_1, the feedback resistor, with a potentiometer. At the minimum setting of the potentiometer, the output voltage would be V_{ref}, or approximately 7.15 V. The maximum output voltage is determined by the total potentiometer resistance. The potentiometer value must be calculated by assuming some value for R_2, typically 10 kΩ, substituting the desired maximum voltage into Equation 8-1, and solving for R_1. Remember that the maximum output voltage is limited by the available supply voltage.

EXAMPLE 8-1

Design a regulator circuit based on the LM723C to provide 12 V at 20 mA from a rectifier circuit supplying 18 V DC with a 2 V peak-to-peak ripple at a current of 20 mA. Do not use current limiting.

Solution

Because we need only 20 mA of current, we can use the basic high-voltage regulator circuit shown in Figure 8-3. This circuit has been redrawn as Figure 8-4 with a parts list included.

First design the voltage divider consisting of R_1 and R_2. Use Equation 8-2 and assume that $R_1 + R_2 = 10{,}000\ \Omega$. From the data sheet for the LM723C, $V_{\text{ref}} = 7.15$ V. Thus,

$$R_2 = (R_1 + R_2) \times \frac{V_{\text{ref}}}{V_{\text{out}}} = 10\ \text{k}\Omega \times \frac{7.15\ \text{V}}{12\ \text{V}} = 5958\ \Omega$$

Now determine R_1 as $10\ \text{k}\Omega - 5958\ \Omega = 4042\ \Omega$. Choose the closest standard values for R_1 and R_2 from the table in Appendix A-1, which are 3.9 kΩ and 5.6 kΩ, respectively. Note that we chose R_1 and R_2 smaller than the calculated values to hold the ratio of $(R_1 + R_2)/R_2$ as close as possible to the calculated value. Alternatively, we could have chosen both resistors larger than the calculated values. If we had chosen the true closest values ($R_1 = 4.3$ kΩ and $R_2 = 5.6$ kΩ), our ratio would deviate considerably more from the calculated value. This is a relatively minor point, but a simple way to improve the accuracy of voltage dividers.

Parts List

Resistors:
- R_1 — 3.9 kΩ
- R_2 — 5.6 kΩ
- R_3 — 2.4 kΩ

Capacitors:
- C_C — 100 pF

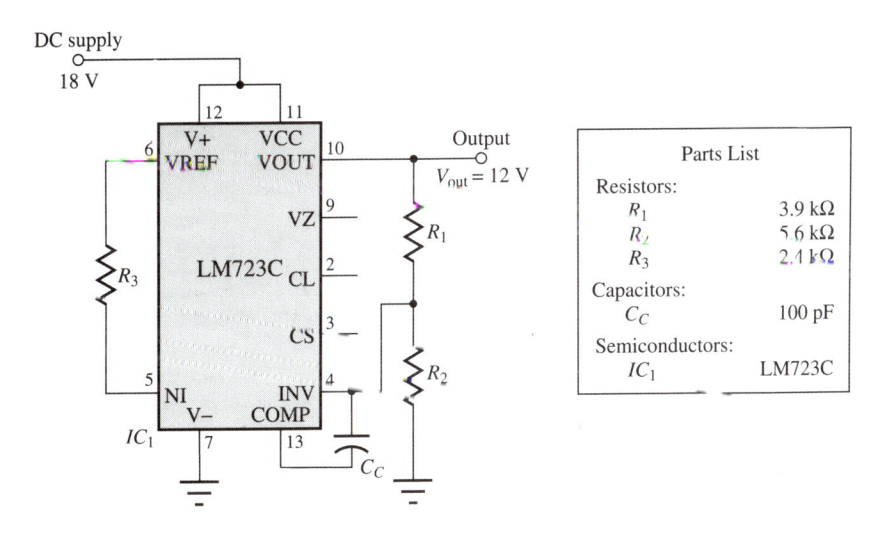

Semiconductors:
- IC_1 — LM723C

Figure 8-4

Next, calculate R_3 by using Equation 8-3:

$$R_3 = \frac{R_1 R_2}{R_1 + R_2} = \frac{3.9 \text{ k}\Omega \times 5.6 \text{ k}\Omega}{3.9 \text{ k}\Omega + 5.6 \text{ k}\Omega} = 2.30 \text{ k}\Omega$$

We will select a value of 2.4 kΩ.

Finally, we must include the compensating capacitor C_C. The manufacturer recommends a value of 100 pF, which we will use.

The design of the regulator circuit is now complete.

EXAMPLE 8-2

Calculate the line regulation, the load regulation, and output ripple voltage for the regulator designed in Example 8-1.

Solution

Refer to the electrical characteristics on the data sheets in Figure 8-1 for these quantities.

(a) Since the supply voltage used in Example 8-1 is 18 V, we must use the line regulation for the LM723C for $V_{in} = 12$ V to $V_{in} = 40$ V listed as 0.1% V_{out}. This is the percent line regulation that we defined in Equation 7-12. If we want the line regulation as given by Equation 7-11, we divide this number by 100 and multiply by V_{out}. Thus,

$$\text{line regulation} = \frac{0.1 \times 12 \text{ V}}{100} = 0.012 \text{ V/V}$$

The line regulation of the regulator designed in Example 8-1 is 12 mV/V.

(b) The load regulation for the LM723C is listed as 0.03% V_{out}. This is the percent load regulation defined in Equation 7-14. If we want the load regulation as defined by Equation 7-13, we divide this number by 100 and multiply it by V_{out}. We have

$$\text{load regulation} = \frac{0.03 \times 12 \text{ V}}{100} = 0.0036 \text{ V/A}$$

The load regulation of the regulator designed in Example 8-1 is 3.6 mV/A.

(c) The data sheets list the ripple rejection as −74 dB. We must convert dB to a voltage ratio. From the definition of dB:

$$\text{dB} = -20 \log \frac{\text{output ripple}}{\text{input ripple}}$$

We solve for output ripple/input ripple

$$\frac{\text{output ripple}}{\text{input ripple}} = 10^{-74/20} = 0.0001995$$

For a 2 V peak-to-peak input ripple, the output ripple is 0.000399 V or 0.399 mV. Hence, the output ripple of the regulator designed in Example 8-1 is 0.399 mV.

Our line regulation, load regulation, and ripple are quite small. We will compare these later to other types of regulators that we will be designing.

Basic Low-Voltage Regulator

Our second option with the LM723C is to configure it as a low-voltage regulator. The circuit for a basic low-voltage regulator is shown in Figure 8-5. In this case, we use a voltage divider consisting of R_1 connected between the reference voltage (pin 6) and the noninverting input (pin 5), and R_2 connected from the noninverting input to ground. This gives a lower reference voltage at the noninverting input. The output voltage is fed back directly from the output (pin 10) to the inverting input (pin 4). This configures the error amplifier as a unity-gain amplifier. In this case, the output voltage is calculated from the equation

$$V_{out} = V_{ref} \times \frac{R_2}{R_1 + R_2} \tag{8-4}$$

As in the case of the high-voltage regulator, a value can be chosen for $R_1 + R_2$, typically 10 kΩ, and Equation 8-4 solved for R_2:

$$R_2 = (R_1 + R_2) \times \frac{V_{out}}{V_{ref}} \tag{8-5}$$

Notice that this is *not* the same as Equation 8-2—the ratio V_{out}/V_{ref} is the inverse of that used in Equation 8-2.

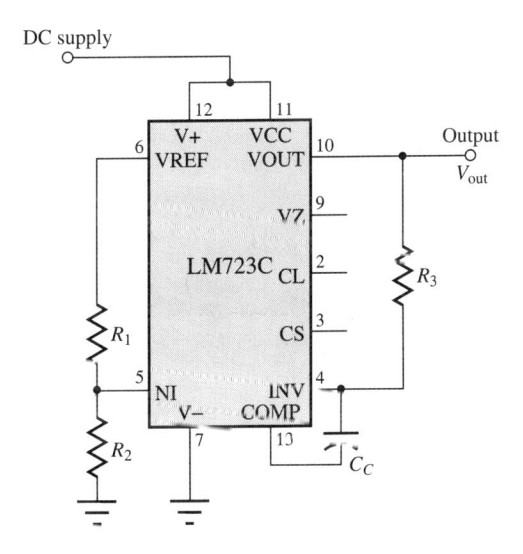

Figure 8-5 Basic low-voltage regulator circuit using an LM723C voltage regulator.

Resistor R_3 is calculated exactly as for the high-voltage regulator, namely by Equation 8-3, and performs the same function.

EXAMPLE 8-3

Design a 5 V, 20 mA regulated power supply using an LM723C regulator. Assume that the rectifier circuit delivers 12 V. Do not use current limiting.

Solution

This problem is similar to that in Example 8-1, except now, because we want an output of 5 V, we must use a low-voltage regulator. Because we need only 20 mA of current, we can use the basic circuit in Figure 8-5. This circuit has been redrawn as Figure 8-6 with a parts list included.

First design the voltage divider consisting of R_1 and R_2. Use Equation 8-4 with $R_1 + R_2 = 10$ kΩ and $V_{ref} = 7.15$ V.

$$R_2 = (R_1 + R_2) \times \frac{V_{out}}{V_{ref}}$$
$$= 10 \text{ k}\Omega \times \frac{5.0 \text{ V}}{7.15 \text{ V}} = 6993 \text{ }\Omega$$

Now determine R_1 as 10 kΩ − 6993 Ω = 3007 Ω. Choose the closest standard values for R_1 and R_2 from Appendix A-1: 3.0 kΩ and 6.8 kΩ, respectively. Note, as in Example 8-1, we choose R_1 and R_2 as the closest smaller value, not necessarily the actual closest value, to give the highest precision in the voltage divider.

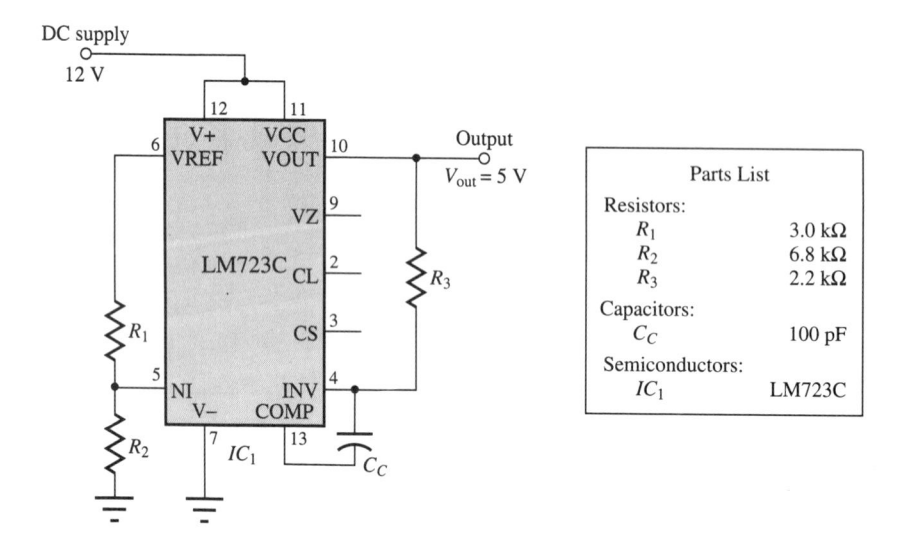

Figure 8-6

Next, calculate R_3 from Equation 8-3.

$$R_3 = \frac{R_1 R_2}{R_1 + R_2}$$

$$= \frac{3.0 \text{ k}\Omega \times 6.8 \text{ k}\Omega}{3.0 \text{ k}\Omega + 6.8 \text{ k}\Omega} = 2.08 \text{ k}\Omega$$

We will select a value of 2.2 kΩ.
 Finally, we will choose 100 pF for the compensating capacitor C_C.
 The design of the regulator circuit is now complete.

The LM723C with a Pass Transistor

The LM723C has an internal pass transistor that allows it to handle output currents as large as 150 mA. If we wish higher current capability, we must use an external pass transistor. The various configurations are shown in Figure 8-7. A high-voltage regulator is shown in Figure 8-7(a), and a low-voltage regulator is shown in Figure 8-7(b). In both cases, the base of the pass transistor Q_1 is driven directly from the output pin of the LM723C (pin 10), and the regulated output is taken from the emitter of the transistor. The resistors to the inverting input (R_1 for the high-voltage regulator, R_3 for the low-voltage regulator) are connected to the power supply output, that is, the emitter of Q_1.

As for the op-amp regulators, no additional design is necessary for the regulator circuit, although some calculations must be performed to verify that the

(a) High voltage regulator

(b) Low voltage regulator

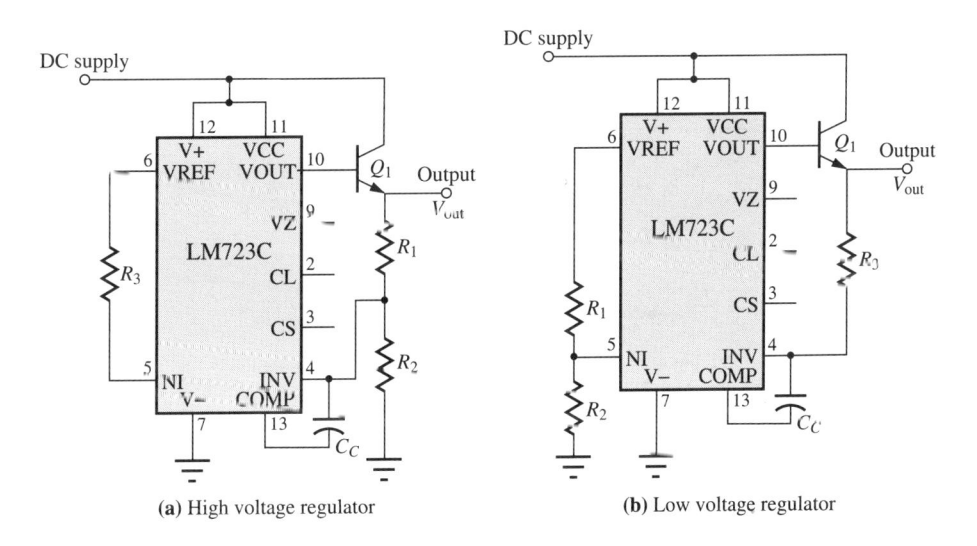

Figure 8-7 LM723C voltage regulator with high-current output.

selected pass transistor can handle the current and power. Transistor Q_1 must be able to pass the desired load current $I_L = I_C$, it must be able to withstand the collector-emitter voltage $V_{CE} = V_{supply} - V_{out}$, and it must be able to dissipate the power $P = I_C \times V_{CE}$.

EXAMPLE 8-4

Design a 5 V, 3 A regulated power supply based on the LM723C regulator chip. Assume that the rectifier circuit can deliver 12 V at 3 A with a 2 V peak-to-peak ripple. Do not use current limiting.

Solution

The circuit will be the same as that in Example 8-3 and shown in Figure 8-6, except that, because of the large current, we must use a pass transistor. The circuit for a low-voltage regulator with pass transistor was shown in Figure 8-7(b). This circuit has been redrawn as Figure 8-8 with a parts list included.

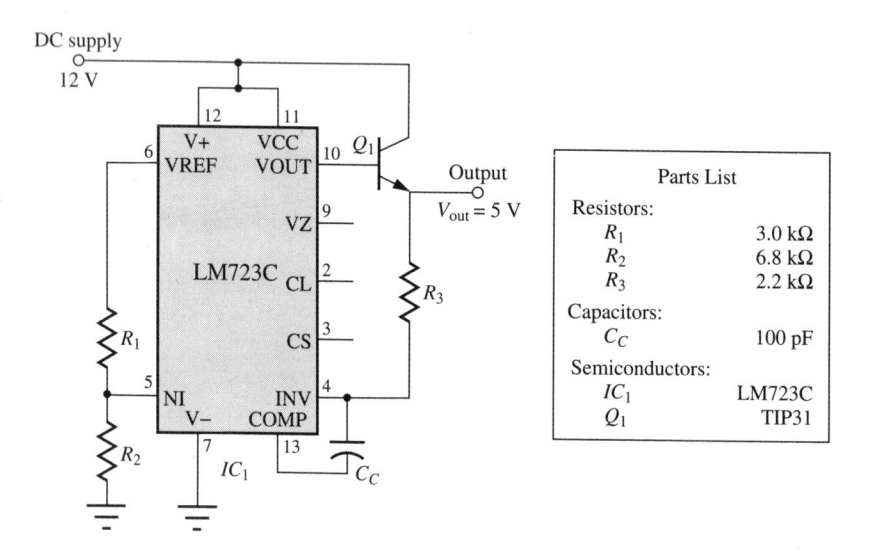

Figure 8-8

The design of the basic regulator circuit is identical to Example 8-3, and we will not repeat it here. The resistor values calculated were $R_1 = 3.0$ kΩ, $R_2 = 6.8$ kΩ, and $R_3 = 2.2$ kΩ. We will also use the same compensating capacitor, $C_C = 100$ pF.

We must select a suitable power transistor. The voltage drop across the transistor is

$$V_{CE} = V_{supply} - V_{out} = 12 \text{ V} - 5 \text{ V} = 7 \text{ V}$$

The current through the transistor I_C is simply the load current of 3 A. Hence, the power dissipation is given by

$$P = V_{CE} \times I_C = 7 \text{ V} \times 3 \text{ A} = 21 \text{ W}$$

These quantities are within the current, voltage, and power ratings of a TIP31 *if suitably heatsinked*. Since the maximum current is 3 A, and the LM723C is capable of outputting 150 mA, the transistor must have a minimum β of 3/0.150 = 20. This is well within the capabilities of the TIP31, so we do not need to use a Darlington configuration.

The design of our high-current regulator is complete.

The LM723C with Current Limiting

The LM723C has an internal current-limiting transistor, and it is quite easy to add either simple current limiting or foldback current limiting. For simple current limiting, we use the circuits in Figure 8-9. This figure shows current limiting for the high- and low-voltage configurations. Current limiting is accomplished by connecting R_{sens} between pin 2 (current limiting) and pin 3 (current sensing). The output from the emitter of the pass transistor (or from pin 10 if current limiting is used without the pass transistor) is connected to pin 2, and the final regulated output is connected to pin 3. The value of the sensing resistor is

$$R_{sens} = \frac{V_{sens}}{I_{max}} \tag{8-6}$$

(a) High voltage regulator

(b) Low voltage regulator

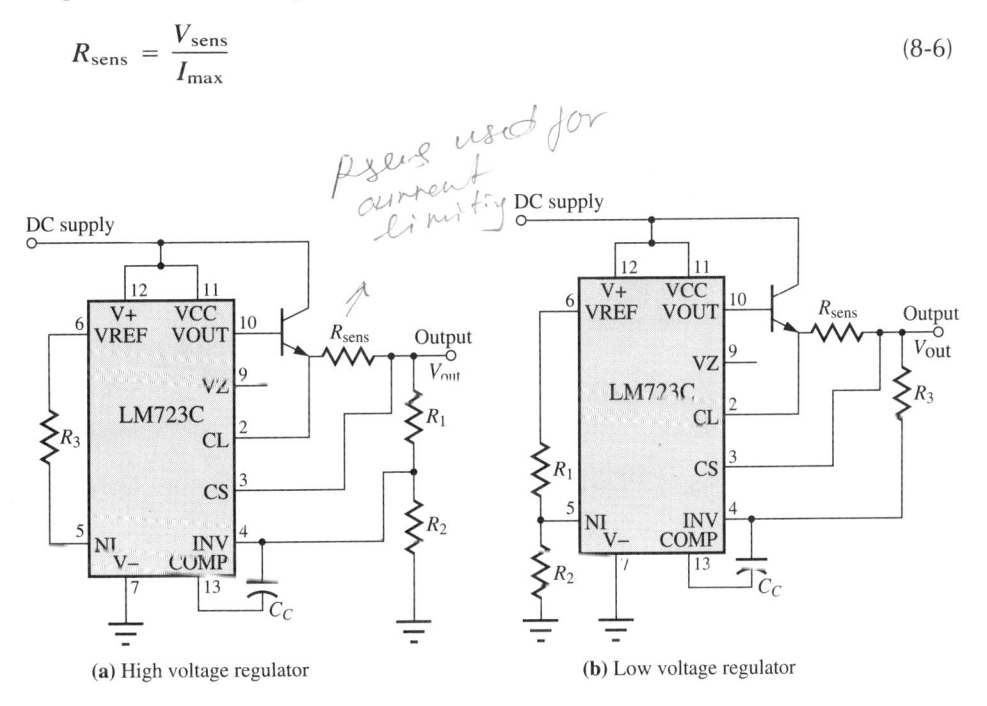

Figure 8-9 LM723C voltage regulator with simple current limiting.

The data sheets for the LM723C list V_{sens} as 0.7 V, the same as for our simple op-amp regulator.

If we wish to incorporate foldback limiting, we use the circuits in Figure 8-10. In this configuration, pin 2 (current limiting) is connected to a voltage divider consisting of R_4 and R_5 connected between the emitter of the pass transistor (or

(a) High voltage regulator

(b) Low voltage regulator

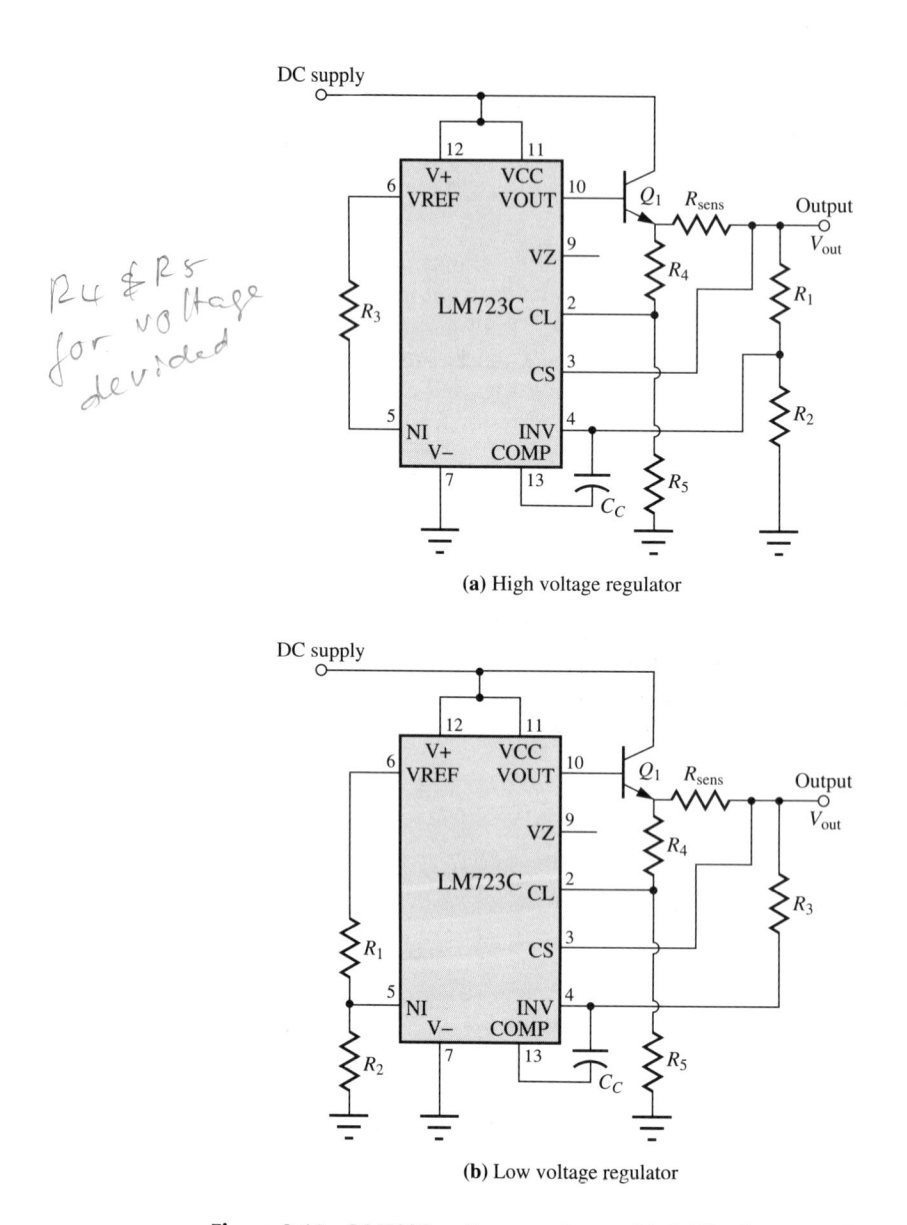

Figure 8-10 LM723C voltage regulator with foldback current limiting.

pin 10 if no pass transistor is used) and ground. Pin 3 (current sensing) is connected directly to the output. Values of R_{sens}, R_4, and R_5 are calculated by the following equations:

$$I_{L\text{max}} = \frac{R_4 + R_5}{R_5 R_{\text{sens}}} \times (V_{\text{out}} + V_{\text{sens}}) - \frac{V_{\text{out}}}{R_{\text{sens}}} \tag{8-7}$$

$$I_{\text{short}} = \frac{R_4 + R_5}{R_5 R_{\text{sens}}} \times V_{\text{sens}} \tag{8-8}$$

If we replace V_{sens} with 0.7 V, R_4 by R_A, and R_5 by R_B, we find that these equations are the same as Equations 7-29 and 7-30 derived for the op-amp regulators designed in the previous chapter.

To design R_{sens}, R_4, and R_5, we first use the following equation to determine a value for R_{sens}:

$$R_{\text{sens}} = \frac{V_{\text{sens}} V_{\text{out}}}{V_{\text{out}} I_{\text{short}} + V_{\text{sens}} I_{\text{short}} - V_{\text{sens}} I_{L\text{max}}} \tag{8-9}$$

Substitute this value of R_{sens} into Equation 8-8, choose a value for $R_4 + R_5$ near 1 kΩ, and solve for R_5:

$$R_5 = \frac{0.7}{I_{\text{short}}} \times \frac{R_4 + R_5}{R_{\text{sens}}} \tag{8-10}$$

Finally solve for R_4. Equations 8-9 and 8-10 are the same as Equations 7-32 and 7-33 for the simple op-amp regulator and are derived in the same manner. The design procedure is the same as for the op-amp regulator.

One word of caution about using the LM723C should be injected here. This regulator, unlike the three-terminal regulators to be discussed later, is not internally protected against overload currents and can easily be blown by drawing too much current. It is advisable to *always use some form of current limiting*, whether a pass transistor is used or not.

EXAMPLE 8-5

Add simple current limiting to the 5 V, 3 A regulated power supply designed in Example 8-4 to limit the maximum current to 3 A.

Solution

To add current limiting to the circuit in Figure 8-8 of Example 8-4, we need to add a sensing resistor R_{sens} to the output of the power transistor and connect pins 2 and 3 of the regulator chip to this resistor, as in the circuit in Figure 8-7(b). This circuit has been redrawn as Figure 8-11, and a parts list is included.

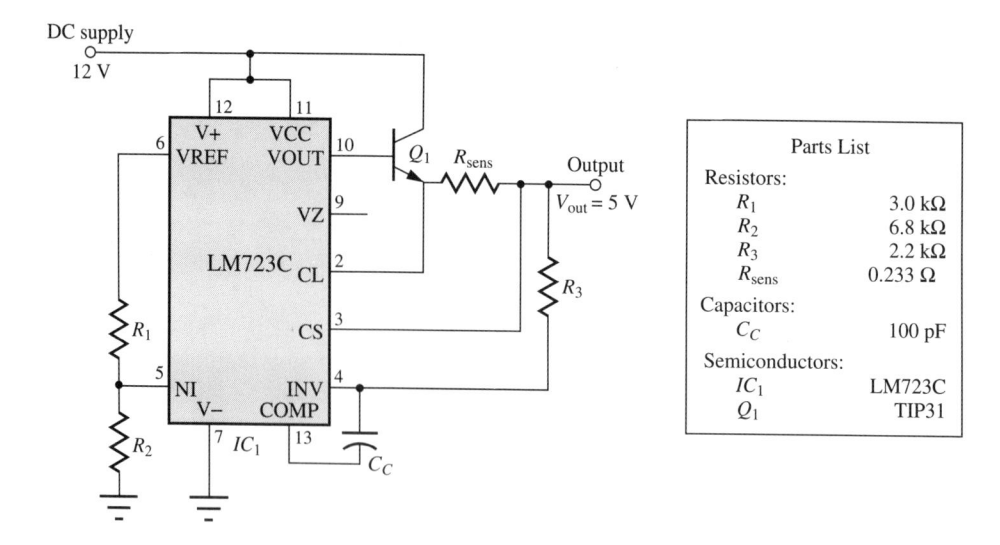

Figure 8-11

Parts List

Resistors:
- R_1 3.0 kΩ
- R_2 6.8 kΩ
- R_3 2.2 kΩ
- R_{sens} 0.233 Ω

Capacitors:
- C_C 100 pF

Semiconductors:
- IC_1 LM723C
- Q_1 TIP31

The regulator circuit was designed in Examples 8-3 and 8-4, and the design will not be repeated here. The resistor values calculated were $R_1 = 3.0$ kΩ, $R_2 = 6.8$ kΩ, and $R_3 = 2.2$ kΩ, with compensating capacitor $C_C = 100$ pF.

To add current limiting, we need to calculate R_{sens}, using Equation 8-6, where $V_{sens} = 0.7$ V and $I_{max} = 3.0$ A:

$$R_{sens} = \frac{V_{sens}}{I_{max}} = \frac{0.7}{3.0} = 0.233 \text{ Ω}$$

This is an extremely small value, and it will have to dissipate approximately 2 W of power. It is probably best to custom-wind this resistor with resistance wire on a suitable form.

Our regulator is now completely designed.

EXAMPLE 8-6

Design a 15 V, 2.5 A regulated power supply using an LM723C regulator. Assume that the rectifier circuit can deliver 20 V at 2.5 A with 2 V RMS ripple. Include foldback current limiting to limit the short-circuit current to 0.6 A.

Solution

The circuit for a high-voltage regulator with pass transistor and foldback current limiting was shown in Figure 8-10(b). This circuit has been redrawn as Figure 8-12, and a parts list is included.

First design the voltage divider consisting of R_1 and R_2 using Equation 8-2. Assume that $R_1 + R_2 = 10{,}000 \ \Omega$.

$$R_2 = (R_1 + R_2) \times \frac{V_{\text{ref}}}{V_{\text{out}}}$$

$$= 10 \ \text{k}\Omega \times \frac{7.15 \ \text{V}}{15 \ \text{V}} = 4767 \ \Omega$$

Now determine R_1 as $10 \ \text{k}\Omega - 4767 \ \Omega = 5233 \ \Omega$. The closest standard values for R_1 and R_2 are 5.1 kΩ and 4.7 kΩ, respectively.

Next, calculate R_3 from Equation 8-3.

$$R_3 = \frac{R_1 R_2}{R_1 + R_2}$$

$$= \frac{5.1 \ \text{k}\Omega \times 4.7 \ \text{k}\Omega}{5.1 \ \text{k}\Omega + 4.7 \ \text{k}\Omega} = 2.44 \ \text{k}\Omega$$

We will select a value of 2.4 kΩ.

We must include the compensating capacitor C_C for which the manufacturer recommends a value of 100 pF.

To design the foldback current-limiting circuit, we first must calculate R_{sens} from Equation 8-9:

$$R_{\text{sens}} = \frac{V_{\text{sens}} V_{\text{out}}}{V_{\text{out}} I_{\text{short}} + V_{\text{sens}} I_{\text{short}} - V_{\text{sens}} I_{L\max}}$$

$$= \frac{0.7 \ \text{V} \times 15 \ \text{V}}{15 \ \text{V} \times 0.6 \ \text{A} + 0.7 \ \text{V} \times 0.6 \ \text{A} - 0.7 \ \text{V} \times 2.5 \ \text{A}} = 1.369 \ \Omega$$

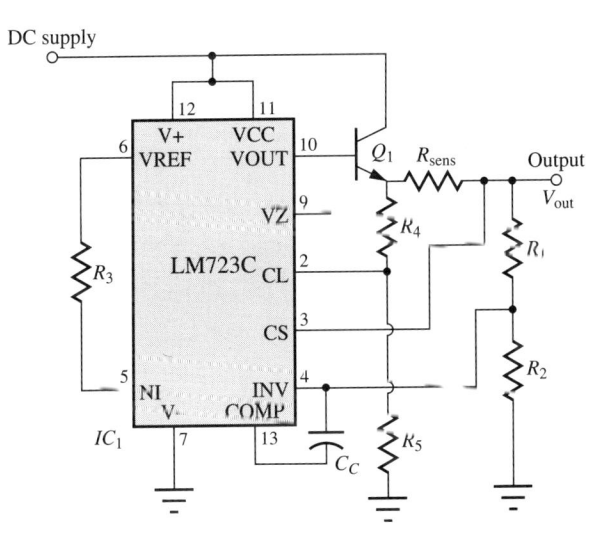

Figure 8-12

This resistor must carry 2.5 A of current at maximum load and dissipate approximately 8.5 W. Hence, we must select a power resistor. The closest standard power resistor is 1.5 Ω, so we use this value for R_{sens}.

Next we assume $R_4 + R_5 = 1.0$ kΩ and calculate R_5 from Equation 8-10.

$$R_5 = \frac{V_{sens}}{I_{short}} \times \frac{R_4 + R_5}{R_{sens}} = \frac{0.7 \text{ V}}{0.6 \text{ A}} \times \frac{1.0 \text{ k}\Omega}{1.369 \text{ }\Omega} = 852 \text{ }\Omega$$

This gives us R_4 as 148 Ω. Choose the closest standard values from the table in Appendix A-1: $R_4 = 150$ Ω and $R_5 = 820$ Ω.

We still have to select the power transistor. The voltage drop across the transistor is

$$V_{CE} = V_{DC} - V_{out} = 20 \text{ V} - 15 \text{ V} = 5 \text{ V}$$

The current through the transistor I_C is simply the load current of 2.5 A. Hence, the power dissipation is

$$P = V_{CE} \times I_C = 5 \text{ V} \times 2.5 \text{ A} = 12.5 \text{ W}$$

These quantities are within the current, voltage, and power ratings of a TIP31 *if used with a suitable heatsink*. We will choose a TIP31C power transistor for Q_1.

To verify the operation of the foldback current limiting, we will calculate the power that would be dissipated in the pass transistor under short-circuit conditions:

$$P = \text{supply voltage} \times \text{foldback limited current}$$
$$= 20 \text{ V} \times 0.6 \text{ A} = 12 \text{ W}$$

This is slightly less than the full-load power and illustrates the effectiveness of foldback limiting. With simple current limiting only, the power under short-circuit conditions would be 20 V × 2.5 A = 50 W, exceeding the maximum power of the TIP31.

Finally, we will check the line regulation, load regulation, and ripple of this regulator.

(a) From the data sheets in Figure 8-1, the line regulation is 0.1% V_{out}. This is 0.015 V/V or 15 mV/V, as defined by Equation 7-11.
(b) The load regulation is 0.03% V_{out}. This is 0.0045 V/A or 4.5 mV/A, as defined by Equation 7-13.
(c) The ripple rejection is −74 dB or a factor of 0.0001995, so the actual ripple is 2 V × 0.0001995 or 0.4 mV.

These values are comparable to the simple regulator as calculated in Example 8-2.

Our regulator design is completed.

EXAMPLE 8-7

A power supply circuit using the LM723C is shown in Figure 8-13. Determine the

(a) output voltage
(b) maximum output current
(c) short-circuit current
(d) line regulation
(e) load regulation
(f) output ripple
(g) full-load power and short-circuit power dissipated in the pass transistor.

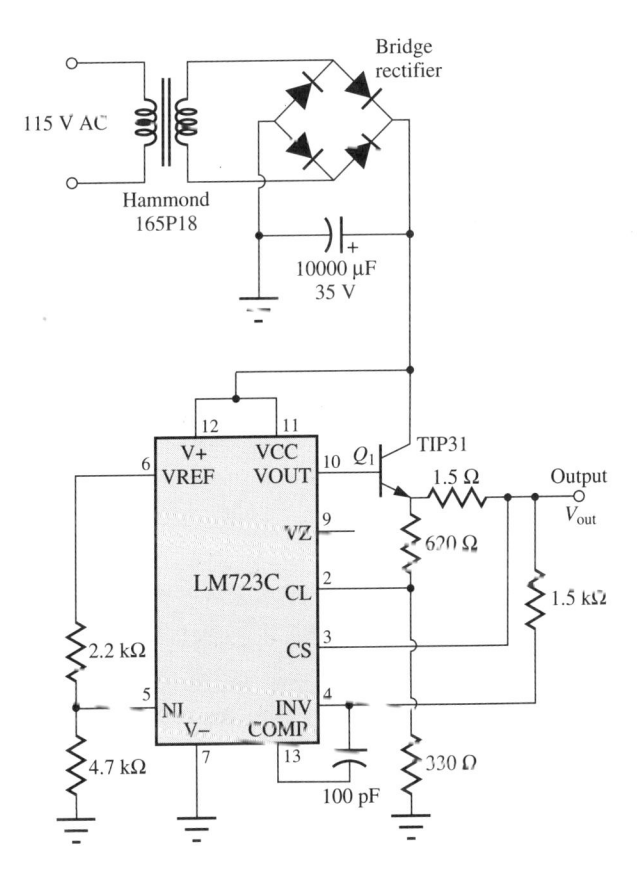

Figure 8-13

8 Integrated Circuit Regulators

Solution

From the circuit in Figure 8-13, it is obvious that we have a low-voltage regulator with pass transistor and foldback current limiting. Compare this figure with Figure 8-10(b) to identify the various resistors.

(a) To determine the output voltage, first locate R_1 and R_2 in Figure 8-13. We find $R_1 = 2.2$ kΩ and $R_2 = 4.7$ kΩ. Calculate output voltage from Equation 8-1:

$$V_{out} = V_{ref} \times \frac{R_2}{R_1 + R_2} = 7.15 \text{ V} \times \frac{4.7 \text{ k}\Omega}{4.7 \text{ k}\Omega + 2.2 \text{ k}\Omega} = 4.87 \text{ V}$$

(b) This circuit uses foldback current limiting. We can identify the various resistor values as $R_4 = 620$ Ω, $R_5 = 330$ Ω, and $R_{sens} = 1.5$ Ω. Use Equation 8-7 to calculate the maximum output current:

$$I_{Lmax} = \frac{R_4 + R_5}{R_5 R_{sens}} \times (V_{out} + V_{sens}) - \frac{V_{out}}{R_{sens}}$$

$$= \frac{620 \text{ }\Omega + 330 \text{ }\Omega}{330 \text{ }\Omega \times 1.5 \text{ }\Omega} \times (4.87 \text{ V} - 0.7 \text{ V}) - \frac{4.87 \text{ V}}{1.5 \text{ }\Omega} = 4.76 \text{ A}$$

(c) Next, use Equation 8-8 to calculate the current under short-circuit conditions:

$$I_{short} = \frac{V_{sens}}{R_{sens}} \times \frac{R_4 + R_5}{R_5} = \frac{0.7 \text{ V}}{1.5 \text{ }\Omega} \times \frac{620 \text{ }\Omega + 330 \text{ }\Omega}{330 \text{ }\Omega} = 1.34 \text{ A}$$

(d) To calculate the line regulation, we will assume that the supply voltage is between 12 and 40 V, and read the regulation from the data sheet in Figure 8-1 as 0.1% V_{out}. Calculate the line regulation from Equation 7-11:

$$\text{line regulation} = \frac{0.1}{100} \times 4.87 \text{ V} = 4.87 \text{ mV/V}$$

(e) The load regulation from the data sheets is 0.03% of V_{out}. From Equation 7-13, the load regulation is

$$\text{load regulation} = \frac{0.03}{100} \times 4.87 \text{ V} = 1.46 \text{ mV/A}$$

(f) To determine the output ripple, we must first determine the input ripple. We are given the rectifier circuit, but if we refer to Equations 7-7 and 7-9, we see that we need a quantity identified as R_L. This R_L is not a physical load resistor which can be calculated from the output voltage and maximum current, but is the effective load resistance as seen by the rectifier itself. This resistance is equal to V_{DC}/I_{Lmax}, but unfortunately we do not know V_{DC}. We can calculate R_L by the following sequence of steps.

First calculate a value of V_P from Equation 7-3. The transformer is a Hammond 165P18, which, from the data sheet in Appendix A-3, has an RMS output voltage of 18 V under a full-load current of 5 A, close to our actual maximum current. Thus,

$$V_P = \sqrt{2}V_{RMS} - 1.4 = \sqrt{2} \times 18 \text{ V} - 1.4 \text{ V} = 24.06 \text{ V}$$

Although we do not know R_L or V_{DC}, we do know that the load current is 4.76 A. Therefore, we can express R_L and V_{DC} in terms of this current. From Ohm's law, we have

$$R_L = \frac{V_{DC}}{I_L} \quad \text{or} \quad V_{DC} = I_L R_L$$

Substitute this into Equation 7-7 and solve for R_L:

$$V_{DC} = V_P - \frac{0.00417}{R_L C} \times V_P$$

$$I_L R_L = V_P - \frac{0.00417}{R_L C} \times V_P$$

Therefore,

$$R_L^2 I_L C - V_P R_L C + 0.00417 V_P = 0$$

$$R_L^2 - \frac{V_P}{I_L} \times R_L + \frac{0.00417}{C} \times \frac{V_P}{I_L} = 0$$

$$R_L^2 - 5.055 R_L + 2.108 = 0$$

This is a quadratic equation in R_L. The roots are $R_L = 4.596 \ \Omega$ or $R_L = 0.459 \ \Omega$. We have to decide which root is correct. We know that the current is about 5 A, and from V_P we can estimate that V_{DC} will be approximately 24 V, so we expect that R_L will be approximately $24/5 = 4.8 \ \Omega$. The first root is correct.

Now calculate V_{DC} from Equation 7-7:

$$V_{DC} = V_P - \frac{0.00417}{R_L C} \times V_P$$

$$= 24.06 \text{ V} - \frac{0.00417}{4.596 \ \Omega \times 0.01 \text{ F}} \times 24.06 \text{ V} = 21.88 \text{ V}$$

Next, calculate the ripple from Equation 7-9:

$$V_{ripple} = \frac{0.0024}{R_L C} \times V_P$$

$$= \frac{0.0024}{4.596 \ \Omega \times 0.01 \text{ F}} \times 24.06 \text{ V} = 1.256 \text{ V}$$

From the data sheets, the ripple reduction is -74 dB, or, as we calculated earlier, a factor of 0.0001995. Hence, the output ripple is 0.0001995×1.256 V = 0.250 mV.

(g) Lastly, we must calculate the power dissipated in the pass transistor to verify that our choice of transistor is adequate. We are not given any data on the heatsink, so we will not be able to determine if the overall design is correct, but at least we can check if the pass transistor is suitable, *assuming an adequate heatsink*. The power dissipated in the pass transistor at maximum load is simply the maximum current times V_{CE} of the transistor:

$$V_{CE} = V_{DC} - V_{out} = 21.88 \text{ V} - 4.87 \text{ V} = 17.0 \text{ V}$$

Hence, transistor power is

$$P = I_{Lmax} \times V_{CE} = 4.76 \text{ A} \times 17.0 \text{ V} = 81.0 \text{ W}$$

This power far *exceeds* the ratings of the TIP31 transistor specified for the design! A much larger pass transistor is required.

8-2 Three-Terminal Fixed Regulators

The regulator circuits discussed thus far, using the LM723C, have been relatively complicated. The regulation and the flexibility of the LM723C are good, but the 14 pin package with a number of external resistors, a pass transistor, and a compensating capacitor makes the circuit unduly complex. There is, however, an alternative to multipin regulators such as the LM723C—the **three-terminal regulators**. These regulators have the voltage reference, sensing circuit, pass transistor, and current limiting built into a standard power transistor package, usually a TO-3 or a TO-220. For many regulator applications, the only connections required are the input lead of the regulator to the output of the rectifier, the common lead to the common from the rectifier, and the output to load. Although we sacrifice some flexibility, we certainly make up for it in convenience. Even so, these devices still have considerable flexibility for a variety of applications. The overall convenience of only three connections makes this type of regulator far more popular than the multipin regulators.

There are a number of three-terminal regulators. The most common are the 7800 series of positive regulators and the 7900 series of negative regulators. The actual devices are numbered in the form 78XX and 79XX, where the XX refers to the regulated output voltage. For example, a 7805 is a positive 5 V regulator, and a 7912 is a negative 12 V regulator.

The data sheets for the 7800 series regulators are shown in Figure 8-14. These are Motorola data sheets, so the regulators are listed with the MC prefix. Most major manufacturers produce these regulators and generally use their own prefix. We will use the MC prefix in this book. The 7900 series regulators have very similar characteristics except that they require a negative input voltage. This difference will be discussed later.

Three-terminal regulators are available in a variety of packages and with a variety of current, voltage, and power ratings. The data sheets for the MC7800 series regulators in Figure 8-14 show only two types: the TO-3 package with a 20 W rating (1.25 A), and the TO-220 package with a 15 W rating (1.0 A). As well, the MC78L00/MC79L00 series and the MC78M00/MC79M00 series regulators are available in the TO-202 package with a power rating of 7.5 W (0.5 A), in the TO-39 package with a 2 W rating, in the TO-92 package with a 500 mW rating, and in surface-mount packages. Even in small quantities, these regulators sell for less than $1. When they were first introduced, they were also available in the TO-3 cases with current ratings up to 10 A, but these have been discontinued. These high-current devices were relatively expensive, and it proved far more economical to use an external pass transistor with a low-power regulator when a large current rating was required. In fact, for many applications, currents greater than 1 A are not required.

One cautionary note should be interjected here. Although a regulator in a TO-220 package is capable of passing approximately 1 A of current, it must have a good heatsink to be able to do this. If a 1 A regulator is used without a heatsink, it will shut down at currents greater than approximately 0.1 A. These three-terminal regulators have full overload protection with a thermal shutdown. If you try to draw too much current from the regulator, it heats up and the thermal shutdown limits the current.

The MC7800 and MC7900 series regulators are available with output voltages of 5, 6, 8, 12, 15, 18, and 24 V. Most standard three-terminal regulators require that the supply voltage be at least 3 V greater than the regulated output (that is, they have a dropout voltage of 3 V), and the total voltage between input terminal and output terminal must not exceed 37 V. Some three-terminal regulators are designed with very low dropout voltages. For example, the LM2931 has a typical dropout voltage of only 0.6 V, and is designed for regulating from battery supplies. Some three-terminal regulators are also designed to withstand very high input/output voltage ranges. For example, the TL783 has a maximum input/output voltage range of 125 V.

There is an equivalent series of regulators to the MC7800/MC7900 series regulators—the LM140/LM340 series positive regulators and the LM120/LM320 series negative regulators. These are numbered in the form LM320-XX and LM340-XX, where the XX refers to the voltage. For example, LM340-18 indicates an 18 V positive regulator. The LM120/LM320 and LM140/LM340 series regulators are direct replacements for the MC7800 and MC7900 series devices. They are more modern and have approximately two times better line and load regulation, with improved ripple rejection. Replacing an LM120/LM320 or LM140/LM340 series regulator with an MC7800/MC7900 series regulator, however, is not advisable because the LM regulators are higher-performance devices.

Regulation in MC7800 Series Devices

The characteristics that describe the quality of any regulator are the line regulation, load regulation, and ripple rejection as defined in Chapter 7. For the MC7800 series regulators, the values of these quantities are found on the data sheets in Figure 8-14.

(a) *Line Regulation:* The line regulation listed for the MC7800 series regulators is equivalent to the line regulation as defined by Equation 7-11, but note the incorrect units on the data sheets. It should be mV/V instead of simply mV.

348   8 Integrated Circuit Regulators

MOTOROLA

MC7800 Series

THREE-TERMINAL POSITIVE VOLTAGE REGULATORS

These voltage regulators are monolithic integrated circuits designed as fixed-voltage regulators for a wide variety of applications including local, on-card regulation. These regulators employ internal current limiting, thermal shutdown, and safe-area compensation. With adequate heatsinking they can deliver output currents in excess of 1.0 ampere. Although designed primarily as a fixed voltage regulator, these devices can be used with external components to obtain adjustable voltages and currents.

- Output Current in Excess of 1.0 Ampere
- No External Components Required
- Internal Thermal Overload Protection
- Internal Short-Circuit Current Limiting
- Output Transistor Safe-Area Compensation
- Output Voltage Offered in 2% and 4% Tolerance

THREE-TERMINAL POSITIVE FIXED VOLTAGE REGULATORS

SILICON MONOLITHIC INTEGRATED CIRCUITS

K SUFFIX
METAL PACKAGE
CASE 1-03

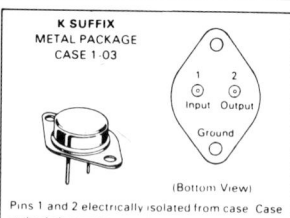

(Bottom View)
Pins 1 and 2 electrically isolated from case. Case is third electrical connection.

T SUFFIX
PLASTIC PACKAGE
CASE 221A-04

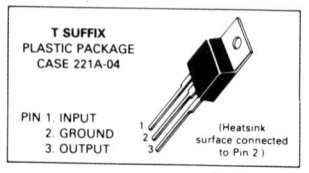

PIN 1. INPUT
     2. GROUND
     3. OUTPUT

(Heatsink surface connected to Pin 2.)

REPRESENTATIVE SCHEMATIC DIAGRAM

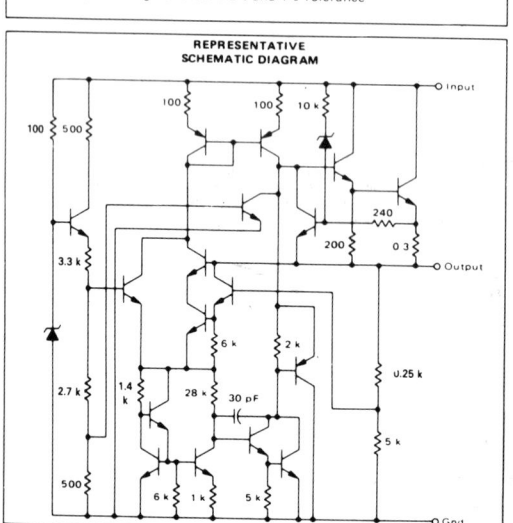

STANDARD APPLICATION

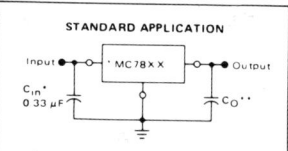

$C_{in}^* = 0.33 \mu F$

A common ground is required between the input and the output voltages. The input voltage must remain typically 2.0 V above the output voltage even during the low point on the input ripple voltage.

XX = these two digits of the type number indicate voltage.

* = C_{in} is required if regulator is located an appreciable distance from power supply filter.

** = C_O is not needed for stability; however, it does improve transient response.

XX indicates nominal voltage.

ORDERING INFORMATION

Device	Output Voltage Tolerance	Tested Operating Junction Temp. Range	Package
MC78XXK	4%	−55 to +150°C	Metal Power
MC78XXAK*	2%		
MC78XXCK	4%	0 to +125°C	
MC78XXACK*	2%		
MC78XXCT	4%		Plastic Power
MC78XXACT	2%		
MC78XXBT	4%	−40 to +125°C	

*2% regulators in Metal Power packages are available in 5, 12 and 15 volt devices.

TYPE NO./VOLTAGE

MC7805	5.0 Volts	MC7812	12 Volts
MC7806	6.0 Volts	MC7815	15 Volts
MC7808	8.0 Volts	MC7818	18 Volts
MC7809	9.0 Volts	MC7824	24 Volts

MOTOROLA LINEAR/INTERFACE DEVICES

Figure 8-14 Data sheets for the 7800 series three-terminal voltage regulator. *(Copyright of Motorola, Inc.)*

MC7800 Series

MAXIMUM RATINGS (T_A = +25°C unless otherwise noted.)

Rating	Symbol	Value	Unit
Input Voltage (5.0 V – 18 V)	V_{in}	35	Vdc
(24 V)		40	
Power Dissipation and Thermal Characteristics			
Plastic Package			
T_A = +25°C	P_D	Internally Limited	Watts
Derate above T_A = +25°C	$1/\theta_{JA}$	15.4	mW/°C
Thermal Resistance, Junction to Air	θ_{JA}	65	°C/W
T_C = +25°C	P_D	Internally Limited	Watts
Derate above T_C = +75°C (See Figure 1)	$1/\theta_{JC}$	200	mW/°C
Thermal Resistance, Junction to Case	θ_{JC}	5.0	°C/W
Metal Package			
T_A = +25°C	P_D	Internally Limited	Watts
Derate above T_A = +25°C	$1/\theta_{JA}$	22.5	mW/°C
Thermal Resistance, Junction to Air	θ_{JA}	45	°C/W
T_C = +25°C	P_D	Internally Limited	Watts
Derate above T_C = +65°C (See Figure 2)	$1/\theta_{JC}$	182	mW/°C
Thermal Resistance, Junction to Case	θ_{JC}	5.5	°C/W
Storage Junction Temperature Range	T_{stg}	–65 to +150	°C
Operating Junction Temperature Range	T_J		°C
MC7800,A		–55 to +150	
MC7800C,AC		0 to +150	
MC7800B		–40 to +150	

DEFINITIONS

Line Regulation — The change in output voltage for a change in the input voltage. The measurement is made under conditions of low dissipation or by using pulse techniques such that the average chip temperature is not significantly affected.

Load Regulation — The change in output voltage for a change in load current at constant chip temperature.

Maximum Power Dissipation — The maximum total device dissipation for which the regulator will operate within specifications.

Quiescent Current — That part of the input current that is not delivered to the load.

Output Noise Voltage — The rms ac voltage at the output with constant load and no input ripple, measured over a specified frequency range.

Long Term Stability — Output voltage stability under accelerated life test conditions with the maximum rated voltage listed in the devices electrical characteristics and maximum power dissipation.

MC7805, B, C
ELECTRICAL CHARACTERISTICS (V_{in} = 10 V, I_O = 500 mA, T_J = T_{low} to T_{high} [Note 1] unless otherwise noted)

Characteristic	Symbol	MC7805 Min	MC7805 Typ	MC7805 Max	MC7805B Min	MC7805B Typ	MC7805B Max	MC7805C Min	MC7805C Typ	MC7805C Max	Unit
Output Voltage (T_J = +25°C)	V_O	4.8	5.0	5.2	4.8	5.0	5.2	4.8	5.0	5.2	Vdc
Output Voltage	V_O										Vdc
(5.0 mA ≤ I_O ≤ 1.0 A, P_O ≤ 15 W)								4.75	5.0	5.25	
7.0 Vdc ≤ V_{in} ≤ 20 Vdc								—	—	—	
8.0 Vdc ≤ V_{in} ≤ 20 Vdc		4.65	5.0	5.35	4.75	5.0	5.25				
Line Regulation (T_J = +25°C, Note 2)	Reg$_{line}$										mV
7.0 Vdc ≤ V_{in} ≤ 25 Vdc		—	2.0	50	—	7.0	100	—	7.0	100	
8.0 Vdc ≤ V_{in} ≤ 12 Vdc		—	1.0	25	—	2.0	50	—	2.0	50	
Load Regulation (T_J = +25°C, Note 2)	Reg$_{load}$										mV
5.0 mA ≤ I_O ≤ 1.5 A		—	25	100	—	40	100	—	40	100	
250 mA ≤ I_O ≤ 750 mA		—	8.0	25	—	15	50	—	15	50	
Quiescent Current (T_J = +25°C)	I_B	—	3.2	6.0	—	4.3	8.0	—	4.3	8.0	mA
Quiescent Current Change	ΔI_B										mA
7.0 Vdc ≤ V_{in} ≤ 25 Vdc		—	0.3	0.8	—	—	1.3	—	—	1.3	
8.0 Vdc ≤ V_{in} ≤ 25 Vdc		—	0.04	0.5	—	—	0.5	—	—	0.5	
5.0 mA ≤ I_O ≤ 1.0 A											
Ripple Rejection	RR	68	75	—	68	—	—	68	—	—	dB
8.0 Vdc ≤ V_{in} ≤ 18 Vdc, f = 120 Hz											
Dropout Voltage (I_O = 1.0 A, T_J = +25°C)	$V_{in} - V_O$	—	2.0	2.5	—	2.0	—	—	2.0	—	Vdc
Output Noise Voltage (T_A = +25°C)	V_n	—	10	40	—	10	—	—	10	—	µV/V_O
10 Hz ≤ f ≤ 100 kHz											
Output Resistance f = 1.0 kHz	r_O	—	17	—	—	17	—	—	17	—	mΩ
Short-Circuit Current Limit (T_A = +25°C)	I_{sc}	—	0.2	1.2	—	0.2	—	—	0.2	—	A
V_{in} = 35 Vdc											
Peak Output Current (T_J = +25°C)	I_{max}	1.3	2.5	3.3	—	2.2	—	—	2.2	—	A
Average Temperature Coefficient of Output Voltage	TCV$_O$	—	±0.6	—	—	1.1	—	—	1.1	—	mV/°C

MOTOROLA LINEAR/INTERFACE DEVICES

Figure 8-14 (*Continued*)

MC7800 Series

MC7806, B, C
ELECTRICAL CHARACTERISTICS (V_{in} = 11 V, I_O = 500 mA, T_J = T_{low} to T_{high} [Note 1] unless otherwise noted).

Characteristic	Symbol	MC7806 Min	MC7806 Typ	MC7806 Max	MC7806B Min	MC7806B Typ	MC7806B Max	MC7806C Min	MC7806C Typ	MC7806C Max	Unit
Output Voltage (T_J = +25°C)	V_O	5.75	6.0	6.25	5.75	6.0	6.25	5.75	6.0	6.25	Vdc
Output Voltage (5.0 mA ≤ I_O ≤ 1.0 A, P_O ≤ 15 W) 8.0 Vdc ≤ V_{in} ≤ 21 Vdc 9.0 Vdc ≤ V_{in} ≤ 21 Vdc	V_O	— 5.65	— 6.0	— 6.35	— 5.7	— 6.0	— 6.3	5.7 —	6.0 —	6.3 —	Vdc
Line Regulation (T_J = +25°C, Note 2) 8.0 Vdc ≤ V_{in} ≤ 25 Vdc 9.0 Vdc ≤ V_{in} ≤ 13 Vdc	Regline	— —	3.0 2.0	60 30	— —	9.0 3.0	120 60	— —	9.0 3.0	120 60	mV
Load Regulation (T_J = +25°C, Note 2) 5.0 mA ≤ I_O ≤ 1.5 A 250 mA ≤ I_O ≤ 750 mA	Regload	— —	27 9.0	100 30	— —	43 16	120 60	— —	43 16	120 60	mV
Quiescent Current (T_J = +25°C)	I_B	—	3.2	6.0	—	4.3	8.0	—	4.3	8.0	mA
Quiescent Current Change 8.0 Vdc ≤ V_{in} ≤ 25 Vdc 9.0 Vdc ≤ V_{in} ≤ 25 Vdc 5.0 mA ≤ I_O ≤ 1.0 A	ΔI_B	— — —	— 0.3 0.04	— 0.8 0.5	— — —	— — —	— 1.3 0.5	— — —	— — —	— 1.3 0.5	mA
Ripple Rejection 9.0 Vdc ≤ V_{in} ≤ 19 Vdc, f = 120 Hz	RR	65	73	—	—	65	—	—	65	—	dB
Dropout Voltage (I_O = 1.0 A, T_J = +25°C)	$V_{in}-V_O$	—	2.0	2.5	—	2.0	—	—	2.0	—	Vdc
Output Noise Voltage (T_A = +25°C) 10 Hz ≤ f ≤ 100 kHz	V_n	—	10	40	—	10	—	—	10	—	μV/V_O
Output Resistance f = 1.0 kHz	r_O	—	17	—	—	17	—	—	17	—	mΩ
Short-Circuit Current Limit (T_A = +25°C) V_{in} = 35 Vdc	I_{sc}	—	0.2	1.2	—	0.2	—	—	0.2	—	A
Peak Output Current (T_J = +25°C)	I_{max}	1.3	2.5	3.3	—	2.2	—	—	2.2	—	A
Average Temperature Coefficient of Output Voltage	TCV_O	—	±0.7	—	—	−0.8	—	—	−0.8	—	mV/°C

MC7800 Series

MC7808, B, C
ELECTRICAL CHARACTERISTICS (V_{in} = 14 V, I_O = 500 mA, T_J = T_{low} to T_{high} [Note 1] unless otherwise noted).

Characteristic	Symbol	MC7808 Min	MC7808 Typ	MC7808 Max	MC7808B Min	MC7808B Typ	MC7808B Max	MC7808C Min	MC7808C Typ	MC7808C Max	Unit
Output Voltage (T_J = +25°C)	V_O	7.7	8.0	8.3	7.7	8.0	8.3	7.7	8.0	8.3	Vdc
Output Voltage (5.0 mA ≤ I_O ≤ 1.0 A, P_O ≤ 15 W) 10.5 Vdc ≤ V_{in} ≤ 23 Vdc 11.5 Vdc ≤ V_{in} ≤ 23 Vdc	V_O	— 7.6	— 8.0	— 8.4	— 7.6	— 8.0	— 8.4	7.6 —	8.0 —	8.4 —	Vdc
Line Regulation (T_J = +25°C, Note 2) 10.5 Vdc ≤ V_{in} ≤ 25 Vdc 11 Vdc ≤ V_{in} ≤ 17 Vdc	Regline	— —	3.0 2.0	80 40	— —	12 5.0	160 80	— —	12 5.0	160 80	mV
Load Regulation (T_J = +25°C, Note 2) 5.0 mA ≤ I_O ≤ 1.5 A 250 mA ≤ I_O ≤ 750 mA	Regload	— —	28 9.0	100 40	— —	45 16	160 80	— —	45 16	160 80	mV
Quiescent Current (T_J = +25°C)	I_B	—	3.2	6.0	—	4.3	8.0	—	4.3	8.0	mA
Quiescent Current Change 10.5 Vdc ≤ V_{in} ≤ 25 Vdc 11.5 Vdc ≤ V_{in} ≤ 25 Vdc 5.0 mA ≤ I_O ≤ 1.0 A	ΔI_B	— — —	— 0.3 0.04	— 0.8 0.5	— — —	— — —	— 1.0 0.5	— — —	— — —	1.0 — 0.5	mA
Ripple Rejection 11.5 Vdc ≤ V_{in} ≤ 21.5 Vdc, f = 120 Hz	RR	62	70	—	—	62	—	—	62	—	dB
Dropout Voltage (I_O = 1.0 A, T_J = +25°C)	$V_{in}-V_O$	—	2.0	2.5	—	2.0	—	—	2.0	—	Vdc
Output Noise Voltage (T_A = +25°C) 10 Hz ≤ f ≤ 100 kHz	V_n	—	10	40	—	10	—	—	10	—	μV/V_O
Output Resistance f = 1.0 kHz	r_O	—	18	—	—	18	—	—	18	—	mΩ
Short-Circuit Current Limit (T_A = +25°C) V_{in} = 35 Vdc	I_{sc}	—	0.2	1.2	—	0.2	—	—	0.2	—	A
Peak Output Current (T_J = +25°C)	I_{max}	1.3	2.5	3.3	—	2.2	—	—	2.2	—	A
Average Temperature Coefficient of Output Voltage	TCV_O	—	±1.0	—	—	−0.8	—	—	−0.8	—	mV/°C

Figure 8-14   (*Continued*)

MC7800 Series

MC7812, B, C
ELECTRICAL CHARACTERISTICS (V_{in} = 19 V, I_O = 500 mA, T_J = T_{low} to T_{high} [Note 1] unless otherwise noted)

Characteristic	Symbol	MC7812 Min	MC7812 Typ	MC7812 Max	MC7812B Min	MC7812B Typ	MC7812B Max	MC7812C Min	MC7812C Typ	MC7812C Max	Unit
Output Voltage (T_J = +25°C)	V_O	11.5	12	12.5	11.5	12	12.5	11.5	12	12.5	Vdc
Output Voltage (5.0 mA ≤ I_O ≤ 1.0 A, P_O ≤ 15 W) 14.5 Vdc ≤ V_{in} ≤ 27 Vdc; 15.5 Vdc ≤ V_{in} ≤ 27 Vdc	V_O	11.4	12	12.6	11.4	12	12.6	11.4	12	12.6	Vdc
Line Regulation (T_J = +25°C, Note 2) 14.5 Vdc ≤ V_{in} ≤ 30 Vdc; 16 Vdc ≤ V_{in} ≤ 22 Vdc	Reg$_{line}$	—	5.0 / 3.0	120 / 60	—	13 / 6.0	240 / 120	—	13 / 6.0	240 / 120	mV
Load Regulation (T_J = +25°C, Note 2) 5.0 mA ≤ I_O ≤ 1.5 A; 250 mA ≤ I_O ≤ 750 mA	Reg$_{load}$	—	30 / 10	120 / 60	—	46 / 17	240 / 120	—	46 / 17	240 / 120	mV
Quiescent Current (T_J = +25°C)	I_B	—	3.4	6.0	—	4.4	8.0	—	4.4	8.0	mA
Quiescent Current Change 14.5 Vdc ≤ V_{in} ≤ 30 Vdc; 15 Vdc ≤ V_{in} ≤ 30 Vdc; 5.0 mA ≤ I_O ≤ 1.0 A	ΔI_B	—	0.3 / 0.04	0.8 / 0.5	—	1.0 / 0.5	—	—	1.0 / 0.5	—	mA
Ripple Rejection 15 Vdc ≤ V_{in} ≤ 25 Vdc, f = 120 Hz	RR	61	68	—	—	60	—	—	60	—	dB
Dropout Voltage (I_O = 1.0 A, T_J = +25°C)	$V_{in} - V_O$	—	2.0	2.5	—	2.0	—	—	2.0	—	Vdc
Output Noise Voltage (T_A = +25°C) 10 Hz ≤ f ≤ 100 kHz	V_n	—	10	40	—	10	—	—	10	—	µV/V_O
Output Resistance f = 1.0 kHz	r_O	—	18	—	—	18	—	—	18	—	mΩ
Short-Circuit Current Limit (T_A = +25°C) V_{in} = 35 Vdc	I_{SC}	—	0.2	1.2	—	0.2	—	—	0.2	—	A
Peak Output Current (T_J = +25°C)	I_{max}	1.3	2.5	3.3	—	2.2	—	—	2.2	—	A
Average Temperature Coefficient of Output Voltage	TCV$_O$	—	±1.5	—	—	-1.0	—	—	-1.0	—	mV/°C

MC7800 Series

MC7815, B, C
ELECTRICAL CHARACTERISTICS (V_{in} = 23 V, I_O = 500 mA, T_J = T_{low} to T_{high} [Note 1] unless otherwise noted)

Characteristic	Symbol	MC7815 Min	MC7815 Typ	MC7815 Max	MC7815B Min	MC7815B Typ	MC7815B Max	MC7815C Min	MC7815C Typ	MC7815C Max	Unit
Output Voltage (T_J = +25°C)	V_O	14.4	15	15.6	14.4	15	15.6	14.4	15	15.6	Vdc
Output Voltage (5.0 mA ≤ I_O ≤ 1.0 A, P_O ≤ 15 W) 17.5 Vdc ≤ V_{in} ≤ 30 Vdc; 18.5 Vdc ≤ V_{in} ≤ 30 Vdc	V_O	14.25	15	15.75	14.25	15	15.75	14.25	15	15.75	Vdc
Line Regulation (T_J = +25°C, Note 2) 17.5 Vdc ≤ V_{in} ≤ 30 Vdc; 20 Vdc ≤ V_{in} ≤ 26 Vdc	Reg$_{line}$	—	6.0 / 3.0	150 / 75	—	13 / 6.0	300 / 150	—	13 / 6.0	300 / 150	mV
Load Regulation (T_J = +25°C, Note 2) 5.0 mA ≤ I_O ≤ 1.5 A; 250 mA ≤ I_O ≤ 750 mA	Reg$_{load}$	—	32 / 10	150 / 75	—	52 / 20	300 / 150	—	52 / 20	300 / 150	mV
Quiescent Current (T_J = +25°C)	I_B	—	3.4	6.0	—	4.4	8.0	—	4.4	8.0	mA
Quiescent Current Change 17.5 Vdc ≤ V_{in} ≤ 30 Vdc; 18.5 Vdc ≤ V_{in} ≤ 30 Vdc; 5.0 mA ≤ I_O ≤ 1.0 A	ΔI_B	—	0.3 / 0.04	0.8 / 0.5	—	1.0 / 0.5	—	—	1.0 / 0.5	—	mA
Ripple Rejection 18.5 Vdc ≤ V_{in} ≤ 28.5 Vdc, f = 120 Hz	RR	60	66	—	—	58	—	—	58	—	dB
Dropout Voltage (I_O = 1.0 A, T_J = +25°C)	$V_{in} - V_O$	—	2.0	2.5	—	2.0	—	—	2.0	—	Vdc
Output Noise Voltage (T_A = +25°C) 10 Hz ≤ f ≤ 100 kHz	V_n	—	10	40	—	10	—	—	10	—	µV/V_O
Output Resistance f = 1.0 kHz	r_O	—	19	—	—	19	—	—	19	—	mΩ
Short-Circuit Current Limit (T_A = +25°C) V_{in} = 35 Vdc	I_{SC}	—	0.2	1.2	—	0.2	—	—	0.2	—	A
Peak Output Current (T_J = +25°C)	I_{max}	1.3	2.5	3.3	—	2.2	—	—	2.2	—	A
Average Temperature Coefficient of Output Voltage	TCV$_O$	—	±1.8	—	—	-1.0	—	—	-1.0	—	mV/°C

Figure 8-14 (*Continued*)

To convert this line regulation into percent line regulation as defined in Equation 7-12, divide the value from the data sheets by the output voltage and multiply by 100.

(b) *Load Regulation:* The load regulation listed for the MC7800 series regulators is equivalent to the load regulation as defined by Equation 7-13, and, like line regulation, the units on the data sheets are incorrect. They should be mV/A instead of mV. To convert the data sheet value into the percent load regulation as introduced in Equation 7-14, again simply divide by the device voltage and multiply by 100.

(c) *Ripple Rejection:* The ripple rejection for the MC7800 series regulators varies slightly from voltage rating to voltage rating, but is typically better than -60 dB. Converting this to a linear factor, we find that any MC7800 series regulator reduces input ripple by a factor of at least 1000.

In Example 8-8, we will compare these quantities to the values that we calculated for the LM723C regulator used in a 12 V power supply in Example 8-1. Basically, the regulation in the MC7800 series regulators is inferior to that of the LM723C. The poorer performance of the MC7800 series regulators is largely due to thermal effects from the built-in pass transistor. Even though the regulation is not as good as the LM723C, the regulation is still more than adequate for most applications. In fact, if we use an MC7800 series regulator to drive an external pass transistor, keeping the operating point of the MC7800 series regulator constant, we can greatly improve the regulation.

EXAMPLE 8-8

Determine the line regulation and load regulation for an MC7812 three-terminal regulator from the data sheets shown in Figure 8-14. Calculate percent line regulation and percent load regulation, and compare these values with the values determined for the LM723C regulator in Example 8-2.

Solution

From the data sheets in Figure 8-14 the MC7812 has a line regulation of 13 mV/V (note corrected units) for a supply voltage range of 14.5 to 30 V and a current of 100 mA. To convert this line regulation into percent line regulation as defined by Equation 7-12, we divide the data sheet value by the output voltage and multiply by 100:

$$\text{percent line regulation} = \frac{0.013 \text{ V/V}}{12.0 \text{ V}} \times 100 = 0.11\%$$

From the data sheets, the MC7812 has a load regulation for currents from 5 mA to 1.5 A of 46 mV/A (note corrected units). Convert this load regulation to percent regulation as defined by Equation 7-14:

$$\text{percent load regulation} = \frac{0.046 \text{ V/A}}{12.0 \text{ V}} \times 100 = 0.38\%$$

Now compare these values with the load and line regulation found in Example 8-2. From that example, the line regulation was 12 mV/V, the percent line regulation was 0.1%, the load regulation was 3.6 mV/A, and the percent load regulation was 0.03%. These regulation values are compared with the MC7812 in Table 8-1.

Table 8-1 Comparison of LM723C regulation with MC7812 regulation

	LM723C	MC7812
Line regulation	12 mV/V	13 mV/V
Percent line regulation	0.1%	0.11%
Load regulation	3.6 mV/A	46 mV/A
Percent load regulation	0.03%	0.38%

The LM723C has only slightly better line regulation than the MC7812, but the load regulation is over ten times better. As explained earlier, the poorer load regulation of the MC7812 is largely due to thermal effects of the large pass transistor incorporated as part of the integrated circuit.

Basic Circuits Using the MC7800 Series Regulators

The standard circuit for using an MC7800 series regulator to regulate a power supply is shown in Figure 8-15. Note the use of capacitors C_1 and C_2. Both capacitors are generally optional. The MC7800/MC7900 series regulators are sensitive to inductance, and the input capacitor C_1 is used to cancel any inductance present at the input to the regulator chip. It is only necessary if there are more than 4 in. of wire connecting the input of the regulator to the filter capacitor C_F.

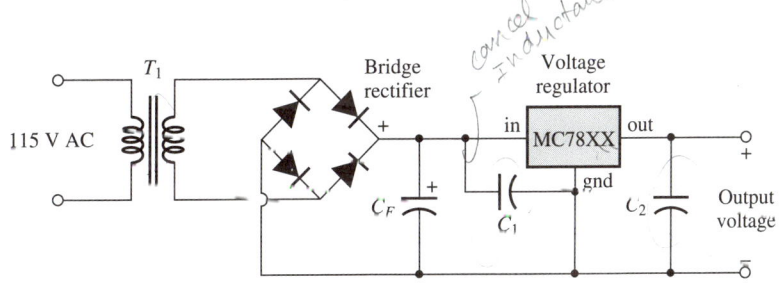

Figure 8-15 Basic power supply using a MC7800 series regulator.

in the rectifier. This capacitor, if used, should be a type with a low internal inductance, preferably a tantalum or mica disk. Typical values are 1.0 μF for a tantalum capacitor or 0.1 μF for a mica disk capacitor. The output capacitor C_2 improves the transient response of the regulator and, if used, should be a 1.0 μF tantalum or a 0.1 μF mica disk. In most cases the capacitors are not required or used.

Although these regulators are available in both positive and negative voltages, for most applications either an MC7800 series regulator or an MC7900 series regulator can be used, regardless of the output polarity. Suppose we have a negative regulator and want a positive regulated output. Simply connect the negative regulator as shown in Figure 8-16. The regulator is connected to the negative lead from the power supply. The general rule for using any MC7800/MC7900 series regulator as a simple fixed-voltage regulator is to connect the regulator input to the matching polarity lead from the rectifier, and to connect the common regulator lead to the other lead from the rectifier.

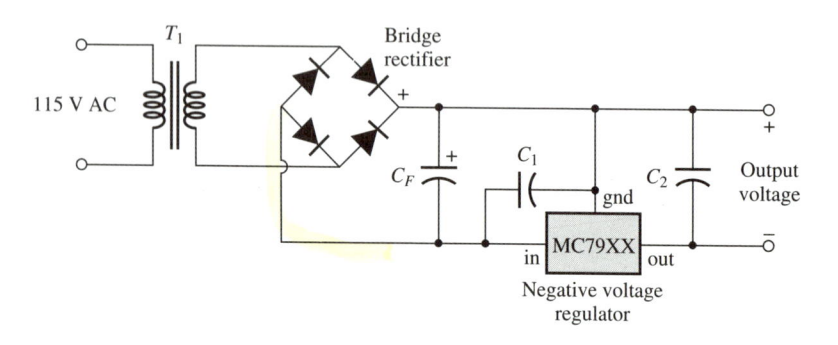

Figure 8-16 Basic power supply using a MC7900 series regulator.

Often in op-amp circuits, a dual-polarity supply is required. For this type of power supply, simply use a single bridge rectifier across a center-tapped transformer capable of supplying more than twice the required single supply voltage. Use two filter capacitors and two regulators: an MC7800 series positive regulator and an MC7900 series negative regulator. The capacitors shown in Figure 8-15 can also be included for each regulator if required. The circuit for a simple dual-polarity supply is shown in Figure 8-17. With proper heatsinks on the regulators, this power supply is capable of supplying typically as much as 1 A. If more output current is required, then pass transistors can be added. The power supply in Figure 8-17 is ideal as a general-purpose op-amp power supply. Use an MC7815 and an MC7915 to give a ±15 V supply.

These three circuits (Figures 8-15, 8-16, and 8-17) show how easy it is to use the MC7800/MC7900 series voltage regulators. They are simply placed in the circuit; no other design is necessary. It is, however, still necessary to design the transformer, rectifier, and filter capacitor as discussed in Chapter 7. This design is illustrated in Example 8-9.

8-2 Three-Terminal Fixed Regulators   355

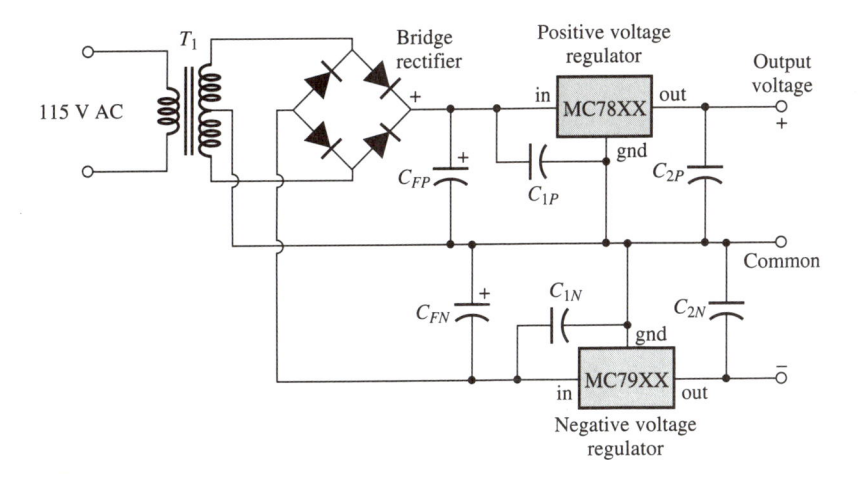

Figure 8-17 Dual-polarity power supply using a MC7800 series positive regulator and a MC7900 series negative regulator.

EXAMPLE 8-9

Design a line-operated 15 V supply to deliver up to 1 A of current. The output ripple must be less than 1 mV.

Solution

For this power supply, we will use a simple bridge rectifier/filter combination feeding an MC7815 voltage regulator. The circuit is shown in Figure 8-18, with a parts list included.

Parts List

Capacitors:
 C_F 3000 µF
 C_1 1.0 µF tantalum
 C_2 1.0 µF tantalum

Transformers:
 T_1 166J16

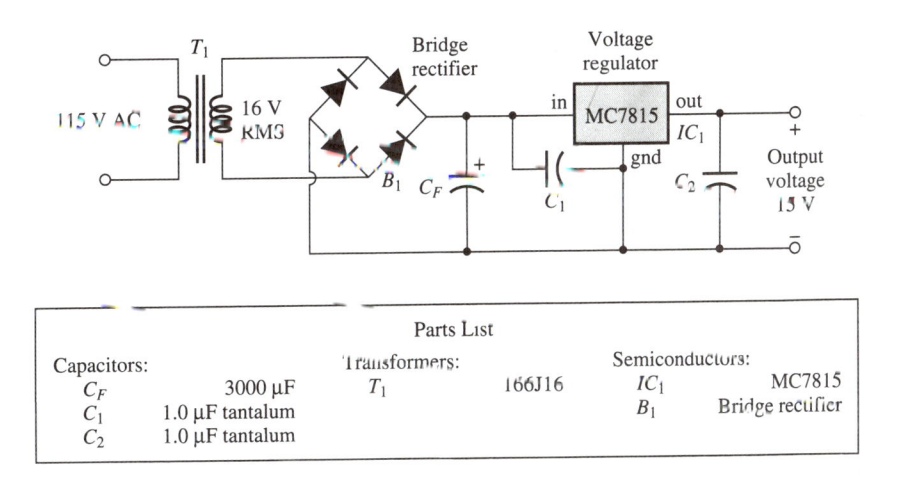

Semiconductors:
 IC_1 MC7815
 B_1 Bridge rectifier

Figure 8-18

The design of rectifiers was discussed in the previous chapter. Our main concern in this chapter is the design of the regulator circuit. However, since we are required to hold the ripple to a prescribed level, we must design the filter capacitor to provide a suitable ripple level.

The design of the regulator portion of the circuit is easy. We simply insert an MC7815 regulator after the filter with capacitors C_1 and C_2 if necessary. These capacitors have been included in the circuit in Figure 8-18.

The most difficult step is to design the rectifier circuit. Our output voltage is 15 V, and we will need at least 3 V more than this for the regulator to operate, so our rectifier must provide a V_{DC} of at least 18 V at 1 A. Use Ohm's law to calculate the effective $R_L = V_{DC}/I_L = 18 \text{ V}/1 \text{ A} = 18 \text{ }\Omega$.

The MC7815 has a typical ripple rejection of 60 dB. This corresponds to a factor of 0.001. If our output ripple is to be less than 1 mV, then our input ripple must be less than 1 V.

Calculate a minimum V_P from Equation 7-10:

$$V_P = V_{DC} + 1.736 V_{ripple} = 18 \text{ V} + 1.736 \times 1.0 \text{ V} = 19.74 \text{ V}$$

This is a minimum value of V_P since V_{DC} is the minimum value. We will have to select a transformer before we calculate the actual V_P. We need V_{RMS} to select a transformer, and we can calculate a minimum V_{RMS} from our minimum V_P value by using Equation 7-3:

$$V_p = \sqrt{2} V_{RMS} - 1.4$$

$$V_{RMS} = \frac{V_P + 1.4}{\sqrt{2}} = \frac{19.74 \text{ V} + 1.4 \text{ V}}{1.414} = 14.95 \text{ V}$$

We need a transformer that provides an RMS secondary voltage of at least 14.95 V at 1 A. Refer to the transformer table in Appendix A-3. Select a 166J16 16 V, 1 A transformer.

Now calculate the actual value of V_P, using the transformer that we selected:

$$V_P = \sqrt{2} \; V_{RMS} - 1.4$$
$$= 1.414 \times 16 \text{ V} - 1.4 \text{ V} = 21.23 \text{ V}$$

Next, use Equation 7-9 and rearrange to solve for the filter capacitor:

$$V_{ripple} = \frac{0.0024}{R_L C_F} \times V_P$$

$$C_F = \frac{0.0024 V_P}{R_L V_{ripple}}$$

$$= \frac{0.0024 \times 21.23 \text{ V}}{18 \text{ }\Omega \times 1.0 \text{ V}} = 0.00283 \text{ F}$$

Refer to Appendix A-2 and select a 3000 μF electrolytic capacitor.

The design of our power supply is complete. Notice that although the design of the regulator portion of the power supply was very easy, the design of a suitable filter capacitor was quite lengthy.

MC7800/MC7900 Series Regulators with Pass Transistors

The current rating of an MC7800 series regulator in a TO-220 plastic power transistor case is approximately 1 A. To regulate larger currents than this, we can use a pass transistor in the configuration shown in Figure 8-19. For low currents, the regulator alone conducts, but for high currents the transistor conducts to provide the larger current. The current at which the pass transistor starts to conduct is determined by the resistor R_P. When the voltage drop across R_P due to the current through the regulator reaches 0.7 V, the pass transistor starts to conduct, thereby supplying the additional current. The formula for designing R_P is

$$R_P = \frac{V_{BE}}{I_{reg}} = \frac{0.7}{I_{reg}} \qquad (8\text{-}11)$$

In this formula, I_{reg} is the regulator current at which transistor conduction is to begin. This could be any value up to the current limit of the regulator, but, typically, make I_{reg} less than or equal to one-tenth of the maximum regulator current so that it will not be necessary to use a heatsink for the regulator and so that a standard resistor can be used for R_P instead of a custom-wound small-value resistor.

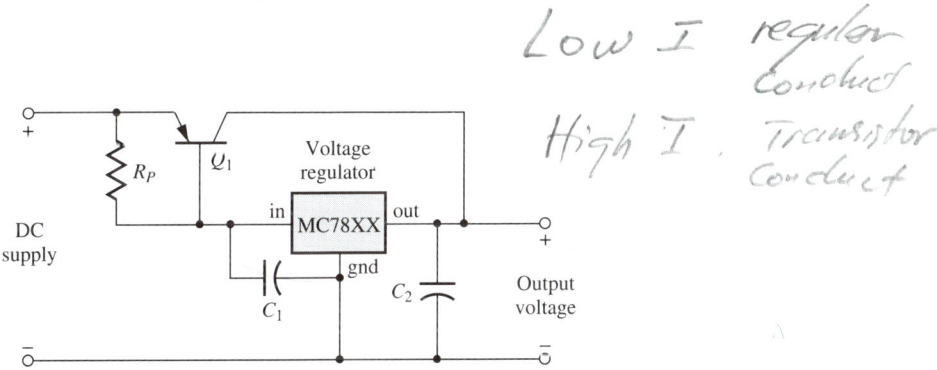

Figure 8-19 Regulator circuit using a MC7800 series regulator with pass transistor.

EXAMPLE 8-10

Design a 12 V power supply with 3 A capability using an 18 V DC supply.

Solution

Since the output current must be 3 A, it is necessary to use a pass transistor with the regulator. The circuit for the MC7800 series regulator with pass transistor was shown in Figure 8-19. This circuit has been redrawn as Figure 8-20, with a parts list included. The first step in designing the regulator circuit is to select a suitable regulator. Since we need an output of 12 V, we will use an MC7812 regulator. We will select 1.0 μF tantalum capacitors for C_1 and C_2.

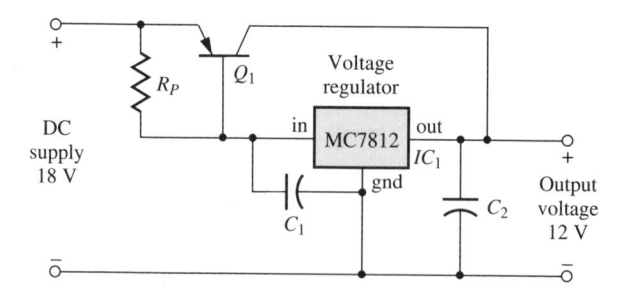

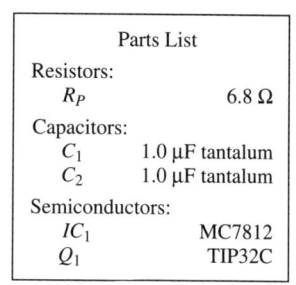

Figure 8-20

For the pass transistor circuit, we have to design the resistor R_P and select a suitable transistor Q_1. The maximum current that the regulator itself can handle is 1 A, so we will choose an operating level of one-tenth of this, or 0.1 A. The value of R_P, given by Equation 8-11, is

$$R_P = \frac{V_{BE}}{I_{reg}} = \frac{0.7 \text{ V}}{0.1 \text{ A}} = 7 \, \Omega$$

From Appendix A-1, the closest standard resistor is 6.8 Ω. The power rating may be important here, so we will calculate the power P in this resistor.

$$P = \text{voltage} \times \text{current} = 0.7 \text{ V} \times 0.1 \text{ A} = 0.07 \text{ W}$$

A standard ¼ W resistor will be adequate.

Now we must choose a suitable pass transistor. The maximum current $I_C = I_L$ through the transistor is 3 A, the maximum voltage across the transistor V_{CE} is $V_{\text{supply}} - V_{\text{out}} = 18 \text{ V} - 12 \text{ V} = 6 \text{ V}$. The maximum power dissipated in the transistor is

$$P = V_{CE} \times I_C = 6 \text{ V} \times 3 \text{ A} = 18 \text{ W}$$

A TIP32 power transistor, as described in Appendix A-5, could be used *if mounted on a suitable heatsink*. Note that this circuit requires a PNP transistor.

The design of our regulator circuit with pass transistor is complete. There is, however, one problem that we did not address in this example: if the output is shorted, the transistor will blow. The solution, which we will now discuss, is to use current limiting.

Current Limiting with MC7800/MC7900 Series Regulators

The MC7800/MC7900 series regulators have built-in short-circuit protection, so no additional circuitry is necessary to protect the regulator itself. If we are using a pass transistor, however, it has no short-circuit protection and will blow under short-circuit conditions. We can provide short-circuit protection for the pass tran-

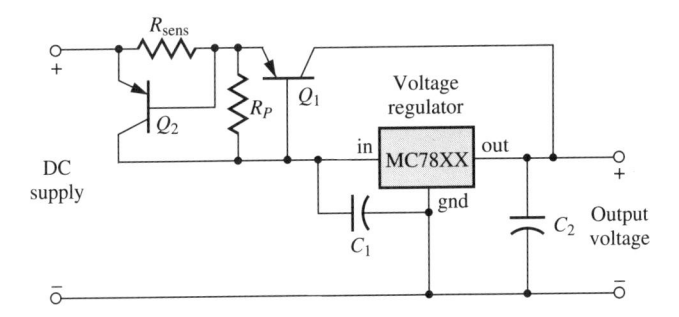

Figure 8-21 Regulator circuit using a MC7800 series regulator with pass transistor protected by current limiting.

sistor by using the regulator short-circuit protection with the circuit in Figure 8-21.

Transistor Q_2 is used to protect pass transistor Q_1. This transistor conducts whenever the voltage drop across R_{sens} due to the current through Q_1 exceeds 0.7 V. The current flows through Q_2 to the regulator where it causes the regulator short-circuit protection to cut in. The design equation for R_{sens} is

$$R_{sens} = \frac{V_{BE}}{I_{max}} = \frac{0.7}{I_{max}} \tag{8-12}$$

Note that we must use a PNP transistor for Q_2. This bypass transistor must be capable of handling the short-circuit cutoff current of the regulator, typically 1 A. The collector-emitter voltage across Q_2 when the output is shorted will only be 1.4 V, from the sum of the voltage drops across R_{sens} and R_P. Hence, the power dissipated in Q_2 will be quite small.

EXAMPLE 8-11

Design the regulator circuit for a 5 V power supply with a 4 A current-limited capability. The DC source is 11.5 V.

Solution

The design of this circuit is similar to that of Example 8-10, except that now we are including current limiting to protect the pass transistor. The circuit used for current limiting was shown in Figure 8-21. This circuit has been redrawn here as Figure 8-22, with a parts list included. Since we require a 5 V output, we must use an MC7805 regulator. We will use 1.0 μF tantalum capacitors for C_1 and C_2.

The design requires the calculation of R_P and R_{sens} and the selection of suitable transistors Q_1 and Q_2.

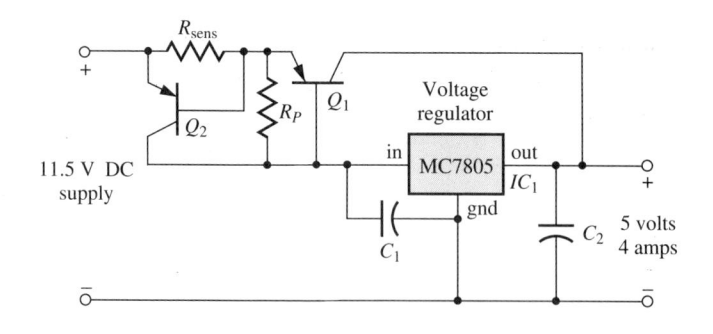

Figure 8-22

First calculate R_P. Assume a regulator current of 0.1 A and use Equation 8-11:

$$R_P = \frac{V_{BE}}{I_{reg}} = \frac{0.7 \text{ V}}{0.1 \text{ A}} = 7.0 \text{ }\Omega$$

We will choose a value of 6.8 Ω for R_P as the closest standard value in Appendix A-1. We found in the previous example that a power rating of $\frac{1}{4}$ W was adequate. Next calculate R_{sens}. Use the full-load current of 4 A in Equation 8-12:

$$R_{sens} = \frac{V_{BE}}{I_{max}} = \frac{0.7 \text{ V}}{4.0 \text{ A}} = 0.175 \text{ }\Omega$$

In this case, it is probably best to custom-wind our sensing resistor from resistance wire onto a suitable form. It will have to dissipate approximately 3 W.

Finally, we must select suitable transistors. Because the pass transistor must carry up to 4 A, we cannot use a TIP32 transistor for Q_1. The maximum voltage V_{CE} across transistor Q_1 is $V_{supply} - V_{out} = 11.5 \text{ V} - 5 \text{ V} = 6.5 \text{ V}$. The maximum power dissipated in transistor Q_1 is

$$P = V_{CE} \times I_C = 6.5 \text{ V} \times 4 \text{ A} = 26 \text{ W}$$

A TIP34 power transistor, with suitable heatsink, would be satisfactory.

The bypass transistor Q_2 must pass the maximum regulator current of approximately 1 A to initiate current limiting, so the maximum current $I_C = I_L$ through Q_2 is 1 A. The maximum voltage V_{CE} across transistor Q_2 is the 0.7 V drop across R_{sens} plus the 0.7 V drop across R_P, or a total of 1.4 V. The maximum power dissipated in transistor Q_2 is

$$P = V_{CE} \times I_C = 1.4 \text{ V} \times 1 \text{ A} = 1.4 \text{ W}$$

A TIP32 power transistor could be used *without a heatsink*.
The design of our high-current regulator with current limiting is complete.

Floating Regulators Using MC7800/MC7900 Series Regulators

The MC7800/MC7900 series regulators are fixed regulators, designed to produce a single output voltage. They can, however, be used as **floating regulators** to give any regulated output voltage larger than the characterized value up to the maximum rated voltage of approximately 37 V. The circuit of an MC7800 series regulator connected as a floating regulator is shown in Figure 8-23.

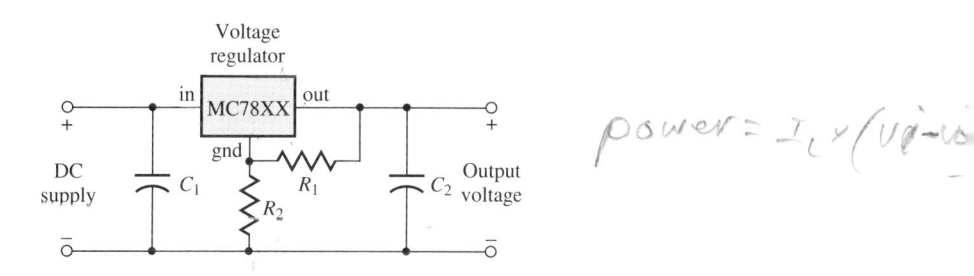

Figure 8-23 MC7800 series regulator used as a floating regulator.

The output voltage of the floating regulator is the sum of the voltage drops across R_1 and R_2. The voltage drop across R_1 is simply the regulator voltage V_{reg}. The voltage drop across R_2 is caused by the current through R_2. This current comprises the current flowing through R_1, equal to V_{reg}/R_1, and the **quiescent current** I_Q of the regulator. The output voltage is

$$V_{\text{out}} = V_{\text{reg}} + \left(\frac{V_{\text{reg}}}{R_1} + I_Q\right) \times R_2 \tag{8-13}$$

Rearranging this equation, we get

$$V_{\text{out}} = \frac{R_1 + R_2}{R_1} \times V_{\text{reg}} + I_Q R_2 \tag{8-14}$$

The quiescent current is listed on the data sheets for the MC7800 series regulators in Figure 8-14. For example, the typical value of I_Q for the MC7805 is 4.3 mA. The quiescent current is somewhat variable and may be different from one regulator to another. Therefore, in the circuit design it is important to minimize the effects of I_Q. We must choose R_1 such that the current V_{ref}/R_1 is much larger than I_Q, and typically we would want V_{ref}/R_1 to be at least ten times the maximum possible I_Q. This condition gives the following design equation for R_1:

$$R_1 \leq \frac{1}{10} \times \frac{V_{reg}}{I_Q} \qquad (8\text{-}15)$$

Rearranging Equation 8-13, we get an equation for designing R_2.

$$R_2 = \frac{V_{out} - V_{reg}}{V_{reg}/R_1 + I_Q} \qquad (8\text{-}16)$$

Theoretically, the circuit we have just discussed is capable of regulating any voltage larger than the characterized value of the regulator—hundreds of volts if necessary—provided that the voltage across the regulator chip ($V_{in} - V_{out}$) never exceeds the 37 V maximum. In practice, however, any input voltage greater than 37 V *may* destroy the regulator chip. If there is a capacitor across the output (for example, C_2), then when voltage is first applied to the regulator, the capacitor must charge. The initial output voltage, with the capacitor uncharged, is zero, and the full input voltage will appear across the regulator. If this voltage is greater than the 37 V limit, the regulator may blow.

EXAMPLE 8-12

Design a regulator circuit capable of supplying 0.5 A at 30 V. Use an MC7815 regulator. The DC power source is 36 V.

Solution

To design a 30 V regulated power supply, we must use the MC7815 as a floating regulator. We shall use the circuit in Figure 8-23. This circuit has been redrawn as Figure 8-24, and a parts list is included.

Because the current is only 0.5 A, we will not need a pass transistor, but the regulator may require a heatsink. Our DC supply is 36 V and our required output is 30 V. This means that we would have 6 V dropped across the regulator. We

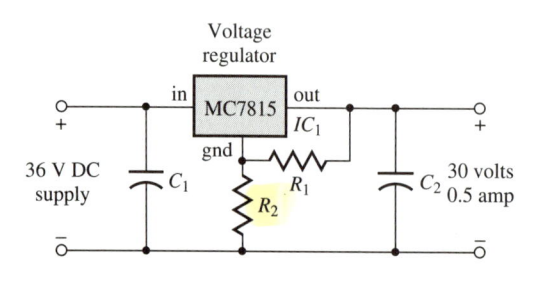

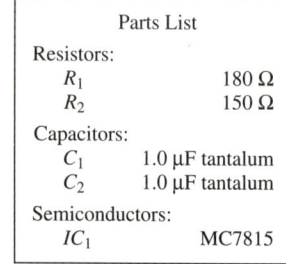

Figure 8-24

have to check for two things: (1) that the supply voltage is at least the regulator voltage plus 3 V, to allow for dropout, and (2) that the supply voltage is less than the maximum value of 37 V, to avoid blowing the regulator. Both conditions are satisfied, so we can proceed with the design.

The data sheet for the MC7815 in Figure 8-14 lists the typical quiescent current as 4.4 mA and the maximum quiescent current as 8.0 mA. To ensure that R_1 is satisfactory for all values of I_Q, we should use the maximum value. Use $I_Q = 8.0$ mA in Equation 8-15 to calculate R_1:

$$R_1 \leq \frac{1}{10} \times \frac{V_{reg}}{I_Q} = \frac{15 \text{ V}}{10 \times 8.0 \text{ mA}} = 187 \text{ }\Omega$$

Refer to Appendix A-1 and choose a smaller-value standard resistor. We will use 180 Ω.

Next calculate R_2 from Equation 8-16:

$$R_2 = \frac{V_{out} - V_{reg}}{V_{reg}/R_1 + I_Q} = \frac{30 \text{ V} - 15 \text{ V}}{15 \text{ V}/180 \text{ }\Omega + 8.0 \text{ mA}} = 164 \text{ }\Omega$$

Again, refer to Appendix A-1 and choose the nearest standard value. We will choose 150 Ω for R_2.

Because we are dealing with relatively high voltages, it is wise to calculate the power dissipated in these resistors.

The power dissipated in R_1 is

$$P = \frac{V^2}{R} = \frac{(15 \text{ V})^2}{180 \text{ }\Omega} = 1.25 \text{ W}$$

The power dissipated in R_2 is

$$P = \frac{V^2}{R} = \frac{(15 \text{ V})^2}{150 \text{ }\Omega} = 1.50 \text{ W}$$

Both resistors should be selected with 2 W ratings.

We should now substitute our real resistor values into Equation 8-14 to verify the output voltage. Use the typical value for quiescent current of 4.4 mA:

$$V_{out} = \frac{R_1 + R_2}{R_1} \times V_{reg} + I_Q R_2$$

$$= \frac{180 \text{ }\Omega + 150 \text{ }\Omega}{180 \text{ }\Omega} \times 15 \text{ V} + 4.4 \text{ mA} \times 150 \text{ }\Omega = 28.2 \text{ V}$$

This is about 6% lower than the design value of 30 V. Now suppose I_Q were at its maximum value of 8.0 mA. Then

$$V_{out} = \frac{180 \text{ }\Omega + 150 \text{ }\Omega}{180 \text{ }\Omega} \times 15 \text{ V} + 8.0 \text{ mA} \times 150 \text{ }\Omega = 28.7 \text{ V}$$

This output value is approximately 4% lower than the design value, but within 2% of the value for $I_Q = 4.4$ mA.

364 8 Integrated Circuit Regulators

The error in the output voltage is due to the choice of value for R_2. The important point to note, however, is that the variation in output due to a change in I_Q from 4.4 to 8.0 mA is only 2%. Our choice of R_1 has stabilized the circuit against I_Q. For most applications, 10% accuracy is adequate, and our regulator is well within this range.

8-3 Three-Terminal Variable Regulators

The floating regulator described in the previous section and designed in Example 8-12 could easily be made into a variable regulator. We need only make R_2 variable, and, providing that the DC supply voltage does not exceed the 37 V limit of the regulator, we have a regulator capable of varying from the regulator voltage up to approximately 3 V less than the DC supply voltage. This circuit works and has been frequently used.

The circuit, however, has several limitations. First, the minimum voltage is equal to the voltage of the regulator chip. For variable supplies, it is usually desirable to adjust the voltage to zero. Second, the quiescent current I_Q is relatively large and varies from regulator chip to regulator chip. Our design constraint on R_1 was an attempt to stabilize the circuit, but it is only good to approximately 5%. Third, the power dissipation in R_2 can in some cases be quite large. Because this would be the variable resistor in a variable power supply, the power supply would be bulky and expensive. All of these points are minor, and the simple MC7800/MC7900 series regulators can be used for variable power supplies without difficulty. There are, however, regulators designed specifically for variable-voltage power supplies, and this section describes some of these devices.

A variety of three-terminal variable regulators are manufactured, the most popular being the LM117/LM217/LM317 positive variable regulators and the LM137/LM237/LM337 negative variable regulators. All of the circuits designed in this section will use the LM317 positive regulator, but the design is basically the same for most three-terminal regulators.

Data sheets for the LM117/LM217/LM317 positive variable regulator are shown in Figure 8-25. Three device numbers are listed (LM117, LM217, and LM317). They simply refer to the same device with different operating temperature ranges. Throughout this section, we will use the LM317, which has an operating range from 0 to +125°C. The LM317 is available in a range of packages and current ratings, including 1.5 A from a TO-3 or TO-220 package and 0.5 A from a TO-39 package. The maximum voltage across the regulator must be less than 40 V, and the regulator has an output range of 1.2 to 37 V. Like the MC7800/MC7900 series regulators, the LM317 has thermal overload protection and internal short-circuit current limiting.

There is an equivalent series of negative regulators, the LM137/LM237/LM337. These are essentially identical to the LM117/LM217/LM317 except for polarity. As for the MC7800/MC7900 series regulators, a positive or negative regulator can be used to regulate a simple power supply as long as the input of a positive regulator is connected to the positive output from the source, or the

Voltage Regulators

LM117/LM217/LM317 3-Terminal Adjustable Regulator

General Description

The LM117/LM217/LM317 are adjustable 3-terminal positive voltage regulators capable of supplying in excess of 1.5A over a 1.2V to 37V output range. They are exceptionally easy to use and require only two external resistors to set the output voltage. Further, both line and load regulation are better than standard fixed regulators. Also, the LM117 is packaged in standard transistor packages which are easily mounted and handled.

In addition to higher performance than fixed regulators, the LM117 series offers full overload protection available only in IC's. Included on the chip are current limit, thermal overload protection and safe area protection. All overload protection circuitry remains fully functional even if the adjustment terminal is disconnected.

Features

- Adjustable output down to 1.2V
- Guaranteed 1.5A output current
- Line regulation typically 0.01%/V
- Load regulation typically 0.1%
- Current limit constant with temperature
- 100% electrical burn-in
- Eliminates the need to stock many voltages
- Standard 3-lead transistor package
- 80 dB ripple rejection

Normally, no capacitors are needed unless the device is situated far from the input filter capacitors in which case an input bypass is needed. An optional output capacitor can be added to improve transient response. The adjustment terminal can be bypassed to achieve very high ripple rejections ratios which are difficult to achieve with standard 3 terminal regulators.

Besides replacing fixed regulators, the LM117 is useful in a wide variety of other applications. Since the regulator is "floating" and sees only the input-to-output differential voltage, supplies of several hundred volts can be regulated as long as the maximum input to output differential is not exceeded.

Also, it makes an especially simple adjustable switching regulator, a programmable output regulator, or by connecting a fixed resistor between the adjustment and output, the LM117 can be used as a precision current regulator. Supplies with electronic shutdown can be achieved by clamping the adjustment terminal to ground which programs the output to 1.2V where most loads draw little current.

The LM117K, LM217K and LM317K are packaged in standard TO-3 transistor packages while the LM117H, LM217H and LM317H are packaged in a solid Kovar base TO-39 transistor package. The LM117 is rated for operation from $-55°C$ to $+150°C$, the LM217 from $-25°C$ to $+150°C$ and the LM317 from $0°C$ to $+125°C$. The LM317T and LM317MP, rated for operation over a $0°C$ to $+125°C$ range, are available in a TO-220 plastic package and a TO-202 package, respectively.

For applications requiring greater output current in excess of 3A and 5A, see LM150 series and LM138 series data sheets, respectively. For the negative complement, see LM137 series data sheet.

LM117 Series Packages and Power Capability

DEVICE	PACKAGE	RATED POWER DISSIPATION	DESIGN LOAD CURRENT
LM117	TO-3	20W	1.5A
LM217 LM317	TO-39	2W	0.5A
LM317T	TO-220	15W	1.5A
LM317M	TO-202	7.5W	0.5A
LM317LZ	TO-92	0.6W	0.1A

Typical Applications

1.2V–25V Adjustable Regulator

†Optional—improves transient response. Output capacitors in the range of 1 µF to 1000 µF of aluminum or tantalum electrolytic are commonly used to provide improved output impedance and rejection of transients.

*Needed if device is far from filter capacitors.

††$V_{OUT} = 1.25V \left(1 + \dfrac{R2}{R1}\right)$

Digitally Selected Outputs

*Sets maximum V_{OUT}

5V Logic Regulator with Electronic Shutdown*

*Min output ≈ 1.2V

1-23

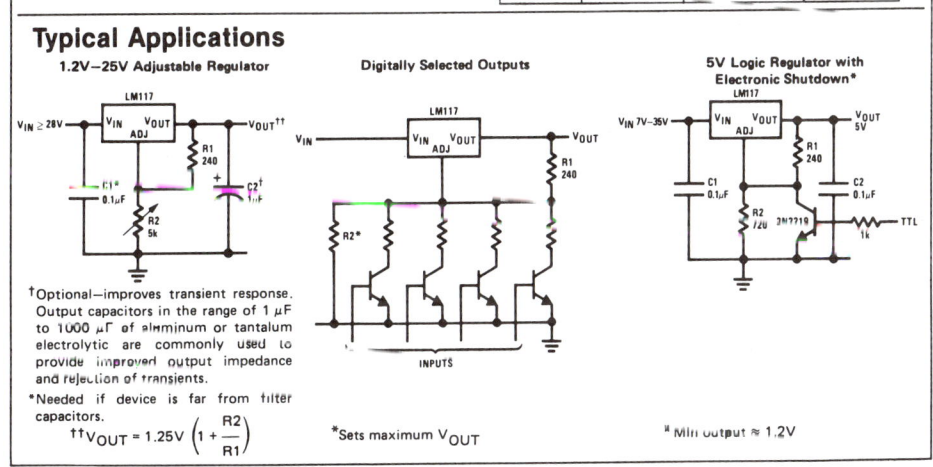

Figure 8-25 Data sheets for the LM317 variable voltage regulator. *(Reprinted with permission of National Semiconductor Corporation)*

LM117/LM217/LM317

Absolute Maximum Ratings

Power Dissipation	Internally limited
Input–Output Voltage Differential	40V
Operating Junction Temperature Range	
LM117	$-55°C$ to $+150°C$
LM217	$-25°C$ to $+150°C$
LM317	$0°C$ to $+125°C$
Storage Temperature	$-65°C$ to $+150°C$
Lead Temperature (Soldering, 10 seconds)	$300°C$

Preconditioning

Burn-In in Thermal Limit	100% All Devices

Electrical Characteristics (Note 1)

PARAMETER	CONDITIONS	LM117/217 MIN	LM117/217 TYP	LM117/217 MAX	LM317 MIN	LM317 TYP	LM317 MAX	UNITS
Line Regulation	$T_A = 25°C$, $3V \leq V_{IN} - V_{OUT} \leq 40V$ (Note 2)		0.01	0.02		0.01	0.04	%/V
Load Regulation	$T_A = 25°C$, $10\,mA \leq I_{OUT} \leq I_{MAX}$							
	$V_{OUT} \leq 5V$, (Note 2)		5	15		5	25	mV
	$V_{OUT} \geq 5V$, (Note 2)		0.1	0.3		0.1	0.5	%
Thermal Regulation	$T_A = 25°C$, 20 ms Pulse		0.03	0.07		0.04	0.07	%/W
Adjustment Pin Current			50	100		50	100	µA
Adjustment Pin Current Change	$10\,mA \leq I_L \leq I_{MAX}$ $3V \leq (V_{IN}-V_{OUT}) \leq 40V$		0.2	5		0.2	5	µA
Reference Voltage	$3V \leq (V_{IN}-V_{OUT}) \leq 40V$, (Note 3) $10\,mA \leq I_{OUT} \leq I_{MAX}$, $P \leq P_{MAX}$	1.20	1.25	1.30	1.20	1.25	1.30	V
Line Regulation	$3V \leq V_{IN} - V_{OUT} \leq 40V$, (Note 2)		0.02	0.05		0.02	0.07	%/V
Load Regulation	$10\,mA \leq I_{OUT} \leq I_{MAX}$, (Note 2)							
	$V_{OUT} \leq 5V$		20	50		20	70	mV
	$V_{OUT} \geq 5V$		0.3	1		0.3	1.5	%
Temperature Stability	$T_{MIN} \leq T_j \leq T_{MAX}$		1			1		%
Minimum Load Current	$V_{IN}-V_{OUT} = 40V$		3.5	5		3.5	10	mA
Current Limit	$V_{IN}-V_{OUT} \leq 15V$							
	K and T Package	1.5	2.2		1.5	2.2		A
	H and P Package	0.5	0.8		0.5	0.8		A
	$V_{IN}-V_{OUT} = 40V$, $T_j = +25°C$							
	K and T Package	0.30	0.4		0.15	0.4		A
	H and P Package	0.15	0.07		0.075	0.07		A
RMS Output Noise, % of V_{OUT}	$T_A = 25°C$, $10\,Hz \leq f \leq 10\,kHz$		0.003			0.003		%
Ripple Rejection Ratio	$V_{OUT} = 10V$, $f = 120\,Hz$		65			65		dB
	$C_{ADJ} = 10\,µF$	66	80		66	80		dB
Long-Term Stability	$T_A = 125°C$		0.3	1		0.3	1	%
Thermal Resistance, Junction to Case	H Package		12	15		12	15	°C/W
	K Package		2.3	3		2.3	3	°C/W
	T Package					4		°C/W
	P Package					12		°C/W

Note 1: Unless otherwise specified, these specifications apply $-55°C \leq T_j \leq +150°C$ for the LM117, $-25°C \leq T_j \leq +150°C$ for the LM217, and $0°C \leq T_j \leq +125°C$ for the LM317; $V_{IN} - V_{OUT} = 5V$; and $I_{OUT} = 0.1A$ for the TO-39 and TO-202 packages and $I_{OUT} = 0.5A$ for the TO-3 and TO-220 packages. Although power dissipation is internally limited, these specifications are applicable for power dissipations of 2W for the TO-39 and TO-202, and 20W for the TO-3 and TO-220. I_{MAX} is 1.5A for the TO-3 and TO-220 packages and 0.5A for the TO-39 and TO-202 packages.

Note 2: Regulation is measured at constant junction temperature, using pulse testing with a low duty cycle. Changes in output voltage due to heating effects are covered under the specification for thermal regulation.

Note 3: Selected devices with tightened tolerance reference voltage available.

Figure 8-25   *(Continued)*

8-3 Three-Terminal Variable Regulators

input of a negative regulator is connected to the negative output from the source. The only time when both types of regulators are required is when a dual-polarity supply is being designed.

Basic Circuits Using the LM317 Variable Regulator

Circuits for a basic variable power supply using an LM317 variable regulator are shown in Figure 8-26. Three circuits are shown. Figure 8-26(a) shows the simplest configuration, which is usable in most applications.

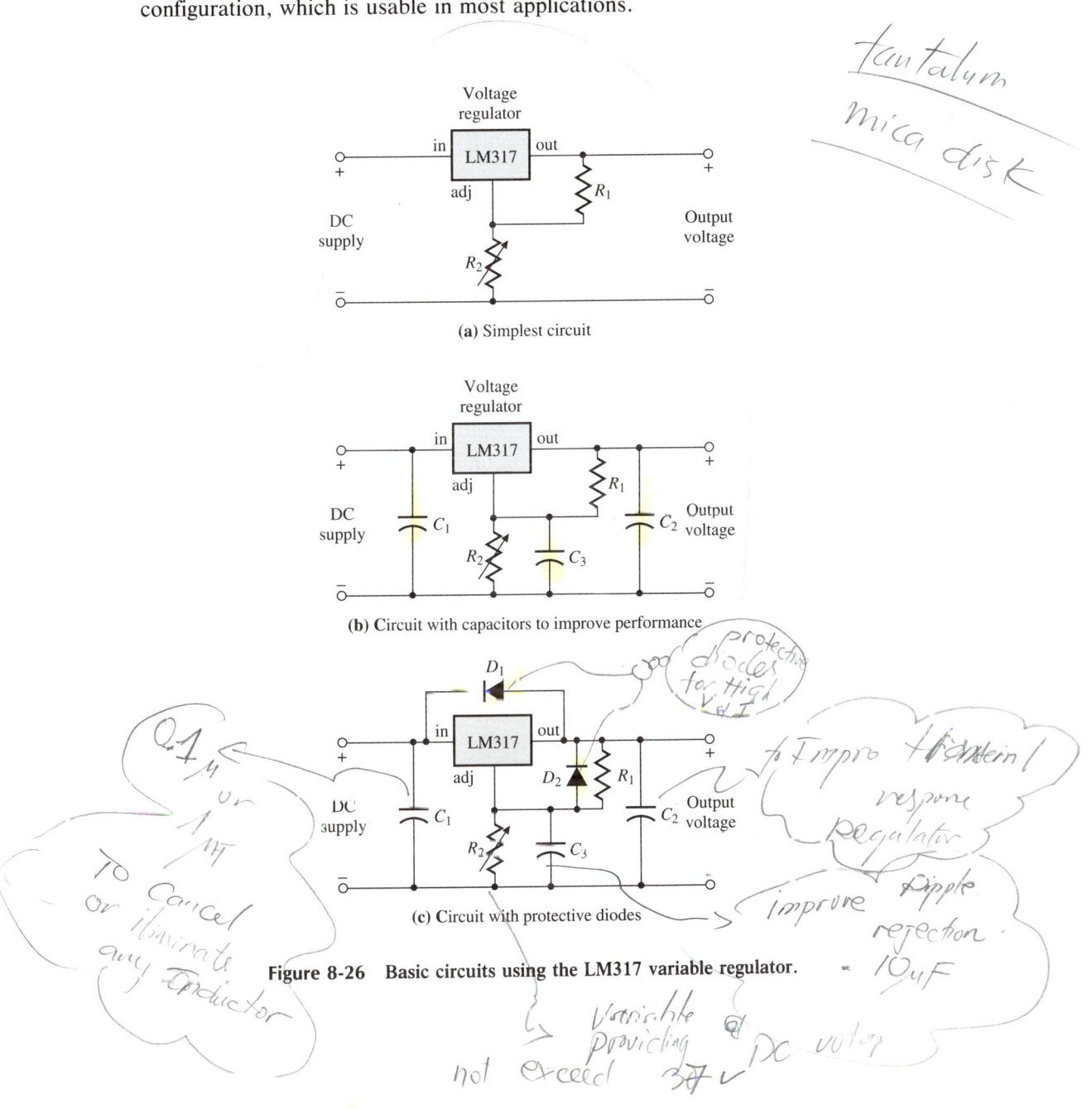

Figure 8-26 Basic circuits using the LM317 variable regulator.

(a) Simplest circuit

(b) Circuit with capacitors to improve performance

(c) Circuit with protective diodes

The circuit in Figure 8-26(b) has capacitors to improve the circuit performance. The function of these capacitors is similar to those used in the MC7800/MC7900 fixed regulators. Capacitor C_1 at the input is used to eliminate any input inductance and is only necessary if there are more than 4 in. of lead between the rectifier filter capacitor C_F and the regulator input terminal. If used, this capacitor should be either a 0.1 µF mica disk capacitor or a 1.0 µF tantalum capacitor. Capacitor C_2 at the output is used to improve the transient response of the regulator and should be a 0.1 µF mica disk capacitor or a 1.0 µF tantalum capacitor. Capacitor C_3 from the adjustment terminal of the regulator is used to improve ripple rejection. A 10 µF tantalum capacitor will reduce the ripple rejection by approximately 15 dB.

The circuit in Figure 8-26(c) has protective diodes for high-current/high-voltage applications. These diodes are necessary only if the capacitors have been used and are required to protect the regulator from the discharge currents of the capacitors when the power is switched off. Protective diodes are only necessary if the capacitors are greater than 10 µF and the voltages are greater than 25 V. Diode D_1 protects the regulator from C_2, and diode D_2 protects the regulator from C_3.

The output voltage of the regulator is

$$V_{out} = \frac{R_1 + R_2}{R_1} \times V_{ref} + I_{adj}R_2 \quad (8\text{-}17)$$

This equation should be familiar. It is the same basic equation as developed for floating regulators in the previous section, except that now we have V_{ref} as the internal reference voltage of the regulator, with a value of 1.25 V, instead of V_{reg}, and I_{adj} as the current from the adjustment terminal instead of the quiescent current I_Q.

Resistor R_1 must be chosen to ensure that a minimum load current flows through the regulator. The voltage across this resistor is the reference voltage 1.25 V, and the minimum current through this resistor must be equal to or greater than the minimum load current given on the data sheets. The data sheets for the LM317 in Figure 8-25 list the typical minimum load current as 3.5 mA, but this may be as large as 10 mA in the worst case. We should design for the worst case. We know the voltage across R_1 and the current through it, so we can calculate R_1 from Ohm's law:

$$R_1 = \frac{V_{ref}}{I_{Lmin}} = \frac{1.25 \text{ V}}{10 \text{ mA}} = 125 \ \Omega \quad (8\text{-}18)$$

The closest value of resistor from Appendix A-1 is 120 Ω for R_1. This is a standard value that we will use in all future designs with the LM317.

To obtain a design equation for R_2, solve Equation 8-17 for R_2, then substitute values for V_{ref}, R_1, and I_{adj}:

$$R_2 = \frac{V_{out} - V_{ref}}{V_{ref}/R_1 + I_{adj}} = \frac{V_{out} - 1.25 \text{ V}}{10.47 \text{ mA}} \quad (8\text{-}19)$$

To calculate R_2, we simply select the maximum voltage for the power supply and substitute this into Equation 8-19.

EXAMPLE 8-13

Design a 1.25 to 25 V variable power supply with a 1 A current capability. Use an LM317 positive variable regulator, and assume that the rectifier circuit can supply 32 V at 1 A with 2 V peak-to-peak ripple.

Solution

Because our voltage is over 25 V, if we use capacitors to improve the performance of the circuit we will have to use protective diodes. We will use the capacitors, and the circuit is shown in Figure 8-26(c). It has been redrawn as Figure 8-27, with a parts list included.

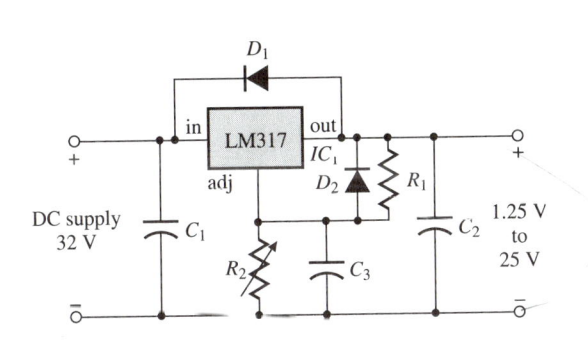

Parts List

Resistors:
R_1 120 Ω
R_2 2.5 kΩ potentiometer

Capacitors:
C_1 1.0 μF tantalum
C_2 1.0 μF tantalum
C_3 10 μF tantalum

Semiconductors:
IC_1 LM317
D_1, D_2 1N4004

Figure 8-27

The only thing that we have to design in this circuit is R_2. We know from Equation 8-18 that R_1 must be 120 Ω. Use Equation 8-19 to calculate R_2 for a voltage of 25 V. This gives the value of the potentiometer that we must use in our variable supply:

$$R_2 = \frac{V_{out} - V_{ref}}{V_{ref}/R_1 + I_{adj}} = \frac{V_{out} - 1.25 \text{ V}}{10.47 \text{ mA}}$$

$$= \frac{25 \text{ V} - 1.25 \text{ V}}{10.47 \text{ mA}} = 2.27 \text{ k}\Omega$$

The closest standard value potentiometer is 2500 Ω. Using this value will give us a maximum output of

$$V_{out} = \frac{R_1 + R_2}{R_1} \times V_{ref} + I_{adj}R_2$$

$$= \frac{120 \text{ }\Omega + 2500 \text{ }\Omega}{120 \text{ }\Omega} \times 1.25 \text{ V} + 50 \text{ }\mu\text{A} \times 2500 \text{ }\Omega$$

$$= 27.4 \text{ V}$$

The circuit design is complete.

EXAMPLE 8-14

Calculate the line regulation, load regulation, and output ripple of the variable regulator designed in Example 8-13 for an output voltage of 12 V.

Solution

(a) *Line regulation:* The line regulation of the LM317 from the data sheets in Figure 8-25 is listed as 0.01%/V, where V is the voltage across the regulator. For an output of 12 V, we have 32 V − 12 V = 20 V across the regulator, and our percent line regulation is $0.0001 \times 20 = 0.2\%$, as defined by Equation 7-12. Converting this to line regulation as defined by Equation 7-11, we get 24 mV/V.

The line regulation becomes worse for smaller output voltages.

(b) *Load regulation:* The load regulation from the data sheets is given as 0.1% of V_{out}. Hence, for an output of 12 V the line regulation is $0.001 \times 12\ V = 0.012$ V/A or 12 mV/A.

The load regulation becomes worse for higher output voltages.

(c) *Output ripple:* The ripple rejection is 65 dB without C_3 and 80 dB if $C_3 = 10$ μF. Since we are using the capacitor in our circuit, we have the 80 dB rejection. Our ripple is reduced by a factor of 0.0001, so we have an output ripple of 0.2 mV.

Table 8-2 Comparison of LM317 regulation with LM723C regulation and MC7812 regulation at 12 V output

	LM317	LM723C	MC7812
Line regulation	24 mV/V	12 mV/V	13 mV/V
Percent line regulation	0.2%	0.1%	0.11%
Load regulation	12 mV/A	3.6 mV/A	46 mV/A
Percent load regulation	0.1%	0.03%	0.38%

Table 8-2 compares the line regulation and load regulation with the corresponding regulation of an MC7812 fixed three-terminal regulator (from Example 8-8) and with an LM723C multipin regulator (from Example 8-2). The line regulation for the LM317 is worse than either of the other two regulators, presumably because of the large voltage dropped across the regulator (20 V). The load regulation is between that of the LM723C and that of the MC7812. It is approximately four times better than the MC7812.

The LM317 Variable Regulator with Pass Transistor

The LM317 variable regulator is capable of handling currents as large as 1.5 A. This is, of course, assuming that the regulator is mounted on a suitable heatsink. Without a heatsink, the maximum current would be less than 0.2 A. To achieve a higher current-handling capability, an external pass transistor can be added. Figure 8-28 shows the regulator circuit with a pass transistor. The use of a pass transistor with the LM317 regulator in this circuit is identical to the MC7800/MC7900 series regulators shown in Figure 8-19. Resistor R_P is designed in exactly the same way as for the MC7800/MC7900 series fixed regulators by using Equation 8-11. The complete design procedure introduced in Example 8-10 is also valid for variable regulators.

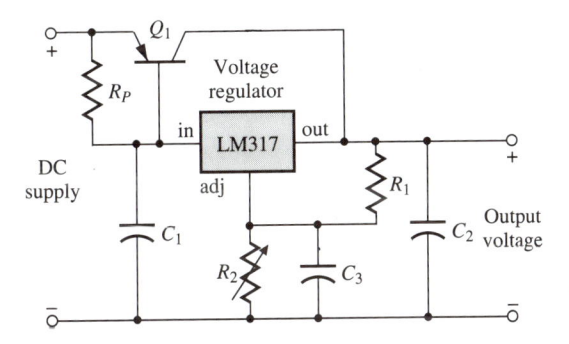

Figure 8-28 High-current power supply using a LM317 variable regulator with a pass transistor.

Current Limiting with the LM317 Variable Regulator

The LM317 variable regulator is fully current-limited and thermally protected. However, an external pass transistor, such as used in Figure 8-28, is not protected and will blow under short-circuit conditions. Figure 8-29 shows the inclusion of

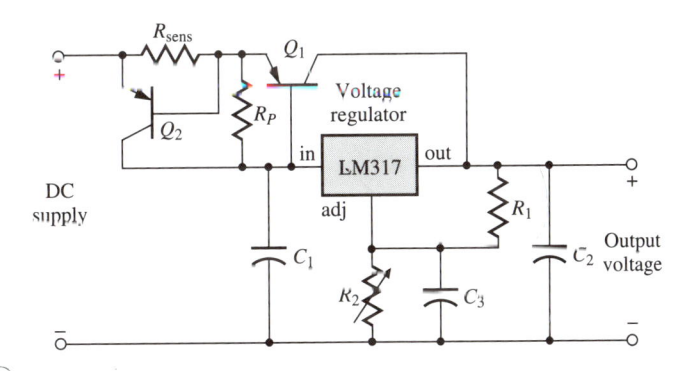

Figure 8-29 The LM317 regulator circuit using a pass transistor with current limiting.

current limiting to protect the pass transistor. This is the same basic circuit as used for the MC7800/MC7900 series fixed regulators and operates on the same principle. The design of R_{sens} in this circuit is exactly the same as for the MC7800/MC7900 series regulators. The design procedure introduced in Example 8-11 for fixed regulators is also valid for variable regulators.

A Variable Power Supply Using the LM317 Regulator

There is one restriction with the LM317 regulator—the minimum output voltage is 1.25 V, not 0 V. For some applications, this poses no problem. However, generally it is desirable that the output voltage in a variable power supply be able to go to zero. This restriction in the LM317 is very easily overcome by offsetting the bottom end of the adjustment potentiometer by 1.22 V by using a LM313 voltage reference as in the circuit in Figure 8-30. Resistor R_S is used to set the bias current from the negative supply to the 1.0 mA required by the LM113. The negative supply for the LM113 must be independent of the main supply, with a typical voltage between -5 and -15 V, and capable of supplying at least 1 mA. It can be built by using a separate transformer, but a more elegant solution is to use a voltage converter. Voltage converters will be introduced in the next section, and a full variable power supply will be designed.

8-4 Voltage Converters

In the variable power supply in Figure 8-30, we required a negative voltage for the LM313 voltage reference, but we had only a positive supply. This problem is often encountered—a positive supply is available but a negative voltage is required. Sometimes this negative voltage can be obtained simply by changing the ground point. For example, if a 12 V battery is available and a ± 6 V supply is required, then a simple 1:1 resistive voltage divider across the battery with the center of the voltage divider defined as ground will be the answer. This procedure

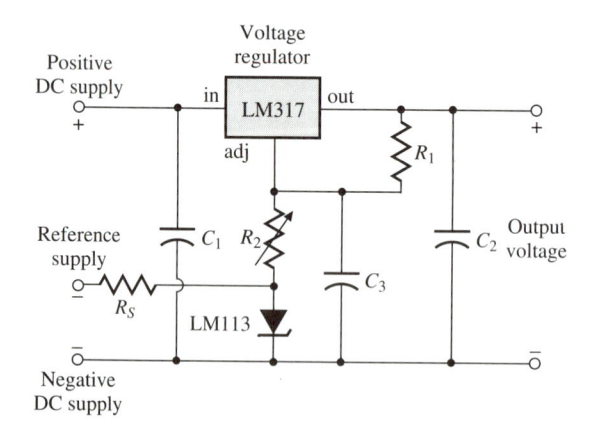

Figure 8-30 Variable regulator with zero minimum output voltage.

HARRIS
HARRIS · RCA · GE · INTERSIL

ICL7660
CMOS Voltage Converter

GENERAL DESCRIPTION

The Harris ICL7660 is a monolithic CMOS power supply circuit which offers unique performance advantages over previously available devices. The ICL7660 performs supply voltage conversion from positive to negative for an input range of +1.5V to +10.0V, resulting in complementary output voltages of −1.5V to −10.0V. Only 2 non-critical external capacitors are needed for the charge pump and charge reservoir functions. The ICL7660 can also be connected to function as a voltage doubler and will generate output voltages up to +18.6V with a +10V input.

Contained on chip are a series DC power supply regulator, RC oscillator, voltage level translator, and four output power MOS switches. A unique logic element senses the most negative voltage in the device and ensures that the output N-channel switch source-substrate junctions are not forward biased. This assures latchup free operation.

The oscillator, when unloaded, oscillates at a nominal frequency of 10kHz for an input supply voltage of 5.0 volts. This frequency can be lowered by the addition of an external capacitor to the "OSC" terminal, or the oscillator may be overdriven by an external clock.

The "LV" terminal may be tied to GROUND to bypass the internal series regulator and improve low voltage (LV) operation. At medium to high voltages (+3.5 to +10.0 volts), the LV pin is left floating to prevent device latchup.

An enhanced direct replacement for this part, the ICL7660S, is now available and should be used for all new designs.

FEATURES
- Simple Conversion of +5V Logic Supply to ±5V Supplies
- Simple Voltage Multiplication ($V_{OUT} = (-)nV_{IN}$)
- 99.9% Typical Open Circuit Voltage Conversion Efficiency
- 98% Typical Power Efficiency
- Wide Operating Voltage Range 1.5V to 10.0V
- Easy to Use — Requires Only 2 External Non-Critical Passive Components
- No External Diode Over Full Temperature and Voltage Range

APPLICATIONS
- On Board Negative Supply for Dynamic RAMs
- Localized μ-Processor (8080 Type) Negative Supplies
- Inexpensive Negative Supplies
- Data Acquisition Systems

ORDERING INFORMATION

Part Number	Temp. Range	Package
ICL7660CTV	0° to +70°C	TO-99
ICL7660CBA	0°C to +70°C	8 PIN SOIC
ICL7660CPA	0° to +70°C	8 PIN MINI DIP
ICL7660MTV*	−55° to +125°C	TO-99

*Add /883B to part number if 883B processing is required.

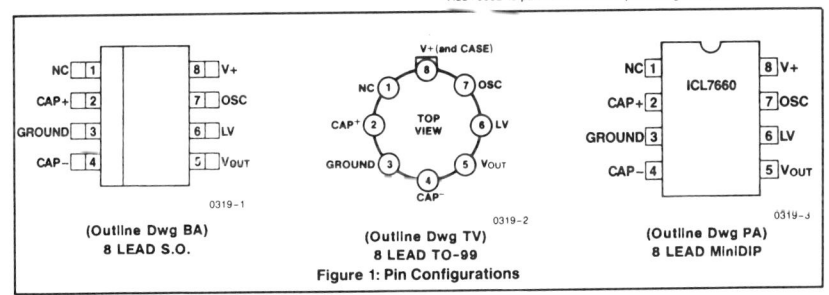

Figure 1: Pin Configurations

Figure 8-31 Data sheets for the ICL7660 voltage converter. *(Reprinted with permission of Harris Semiconductor Corporation)*

ICL7660

ABSOLUTE MAXIMUM RATINGS

Supply Voltage 10.5V
LV and OSC Input Voltage
(Note 1) .... −0.3V to (V+ +0.3V) for V+ <5.5V
            (V+ −5.5V) to (V+ +0.3V) for V+ >5.5V
Current into LV (Note 1) .... 20µA for V+ >3.5V
Output Short Duration (V$_{SUPPLY}$ ≤ 5.5V) ...... Continuous
Power Dissipation (Note 2)
  ICL7660CTV 500mW
  ICL7660CPA 300mW
  ICL7660MTV 500mW

Operating Temperature Range
  ICL7660M .... −55°C to +125°C
  ICL7660C .... 0°C to +70°C
Storage Temperature Range .... −65°C to +150°C
Lead Temperature
  (Soldering, 10sec) .... 300°C

NOTE: Stresses above those listed under "Absolute Maximum Ratings" may cause permanent damage to the device. These are stress ratings only and functional operation of the device at these or any other conditions above those indicated in the operational sections of the specifications is not implied. Exposure to absolute maximum rating conditions for extended periods may affect device reliability.

ELECTRICAL CHARACTERISTICS

V+ = 5V, T$_A$ = 25°C, C$_{OSC}$ = 0, Test Circuit Figure 3 (unless otherwise specified)

Symbol	Parameter	Test Conditions	Min	Typ	Max	Units
I+	Supply Current	R$_L$ = ∞		170	500	µA
V$_L^+$	Supply Voltage Range – Lo	MIN ≤ T$_A$ ≤ MAX, R$_L$ = 10kΩ, LV to GROUND	1.5		3.5	V
V$_H^+$	Supply Voltage Range – Hi	MIN ≤ T$_A$ ≤ MAX, R$_L$ = 10kΩ, LV Open	3.0		10.0	V

ELECTRICAL CHARACTERISTICS

V+ = 5V, T$_A$ = 25°C, C$_{OSC}$ = 0, Test Circuit Figure 3 (unless otherwise specified) (Continued)

Symbol	Parameter	Test Conditions	Min	Typ	Max	Units
R$_{OUT}$	Output Source Resistance	I$_{OUT}$ = 20mA, T$_A$ = 25°C		55	100	Ω
		I$_{OUT}$ = 20mA, 0°C ≤ T$_A$ ≤ +70°C			120	Ω
		I$_{OUT}$ = 20mA, −55°C ≤ T$_A$ ≤ +125°C			150	Ω
		V+ = 2V, I$_{OUT}$ = 3mA, LV to GROUND, 0°C ≤ T$_A$ ≤ +70°C			300	Ω
		V+ = 2V, I$_{OUT}$ = 3mA, LV to GROUND, −55°C ≤ T$_A$ ≤ +125°C			400	Ω
f$_{OSC}$	Oscillator Frequency			10		kHz
P$_{Ef}$	Power Efficiency	R$_L$ = 5kΩ	95	98		%
V$_{OUT\ Ef}$	Voltage Conversion Efficiency	R$_L$ = ∞	97	99.9		%
Z$_{OSC}$	Oscillator Impedance	V+ = 2 Volts		1.0		MΩ
		V+ = 5 Volts		100		kΩ

Notes: 1. Connecting any input terminal to voltages greater than V+ or less than GROUND may cause destructive latchup. It is recommended that no inputs from sources operating from external supplies be applied prior to "power up" of the ICL7660.
2. Derate linearly above 50°C by 5.5mW/°C.

TYPICAL PERFORMANCE CHARACTERISTICS (Circuit of Figure 3)

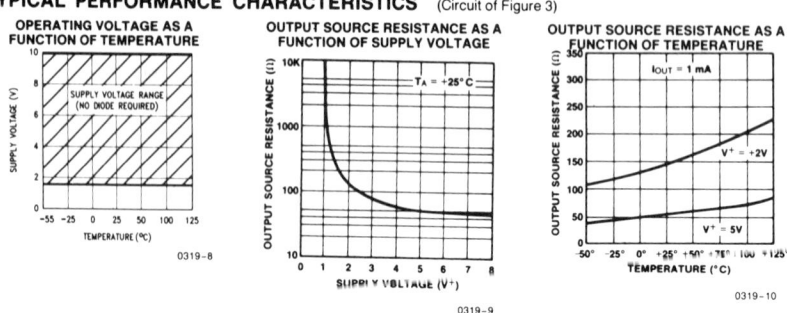

NOTE: All typical values have been characterized but are not tested.

Figure 8-31 (*Continued*)

8-4 Voltage Converters

was used to allow standard op-amps to operate from a single supply. In the case of the variable regulator, however, this simple procedure will not work because the full regulator current will flow through one of the voltage divider resistors.

A problem also arises with the 12 V battery example when a ±6 V supply *and* a +12 V supply are both required. The definition of a +12 V supply means that the negative terminal of the battery must be defined as ground. A +6 V point can be defined with respect to this ground, but a −6 V point cannot. If we were to include a simple voltage divider with the center defined as ground, we would have two grounds defined in the circuit.

This type of problem is frequently encountered in the design of computer backup power supplies where the computer requires +12 V, +5 V, and −5 V. The obvious solution is to use separate 5 and 12 V batteries. Another alternative is to use a switch-mode power supply. This alternative will be described in the next chapter. If considerable current must be supplied by the negative source, then one of these solutions must be used.

Often, however, the negative source is not required to supply very much current. For example, the LM313 voltage reference used in Figure 8-30 to provide a negative offset to the variable regulator requires only 1 mA of current but its own negative supply. We could build a separate rectified supply, but there is a better solution—use a **voltage converter.**

The ICL7660 is a voltage converter designed exactly for this type of problem. The ICL7660 takes any positive voltage from +1.5 to +10 V and outputs the negative complementary voltage. For example, the ICL7660 can take a standard +5 V digital supply and produce a −5 V supply. Data sheets for this voltage converter are shown in Figure 8-31. The chip is basically a square-wave oscillator converting the DC to high frequency AC (about 10 kHz). This AC signal is rectified as described in Section 7-1 and filtered to produce a negative output signal.

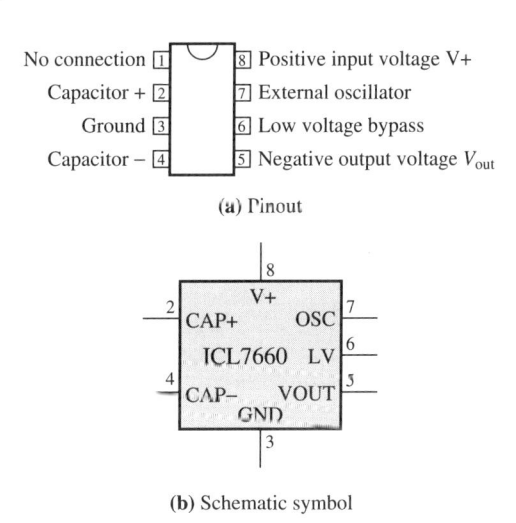

```
No connection   1       8  Positive input voltage V+
Capacitor +     2       7  External oscillator
Ground          3       6  Low voltage bypass
Capacitor −     4       5  Negative output voltage V_out
```

(a) Pinout

```
         8
         V+
   2 ─┤CAP+    OSC├─ 7
        ICL7660 LV ├─ 6
   4 ─┤CAP−   VOUT├─ 5
         GND
         3
```

(b) Schematic symbol

Figure 8-32 The ICL7660 voltage converter.

The ICL7660 is only one of a family of voltage converters. This family also includes the ICL7662, which produces negative output voltages for input voltages from 4.5 to 20 V, and the ICL7663/7664, which is programmable for 1.6 to 16 V output.

The pinout of the ICL7660 is given on the data sheets in Figure 8-31 and has been redrawn as Figure 8-32(a) to a larger scale. Figure 8-32(b) shows the schematic symbol for the ICL7660. The circuit of a simple voltage converter using this chip is shown in Figure 8-33. There are actually two circuits in Figure 8-33. Figure 8-33(a) is for supply voltages less than 6.5 V, and Figure 8-33(b) is for supply voltages greater than 6.5 V. Both capacitors should be 10 μF tantalum electrolytics. Pin 7, labelled OSC, is for connecting an external oscillator. Sometimes because of noise problems or interference, it is necessary to operate the chip at a frequency different from the built-in frequency. In this case, an external oscillator is connected to pin 7.

The voltage converter acts as if it were an ideal voltage source, equal to but opposite in polarity to the input voltage, in series with a 55 Ω resistance. Any current drawn will reduce the effective output voltage of the circuit. The output voltage is

$$V_{out} = -(V_{in} - I_L \times 55 \text{ Ω})$$

(a) Supply voltages less than 6.5 volts

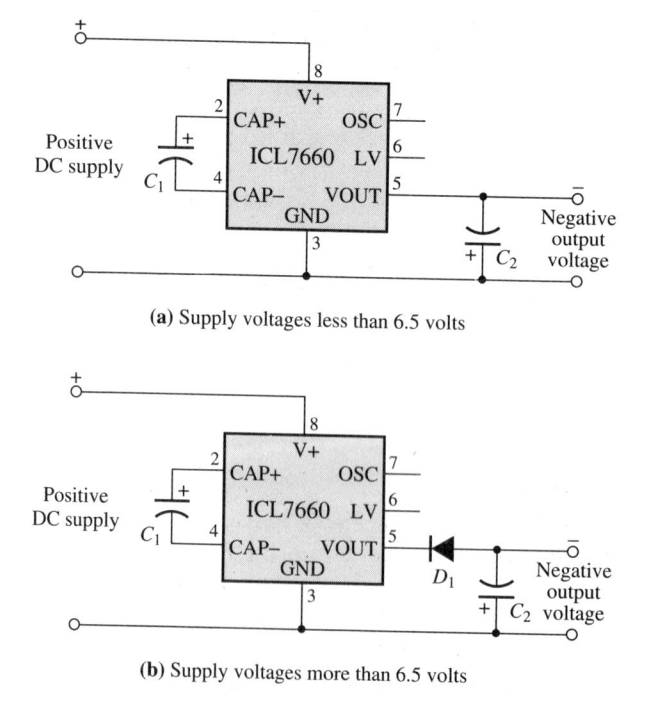

(b) Supply voltages more than 6.5 volts

Figure 8-33 Voltage converter circuits using the ICL7660.

For example, the output from a converter operating from a +5 V supply and delivering 10 mA to a load would be

$$V_{out} = -(5 \text{ V} - 0.01 \text{ A} \times 55 \text{ }\Omega) = -4.45 \text{ V}$$

Although the ICL7660 voltage converter is basically limited to a few milliamps of output current, it nevertheless has a variety of applications. In Example 8-15, we will demonstrate how it can be used in a power supply circuit where the adjustment voltage is offset with a LM313 voltage reference. It can also be used as an alternative method for operating op-amps from a single supply.

If we need more current than the ICL7660 (or similar device) can supply, then we must use a more complicated solution. We will introduce the switch-mode power supply in the next chapter to solve this type of problem.

EXAMPLE 8-15

Design a circuit for a 0 to 25 V, 1 A variable power supply using a LM317 variable regulator with an ICL7660 voltage converter and LM313 voltage reference. The DC supply is 32 V.

Solution

This is actually a continuation of the power supply circuit that we started to design in Example 8-13. The basic circuit, including the LM313 voltage reference was originally drawn in Figure 8-30. It has been redrawn as Figure 8-34, including the ICL7660 voltage converter. The ICL7660 serves as a negative supply for the LM313. The ICL7660 simply inverts the positive voltage obtained from a voltage divider consisting of R_A and R_B. Series resistor R_S limits the current to the LM313. Consequently, we have three components that must be designed: R_A, R_B, and R_S.

We must first select the voltage that we want supplied by the voltage converter. As a general rule, this should be at least twice the voltage of the voltage reference. We will arbitrarily choose 5 V.

Now we can design R_S. Use Equation 7-18 where the input voltage V_{supply} is 5 V, the reference voltage V_Z is 1.22 V, and the current I_{ZT} is 1 mA. This is similar to Example 7-4.

$$R_S = \frac{V_L - V_Z}{I_{ZT}} = \frac{5 \text{ V} - 1.22 \text{ V}}{1.0 \text{ mA}} = 3.78 \text{ k}\Omega$$

The closest standard value from Appendix A-1 is 3.9 kΩ.

Next, design the voltage divider. From the data sheet for the ICL7660, we see that this device needs a maximum internal supply current of 500 μA (0.5 mA). Because the divider must also provide an output current of 1 mA to the LM113, the total current drawn from the voltage divider is 1.5 mA. Design the voltage divider so that a bleeder current of approximately ten times this total current flows through the divider; hence, make the bleeder current 15 mA.

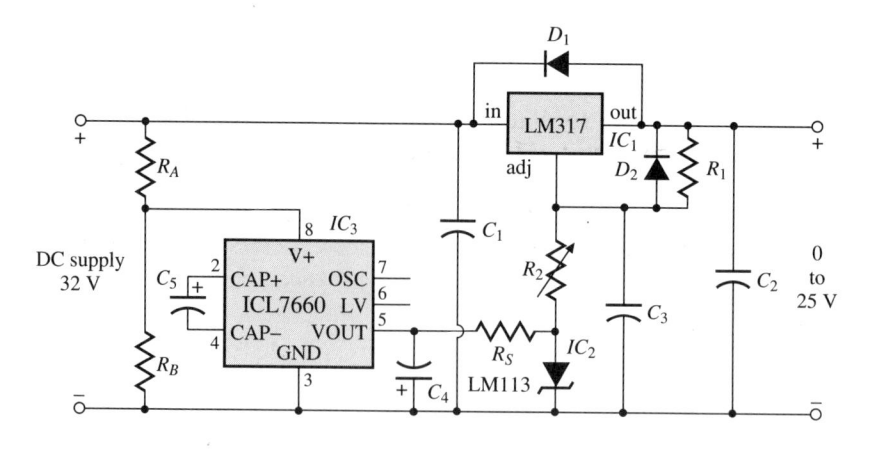

Figure 8-34

Calculate R_A. The voltage drop across this resistor will be the supply voltage minus the 5 V supplied to the ICL7660: $V_{R_A} = 32 \text{ V} - 5 \text{ V} = 27 \text{ V}$. The current through it will be the bleeder current plus the device current: $I_{R_A} = 15 \text{ mA} + 1.5 \text{ mA} = 16.5 \text{ mA}$.

$$R_A = \frac{V_{R_A}}{I_{R_A}} = \frac{27 \text{ V}}{16.5 \text{ mA}} = 1636 \text{ }\Omega$$

$$P = V_{R_A} \times I_{R_A} = 0.4455 \text{ W}$$

Choose a 1.6 kΩ, 1 W resistor.

Calculate R_B. The voltage drop V_{R_B} is simply the 5 V supplied to the ICL7660. The current I_{R_B} is simply the bleeder current of 15 mA. Thus,

$$R_B = \frac{V_{R_B}}{I_{R_B}} = \frac{5 \text{ V}}{15 \text{ mA}} = 333 \text{ }\Omega$$

$$P = V_{R_B} \times I_{R_B} = 0.075 \text{ W}$$

Choose a 330 Ω, $\frac{1}{4}$ W resistor.

Our design is now complete. We could go back and verify the operation of the voltage divider. The reader is encouraged to do this. Our design is flexible enough such that none of the components is critical.

Summary

This chapter discussed practical regulator circuits. Designs for high- and low-voltage output regulators using the LM723C multipin regulator were introduced. A pass transistor, to increase the current-handling capability, and current limiting, to protect the pass transistor, extended the usefulness of the LM723C. Although the LM723C is very flexible and gives good regulation, it requires relatively complex circuits, very similar to those introduced in the previous chapter for simple op-amp regulators.

A simpler and much more popular form of regulator is the three-terminal regulator. The MC7800/MC7900 series fixed regulators and the LM317/LM337 variable regulators are the most common forms of this type of regulator. Applications were given to demonstrate the simplicity of this type of regulator. The use of pass transistors to boost the current output and current limiting to protect the pass transistor were described.

Lastly, this chapter introduced the ICL7660 voltage inverter. Voltage inverters are little known but very useful devices that take a positive input voltage and generate an equivalent negative output voltage. Although their output current is limited, they have many applications.

Review Questions

8-1 Multipin voltage regulators

1. In what type of packages is the LM723C available?
2. What does the LM723C require for frequency compensation?
3. What is the purpose of R_3 in the LM723C circuit?
4. What is the difference in circuit configuration between a high-voltage regulator and a low-voltage regulator using an LM723C?
5. What resistors must be designed for an LM723C regulator circuit using foldback current limiting?
6. What is the maximum current that an LM723C is capable of delivering, and the maximum power that it can dissipate, without the addition of an external pass transistor?
7. What is the maximum voltage that can be dropped across an LM723C voltage regulator?
8. What is the percent line regulation and the percent load regulation for an LM723C acting as a high-voltage regulator?

8-2 Three-terminal fixed regulators

9. List three improvements that the MC7800 series regulators have over the LM723C regulator.
10. In what way is the LM723C superior to the MC7800 series regulators?
11. What are the three connections required for a three-terminal regulator?

12. In what case styles and power ratings are the MC7800 series regulators available?
13. In what voltage outputs are the MC7800 series regulators available?
14. What are the requirements or limitations on the DC supply voltage of the MC7800 series regulators?
15. What is the difference between an LM140-12 and an MC7812 regulator? Are they fully interchangeable?
16. Two capacitors are sometimes used with the MC7800 series regulators. What are their purposes, and what are typical values?
17. What is the difference between an MC7808 and an MC7908 regulator? What is the difference in the circuit connections if they are both used to produce an 8 V regulated supply?
18. If we are using an external pass transistor with an MC7800 series regulator, what is the typical regulator current?
19. What procedure is used to add current limiting to an MC7800 series regulator with a pass transistor?
20. What change in the regulator circuit is necessary to make the output voltage of an MC7800 series regulator variable?
21. What are the three limitations in using an MC7800 series regulator as a variable regulator?

8-3 Three-terminal variable regulators

22. What is the difference between an LM117, an LM217, and an LM317 variable-voltage regulator?
23. What is the reference voltage of an LM317 regulator?
24. What is the typical current required by the LM317 internal circuitry?
25. What is the minimum load current required by the LM317 regulator? What value of R_1 does this require?
26. What is the function of the capacitor between the adjustment terminal and ground? What is its typical value?
27. When are protective diodes required?
28. What is the minimum output voltage of an LM317 regulator?
29. How can the circuit be modified to give zero volts as the minimum output voltage?
30. Compare the line regulation, load regulation, and ripple rejection of the LM317 regulator to an MC7800 series regulator.

8-4 Voltage converters

31. What is the function of a voltage converter?
32. Give three specific applications of a voltage converter.
33. What is the principle of operation of a voltage converter?
34. What is the circuit difference for the ICL7660 used as a low-voltage converter and used as a high-voltage converter?
35. What is the output impedance of the ICL7660 voltage converter? What effect does this have on the output voltage and current?
36. If a high output current at negative voltage was required from a positive voltage source, what is the best circuit to use?

Design and Analysis Problems

8-1 Multipin voltage regulators

1. Design a regulator circuit using an LM723C voltage regulator to give an output of 9.0 V at a current of 1.5 A from a rectifier supplying a voltage of 16.3 V at a current of 1.5 A. Include simple current limiting.
   (a) Sketch the circuit you are going to use.
   (b) Calculate values for R_1, R_2, R_3, and R_{sens}. Select the nearest standard values for each resistor.
   (c) Calculate the power dissipated by the pass transistor.
2. Analyze the regulator circuit in Figure 8-35 to determine the
   (a) output voltage
   (b) maximum output current
   (c) power dissipated in the pass transistor at maximum load
   (d) output ripple if the supply has a ripple of 2.85 V
   (e) line regulation and the percent line regulation
   (f) load regulation and the percent load regulation.

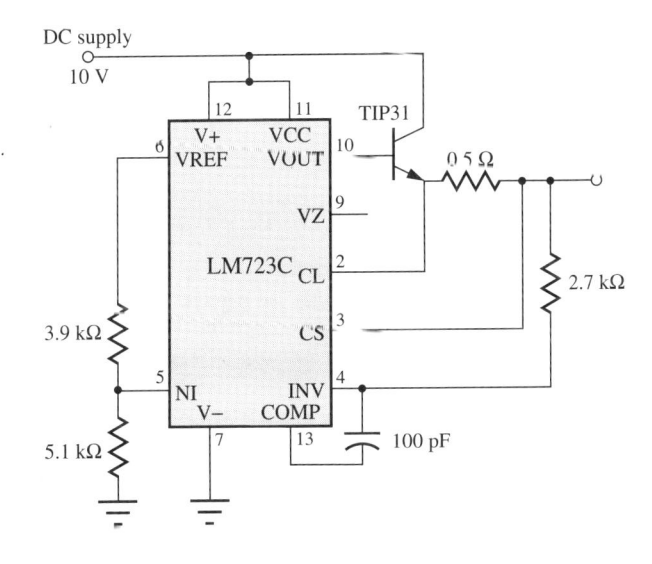

Figure 8-35

3. Design a 15 V, 2 A regulated power supply using an LM723C voltage regulator. Foldback current limiting should limit the short-circuit current to 0.3 A. The output ripple should be less than 0.5 mV.
   (a) Sketch the circuit you are going to use.
   (b) Determine the minimum DC voltage that must be provided by the rectifier circuit, and the maximum ripple voltage. Calculate the effective load resistance seen by the rectifier, and select a suitable transformer, filter capacitor, and bridge rectifier.
   (c) Design the voltage divider R_1 and R_2 of the regulator circuit, and calculate a value for R_3. Select the nearest values of standard resistors.

382   8 Integrated Circuit Regulators

(d) Design the value of the sensing resistor, and calculate values of R_4 and R_5 for the foldback current limiting. Select the nearest values of standard resistors.
(e) Calculate the voltage, current, and power for the pass transistor, and select a suitable pass transistor.

4. Analyze the regulator circuit shown as Figure 8-36 to determine the
   (a) output voltage
   (b) maximum current
   (c) DC output voltage from the rectifier circuit
   (d) ripple from the rectifier circuit, and the final output ripple voltage
   (e) line regulation and percent line regulation
   (f) load regulation and percent load regulation

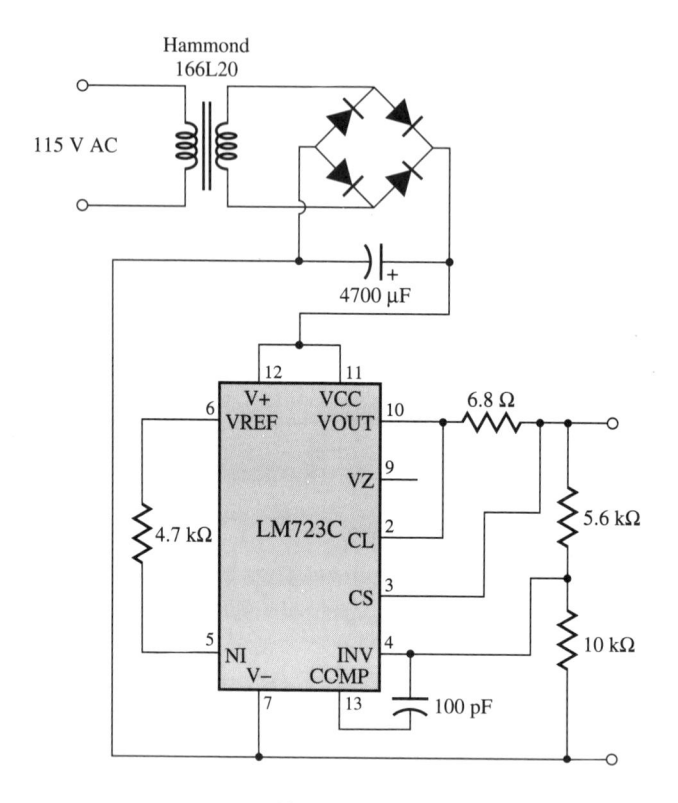

Figure 8-36

8-2 Three-terminal fixed regulators

5. Design a regulator circuit using an MC7805 regulator to give a regulated output of 30 V at 1 A from a DC supply which is capable of providing 35 V.
   (a) Sketch the circuit you are going to use.
   (b) Calculate values of R_1 and R_2 to give the required output. Calculate power ratings for each of these resistors, and select nearest standard values.
   (c) Calculate the power dissipated in the MC7805.

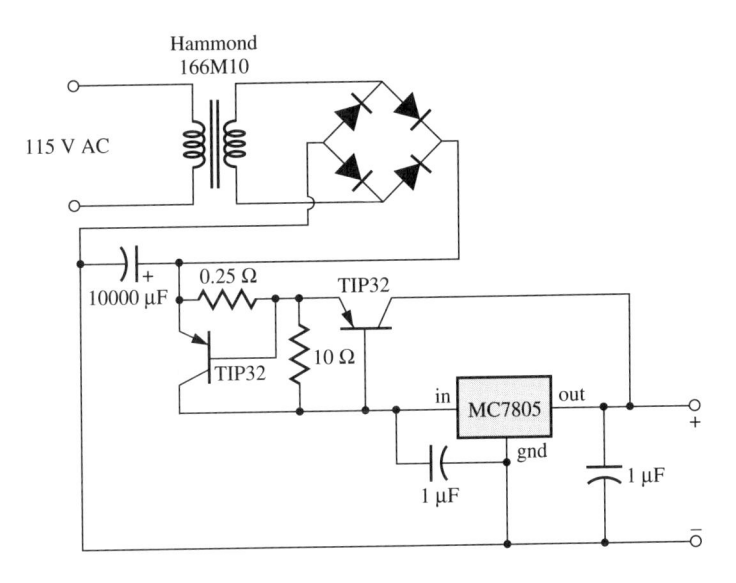

Figure 8-37

6. Analyze the regulator circuit in Figure 8-37 to determine the
   (a) maximum current
   (b) current through the regulator
   (c) line regulation and percent line regulation
   (d) load regulation and percent load regulation
   (e) output ripple voltage
   (f) voltage drop across the regulator
   (g) power dissipated by the regulator and power dissipated by the pass transistor under full-load conditions.
7. Design a ± 15 V, 3 A DC power supply using MC7815 and MC7915 voltage regulators. Assume that the MC7915 regulator has characteristics identical to the MC7815. Use a Hammond 167M36 36 V, 3 A power transformer which has a no-load output voltage of 37.9 V. The circuit should use current limiting. The maximum output ripple should be less than 1.0 mV.
   (a) Sketch the circuit you are going to use.
   (b) Calculate the maximum allowable ripple from the rectifier circuit, and determine suitable values for the filter capacitors.
   (c) Calculate suitable values for R_P and R_{sens}
   (d) Calculate the DC output voltage of the rectifier, and determine the voltage drop across the power transistor and the regulator. Calculate the power dissipated in each under maximum load.

8-3 Three-terminal variable regulators

8. Analyze the regulator circuit in Figure 8-38 to determine the
   (a) maximum output voltage
   (b) maximum output current
   (c) line regulation and percent line regulation
   (d) load regulation and percent load regulation.

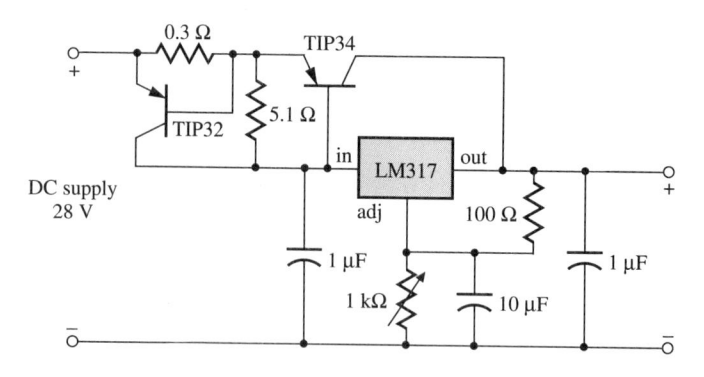

Figure 8-38

8-4 Voltage converters

9. Design a 0 to 15 V, 3 A variable power supply using an LM317 variable-voltage regulator, an LM313 voltage reference, and an ICL7660 voltage converter. The output ripple should be less than 0.1 mV. The circuit should have current limiting.
   (a) Sketch the circuit you are going to use.
   (b) Calculate the maximum allowable output ripple from the rectifier and the minimum DC output from the rectifier. Use these values to select a power transformer, a filter capacitor, and a rectifier.
   (c) Calculate the actual values of rectifier DC output and rectifier ripple. Use these to determine the output ripple and the voltage drop across the pass transistor and voltage regulator.
   (d) Calculate values for R_P and R_{sens}.
   (e) Calculate values for R_1 and R_2 to give the required range of voltage variation. Select nearest standard values.
   (f) Design a voltage divider to feed the ICL7660. Design and select values for R_A and R_B.
   (g) Design and select R_S for the voltage reference.

Switch-Mode Power Supplies 9

OUTLINE

9-1 Principles of Switch-Mode Power Supplies
9-2 Pulse-Width Modulators
9-3 Single-Package Switching Regulators

KEY TERMS

Switch-mode power supply
Pulse-width modulator
Duty cycle
Error amplifier
Modulator
Transistor switch
Effective series resistance (ESR)
Catch diode
Toroidal core
Step-up regulator
Step-down regulator
Voltage inverting regulator

OBJECTIVES

On completion of this chapter, the reader should be able to:

- explain the advantages of a switched power supply over a linear power supply.
- list the five basic components of a switch-mode power supply, describe what the function of each is, and connect the basic components together to produce a step-down regulator, a step-up regulator, or a voltage-inverting regulator.
- calculate the capacitor and inductor values for a step-down regulator, a step-up regulator, and a voltage-inverting regulator.
- design a variety of switch-mode regulator circuits using the LM3524 pulse-width modulator, including step-down regulators, step-up regulators, voltage-inverting regulators, and multiple-output regulators.
- design step-down switch-mode regulators using the LH1605.

9 Switch-Mode Power Supplies

Introduction

The previous chapter introduced a number of regulator circuits. These regulators were linear; that is, they regulated by continuously varying the effective circuit resistance to compensate for changes in load. For many applications, this type of regulator is the simplest and most convenient to use.

There is, however, a potential problem with these linear regulators. The difference in voltage between the input and output is dropped across the regulator, and the full-load current flows through it. This produces a large power dissipation in the regulator. Power dissipated in the regulator is waste power—the more power wasted, the lower the efficiency of the regulator. Power dissipated in the regulator also means that the regulator or pass transistor will probably require a large heatsink, which in turn means that the packaging of the power supply will be large and bulky. In some areas, such as military and aerospace applications, size and efficiency are critical.

Switch-mode regulators are designed to overcome this problem of low efficiency. Instead of controlling power by resistively dissipating excess power, switched power supplies switch the power only as required by the load. The switching operation wastes very little power. Switch-mode supplies can operate at 80% or better efficiency and can be packaged in a fraction of the size of linear regulated supplies. Unlike linear regulators, switch-mode regulators can also be used to step up the output voltage or to invert the polarity. As well, the output of the pulse-width modulator in a switch-mode regulator is effectively an AC waveform, which can be put through a transformer to generate more than one regulated output voltage from a single regulator.

Conventional switch-mode regulators use a separate pulse-width modulator circuit to control the power switching, and the corresponding circuit requires a number of external components, in some respects similar to the LM723C linear regulator circuits. Modern switch-mode regulators, however, are being designed with most of the external components enclosed in a single power transistor package, similar to the three-terminal linear regulators, and are starting to replace the older pulse-width modulator circuits.

9-1 Principles of Switch-Mode Power Supplies

The linear regulator, as developed in the previous chapter, maintains a constant output voltage by continuously monitoring the load demand and varying the effective internal regulator resistance to compensate for load changes. This means that there is always power being dissipated in the regulator, and sometimes this power can be far greater than the power supplied to the load. For example, if a regulated 5 V, 1 A output is required from a 24 V DC supply, the voltage drop across the regulator is 19 V, the current through the regulator is 1 A, and the power dissipated in the regulator is

$$P_R = \text{regulator voltage} \times \text{current} = 19 \text{ V} \times 1 \text{ A} = 19 \text{ W}$$

By a similar calculation the load power is

$$P_L = \text{load voltage} \times \text{current} = 5 \text{ V} \times 1 \text{ A} = 5 \text{ W}$$

The efficiency is about 21%. Furthermore, the power dissipated in the regulator heats the regulator, necessitating a large heatsink and a correspondingly bulky package for the complete power supply.

If we were to switch power directly from the supply to the load only as required, we would reduce the power dissipated in the regulator. The power dissipated in an open switch is zero, and the power dissipated in a closed switch is typically quite small. When the power is switched, a pulsating DC output voltage is produced. Filtering is necessary. However, unlike the simple rectifier circuits discussed in Chapter 7, where the output voltage was pulsating at AC line frequency, in switch-mode supplies we can select the switching frequency. Normally, a relatively high switching frequency is used, since at high frequencies capacitive reactance is low, inductive reactance is high, and effective filtering can be accomplished by relatively low-value, and hence physically small, capacitors and inductors.

The linear regulators discussed in the previous two chapters took an input voltage and dropped it to a lower regulated output of the same polarity. This is a step-down regulator. The standard switch-mode regulator is also generally used as a step-down regulator. However, in switch-mode supplies, it is relatively easy to reconfigure the circuit to act as a step-up regulator to produce an output voltage higher than the supply voltage, or to reconfigure the circuit in another way to act as a voltage inverter to produce a negative output for a positive supply voltage. In our discussion of switching regulators we will describe the operation of all three configurations.

Components of a Switch-Mode Regulator

Any **switch-mode power supply** consists of five standard components:

- pulse-width modulator
- transistor switch
- inductor
- capacitor
- diode

The arrangement of these components determines whether we have a step-down regulator, a step-up regulator, or an inverting regulator. All of these components are necessary, and each performs a function vital to the operation of the power supply. To understand the operation of a switch-mode regulator, we must understand the function and characteristics of each component.

The Pulse-Width Modulator

The **pulse-width modulator** controls the switching of the transistor and is the most complicated part of a switch-mode regulator. Although it is possible to design a pulse-width modulator circuit using simple op-amps, the circuit is rather involved

388 9 Switch-Mode Power Supplies

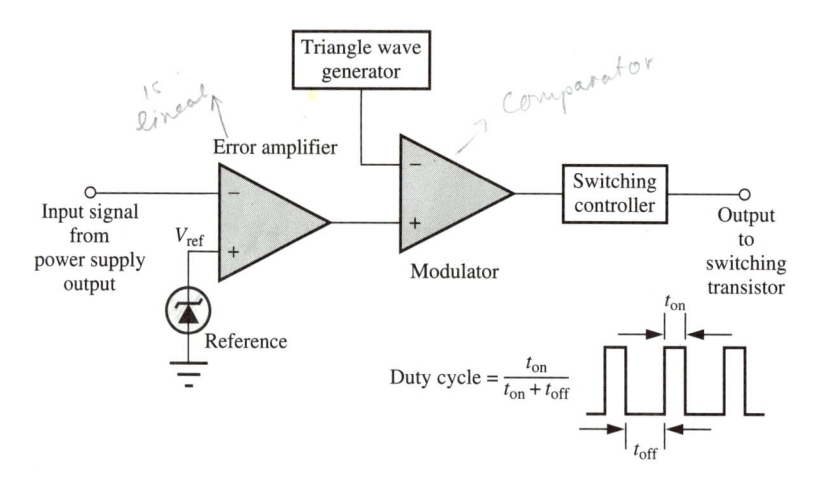

Figure 9-1 Pulse-width modulator.

and we will not attempt to design one. Generally, switch-mode regulators use pulse-width modulator integrated circuits. The simplified functional diagram of a typical pulse-width modulator is shown in Figure 9-1.

The input signal is a feedback signal from the output of the power supply. The pulse-width modulator generates output pulses with variable duty cycle to hold the output voltage equal to an internal reference voltage. The duty cycle is the pulse duration t_{on} divided by the pulsation period. The length of time that the transistor is switched on is equal to the pulse duration.

The **error amplifier** compares the signal fed back from the regulator output to an internal reference voltage. It is similar to a level comparator, such as introduced in Section 5-3, with a reference voltage applied to one input and the output voltage of the regulator applied to the other input. Unlike a comparator, however, the output voltage of the error amplifier varies linearly with the difference between the regulator output voltage and the reference voltage. The output signal from the error amplifier is used to control a modulator.

The **modulator** is a comparator, comparing the error amplifier signal to the signal produced by a triangle-wave oscillator. The comparator produces a pulse-wave output with a duty cycle determined by the error amplifier output level. The modulator output, through a switch controller, produces the actual pulse-width modulator output signal.

The Transistor Switch

The **transistor switch** is the heart of the switch-mode supply, actually controlling the power supplied to the load. TMOS power transistors are in some ways superior to bipolar transistors, but many readers may be unfamiliar with this type of transistor, so for all of the designs in this book we will use bipolar power transistors. Transistors chosen for use in switching power supplies must have fast switching times and be able to withstand the voltage spikes produced by the inductor.

The bipolar transistor is not a perfect switch. When it is switched on, there is always a voltage drop developed between the collector and emitter. This voltage drop, V_{CEsat}, or more simply V_{sat}, depends on the specific transistor and on the collector current flowing in the transistor, but is typically less than 1 V. Unless otherwise stated, we shall assume a value of V_{sat} equal to 1.0 V. The power dissipated in the transistor switch is simply the collector current times V_{sat}. When the transistor switches off, the only current flowing in the transistor is a very small leakage current, which we will assume is zero. Because the current is zero, the power is zero. The total power dissipated in the transistor over the full switching cycle, even for large currents, is quite small.

We will see shortly that step-down regulators and voltage-inverting regulators use PNP switching transistors, whereas step-up regulators use NPN switching transistors. The switching transistor's polarity and its biasing circuit are important.

The Inductor

The inductor is the device that ultimately produces the output voltage, and, as such, is *the most important element in a switch-mode regulator*. To understand the operation of a switch-mode regulator, we need to understand the operation of an inductor. An inductor is a device that stores magnetic energy. Current flowing in the coils of an inductor generates a magnetic field, which stores energy. Current cannot change instantaneously in an inductor, and the voltage developed across an inductor is proportional to the rate of change of current. When a voltage is suddenly applied across an inductor, the initial current is zero, but starts immediately to increase to a value determined by the external circuit resistance. As the current increases, the magnetic field increases, the energy stored in the magnetic field increases, and the voltage drop across the windings decreases toward zero. If the voltage source is now suddenly removed, the current cannot change instantaneously to zero. The magnetic field starts to collapse, producing a negative voltage across the windings in an attempt to maintain current flow. This negative voltage and the current (unchanged in direction) decrease exponentially toward zero. The energy stored in the magnetic field is converted back to current flow.

Inductors used in switch-mode supplies are usually wound on toroidal cores made of ferrite or powdered iron to minimize core losses at high frequencies. Often, in an experimental design, it is necessary to design and construct the actual inductor. The procedure for designing inductors (and transformers) using toroidal cores is outlined in Appendix B. Portions of a catalog listing toroidal cores are shown in Appendix B.

The Capacitor

The capacitor, as in a linear power supply, is used to store charge and deliver it as necessary to the load. It is always connected directly across the regulator output, and its main purpose is to reduce ripple. Since switch-mode regulators are usually employed in high-current, high-performance power supplies, the capacitor should be chosen for minimum loss. Loss in capacitors occurs because of internal series resistance and inductance. Capacitors for switch-mode regulators are chosen on the basis of low **effective series resistance** (ESR). Solid tantalum capacitors are best in this respect. For very high performance power supplies, sometimes it is necessary to parallel capacitors to get a low enough effective series resistance. To minimize ground-loop resistance, the capacitor ground connection and all output ground connections should be made to one point, with lead length kept to a minimum.

The Diode

The diode used in switch-mode regulators is sometimes referred to as a **catch diode**. The purpose of the catch diode is not rectification, but to direct current flow in the circuit and to ensure that there is always a path for current to flow into the inductor. Although the transistor switches the power to the load, it does not produce an AC signal, but pulsating DC, in some respects similar to the pulsating output from the rectifier of a linear power supply.

The Load Resistance

One more important component, although not part of the regulator, must be in the power supply circuit—*a load resistance*. For switch-mode regulators to perform properly, current must always flow through the inductor; hence, a load must always be attached to the power supply.

Step-Down Switch-Mode Supply

The first configuration for the various switch-mode regulator components that we have just discussed is shown in Figure 9-2. This is the basic circuit for a **step-down switch-mode regulator**. The PNP switching transistor is in series with the inductor and switches current directly through the inductor. The pulse-width modulator controls the switching of the transistor. At this point we will not be concerned with the exact mechanics of the switching, and hence the bias circuitry for the transistor has been omitted to simplify the discussion.

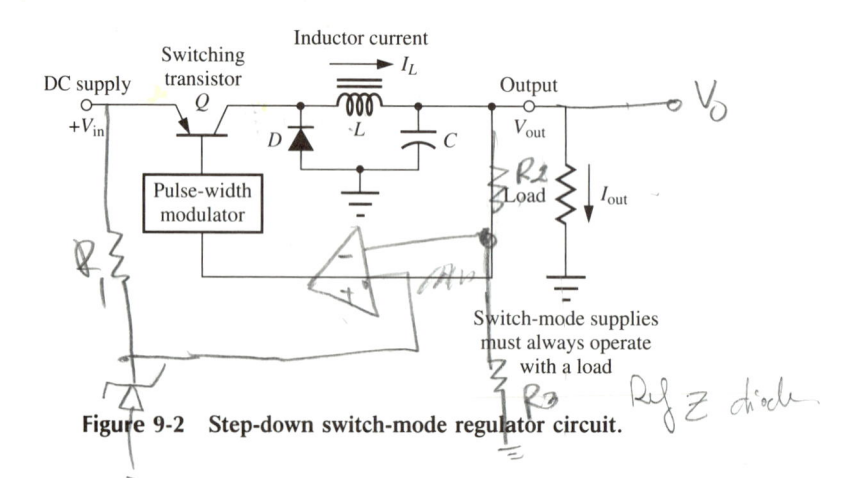

Figure 9-2 Step-down switch-mode regulator circuit.

9-1 Principles of Switch-Mode Power Supplies

To explain the operation of this circuit, we shall assume that there is some steady-state current flowing through the inductor to the load and that the pulse-width modulator is producing the output shown in Figure 9-3(a). When the switch is closed, the power supply is switched to the inductor, and additional current tries to flow from the supply through the inductor to the capacitor and the load. Because the inductor will oppose any instantaneous change in current, it acts initially as a very high impedance to the power supply, and the entire supply voltage appears across the inductor. The current through the inductor starts to increase because of the voltage across it. Some of this additional current flows into the capacitor, increasing the capacitor charge and, hence, the output voltage. The positive voltage initially developed across the inductor drops exponentially as the current increases.

When the switch opens, the inductor again opposes the change in current, and the magnetic field stored in the inductor starts to collapse as it tries to keep the same current flowing. The voltage across the inductor after the transistor switches off is initially the same magnitude as immediately before the transistor

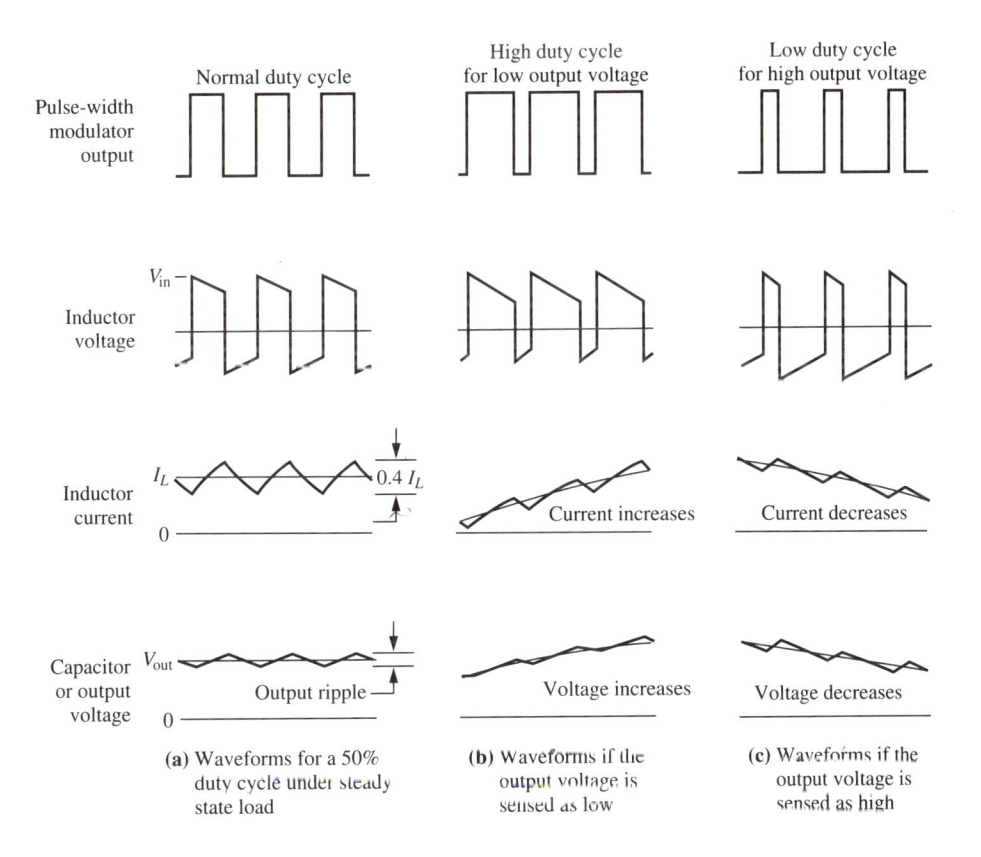

Figure 9-3 Pulse-width modulator, inductor voltage and current, and output voltage waveforms for a step-down switching regulator.

switched, but is of opposite polarity. The same current initially flows, only now from ground through the catch diode, which has become forward-biased. The current in the inductor immediately starts to decrease because there is no voltage source to sustain it. Both the inductor voltage and current will decrease in magnitude until the transistor switches on again. The inductor voltage and current are shown in Figure 9-3(a).

Now suppose that the output voltage detected at the voltage divider is less than the reference voltage. The error detector amplifier sends a signal to the pulse-width modulator, causing it to increase the duty cycle of the pulse as shown in Figure 9-3(b). The pulse will now be on longer than it is off, switching the transistor on for a longer period of time. Because the transistor is on longer, the current through the inductor has more time to increase, so it is larger than under normal load conditions. This causes the capacitor voltage to increase until it balances the reference voltage. The resultant inductor voltage and current waveforms and the output voltage are shown in Figure 9-3(b).

If the output is higher than the reference voltage, the error amplifier will send a signal to the pulse-width modulator and cause it to decrease the duty cycle. The pulse will be off longer than it is on, switching the transistor on for a shorter period of time, as shown in Figure 9-3(c). The net current through the inductor will decrease, and the output voltage across the capacitor will also decrease. The voltage and current waveforms are shown in Figure 9-3(c). At no time does the output voltage reach the level of the input voltage; thus, we have a step-down regulator.

We need to derive equations for designing the inductor and capacitor of a step-down regulator. First, we will develop an equation for designing the inductor. From basic circuit theory, the equation for the voltage drop across an inductor is

$$V_L = L\frac{dI}{dt} \approx L\frac{\Delta I_L}{\Delta t} \tag{9-1}$$

Rearrange Equation 9-1 to solve for ΔI_L:

$$\Delta I_L = \frac{V_L \, \Delta t}{L} \tag{9-2}$$

When the transistor switch is on, for $\Delta t = t_{on}$, the voltage across the inductor is $V_{in} - V_{out}$:

$$\Delta I_{Lon} = \frac{(V_{in} - V_{out}) \times t_{on}}{L} \tag{9-3}$$

When the transistor switch is off, for $\Delta t = t_{off}$, the voltage across the inductor is V_{out}:

$$\Delta I_{Loff} = \frac{V_{out} \times t_{off}}{L} \tag{9-4}$$

For steady-state operation, the change in current when the switch is off, ΔI_{Loff}, must be equal to the change in current when the switch is on, ΔI_{Lon}. Equating

Equations 9-3 and 9-4 and cancelling the inductance L common to each term, we get

$$V_{out} \times t_{off} = (V_{out} - V_{in}) \times t_{on} \qquad (9\text{-}5)$$

Solve this equation for V_{out}/V_{in}:

$$\frac{V_{out}}{V_{in}} = \frac{t_{on}}{t_{on} + t_{off}} = \frac{t_{on}}{T} \qquad (9\text{-}6)$$

where the quantity $T = t_{on} + t_{off}$ is the pulse period. Now rearrange this equation in terms of the duty cycle t_{on}/T:

$$\frac{t_{on}}{T} = \frac{V_{out}}{V_{in}} \qquad (9\text{-}7)$$

Notice that the duty cycle is equal to the ratio of the output voltage to the input voltage.

Normally, the inductor is designed so that the peak-to-peak inductor current ΔI_L varies by 40% of the maximum output current I_{out} as shown in Figure 9-4(a). Substitute $0.4 I_{out}$ for ΔI in Equation 9-3 and solve for the inductance L:

$$L = \frac{(V_{in} - V_{out}) \times t_{on}}{0.4 I_{out}} \qquad (9\text{-}8)$$

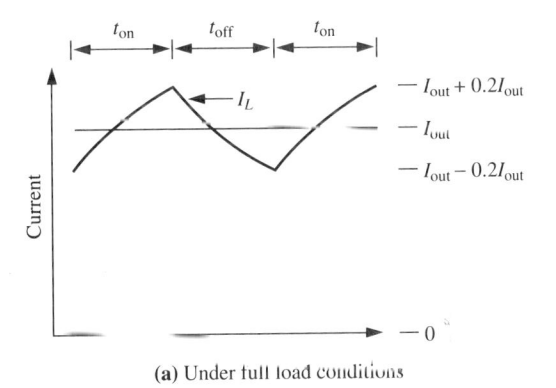

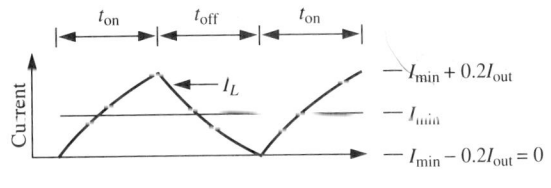

(a) Under full load conditions

(b) Under minimum load conditions

Figure 9-4 Inductor current for step-down regulator.

Now solve Equation 9-7 for t_{on}:

$$t_{on} = T \times \frac{V_{out}}{V_{in}} \tag{9-9}$$

Substitute this into Equation 9-8:

$$L = \frac{T \times V_{out} \times (V_{in} - V_{out})}{0.4 I_{out} V_{in}} \tag{9-10}$$

The oscillator pulse frequency f_{osc} is simply $1/T$. Replace the period T in Equation 9-10 with $1/f_{osc}$ to get

$$L = \frac{2.5 V_{out}(V_{in} - V_{out})}{I_{out} V_{in} f_{osc}} \tag{9-11}$$

This is the equation for designing the inductor in a step-down switch-mode power supply. The inductance value determined by this equation is in henries.

Note that the factor 2.5 comes from the requirement that the variation in the inductor current be $0.4 I_{out}$, where I_{out} is the design output current. This places a severe limit on the minimum current that the supply can provide while maintaining proper regulation. The minimum current I_{min} must be greater than $0.2 I_{out}$ to ensure that the inductor current, which is varying from $-0.2 I_{out}$ to $+0.2 I_{out}$ by design, never goes less than zero, as shown in Figure 9-4(b). This point has been made before and is important—*Switch-mode supplies must always be operated such that they supply at least the minimum current.*

If the regulator is to provide current to a large range of loads, it may be necessary to limit the variation in inductor current to a smaller fraction of the total current. If, for example, we restricted the variation in inductor current to 4% instead of 40%, the factor 2.5 in Equation 9-11 would be replaced by a factor of 25 (1/0.04), and the minimum current would be $0.02 I_{out}$. In this case the inductor would have ten times as much inductance and would be proportionally larger physically.

The capacitor is chosen to reduce the output ripple to some acceptable level. The peak-to-peak ripple voltage is simply the change in output voltage across the capacitor ΔV. The voltage across a capacitor is $V = Q/C$. The change in voltage across the capacitor is

$$\Delta V = \frac{1}{C} \times \Delta Q \tag{9-12}$$

where the change in charge on the capacitor ΔQ is equal to the average charging current I_{avg} times the charging time t.

The current charging the capacitor is the difference between the inductor current I_L and the output current I_{out}. Referring to the inductor current in Figure 9-4(a), $I_L = I_{out}$ halfway through the on cycle. Charging current starts to flow into the capacitor and reaches a maximum of $0.2 I_{out}$ at the end of the on cycle. The charging current then drops, reaching zero when $I_L = I_{out}$ at the midpoint of the off cycle. The average charging current is $I_{avg} = \frac{1}{2} \times 0.2I_{out}$, which flows for $t = \frac{1}{2}t_{on} + \frac{1}{2}t_{off} = \frac{1}{2}T$. The total change of charge on the capacitor is

$$\Delta Q = \frac{1}{2} \times 0.2 \times I_L \times \frac{1}{2}T \tag{9-13}$$

Substituting this into Equation 9-12 and replacing T with $1/f_{osc}$, we get

$$\Delta V = \frac{1}{C} \times \frac{1}{2} \times 0.2 \times I_L \times \frac{1}{2} \times T = \frac{0.05}{Cf_{osc}} \times I_L \tag{9-14}$$

Now solve Equation 9-14 for C:

$$C = \frac{0.05 I_{out}}{\Delta V f_{osc}} \tag{9-15}$$

Normally, the ripple voltage is given in RMS volts; hence, $V_{ripple} = \Delta V/2\sqrt{2}$, and our final design equation becomes

$$C = \frac{0.01768 I_{out}}{V_{ripple} f_{osc}} \tag{9-16}$$

Note that both the capacitor and inductor values are inversely dependent on the frequency—the higher the frequency, the smaller L and C, and the more compact the physical size of the circuit. There are, however, several practical constraints on the selection of the oscillator frequency. The minimum frequency must be greater than 20 kHz, or the power supply will produce an audible whistle from the vibration of the inductor windings. The maximum frequency is generally limited by the ferromagnetic material of the inductor. Many powdered-iron cores have a maximum frequency of about 50 kHz, although premium cores may be good to frequencies several times as high. At high frequencies, stray inductance and capacitance may make the operation of the circuit unpredictable, and there may be considerable energy radiated from the circuit. The maximum upper frequency is typically 50 to 100 kHz.

Step-Up Regulators

The second configuration for the switch-mode regulator components is the **step-up regulator** as shown in Figure 9-5. Compare the configuration of this step-up regulator with the step-down regulator in Figure 9-2. In the present circuit, the switching transistor is in shunt with the load and, when switched on, tends to bypass some of the current to ground. As for the step-down regulator, we will assume that the pulse-width modulator properly switches the transistor and won't worry about proper biasing.

To describe the operation of this circuit, we shall again assume a steady-state current flowing to the load. If the transistor switch was not included in the circuit, the full steady-state load current would be flowing through the inductor and the full supply voltage would appear across the capacitor. With the transistor switch in the circuit, however, additional current is forced to flow through the inductor, increasing the capacitor charge and, hence, the output voltage. When the transistor is switched on, it essentially shorts the inductor to ground, the full supply

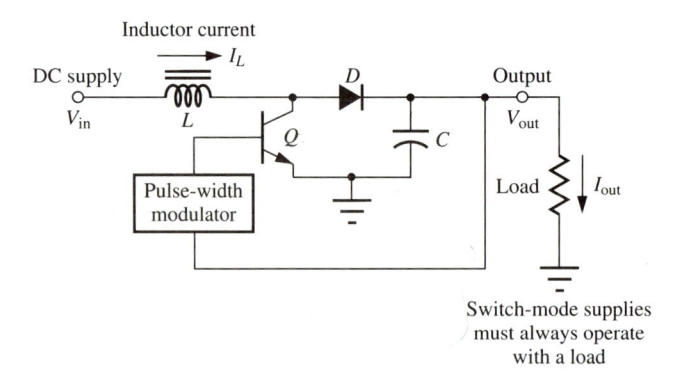

Figure 9-5 Step-up switch-mode regulator circuit.

voltage is dropped across the inductor, and the inductor current increases. The current will continue to increase for the duration of the on cycle. When the transistor switches off, whatever current is flowing in the inductor will continue to flow. The only place it can flow is through the diode to the capacitor, where the additional charge will increase the capacitor voltage. The variation in the inductor voltage and current is identical to that shown in Figure 9-3(a) for the step-down regulator.

Because the step-up regulator operates somewhat differently from the step-down regulator, it is necessary to develop new design equations for the inductor and capacitor. To develop the design equation for the inductor, we will start with Equation 9-2, giving the inductance in terms of the voltage and current:

$$\Delta I_L = \frac{V_L \, \Delta t}{L} \tag{9-2}$$

When the transistor switch is on, for $\Delta t = t_{on}$, the voltage across the inductor is V_{in}:

$$\Delta I_{Lon} = \frac{V_{in} \times t_{on}}{L} \tag{9-17}$$

When the transistor switch is off, for $\Delta t = t_{off}$, the voltage across the inductor is $V_{out} - V_{in}$:

$$\Delta I_{Loff} = \frac{(V_{out} - V_{in}) \times t_{off}}{L} \tag{9-18}$$

For steady-state operation, the change in current when the switch is off, ΔI_{Loff}, must be equal to the change in current when the switch is on, ΔI_{Lon}. Equating Equations 9-17 and 9-18 and cancelling the inductance L common to each term, we get

$$V_{in} \times t_{on} = (V_{out} - V_{in}) \times t_{off} \tag{9-19}$$

9-1 Principles of Switch-Mode Power Supplies

This is similar to the step-down regulator except that t_{on} and t_{off} are reversed. Solve this equation for V_{out}/V_{in}:

$$\frac{V_{out}}{V_{in}} = \frac{t_{on} + t_{off}}{t_{off}} = \frac{T}{t_{off}} \tag{9-20}$$

Now rearrange this equation in terms of the duty cycle t_{on}/T by inverting and subtracting each side from 1:

$$\frac{t_{on}}{T} = \frac{V_{out} - V_{in}}{V_{out}} \times 100 \quad \leftarrow \text{step up} \tag{9-21}$$

As for the step-down regulator, we will assume a peak-to-peak variation in the inductor current of 40%. In this case, however, the inductor current I_L is not equal to the output current I_{out}. However, the product of the inductor current and the input voltage is equal to the product of the output current and the output voltage: $I_L \times V_{in} = I_{out} \times V_{out}$. Solve for I_L:

$$I_L = I_{out} \times \frac{V_{out}}{V_{in}} \tag{9-22}$$

Substitute $0.4 I_{out} \times V_{out}/V_{in}$ for ΔI in Equation 9-17 and solve for the inductance L:

$$L = \frac{V_{in}^2 \times t_{on}}{0.4 I_{out} V_{out}} \tag{9-23}$$

Now solve Equation 9-21 for t_{on}:

$$t_{on} = T \times \frac{V_{out} - V_{in}}{V_{out}} \tag{9-24}$$

Substitute this into Equation 9-23:

$$L = \frac{T \times V_{in}^2 \times (V_{out} - V_{in})}{0.4 I_{out} V_{out}^2} \tag{9-25}$$

The oscillator pulse frequency f_{osc} is simply $1/T$. Replace the period T in Equation 9-25 with $1/f_{osc}$ to get

$$L = \frac{2.5 V_{in}^2 (V_{out} - V_{in})}{f_{osc} I_{out} V_{out}^2} \tag{9-26}$$

This is the equation for designing the inductor in a step-up switch-mode power supply. The inductance value is in henries.

Next we must determine an equation for designing the capacitor. As for the step-down regulator, the change in voltage across the capacitor is given by Equation 9-12:

$$\Delta V = \frac{1}{C} \times \Delta Q \tag{9-12}$$

where the change in charge on the capacitor ΔQ is equal to the average charging current I_{avg} times the charging time t. The capacitor in this case supplies current when the transistor is switched on at an approximately constant rate of I_{out}. The change in charge $\Delta Q = I_{out} \times t_{on}$. Substituting this into Equation 9-12, we get

$$\Delta V = \frac{1}{C} \times I_{out} \times t_{on} \qquad (9\text{-}27)$$

Substitute for t_{on} from Equation 9-24 to get

$$\Delta V = \frac{1}{C} \times I_{out} \times \frac{V_{out} - V_{in}}{V_{out}} \times T \qquad (9\text{-}28)$$

Solve Equation 9-28 for C and substitute $1/f_{osc}$ for T:

$$C = \frac{(V_{out} - V_{in}) \times I_{out}}{(\Delta V) V_{in} f_{osc}} \qquad (9\text{-}29)$$

wrong

Converting ΔV to RMS ripple, we get the final design equation:

$$C = \frac{0.3536 I_{out}(V_{out} - V_{in})}{f_{osc} V_{ripple} V_{out}} \qquad (9\text{-}30)$$

Both the inductor and capacitor must be chosen subject to the constraints discussed earlier.

Voltage-Inverting Regulator

The third configuration that we can use for a switching power supply is the **voltage-inverting regulator,** or simply, **voltage inverter.** This circuit, shown in Figure 9-6, is capable of providing a negative output voltage for a positive supply. Compare the position of the inductor and transistor switch with the step-down and step-up circuits. Here we have the inductor shunting current to ground. As for the other regulators, we will assume that the pulse-width modulator properly switches the transistor without worrying about proper biasing.

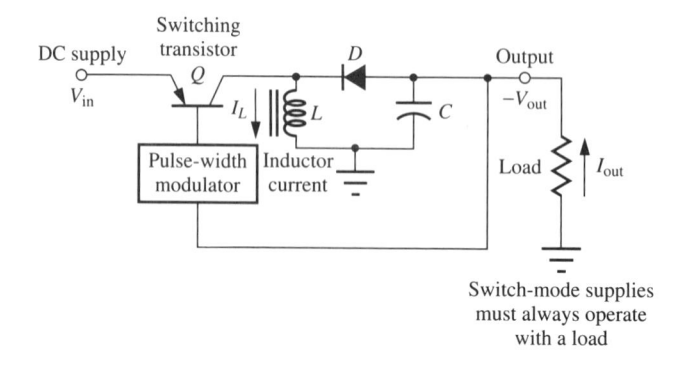

Figure 9-6 Voltage-inverting switch-mode regulator circuit.

Again, to describe the operation of this circuit, we shall assume a steady-state current is flowing. In this case, the output is negative, so the current is flowing from the load into the inductor. When the switching transistor turns on, the diode is reverse-biased, and current will attempt to flow from the supply through the inductor. Because the inductor opposes instantaneous change in current, the full supply voltage will appear across the inductor. The current through the inductor will start to increase. When the transistor switches off, the inductor voltage will change polarity to keep the current flowing, forward-biasing the diode and now allowing current to flow from the load and capacitor through the inductor to ground. Since charge is being withdrawn from the capacitor, the capacitor voltage will become negative and the net current flow will be into the regulator from the load. The voltage and current waveforms, although opposite in polarity, are identical to those shown in Figure 9-3(a) for the step-down regulator.

As for the previous types of regulators, we must develop design equations for the inductor and capacitor. To derive the equation for the inductor, we will use Equation 9-2:

$$\Delta I_L = \frac{V_L \, \Delta t}{L} \tag{9-2}$$

When the transistor switch is on, for $\Delta t = t_{on}$, the voltage across the inductor is V_{in}:

$$\Delta I_{Lon} = \frac{V_{in} \times t_{on}}{L} \tag{9-31}$$

When the transistor switch is off, for $\Delta t = t_{off}$, the voltage across the inductor is V_{out}:

$$\Delta I_{Loff} = \frac{V_{out} \times t_{off}}{L} \tag{9-32}$$

For steady-state operation, the change in current when the switch is off, ΔI_{Loff}, must be equal to the change in current when the switch is on, ΔI_{Lon}. Equating Equation 9-31 to Equation 9-32 and cancelling the inductance L common to each term, we get

$$V_{out} \times t_{off} = V_{in} \times t_{on} \tag{9-33}$$

Solve this equation for V_{in}/V_{out}:

$$\frac{V_{in}}{V_{out}} = \frac{t_{off}}{t_{on}} \tag{9-34}$$

Now add 1 to each side and rearrange this equation to solve for the duty cycle t_{on}/T:

$$\frac{t_{on}}{T} = \frac{V_{out}}{V_{in} + V_{out}} \tag{9-35}$$

As in the previous designs, assume a peak-to-peak inductor current ΔI_L that varies by 40% of the maximum output current I_{out}. Substitute $0.4I_{out}$ for ΔI in Equation 9-31 and solve for the inductance L:

$$L = \frac{V_{in} \times t_{on}}{0.4I_{out}} \qquad (9\text{-}36)$$

Now solve Equation 9-35 for t_{on}:

$$t_{on} = T \times \frac{V_{out}}{V_{in} + V_{out}} \qquad (9\text{-}37)$$

Substitute this into Equation 9-36:

$$L = \frac{V_{in} V_{out} T}{0.4I_{out}(V_{out} + V_{in})} \qquad (9\text{-}38)$$

The oscillator pulse frequency f_{osc} is simply $1/T$. Replace the period T in Equation 9-38 with $1/f_{osc}$ to get

$$L = \frac{2.5 V_{in} V_{out}}{I_{out}(V_{out} + V_{in})f_{osc}} \qquad (9\text{-}39)$$

This is the equation for designing the inductor in a voltage-inverting switch-mode power supply. The inductance is in henries.

Next we will develop an equation for designing the capacitor. As for the step-down and step-up regulators, the change in voltage across the capacitor is given by Equation 9-12:

$$\Delta V = \frac{1}{C} \times \Delta Q \qquad (9\text{-}12)$$

where the change in charge on the capacitor ΔQ is equal to the average charging current I_{avg} times the charging time t. Like the step-up regulator, the capacitor supplies current only when the transistor is switched on. The current I_{out} is approximately constant, so the change in charge is simply $\Delta Q = I_{out} \times t_{on}$. Substituting this into Equation 9-12, we get

$$\Delta V = \frac{1}{C} \times I_{out} \times t_{on} \qquad (9\text{-}40)$$

Substitute for t_{on} from Equation 9-37 to get

$$\Delta V = \frac{1}{C} \times I_{out} \times \frac{V_{out}}{V_{in} + V_{out}} \times T \qquad (9\text{-}41)$$

Solve Equation 9-41 for C and substitute $1/f_{osc}$ for T:

$$C = \frac{I_{out} \times V_{out}}{\Delta V(V_{in} + V_{out})f_{osc}} \qquad (9\text{-}42)$$

Converting ΔV to RMS ripple, we get the final design equation:

$$C = \frac{0.3536 I_{out} V_{out}}{V_{ripple} f_{osc}(V_{out} + V_{in})} \qquad (9\text{-}43)$$

Both the inductor and capacitor must be chosen subject to the constraints discussed earlier.

Multiple-Output Switch-Mode Regulator

There is actually another configuration that we can use for our switch-mode regulator components—a multiple-output regulator. The output from the transistor switch, although strictly not an AC waveform, is nevertheless varying with time and can be stepped up or down by a transformer. By using a transformer, we can generate as many output voltages as we want. A typical circuit is shown in Figure 9-7. The switching transistor and primary coil of the transformer are connected as a standard step-up regulator (see Figure 9-5), and the circuits on each secondary are step-down regulators. Note the extra diode in the secondary circuits. This is to prevent current from flowing back through the secondary of the transformer—although the primary current is switched DC, in the absence of a ground for the secondary we have true AC.

The design equations for the inductors and capacitors are identical to those developed for the step-down regulator. Only one output is fully regulated—the one connected to the error amplifier. If the load on this secondary changes, the pulse width from the pulse-width modulator will change to maintain a constant output voltage *for this secondary*, but, in so doing, will change the output voltages of the other secondaries. If this is a problem, then linear regulators can be used on all the unregulated secondaries, as shown in Figure 9-8.

The circuits in Figure 9-7 and Figure 9-8 have one additional element that must be designed—the transformer. Our basic design will simply consist of determining the turns ratio for each secondary. A detailed design procedure for toroidal transformers is described in Appendix B.

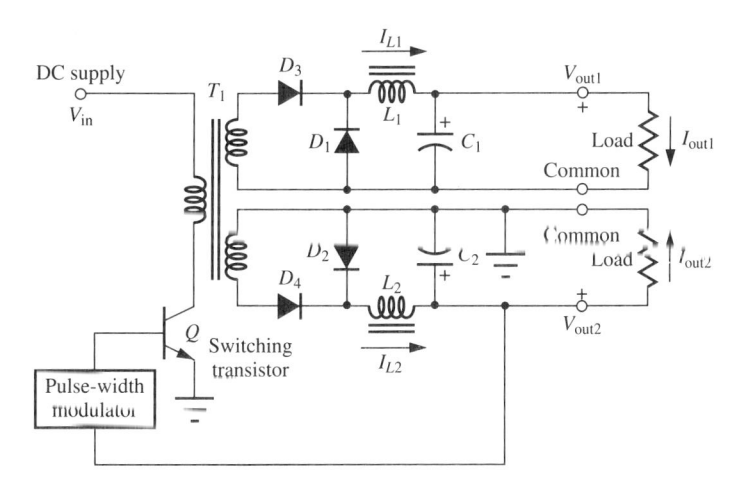

Figure 9-7 Multiple-output switch-mode regulator. Current is switched through the primary of a transformer. Each secondary can be used for a separate output.

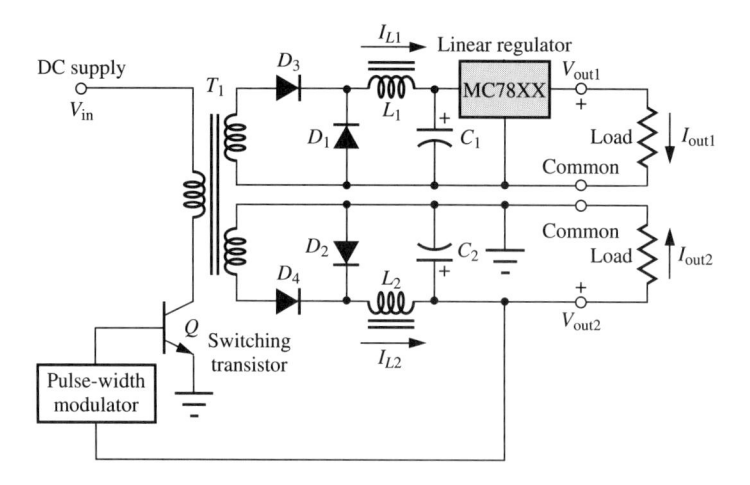

Figure 9-8 Double regulation using a linear regulator on the secondary of a multiple output switch-mode supply.

One secondary winding must be used by the error amplifier to sense the output voltage. Generally, the output winding that must provide the highest output current at a step-down voltage is used. This winding is made as a 1:1 winding, that is, the secondary has the same number of turns as the primary. For the other secondaries, turns ratios must be determined. Because of the high frequencies involved in switch-mode regulators, the transformer can be assumed to be operating almost ideally, so we use simple turns ratios to determine the output voltage:

$$\frac{N_{\text{sec}}}{N_{\text{pri}}} = \frac{V_{\text{out}}}{V_{\text{in}}} \tag{9-44}$$

Here, V_{out} is the output voltage of the secondary of the transformer being designed; it is not the output voltage of the regulator connected to this secondary. Because we will always use a step-down regulator on the secondary, it is only necessary to choose a voltage for V_{out} several volts larger than the desired regulated output. If we wish to employ double regulation using a linear regulator in conjunction with the switched regulator, then values several volts higher than the desired output should be used to allow for the dropout voltage of the linear regulator.

9-2 Pulse-Width Modulators

In the previous section, we discussed the various types of switch-mode regulator circuits without looking at specific devices. In this section, we will introduce a specific device, the LM3524 pulse-width modulator, and design actual working circuits.

National Semiconductor

Voltage Regulators

LM1524/LM2524/LM3524
Regulating Pulse Width Modulator

General Description

The LM1524 series of regulating pulse width modulators contains all of the control circuitry necessary to implement switching regulators of either polarity, transformer coupled DC to DC converters, transformerless polarity converters and voltage doublers, as well as other power control applications. This device includes a 5V voltage regulator capable of supplying up to 50 mA to external circuitry, a control amplifier, an oscillator, a pulse width modulator, a phase splitting flip-flop, dual alternating output switch transistors, and current limiting and shutdown circuitry. Both the regulator output transistor and each output switch are internally current limited and, to limit junction temperature, an internal thermal shutdown circuit is employed. The LM1524 is rated for operation from −55°C to +125°C and is packaged in a hermetic 16-lead DIP (J). The LM2524 and LM3524 are rated for operation from 0°C to +70°C and are packaged in either a hermetic 16-lead DIP (J) or a 16-lead molded DIP (N).

Features

- Complete PWM power control circuitry
- Frequency adjustable to greater than 100 kHz
- 2% frequency stability with temperature
- Total quiescent current less than 10 mA
- Dual alternating output switches for both push-pull or single-ended applications
- Current limit amplifier provides external component protection
- On-chip protection against excessive junction temperature and output current
- 5V, 50 mA linear regulator output available to user

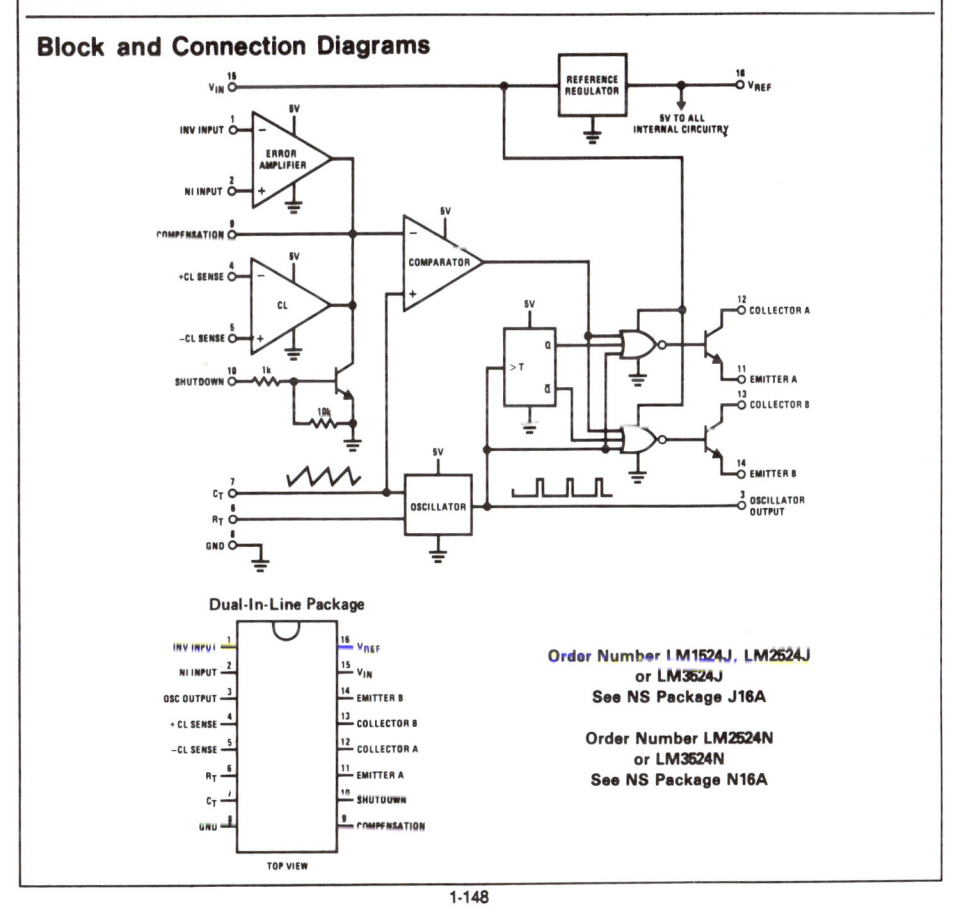

Block and Connection Diagrams

Order Number LM1524J, LM2524J or LM3524J
See NS Package J16A

Order Number LM2524N or LM3524N
See NS Package N16A

1-148

Figure 9-9 Data sheets for the LM3524 pulse-width modulator. *(Reprinted with permission of National Semiconductor Corporation)*

Absolute Maximum Ratings

Input Voltage	40V	Maximum Junction Temperature	
Reference Voltage, Forced	6V	(J Package)	150°C
Reference Output Current	50 mA	(N Package)	125°C
Output Current (Each Output)	100 mA	Storage Temperature Range	−65°C to +150°C
Oscillator Charging Current (Pin 6 or 7)	5 mA	Lead Temperature (Soldering, 10 seconds)	300°C
Internal Power Dissipation (Note 1)	1W		
Operating Temperature Range			
LM1524	−55°C to +125°C		
LM2524/LM3524	0°C to +70°C		

Electrical Characteristics

Unless otherwise stated, these specifications apply for $T_A = -55°C$ to $+125°C$ for the LM1524 and $0°C$ to $+70°C$ for the LM2524 and LM3524, $V_{IN} = 20V$, and $f = 20$ kHz. Typical values other than temperature coefficients, are at $T_A = 25°C$.

PARAMETER	CONDITIONS	LM1524/LM2524 MIN	TYP	MAX	LM3524 MIN	TYP	MAX	UNITS
Reference Section								
Output Voltage		4.8	5.0	5.2	4.6	5.0	5.4	V
Line Regulation	$V_{IN} = 8$–$40V$		10	20		10	30	mV
Load Regulation	$I_L = 0$–20 mA		20	50		20	50	mV
Ripple Rejection	$f = 120$ Hz, $T_A = 25°C$		66			66		dB
Short-Circuit Output Current	$V_{REF} = 0$, $T_A = 25°C$		100			100		mA
Temperature Stability	Over Operating Temperature Range		0.3	1		0.3	1	%
Long Term Stability	$T_A = 25°C$		20			20		mV/khr
Oscillator Section								
Maximum Frequency	$C_T = 0.001$ μF, $R_T = 2$ kΩ		350			350		kHz
Initial Accuracy	R_T and C_T constant		5			5		%
Frequency Change with Voltage	$V_{IN} = 8$–$40V$, $T_A = 25°C$			1			1	%
Frequency Change with Temperature	Over Operating Temperature Range			2			2	%
Output Amplitude (Pin 3)	$T_A = 25°C$		3.5			3.5		V
Output Pulse Width (Pin 3)	$C_T = 0.01$ μF, $T_A = 25°C$		0.5			0.5		μs
Error Amplifier Section								
Input Offset Voltage	$V_{CM} = 2.5V$		0.5	5		2	10	mV
Input Bias Current	$V_{CM} = 2.5V$		2	10		2	10	μA
Open Loop Voltage Gain		72	80		60	80		dB
Common-Mode Input Voltage Range	$T_A = 25°C$	1.8		3.4	1.8		3.4	V
Common-Mode Rejection Ratio	$T_A = 25°C$		70			70		dB
Small Signal Bandwidth	$A_V = 0$ dB, $T_A = 25°C$		3			3		MHz
Output Voltage Swing	$T_A = 25°C$	0.5		3.8	0.5		3.8	V
Comparator Section								
Maximum Duty Cycle	% Each Output ON	45			45			%
Input Threshold (Pin 9)	Zero Duty Cycle		1			1		V
Input Threshold (Pin 9)	Maximum Duty Cycle		3.5			3.5		V
Input Bias Current			−1			−1		μA
Current Limiting Section								
Sense Voltage	V(Pin 2) $- V$(Pin 1) ≥ 50 mV, Pin 9 = 2V, $T_A = 25°C$	190	200	210	180	200	220	mV
Sense Voltage T.C.			0.2			0.2		mV/°C
Common-Mode Voltage		−0.7		1	−0.7		1	V
Output Section (Each Output)								
Collector-Emitter Voltage		40			40			V
Collector Leakage Current	$V_{CE} = 40V$		0.1	50		0.1	50	μA
Saturation Voltage	$I_C = 50$ mA		1	2		1	2	V
Emitter Output Voltage	$V_{IN} = 20V$, $I_E = -250$ μA	17	18		17	18		V
Rise Time (10% to 90%)	$R_C = 2$ kΩ, $T_A = 25°C$		0.2			0.2		μs
Fall Time (90% to 10%)	$R_C = 2$ kΩ, $T_A = 25°C$		0.1			0.1		μs
Total Standby Current	$V_{IN} = 40V$, Pins 1, 4, 7, 8, 11 and 14 are grounded, Pin 2 = 2V All Other Inputs and Outputs Open		5	10		5	10	mA

Note 1: For operation at elevated temperatures, devices in the J package must be derated based on a thermal resistance of 100°C/W, junction to ambient, and devices in the N package must be derated based on a thermal resistance of 150°C/W junction to ambient.

Figure 9-9 (*Continued*)

9-2 Pulse-Width Modulators

Although the pulse-width modulator circuit is relatively complex, it can easily be integrated onto a single chip in much the same way that the linear op-amp regulator circuitry was integrated into a single package to produce the LM723C. There are a number of pulse-width modulator integrated circuits on the market, and we will use the National Semiconductor LM1524/LM2524/LM3524 pulse-width modulator as a representative example.

Data sheets for the LM1524/LM2524/LM3524 are shown in Figure 9-9. The designations LM1524, LM2524, and LM3524 all refer to the same integrated circuit, but with different temperature ratings or in slightly different package styles. We will only consider the LM3524. The LM3524 has an operating temperature range of 0 to +70°C and is packaged in a 16-lead molded DIP. Although the data sheets in Figure 9-9 include the pinout of the LM3524, it is redrawn as Figure 9-10(a), and Figure 9-10(b) shows the schematic symbol.

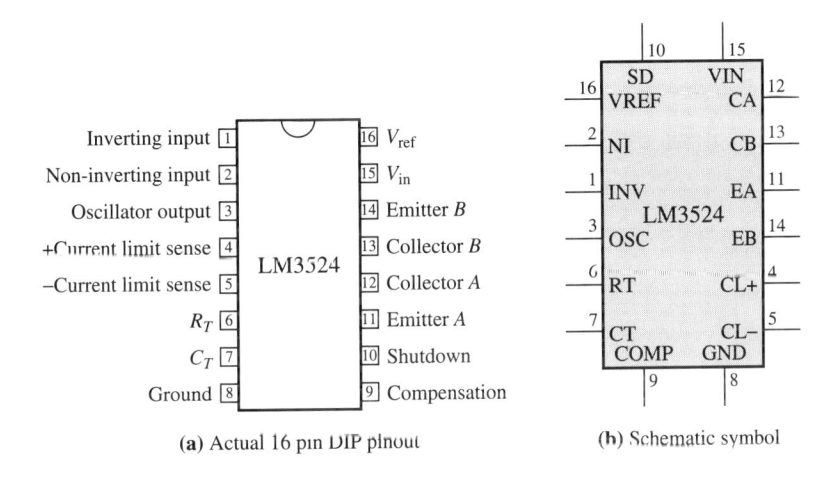

(a) Actual 16 pin DIP pinout

(b) Schematic symbol

Figure 9-10 Pinout of the LM3524 pulse-width modulator.

The LM3524 requires a supply voltage between 8 and 40 V, with pin 15 connected to the positive input and pin 8 to ground. Normally, it is powered from the DC supply being regulated. There are two output driver transistors with uncommitted collectors (pins 12 and 13) and uncommitted emitters (pins 11 and 14) which also must be supplied with power. These output transistors switch out of phase and may be connected in parallel. They are each capable of handling 100 mA, for a total of 200 mA when connected in parallel. For larger currents, the internal transistors can be used to drive external switching transistors.

The LM3524 can be used as a step-down regulator, as a step-up regulator, or as a voltage-inverting regulator, either as a stand-alone regulator or with external switching transistors. The basic configuration for the LM3524 pulse-width modulator is the same for all regulator circuits. We will use a step-down regulator to illustrate the design of this part of the circuit. The configuration of the inductor,

Basic Step-Down Regulator Using the LM3524

The simplest regulator circuit using the LM3524 is the basic step-down regulator in Figure 9-11. This is a low-current regulator using internal switching transistors only. Note that the two internal transistors are connected in parallel with both collectors (pins 12 and 13) connected to the DC supply and both emitters (pins 11 and 14) connected to the inductor. This regulator can provide a maximum output current of 200 mA.

A voltage divider consisting of resistors R_1 and R_2 is used to apply a feedback signal from the output of the power supply to pin 1, the inverting input of the error amplifier. The LM3524 will control the pulse width to hold this feedback voltage equal to a reference voltage V_{ref} applied to pin 2, the noninverting input of the error amplifier. Using the standard voltage divider equation, we see that

$$V_{out} \times \frac{R_2}{R_1 + R_2} = V_{ref} \tag{9-45}$$

Solving for V_{out}, we obtain the equation for designing V_{out}:

$$V_{out} = \frac{R_1 + R_2}{R_2} \times V_{ref} \tag{9-46}$$

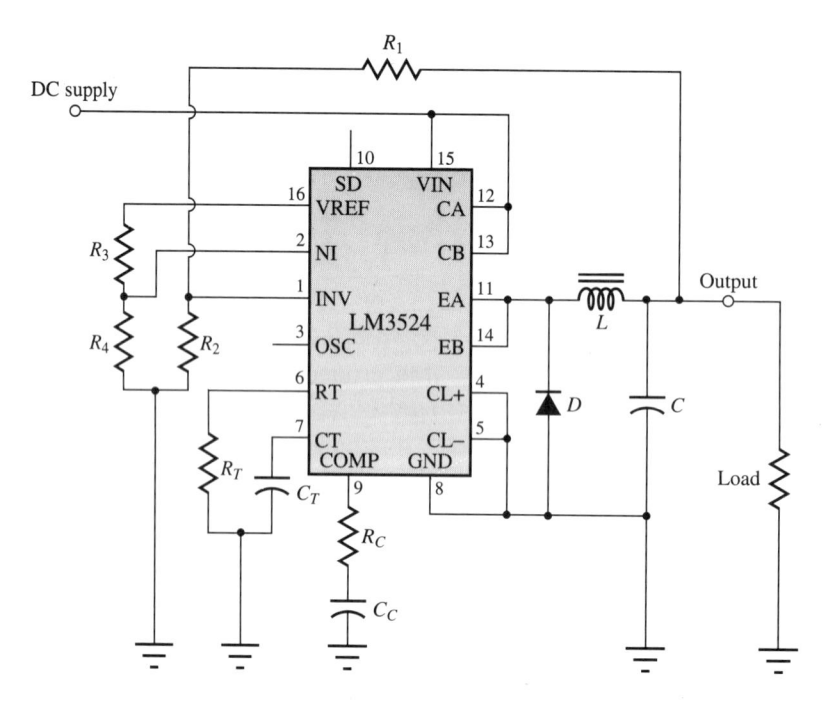

Figure 9-11 Step-down switching regulator using an LM3524 pulse-width modulator.

The reference voltage V_{ref} can be obtained from an external reference, or from the LM3524 internal 5 V reference, available from pin 16. Normally, we will use the internal reference, which we will call V_{int}. This must be connected externally to the noninverting input (pin 2) of the error amplifier by a voltage divider consisting of resistors R_3 and R_4. The effective reference voltage V_{ref} applied to pin 2 is

$$V_{ref} = \frac{R_4}{R_3 + R_4} \times V_{int} \qquad (9\text{-}47)$$

The internal reference requires a current path to ground for stability, consequently the voltage divider configuration should always be used. Usually, R_3 is chosen equal to R_4 to give a reference voltage of 2.5 V, and both resistors are normally selected as 10 kΩ. If any other reference voltage is to be used, choose R_4 as 10 kΩ and solve Equation 9-47 for R_3.

Since the error amplifier is essentially a bipolar op-amp, it should have equal bias currents into each input. To achieve this, make $R_1 \| R_2$ (the resistors connected to the inverting input) equal to $R_3 \| R_4$ (the resistors connected to the noninverting input). This gives the equation

$$\frac{R_1 R_2}{R_1 + R_2} = R_3 \| R_4 \qquad (9\text{-}48)$$

If we choose $R_3 = R_4 = 10$ kΩ, then $R_3 \| R_4 = 5$ kΩ. Solve Equation 9-48 for $(R_1 + R_2)/R_2$, substitute this into Equation 9-46, and solve for R_1:

$$R_1 = \frac{V_{out}}{V_{ref}} \times R_3 \| R_4 \qquad (9\text{-}49)$$

Finally, substitute the value of R_1 from Equation 9-49 back into Equation 9-48 to obtain R_2.

The switching pulses are provided by an internal oscillator whose frequency is determined by the combination of timing resistor R_T and timing capacitor C_T connected from pins 6 and 7, respectively, to ground. Typical switching frequencies for the LM3524 range from 10 to 100 kHz. The oscillator frequency is determined by the formula

$$f_{osc} = \frac{1}{R_T C_T} \qquad (9\text{-}50)$$

The value of the timing resistor should be around 10 kΩ. Suggested capacitor values range from 0.001 to 0.1 μF. The R_T and C_T values are not critical, and neither, generally, is the exact switching frequency. The oscillator output is available from pin 3 in the form of a pulse wave. This output can be used for driving other regulators as slaves to the master regulator, or it can be used for receiving driving pulses from a master regulator.

The LM3524 has provision for current limiting. A sensing resistor R_{sens} is connected in series with the output of the regulator, and the voltage drop across this resistor is connected to the internal current-limiting circuitry of the LM3524 through pins 4 and 5. Pin 4 is connected to the positive end of the sensing resistor, while pin 5 is connected to the negative end. The value of the sensing resistor is

$$R_{sens} = \frac{\text{sense voltage}}{I_{out(max)}} \quad (9\text{-}51)$$

The sense voltage is 200 mV. Current limiting is optional. If it is not used, the current-limiting terminals should be tied together and connected to ground.

Pin 9 is used for compensation. The recommended compensation network for all applications is a 47 kΩ resistor in series with a 0.001 μF capacitor, connected from pin 9 to ground.

There is one last pin on the LM3524 to consider. This is pin 10, labelled "shutdown." If this pin is connected to ground, then the switching regulator is effectively shut off. This provides for low-power control of the switching power supply.

The inductor and capacitor are calculated by Equations 9-11 and 9-16, respectively, as derived earlier for the basic step-down regulator discussed in Section 9-1. These equations are repeated here for the sake of completeness. The inductor is calculated from the equation

$$L = \frac{2.5 V_{out}(V_{in} - V_{out})}{I_{out} V_{in} f_{osc}} \quad (9\text{-}11)$$

Often, it is not possible to purchase a suitable inductor. In this case, it is necessary to design and construct the actual device. Detailed instructions for the design and construction of toroidal inductors are given in Appendix B.

The capacitor is calculated from the equation

$$C = \frac{0.01768 I_{out}}{V_{ripple} f_{osc}} \quad (9\text{-}16)$$

The capacitor should always be chosen larger than the calculated value.

EXAMPLE 9-1

Design a step-down switching regulator using an LM3524 to give an output voltage of 9 V at 200 mA from a 30 V supply. Use a switching frequency of 45 kHz. The output ripple should be less than 5.0 mV. Include current limiting.

Solution

We require a simple step-down regulator circuit similar to that illustrated in Figure 9-11. This circuit has been redrawn as Figure 9-12, and a parts list is included.

First, calculate the values for the timing capacitor C_T and timing resistor R_T, using Equation 9-50. The manufacturer recommends a value near 10 kΩ for R_T, so we will use it to calculate a value for C_T. Solve Equation 9-50 for C_T:

$$C_T = \frac{1}{f_{osc}R_T}$$

$$= \frac{1}{45 \text{ kHz} \times 10 \text{ k}\Omega} = 0.002222 \text{ }\mu\text{F}$$

Choose C_T as 0.0022 μF, a standard value from Appendix A-2.

Next, we will design the voltage divider consisting of R_1 and R_2. First we must select a reference voltage and determine values for R_3 and R_4. We will use a reference voltage of 2.5 V, and choose $R_3 = R_4 = 10$ kΩ. Now use Equation 9-49 to calculate R_1:

$$R_1 = \frac{V_{out}}{V_{ref}} \times R_3 \parallel R_4 = \frac{9.0 \text{ V}}{2.5 \text{ V}} \times 5.0 \text{ k}\Omega = 18 \text{ k}\Omega$$

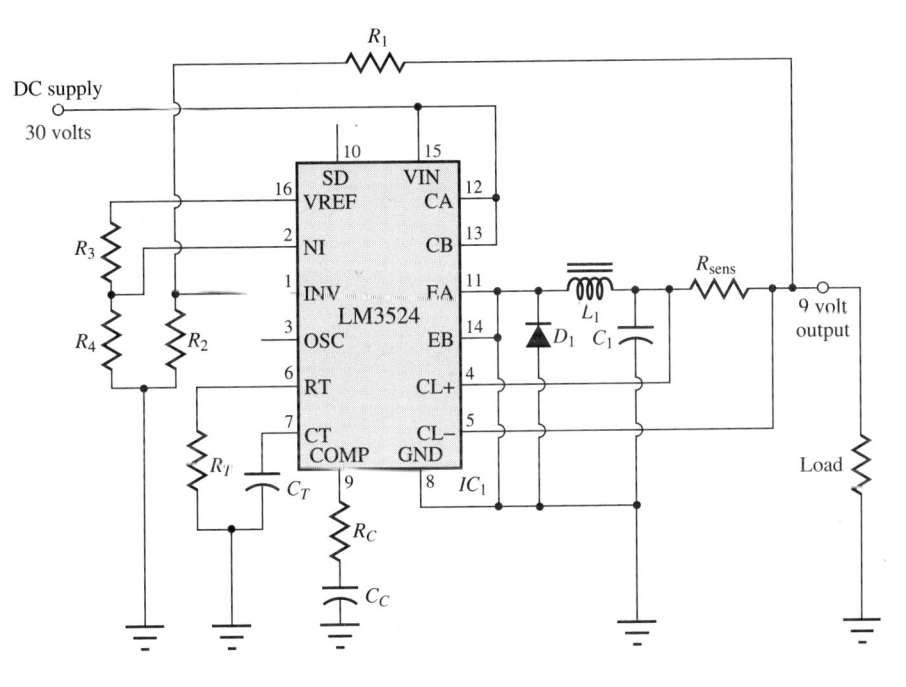

Parts List

Resistors:
R_1 18 kΩ
R_2 6.8 kΩ
R_3 10 kΩ
R_4 10 kΩ
R_T 10 kΩ
R_C 47 kΩ
R_{sens} 1.0 Ω

Inductors:
L_1 1.75 mH

Capacitors:
C_1 18 μF
C_I 0.0022 μF
C_C 0.001 μF

Semiconductors:
IC_1 LM3524

Figure 9-12

This is a standard value, so we will make $R_1 = 18$ kΩ. Substitute this value into Equation 9-48 to solve for R_2. The easiest procedure here is to note that

$$\frac{1}{R_1} + \frac{1}{R_2} = \frac{1}{5 \text{ k}\Omega} \quad \text{or} \quad \frac{1}{R_2} = \frac{1}{5 \text{ k}\Omega} - \frac{1}{18 \text{ k}\Omega}$$

From this, we calculate a value for R_2 of 6.92 kΩ. The closest standard value is 6.8 kΩ.

We will use the standard compensating network of $R_C = 47$ kΩ and $C_C = 0.001$ μF.

Because we want to use current limiting, we will have to design a sensing resistor R_{sens}. The maximum output current is to be 200 mA, and the sense voltage is 0.200 V. Use Equation 9-51:

$$R_{\text{sens}} = \frac{\text{sense voltage}}{I_{\text{out(max)}}} = \frac{0.20 \text{ V}}{0.20 \text{ A}} = 1.0 \text{ }\Omega$$

Since this is a standard value, we will make $R_{\text{sens}} = 1.0$ Ω.

Lastly, we must design the inductor and capacitor. Calculate the inductor from Equation 9-11:

$$L = \frac{2.5 V_{\text{out}}(V_{\text{in}} - V_{\text{out}})}{I_{\text{out}} V_{\text{in}} f_{\text{osc}}}$$

$$= \frac{2.5 \times 9.0 \text{ V} \times (30.0 \text{ V} - 9.0 \text{ V})}{200 \text{ mA} \times 30.0 \text{ V} \times 45 \text{ kHz}} = 1.75 \text{ mH}$$

A suitable inductor can be designed by using the procedures described in Appendix B, if necessary.

Next design the capacitor using Equation 9-16:

$$C = \frac{0.01768 I_{\text{out}}}{V_{\text{ripple}} f_{\text{osc}}} = \frac{0.01768 \times 200 \text{ mA}}{5 \text{ mV} \times 45 \text{ kHz}} = 15.7 \text{ }\mu\text{F}$$

Choose an 18 μF capacitor from Appendix A-2.

The design of the regulator is now complete.

Step-Down Regulator Using the LM3524 with External Transistor

Generally switch-mode regulators are designed to handle currents larger than 200 mA. In this case, an external switching transistor is used, as shown in Figure 9-13. Note that the switching transistor must be a negative-polarity PNP, and must be biased with resistors R_{B_1} and R_{B_2}. The internal transistors of the LM3524 act as switches, alternately connecting the base of the external switching transistor to ground and opening the circuit.

If the internal switch is open, no current flows through the biasing resistors, and the external transistor is off. If the internal switch is closed, the bottom end of resistor R_{B_2} is close to ground potential. Actually it will be V_{sat} volts above ground, typically between 0 and 1 V. The voltage across R_{B_1} and R_{B_2} is the DC

9-2 Pulse-Width Modulators

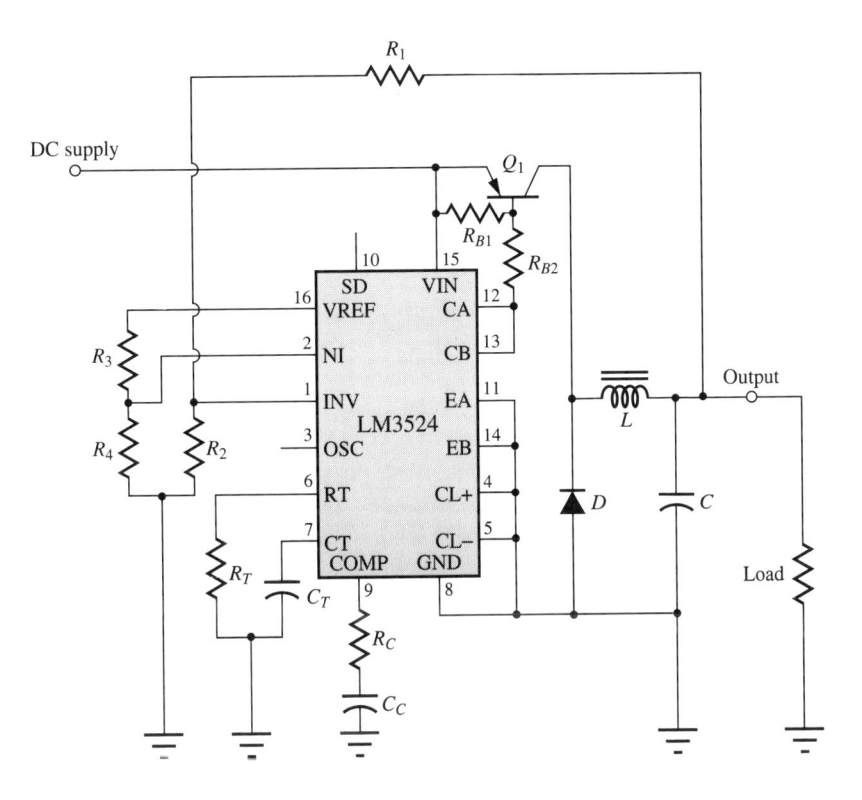

Figure 9-13 High-current step-down switching regulator using an LM3524 pulse-width modulator with external transistor.

supply voltage minus the transistor saturation voltage. When the internal switch is closed, current flows through the resistors which effectively form a voltage divider. The voltage at the base of the external switching transistor is 0.7 V less than the voltage at its emitter (the DC supply voltage), and hence the transistor switches on.

To design R_{B_1} and R_{B_2}, assume that a bleeder current I_{bleeder} flows through R_{B_1}. This same bleeder current plus the base current to the switching transistor flows through R_{B_2}. We can arbitrarily choose this bleeder current—typically a value about one-tenth of the transistor base current. To determine the transistor base current, we must use the maximum collector current (that is, the maximum current that will be delivered to the load) $I_{C\text{max}}$ and the minimum β of the transistor obtained from the data sheets, β_{min}. The equation for calculating the base current is

$$I_{B\text{min}} = \frac{I_{C\text{max}}}{\beta_{\text{min}}} \qquad (9\text{-}52)$$

The voltage drop across resistor R_{B_1} (from emitter to base of the switching transistor) must be 0.7 V (the transistor V_{BE}) to ensure that the transistor turns on. The current through R_{B_1} is $I_{bleeder}$. We calculate R_{B_1} from Ohm's law:

$$R_{B_1} = \frac{V_{BE}}{I_{bleeder}} \qquad (9\text{-}53)$$

Always select a value larger than this calculated value to ensure that there is at least 0.7 V between the base and emitter of the transistor.

The voltage drop across resistor R_{B_2} is the DC supply voltage minus the saturation voltage of the internal switching transistor V_{sat} minus the V_{BE} voltage drop across R_{B_1}. The current through R_{B_2} is I_{Bmin} plus $I_{bleeder}$. We calculate R_{B_2} from Ohm's law:

$$R_{B_2} = \frac{V_{DC\ supply} - V_{BE} - V_{sat}}{I_{bleeder} + I_{Bmin}} \qquad (9\text{-}54)$$

The maximum collector current that can be passed by either internal switching transistor is 100 mA. It is important to check that the current that would flow through the internal transistor when it is switched on, equal to the external transistor base current I_{Bmin} plus $I_{bleeder}$, does not exceed this value. If it does, it will probably be necessary to use a Darlington transistor for the external switching transistor.

When a single external switching transistor is used, as in Figure 9-13, the collectors of the internal switching transistors are normally connected together. The internal transistors switch 180° out of phase, each at half the frequency f_{osc} determined by Equation 9-49. When they are connected in parallel, the net switching frequency is f_{osc}.

EXAMPLE 9-2

Design a step-down switching regulator using an LM3524 to give an output voltage of 9 V at 5 A from a 30 V supply. Use a switching frequency of 45 kHz. The output ripple should be less than 5.0 mV. Include current limiting. Assume that the switching transistor has a minimum β of 60.

Solution

This is the same problem solved in Example 9-1, except that now we want an output current of 5.0 A and must use an external switching transistor. We will use the step-down regulator circuit shown in Figure 9-13. This circuit is redrawn as Figure 9-14, and a parts list is included.

The design of the LM3524 regulator portion of this circuit is identical to Example 9-1. We calculated values for the timing capacitor C_T and timing resistor R_T of 0.0022 μF and 10 kΩ, respectively. These will be unchanged in this circuit. We will use a reference voltage of 2.5 V, so we make $R_3 = R_4 = 10$ kΩ. The feedback voltage divider consisting of R_1 and R_2 will also be the same as in Example 9-1, namely $R_1 = 18$ kΩ and $R_2 = 6.8$ kΩ. We will use the standard compensating network with $R_C = 47$ kΩ and $C_C = 0.001$ μF.

9-2 Pulse-Width Modulators

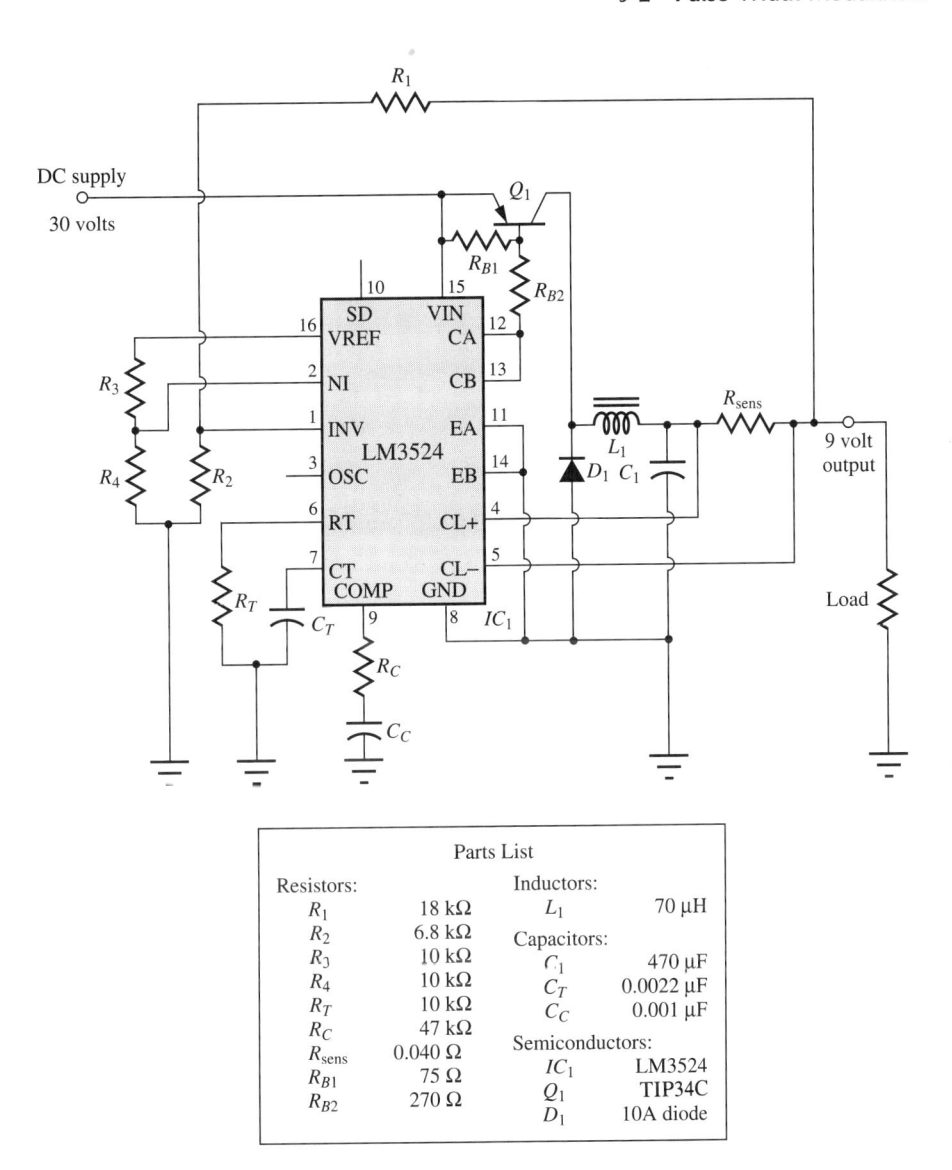

Figure 9-14

Parts List			
Resistors:		Inductors:	
R_1	18 kΩ	L_1	70 µH
R_2	6.8 kΩ	Capacitors:	
R_3	10 kΩ	C_1	470 µF
R_4	10 kΩ	C_T	0.0022 µF
R_T	10 kΩ	C_C	0.001 µF
R_C	47 kΩ	Semiconductors:	
R_{sens}	0.040 Ω	IC_1	LM3524
R_{B1}	75 Ω	Q_1	TIP34C
R_{B2}	270 Ω	D_1	10A diode

The sensing resistor R_{sens} for current limiting will not be the same as in Example 9-1 because we now have a current of 5 A. Use Equation 9-51 to calculate R_{sens}:

$$R_{sens} = \frac{\text{sense voltage}}{I_{out(max)}}$$

$$= \frac{200 \text{ mV}}{5.0 \text{ A}} = 0.040 \text{ Ω}$$

This resistor will have to be custom-wound from resistance wire on a standard form.

Next, calculate the value of the biasing resistors R_{B_1} and R_{B_2}. Since the current is 5 A and the minimum β of the transistor is 60, the minimum base current from Equation 9-52 is

$$I_{Bmin} = \frac{I_{Cmax}}{\beta_{min}} = \frac{5.0 \text{ A}}{60} = 0.0833 \text{ A}$$

The bleeder current should be approximately one-tenth of I_{Bmin}, so we will choose a bleeder current of 10 mA. Calculate R_{B_1} from Equation 9-53:

$$R_{B_1} = \frac{V_{BE}}{I_{bleeder}} = \frac{0.7 \text{ V}}{10 \text{ mA}} = 70 \text{ }\Omega$$

To ensure that the transistor always turns on, choose a *larger* standard value. From the table in Appendix A-1, choose 75 Ω.

We will assume the saturation voltage V_{sat} of the internal switching transistor is 1.0 V and the base-emitter voltage V_{BE} for the external transistor is 0.7 V. Calculate R_{B_2} from Equation 9-54:

$$R_{B_2} = \frac{V_{DC\ supply} - V_{BE} - V_{sat}}{I_{bleeder} + I_{Bmin}} = \frac{30 \text{ V} - 0.7 \text{ V} - 1.0 \text{ V}}{10 \text{ mA} + 83.3 \text{ mA}} = 303 \text{ }\Omega$$

To ensure that the full current can flow in this resistor, we must choose a smaller value. From Appendix A-1, we will choose a 270 Ω resistor. Because there will be 28.3 V across the resistor and a current of 0.0833 A through it, we should check the power rating. The power is 28.3 V × 0.0833 A = 2.36 W. Even though this power is not continuous, we still should select a 5 W resistor.

Lastly, we must design the inductor and capacitor. Use Equation 9-11 to design the inductor:

$$L = \frac{2.5 V_{out}(V_{in} - V_{out})}{I_{out} V_{in} f_{osc}}$$

$$= \frac{2.5 \times 9.0 \text{ V} \times (30.0 \text{ V} - 9.0 \text{ V})}{5.00 \text{ A} \times 30.0 \text{ V} \times 45 \text{ kHz}} = 70.0 \text{ }\mu\text{H}$$

Use Equation 9-16 to design the capacitor:

$$C = \frac{0.01768 I_{out}}{V_{ripple} f_{osc}}$$

$$= \frac{0.01768 \times 5.0 \text{ A}}{0.005 \text{ V} \times 45 \text{ kHz}} = 393 \text{ }\mu\text{F}$$

Select a 470 μF capacitor from Appendix A-2.

Our design is now complete.

Step-Up Regulators Using the LM3524

The LM3524 can also be used to make a step-up regulator. The basic circuit is shown in Figure 9-15. This is a low-current regulator using the internal switching transistors only. Note that the connections to the LM3524 in this step-up regulator circuit are identical to the step-down regulator circuit discussed earlier. The design procedure is identical.

Since we have a step-up regulator, however, the connections to the inductor, diode, and capacitor are different. The inductor is connected from the DC supply to the collectors of the internal transistors. The catch diode is connected to the emitters of the internal transistors. The inductor is calculated from Equation 9-26:

$$L = \frac{2.5 V_{in}^2 (V_{out} - V_{in})}{f_{osc} I_{out} V_{out}^2} \qquad (9\text{-}26)$$

The capacitor is calculated from Equation 9-30:

$$C = \frac{0.3536 I_{out} (V_{out} - V_{in})}{f_{osc} V_{ripple} V_{out}} \qquad (9\text{-}30)$$

Both of these equations were derived in the previous section.

If an output current larger than the 200 mA limit of the basic step-up regulator is required, an external switching transistor can be added as shown in Figure 9-

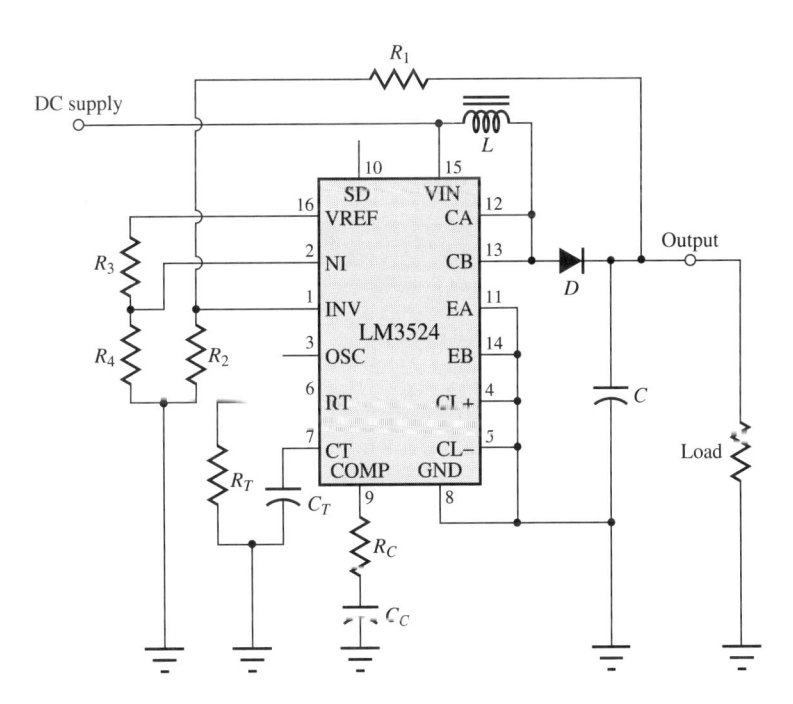

Figure 9-15 Step-up switching regulator using an LM3524 pulse-width modulator.

16. Although the basic regulator circuit using the LM3524 remains unchanged, the external switching circuit is quite different from that of a step-down regulator and will require some explanation.

The external switching circuit requires two positive-polarity NPN transistors. Transistor Q_1 acts as an inverter and is biased with resistors R_{B_1} and R_{B_2}, separated with a diode. Transistor Q_2 is the actual switching transistor and is biased with resistor R_{B_3}. The internal switching transistors are connected in parallel, with their collectors connected to the DC supply by resistor R_{B_1} and their emitters connected to ground. To design the biasing resistors, first consider the switching transistor bias resistor R_{B_3}. This resistor limits the base current flowing into Q_2 but must be able to pass sufficient current to ensure that Q_2 is fully saturated when switched on. If I_{C2max} is the maximum collector current that would ever flow through Q_2, and β_{2min} is the minimum β of this transistor, then $I_{R_{B3}}$, the minimum current that must flow through the biasing resistor R_{B_3}, is

$$I_{R_{B3}} = I_{B2min} = \frac{I_{C2max}}{\beta_{2min}} \tag{9-55}$$

This base current will always ensure that transistor Q_2 is saturated when it is turned on.

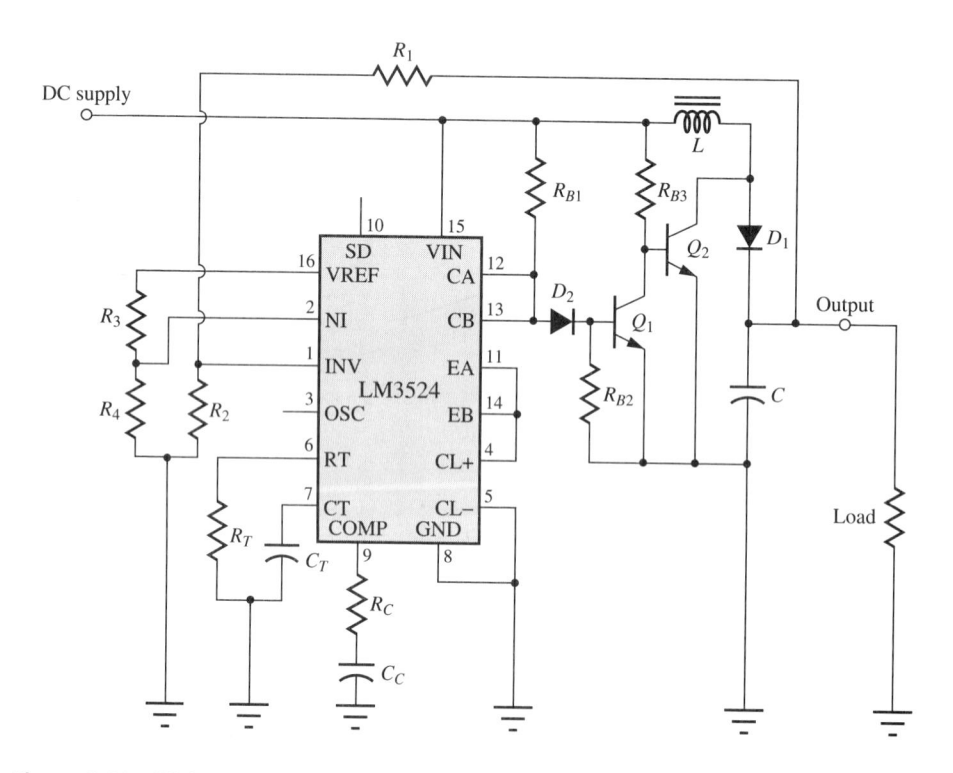

Figure 9-16 High-current step-up switching regulator using an LM3524 pulse-width modulator with an external switching transistor.

9-2 Pulse-Width Modulators

The voltage drop across R_{B_3} is the DC supply voltage minus the base-emitter junction voltage of Q_2 (typically 0.7 V).

$$V_{R_{B_3}} = V_{\text{DC supply}} - V_{BE_2} \tag{9-56}$$

This allows us to calculate a value for R_{B_3} by Ohm's law:

$$R_{B_3} = \frac{V_{\text{DC supply}} - V_{BE_2}}{I_{C_2\text{max}}} \times \beta_{2\text{min}} \tag{9-57}$$

This equation gives the maximum value of R_{B_3}. The actual value chosen must always be less than this to ensure that sufficient current will flow to saturate the transistor.

Next consider transistor Q_1. The external switching transistor must turn on when the internal transistors switch on. Without Q_1, the external transistor would switch *off* when the internal transistors switch on. With this inverter circuit, Q_1 switches off when the internal transistors switch on, in turn causing Q_2 to switch on as required.

The inverter transistor Q_1 is biased with resistors R_{B_1} and R_{B_2}. Resistor R_{B_2} is connected from the base of Q_1 to ground to provide a path for the charge stored in the base of Q_1 to flow to ground when the transistor is switched off. Typically a value of 1.0 kΩ is chosen for this resistor. We shall define R_{B_2} as

$$R_{B_2} = 1.0 \text{ k}\Omega \tag{9-58}$$

The other biasing resistor R_{B_1} limits the base current flowing into Q_1, yet must pass sufficient base current to ensure that Q_1 is always saturated when switched on. The maximum collector current of Q_1 is equal to the base current of Q_2 as given by Equation 9-55. If $\beta_{1\text{min}}$ is the minimum β of Q_1, the minimum base current for Q_1 is

$$I_{B_1\text{min}} = \frac{I_{C_1\text{max}}}{\beta_{1\text{min}}} = \frac{I_{C_2\text{max}}}{\beta_{1\text{min}}\beta_{2\text{min}}} \tag{9-59}$$

The current through R_{B_1} is the transistor base current $I_{B_1\text{min}}$ plus the current flowing through R_{B_2}. Since R_{B_2} is connected between the base and emitter of Q_1, it has a voltage drop of V_{BE_1} (approximately 0.7 V) across it. The current through resistor R_{B_2} is given by Ohm's law as

$$I_{R_{B_2}} = \frac{V_{BE_1}}{R_{B_2}} \tag{9-60}$$

The current through R_{B_1} is

$$I_{R_{B_1}} = I_{B_{Q_1}} + I_{R_{B_2}} = \frac{I_{C_2\text{max}}}{\beta_{1\text{min}}\beta_{2\text{min}}} + \frac{V_{BE_1}}{R_{B_2}} \tag{9-61}$$

The voltage across R_{B_1} is the DC supply voltage minus the forward voltage drop across the diode minus the base-emitter voltage drop of Q_1:

$$V_{R_{B_1}} = V_{\text{DC supply}} - V_D - V_{BE_1} \tag{9-62}$$

418   9 Switch-Mode Power Supplies

By Ohm's law, the value of R_{B_1} is

$$R_{B_1} = \frac{V_{\text{DC supply}} - V_D - V_{BE_1}}{\dfrac{I_{C_2\text{max}}}{\beta_{1\text{min}}\beta_{2\text{min}}} + \dfrac{V_{BE_1}}{R_{B_2}}} \tag{9-63}$$

This value of R_{B_1} is a maximum value. Always select a smaller value of R_{B_1}. Note that R_{B_1} must be large enough to limit the current through the internal switching transistors to less than 100 mA. This should not be a problem.

Diode D_2 is required to ensure that transistor Q_1 is switched off when either of the internal switching transistors are conducting. The internal transistors can have a value of V_{sat} as large as 1 V. Without the diode, if V_{sat} were greater than 0.7 V, transistor Q_1 would remain conducting at all times, and not switch off. The addition of the diode imposes a second 0.7 V drop. Now V_{sat} has to be larger than 1.4 V to interfere with the switching. Transistor Q_2 must be a power transistor capable of passing full-load current. Transistor Q_1 can be a considerably smaller transistor, since it must pass only the base current of Q_2.

The inductor and capacitor for this regulator are calculated by Equations 9-26 and 9-30, exactly as for the basic step-up regulator.

EXAMPLE 9-3

Design a step-up switching regulator using an LM3524 pulse-width modulator to give an output voltage of 36 V at 2.5 A from a 12.0 V supply. Use a switching frequency of 50 kHz, and make the output ripple less than 20 mV RMS. Current limiting is not required. Assume that the switching transistor Q_2 has a minimum β of 20 and that the inverting transistor Q_1 has a minimum β of 100.

Solution

Since we require a high-current step-up regulator, we must use a circuit similar to that shown in Figure 9-16. This circuit has been redrawn as Figure 9-17, and a parts list is included.

The switching frequency is determined by Equation 9-50. Use a value of 10 kΩ for R_T, as recommended by the manufacturer, and solve Equation 9-50 for C_T:

$$C_T = \frac{1}{f_{\text{osc}} R_T} = \frac{1}{50\ \text{kHz} \times 10\ \text{k}\Omega} = 0.00200\ \mu\text{F}$$

Choose C_T as 0.0022 μF from Appendix A-2.

Choose $R_3 = R_4 = 10$ kΩ to set the reference voltage V_{ref} at pin 2 to 2.5 V. Now design the feedback voltage divider. Use Equation 9-49 to calculate R_1:

$$R_1 = \frac{V_{\text{out}}}{V_{\text{ref}}} \times R_3 \parallel R_4$$

$$= \frac{36\ \text{V}}{2.5\ \text{V}} \times 5.0\ \text{k}\Omega = 72.0\ \text{k}\Omega$$

9-2 Pulse-Width Modulators

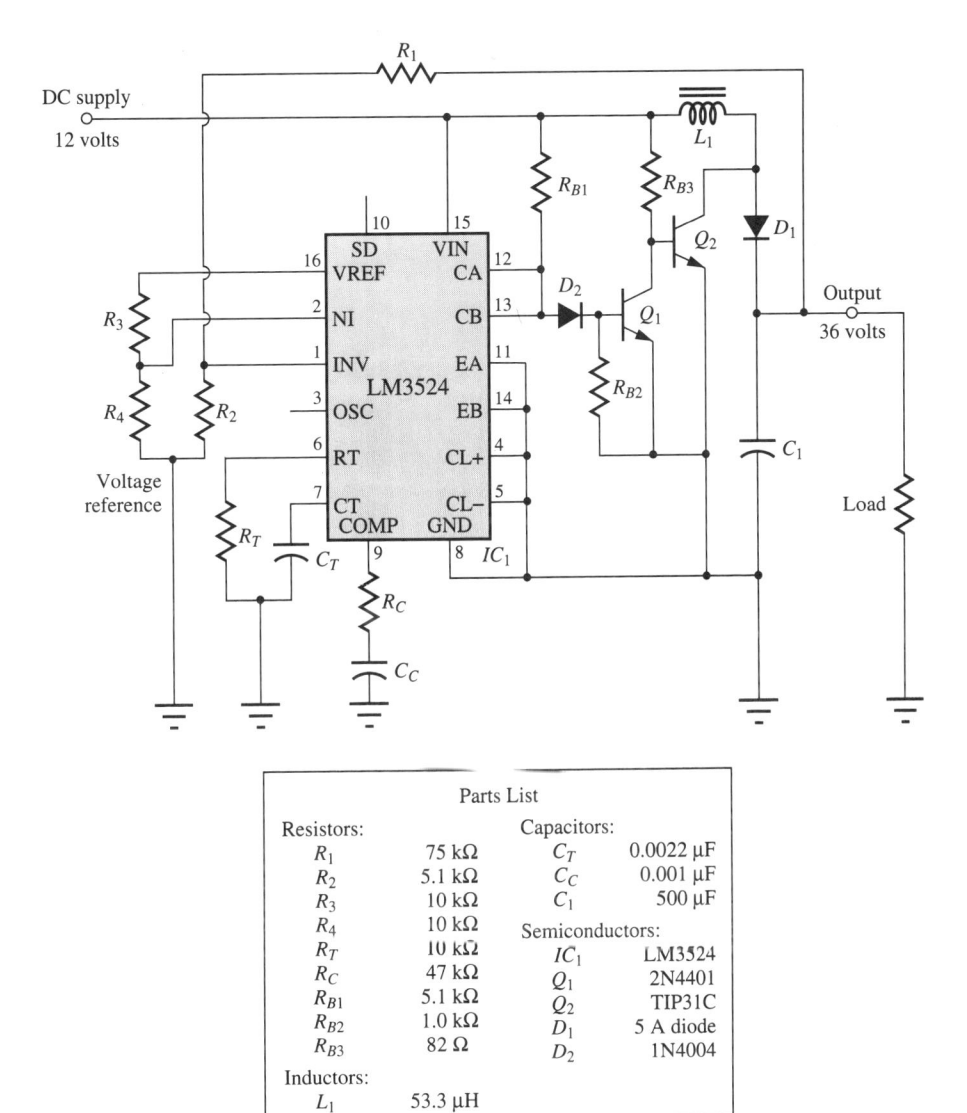

Figure 9-17

Parts List			
Resistors:		Capacitors:	
R_1	75 kΩ	C_T	0.0022 µF
R_2	5.1 kΩ	C_C	0.001 µF
R_3	10 kΩ	C_1	500 µF
R_4	10 kΩ		
R_T	10 kΩ	Semiconductors:	
R_C	47 kΩ	IC_1	LM3524
R_{B1}	5.1 kΩ	Q_1	2N4401
R_{B2}	1.0 kΩ	Q_2	TIP31C
R_{B3}	82 Ω	D_1	5 A diode
Inductors:		D_2	1N4004
L_1	53.3 µH		

The closest standard value is 75 kΩ. We will use this value. Now solve for R_2 as in Example 9-1:

$$\frac{1}{R_1} + \frac{1}{R_2} = \frac{1}{5 \text{ k}\Omega} \quad \text{or} \quad \frac{1}{R_2} = \frac{1}{5 \text{ k}\Omega} - \frac{1}{75 \text{ k}\Omega}$$

From this, we calculate a value for R_2 of 5.36 kΩ. The closest standard value is 5.1 kΩ.

Use the standard compensating network with $R_C = 47$ kΩ and $C_C = 0.001$ µF.

Next, calculate the values of the biasing resistors R_{B_1}, R_{B_2}, and R_{B_3}. First use Equation 9-57 to calculate bias resistor R_{B_3}:

$$R_{B_3} = \frac{V_{DC\ supply} - V_{BE_2}}{I_{C_2 max}} \times \beta_{2min}$$

$$= \frac{12.0\ V - 0.7\ V}{2.5\ A} \times 20 = 90.4\ \Omega$$

Choose the smaller standard value of 82 Ω from Appendix A-1. Because there is a relatively large voltage across this resistor, we should check the power. With 11.3 V across an 82 Ω resistor, the power is 1.56 W. We should use at least a 2 W resistor.

Bias resistor R_{B_2} is calculated from Equation 9-58:

$$R_{B_2} = 1.0\ k\Omega$$

This is a standard value. Since the resistor has only 0.7 V across it, a $\frac{1}{2}$ W resistor will be satisfactory.

Finally, use Equation 9-63 to calculate R_{B_1}.

$$R_{B_1} = \frac{V_{DC\ supply} - V_D - V_{BE_1}}{I_{C_2 max}/\beta_{1min}\beta_{2min} + V_{BE_1}/R_{B_2}}$$

$$= \frac{12.0\ V - 0.7\ V - 0.7\ V}{2.5\ A/(100 \times 20) + 0.7\ V/1000\ \Omega} = \frac{10.6\ V}{0.00125\ A + 0.0007\ A} = 5.44\ k\Omega$$

Choose a standard 5.1 kΩ resistor from Appendix A-1. Because we have a relatively large voltage across this resistor, we should check the power. There are 10.6 V across a 5.1 kΩ resistor, so the resistor power is 0.0239 W. A standard $\frac{1}{2}$ W resistor will be satisfactory.

Now calculate the inductor and capacitor. From Equation 9-26:

$$L = \frac{2.5\ V_{in}^2 (V_{out} - V_{in})}{f_{osc} I_{out} V_{out}^2}$$

$$= \frac{2.5 \times (12\ V)^2 \times (36.0\ V - 12.0\ V)}{50\ kHz \times 2.5\ A \times (36\ V)^2} = 53.3\ \mu H$$

From Equation 9-30,

$$C = \frac{0.3536 I_{out}(V_{out} - V_{in})}{f_{osc} V_{ripple} V_{out}}$$

$$= \frac{0.3536 \times 2.0\ A \times (36.0\ V - 12.0\ V)}{50\ kHz \times 20\ mA \times 36.0\ V} = 471\ \mu F$$

Choose a 500 µF capacitor from Appendix A-2.

Our design is complete.

Voltage-Inverter Regulators Using the LM3524

The circuit for a low-current voltage-inverting regulator using the LM3524 is shown in Figure 9-18. Notice that the feedback through R_1 is connected to the *noninverting input* of the error amplifier, and the reference voltage through R_3 is connected to the *inverting input*. This is necessary because the output voltage is negative. Otherwise, the connections to the LM3524 in this voltage-inverting regulator circuit are the same as the step-down and step-up regulators discussed earlier, and the design procedure is identical.

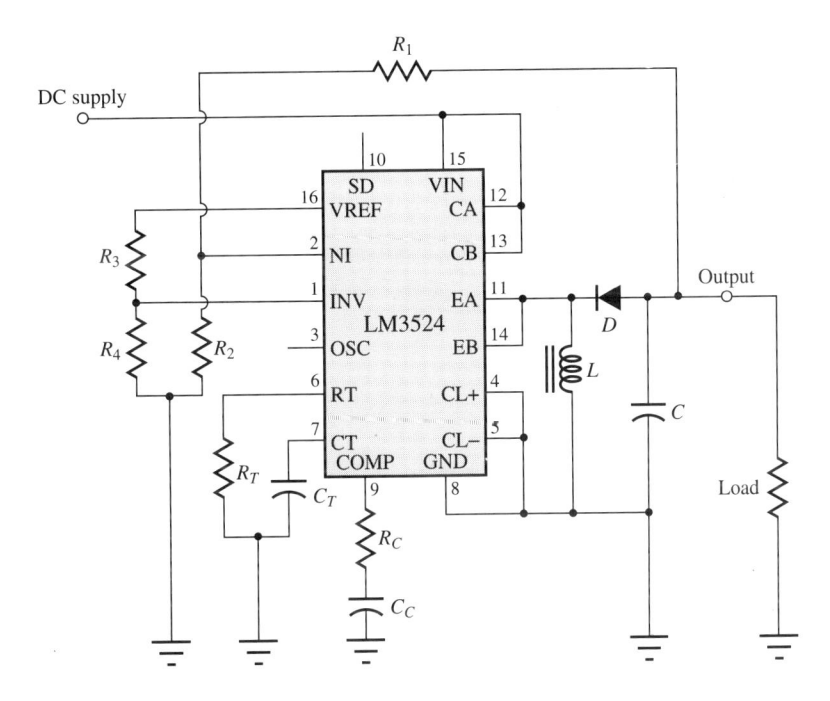

Figure 9-18 Voltage-inverting regulator using an LM3524 pulse-width modulator.

Since we have a voltage-inverting regulator, however, the connections to the inductor, diode, and capacitor are different. The inductor is connected from the emitters of the internal transistors to ground. The catch diode is connected with the negative end to the emitters of the internal transistors. The capacitor, as in all types of regulators, is connected directly across the output.

The inductor and capacitor values for an inverting regulator are calculated by Equations 9-39 and 9-43, respectively, as derived in Section 9-1. The inductor is calculated from the equation

$$L = \frac{2.5 V_{in} V_{out}}{f_{osc}(V_{out} + V_{in})I_{out}} \tag{9-39}$$

The capacitor is calculated from the equation

$$C = \frac{0.3536 I_{\text{out}} V_{\text{out}}}{V_{\text{ripple}} f_{\text{osc}}(V_{\text{out}} + V_{\text{in}})} \qquad (9\text{-}43)$$

If an output current larger than the 200 mA limit of the inverting regulator shown in Figure 9-18 is required, an external switching transistor can be added, as shown in Figure 9-19. The biasing circuit for the external switching transistor is similar to that used for a step-down regulator as shown in Figure 9-13. A PNP transistor must be used and must be biased with resistors R_{B_1} and R_{B_2}. These resistors are calculated in exactly the same manner as for the step-down regulator, namely by using Equations 9-53 and 9-54. Biasing resistor R_{B_1} is calculated as

$$R_{B_1} = \frac{V_{BE}}{I_{\text{bleeder}}} \qquad (9\text{-}53)$$

Biasing resistor R_{B_2} is calculated as

$$R_{B_2} = \frac{V_{\text{DC supply}} - V_{BE} - V_{\text{sat}}}{I_{\text{bleeder}} + I_{B\text{min}}} \qquad (9\text{-}54)$$

The rest of the circuit is designed in the same way as for a simple regulator.

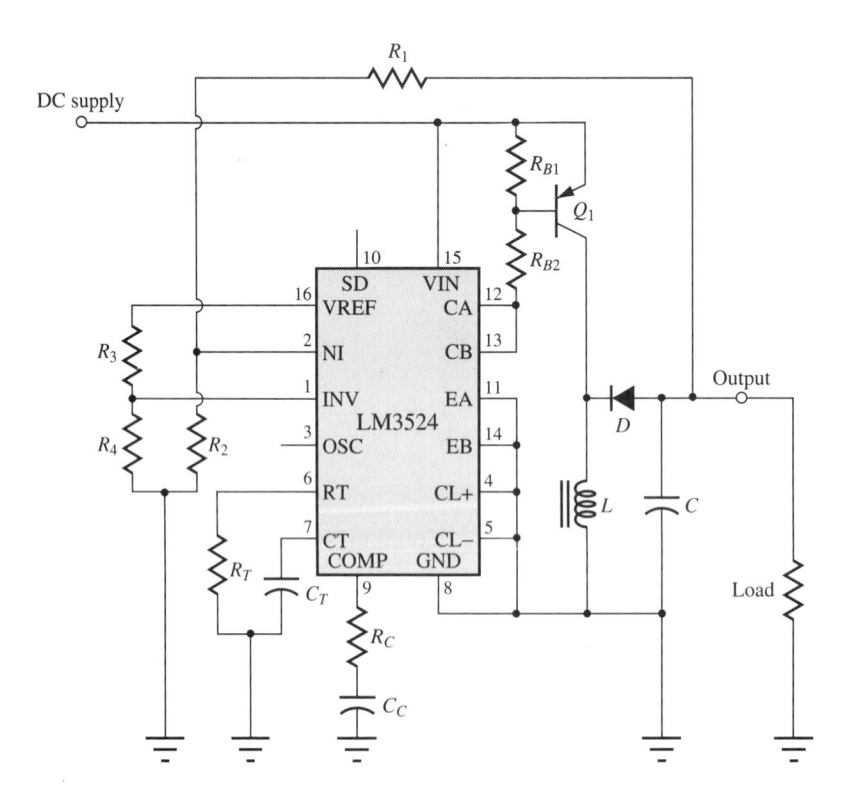

Figure 9-19 High-current voltage-inverting regulator using an LM3524 pulse-width modulator with external transistor.

EXAMPLE 9-4

Design a voltage-inverting switch-mode regulator using an LM3524 to give an output voltage of −5 V at 3 A from a +5 V supply. Use a switching frequency of 30 kHz. The output ripple should be less than 5.0 mV. Include current limiting. Assume that the switching transistor has a minimum β of 60. Operate the LM3524 from a separate 12 V supply.

Solution

In Chapter 8, we introduced the ICL7660 voltage converter to convert a +5 V supply to −5 V. In this example, however, we require a large current, far beyond the capabilities of the ICL7660, so we must use a voltage-inverting switch-mode regulator. The circuit is shown in Figure 9-20, with a parts list included. Note that we must power the LM3524 from a separate supply since it requires a minimum of 8 V to operate.

The switching frequency is determined by Equation 9-50. Use a value of 10 kΩ for R_T, as recommended by the manufacturer, and solve Equation 9-50 for C_T:

$$C_T = \frac{1}{f_{osc} R_T} = \frac{1}{30 \text{ kHz} \times 10 \text{ k}\Omega} = 0.00333 \text{ μF}$$

Choose C_T as 0.0033 μF from Appendix A-2.

Choose $R_3 = R_4 = 10$ kΩ to set the reference voltage V_{ref} at pin 1 to 2.5 V (note that the reference goes to the inverting input in this regulator). Now design the feedback voltage divider. Use Equation 9-49 to calculate R_1:

$$R_1 = \frac{V_{out}}{V_{ref}} \times R_3 \parallel R_4 = \frac{5.0 \text{ V}}{2.5 \text{ V}} \times 5.0 \text{ k}\Omega = 10 \text{ k}\Omega$$

This is a standard value. Now solve for R_2 as in Example 9-1.

$$\frac{1}{R_1} + \frac{1}{R_2} = \frac{1}{5 \text{ k}\Omega} \quad \text{or} \quad \frac{1}{R_2} = \frac{1}{5 \text{ k}\Omega} - \frac{1}{10 \text{ k}\Omega}$$

From this, we calculate R_2 as 10 kΩ, a standard value.

We will use the standard compensating network with $R_C = 47$ kΩ and $C_C = 0.001$ μF.

Use Equation 9-51 to calculate R_{sens}:

$$R_{sens} = \frac{\text{sense voltage}}{I_{out(max)}} = \frac{200 \text{ mV}}{3.0 \text{ A}} = 0.0667 \text{ }\Omega$$

As in the previous examples, this resistor will have to be custom-wound from resistance wire on a standard form.

Next, calculate the value of the biasing resistors R_{B_1} and R_{B_2}. Since the current is 3 A and the minimum β of the transistor is 60, the minimum base current from

424   **9 Switch-Mode Power Supplies**

Figure 9-20

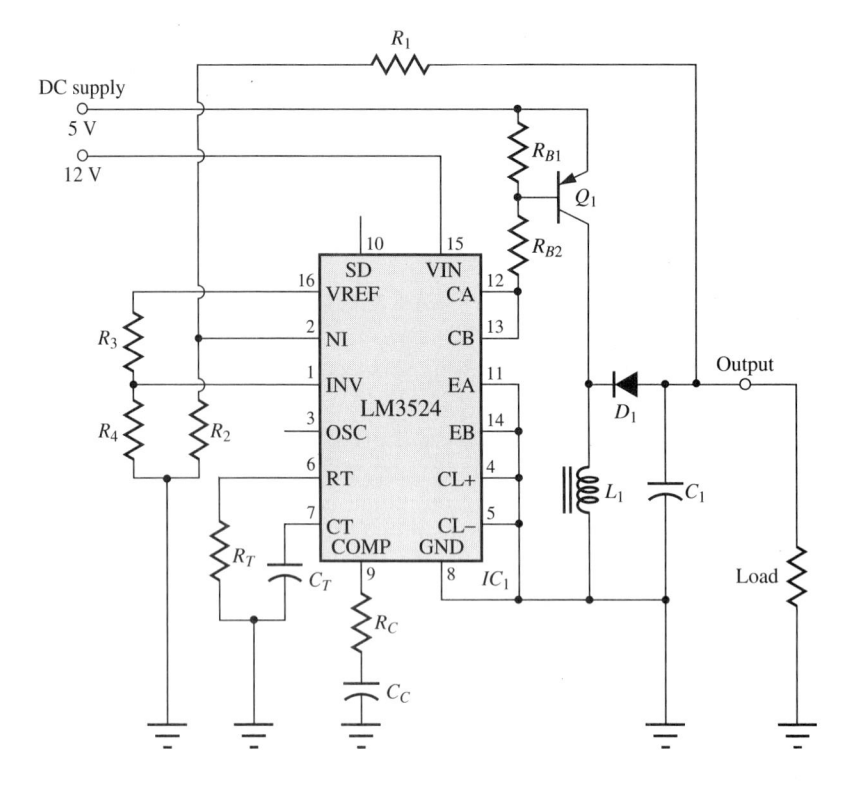

Parts List			
Resistors:		Inductors:	
R_1	10 kΩ	L_1	69.4 μH
R_2	10 kΩ	Capacitors:	
R_3	10 kΩ	C_1	4000 μF
R_4	10 kΩ	C_C	0.001 μF
R_T	10 kΩ	C_T	0.0033 μF
R_C	47 kΩ	Semiconductors:	
R_{sens}	0.0667 Ω	IC_1	LM3524
R_{B1}	150 Ω	Q_1	TIP32C
R_{B2}	56 Ω	D_1	5 A diode

Equation 9-52 is

$$I_{B\text{min}} = \frac{I_{C\text{max}}}{\beta_{\text{min}}} = \frac{3.0 \text{ A}}{60} = 0.050 \text{ A}$$

The bleeder current should be approximately one-tenth of $I_{B\text{min}}$, so we will choose a bleeder current of 5 mA. Calculate R_{B_1} from Equation 9-53:

$$R_{B_1} = \frac{V_{BE}}{I_{\text{bleeder}}} = \frac{0.7 \text{ V}}{5 \text{ mA}} = 140 \text{ Ω}$$

To ensure that the transistor always turns on, choose a *larger* standard value. We will use 150 Ω.

Assume the saturation voltage V_{sat} of the internal switching transistors is 1.0 V and the base-emitter voltage V_{BE} for the external transistor is 0.7 V. Calculate R_{B_2} from Equation 9-54:

$$R_{B_2} = \frac{V_{DC\ supply} - V_{BE} - V_{sat}}{I_{bleeder} + I_{Bmin}} = \frac{5\ V - 0.7\ V - 1.0\ V}{5\ mA + 50\ mA} = 60\ \Omega$$

Choose the closest smaller value of 56 Ω. It is left to the reader to verify that a $\frac{1}{2}$ W rating is adequate.

Lastly, we must design the inductor and capacitor. Design the inductor from Equation 9-39:

$$L = \frac{2.5 V_{in} V_{out}}{f_{osc}(V_{out} + V_{in}) I_{out}} = \frac{2.5 \times 5.0\ V \times 5.0\ V}{30\ kHz \times (5.0\ V + 5.0\ V) \times 3.0\ A} = 69.4\ \mu H$$

Next design the capacitor from Equation 9-43:

$$C = \frac{0.3536 I_{out} V_{out}}{V_{ripple} f_{osc}(V_{out} + V_{in})} = \frac{0.3536 \times 3.0\ A \times 5.0\ V}{0.005\ V \times 30\ kHz \times (5.0\ V + 5.0\ V)} = 3536\ \mu F$$

Select a 4000 μF capacitor from Appendix A-2.

Our design is now complete. Although we designed for an output voltage of −5 V from an input supply voltage of +5 V, we are not limited to a single output voltage as we were with the ICL7660 voltage converter. We could have designed the regulator for any output voltage greater than −2.5 V.

Multiple-Output Regulators Using the LM3524

We saw in Section 9-1 that it was possible to obtain multiple output voltages from a switch-mode regulator by using the switching transistor to drive a transformer and taking the outputs from separate secondaries of the transformer. There are four possible configurations in which we can connect an LM3524 to drive a transformer with multiple outputs. These are shown in Figures 9-21 to 9-24.

In all cases the connections to the LM3524 are identical to the regulator circuits discussed earlier, and the design procedure is identical. The feedback resistor R_1 is always connected to the output supplying the largest output current. The other outputs normally require additional regulation from a linear regulator. The transformer is always connected in the configuration of a *step-up* regulator; that is, the primary is connected between the DC supply and the collector of the switching transistor. The secondary circuits are always configured as *step-down* regulators, except that a second diode is placed in series with the transformer secondary to prevent inductor current flowing back through the secondary winding. If a high output voltage is required from one of the secondaries (that is, an output voltage where we would normally require a step-up regulator), we step up the voltage by using the transformer, and then regulate this down to the required

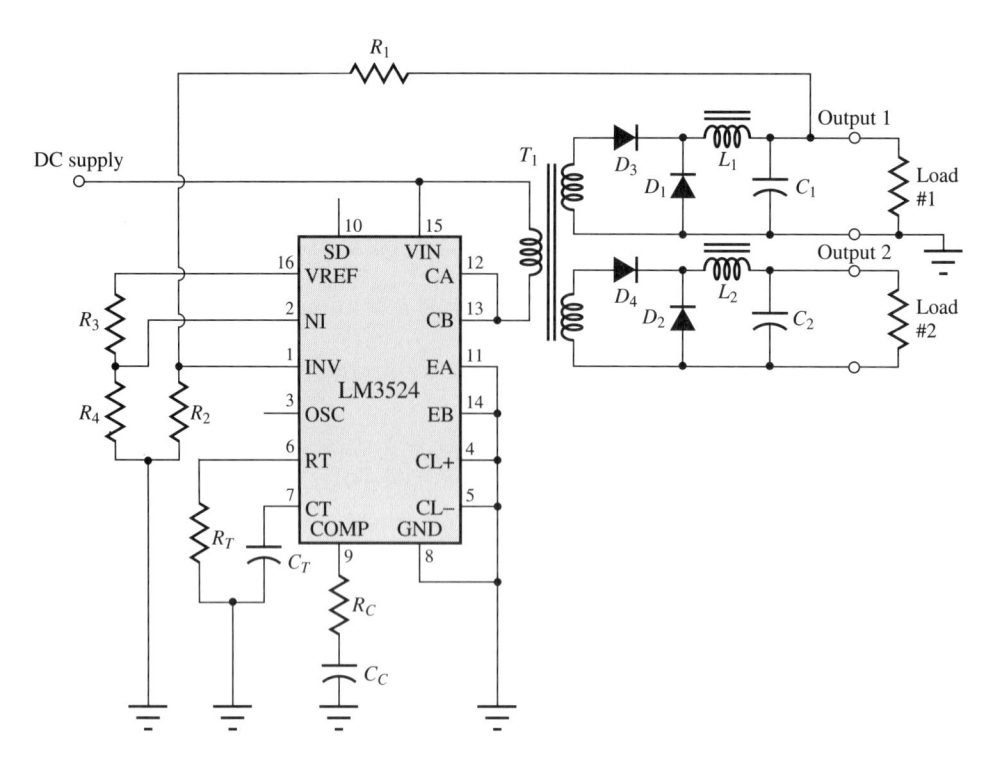

Figure 9-21 Simple multiple-output switch-mode regulator using an LM3524 pulse-width modulator.

voltage by using a normal step-down regulator. If a negative voltage is required from one of the secondaries, we simply reverse the direction of the windings. The inductors and capacitors for each secondary are calculated in exactly the same way as for a normal *step-down* regulator, by using Equations 9-11 and 9-16, respectively.

The simplest circuit, shown in Figure 9-21, is used where only a low output current is required. The transformer is driven directly by using the two internal switching transistors connected in parallel.

If we require a larger output current than we can obtain from the LM3524 directly, we can use the external switching transistor configuration shown in Figure 9-22. Both the external switching transistor and the inverter driver must be NPN transistors, and the bias circuitry is exactly the same as for a *step-up* regulator. Bias resistor R_{B_1} is calculated from Equation 9-61:

$$R_{B_1} = \frac{V_{\text{DC supply}} - V_D - V_{BE_1}}{I_{C_2\text{max}}/\beta_{1\text{min}}\beta_{2\text{min}} + V_{BE_1}/R_{B_2}} \tag{9-61}$$

Bias resistor R_{B_2} is calculated from Equation 9-58:

$$R_{B_2} = 1.0 \text{ k}\Omega \tag{9-58}$$

9-2 Pulse-Width Modulators

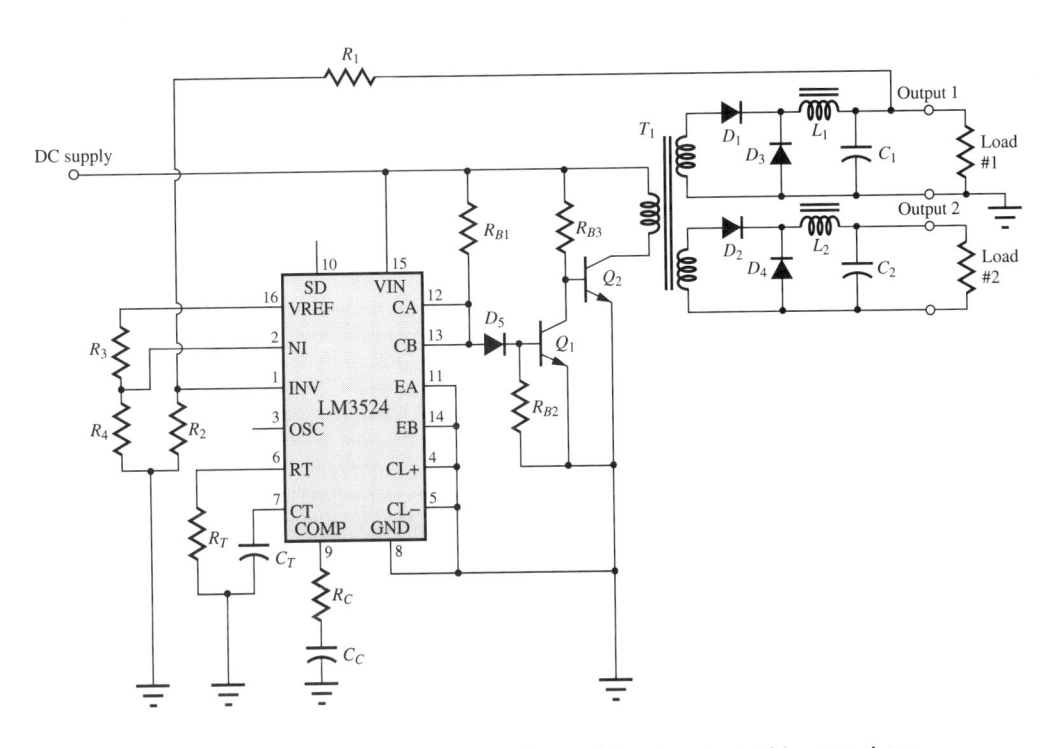

Figure 9-22 Multiple-output switch-mode regulator with external switching transistor.

Bias resistor R_{B_3} is calculated from Equation 9-57:

$$R_{B_3} = \frac{V_{\text{DC supply}} - V_{BE_2}}{I_{C_{2\text{max}}}} \times \beta_{2\text{min}} \tag{9-57}$$

We saw earlier that the output of the LM3524 pulse-width modulator is separately available from the two internal driver transistors, where one output is 180° out of phase with the other. In the previous discussion, we made use of only a single primary transformer winding driven by the two internal switching transistors connected in parallel (either directly or with an external switching transistor). It is more efficient to use the internal transistors separately to drive a center-tapped transformer, as shown in Figure 9-23. The LM3524 was actually designed for this type of application.

The design of this circuit is almost the same as for the circuit with the two switching transistors connected in parallel. The only difference is that the transformer is center-tapped, with each transistor driving half of the transformer. The number of windings on each side of the center tap is the same as for a single-transistor circuit; hence, the center-tapped transformer has twice the number of windings as the simple transformer. The secondary windings are also center-tapped in exactly the same way as the primary windings. Note how the two diodes are used in the secondary circuit. The primary transformer windings must always be placed in the collector circuits of the switching transistors, and the emitters

428   9   Switch-Mode Power Supplies

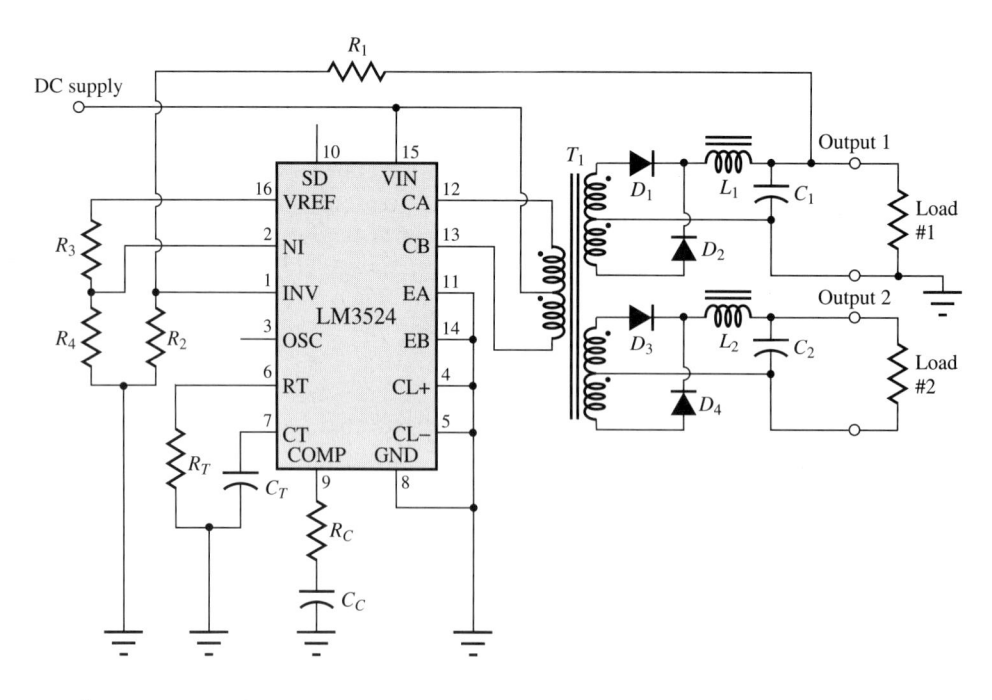

Figure 9-23   Multiple-output switch-mode regulator with center-tapped transformer.

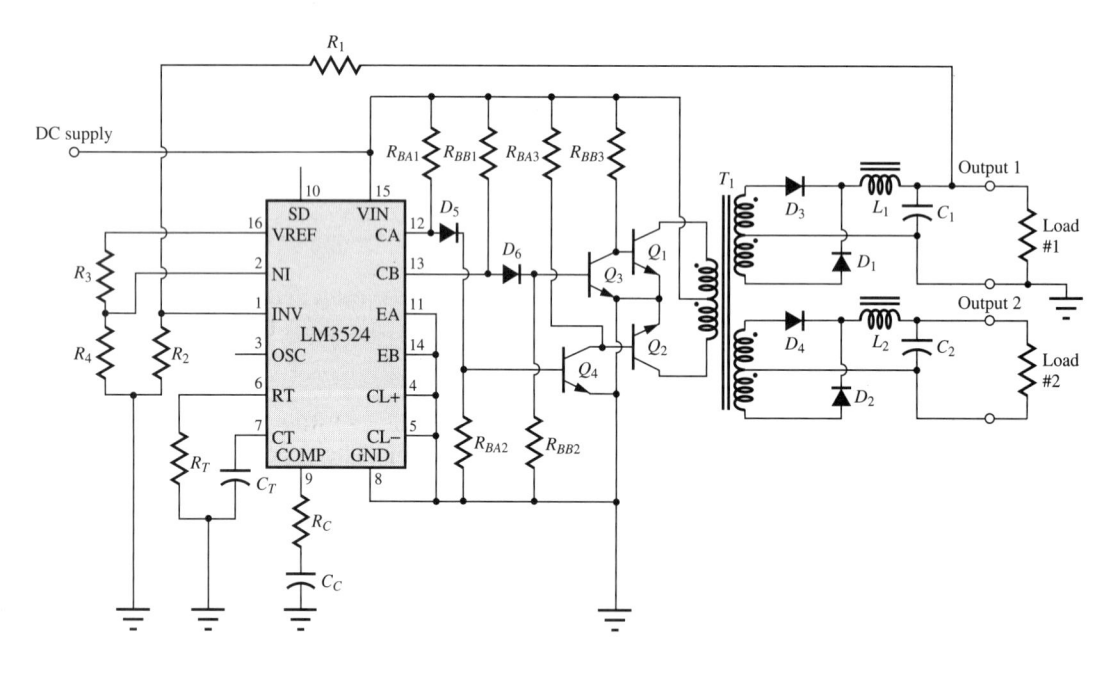

Figure 9-24   Multiple-output switch-mode regulator with external switching transistors driving a center-tapped transformer.

must be grounded. The outputs of the transformer are always connected to standard step-down regulator circuits.

When internal switching transistor *A* is triggered on by the pulse-width modulator, current flows from the DC supply through the top half of the transformer. The pulse is transferred through the transformer to the secondary winding where current flows through the upper diode to the inductor and capacitor. Internal transistor *A* is only triggered on during a portion of the first 180° of the switching cycle. Internal transistor *B* is triggered on by the pulse-width modulator only during a portion of the second 180° of the switching cycle. When transistor *B* is triggered on, current flows through the bottom half of the transformer. The net output is similar to the case when there was only a single transistor driving a single primary winding.

If a higher current output is required, then external switching transistors can be used, as in the circuit in Figure 9-24. The biasing for each of the transistors is exactly the same as for the single-transistor regulator shown in Figure 9-22. The transformer windings are connected to the collector circuits of the external switching transistors.

EXAMPLE 9-5

Design a dual-output switching power supply using an LM3524 pulse-width modulator to provide 12 V at 1 A and 5 V at 3 A. The DC supply is 10.0 V. Use a switching frequency of 50 kHz and design the capacitors to make the output ripple less than 12 mV. Do not use current limiting. Assume the switching transistors have a minimum β of 20 and the inverting transistors have a minimum β of 100.

Solution

Because of the large current required for both outputs, we will use the LM3524 with two external switching transistors driving a center-tapped transformer. The circuit is shown in Figure 9-25, with a parts list included.

First, we must design the transformer. This is relatively easy. The 5 V secondary must deliver the largest current, so we will make it with a 1:1 turns ratio. The input to the transformer is the 10 V of the supply, so the output of the 5 V secondary will also be 10 V. We will arbitrarily choose 18 V for the output of the 12 V secondary. The turns ratio for this secondary is 18:10 or 1.8:1.

Next, we will design the pulse-width modulator circuit. The procedure will be similar to that used in the previous examples.

The switching frequency is determined by R_T and C_T. We will choose $R_T = 10 \text{ k}\Omega$ according to the manufacturer's recommendation, and solve Equation 9-50 for C_T:

$$C_T = \frac{1}{f_{osc}R_T} = \frac{1}{50 \text{ kHz} \times 10 \text{ k}\Omega} = 0.0020 \text{ }\mu\text{F}$$

Choose $C_T = 0.0022 \text{ }\mu\text{F}$ from Appendix A-2.

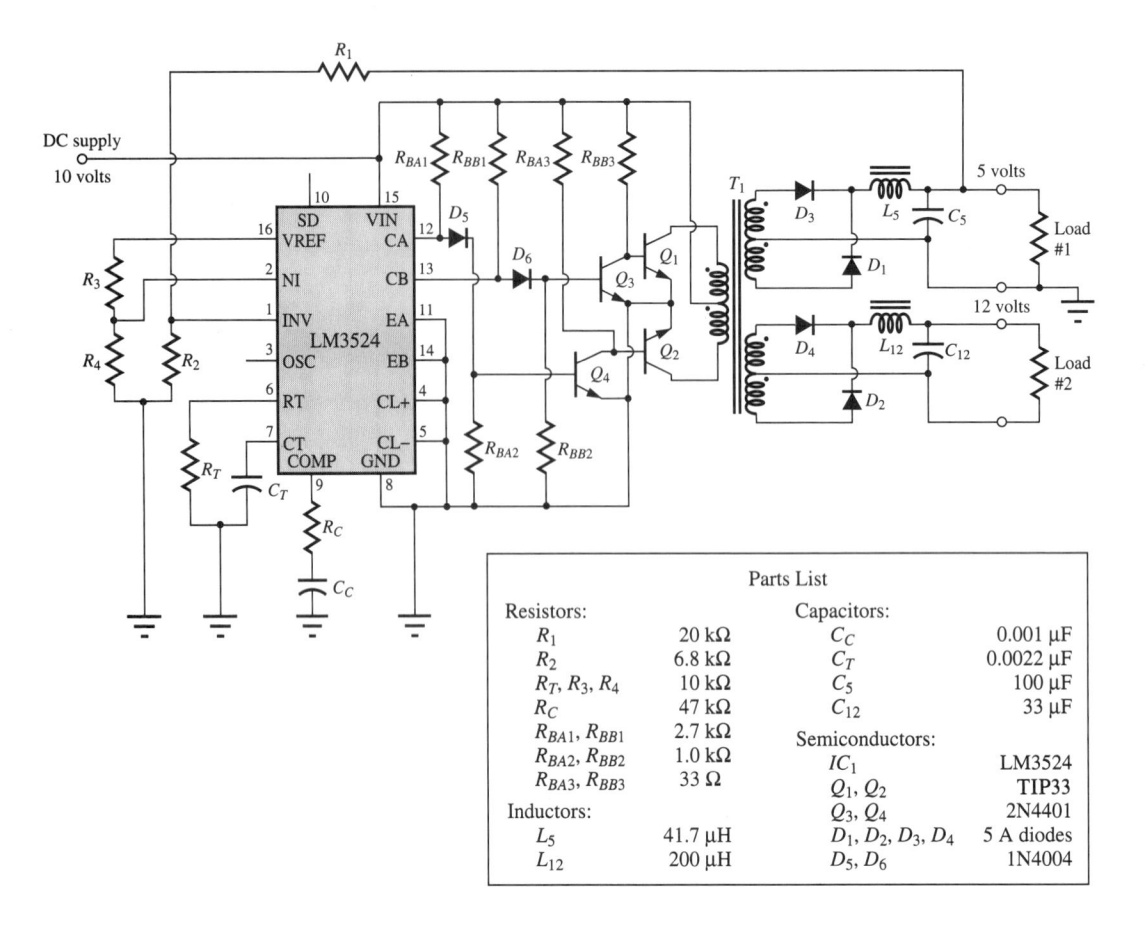

Figure 9-25

Parts List

Resistors:		Capacitors:	
R_1	20 kΩ	C_C	0.001 μF
R_2	6.8 kΩ	C_T	0.0022 μF
R_T, R_3, R_4	10 kΩ	C_5	100 μF
R_C	47 kΩ	C_{12}	33 μF
R_{BA1}, R_{BB1}	2.7 kΩ		
R_{BA2}, R_{BB2}	1.0 kΩ	Semiconductors:	
R_{BA3}, R_{BB3}	33 Ω	IC_1	LM3524
Inductors:		Q_1, Q_2	TIP33
L_5	41.7 μH	Q_3, Q_4	2N4401
L_{12}	200 μH	D_1, D_2, D_3, D_4	5 A diodes
		D_5, D_6	1N4004

Choose $R_3 = R_4 = 10$ kΩ to set the reference voltage V_{ref} at pin 2 to 2.5 V. Now design the feedback voltage divider. Use Equation 9-49 to calculate R_1:

$$R_1 = \frac{V_{out}}{V_{ref}} \times R_3 \| R_4 = \frac{10 \text{ V}}{2.5 \text{ V}} \times 5.0 \text{ k}\Omega = 20 \text{ k}\Omega$$

This is a standard value. Now solve for R_2 as in Example 9-1.

$$\frac{1}{R_1} + \frac{1}{R_2} = \frac{1}{5 \text{ k}\Omega} \quad \text{or} \quad \frac{1}{R_2} = \frac{1}{5 \text{ k}\Omega} - \frac{1}{20 \text{ k}\Omega}$$

From this, we calculate a value of 6.67 kΩ for R_2. The closest standard value is 6.8 kΩ.

We will use the standard compensating network with $R_C = 47$ kΩ and $C_C = 0.001$ μF.

Next, we must calculate the values of the biasing resistors R_{B_1}, R_{B_2}, and R_{B_3}. Since there are two external switching transistors, we will require two biasing networks, but both will be the same. The procedure for calculating values for these biasing resistors is the same as for the step-up regulator in Example 9-3. First, however, we must determine the primary current. This is equal to the sums of the output currents as transformed by the transformer. The 5 V secondary has a 1:1 turns ratio; hence, the primary current due to this winding is 3.0 A. The 12 V secondary has a 1.8:1 turns ratio; hence, the primary current due to this winding is 1 × 1.8 or 1.8 A. The total current is 3.0 + 1.8 = 4.8 A.

Now we can determine the values of the biasing resistors R_{B_1}, R_{B_2}, and R_{B_3}. First use Equation 9-57 to calculate R_{B_3}:

$$R_{B_3} = \frac{V_{\text{DC supply}} - V_{BE_2}}{I_{C_2\text{max}}} \times \beta_{2\text{min}} = \frac{10.0 \text{ V} - 0.7 \text{ V}}{4.8 \text{ A}} \times 20 = 38.75 \text{ }\Omega$$

Choose the smaller standard value of 33 Ω from Appendix A-1. Because there is a relatively large voltage across this resistor, we should check the power. With 9.3 V across a 33 Ω resistor, the power is 2.62 W. Use a 5 W resistor.

Bias resistor R_{B_2} is calculated from Equation 9-58:

$$R_{B_2} = 1.0 \text{ k}\Omega$$

This is a standard value. Since the resistor has only 0.7 V across it, a $\frac{1}{2}$ W resistor will be satisfactory.

Finally, use Equation 9-63 to calculate R_{B_1}.

$$R_{B_1} = \frac{V_{\text{DC supply}} - V_D - V_{BE_1}}{I_{C_2\text{max}}/\beta_{1\text{min}}\beta_{2\text{min}} + V_{BE_1}/R_{B_2}}$$

$$= \frac{10.0 \text{ V} - 0.7 \text{ V} - 0.7 \text{ V}}{4.8 \text{ A}/(100 \times 20) + 0.7 \text{ V}/1000 \text{ }\Omega}$$

$$= \frac{8.6 \text{ V}}{0.0024 \text{ A} + 0.0007 \text{ A}} = 2.77 \text{ k}\Omega$$

Choose a standard 2.7 kΩ resistor from Appendix A-1. As we saw in Example 9-3, the power dissipated in this resistor will be quite small, so a standard $\frac{1}{2}$ W resistor will be satisfactory.

Next, design the inductors and capacitors for each secondary. For the 5 V secondary, the input voltage V_{in} is 10 V. The output voltage is 5.0 V. To calculate the inductance, use Equation 9-11:

$$L = \frac{2.5 V_{\text{out}}(V_{\text{in}} - V_{\text{out}})}{I_{\text{out}} V_{\text{in}} f_{\text{osc}}}$$

$$= \frac{2.5 \times 5.0 \text{ V} \times (10.0 \text{ V} - 5.0 \text{ V})}{3.0 \text{ A} \times 10.0 \text{ V} \times 50 \text{ kHz}} = 41.67 \text{ }\mu\text{H}$$

To calculate the capacitance, use Equation 9-16:

$$C = \frac{0.01768 I_{\text{out}}}{V_{\text{ripple}} f_{\text{osc}}} = \frac{0.01768 \times 3.0 \text{ A}}{0.012 \text{ V} \times 50 \text{ kHz}} = 88.4 \text{ }\mu\text{F}$$

From Appendix A-2, we will select a 100 μF capacitor.

Now design the capacitor and inductor for the 12 V secondary. By choosing a turns ratio of 1.8:1, we have set V_{in} as 18 V. The output voltage is 12.0 V. To calculate the inductance, use Equation 9-11:

$$L = \frac{2.5 V_{out}(V_{in} - V_{out})}{I_{out} V_{in} f_{osc}}$$

$$= \frac{2.5 \times 12.0 \text{ V} \times (18.0 \text{ V} - 12.0 \text{ V})}{1.0 \text{ A} \times 18.0 \text{ V} \times 50 \text{ kHz}} = 200 \text{ μH}$$

To calculate the capacitance, use Equation 9-16:

$$C = \frac{0.01768 I_{out}}{V_{ripple} f_{osc}}$$

$$= \frac{0.01768 \times 1.0 \text{ A}}{0.012 \text{ V} \times 50 \text{ kHz}} = 29.5 \text{ μF}$$

From Appendix A-2, we will select a 33 μF capacitor.

Our dual-output regulator is designed. Notice that the 5 V output is fully regulated, with feedback from this output controlling the pulse-width modulator. The 12 V output, however, is regulated only if the 5 V load remains constant. If the 5 V output changes, then the pulse width will change, and the 12 V output will change. The normal procedure is to use a linear regulator on this output. In this case, the output voltage would be designed as at least 15 V, and an MC7812 linear regulator would be used.

9-3 Single-Package Switching Regulators

In the previous chapter, we saw that three-terminal linear regulators are generally preferred to multipin regulators such as the LM723C because of the design simplicity of the three-terminal devices. The three-pin regulators incorporated the pass transistor and current limiting into a single package. The same process is currently happening to switch-mode regulators. Pulse-width modulator chips such as the LM3524 presented in the previous section work well and are quite satisfactory. Unfortunately, the circuit is relatively complex with compensating resistor and capacitor, external timing resistor and capacitor, two external voltage dividers, and an external switching transistor. Single-chip switch-mode regulators incorporating all of the above components are becoming popular and are available from a number of manufacturers in both the TO-3 metal case and the TO-220 plastic package. Most of these circuits are available in step-down configuration only, but this is seldom a problem. In this section, we will look at the applications of the LH1605 switching regulator chip produced by National Semiconductor. This is typical of the single-package regulators produced by a number of manufacturers.

The data sheets for the LH1605 are shown in Figure 9-26. This regulator is available in an 8 pin TO-3 metal case. The figure shows a block diagram of the

National Semiconductor

Voltage Regulators

LH1605/LH1605C
5 Amp, High Efficiency Switching Regulator

General Description

The LH1605 is a hybrid switching regulator with high output current capability. It incorporates a temperature-compensated voltage reference, a duty cycle modulator with the oscillator frequency programmable, error amplifier, high current-high voltage output switch, and a power diode. The LH1605 can supply up to 5 A of output current over a wide range of regulated output voltages.

Features

- Step down switching regulator
- Output adjustable from 3.0 to 30V
- 5 A output current
- High efficiency
- Frequency adjustable to 100 kHz
- Standard 8-pin TO-3 package

Block Diagram and Connection Diagram

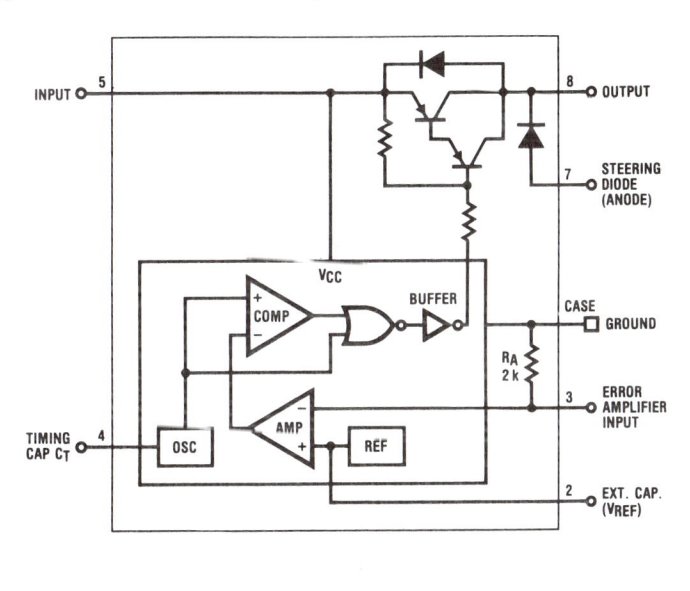

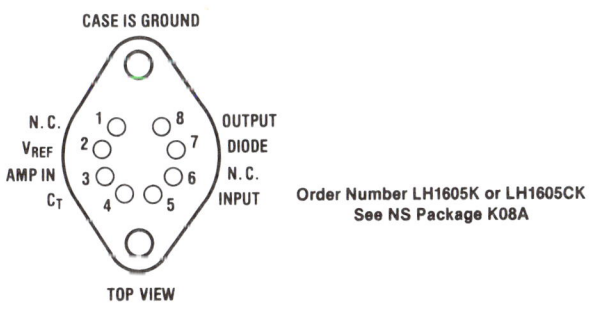

Order Number LH1605K or LH1605CK
See NS Package K08A

Figure 9-26 Data sheets for the LH1605 switch-mode regulator. *(Reprinted with permission of National Semiconductor Corporation)*

Absolute Maximum Ratings

V_{IN}	Input Voltage	35V Max.
I_{OUT}	Output Current	6A
T_J	Operating Temperature	150°C
P_D	Internal Power Dissipation	20W
T_A	Operating Temperature Range	
	LH1605C	−25°C to +85°C
	LH1605	−55°C to +125°C
T_{STG}	Storage Temperature Range	−65°C to +150°C
V_R (V_{8-7})	Steering Diode Reverse Voltage	60V
I_D (I_{7-8})	Steering Diode Forward Current	6A

Electrical Characteristics

$T_C = 25°C$, $V_{IN} = 15V$ unless otherwise specified.

Symbol	Characteristics	Conditions	LH1605 Min.	LH1605 Typ.	LH1605 Max.	LH1605C Min.	LH1605C Typ.	LH1605C Max.	Units
V_{OUT}	Output Voltage Range	$V_{IN} \geq V_{OUT} + 5V$ $I_{OUT} = 2A$ (Note 2)	3.0		30	3.0		30	V
V_S	Switch Saturation Voltage	$I_C = 5.0A$ $I_C = 2.0A$		1.5 1.0	2.0 1.2		1.5 1.0	2.0 1.2	V
V_F	Steering Diode On Voltage	$I_D = 5.0A$ $I_D = 2.0A$		2.0 1.6	2.8 2.0		2.0 1.6	2.8 2.0	V
V_{IN}	Supply Voltage Range		10		35	10		35	V
I_R	Steering Diode Reverse Current	$V_R = 25V$		0.1	10.0		0.1	10.0	µA
I_Q	Quiescent Current	$I_{OUT} = 0.2A$ (Note 3) 50% Duty Cycle		30			30		mA
		0% Duty Cycle ($V_3 = 3.0V$)		6			6		
		100% Duty Cycle ($V_3 = 0V$)		46			46		
V_2	Reference Voltage on Pin 2		2.42	2.50	2.58		2.50		V
		$T_{MIN} \leq T_A \leq T_{MAX}$	2.40	2.50	2.60		2.50		
$\Delta V_2/\Delta T$	V_2 Temperature Coefficient			100			100		ppm/°C
ΔV_2	Line Regulation of Reference Voltage on Pin 2	$10V \leq V_{IN} \leq 35V$ $T_{MIN} \leq T_A \leq T_{MAX}$		20	30		20		mV
V_3	Voltage on Pin 3	(Note 4)	2.45	2.50	2.55		2.50		V
		$T_{MIN} \leq T_C \leq T_{MAX}$	2.42	2.50	2.58		2.50		
V_4	Voltage Swing — Pin 4			3.0			3.0		V
I_4	Charging Current — Pin 4			70			70		µA
$\Delta R_A/\Delta T$	Resistance Temp. Coeff.			75			75		ppm/°C
t_r	Voltage Rise Time	$V_{OUT} = 10V$ $I_{OUT} = 2.0A$ $I_{OUT} = 5.0A$		350 500			350 500		ns
t_f	Voltage Fall Time	$V_{OUT} = 10V$ $I_{OUT} = 2.0A$ $I_{OUT} = 5.0A$		300 400			300 400		ns
t_s	Storage Time	$V_{OUT} = 10V$		1.5			1.5		µs
t_d	Delay Time	$I_{OUT} = 5.0A$		100			100		ns
P_D	Power Dissipation	$V_{OUT} = 10V$		16			16		W
η	Efficiency	$I_{OUT} = 5.0A$		75			75		%
θ_{JC}	Thermal Resistance			5.0			5.0		°C/W

Note 1: θ_{JA} is typically 30°C/W for natural convection cooling.
Note 2: V_{OUT} and I_{OUT} refer to the output DC voltage and output current of a switching supply after the output LC filter as shown in the Typical Application circuit.
Note 3: Quiescent current depends on the duty cycle of the switching transistor. The average quiescent current may be calculated from known operating parameters.
Note 4: Voltage on pin 3 is tested by applying a +5.0V_{DC} voltage through a precision 2.0kΩ resistor to pin 3. This method combines the error due to the input bias current of the error amplifier, and the tolerance of the 2kΩ resistor from pin 3 to ground.
Note 5: The input offset voltage of the error amplifier is wafer tested to a maximum of 10mV.

Figure 9-26 (*Continued*)

internal components. This diagram should be compared with the circuits developed in the previous section. The LH1605 is a 5 A step-down switching regulator with output adjustable from 3.0 to 30 V. External components required include the standard step-down inductor and capacitor, a timing capacitor, a bypass capacitor for the voltage reference, and a feedback resistor. An optional current-limiting circuit can be added if desired.

The basic circuit using the LH1605 as a step-down regulator is shown in Figure 9-27. Note the simplicity of this circuit compared with the pulse-width modulator circuits developed in the previous section (for example, see Figure 9-13). The output voltage is given by the same formula that was used in previous sections, namely Equation 9-46:

$$V_{out} = \frac{R_1 + R_2}{R_2} \times V_{ref} \qquad (9\text{-}46)$$

The feedback resistor R_1 can be designed by rearranging this equation:

$$R_1 = R_2 \times \left(\frac{V_{out}}{V_{ref}} - 1\right) \qquad (9\text{-}64)$$

For the LH1605, $V_{ref} = 2.5$ V and $R_2 = 2000\ \Omega$. Equation 9-46 can be rewritten as

$$V_{out} = 2.50 + 0.00125 R_1 \qquad (9\text{-}65)$$

The design equation for R_1 is

$$R_1 = 800 V_{out} - 2000 \qquad (9\text{-}66)$$

To design the circuit, we calculate resistor R_1 to give the desired output voltage. This resistor is connected from the output to pin 3. For a variable regulator, simply replace R_1 with a potentiometer. Note that the minimum output voltage is approximately 2.50 V.

The switching frequency is

$$f_{osc} = \frac{1}{R_T C_T} \qquad (9\text{-}67)$$

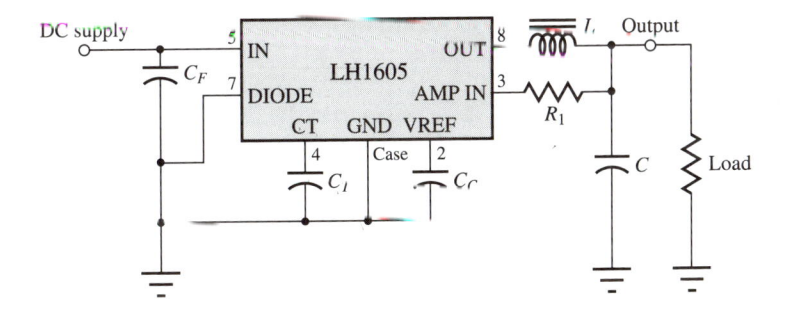

Figure 9-27 LH1605 step-down regulator circuit.

436   9 Switch-Mode Power Supplies

The timing resistor is internal to the LH1605 chip and has a value of approximately 40,000 Ω. We calculate C_T to give the desired switching frequency by rearranging the above equation:

$$C_T = \frac{1}{40,000 f_{osc}} \tag{9-68}$$

Timing capacitor C_T is connected between pin 4 and ground.

The manufacturer recommends that a bypass capacitor be connected to the voltage reference to eliminate switching voltage spikes. A 10 μF solid tantalum capacitor with a voltage rating at least 10 V higher than the highest operating voltage is suggested. This bypass capacitor is connected from pin 2 to ground.

The inductor and capacitor are calculated by using the standard equations for a step-down regulator, namely Equations 9-11 and 9-16, respectively. The inductor is calculated from the equation

$$L = \frac{2.5 V_{out}(V_{in} - V_{out})}{I_{out} V_{in} f_{osc}} \tag{9-11}$$

The capacitor is calculated from the equation

$$C = \frac{0.01768 I_{out}}{V_{ripple} f_{osc}} \tag{9-16}$$

The capacitor should always be chosen larger than the calculated value.

It is possible to add current limiting to the basic LH1605 regulator by using the circuit in Figure 9-28. The current-limiting process is quite simple. The current is sensed by resistor R_{sens}. When the voltage drop across R_{sens} equals 0.7 V, transistor Q_1 turns on. This, in turn, causes transistor Q_2 to turn on, shorting the reference voltage to ground and driving the output voltage to zero. The sensing resistor is calculated in exactly the same way as for the linear regulators in Chapter 8:

$$R_{sens} = \frac{V_{BE}}{I_{max}} \tag{9-69}$$

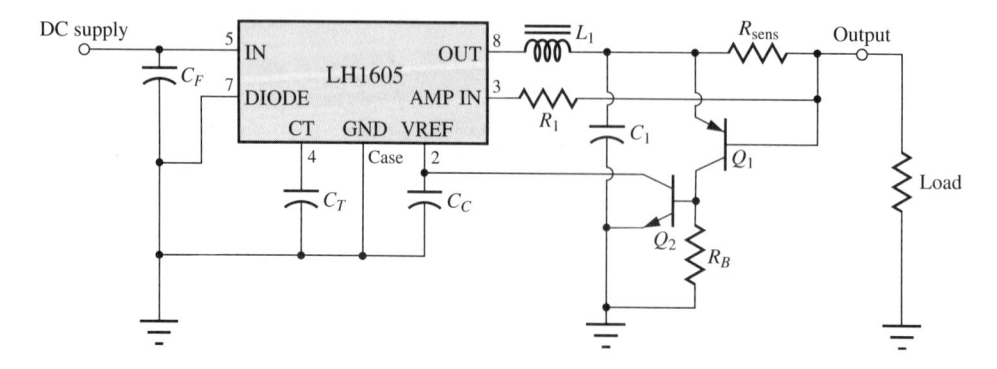

Figure 9-28 LH1605 step-down regulator with current limiting.

The value of V_{BE} is normally 0.7 V. Neither transistor handles much power, so almost any small-signal transistor can be used. A typical value for the base resistor R_B of Q_2 is 10 kΩ. It is used to limit the current through Q_1.

EXAMPLE 9-6

Design a 5.0 V, 2.0 A step-down regulator using an LH1605 single-chip switching regulator. The supply voltage is 12 V. Use a switching frequency of 25 kHz. The output ripple should be less than 10.0 mV RMS. Current limiting is not required.

Solution

We will use the regulator circuit in Figure 9-27. This circuit has been redrawn as Figure 9-29, and a parts list is included.

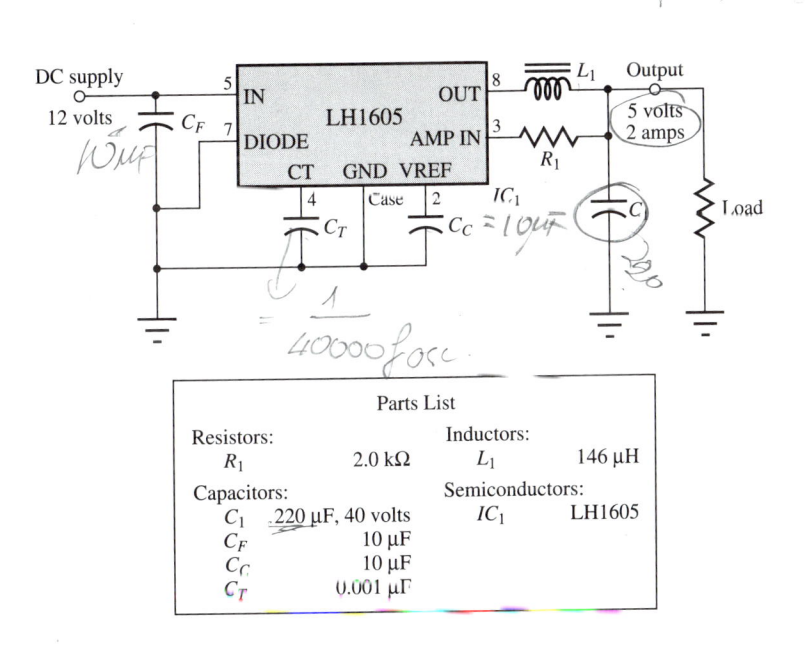

Figure 9-29

Parts List

Resistors:
R_1 2.0 kΩ

Capacitors:
C_1 220 μF, 40 volts
C_F 10 μF
C_C 10 μF
C_T 0.001 μF

Inductors:
L_1 146 μH

Semiconductors:
IC_1 LH1605

First, calculate the value of the feedback resistor R_1 from Equation 9-66:

$R_1 = 800 V_{out} - 2000$
     $= 800 \times 5.0\text{ V} - 2000\text{ Ω} = 2.0\text{ kΩ}$

This is a standard value of resistor.

Now we will calculate the value of the timing capacitor C_T by using Equation 9-68:

$$C_T = \frac{1}{40,000 f_{osc}}$$
$$= \frac{1}{40,000 \times 25 \text{ kHz}} = 0.001 \text{ μF}$$

This is also a standard value.

We will have to select a bypass capacitor for the voltage reference. The recommended capacitor is a 10 μF solid core tantalum. We will use this.

The inductor is calculated by using Equation 9-11:

$$L = \frac{2.5 V_{out}(V_{in} - V_{out})}{I_{out} V_{in} f_{osc}}$$
$$= \frac{2.5 \times 5.0 \text{ V} \times (12 \text{ V} - 5 \text{ V})}{2.0 \text{ A} \times 12.0 \text{ V} \times 25 \text{ kHz}} = 146 \text{ μH}$$

The capacitor is calculated from Equation 9-16:

$$C = \frac{0.01768 I_{out}}{V_{ripple} f_{osc}}$$
$$= \frac{0.01768 \times 2.0 \text{ A}}{0.010 \text{ V} \times 25 \text{ kHz}} = 141 \text{ μF}$$

We will choose a 220 μF capacitor from Appendix A-2.

Our design is now complete.

EXAMPLE 9-7

Design a 12.0 V, 5.0 A step-down regulator using an LH1605 single-chip switching regulator. The supply voltage is 24 V. Use a switching frequency of 60 kHz. The output ripple should be less than 2.0 mV RMS. The circuit should have current limiting.

Solution

The circuit for the LH1605 using current limiting was shown in Figure 9-28. This circuit has been redrawn as Figure 9-30, with a parts list included.

The feedback resistor R_1 is calculated from Equation 9-66:

$$R_1 = 800 V_{out} - 2000$$
$$= 800 \times 12 \text{ V} - 2000 = 7.6 \text{ k}\Omega$$

Choose $R_1 = 7.5$ kΩ as the nearest standard value from Appendix A-1.

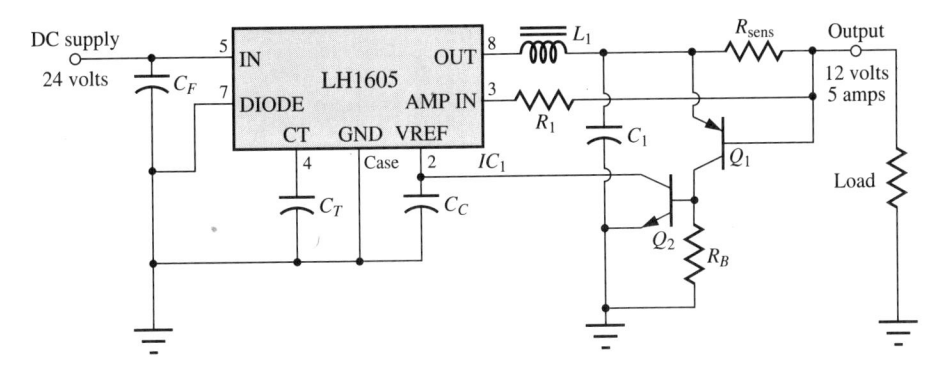

9-3 Single-Package Switching Regulators

Figure 9-30

Parts List

Resistors:		Capacitors:	
R_1	7.5 kΩ	C_1	1000 μF
R_B	10 kΩ	C_F	10 μF
R_{sens}	0.14 Ω	C_C	10 μF
Semiconductors:		C_T	390 pF
IC_1	LH1605	Inductors:	
Q_1	2N4403	L_1	50 μH
Q_2	2N4401		

Next calculate the value of the timing capacitor C_T from Equation 9-68:

$$C_T = \frac{1}{40{,}000 f_{osc}}$$

$$= \frac{1}{40{,}000 \times 60 \text{ kHz}} = 417 \text{ pF}$$

We will choose $C_T = 390$ pF as the closest standard value from Appendix A-2.

We will have to select a bypass capacitor for the voltage reference. The recommended capacitor is a 10 μF solid core tantalum. We will use this.

The inductor is calculated from Equation 9-11:

$$L = \frac{2.5 V_{out}(V_{in} - V_{out})}{I_{out} V_{in} f_{osc}}$$

$$= \frac{2.5 \times 12.0 \text{ V} \times (24 \text{ V} - 12 \text{ V})}{5.0 \text{ A} \times 24.0 \text{ V} \times 60 \text{ kHz}} = 50.0 \text{ μH}$$

The capacitor is calculated from Equation 9-16:

$$C = \frac{0.01768 I_{out}}{V_{ripple} f_{osc}} = \frac{0.01768 \times 5.0 \text{ A}}{0.002 \text{ V} \times 60 \text{ kHz}} = 737 \text{ μF}$$

We will choose a 1000 μF capacitor from Appendix A-2.

Lastly, we must design the current-limiting circuitry. We will use a 2N4403 for Q_1 and a 2N4401 for Q_2. These are typical small-signal transistors suitable for this application. Data sheets for these transistors are shown in Appendix A-5. We shall use a current-limiting base resistor for Q_2 of 10 kΩ. This value is not critical. The sensing resistor R_{sens} is calculated from Equation 9-69:

$$R_{sens} = \frac{V_{BE}}{I_{max}} = \frac{0.7 \text{ V}}{5.0 \text{ A}} = 0.14 \text{ }\Omega$$

This resistor will dissipate 3.5 W and will probably have to be custom-wound from resistance wire.

Our design of the step-down regulator using the LH1605 is complete. Note how easy the design was and how simple the circuit is compared with the regulators designed in the previous section.

Summary

This chapter described the design and applications of switch-mode regulators. Switch-mode regulators switch power as required by the load. Through power switching, high regulator efficiencies can be attained, and, consequently, very compact power supplies can be produced. Switch-mode regulators can be used in different configurations, including step-down regulators, step-up regulators, voltage-inverting regulators, and multiple-output regulators.

We saw that all switch-mode regulators consist of five components: a transistor switch, a pulse-width modulator, an inductor, a capacitor, and a diode. The pulse-width modulator which controls the actual switching is a relatively complex circuit, usually produced as an integrated circuit. The LM3524 pulse-width modulator was introduced as a typical example. Pulse-width modulators such as the LM3524 are very flexible, and their use in different regulators was demonstrated. The regulator circuits, however, were relatively complex. The LH1605 was introduced as an example of a single-chip switch-mode regulator. While lacking the flexibility of the simple pulse-width modulators, these devices are far easier to use.

Review Questions

9-1 Principles of switch-mode power supplies

1. What is the major disadvantage of a linear voltage regulator?
2. Determine the efficiency of a linear voltage regulator if the input voltage is 35.5 V, the regulated output voltage is 8.0 V, and the current is 1.2 A.
3. How does a switching regulator overcome the disadvantages of a linear regulator?
4. What is the advantage of operating a switching regulator at high frequency?
5. What limits the maximum frequency at which a switching regulator can be operated?

6. Distinguish among a step-down regulator, a step-up regulator, and a voltage-inverting regulator.
7. List the functional components of a switching regulator.
8. Describe the basic principle of operation of a switching regulator.
9. What device performs the switching action in a switching power supply? What is the best type of switching device?
10. What is V_{sat}, and what is its typical value?
11. What device controls the switching?
12. What is the most important element in a switching regulator?
13. Describe what happens when voltage is suddenly applied across an inductor. Describe what happens when current is flowing through an inductor and the circuit opens.
14. Describe the configuration of the switch-mode components in a step-down regulator.
15. What is the difference in circuit configuration between a step-up regulator and a step-down regulator?
16. What is the difference in circuit configuration between a voltage-inverting regulator and a step-down regulator?
17. Describe what happens in a step-down regulator when the output voltage increases above the reference voltage. What happens when the output voltage decreases below the reference voltage?
18. What restriction is placed on the load of a switching power supply if the regulator is to function properly?
19. What part of the circuit determines the output voltage? How can the output be made variable?
20. Define duty cycle.
21. If the output voltage is too high, what happens to the duty cycle? What happens if the output is too low?
22. What normal variation is designed for the inductor current? What limitation does this impose?
23. What is the purpose of the capacitor in a switching power supply? What characteristic other than the capacitance is important in selecting a capacitor? What is the best type of capacitor?
24. Describe the differences in the circuits of a step-down regulator, a step-up regulator, and a voltage-inverting regulator.
25. What is the advantage of combining a linear regulator with a switching regulator?
26. What is the requirement for the output voltage of the switching regulator if it is supplying a linear regulator?
27. Describe how to generate multiple output voltages from a single switching regulator.

9-2 Pulse-width modulators

28. What is the LM3524?
29. From where does the LM3524 usually receive its power? What is the minimum voltage that it requires to operate?
30. What is the maximum output current of an LM3524 acting without an external transistor?
31. How is the oscillator frequency determined in an LM3524 pulse-width modulator?
32. What is the purpose of the external oscillator connection (pin 3)?
33. What is the value of the internal reference voltage of an LM3524?
34. In what configurations can an LM3524 pulse-width modulator be used?
35. Why are biasing circuits necessary when an external switching transistor is used?

442   9 Switch-Mode Power Supplies

36. What is the difference in biasing circuits for an external switching transistor between a step-down regulator and a voltage-inverting regulator?
37. Why are two transistors required for the step-up regulator using an external switching transistor?
38. What is the purpose of the diode between the biasing resistors of a step-up regulator with external switching transistor?
39. What type of configuration is used when the switching transistor is driving a transformer?
40. In what configuration are the inductor, capacitor, and diode always placed in the secondary of a transformer-coupled regulator?
41. Why are two diodes required in each secondary circuit of a multiple-output regulator?
42. Why are two output driver transistors provided in the LM3524?
43. How does the primary of a transformer intended for using both transistors in the LM3524 differ from a transformer intended for using a single transistor?
44. What is the major disadvantage of a pulse-width modulator chip such as the LM3524?

9-3 Single-package switching regulators

45. How does the LH1605 overcome the disadvantage of a pulse-width modulator such as the LM3524?
46. What is the only configuration in which the LH1605 can be used?
47. What is the maximum output current of the LH1605?
48. What external connections must be made to the LH1605 so that it can function as a switching power supply regulator?
49. What component(s) determine the oscillator frequency?
50. What is the value of the internal feedback resistor?
51. What is the internal reference voltage of the LH1605? Can an external reference be used instead?
52. In what type of package is the LH1605 available?
53. Explain how current limiting can be obtained when using an LH1605 regulator.
54. What is the limitation of the LH1605 regulator?

Design and Analysis Problems

9-2 Pulse-width modulators

1. Design a step-down switch-mode regulator using an LM3524 to give an output voltage of 5 V at 3 A from a 15 V supply. Use a switching frequency of 54 kHz and make the output ripple less than 3.0 mV RMS. Include current limiting. Assume that the switching transistor has a minimum β of 75.
   (a) Sketch the circuit you are going to use. Prepare a parts list to be completed during the design.
   (b) Calculate values for the timing capacitor and resistor.
   (c) Design the feedback voltage divider.
   (d) Design the current-limiting sensing resistor.
   (e) Design the inductor and capacitor values.
   (f) Calculate biasing resistors R_{B_1} and R_{B_2} for the external switching transistor.
2. Design a three-output power supply using the LM3524 pulse-width modulator. The outputs are 5 V, 8 V, and 15 V all at 1 A. The error sensor should be connected to the

8 V output. Use a switching frequency of 35 kHz and make the ripple voltage less than 12 mV for each output. The DC supply is 13.5 V. Use the internal reference, and do not include current limiting. Assume that the switching transistors have a minimum β of 65.

(a) Sketch the circuit for the regulator. Include a parts list that you will complete as you do the design.
(b) Design inductors and capacitors for each output.
(c) Determine the timing capacitor and resistor.
(d) Design the feedback voltage divider.
(e) Design the biasing resistors for the external switching transistors.

3. Repeat Problem 2 except use double regulation for the 5 and 15 V outputs. Use an MC7805 for the 5 V output and an MC7815 for the 15 V output. Design the switching supply to give 3 V more than the desired output voltages on the doubly regulated outputs.

4. Analyze the step-up switch-mode power supply using an LM3524 pulse-width modulator shown in Figure 9-31 to determine the
   (a) switching frequency
   (b) output voltage.

5. Design a ±12 V 1A doubly regulated power supply using an LM3524 pulse-width modulator, with MC7812 and MC7912 linear regulators. The switching supply is powered by a 12 V battery and should operate at a frequency of 33 kHz. The maximum input ripple to the linear regulators must be 25 mV RMS. Assume that the switching transistors have a minimum β of 50.

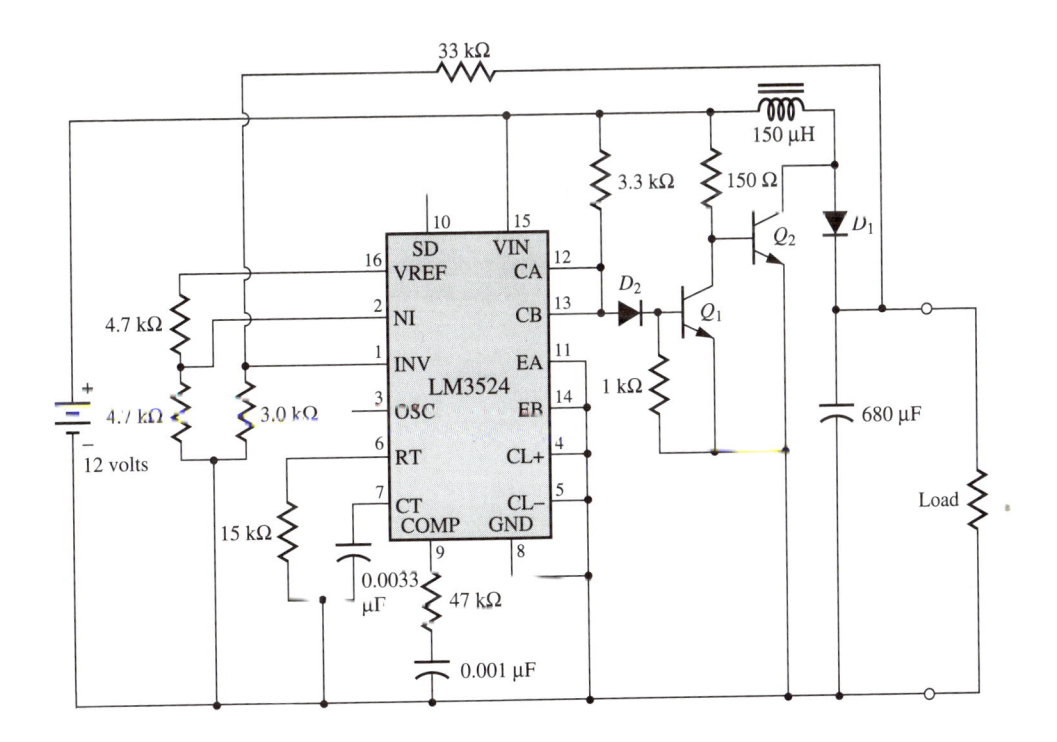

Figure 9-31

9-3 Single-package switching regulators

6. Analyze the switch-mode power supply using an LH1605 single-chip regulator shown in Figure 9-32 to determine the
   (a) switching frequency f_{osc}
   (b) output voltage. V_o

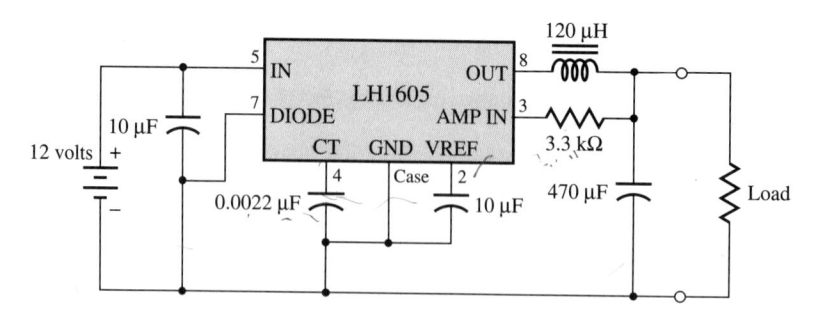

Figure 9-32

7. Design a 5 V, 3 A step-down regulator using an LH1605 single-chip switching regulator. Include current limiting. The supply voltage is 15 V. Use a switching frequency of 45 kHz and make the output ripple less than 10 mV RMS.
   (a) Sketch the circuit. Include a parts list that will be completed as you do the design.
   (b) Calculate the value of the feedback resistor.
   (c) Calculate the value of the timing capacitor.
   (d) Calculate the value of the current-limiting resistor.
   (e) Calculate the values of the capacitor and inductor.

Oscillator Circuits 10

OUTLINE

10-1 Theory of Oscillation
10-2 Relaxation Oscillators
10-3 Bootstrap Oscillators
10-4 Sine-Wave Oscillators
10-5 Other Oscillator Circuits

KEY TERMS

Oscillation
Periodic waveform
Period
Cycle
Frequency
Phase

Amplitude
Sine wave
Square wave
Pulse wave
Triangle wave
Sawtooth wave (or ramp wave)

Barkhausen criteria
Phase-shift network
Wien bridge network
Programmable unijunction transistor (PUT)

OBJECTIVES

On completion of this chapter, the reader should be able to:

- define the various terms describing oscillation and periodic waveforms.
- describe the most common types of periodic waveforms.
- list the conditions for a circuit to oscillate, and recognize these conditions in a circuit.
- design and analyze relaxation oscillator circuits to produce square waves and triangle waves.
- design and analyze bootstrap oscillator circuits.
- explain the Barkhausen criteria as they relate to sine-wave oscillators.
- design phase-shift and Wien bridge sine-wave oscillators, describing how to set the proper gain and how to make the circuit self starting.
- design a bootstrap sine-wave oscillator.
- design a ramp generator using a programmable unijunction transistor.

10 Oscillator Circuits

Introduction

An oscillator circuit generates some form of time-varying output. There are many types of oscillator circuits, and many types of output waveforms are generated. Most electronic systems, other than basic amplifiers and linear power supplies, use an oscillator of some form. We have already encountered oscillators in our discussion of switch-mode power supplies. There, an oscillator was used to produce a variable-duty-cycle pulse wave.

The simplest form of oscillator is the square-wave generator. Square waves, and the related pulse waves, have many applications. Their use in switch-mode power supplies has already been mentioned. They are also used as time-base signals for most digital systems, as chopping signals for multiplexers, and for various types of modulation. Most function generators produce square and pulse waves as test signals.

Probably the most frequently encountered type of oscillator, however, is the sine-wave generator. Audio test signals, radio frequency carriers, and communication signals of many types are sine waves. Sine-wave generator circuits are generally more complex than square-wave generating circuits, yet the basic principles of oscillation are the same.

Other types of oscillators and waveforms will be introduced. Triangle waves are widely used as test signals and as a base for producing sine waves in some function generators. Sawtooth waves are used in oscilloscopes and television receivers to generate the sweep or deflection signals. Pulse waves are used for various types of triggering circuits, such as SCR motor speed controls.

In this chapter we will develop op-amp oscillator circuits for generating a variety of waveforms. The circuits are relatively simple but suitable for many applications. A number of integrated circuit oscillators are available for more complex applications. We will defer the study of these until the next chapter.

10-1 Theory of Oscillation

Oscillation is the production of some form of periodic output. Although we will be studying electronic oscillations, many other types exist. The waves on a lake or ocean are oscillations, the motion of a pendulum in an old-style clock is an oscillation, the motion of the tines of a tuning fork is an oscillation. These examples are of mechanical oscillations. Sometimes mechanical oscillations are used to produce electrical oscillations. In alternators and generators, the mechanical rotation of a conductor in a magnetic field produces a time-varying electrical output; in quartz crystals, an electrical oscillation is transferred through the crystal by the mechanical vibration of the crystal; and the vibrators, used many years ago in the era of tube-operated radios, employed a vibrating reed to convert DC into AC. In our discussion of oscillation and oscillator circuits, we will consider only the electronic generation of periodic electrical waveforms.

Periodic Waveforms

A **waveform** is the shape or pattern of the voltage or current variation, plotted against time, or as portrayed on the screen of an oscilloscope. A **periodic** waveform is a time-varying pattern that keeps repeating itself, exactly, over an indefinite period of time. A general periodic waveform is shown in Figure 10-1. We will use this waveform to define several properties of waves.

The **period** T of a waveform is the time for the wave to trace out one complete cycle, where a **cycle** is simply the repeating waveform shape. In Figure 10-1 the period is shown as the interval from one peak to the next peak of the waveform. This is not necessary to the definition of the period. In general, the period is the time from any point in one cycle to the identical point in the next cycle.

The **frequency** of the wave is the reciprocal of the period and is given in cycles per second or hertz:

$$\text{frequency} = \frac{1}{\text{period}} \quad \text{or} \quad f = \frac{1}{T} \tag{10-1}$$

The **phase** of a wave at any point in a cycle is the fraction of the period elapsed from the start of a cycle to the selected point in the cycle. Phase is usually expressed in degrees, where 360° is one complete cycle.

$$\text{phase} = 360° \times \frac{\text{ellapsed time in cycle}}{\text{period}} \tag{10-2}$$

In Figure 10-1, the phase of point A is approximately 90° from where we define the start of the cycle.

The **amplitude** of a waveform can be defined in a number of ways. The *instantaneous amplitude* of the waveform in Figure 10-1 is its value above (or below) zero at any time. The *peak-to-peak amplitude* of the waveform is the difference between the most positive level of the wave and the most negative level of the wave. The *peak amplitude* is the difference between either the positive maximum

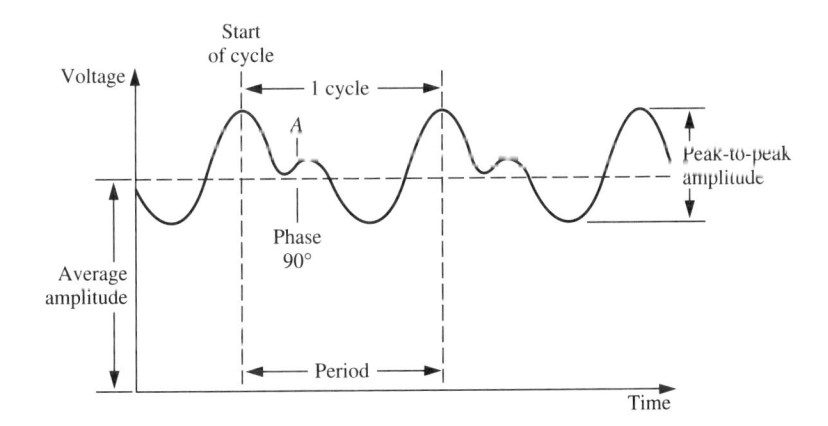

Figure 10-1 General periodic waveform.

and zero (*positive peak amplitude*) or the negative maximum and zero (*negative peak amplitude*). Note that the positive peak amplitude does not necessarily equal the negative peak amplitude. The *average amplitude* of the waveform is the level at which there is an equal area of the waveform above and below the average. For a sine wave, such as is supplied from the power line, the average value is zero.

Standard Electronic Waveforms

The waveform in Figure 10-1 is arbitrary. As such, it is not a particularly useful waveform, although it could certainly arise from some electronic circuit. In this chapter we will be concerned with the generation of waveforms that have application in electronic systems. Some of these are shown in Figure 10-2.

(a) Sine wave

(b) Square wave

(c) Pulse wave

(d) Triangle wave

(e) Ramp or sawtooth wave

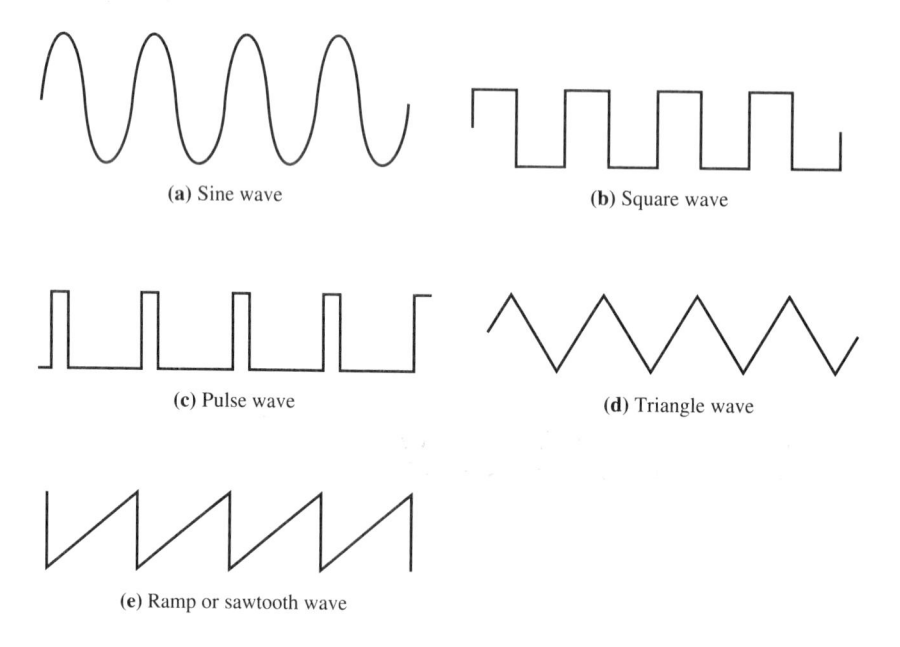

Figure 10-2 Common waveforms used in electronic systems.

Figure 10-2(a) shows a **sine wave**. This is the most common waveform used in electronics. It is the AC power line waveform, a common test waveform, and the carrier waveform used in communication systems. Figure 10-2(b) shows a **square wave**. This is often used as a test waveform, but is most widely used as a timing signal in digital systems, multiplexers, and various instrument applications. The **pulse wave** in Figure 10-2(c) is used for triggering in digital circuits and in analog switching systems. The **triangle wave** in Figure 10-2(d) is commonly used as a test waveform. The **sawtooth wave** (or **ramp wave**) shown in Figure 10-2(e) is used in oscilloscopes and function generators for sweep circuits. Throughout this chapter we will be introducing circuits for generating these waveforms.

Conditions for Oscillation

An oscillator can be considered as a DC to AC converter. As explained in Chapter 7, virtually all electronic circuits and systems require DC power to operate. An oscillator is no different. It takes DC power and converts it to a time-varying signal. Obviously, not all circuits act as oscillators. What are the requirements for a circuit to oscillate?

The first, and most important, requirement is *positive feedback*. We discussed positive feedback briefly in Chapter 2. Positive feedback occurs in an amplifier circuit when some of the output signal is routed back to the input in phase with the input signal. Although there is no input in an oscillator circuit, part of the output signal is routed back through the circuit with zero phase shift.

The second requirement is circuit gain. Thus, we need an amplifier. For many types of oscillators, the amount of gain is not important, and, in general, the more gain the better. For sine-wave oscillators, however, the amount of gain is critical. Sine-wave oscillators will be discussed later, and the special requirements for sine-wave oscillation, known as the Barkhausen criteria, will be introduced at that time.

The third requirement is a system for determining the oscillator operating frequency. This may simply be a resistor and capacitor where the frequency is determined by the *R-C* time constant, or it may be some form of filter circuit that allows positive feedback at only one frequency. Even if there is no frequency-sensitive element in the circuit, some circuits may still oscillate at a high frequency due to stray inductance and capacitance forming an effective frequency-sensitive network.

In the following sections we will be looking at a variety of oscillator circuits. In some, the elements necessary for producing oscillations will be easy to identify. In others, they may be quite difficult to identify. But always, in any oscillator circuit, the above three requirements must be fulfilled.

10-2 Relaxation Oscillators

The first oscillator type we will introduce is the relaxation oscillator. This is the simplest form of oscillator, consisting of a comparator controlled by the charge and discharge of a capacitor. Although the basic relaxation oscillator produces a square-wave output, we will see how it can be modified to produce a triangle wave, a pulse wave, or a sawtooth wave.

Square-Wave Relaxation Oscillator

The circuit of a simple square-wave relaxation oscillator is shown in Figure 10-3. The three requirements for oscillation are easily seen in this circuit. The *R-C* time constant of capacitor C_1 and resistor R_1 determines the frequency of oscillation. Amplification is provided by the op-amp. Positive feedback is accomplished through the voltage divider consisting of R_2 and R_3 connected to the noninverting input.

10 Oscillator Circuits

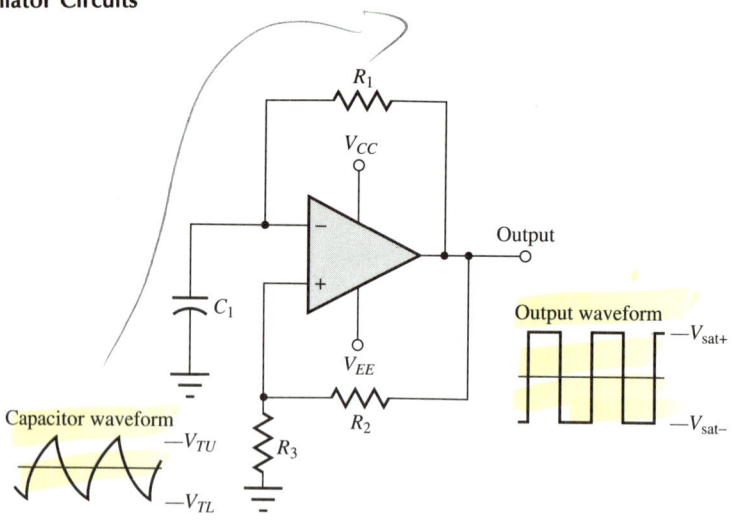

Figure 10-3 Square-wave relaxation oscillator.

The op-amp in this circuit acts as a comparator, comparing the voltage across the capacitor at the inverting input with the fraction of the output voltage produced by the voltage divider at the noninverting input. This is a Schmitt trigger circuit as introduced in Chapter 5. Assume that when the power is first turned on the output voltage is V_{sat+}, the positive saturation voltage of the op-amp, typically 1 V less than the positive supply voltage. The voltage divider consisting of R_2 and R_3 applies the upper trigger voltage V_{TU} to the noninverting input:

$$V_{TU} = V_{sat+} \times \frac{R_3}{R_2 + R_3} \qquad (10\text{-}3)$$

Capacitor C_1 is initially charged to less than V_{TU}. Since the voltage at the inverting input is less than the voltage at the noninverting input, the op-amp will produce the maximum positive output voltage V_{sat+} to try to make the inverting input voltage increase to the noninverting input voltage. This clamps the output at V_{sat+}. Capacitor C_1 starts to charge through R_1 and will try to charge until its voltage is equal to V_{sat+}. The instant the capacitor voltage exceeds V_{TU}, however, the voltage at the inverting input becomes more positive than the voltage at the noninverting input, and the output will become V_{sat-} to try to drive the inverting input voltage back to the noninverting input voltage level. This voltage clamps the output at V_{sat-}. Now the voltage applied to the noninverting input of the op-amp will be

$$V_{TL} = V_{sat-} \times \frac{R_3}{R_2 + R_3} \qquad (10\text{-}4)$$

where V_{TL} is the lower trigger voltage. If the op-amp is operated from a balanced supply, then $V_{TL} = -V_{TU}$. A little later, we will develop the case where these are not equal.

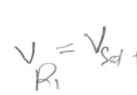

10-2 Relaxation Oscillators

With the op-amp output clamped to V_{sat-}, capacitor C_1 starts to discharge through R_1 and will try to discharge until its voltage is equal to V_{sat-}. The instant the capacitor voltage becomes less than V_{TL}, however, the inverting input voltage becomes less than the noninverting input voltage, and the op-amp output will switch back again to V_{sat+}.

The period of the square-wave output is simply the time for the capacitor to charge from V_{TL} to V_{TU}, then discharge from V_{TU} to V_{TL}. The charge time (or discharge time since the R-C time constant is the same for both charge and discharge) is given by the standard equation for charging a capacitor:

$$V(t) = V_0(1 - \epsilon^{-t/R_1C_1}) \tag{10-5}$$

Here, $V(t)$ is the change in voltage on the capacitor after t seconds and is equal to $V_{TU} - V_{TL}$, and V_0 is the total voltage range through which the capacitor will try to charge and is equal to $V_{TU} - V_{sat-}$. Rearrange Equation 10-5 to solve for $1 - V(t)/V_0$, take the natural logarithm of each side, and solve for t:

$$t = R_1C_1 \ln\left(\frac{V_0}{V_0 - V(t)}\right) \tag{10-6}$$

The wave will always be symmetrical (a pure square wave) with period $2t$. Consequently, the frequency f_{osc} will be $1/2t$ or

$$f_{osc} = \frac{1}{2R_1C_1 \ln(V_0/[V_0 - V(t)])} \tag{10-7}$$

The logarithmic portion of this expression is a nuisance to calculate and makes the design extremely difficult. We can eliminate the logarithm term by setting it to a constant and selecting values of R_2 and R_3 to give this value. We can set it equal to any value, but setting it equal to 1 is the best choice:

$$\ln\left(\frac{V_0}{V_0 - V(t)}\right) = 1 \tag{10-8}$$

By taking the exponential, we get

$$\frac{V_0}{V_0 - V(t)} = 2.71828 \tag{10-9}$$

Under steady-state operation, the capacitor is charging and discharging between V_{TU} and V_{TL}. Thus,

$$V(t) = V_{TU} - V_{TL}$$
$$= \left(V_{sat+} \times \frac{R_3}{R_2 + R_3}\right) - \left(V_{sat-} \times \frac{R_3}{R_2 + R_3}\right) \tag{10-10}$$

If $V_{sat+} = V_{sat-}$, that is, the op-amp is operating from a balanced dual supply, then

$$V(t) = 2V_{sat+} \times \frac{R_3}{R_2 + R_3} \tag{10-11}$$

The total range that the capacitor will try to charge through is

$$V_0 = V_{sat+} - V_{TL}$$
$$= V_{sat+} - V_{sat-} \times \frac{R_3}{R_2 + R_3} \qquad (10\text{-}12)$$

If $V_{sat+} = V_{sat-}$, as before, then

$$V_0 = V_{sat+} \times \left(1 + \frac{R_3}{R_2 + R_3}\right)$$
$$= V_{sat+} \times \frac{R_2 + 2R_3}{R_2 + R_3} \qquad (10\text{-}13)$$

Substitute Equations 10-11 and 10-13 into Equation 10-9 to get

$$\frac{[(R_2 + 2R_3)/(R_2 + R_3)] \times V_{sat+}}{[(R_2 + 2R_3)/(R_2 + R_3)] \times V_{sat+} - [2R_3/(R_2 + R_3)] \times V_{sat+}} = 2.71828 \qquad (10\text{-}14)$$

Cancelling the common V_{sat+} and $R_2 + R_3$ factors gives

$$\frac{R_2 + 2R_3}{R_2} = 2.71828 \qquad (10\text{-}15)$$

Finally, solving for R_3 gives

$$R_3 = 0.859 R_2 \qquad (10\text{-}16)$$

When R_2 and R_3 are chosen to satisfy Equation 10-16, the logarithmic term in Equation 10-7 becomes 1 and the frequency is

Final

$$f_{osc} = \frac{1}{2R_1 C_1} \qquad (10\text{-}17)$$

If we choose $R_2 = 10$ kΩ, then $R_3 = 8.59$ kΩ. Selecting the nearest standard value for R_3 makes it 8.2 kΩ. Assuming precise resistor values, this choice will give a 3% error in the final frequency. A better choice is to make $R_2 = 13$ kΩ and $R_3 = 11$ kΩ, giving a final frequency accurate to 1%.

Normally, in the design of this circuit, we will select capacitor C_1 and calculate R_1 from Equation 10-17 to give the desired frequency. The design equation for R_1 is

$$R_1 = \frac{1}{2 f_{osc} C_1} \qquad (10\text{-}18)$$

We will use Equations 10-16 and 10-18 as the design equations for the relaxation oscillator.

The frequency of a relaxation oscillator can be made variable by either varying R_1 or C_1. Because the capacitance values used are normally quite large (0.001 to 1.0 μF), variable capacitors are not available, and it is necessary to use a potentiometer for R_1 to provide continuous control of the frequency. Often, different

C_1 capacitors are switched to provide decade ranging, and the potentiometer R_1 is used for setting the frequency within the decade. There is one disadvantage with using a variable resistance to control the frequency—the frequency varies as $1/R_1$, not linearly. A change in resistance at the low end of the potentiometer produces a much larger change in frequency than does a change at the high end.

If we want to make the oscillator operate from a single supply, such as a digital timing circuit operating from a single 5 V supply, then we use the circuit shown in Figure 10-4. The design of C_1, R_1, R_2, and R_3 is exactly as discussed above. The frequency is determined from Equation 10-17, provided that the ratio of R_3 to R_2 is given by Equation 10-16.

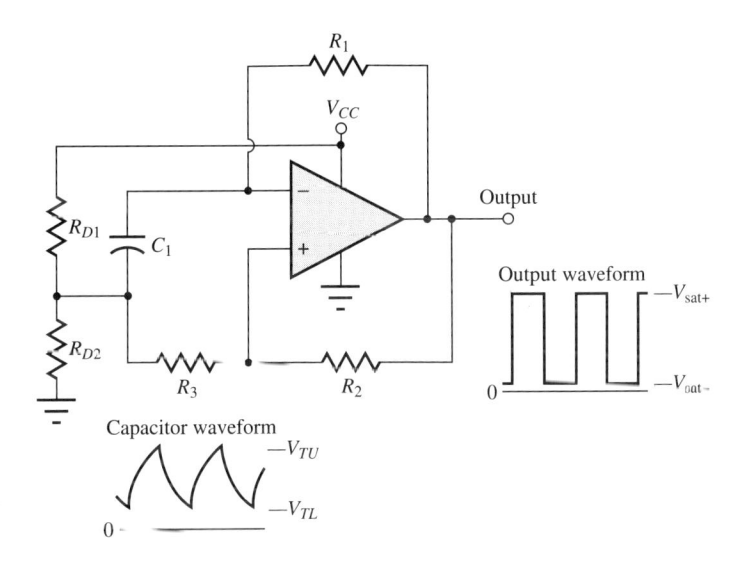

Figure 10-4 Relaxation oscillator operated from a single supply.

The voltage divider across the op-amp power supply inputs is used to establish a floating op-amp ground. Since we normally want symmetric operation, we make the divider resistors R_{D_1} and R_{D_2} equal and choose relatively large values so as not to load the DC supply. Typically, we would set $R_{D_1} = R_{D_2} = 10\ \text{k}\Omega$.

EXAMPLE 10-1

Design a relaxation oscillator to produce a square wave with a frequency of 2500 Hz. Use standard BiFET op-amps operating from a ± 15 V supply.

Solution

First we will draw the circuit of the relaxation oscillator as Figure 10-5. A parts list is included.

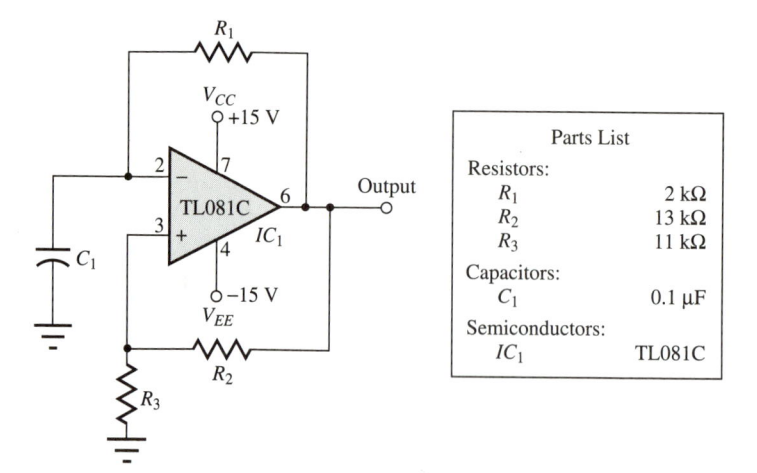

Figure 10-5

Parts List

Resistors:
- R_1 2 kΩ
- R_2 13 kΩ
- R_3 11 kΩ

Capacitors:
- C_1 0.1 μF

Semiconductors:
- IC_1 TL081C

To design this circuit, we will have to choose a value for C_1 and calculate values for resistors R_1, R_2, and R_3. We will use a 0.1 μF capacitor for C_1. Use Equation 10-18 to design R_1 for a frequency of 2.5 kHz:

$$R_1 = \frac{1}{2 f_{osc} C_1} = \frac{1}{2 \times 2.5 \text{ kHz} \times 0.1 \text{ μF}} = 2.0 \text{ kΩ}$$

This is a standard resistor.

The above design equation is valid only if $R_3 = 0.859 R_2$, as given in Equation 10-16. For maximum accuracy, we will choose R_2 as 13 kΩ, making R_3 equal to 11 kΩ.

The amplitude of the output square wave will be $V_{sat+} - V_{sat-}$, or approximately 28 V, since V_{sat+} is approximately +14 V and V_{sat-} is approximately −14 V for a ±15 V supply.

Our design is now complete.

Relaxation Oscillator for Generating Triangle Waves

The relaxation oscillator can also be used to generate a triangle-wave output—simply add an integrating circuit to the basic relaxation oscillator in Figure 10-3. The circuit of a combined square-wave/triangle-wave relaxation oscillator is shown in Figure 10-6. Recall that integrating circuits were discussed in Chapter 4. We must use the circuit that we introduced for integrating periodic waveforms, namely the circuit in Figure 4-11. To design the integrating circuit, we generally select a value for C_F and calculate values for R_F and R_I. The input resistor R_I is calculated from Equation 10-19. This equation is the same as Equation 4-14:

$$R_I = \frac{V_{in}}{V_{out}} \times \frac{1}{4 f_{osc} C_F} \tag{10-19}$$

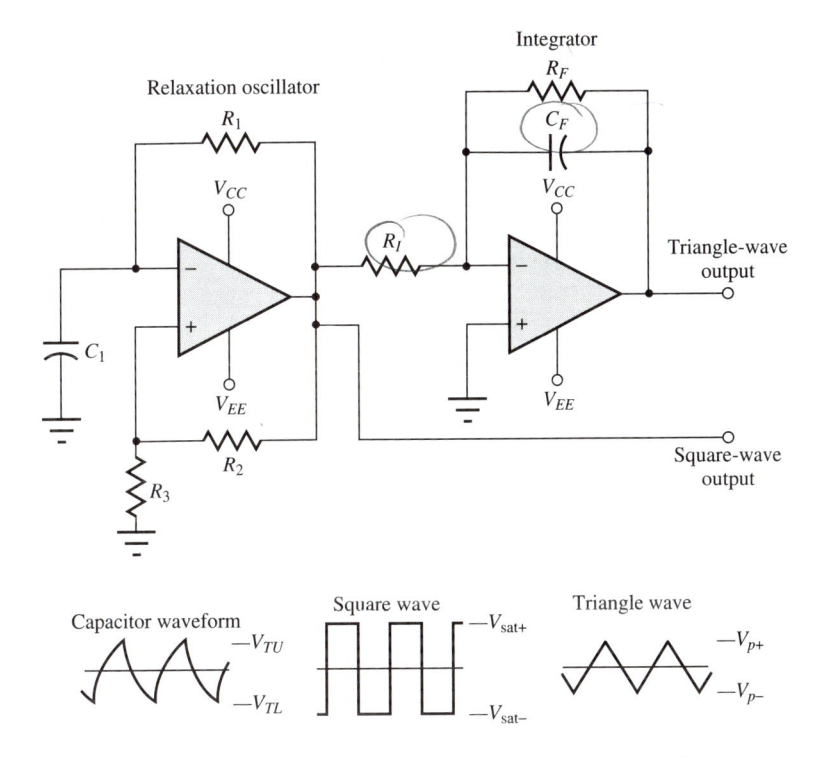

Figure 10-6 Relaxation oscillator with integrator for generation of triangle wave.

Here, V_{in} is the peak-to-peak voltage of the square wave produced by the relaxation oscillator, and V_{out} is the peak-to-peak voltage of the triangle wave. Typically, make V_{out} approximately half of V_{in}, to avoid distortion. As described in Chapter 4, resistor R_F must be included to prevent the integrator op-amp from saturating at high frequencies. Normally make $R_F = 25R_I$.

Making the triangle-wave oscillator variable in frequency is quite difficult. Resistors R_1 and R_I must be made variable and adjusted together. It is generally easier to use some other type of oscillator circuit.

EXAMPLE 10-2

Modify the relaxation oscillator designed in Example 10-1 to produce a triangle wave with a peak-to-peak amplitude of 10 V.

Solution

To make a relaxation oscillator generate a triangle wave, we simply add an integrating amplifier. The complete circuit is shown in Figure 10-7. A parts list is included.

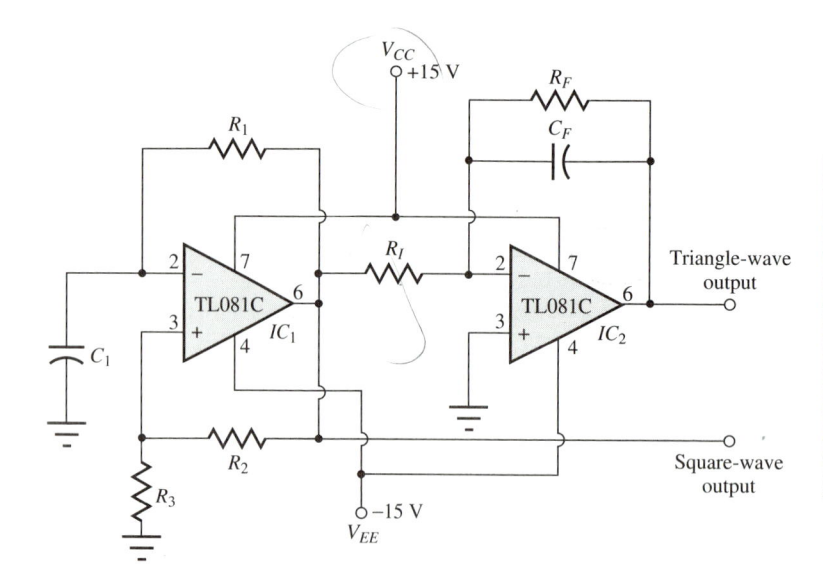

Figure 10-7

The design of the relaxation square-wave oscillator was shown in Example 10-1 and will not be repeated here.

To design the integrating circuit, we must choose a value for C_F and calculate values for R_I and R_F. Normally, when adding an integrating circuit to a relaxation oscillator, choose $C_F = C_1$. Therefore, we will use a value of 0.1 μF for C_F. The input resistor R_I is calculated from Equation 10-19. This equation requires V_{in}, the peak-to-peak voltage of the square wave, which we determined in Example 10-1 to be 28 V, and V_{out}, the peak-to-peak output voltage of the triangle wave, which was specified as 10 V:

$$R_I = \frac{V_{in}}{V_{out}} \times \frac{1}{4 f_{osc} C_F}$$

$$= \frac{28 \text{ V}}{10 \text{ V}} \times \frac{1}{4 \times 2.5 \text{ kHz} \times 0.1 \text{ μF}} = 2.80 \text{ k}\Omega$$

The closest standard resistor value is 2.7 kΩ.

Finally, choose resistor R_F to be $25 \times R_I$. This gives R_F as 67.5 kΩ. The closest standard value is 68 kΩ, which we will use.

Our design is now complete.

Variable-Duty-Cycle Relaxation Oscillator

The basic relaxation oscillator produces a symmetric square wave. The duty cycle of this oscillator is 0.5, where duty cycle D is defined as the ratio of the on time

10-2 Relaxation Oscillators

t_{on} to the period T:

$$D = \frac{t_{on}}{t_{on} + t_{off}} = \frac{t_{on}}{T} \tag{10-20}$$

This definition is the same as for the pulse-width modulator described in Chapter 9. The time t_{on} for the relaxation oscillator is simply the time when the square-wave output is V_{sat+}, whereas the time t_{off} is the time when the output is V_{sat-}. In the basic relaxation oscillator, the charge time for the capacitor is equal to the discharge time, the output is V_{sat+} for exactly half of the complete cycle, and the duty cycle is 0.5.

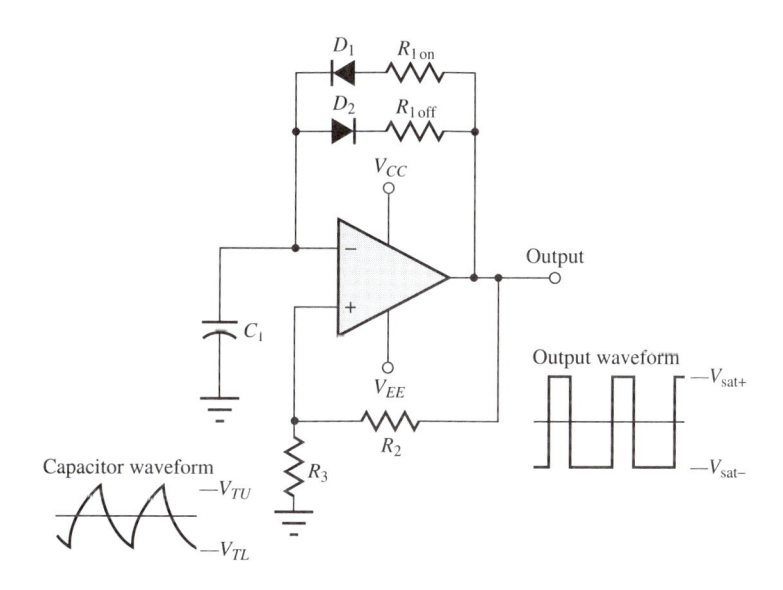

Figure 10-8 Relaxation oscillator circuit for producing a pulse wave.

Sometimes, instead of a symmetric square wave, a short-duration pulse output is required. In this case, the duty cycle is less than 0.5, and t_{on} with an output of V_{sat+} must be less than t_{off} with an output of V_{sat-}. The charge time of the capacitor, which determines t_{on}, must be shorter than the discharge time. This can be accomplished by the circuit in Figure 10-8. Resistor R_1 is replaced by resistors R_{1on} and R_{1off} in series with diodes D_1 and D_2, respectively. When the output is positive, diode D_1 is forward biased, and current flows through R_{1on} to charge C_1. When the output is negative, diode D_2 is forward-biased, and C_1 discharges through R_{1off}. The duty cycle is simply

$$D = \frac{R_{1on}}{R_{1on} + R_{1off}} \tag{10-21}$$

10 Oscillator Circuits

Provided that R_2 and R_3 are chosen according to Equation 10-16, the frequency is

$$f_{osc} = \frac{1}{(R_{1on} + R_{1off})C_1} \tag{10-22}$$

If a continuously variable duty cycle is desired, resistors R_{1on} and R_{1off} can be replaced by a single potentiometer with one end connected to D_1, the other end to D_2, and the wiper arm to the op-amp output. When the wiper arm is near D_1, the duty cycle is less than 0.5. When the wiper arm is near D_2, the duty cycle is greater than 0.5. Since the total resistance is constant, by Equation 10-22, the oscillator frequency remains constant as the duty cycle is changed.

If an integrating circuit is added to the variable-duty-cycle relaxation oscillator, as shown in Figure 10-9, then a sawtooth (or ramp) waveform is produced.

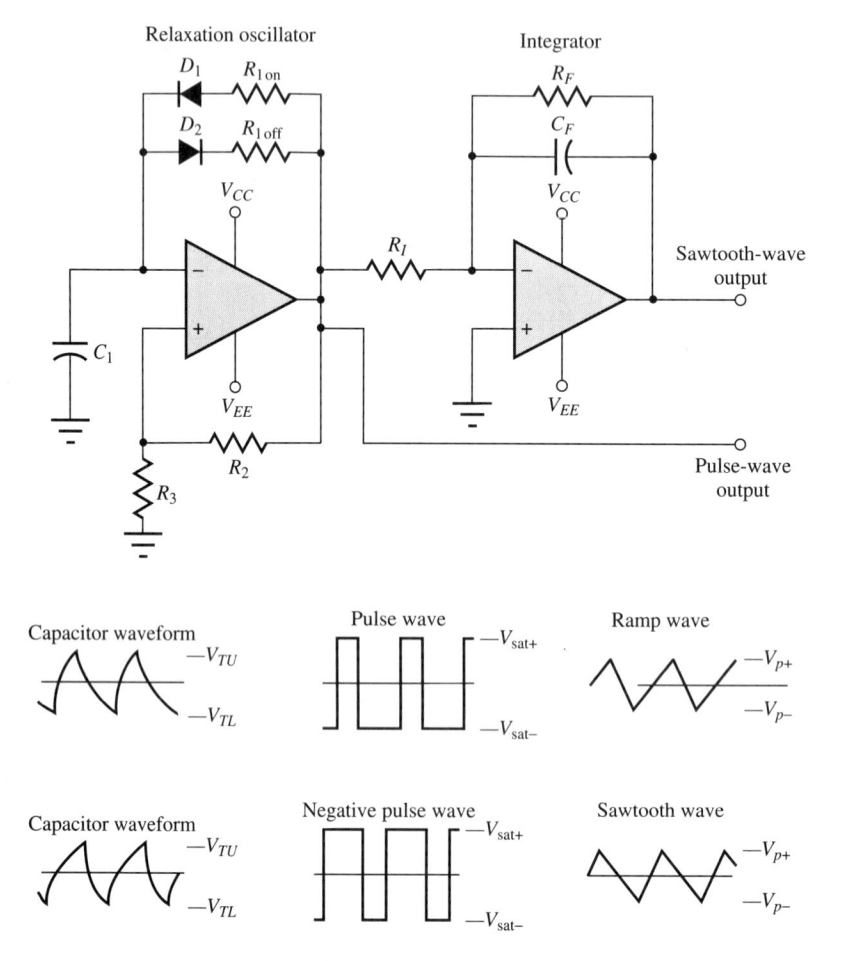

Figure 10-9 Relaxation oscillator with integrator for producing ramp and sawtooth waveforms.

If the duty cycle is less than 0.5, the output linearly increases relatively slowly, then returns rapidly to its initial value. Such waveforms are used in the sweep circuits of television receivers, video monitors, and oscilloscopes. Alternatively, if the duty cycle is more than 0.5, the output increases rapidly, then falls more slowly.

10-3 Bootstrap Oscillators

Another form of oscillator circuit can be used to produce square-wave and triangle-wave outputs. This is the bootstrap oscillator, shown in Figure 10-10. Although this circuit bears some resemblance to the relaxation oscillator with integrator shown in Figure 10-6, it is actually quite a different circuit, with only one capacitor and two fewer resistors than the relaxation oscillator with integrator.

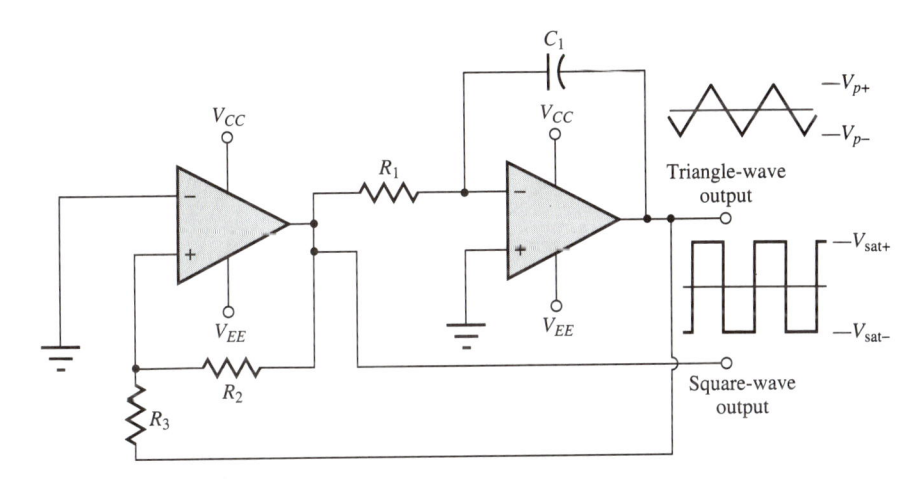

Figure 10-10 Bootstrap oscillator for producing square and triangle waves.

To understand the operation of this circuit, assume that the output of the first op-amp, acting as a comparator, is V_{sat+}. Since the noninverting input of the second op-amp is grounded, its inverting input will be a virtual ground, and the entire output voltage of the first op-amp will appear across resistor R_1. From Ohm's law, the current flowing through R_1 is

$$I_{in} = \frac{V_{sat+}}{R_1} \tag{10-23}$$

Since no actual current flows into the op-amp, this current must flow into the capacitor C_1. The voltage across a capacitor is $V = Q/C$. If the charge changes due to the flow of current I_{in} into the capacitor over a period of t seconds, the voltage will change by

$$\Delta V = \frac{\Delta Q}{C_1} = \frac{I_{in}t}{C_1} = \frac{V_{sat+}t}{R_1 C_1} \qquad (10\text{-}24)$$

Note that ΔV is increasing in a negative direction if the input voltage is positive. Since the voltage across the resistor is constant, the charging current is constant. Therefore, the voltage ramps down at a uniform rate.

Resistors R_2 and R_3 form a voltage divider. The voltage at the output of the first op-amp, applied to the R_2 end of the voltage divider, is V_{sat+}. The voltage at the output of the second op-amp, applied to the R_3 end of the voltage divider, is V_{out}, which would be ramping in a negative direction initially. The voltage at the center of the voltage divider, where it is connected to the noninverting input of the first op-amp V_{NI}, is

$$V_{NI} = V_{out} + (V_{sat+} - V_{out}) \times \frac{R_3}{R_2 + R_3} \qquad (10\text{-}25)$$

As long as this voltage is positive, the first op-amp output will remain positive, and the output voltage from the second op-amp will continue to decrease. When the output voltage of the second op-amp is sufficiently negative, V_{NI} passes zero, becoming negative, and the first op-amp output switches to V_{sat-}. Set V_{NI} to zero in Equation 10-25 and multiply out the remaining coefficients:

$$0 = V_{out} + (V_{sat+} - V_{out}) \times \frac{R_3}{R_2 + R_3}$$
$$R_2 V_{out} + R_3 V_{out} + R_3 V_{sat+} - R_3 V_{out} = 0 \qquad (10\text{-}26)$$

From this equation we can solve for V_{out}:

$$V_{out} = -V_{sat+} \times \frac{R_3}{R_2} \qquad (10\text{-}27)$$

This voltage will be the negative peak voltage of the triangle wave.

When the first op-amp output voltage becomes V_{sat-}, the current flow through the resistor R_1 is in the opposite direction, charge flows out of the capacitor through R_1, and the output voltage starts to ramp up until it reaches its positive peak. The comparator switches again, and the cycle starts over. This generates a triangle-wave output from the second op-amp and a square-wave output from the first op-amp.

If we assume that we are operating the oscillator from a balanced dual supply, the square-wave peak-to-peak output voltage is $+V_{sat}$ to $-V_{sat}$ or $2V_{sat}$:

$$V_{square} = 2V_{sat} \qquad (10\text{-}28)$$

The peak-to-peak triangle-wave output voltage is

$$V_{triangle} = 2V_{sat} \times \frac{R_3}{R_2} \qquad (10\text{-}29)$$

Note that R_2 must always be larger than R_3.

The period is the time for the capacitor C_1 to charge from the negative peak triangle-wave voltage to the positive peak triangle-wave voltage, then to discharge by the same amount. From Equation 10-24,

$$\Delta V = \frac{V_{\text{sat}+} t}{R_1 C_1} \quad \text{or} \quad t = R_1 C_1 \times \frac{\Delta V}{V_{\text{sat}+}} \tag{10-30}$$

where ΔV is the peak-to-peak triangle-wave output voltage, V_{triangle}, from Equation 10-29, and t is the time for one charge cycle or one discharge cycle:

$$t = R_1 C_1 \times \frac{V_{\text{triangle}}}{V_{\text{sat}+}} = R_1 C_1 \times \frac{2 V_{\text{sat}}}{V_{\text{sat}+}} \times \frac{R_3}{R_2}$$

$$= 2 R_1 C_1 \times \frac{R_3}{R_2} \tag{10-31}$$

The actual period is $2t$. Hence, the frequency f_{osc} is $1/2t$ and is

$$f_{\text{osc}} = \frac{1}{2t} = \frac{1}{4 R_1 C_1} \times \frac{R_2}{R_3} \tag{10-32}$$

Note that to design the bootstrap oscillator frequency, it is necessary to know or select values for C_1 and resistors R_1, R_2, and R_3. This makes the design somewhat complicated. Frequently, to simplify the design, the triangle-wave output amplitude is set equal to half of the saturation voltage so that the peak-to-peak triangle-wave output voltage is

$$V_{\text{triangle}} = V_{\text{sat}} \tag{10-33}$$

This requires that $R_2 = 2 R_3$, or

$$R_3 = 0.5 R_2 \tag{10-34}$$

In this case, the frequency is

$$f_{\text{osc}} = \frac{1}{2 R_1 C_1} \tag{10-35}$$

This is the same equation as for the relaxation oscillator. Normally, we will select capacitor C_1 and calculate R_1. The design equation for R_1 is

$$R_1 = \frac{1}{2 f_{\text{osc}} C_1} \tag{10-36}$$

Using Equations 10-34 and 10-36 as the design equations for the relaxation oscillator, we need only choose a value for C_1 and a value for R_3 (typically 10 kΩ) and calculate R_1 from Equation 10-36 and R_2 from Equation 10-34. The square-wave output amplitude is given by Equation 10-28, the triangle-wave output amplitude by Equation 10-29.

For single-supply operation of the bootstrap oscillator, it is necessary to set the floating ground as seen by the inverting input of the first op-amp, and by the noninverting input of the second op-amp, to half of the supply voltage. This is done with a voltage divider circuit as shown in Figure 10-11. The voltage divider

462 10 Oscillator Circuits

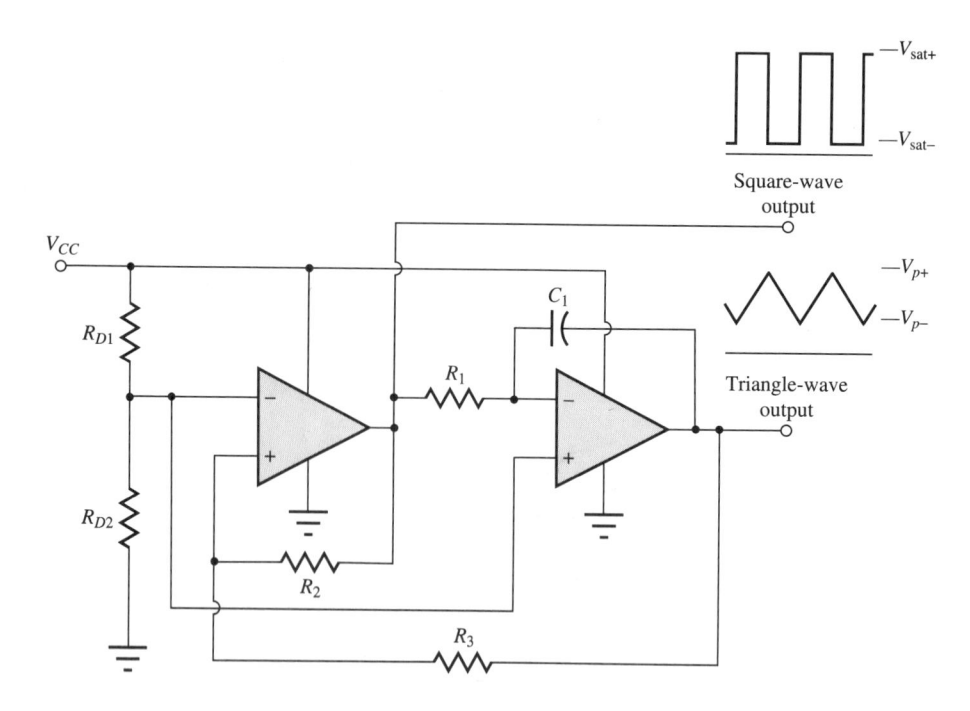

Figure 10-11 Bootstrap oscillator operated from a single power supply.

consisting of divider resistors R_{D_1} and R_{D_2} is placed across the power supply terminals. The inverting input of the first op-amp and the noninverting input of the second op-amp are connected to the junction of these two resistors. For a symmetrical square wave or triangle wave, both divider resistors should be equal in value. Any matched pair of resistors with values between 10 and 100 kΩ would be suitable. Both the triangle wave and square wave would be offset by $V_{CC}/2$ from the power supply ground.

To make the frequency of the bootstrap oscillator variable, normally R_1 is replaced by a potentiometer. Like the relaxation oscillator, the frequency varies as $1/R_1$ instead of linearly. If a linear variation in frequency is desired, resistor R_2 could be replaced by a potentiometer instead. In this case, however, the amplitude would change with frequency.

EXAMPLE 10-3

Design a bootstrap oscillator to produce both a square wave and a triangle wave at a frequency of 2.5 kHz. Use standard BiFET op-amps operating from a ±15 V supply. Estimate the square-wave amplitude, and calculate the triangle-wave amplitude.

Solution

The previous two examples developed a circuit for generating a triangle wave and a square wave by using a relaxation oscillator. The present example uses a bootstrap oscillator to produce the same output. When a triangle wave is required, the bootstrap oscillator is the simpler circuit to use because it does not require the design of a separate integrator. The circuit is shown in Figure 10-12, along with a parts list.

To design this circuit, we will have to choose a value for C_1 and calculate values for resistors R_1, R_2, and R_3. As in Example 10-1, we will use a 0.1 μF capacitor for C_1. We use Equation 10-36 to find R_1:

$$R_1 = \frac{1}{2 f_{osc} C_1} = \frac{1}{2 \times 2.5\ \text{kHz} \times 0.1\ \mu\text{F}} = 2.0\ \text{k}\Omega$$

This is a standard resistor. Note that it is exactly the same value as we found for the relaxation oscillator, since the equation is the same.

Equation 10-36 is only valid if $R_3 = 0.5R_2$ from Equation 10-34. We will choose R_2 as 20 kΩ, so R_3 will be 10 kΩ.

The peak-to-peak amplitude of the square-wave output, as given by Equation 10-28, is $2V_{sat}$, or approximately 28 V. The design equation for R_1 assumes that the peak-to-peak amplitude of the triangle wave is half that of the square wave; thus the triangle-wave amplitude is 14 V.

Our design is complete.

Choose $C_1 \Rightarrow R_1 = \frac{1}{2 f_{osc} C}$
Choose $R_2 \Rightarrow R_3 = 0.5 R_2$

$V_{square} = 24 V_{p-p} = 2V_{sat} = 2(15-1)$
$V_{triang} = 14 V_{p-p}$

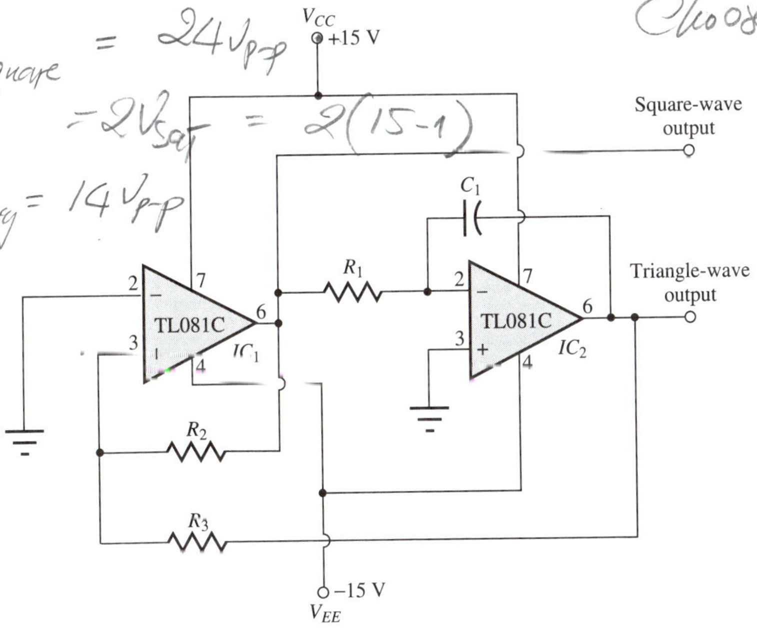

Parts List	
Resistors:	
R_1	2.0 kΩ
R_2	20 kΩ
R_3	10 kΩ
Capacitors:	
C_1	0.1 μF
Semiconductors:	
IC_1, IC_2	TL081C

Figure 10-12

EXAMPLE 10-4

Design a bootstrap oscillator to produce a 20 kHz triangle wave with an amplitude of 3 V. Use standard BiFET op-amps operating from a single 12 V supply.

Solution

In this problem, we have a single supply and must use a voltage divider to establish an effective ground for the oscillator circuit. The circuit is shown in Figure 10-13, with a parts list included.

Parts List

Resistors:
- R_1 — 4.3 kΩ
- R_2 — 10 kΩ
- R_3 — 3.0 kΩ
- R_{D1} — 10 kΩ
- R_{D2} — 10 kΩ

Capacitors:
- C_1 — 0.01 μF

Semiconductors:
- IC_1, IC_2 — TL081C

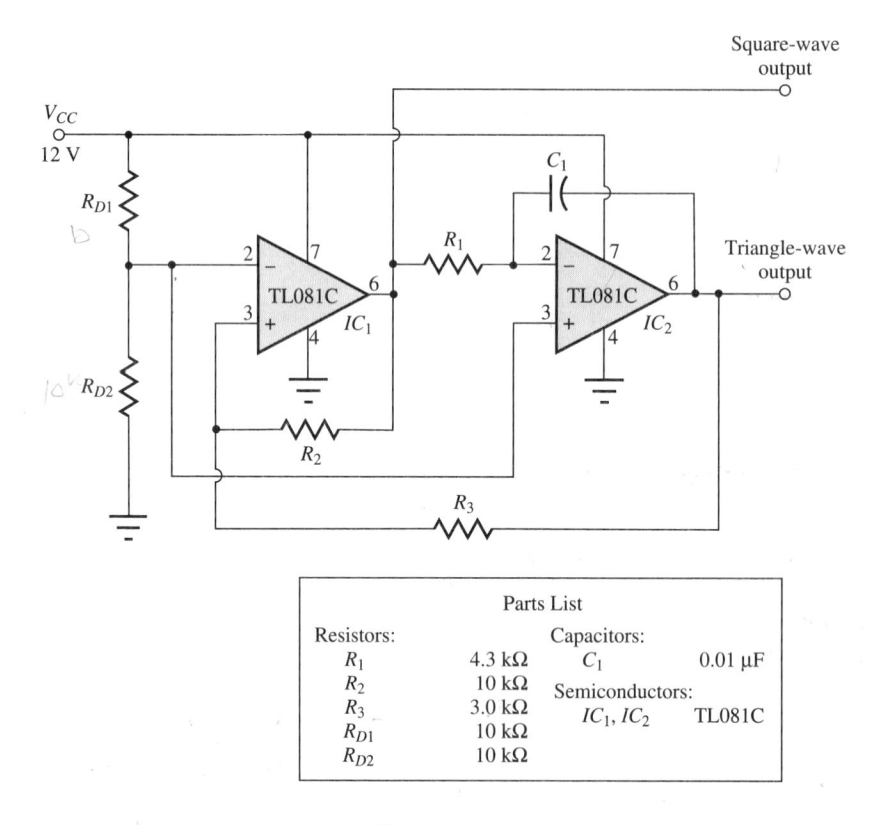

Figure 10-13

Since the peak-to-peak output voltage has been stipulated for this oscillator, we cannot use the simple design equations as in the previous example.

The peak-to-peak triangle-wave output voltage is given by Equation 10-29:

$$V_{\text{triangle}} = 2V_{\text{sat}} \times \frac{R_3}{R_2}$$

The quantity $2V_{\text{sat}}$ corresponds to the peak-to-peak voltage of the square wave. This will be approximately 2 V less than the supply voltage, or approximately

10 V. We will choose R_2 as 10 kΩ. Solve Equation 10-29 for R_3:

$$R_3 = \frac{V_{\text{triangle}}}{V_{\text{square}}} \times R_2 = \frac{3 \text{ V}}{10 \text{ V}} \times 10 \text{ k}\Omega = 3.0 \text{ k}\Omega$$

This is a standard value.

Since we did not choose R_2 and R_3 according to Equation 10-34, we cannot use Equation 10-36 to calculate R_1. Instead, we must use Equation 10-32. This equation defined f_{osc} as

$$f_{\text{osc}} = \frac{1}{4R_1C_1} \times \frac{R_2}{R_3}$$

We know R_2, R_3, and f_{osc}. Choose $C_1 = 0.01$ μF and solve for R_1.

$$R_1 = \frac{1}{4f_{\text{osc}}C_1} \times \frac{R_2}{R_3}$$

$$= \frac{10 \text{ k}\Omega}{4 \times 20 \text{ kHz} \times 0.01 \text{ }\mu\text{F} \times 3.0 \text{ k}\Omega} = 4.17 \text{ k}\Omega$$

The closest standard value from the table in Appendix A-1 is 4.3 kΩ. We will choose this for R_1.

Lastly, we must select the divider resistors. For a symmetric triangle wave, we must have $R_{D_1} = R_{D_2}$. Choose both resistors to be 10 kΩ.

Our design is complete.

Variable-Duty-Cycle Bootstrap Oscillator

In the previous section, we saw how the duty cycle of a relaxation oscillator could be made variable by replacing resistor R_1 with parallel resistors $R_{1\text{on}}$ and $R_{1\text{off}}$ in series with oppositely biased diodes. The duty cycle of a bootstrap oscillator can be changed in a similar manner. The circuit is shown in Figure 10-14. The analysis of this circuit is the same as for the relaxation oscillator, and the same design equations apply. The duty cycle of the pulse wave is

$$D = \frac{R_{1\text{on}}}{R_{1\text{on}} + R_{1\text{off}}} \quad (10\text{-}37)$$

This equation is the same as Equation 10-21 for the relaxation oscillator. Provided that R_2 and R_3 are chosen according to Equation 10-34, the frequency is

$$f_{\text{osc}} = \frac{1}{(R_{1\text{on}} + R_{1\text{off}})C_1} \quad (10\text{-}38)$$

This equation is the same as Equation 10-22 for the relaxation oscillator. If R_2 and R_3 are chosen to set the amplitude of the triangle (or sawtooth) wave, and not chosen according to Equation 10-34, then the frequency is

$$f_{\text{osc}} = \frac{1}{2(R_{1\text{on}} + R_{1\text{off}})C_1} \times \frac{R_2}{R_3} \quad (10\text{-}39)$$

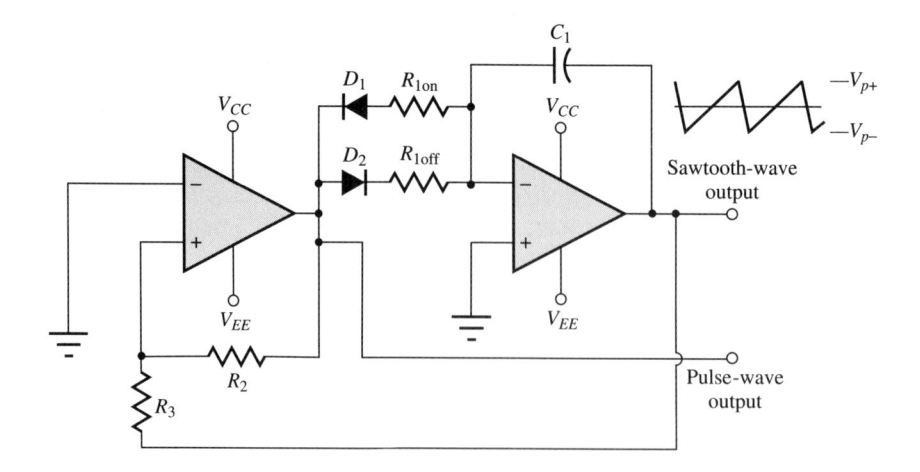

Figure 10-14 Bootstrap oscillator for producing pulse and ramp waves.

To design this oscillator, select a value for C_1 and solve for $R_{1on} + R_{1off}$ in either Equation 10-38 or 10-39. Substitute this value into Equation 10-37 and solve for R_{1on}.

$$R_{1on} = D \times (R_{1on} + R_{1off}) \tag{10-40}$$

Finally, calculate R_{1off} by subtracting R_{1on} from the $R_{1on} + R_{1off}$ value calculated earlier. For a continuously variable duty cycle, replace R_{1on} and R_{1off} by a single potentiometer.

EXAMPLE 10-5

Design a sawtooth-wave oscillator that has a rise time of 0.01 s and a fall time of 0.001 s. Use a TL081C op-amp operating from a ± 15 V supply.

Solution

The circuit is shown in Figure 10-15, with a parts list included.
  First we must determine the duty cycle. The duty cycle is defined as the fall time divided by the total period. The fall time is 0.001 s, and the total period is the rise time plus the fall time, or 0.011 s. Consequently, the duty cycle is 0.001 s/0.011 s = 0.09091.
  The frequency is simply the reciprocal of the period, 1/0.011 s = 90.91 Hz. We will assume that $R_2 = 2R_3$. In this case, the frequency is given by Equation 10-38:

$$f_{osc} = \frac{1}{(R_{1on} + R_{1off})C_1}$$

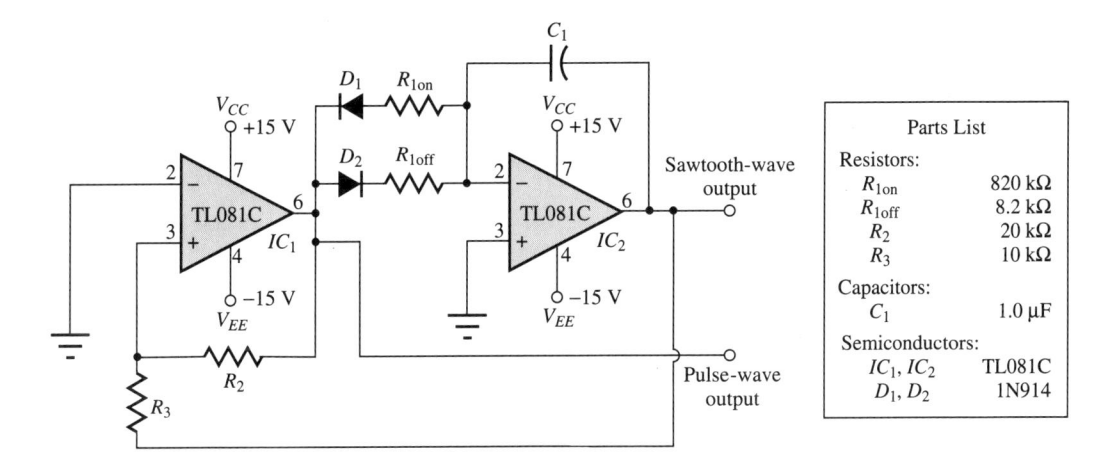

Figure 10-15

Choose a value of 1.0 μF for C_1. Solve for $R_{1on} + R_{1off}$.

$$R_{1on} + R_{1off} = \frac{1}{f_{osc}C_1} = \frac{1}{90.9 \text{ Hz} \times 1.0 \text{ μF}} = 9.31 \text{ k}\Omega$$

Substitute this value into Equation 10-40 to determine R_{1on}.

$$R_{1on} = D \times (R_{1on} + R_{1off}) = 0.09091 \times 9.31 \text{ k}\Omega = 846 \text{ }\Omega$$

Calculate R_{1off} by subtracting R_{1on} from $R_{1on} + R_{1off}$ to obtain 9.31 kΩ − 846 Ω = 8.46 kΩ. Choose R_{1on} = 820 Ω and R_{1off} = 8.2 kΩ.

Finally, we must choose R_2 and R_3. From Equation 10-34 we know that $R_3 = 0.5R_2$. Choose R_2 = 20 kΩ, then R_3 = 10 kΩ.

Our design is now complete.

10-4 Sine-Wave Oscillators

Thus far, we have discussed a number of square-wave oscillator circuits. Square waves are easy to produce, since by using sufficient positive feedback to drive an op-amp into positive or negative saturation, two alternating levels are precisely defined. Triangle waves are also easy to produce by integrating the square wave. Provided that suitable components are used, both the square waves and the triangle waves are essentially perfect in shape.

It is possible to produce a sine wave from a triangle wave by using a special wave-shaping circuit. A typical wave-shaping circuit is developed in Appendix C and will be mentioned later. The sine wave produced by this method is not perfect and may have several percent distortion. The production of a true sine wave requires a special, relatively complex oscillator circuit that satisfies a set of requirements called the **Barkhausen criteria**.

The Barkhausen Criteria

For any circuit to oscillate and produce a true or perfect sine-wave output, three requirements must be satisfied.

1. There must be positive feedback in the circuit.
2. The product of the feedback factor B and the circuit open-loop gain must be exactly unity.
3. There must be some frequency-selective network in the circuit such that a $0°$ (or $360°$) phase shift occurs for only one frequency.

These conditions are known as the Barkhausen criteria. There are various other ways of expressing these criteria, but whatever the form they will always express the same basic requirements.

The first Barkhausen criterion is easily satisfied. Using an op-amp with a noninverting input, a positive feedback path can be achieved simply by connecting the output back to the noninverting input.

The second Barkhausen criterion needs some elaboration. The gain of an amplifier with negative feedback was derived in Chapter 2 and is described by Equation 2-16:

$$A_{VF} = \frac{A_{VOL}}{1 + BA_{VOL}} \qquad (2\text{-}16)$$

The quantity A_{VOL} is the open-loop gain of the amplifier, and B is the feedback factor or fraction of the output that is fed back to the input. The gain of an amplifier with positive feedback is given by a similar equation:

$$A_{VF} = \frac{A_{VOL}}{1 - BA_{VOL}} \qquad (10\text{-}41)$$

The gain of the circuit is always larger than the open-loop gain. If, for example, $BA_{VOL} = 0.9$, then the amplifier gain would be ten times the open-loop gain. As an amplifier, the system could provide stable gain for an input signal. An oscillator, however, does not have an input signal; oscillations must start from zero. With BA_{VOL} equal to 1, the gain becomes infinite, and oscillations starting at zero can build up to a stable level, without the application of an input signal. If BA_{VOL} is greater than 1, more energy is fed back than necessary to build and sustain oscillations, and the output is distorted. The square-wave oscillator circuits that we discussed earlier used the op-amp as a high-gain amplifier with a large value of B. The quantity BA_{VOL} was much larger than 1, causing the output to saturate and produce a square wave. In the case of a sine-wave oscillator, we want an undistorted wave, so we must make BA_{VOL} exactly 1. Normally, this is not a hard criterion to satisfy.

The third Barkhausen criterion has the most complex requirements. In the oscillator circuits discussed earlier, the feedback path was a resistive voltage divider from the output to the noninverting input, providing a $0°$ or $360°$ phase shift that was the same for all frequencies. Fourier analysis of square waves and triangle waves shows that these waveforms can be described as the sum of a series of sine waves of different harmonic frequencies. All of these harmonic components must have the same positive feedback. A sine wave, however, has only one Fourier component—the sine wave itself. The presence of any other frequency components would add distortion to the output waveform. The feedback network must be such that only one frequency receives a 0° or 360° phase shift. Two types of networks are commonly used: the **phase-shift network** and the **Wien bridge network**. We will develop sine-wave oscillators using both of these.

Phase-Shift Oscillator

The phase-shift oscillator uses an op-amp configured as an inverting amplifier, with an additional feedback loop consisting of a phase-shift network. The basic circuit is shown in Figure 10-16. The phase-shift network is simply a series of three capacitor/resistor or resistor/capacitor voltage dividers. The capacitive reactance and, consequently, the phase shift introduced by each divider varies with frequency between 0° and 90°. At the desired operating frequency, *and only at this one frequency*, the combined phase shift from the three dividers is 180°. This feedback signal is returned to the inverting input of the amplifier, introducing an additional 180° phase shift and providing the positive feedback required for oscillation.

In a practical oscillator circuit, all three resistors in the phase-shift network are chosen to have the same value, as are the three capacitors. If this is the case, then the phase shift is exactly 180° for a frequency f_{osc} given by

$$f_{osc} = \frac{1}{2\pi RC \sqrt{6}} \tag{10-42}$$

This equation can be derived from basic circuit theory. In the design of the phase-

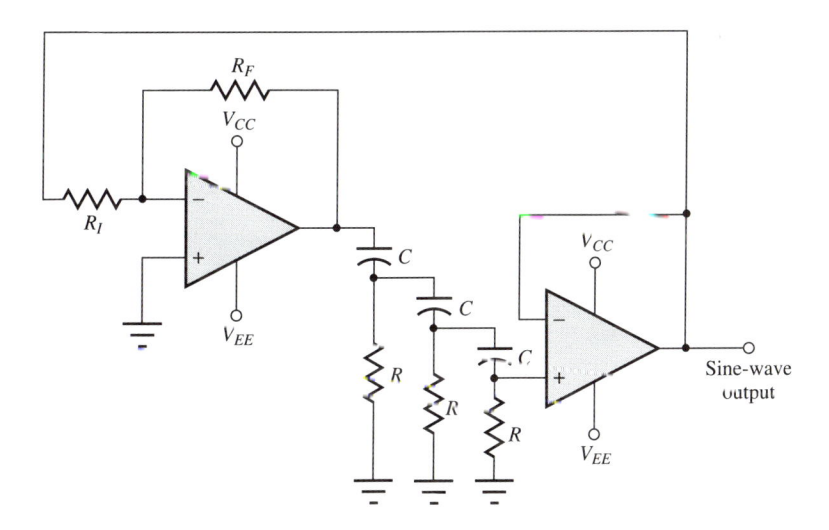

Figure 10-16 Phase-shift oscillator circuit.

shift network for some specific frequency f_{osc}, C is usually chosen and R is calculated from Equation 10-42.

Since the phase-shift network is actually three voltage dividers, it attenuates the input signal. The phase-shift network can be analyzed to show that the input signal is attenuated by a factor of 29 at the frequency given by Equation 10-42. In the absence of any other attenuation, the feedback factor B is $\frac{1}{29}$. Hence, the amplifier must have a gain of 29 to satisfy the requirement $BA_{VOL} = 1$.

The second op-amp acts as a buffer to prevent loading of the phase-shift network. The oscillator shown in Figure 10-16 can be made to work without this second op-amp by simply connecting the output of the phase-shift network back to R_I. In this case, the input resistor R_I must be chosen large enough so that it does not load the phase-shift network. However, as we will see shortly, this buffer amplifier is usually necessary.

If the circuit described in Figure 10-16 were designed and constructed, it probably would not work (either with or without the buffer op-amp). It would not oscillate, or it would oscillate producing a severely distorted sine wave. The gain is critical. If the gain is not large enough to compensate for the loss introduced by the phase-shift network, then the circuit will not oscillate. If the gain is larger than the loss, the output will saturate and the wave will be distorted. If the gain is exactly correct, the oscillations may take a long time to stabilize.

Figure 10-17 shows a practical circuit. The negative feedback loop now has two resistors, R_{F_1} and R_{F_2}, in series forming the effective feedback resistor R_F. Resistor R_{F_2} is in parallel with two back-to-back zener diodes. The zener diodes act to give the amplifier a variable feedback resistance. Any nonlinear resistive element could be used; an incandescent light bulb or a junction field-effect transistor (JFET) are frequently used in place of the zener diodes. When the power

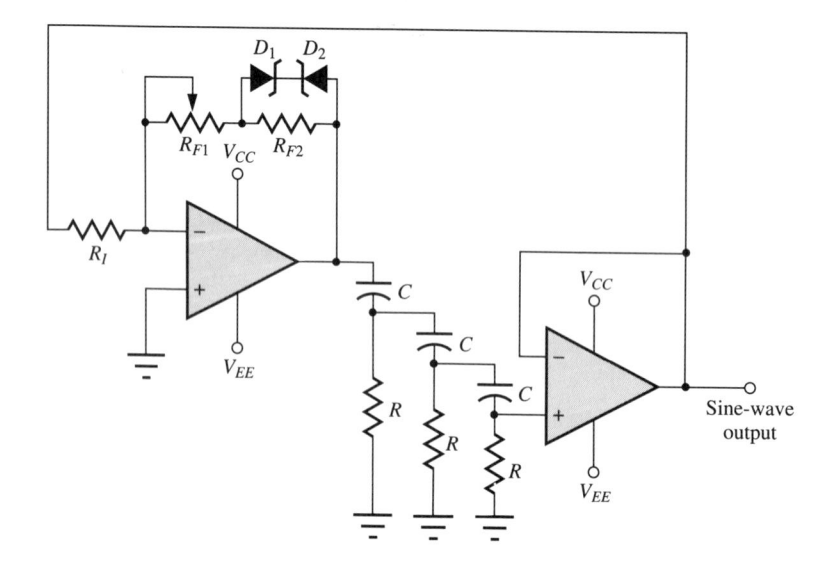

Figure 10-17 Practical phase-shift oscillator circuit.

is first turned on, the oscillations have a very low amplitude and the zener diodes do not conduct. The feedback resistance is equal to $R_{F_1} + R_{F_2}$, giving a high gain. The oscillations build up rapidly. When the peak amplitude of the oscillations reaches the zener voltage, the zeners short out R_{F_2}, the feedback resistance is now R_{F_1}, and the gain drops to the optimal value for sustaining the oscillations.

There is one major constraint on the value of R_{F_1}, however. This resistor must pass sufficient bias current so that the zener diodes have a low impedance when conducting. There is no simple equation for determining a suitable value, but, as a rule, always make R_{F_1} less than or equal to 1.0 kΩ. Now calculate R_I for a normal inverting amplifier with a gain of 29:

$$R_I = \frac{R_{F_1}}{29} \tag{10-43}$$

If we use R_{F_1} as 1.0 kΩ, R_I has a typical value of 33 Ω. This is a very small value and would seriously load the feedback phase-shift loop. The reason for including the buffer amplifier should now be apparent.

The value of R_{F_2} is calculated to give a starting gain larger than 29. The start-up gain $A_{V\text{start}}$ is

$$A_{V\text{start}} = \frac{R_{F_1} + R_{F_2}}{R_I} \tag{10-44}$$

Solving for R_{F_2}, we get

$$R_{F_2} = (A_{V\text{start}} \times R_I) - R_{F_1} \tag{10-45}$$

The value of the start-up gain is not critical; any value larger than 29 is suitable.

Often a potentiometer is used for R_{F_1}, with a value somewhat larger than the necessary resistance. This allows the gain to be adjusted to correct for additional circuit attenuation and for nonzero zener resistance.

The amplitude of the output sine wave is determined by the zener diodes. One zener diode is always forward-biased. The other conducts when the voltage at the output exceeds $(V_Z + 0.7)$ V. This will be approximately the peak amplitude of the sine-wave output.

This phase-shift oscillator can be made variable in frequency by replacing one of the resistors in the phase-shift network with a potentiometer. Unfortunately, this will change the attenuation of the network, so the amplifier gain will no longer be correct and the output sine wave may distort. It is best to replace all three phase-shift resistors with potentiometers and adjust them simultaneously.

EXAMPLE 10-6

Design a phase-shift oscillator to produce a sine wave at a frequency of 300 Hz with an amplitude of at least 10 V peak-to-peak. Use two TL081C op-amps powered from a ±15 V supply

Solution

The phase-shift oscillator circuit is shown in Figure 10-18. Note that we have included the zener diode starting circuit and have used a potentiometer for R_{F_1} to allow the gain to be precisely set. Two parts of the circuit require designing: the phase-shift network and the amplifier.

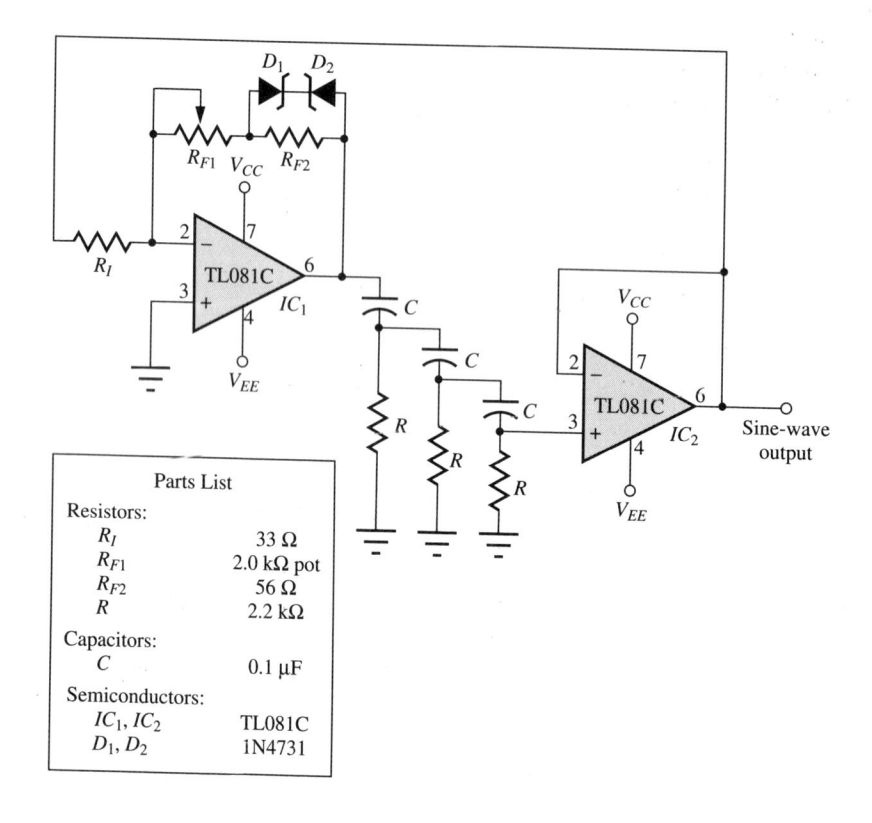

Figure 10-18

Parts List

Resistors:
- R_I — 33 Ω
- R_{F1} — 2.0 kΩ pot
- R_{F2} — 56 Ω
- R — 2.2 kΩ

Capacitors:
- C — 0.1 μF

Semiconductors:
- IC_1, IC_2 — TL081C
- D_1, D_2 — 1N4731

First, design the phase-shift network. Choose $C = 0.1$ μF and rearrange Equation 10-42 to solve for R:

$$R = \frac{1}{2\pi f_{osc} C \sqrt{6}}$$

$$= \frac{1}{2\pi \times 300 \text{ Hz} \times 0.1 \text{ μF} \times 2.44949} = 2.17 \text{ kΩ}$$

Choose a value of 2.2 kΩ for R from Appendix A-1.

Next, design the amplifier stage. We will use a 2.0 kΩ potentiometer for R_{F_1}, but will assume that it has a nominal value of 1.0 kΩ for calculating R_I.

Determine R_I from Equation 10-43:

$$R_I = \frac{R_{F_1}}{29} = \frac{1.0 \text{ k}\Omega}{29} = 34.5 \text{ }\Omega$$

We will use a value of 33 Ω.

Now calculate a value for R_{F_2} from Equation 10-45. We will assume a starting gain of 32:

$$R_{F_2} = 32R_I - R_{F_1} = 32 \times 33 \text{ }\Omega - 1.0 \text{ k}\Omega = 56 \text{ }\Omega$$

This is a standard value, so we will make $R_{F_2} = 56 \text{ }\Omega$.

Lastly, we must select suitable zener diodes. Since we require a minimum 10 V peak-to-peak output, we want $V_Z + 0.7 \geq 5.0$, or $V_Z \geq 4.3$ V. From Appendix A-4, we see that the 1N4731 is a 4.3 V zener. We will use two of these.

Our design is now complete.

Wien Bridge Oscillator

The second frequency-selective network used in oscillator circuits is the Wien bridge. The Wien bridge is an AC version of the Wheatstone bridge. Where the Wheatstone bridge is used for precision resistance measurement, the Wien bridge is used for the precision measurement of capacitance. Two arms of the Wien bridge act as a simple resistive voltage divider; the other two arms are a series reactive circuit and a parallel reactive circuit, respectively. The bridge is balanced when the series reactance equals the parallel reactance. At balance, the phase shift introduced by the series reactance is equal and opposite to the phase shift introduced by the parallel reactance.

Two versions of an oscillator circuit designed using a Wien bridge are shown in Figure 10-19. Figure 10-19(a) shows the bridge circuit with the differential inputs of the op-amp connected between the two diagonal corners of the bridge, with the output and ground connected to the other two corners. Figure 10-19(b) shows the same circuit, only redrawn to emphasize the separate negative and positive feedback loops. Note the zener diode starting circuit, used in exactly the same way as for the phase-shift oscillator.

Positive feedback is through a network consisting of a series R-C circuit and a parallel R-C circuit. The parallel R-C circuit produces a phase lag, and the series R-C circuit produces a phase lead. At only one frequency is the phase lag equal to the phase lead. At this frequency, the total phase shift is zero, defining the operating frequency of the oscillator. Normally both resistors are chosen to be the same value R, and both capacitors the same value C. The phase of the feedback signal at the noninverting input is zero only for the oscillator frequency

$$f_{osc} = \frac{1}{2\pi RC} \tag{10-46}$$

At this frequency, the phase-shift network introduces an attenuation of 3, so the amplifier must be designed with a gain of 3 to make $BA_{VOL} = 1$.

(a) Oscillator drawn to show the Wien Bridge

(b) Standard representation of the Wien Bridge oscillator

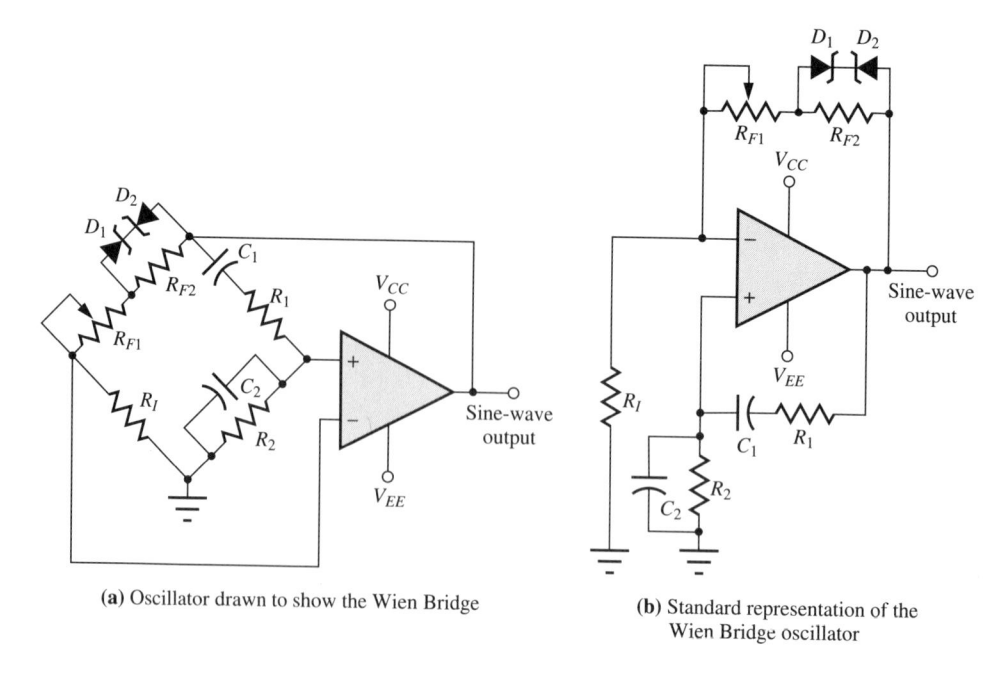

Figure 10-19 Wien Bridge oscillator.

The amplifier gain is determined by the negative feedback loop consisting of R_{F_1}, R_{F_2}, and the zener diode starting circuit, connected between the output and the inverting input, and input resistor R_I connected from the inverting input to ground. This is the circuit of a noninverting amplifier. As with the phase-shift oscillator discussed earlier, R_{F_1} must be kept less than or equal to 1.0 kΩ in order to pass sufficient bias current so that the zener diodes have a low impedance when conducting. Calculate R_I for a noninverting amplifier with a gain of 3:

$$R_I = \frac{R_{F_1}}{A_V - 1} = \frac{R_{F_1}}{2} \tag{10-47}$$

The value of R_{F_2} is calculated to give a starting gain larger than 3. The start-up gain $A_{V\text{start}}$ is

$$A_{V\text{start}} = \frac{R_{F_1} + R_{F_2}}{R_I} + 1 \tag{10-48}$$

Solving for R_{F_2}, we get

$$R_{F_2} = R_I \times (A_{V\text{start}} - 1) - R_{F_1} \tag{10-49}$$

The value of the start-up gain is not critical; any value larger than 3 is suitable.

As for the phase-shift oscillator, a potentiometer is generally used for R_{F_1} to allow the gain to be adjusted to correct for additional circuit attenuation and for nonzero zener resistance. The peak amplitude of the output sine wave will be approximately $(V_Z + 0.7)$ V.

The Wien bridge oscillator can be made variable in frequency by replacing both of the resistors in the phase-shift network with potentiometers. Like the phase-shift oscillator, replacing only one resistor with a potentiometer will make the frequency variable but will change the attenuation and, hence, affect the quality of the output waveform.

EXAMPLE 10-7

Design a Wien bridge oscillator to produce a sine wave at a frequency of 2400 Hz with an amplitude of at least 12 V peak-to-peak. Use a TL081C op-amp powered from a ±15 V supply.

Solution

The Wien bridge oscillator circuit is shown in Figure 10-20. As in the previous example, we have included the zener diode starting circuit and have made R_{F_1} a potentiometer to allow the gain to be precisely set. The design of the Wien bridge oscillator is very similar to the design of the phase-shift oscillator in the previous example.

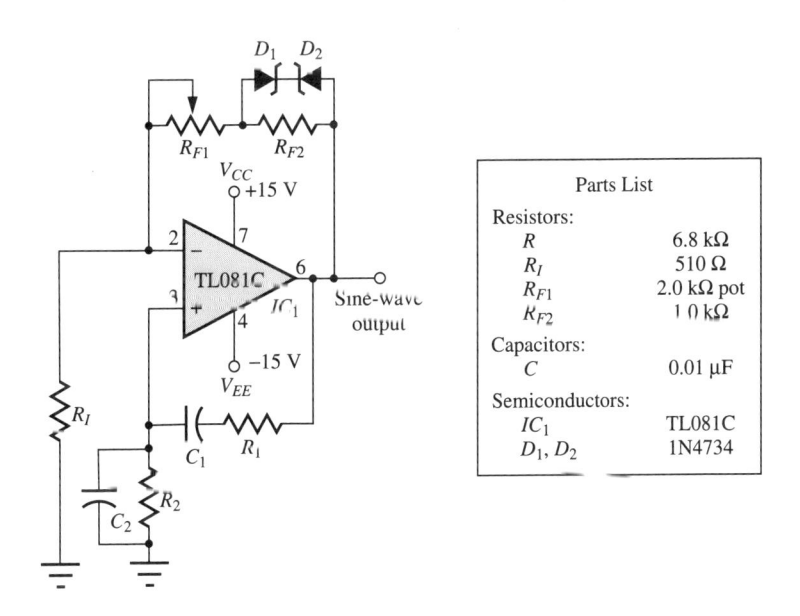

Parts List

Resistors:
R	6.8 kΩ
R_I	510 Ω
R_{F1}	2.0 kΩ pot
R_{F2}	10 kΩ

Capacitors:
C	0.01 μF

Semiconductors:
IC_1	TL081C
D_1, D_2	1N4734

Figure 10-20

First, design the positive feedback network. Choose $C = 0.01 \, \mu F$ and rearrange Equation 10-46 to solve for R:

$$R = \frac{1}{2\pi f_{osc} C}$$

$$= \frac{1}{2\pi \times 2.4 \text{ kHz} \times 0.01 \, \mu F} = 6.63 \text{ k}\Omega$$

The closest standard value is 6.8 kΩ.

Next, design the amplifier stage. We will use a 2.0 kΩ potentiometer for R_{F_1}, but will assume a nominal value of 1.0 kΩ for R_{F_1} for calculating R_I. Calculate R_I from Equation 10-47:

$$R_I = \frac{R_{F_1}}{2} = \frac{1.0 \text{ k}\Omega}{2} = 500 \, \Omega$$

Use the closest standard value of 510 Ω.

Next calculate R_{F_2} from Equation 10-49. We will assume a starting gain of 4. This value is not critical:

$$R_{F_2} = R_I \times (A_{V\text{start}} - 1) - R_{F_1} = 510 \, \Omega \times 3 - 510 \, \Omega = 1.02 \text{ k}\Omega$$

We will use a value of 1.0 kΩ for R_{F_2}.

Lastly, select suitable zener diodes. Since we want a minimum 12 V peak-to-peak output, $V_Z + 0.7 \geq 6.0$, or $V_Z \geq 5.3$ V. From Appendix A-4, we see that the 1N4734, a 5.6 V zener diode, is suitable. We will use two of these.

Our design is now complete.

10-5 Other Oscillator Circuits

Previously we developed circuits for generating square waves, triangle waves, and sine waves. Although these are the most important oscillator circuits, a few other oscillator circuits deserve mention.

Bootstrap Sine-Wave Oscillator

The sine-wave generators that we designed can yield very accurate waveforms, but are quite sensitive to the actual component values. Often a precise sine wave is not required, and a simple sine-wave oscillator that is insensitive to component values is satisfactory. Such a circuit is shown in Figure 10-21. This circuit can best be considered as a bootstrap sine-wave oscillator.

To understand the operation of the circuit, assume that the output of the second op-amp is a sine wave. This signal is fed back to the input of the first stage, and, since the first op-amp acts as a simple zero-crossing comparator, the output of this stage will be a square wave. This square wave is fed into the second stage, a narrow-band bandpass filter identical to those discussed in Chapter 6.

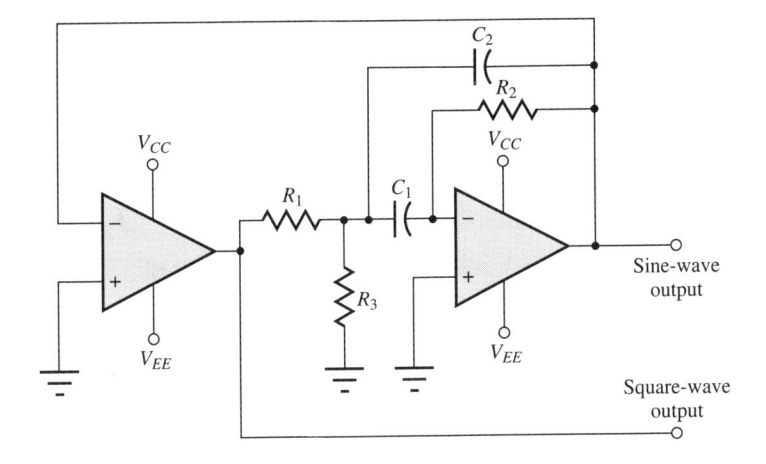

Figure 10-21 Bootstrap sine-wave oscillator.

Recall that, by the Fourier theorem, a square wave can be considered as composed of the sum of a series of sine waves. The bandpass filter passes only one of these sine waves, the one closest in frequency to the center frequency of the filter. The comparator converts this into a square wave of the same frequency and feeds it back into the filter. The final result is a square wave and a sine wave with frequencies equal to the center frequency of the bandpass filter. Because the bandpass filter does not have infinite attenuation for frequencies outside the bandpass, it will pass some of the higher-frequency Fourier components of the square wave, thus introducing some distortion. A more complicated bandpass filter can be used if a better waveform is required.

The design of this circuit involves the design of the bandpass filter. The frequency of the sine wave is simply the center frequency of the filter and is

$$f_{osc} = \frac{1}{2\pi} \times \sqrt{\frac{R_1 + R_3}{C_1 C_2 R_1 R_2 R_3}} \tag{10-50}$$

This is Equation 6-17. Normally we choose $C_1 = C_2$ to simplify the design.

For the best output waveform, we require the highest Q (quality factor) possible, which is

$$Q_{max} = \sqrt{\frac{f_{GB}}{20 f_{center}}} \tag{10-51}$$

This is Equation 6-19. The quantity f_{GB} is simply the unity-gain bandwidth of the op-amp, which for a TL081C BiFET op-amp is approximately 3 MHz.

Once values for capacitors C_1 and C_2 have been chosen, resistor R_2 can be determined from the equation

$$R_2 = \frac{Q}{\pi f_{osc} C_2} \tag{10-52}$$

This is Equation 6-24. Next, resistor R_1 can be calculated from the equation

$$R_1 = \frac{R_2}{4Q^2} \qquad (10\text{-}53)$$

This is Equation 6-25. Finally, resistor R_3 can be calculated from the equation

$$R_3 = \frac{R_1}{4\pi^2 f_{osc} R_1 C_1 R_2 C_2 - 1} \qquad (10\text{-}54)$$

This is Equation 6-26. In practice, it is possible to simply select any value for R_3 much larger than R_1. This value can be substituted into Equation 10-50 to verify that the frequency is correct to within a few percent. The accuracy will be limited by the tolerance of the components to an error larger than this.

On the basis of the preceding discussion, this type of sine-wave oscillator may seem more complicated than the true sine-wave oscillators discussed previously. In fact, it is a much easier circuit to get working and is often preferred over the true sine-wave oscillator circuits. Although precise resistance and capacitance values must be used in the filter network for optimal performance, the circuit will oscillate and produce some form of sine wave with almost any reasonable component values.

EXAMPLE 10-8

Design a simple bootstrap oscillator to generate a sine wave with a frequency of 500 Hz. Use a TL081C op-amp operating from a ±15 V supply.

Solution

The circuit is shown in Figure 10-22.

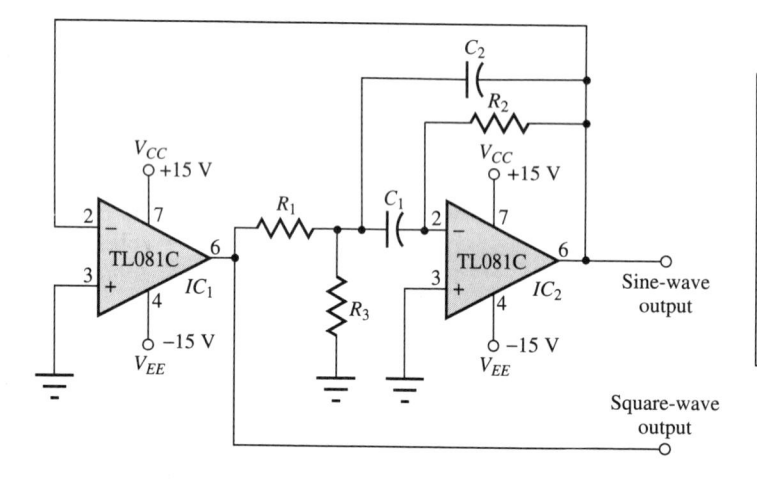

Figure 10-22

First, we must calculate the maximum Q from Equation 10-51, assuming a unity-gain frequency of 3 MHz for a TL081C op-amp:

$$Q_{max} = \sqrt{\frac{f_{GB}}{20 f_{center}}} = \sqrt{\frac{3 \text{ MHz}}{20 \times 500 \text{ Hz}}} = 17.32$$

Select $C_1 = C_2 = 0.01 \; \mu F$. We can now calculate R_2 from Equation 10-52:

$$R_2 = \frac{Q}{\pi f_{osc} C_2} = \frac{17.32}{\pi \times 500 \text{ Hz} \times 0.01 \; \mu F} = 1.10 \text{ M}\Omega$$

Select a value of 1.1 MΩ for R_2 from Appendix A-1.

Now calculate R_1 from Equation 10-53:

$$R_1 = \frac{R_2}{4Q^2} = \frac{1.1 \text{ M}\Omega}{4 \times 17.32^2} = 917 \; \Omega$$

Always choose R_1 larger than this value to ensure that Q does not exceed its maximum value. Select a value of 1.0 kΩ.

Finally, calculate R_3 from Equation 10-54:

$$R_3 = \frac{R_1}{4\pi^2 f_{osc}^2 R_1 C_1 R_2 C_2 - 1}$$

$$= \frac{1.0 \text{ k}\Omega}{4\pi^2 (500 \text{ Hz})^2 \times 1.0 \text{ k}\Omega \times 0.01 \; \mu F \times 1.1 \text{ M}\Omega \times 0.01 \; \mu F - 1} = 11.7 \text{ k}\Omega$$

Use a value of 12 kΩ for R_3.

The design of the oscillator is complete.

Sawtooth Oscillator

We saw earlier how to generate a sawtooth waveform by using either a relaxation oscillator or a bootstrap oscillator. Why use another circuit? Frequently, sawtooth waveforms are required in which the voltage increases linearly, then drops *immediately* back to the starting voltage. The waveforms generated by the relaxation or bootstrap oscillator cannot reset rapidly.

The sawtooth generator in Figure 10-23 can reset very rapidly. The device in parallel with the capacitor in this circuit is a **programmable unijunction transistor (PUT)**. The PUT is programmed by a trigger voltage determined by the voltage divider R_2 and R_3. The PUT is nonconducting until the voltage across it (determined by the voltage across capacitor C_1) reaches the trigger voltage. At this point, the PUT conducts, discharging the capacitor very rapidly. When the voltage across the PUT (that is, the capacitor voltage) has decreased below the forward conducting voltage of the PUT, it stops conducting and the next cycle begins. Because the PUT switches very rapidly, the output voltage drop is very rapid.

The PUT determines the maximum and minimum levels of the sawtooth-wave voltage. The maximum voltage is simply the PUT trigger voltage, determined by

480   10 Oscillator Circuits

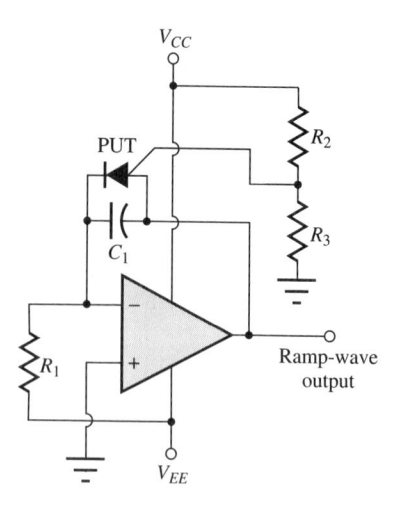

Figure 10-23 Ramp generator using a programmable unijunction transistor (PUT).

the voltage divider R_2 and R_3 and the positive supply voltage:

$$V_{max} = \frac{R_3}{R_2 + R_3} \times V_{CC} \qquad (10\text{-}55)$$

The minimum voltage V_{min} is the foward conducting voltage of the PUT, which is approximately 1 V.

The period is determined by the time for capacitor C_1 to charge from V_{min} to V_{max} through R_1. The charging current is V_{EE}/R_1. The period is

$$T = \frac{R_1 C_1 (V_{max} - V_{min})}{V_{EE}} \qquad (10\text{-}56)$$

The frequency is simply the reciprocal of the period:

$$f_{osc} = \frac{V_{EE}}{R_1 C_1 (V_{max} - V_{min})} \qquad (10\text{-}57)$$

Notice that the period and frequency depend on the amplitude. If the amplitude V_{max} is changed, the frequency will change, so it is best to operate this oscillator at fixed amplitude. The frequency can be changed easily by varying either R_1 or C_1, and this does not affect the amplitude.

EXAMPLE 10-9

Design a sawtooth oscillator that has a rise time of 0.001 s and a maximum amplitude of 10 V. Use a TL081C op-amp operating from a ± 15 V supply.

10-5 Other Oscillator Circuits

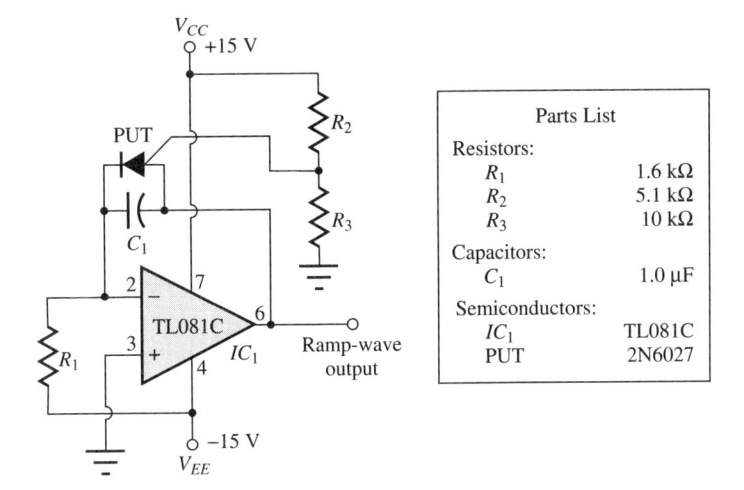

Figure 10-24

Solution

The sawtooth generator circuit is shown in Figure 10-24, with a parts list included.

Note that we must use a PUT. For information on PUTs refer to any text on electronic devices. For this example, we will simply choose a suitable PUT, a 2N6027.

Since the maximum amplitude is specified, we can design the voltage divider consisting of R_2 and R_3 by using Equation 10-55:

$$V_{max} = \frac{R_3}{R_2 + R_3} \times V_{CC}$$

Choose R_3 as 10 kΩ and solve Equation 10-55 for R_2:

$$R_2 = \left(\frac{V_{CC}}{V_{max}} - 1\right) \times R_3 = \left(\frac{15 \text{ V}}{10 \text{ V}} - 1\right) \times 10 \text{ k}\Omega = 5.0 \text{ k}\Omega$$

The closest standard value is 5.1 kΩ. We will use this for R_2.

Now design R_1 and C_1 to give the desired period. The period is given by Equation 10-56:

$$T = \frac{R_1 C_1 (V_{max} - V_{min})}{V_{EE}}$$

Choose $C_1 = 1.0$ μF and solve this equation for R_1:

$$R_1 = \frac{T V_{EE}}{C_1(V_{max} - V_{min})} = \frac{0.001 \text{ s} \times 15 \text{ V}}{1.0 \text{ μF} \times (10 \text{ V} - 1 \text{ V})} = 1.67 \text{ k}\Omega$$

The closest standard value is 1.6 kΩ. We will use this for R_1.

The design of the sawtooth generator is complete.

Summary

This chapter presented the concepts of oscillation and periodic waves, described the conditions for oscillation, and introduced a variety of circuits used as oscillators.

The simplest form of oscillator circuit introduced was the relaxation oscillator. This is a square-wave oscillator with the frequency determined by the charging and discharging time of a capacitor. We saw how to convert the basic relaxation oscillator into a combined square-wave/triangle-wave generator with the addition of an integrating circuit. The same circuit can be used to generate pulse and sawtooth waves by placing diodes in series with the timing resistor.

The bootstrap oscillator, in many ways resembling the relaxation oscillator, was introduced as an alternative generator for square and triangle waves. For generating square waves, the relaxation oscillator is the simpler circuit. For generating triangle waves, however, the bootstrap oscillator is the simpler circuit, using only three resistors and a single capacitor.

Pure sine-wave oscillators were also introduced. The two most common forms, the phase-shift oscillator and the Wien bridge oscillator were compared and design procedures were developed. These circuits are considerably more critical to design than the relaxation oscillator or bootstrap oscillator.

Finally, introducing a few miscellaneous circuits, we saw how a sine-wave oscillator could be produced with a bootstrap type of circuit by placing a bandpass filter in the feedback loop. This is actually an easier way to produce a sine wave if some distortion can be tolerated. We also introduced a sawtooth generator using a programmable unijunction transistor that could produce a sawtooth wave with very short reset time, a much better circuit for producing sawtooth waves than the bootstrap or relaxation oscillators.

Review Questions

10-1 Theory of oscillation

1. Define oscillation.
2. Give three examples of mechanical oscillations.
3. Define waveform.
4. Define a periodic electronic signal.
5. Define the period of an electronic waveform.
6. Define the frequency of an electronic waveform.
7. Define the phase of a periodic waveform.
8. Define the instantaneous peak amplitude of a periodic waveform.
9. Define the peak-to-peak amplitude of a periodic waveform.
10. Name and describe the five most common periodic waveforms found in electronics.
11. Give the three conditions required for oscillation.

10-2 Relaxation oscillators

12. Indicate which part of the relaxation oscillator circuit provides each of the three conditions for oscillation.
13. Describe how oscillations are produced in a relaxation oscillator.
14. How is the logarithmic term in the frequency equation for a relaxation oscillator eliminated?
15. What components must be changed to alter the frequency of a simple relaxation oscillator?
16. What modification is necessary to make a relaxation oscillator operate from a single supply?
17. How can a relaxation oscillator be modified to produce a triangle wave?
18. What components must be changed to alter the frequency of a triangle-wave relaxation oscillator? What problems are encountered?
19. How can a relaxation oscillator be modified to produce a pulse wave?

10-3 Bootstrap oscillators

20. How does the bootstrap oscillator differ from the relaxation oscillator?
21. Indicate which part of the bootstrap oscillator circuit provides each of the three conditions for oscillation.
22. Describe how oscillations are produced in a bootstrap oscillator.
23. How is the triangle-wave output amplitude related to the square-wave output amplitude?
24. How is the frequency made independent of the voltage divider resistors R_2 and R_3?
25. What modification is necessary to make a bootstrap oscillator operate from a single supply?
26. What modification is necessary to make a bootstrap oscillator produce a pulse or sawtooth output? How does this modification differ from the relaxation oscillator?
27. How can the frequency of a bootstrap oscillator be made variable? What advantage does the bootstrap oscillator have over the relaxation oscillator in this case?

10-4 Sine-wave oscillators

28. List the Barkhausen criteria. Describe the meaning of each.
29. How is the frequency determined in a phase-shift oscillator?
30. How much attenuation is introduced by the phase-shift network?
31. What are the amplifier gain requirements for a phase shift oscillator?
32. Describe how to make a sine-wave oscillator self-starting yet essentially distortion-free.
33. What restrictions does the use of zener diodes in the starter circuit place on the feedback resistor?
34. Name two other nonlinear devices frequently used to prevent distortion in sine-wave oscillators
35. How is the frequency determined in a Wien bridge oscillator?
36. What gain is required for a Wien bridge oscillator?
37. How can the frequency be varied in a phase-shift oscillator? In a Wien bridge oscillator?

10-5 Other oscillator circuits

38. What is the difference in output between a bootstrap sine-wave oscillator and either a phase-shift oscillator or a Wien bridge oscillator?
39. What is the advantage of the bootstrap sine-wave oscillator over either the Wien bridge oscillator or the phase-shift oscillator?
40. What is the main weakness of the sine-wave bootstrap oscillator?
41. What determines the frequency of a sine-wave bootstrap oscillator?
42. Describe how a bootstrap oscillator produces a sine wave.
43. What is a PUT?
44. Describe how to make a sawtooth generator by using a PUT.
45. What is the advantage of a PUT sawtooth generator over a relaxation oscillator sawtooth generator or a bootstrap oscillator sawtooth generator?

Design and Analysis Problems

10-2 Relaxation oscillators

1. Design a relaxation oscillator to produce a square wave with a frequency of 1600 Hz. Use standard BiFET op-amps operating from a ±12 V supply.
2. Design a relaxation oscillator to produce a triangle wave with a frequency of 5500 Hz. Use standard BiFET op-amps operating from a single 18 V supply.
3. Analyze the circuit in Figure 10-25 to determine the
   (a) type of oscillator circuit
   (b) frequency of oscillation
   (c) output amplitudes of the square wave and triangle wave.

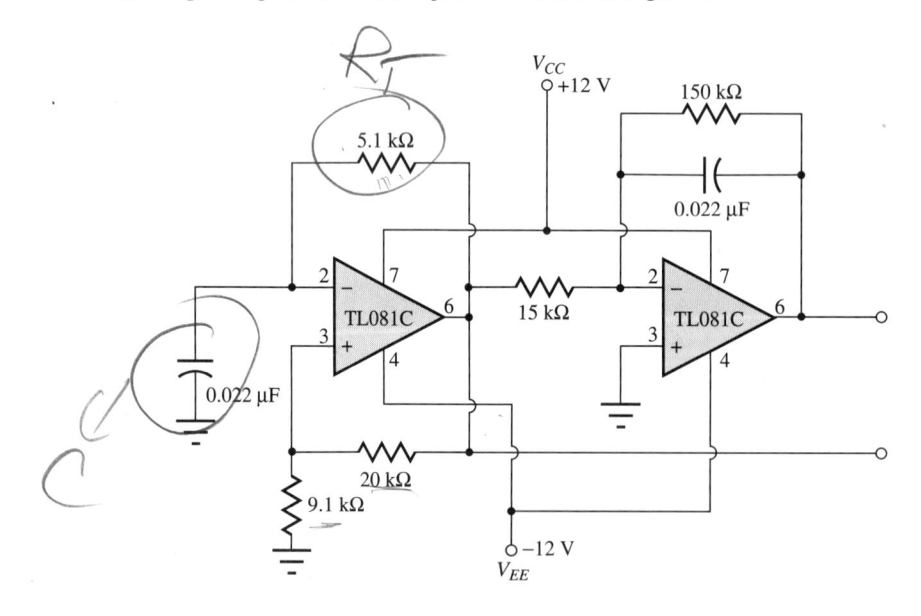

Figure 10-25

10-3 Bootstrap oscillators

4. Design a bootstrap oscillator to produce both a square wave and a triangle wave at a frequency of 4800 Hz. Use standard BiFET op-amps operating from a ± 12 V supply. Estimate the amplitude of the square wave, and calculate the amplitude of the triangle wave.

5. Design a bootstrap oscillator to produce a 5 V peak-to-peak triangle wave at a frequency of 7500 Hz. Use standard BiFET op-amps operating from a single 18 V supply.

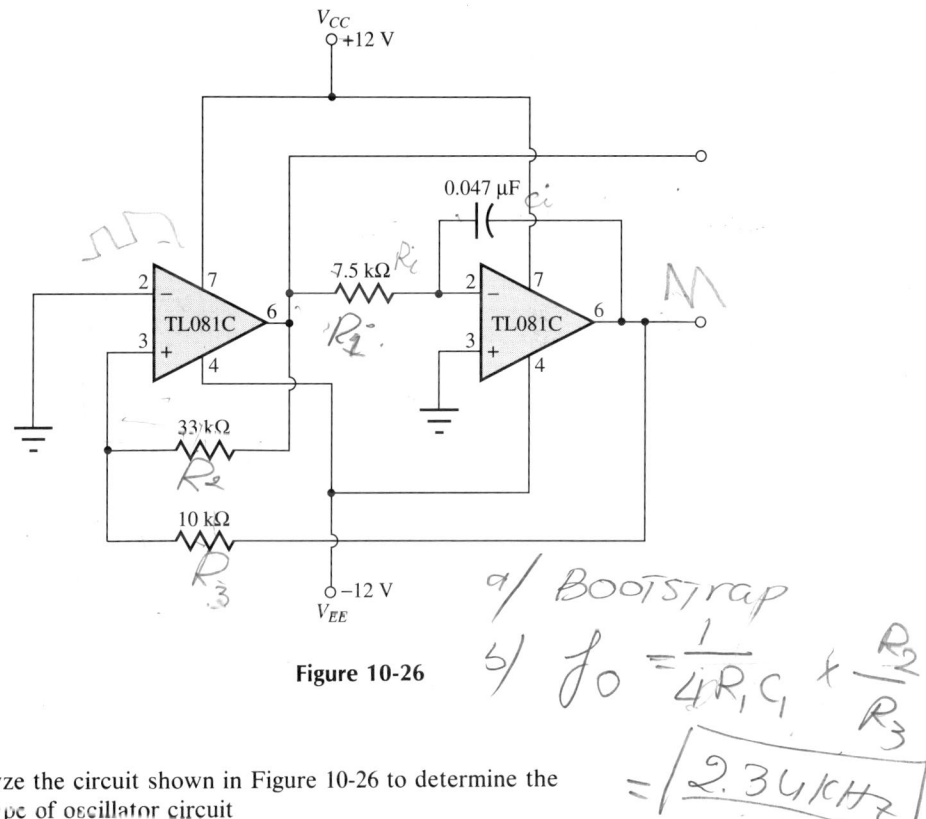

Figure 10-26

6. Analyze the circuit shown in Figure 10-26 to determine the
   (a) type of oscillator circuit
   (b) frequency of oscillation
   (c) output amplitudes of the square wave and triangle wave.

7. Design an oscillator to produce a sawtooth wave that increases linearly to its maximum in 0.005 s, then drops to its minimum value in 0.001 s. It should operate from a ± 15 V supply. Select C_1 as 0.47 µF.

8. An oscillator circuit is shown in Figure 10-27.
   (a) What type of oscillator is it?
   (b) What is the duty cycle of the output waveforms?
   (c) What is the frequency of the output waveforms?
   (d) What is the amplitude of the output waveforms?
   (e) Accurately sketch the two output waveforms.

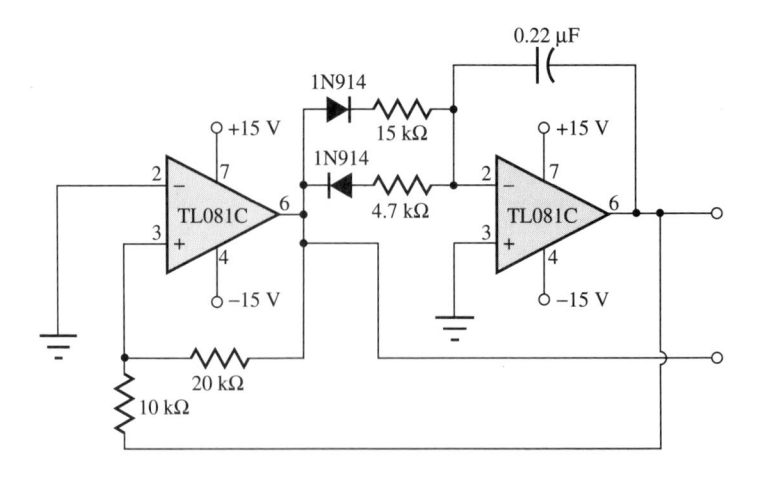

Figure 10-27

10-4 Sine-wave oscillators

9. Design a phase-shift oscillator to operate at a frequency of 450 Hz from a ±12 V supply. Use 0.022 µF capacitors in the phase-shift network. The output sine wave should have a peak-to-peak amplitude of at least 15 V.
10. Design a Wien bridge oscillator to operate at a frequency of 25,000 Hz from a ±15 V supply. Use 680 pF capacitors in the bridge. The output sine wave should have a peak-to-peak amplitude of at least 20 V.

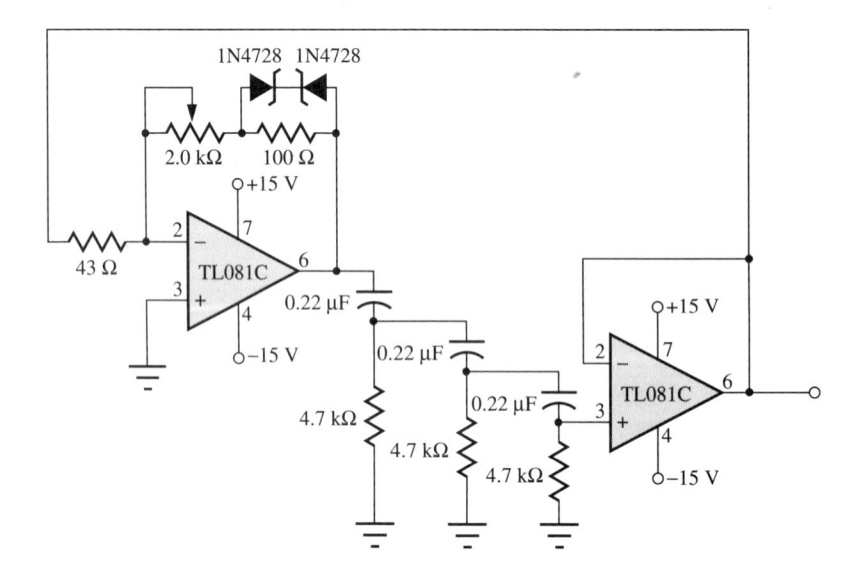

Figure 10-28

11. Analyze the oscillator circuit in Figure 10-28 to determine the
    (a) type of oscillator circuit.
    (b) starting gain. Is it sufficient for oscillations to start?
    (c) maximum operating gain. Is it large enough to sustain oscillations?
    (d) operating frequency.
    (e) sine-wave amplitude.

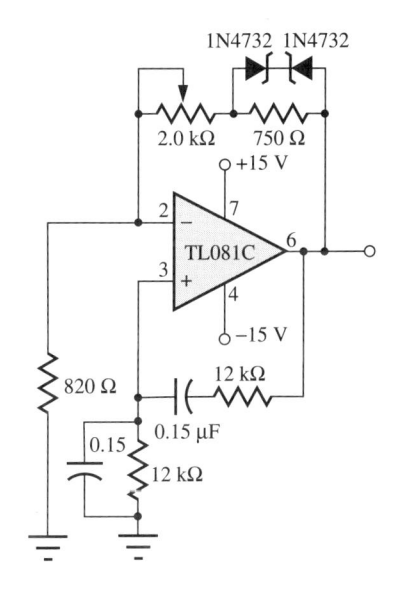

Figure 10-29

12. Analyze the oscillator circuit in Figure 10-29 to determine the
    (a) type of oscillator circuit.
    (b) starting gain. Is it sufficient for oscillations to start?
    (c) maximum operating gain. Is it large enough to sustain oscillations?
    (d) operating frequency.
    (e) sine-wave amplitude.

10-5 Other oscillator circuits

13. Design a sine-wave bootstrap oscillator with a frequency of 1200 Hz and operating from a ±15 V power supply. Use 0.0033 µF capacitors in your circuit.
14. Design a sawtooth generator that has a rise time of 1 s and a maximum operating voltage of 5 V. Use a ±15 V supply and a 2N6027 programmable unijunction transistor.

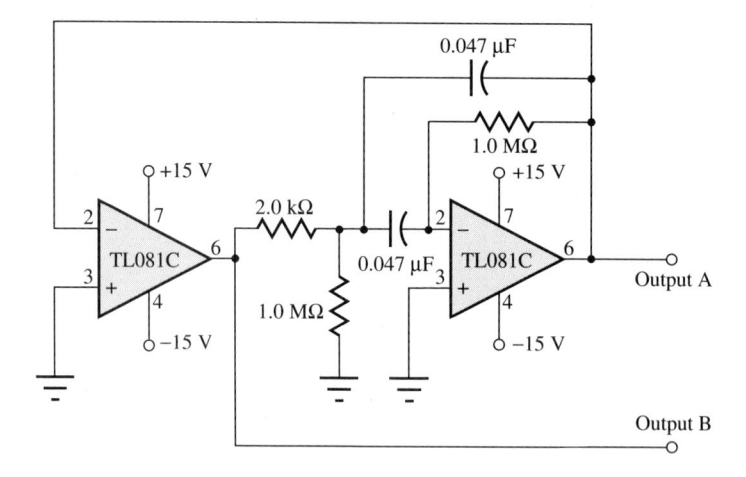

Figure 10-30

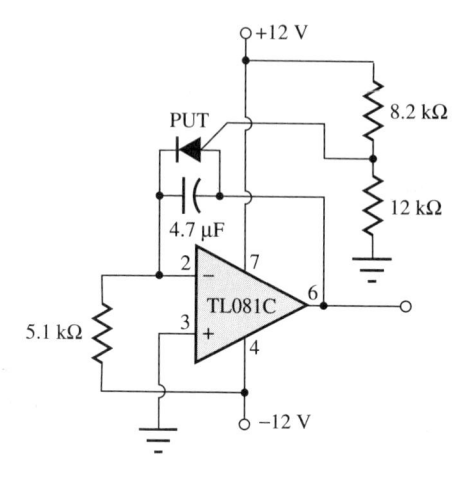

Figure 10-31

15. An oscillator circuit is shown in Figure 10-30.
    (a) What type of oscillator is it? What are the waveforms at output A and output B?
    (b) What is the frequency of oscillation?
16. Determine the rise time and the peak height of the sawtooth wave produced by the circuit in Figure 10-31.

Integrated Circuit Oscillators 11

OUTLINE

11-1 The LM566C Voltage Controlled Oscillator
11-2 The LM555C Timer/Oscillator
11-3 The ICL8038 Single-Chip Function Generator
11-4 The LM565C Phase-Locked Loop

KEY TERMS

Voltage controlled oscillator
Voltage-to-frequency converter
Tone generator
Frequency modulation
Timer
Function generator
Phase-locked loop
Phase comparator (phase detector)
Error signal
Lock range
Capture range
Free-running frequency
FM demodulator
Frequency converter

OBJECTIVES

On completion of this chapter, the reader should be able to:

- design and analyze oscillator and modulator circuits using the LM566C voltage controlled oscillator.
- design pulse-wave and square-wave oscillator circuits using the LM555C timer.
- use the LM555C as a timer.
- use the ICL8038 function generator integrated circuit to produce a function generator.
- explain the operation and applications of phase-locked loops.
- design phase-locked loop circuits using the LM565C phase-locked loop integrated circuit, including an FM demodulator circuit, and a frequency converter.

Introduction

The previous chapter introduced oscillator circuits using op-amps. If a simple square-wave oscillator is required for a time-base generator, then a relaxation oscillator is probably the best circuit to use. However, there are a number of integrated circuit oscillators produced, and if more than just a simple square-wave oscillator is required, then one of these may be the simpler solution.

Integrated circuit oscillators range from the simple LM566C voltage controlled oscillator, to the LM555C oscillator/timer, to the ICL8038 function generator. The ICL8038 is a full-function signal generator capable of producing sine, square, and triangle waveforms, with variable duty cycle, at any frequency from less than 0.01 Hz to over 200 kHz.

In this chapter, we will also introduce a special-purpose oscillator circuit—the phase-locked loop. The phase-locked loop contains a voltage controlled oscillator, but, unlike the standard voltage controlled oscillator, the frequency of oscillation is controlled by an input frequency instead of a control voltage. These are very important oscillator circuits which find many applications in electronic systems. We will look at only one specific device, the LM565C phase-locked loop, and two simple applications.

11-1 The LM566C Voltage Controlled Oscillator

The previous chapter showed how to build a variety of oscillators from simple op-amps. As we have seen a number of times, any frequently used op-amp circuit is usually available in the form of a single integrated circuit. This is true for bootstrap square-wave/triangle-wave oscillator circuits. The bootstrap oscillator introduced in the previous chapter is the basis of **voltage controlled oscillator (VCO)** integrated circuits.

In a voltage controlled oscillator, the output frequency is proportional to an input control voltage. Why control the frequency with an external variable-voltage source? A voltage controlled oscillator can be used as a **voltage-to-frequency converter.** As we will see shortly, in applications such as frequency modulation we have a voltage that we wish to convert to a frequency. A voltage controlled oscillator is ideal for this purpose. If required, however, a voltage controlled oscillator can still be used as a simple oscillator, using a potentiometer or resistor voltage divider to produce the control voltage from the DC supply.

In the bootstrap oscillator discussed in the previous chapter, the frequency could be controlled by varying either R_1, C_1, or R_2. When R_2 is used to control the frequency, the variation of frequency with resistance is linear. The circuit used in a voltage controlled oscillator integrated circuit is similar to the basic bootstrap oscillator, but with one important modification—resistor R_2 is replaced with a voltage-variable resistor circuit. The center frequency of the oscillator is determined by resistor R_1 and capacitor C_1, and the frequency can be varied by changing the input voltage controlling the voltage-variable resistor.

The LM566 voltage controlled oscillator is typical of a bootstrap oscillator integrated circuit. Data sheets for the LM566 voltage controlled oscillator are shown in Figure 11-1. The pinout is reproduced in Figure 11-2, along with the

LM566/LM566C Voltage Controlled Oscillator

Industrial Blocks

General Description

The LM566/LM566C are general purpose voltage controlled oscillators which may be used to generate square and triangular waves, the frequency of which is a very linear function of a control voltage. The frequency is also a function of an external resistor and capacitor.

The LM566 is specified for operation over the −55°C to +125°C military temperature range. The LM566C is specified for operation over the 0°C to +70°C temperature range.

Features

- Wide supply voltage range: 10 to 24 volts
- Very linear modulation characteristics
- High temperature stability
- Excellent supply voltage rejection
- 10 to 1 frequency range with fixed capacitor
- Frequency programmable by means of current, voltage, resistor or capacitor.

Applications

- FM modulation
- Signal generation
- Function generation
- Frequency shift keying
- Tone generation

Schematic and Connection Diagrams

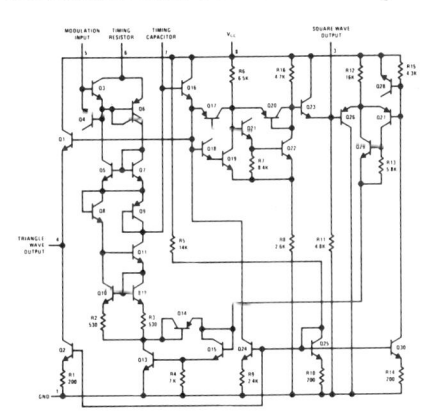

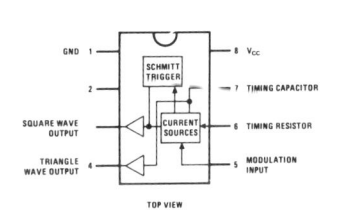

Order Number LM566CN
See NS Package N08B

Typical Application

1 kHz and 10 kHz TTL Compatible Voltage Controlled Oscillator

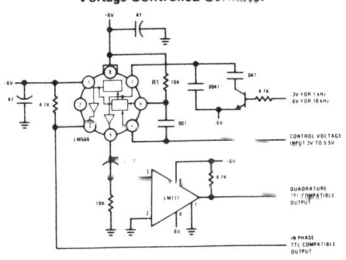

Applications Information

The LM566 may be operated from either a single supply as shown in this test circuit, or from a split (±) power supply. When operating from a split supply, the square wave output (pin 4) is TTL compatible (2 mA current sink) with the addition of a 4.7 kΩ resistor from pin 3 to ground.

A .001 µF capacitor is connected between pins 5 and 6 to prevent parasitic oscillations that may occur during VCO switching.

$$f_O = \frac{2(V^+ - V_5)}{R_1 C_1 V^+}$$

where

$2K < R_1 < 20K$

and V_5 is voltage between pin 5 and pin 1

Figure 11-1 Data sheets for the LM566/LM566C voltage controlled oscillator. *(Reprinted with permission of National Semiconductor Corporation)*

Absolute Maximum Ratings

Power Supply Voltage	26V
Power Dissipation (Note 1)	300 mW
Operating Temperature Range　LM566	$-55°C$ to $+125°C$
LM566C	$0°C$ to $70°C$
Lead Temperature (Soldering, 10 sec)	$300°C$

Electrical Characteristics $V_{CC} = 12V$, $T_A = 25°C$, AC Test Circuit

PARAMETER	CONDITIONS	LM566 MIN	LM566 TYP	LM566 MAX	LM566C MIN	LM566C TYP	LM566C MAX	UNITS
Maximum Operating Frequency	$R_0 = 2k$, $C_0 = 2.7$ pF		1			1		MHz
Input Voltage Range Pin 5		$3/4\ V_{CC}$		V_{CC}	$3/4\ V_{CC}$		V_{CC}	
Average Temperature Coefficient of Operating Frequency			100			200		ppm/°C
Supply Voltage Rejection	10–20V		0.1	1		0.1	2	%/V
Input Impedance Pin 5		0.5	1		0.5	1		MΩ
VCO Sensitivity	For Pin 5, From 8–10V, $f_O = 10$ kHz	6.4	6.6	6.8	6.0	6.6	7.2	kHz/V
FM Distortion	±10% Deviation		0.2	0.75		0.2	1.5	%
Maximum Sweep Rate		800	1		500	1		MHz
Sweep Range			10:1			10:1		
Output Impedance Pin 3			50			50		Ω
Pin 4			50			50		Ω
Square Wave Output Level	$R_{L1} = 10k$	5.0	5.4		5.0	5.4		Vp-p
Triangle Wave Output Level	$R_{L2} = 10k$	2.0	2.4		2.0	2.4		Vp-p
Square Wave Duty Cycle		45	50	55	40	50	60	%
Square Wave Rise Time			20			20		ns
Square Wave Fall Time			50			50		ns
Triangle Wave Linearity	+1V Segment at $1/2\ V_{CC}$		0.2	0.75		0.5	1	%

Note 1: The maximum junction temperature of the LM566 is 150°C, while that of the LM566C is 100°C. For operating at elevated junction temperatures, devices in the TO-5 package must be derated based on a thermal resistance of 150°C/W. The thermal resistance of the dual-in-line package is 100°C/W.

Figure 11-1 *(Continued)*

11-1 The LM566C Voltage Controlled Oscillator

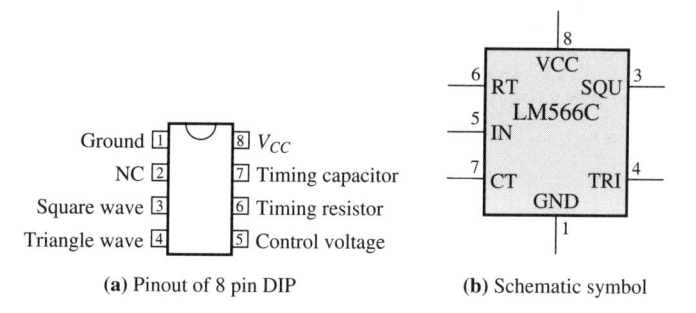

Figure 11-2 Pinout and schematic symbol for the LM566C voltage controlled oscillator.

(a) Pinout of 8 pin DIP

(b) Schematic symbol

schematic symbol. There are two versions of the LM566 listed on the data sheets in Figure 11-1: the LM566 and the LM566C. The LM566C is the commercial version with less stringent specifications; otherwise the two versions are identical. We will use the LM566C throughout this discussion. The LM566C will operate from any single supply voltage from 10 to 24 V, or from a dual-polarity supply as long as the total voltage difference does not exceed 26 V. It generates a triangle-wave output and a square-wave output up to a frequency of about 1 MHz. The center or base frequency is set by a timing resistor R_1 and a timing capacitor C_1. The frequency can be varied by adjusting the control voltage at pin 5 over a range of $0.75V_{CC}$ to V_{CC}.

The LM566 as a Tone Generator

The simplest application of the LM566C is as a fixed-frequency oscillator or **tone generator.** The circuit of the LM566C used as a tone generator is shown in Figure 11-3. For this circuit, it is only necessary to connect a timing capacitor C_1 between pin 7 and ground (pin 1), a timing resistor R_1 between pin 6 and the supply (pin 8), and bias pin 5 to $\frac{7}{8}V_{CC}$, using a voltage divider consisting of resistors R_2 and R_3. A square-wave output is available from pin 3, a triangle-wave output from pin 4.

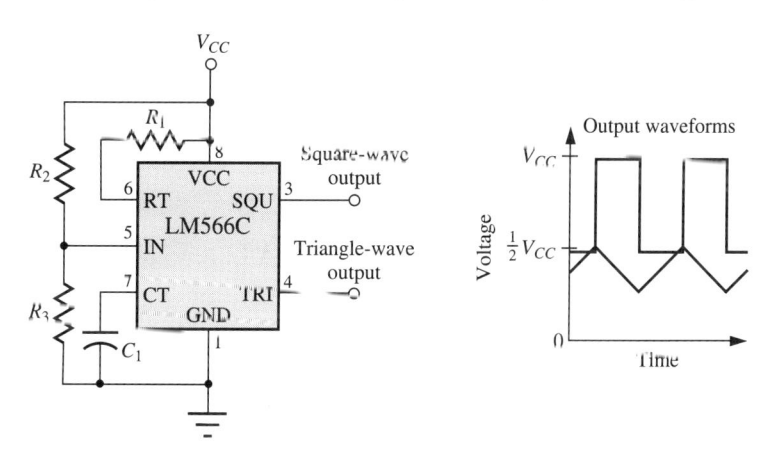

Figure 11-3 Circuit for a square-wave/triangle-wave generator using a LM566C voltage controlled oscillator.

The operating frequency of the oscillator f_{osc} is given by

$$f_{osc} = \frac{2(V_{CC} - V_{pin5})}{C_1 R_1 V_{CC}} \quad (11\text{-}1)$$

Since the voltage control at pin 5 operates only over the range of $\frac{3}{4}V_{CC}$ to V_{CC}, pin 5 is normally biased to the middle of this range, or $\frac{7}{8}V_{CC}$. This biasing is done with a voltage divider consisting of R_2 and R_3 connected between the positive supply and ground. The input impedance at pin 5 is typically 1 MΩ, and the effective resistance of the divider $R_2 \parallel R_3$ should be much less than this. The bias voltage at pin 5 is given by the basic voltage divider formula

$$V_{bias} = 0.875 V_{CC} = V_{CC} \times \frac{R_3}{R_2 + R_3} \quad (11\text{-}2)$$

or, solving for R_2 in terms of R_3,

$$R_2 = R_3 \times \frac{1 - V_{bias}/V_{CC}}{V_{bias}/V_{CC}} = 0.14286 R_3 \quad (11\text{-}3)$$

Any resistor values satisfying this relation could be used, but $R_3 = 110$ kΩ and $R_2 = 16$ kΩ give a ratio of 0.873, very close to $\frac{7}{8}$. For these values, the effective resistance of the divider, equal to R_2 in parallel with R_3, is 14 kΩ, much less than 1 MΩ, as required.

If we bias pin 5 to $\frac{7}{8}V_{CC}$ then the equation for oscillator frequency simplifies to

$$f_{osc} = \frac{1}{4 R_1 C_1} \quad (11\text{-}4)$$

Normally C_1 is chosen. Solve Equation 11-4 for R_1:

$$R_1 = \frac{1}{4 f_{osc} C_1} \quad (11\text{-}5)$$

This is our design equation for R_1. Capacitor C_1 should be chosen so that resistor R_1 has a value between 2 and 20 kΩ.

Both the square-wave and triangle-wave outputs have DC offsets. The square-wave output from pin 4 varies between approximately $\frac{1}{2}V_{CC}$ and V_{CC}, and the triangle-wave output from pin 4 varies between approximately $\frac{1}{3}V_{CC}$ and $\frac{1}{2}V_{CC}$. The relative output levels are shown in Figure 11-3. If a pure AC waveform without DC offset is required, then the outputs must be capacitively coupled.

EXAMPLE 11-1

Design a tone generator using an LM566C voltage controlled oscillator operating from a single 15 V supply to produce an 800 Hz output signal. Describe the output signals produced by this oscillator.

Solution

This is one of the basic applications for which the LM566C tone generator integrated circuit was designed. The circuit is quite simple and is shown in Figure 11-4, with a parts list included.

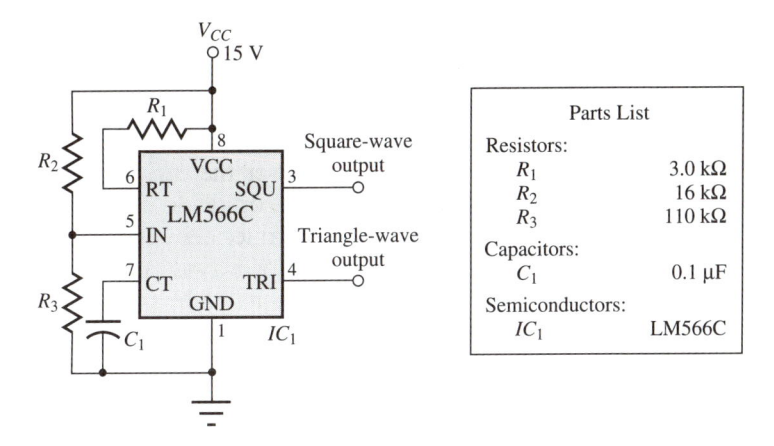

Figure 11-4

First, set the bias point for pin 5 to $\frac{7}{8}V_{CC}$ by using a voltage divider consisting of R_2 and R_3. The optimal resistor values calculated in the text from Equation 11-3 were $R_2 = 16$ kΩ and $R_3 = 110$ kΩ. We will use these values.

We want a frequency of 800 Hz. Choose a value of 0.1 µF for C_1 and calculate R_1 from Equation 11-5:

$$R_1 = \frac{1}{4 f_{osc} C_1} = \frac{1}{4 \times 800 \text{ Hz} \times 0.1 \text{ µF}} = 3.13 \text{ kΩ}$$

The closest standard value from Appendix A-1 is 3.0 kΩ.

Two outputs are available from this oscillator circuit: a square wave from pin 3, and a triangle wave from pin 4. The square wave will vary between approximately 7 and 14 V; the triangle wave will vary between approximately 5 and 7 V. Both obviously have a DC offset. To remove this DC offset, both outputs must be capacitively coupled to their respective loads.

The LM566C as a TTL Time-Base Generator

Most digital systems require a square-wave clocking signal. Several logic levels are used in digital systems, the most common being TTL (transistor-transistor logic), which uses a nominal 5 V for a high logic level and 0 V for a low logic level. Actual levels are typically 3.4 and 0.2 V, respectively. The time-base generator or clocking circuit for TTL digital circuits must produce a square wave varying between these two levels.

The LM566C is ideal for this application, and the circuit of a time-base generator using the LM566C is shown in Figure 11-5. There are two differences from the basic tone generator circuit shown in Figure 11-3. First, the LM566C must be operated from a dual ±5 V power supply so that the output square wave varies from approximately 0 to 5 V, to give TTL logic levels. The negative supply is connected to pin 1. Second, for interfacing to TTL circuits, a 4.7 kΩ resistor must

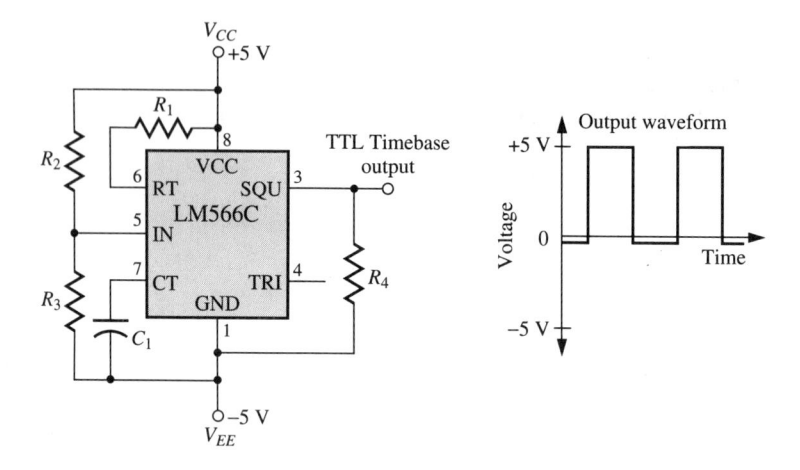

Figure 11-5 Circuit for a TTL compatible timebase generator using a LM566C voltage controlled oscillator.

be connected from the square-wave output to the negative supply (pin 1). The biasing voltage divider, connected across the dual supply, is designed exactly as for the simple tone generator, with R_2 normally chosen as 16 kΩ and R_3 as 110 kΩ.

EXAMPLE 11-2

Design a TTL time-base generator using an LM566C voltage controlled oscillator to produce a TTL square wave with a frequency of 100 kHz. Use a ±5 V power supply.

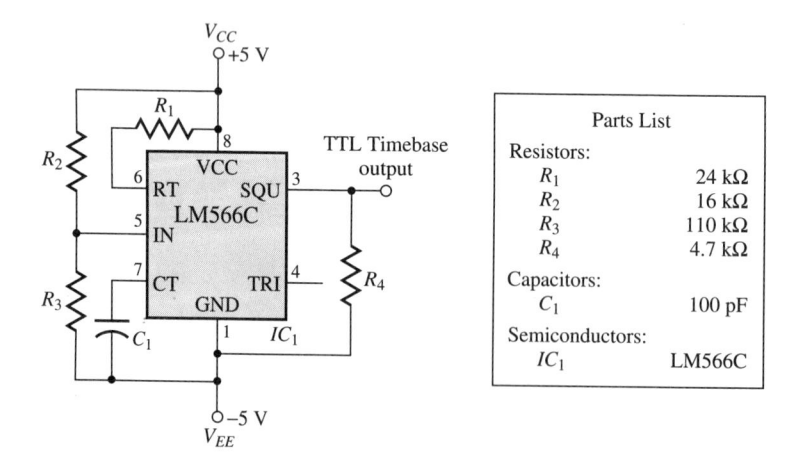

Parts List	
Resistors:	
R_1	24 kΩ
R_2	16 kΩ
R_3	110 kΩ
R_4	4.7 kΩ
Capacitors:	
C_1	100 pF
Semiconductors:	
IC_1	LM566C

Figure 11-6

Solution

The TTL time-base generator circuit is shown in Figure 11-6.

First, set the bias point for pin 5 to $\frac{7}{8}V_{CC}$ by choosing values for R_2 of 16 kΩ and R_3 of 110 kΩ.

Next, calculate a value of R_1 to give a frequency of 100 kHz. Choose a value of 100 pF for C_1 and use Equation 11-5:

$$R_1 = \frac{1}{4f_{osc}C_1} = \frac{1}{4 \times 100 \text{ kHz} \times 100 \text{ pF}} = 25 \text{ k}\Omega$$

The closest standard value from Appendix A-1 is 24 kΩ.

The design is complete.

The LM566C as a Simple Function Generator

Although we will be introducing a single-chip function generator later in this chapter, we can also construct a simple function generator using the LM566C. A **function generator** is simply a circuit that can produce different waveforms over a range of frequencies. The LM566C produces a triangle-wave output and a square-wave output. The frequency can be varied by changing the voltage at pin 5, and different ranges can be selected by changing capacitor C_1.

Figure 11-7 shows the LM566C in a simple function generator circuit. This circuit is the basic tone generator from Figure 11-3, operating from a dual supply, with resistor R_2 in the biasing voltage divider replaced by a potentiometer for

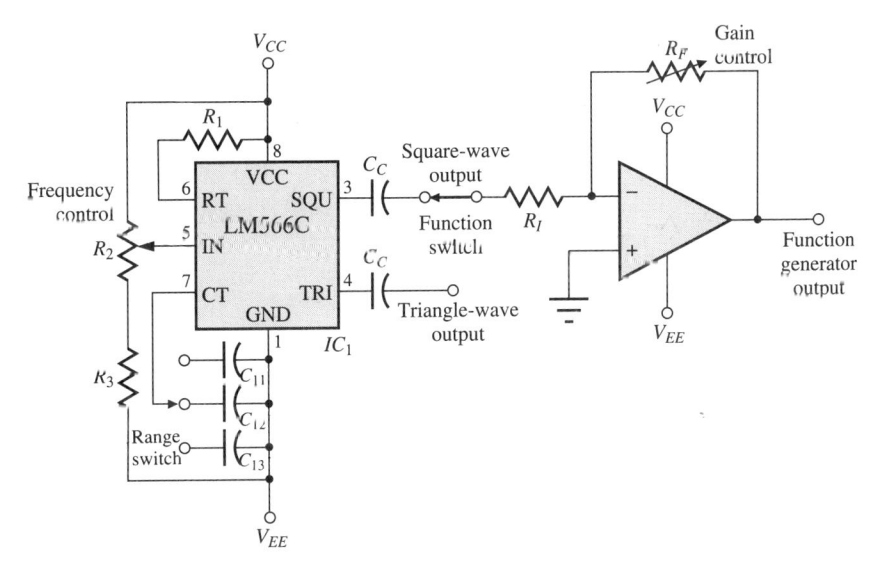

Figure 11-7 A simple function generator using the LM566C voltage controlled oscillator.

varying the frequency. The dual supply is necessary for the op-amp, which is incorporated into the circuit as a gain control.

We must use a slightly different procedure to design the resistors. First, design R_1 by assuming that the voltage at pin 5 is $\frac{3}{4}V_{CC}$ instead of $\frac{7}{8}V_{CC}$. Substituting $V_{\text{pin5}} = \frac{3}{4}V_{CC}$ into Equation 11-1, we get

$$f_{\text{osc}} = \frac{1}{2R_1C_1} \qquad (11\text{-}6)$$

This gives the maximum frequency of the oscillator. Solve Equation 11-6 for R_1:

$$R_1 = \frac{1}{2f_{\text{osc}}C_1} \qquad (11\text{-}7)$$

The voltage divider formula relating R_2 to R_3 is now

$$V_{\text{bias}} = 0.75V_{CC} = V_{CC} \times \frac{R_3}{R_2 + R_3} \qquad (11\text{-}8)$$

Solving for R_3 in terms of R_2 produces

$$R_3 = 3R_2 \qquad (11\text{-}9)$$

To design the function generator, choose potentiometer R_2 and calculate R_3 from Equation 11-9, always selecting the real value of R_3 greater than or equal to the calculated value to ensure that V_{pin5} is at least $\frac{3}{4}V_{CC}$. When calculating R_1, use the frequency of the top of the range. Since we normally switch ranges in a function generator in decades, we can switch decade values of C_1.

One problem with using the LM566 as a function generator is that the square and triangle waves will have DC offset voltages, even when operated from a dual supply. To remove the offset, the easiest procedure is to capacitively couple the outputs. Capacitive coupling will limit the low frequency response of the function generator. The coupling capacitors C_C shown in Figure 11-7 must be chosen such that the capacitive reactance is much less than the sum of the output impedance of the oscillator Z_{out} and the input impedance of the amplifier R_I at the lowest frequency desired from the function generator. The equation for calculating C_C is the same as Equation 3-8, for calculating C_{out} for AC-coupled amplifiers:

$$C_C = \frac{1}{2\pi f_{\text{osc}}(Z_{\text{out}} + R_I)} \qquad (11\text{-}10)$$

Impedance Z_{out} is listed on the LM566C data sheets in Figure 11-1 as 50 Ω.

The circuit in Figure 11-7 includes an op-amp. This op-amp serves two purposes. First, it buffers the output of the oscillator to prevent loading. Second, by using a potentiometer for R_F, we can make the gain variable. The op-amp is configured as a standard inverting amplifier with either of the two waveforms from the oscillator selected by the function switch. Because the two output signals from the oscillator have different peak-to-peak levels, it may be desirable to move R_I to the generator side of the switch and use different values of R_I for the different waveforms to equalize output levels.

This circuit can be improved further, if desired, by adding a wave-shaping circuit to the triangle-wave output to generate a sine-wave output. Although this will not be done here, wave-shaping circuits are discussed in Appendix C, and the reader is encouraged to experiment with one for this function generator.

EXAMPLE 11-3

Design a function generator circuit using an LM566C to generate square waves and triangle waves with frequencies from 100 Hz to 100 kHz. Use three decade frequency ranges; namely, 100 to 1000 Hz, 1000 Hz to 10 kHz, and 10 to 100 kHz. The power supply is ±12 V.

Solution

This example demonstrates how to design a simple function generator with the LM566C voltage controlled oscillator. The circuit, including a parts list, is shown in Figure 11-8.

First, choose a value of C_1 and calculate R_1 to set the maximum frequency of the lowest range. Since the range is from 100 to 1000 Hz, the maximum frequency is 1000 Hz. Choose C_{11} as 0.1 μF and use Equation 11-7 to calculate R_1:

$$R_1 = \frac{1}{2 f_{osc} C_1} = \frac{1}{2 \times 1.0 \text{ kHz} \times 0.1 \text{ μF}} = 5.00 \text{ k}\Omega$$

The closest standard value from Appendix A-1 is 5.1 kΩ.

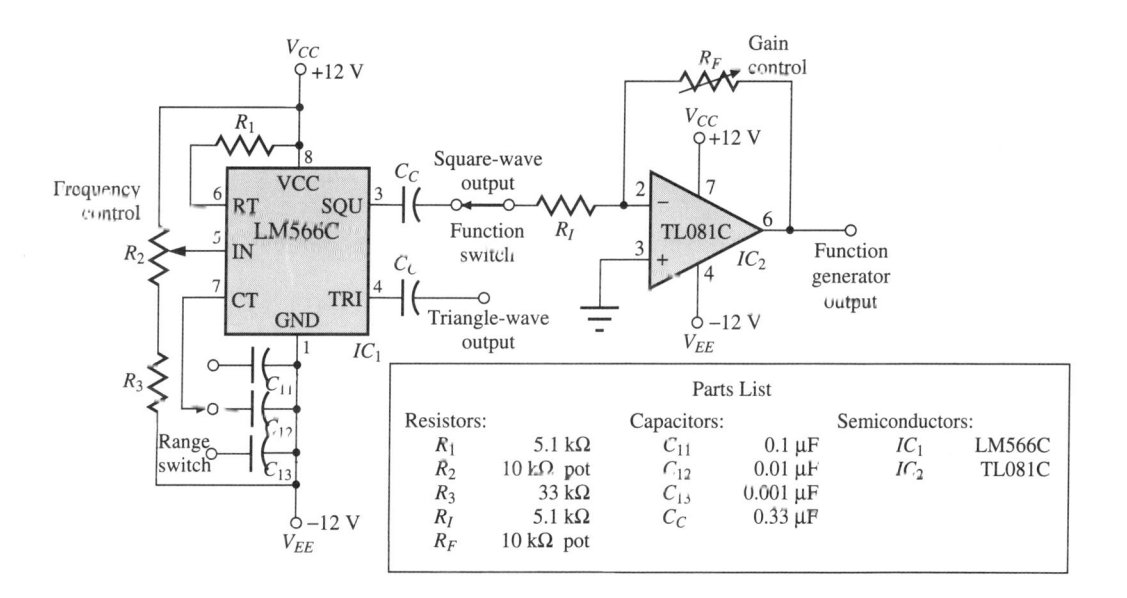

Figure 11-8

We require a voltage divider with potentiometer R_2 and resistor R_3. Choose a 10 kΩ potentiometer for R_2, and calculate R_3 from Equation 11-9 as 30 kΩ. This is a standard value, but we should choose R_3 larger to ensure $V_{pin5} \geq \frac{3}{4}V_{CC}$. Make $R_3 = 33$ kΩ.

For the other two ranges, it is merely necessary to switch in a different capacitor for C_{11}. Because the next frequency range is ten times higher than the first, capacitor C_{12} should be one-tenth as large. In other words, choose $C_{12} = 0.01$ μF. Similarly, choose $C_{13} = 0.001$ μF.

Next, we will design the amplifier/gain control. Since no gain was specified, we will choose a maximum gain of 2, because the square wave from the oscillator has a peak-to-peak amplitude of approximately half the total supply voltage. Choose R_F to be a 10 kΩ potentiometer, and calculate R_I as 5.0 kΩ. The closest standard value is 5.1 kΩ.

Finally, we must design the coupling capacitor to pass the lowest frequency of 100 Hz: $R_I = 5.1$ kΩ and $Z_{out} = 50$ Ω. From Equation 11-10,

$$C_C = \frac{1}{2\pi f_{osc}(Z_{out} + R_I)}$$
$$= \frac{1}{2\pi \times 100 \text{ Hz} \times (50 \text{ Ω} + 5.1 \text{ kΩ})} = 0.309 \text{ μF}$$

Choose a larger value for C_C to ensure that all frequencies greater than 100 Hz will pass unattenuated. We will use 0.33 μF.

The function generator is designed. If a sine wave is also desired, the wave-shaping circuit discussed in Appendix C can be used. Note that it will be necessary to amplify the signal before using the wave-shaping circuit.

The LM566C as a Frequency Modulator

A voltage controlled oscillator such as the LM566C is basically a voltage-to-frequency converter. As such, it can be used for **frequency modulation.** To many readers, the term "frequency modulation" probably implies FM radio. Although voltage-to-frequency converters are used for FM communication systems, the LM566C, with a maximum oscillator frequency of only 1 MHz, is definitely not used for such applications. There are, however, many other applications of frequency modulation. Slowly varying AC instrument signals are often frequency-modulated to reduce noise in transmission from remote sensors. Control signals can be frequency-modulated and transmitted along power lines. Several instrument or control signals can be transmitted along a single data line by frequency-modulating each with a different carrier frequency.

A frequency modulator converts any input voltage to a corresponding output frequency. Usually a base or *carrier* frequency is established for zero input. As the input voltage changes with respect to this zero, the output frequency changes, with the change in frequency proportional to the change in amplitude of the input signal.

11-1 The LM566C Voltage Controlled Oscillator

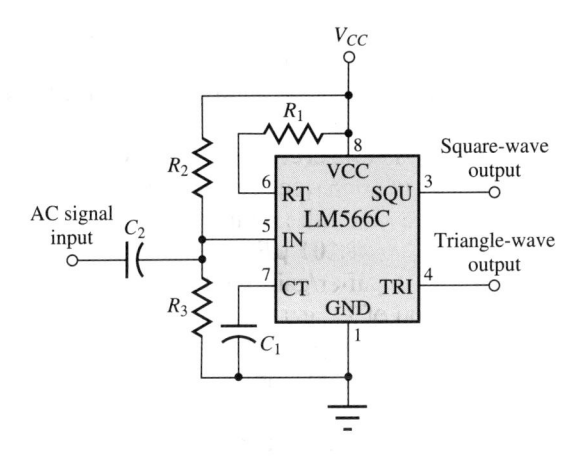

Figure 11-9 The LM566C voltage controlled oscillator used as a frequency modulator for AC signals.

For frequency modulating an AC signal, a circuit identical to the basic tone generator shown in Figure 11-3 can be used. The circuit in Figure 11-9 has the AC signal capacitively coupled to pin 5. The AC signal adds to the biasing voltage, and the output frequency can be calculated by Equation 11-1, where $V_{pin5} = \frac{7}{8}V_{CC} + V_{AC}$. Here, V_{AC} is the instantaneous AC voltage and must be less than $\pm\frac{1}{8}V_{CC}$. Unfortunately, this circuit only works for AC signals.

A better circuit, which works for AC and DC signals is shown in Figure 11-10. An op-amp summing circuit is used to add the modulating signal to the biasing signal. The biasing voltage divider is built into this circuit. Resistors R_F and R_{IB} are chosen to amplify the *negative* supply voltage by $\frac{7}{8}$ to give the bias voltage.

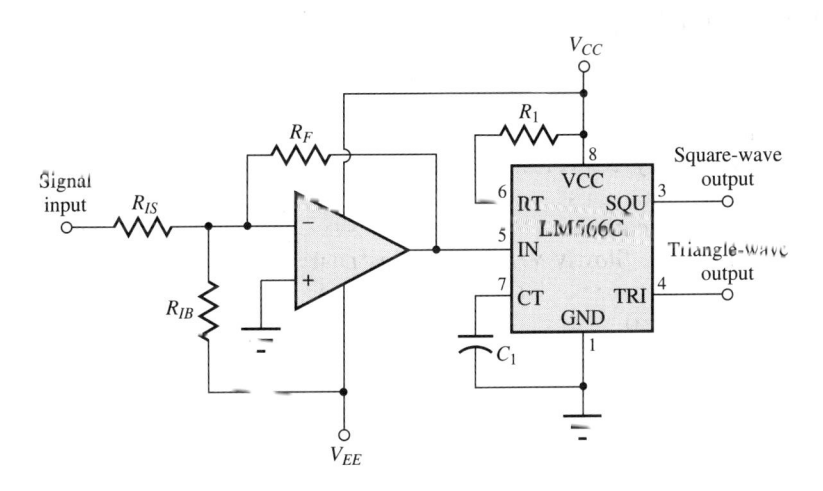

Figure 11-10 General purpose frequency modulating circuit using a LM566C voltage controlled oscillator.

Using the gain equation for an inverting amplifier developed in Chapter 3, we find R_{IB}:

$$R_{IB} = \tfrac{8}{7} \times R_F \qquad (11\text{-}11)$$

The closest ratio is obtained by choosing $R_F = 13$ kΩ and $R_{IB} = 15$ kΩ. Note that the negative supply voltage is used because we have an inverting amplifier and need a positive bias voltage. For the signal input, make $R_{IS} = R_F$ to give a gain of 1.

The carrier frequency f_{osc1} is given by Equation 11-1:

$$f_{osc1} = \frac{2(V_{CC} - V_{pin5})}{C_1 R_1 V_{CC}} \qquad (11\text{-}1)$$

An increase in the signal voltage causes an increase in the output frequency. Increase the voltage at pin 5 by ΔV. The frequency now becomes

$$f_{osc2} = \frac{2(V_{CC} - V_{pin5} + \Delta V)}{C_1 R_1 V_{CC}} \qquad (11\text{-}12)$$

The difference between these is Δf:

$$\Delta f = f_{osc2} - f_{osc1} = \frac{2\,\Delta V}{C_1 R_1 V_{CC}} \qquad (11\text{-}13)$$

Equation 11-13 gives the frequency deviation, that is, the change in frequency Δf for an input voltage change ΔV. This equation can be rearranged to solve for ΔV:

$$\Delta V = \frac{\Delta f\, C_1 R_1 V_{CC}}{2} \qquad (11\text{-}14)$$

In this form, the equation gives the output voltage corresponding to a change in the transmitted frequency, and is useful for determining the amplitude of demodulated signals.

EXAMPLE 11-4

Design a circuit using the LM566C voltage controlled oscillator to modulate an instrument signal onto a carrier with a frequency of 20 kHz. The power supply is ± 15 V. Determine the maximum frequency deviation if the maximum amplitude of the modulating signal is 0.20 V.

Solution

We will design our modulator circuit using the general-purpose circuit in Figure 11-10. This circuit has been redrawn in Figure 11-11 with a parts list.

We will first design the op-amp summing circuit. The bias voltage is set by choosing R_F, then calculating R_{IB} from Equation 11-11. We will choose the resistance values $R_F = 13$ kΩ and $R_{IB} = 15$ kΩ to give the ratio closest to $\tfrac{7}{8}$. Since the signal input is to have unity gain, make $R_{IS} = R_F = 13$ kΩ.

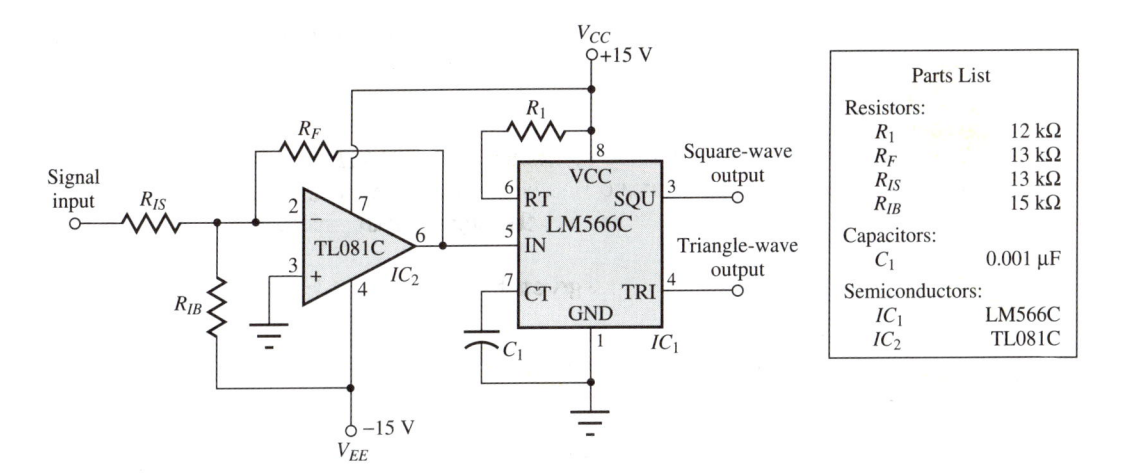

Figure 11-11

Next we must design the LM566C for a carrier or center frequency of 20 kHz. Choose a value of 0.001 µF for C_1, and calculate R_1 from Equation 11-5:

$$R_1 = \frac{1}{4f_{osc}C_1} = \frac{1}{4 \times 20 \text{ kHz} \times 0.001 \text{ µF}} = 12.5 \text{ k}\Omega$$

The closest standard value is 12 kΩ.

Finally, calculate the frequency deviation for the maximum amplitude of the input signal of 0.2 V. Use Equation 11-13:

$$\Delta f = \frac{2\, \Delta V}{C_1 R_1 V_{CC}}$$

$$= \frac{2 \times 0.2 \text{ V}}{0.001 \text{ µF} \times 12 \text{ k}\Omega \times 15 \text{ V}} = 2222 \text{ Hz}$$

This calculation tells us that a change in input voltage of 0.2 V will produce a change in output frequency of 2222 Hz (or 11.1%). In frequency modulation, the change in frequency, that is, the frequency deviation, is proportional to the amplitude of the modulating signal. If an AC signal is used as the modulating signal, the frequency of the modulating signal appears as a periodic change in the frequency deviation.

The design is complete.

11-2 The LM555C Timer/Oscillator

The second integrated circuit widely used for oscillator applications is the LM555C timer/oscillator. Next to the LM741C op-amp, the LM555C timer is probably the best known of all integrated circuits. Entire books have been written on applications of this integrated circuit. Although this is a versatile chip, capable

504 11 Integrated Circuit Oscillators

of performing many functions, it is badly overused, largely by designers who are not aware of the variety of other chips available. For example, the LM555C timer can be used to generate square waves, yet the circuit is more complex than that of an LM566C voltage controlled oscillator and the usable frequency range is much less. The LM555C was designed as a simple timer circuit and is best used for this purpose.

The data sheets for the LM555C are shown in Figure 11-12. There are two versions listed: the LM555 and the LM555C. The only difference is that the LM555 is the military/industrial version capable of operating over a wider temperature range. We will use the LM555C for all discussions in this text. The LM555C can operate from any supply voltage between 4.5 and 16 V, with the output voltage approximately 2 V less than the supply voltage. Figure 11-13 shows the pinout and the schematic symbol of the LM555C.

Although the complete circuit diagram for the LM555C is shown on the data sheets in Figure 11-12, the circuit is so complex that the operation of the device is not apparent. A simplified functional diagram has been drawn in Figure 11-14. This figure shows that the LM555C timer consists of two comparators configured as a window comparator, biased such that the upper trigger point is two-thirds of the supply voltage and the lower trigger point is one-third of the supply voltage. These trigger points are chosen so that the window comparator operates in the center portion of the timing capacitor charge/discharge curve to provide the most linear response. Because the trigger points are determined by the internal voltage divider, the period of oscillation is relatively independent of the supply voltage.

The upper trigger level can also be externally set through the control input (pin 5). The input for the upper comparator (pin 6) is labelled "threshold," that for the lower comparator (pin 2) is labelled "trigger." These two inputs are normally connected together and to the external timing network. The output of the window comparator is connected to a memory which changes level (high or low) whenever the comparator is triggered and provides the output to pin 3. The memory can be reset from an input to pin 4 and provides a control signal to the internal discharge switch. The discharge switch, connected between pin 7 and ground, is used to provide a discharge path for the external capacitor.

The LM555C as a Simple Oscillator

The circuit for using the LM555C as a simple oscillator is shown in Figure 11-15. The timing capacitor C_1 is alternately charged and discharged through the trigger points of the window comparator. The two inputs to the window comparator, pin 2 (trigger) and pin 6 (threshold), are connected together and to the capacitor. The capacitor is charged through two resistors, R_1 and R_2, from the supply, and is discharged through resistor R_2 to pin 7 and ground. For free-running oscillation, the reset (pin 4) must be connected to the supply, and the control (pin 5) should be bypassed to ground through a 0.01 µF capacitor.

In normal operation, the capacitor charges from $\frac{1}{3}V_{CC}$ to $\frac{2}{3}V_{CC}$. The equation for the charge of capacitor C_1 is

$$V(t) = \tfrac{2}{3}V_{CC}(1 - \epsilon^{-t/(R_1+R_2)C_1}) \tag{11-15}$$

National Semiconductor

Industrial Blocks

LM555/LM555C Timer

General Description

The LM555 is a highly stable device for generating accurate time delays or oscillation. Additional terminals are provided for triggering or resetting if desired. In the time delay mode of operation, the time is precisely controlled by one external resistor and capacitor. For astable operation as an oscillator, the free running frequency and duty cycle are accurately controlled with two external resistors and one capacitor. The circuit may be triggered and reset on falling waveforms, and the output circuit can source or sink up to 200 mA or drive TTL circuits.

Features

- Direct replacement for SE555/NE555
- Timing from microseconds through hours
- Operates in both astable and monostable modes
- Adjustable duty cycle
- Output can source or sink 200 mA
- Output and supply TTL compatible
- Temperature stability better than 0.005% per °C
- Normally on and normally off output

Applications

- Precision timing
- Pulse generation
- Sequential timing
- Time delay generation
- Pulse width modulation
- Pulse position modulation
- Linear ramp generator

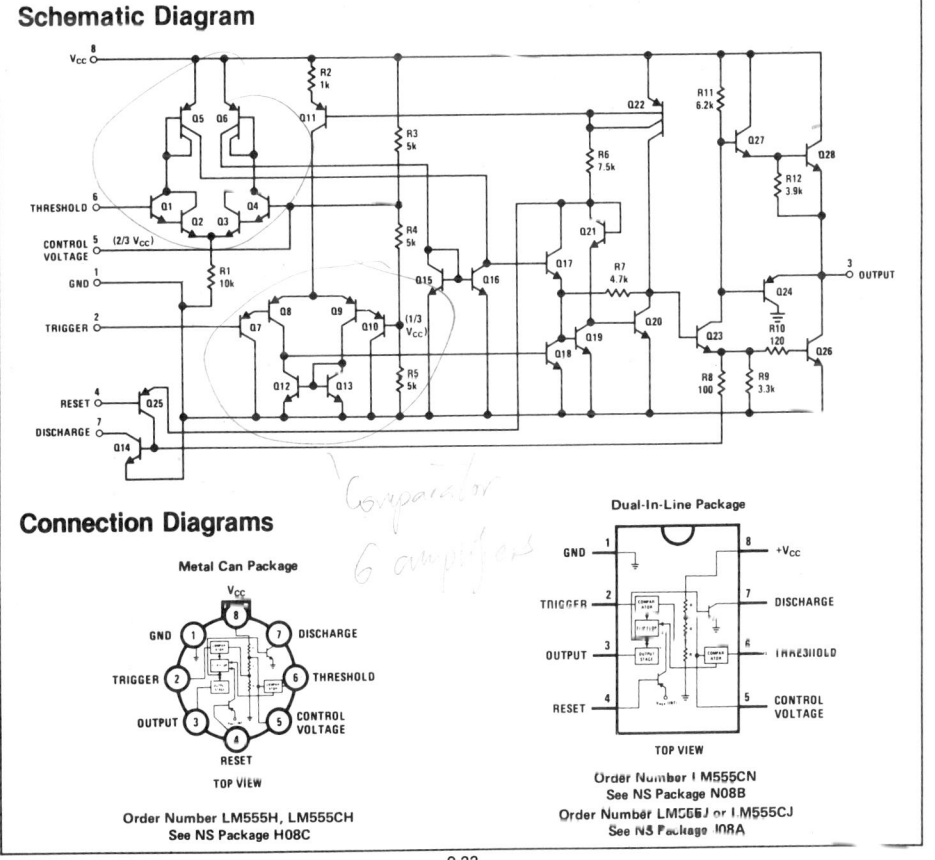

Schematic Diagram

Connection Diagrams

Metal Can Package

Order Number LM555H, LM555CH
See NS Package H08C

Dual-In-Line Package

Order Number LM555CN
See NS Package N08B
Order Number LM555J or LM555CJ
See NS Package J08A

9-33

Figure 11-12 Data sheets for the LM555/LM555C timer. *(Reprinted with permission of National Semiconductor Corporation)*

Absolute Maximum Ratings

Supply Voltage	+18V
Power Dissipation (Note 1)	600 mW
Operating Temperature Ranges	
LM555C	0°C to +70°C
LM555	−55°C to +125°C
Storage Temperature Range	−65°C to +150°C
Lead Temperature (Soldering, 10 seconds)	300°C

Electrical Characteristics (T_A = 25°C, V_{CC} = +5V to +15V, unless otherwise specified)

PARAMETER	CONDITIONS	LM555 MIN	LM555 TYP	LM555 MAX	LM555C MIN	LM555C TYP	LM555C MAX	UNITS
Supply Voltage		4.5		18	4.5		16	V
Supply Current	V_{CC} = 5V, R_L = ∞		3	5		3	6	mA
	V_{CC} = 15V, R_L = ∞ (Low State) (Note 2)		10	12		10	15	mA
Timing Error, Monostable								
Initial Accuracy			0.5			1		%
Drift with Temperature	R_A, R_B = 1k to 100 k, C = 0.1μF, (Note 3)		30			50		ppm/°C
Accuracy over Temperature			1.5			1.5		%
Drift with Supply			0.05			0.1		%/V
Timing Error, Astable								
Initial Accuracy			1.5			2.25		%
Drift with Temperature			90			150		ppm/°C
Accuracy over Temperature			2.5			3.0		%
Drift with Supply			0.15			0.30		%/V
Threshold Voltage			0.667			0.667		x V_{CC}
Trigger Voltage	V_{CC} = 15V	4.8	5	5.2		5		V
	V_{CC} = 5V	1.45	1.67	1.9		1.67		V
Trigger Current			0.01	0.5		0.5	0.9	μA
Reset Voltage		0.4	0.5	1	0.4	0.5	1	V
Reset Current			0.1	0.4		0.1	0.4	mA
Threshold Current	(Note 4)		0.1	0.25		0.1	0.25	μA
Control Voltage Level	V_{CC} = 15V	9.6	10	10.4	9	10	11	V
	V_{CC} = 5V	2.9	3.33	3.8	2.6	3.33	4	V
Pin 7 Leakage Output High			1	100		1	100	nA
Pin 7 Sat (Note 5)								
Output Low	V_{CC} = 15V, I_7 = 15 mA		150			180		mV
Output Low	V_{CC} = 4.5V, I_7 = 4.5 mA		70	100		80	200	mV
Output Voltage Drop (Low)	V_{CC} = 15V							
	I_{SINK} = 10 mA		0.1	0.15		0.1	0.25	V
	I_{SINK} = 50 mA		0.4	0.5		0.4	0.75	V
	I_{SINK} = 100 mA		2	2.2		2	2.5	V
	I_{SINK} = 200 mA		2.5			2.5		V
	V_{CC} = 5V							
	I_{SINK} = 8 mA		0.1	0.25				V
	I_{SINK} = 5 mA					0.25	0.35	V
Output Voltage Drop (High)	I_{SOURCE} = 200 mA, V_{CC} = 15V		12.5			12.5		V
	I_{SOURCE} = 100 mA, V_{CC} = 15V	13	13.3		12.75	13.3		V
	V_{CC} = 5V	3	3.3		2.75	3.3		V
Rise Time of Output			100			100		ns
Fall Time of Output			100			100		ns

Note 1: For operating at elevated temperatures the device must be derated based on a +150°C maximum junction temperature and a thermal resistance of +45°C/W junction to case for TO-5 and +150°C/W junction to ambient for both packages.
Note 2: Supply current when output high typically 1 mA less at V_{CC} = 5V.
Note 3: Tested at V_{CC} = 5V and V_{CC} = 15V.
Note 4: This will determine the maximum value of $R_A + R_B$ for 15V operation. The maximum total ($R_A + R_B$) is 20 MΩ.
Note 5: No protection against excessive pin 7 current is necessary providing the package dissipation rating will not be exceeded.

9-34

Figure 11-12 (*Continued*)

11-2 The LM555C Timer/Oscillator

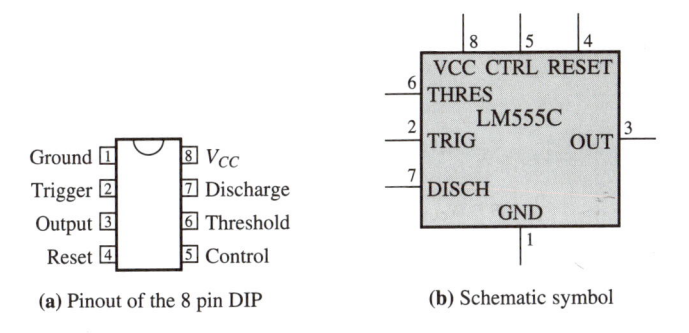

Figure 11-13 Pinout and schematic symbol for the LM555C timer/oscillator.

(a) Pinout of the 8 pin DIP

(b) Schematic symbol

Substitute $\tfrac{1}{3}V_{CC}$ for $V(t)$ and solve for t_{charge}:

$$t_{charge} = 0.693 \times (R_1 + R_2) \times C_1 \tag{11-16}$$

This charge time corresponds to time when the output signal from the memory is high (approximately 2 V less than V_{CC}). When the capacitor voltage reaches the upper trigger point voltage ($\tfrac{2}{3}V_{CC}$), the comparator triggers off, the memory resets to zero, and the discharge path is connected to ground.

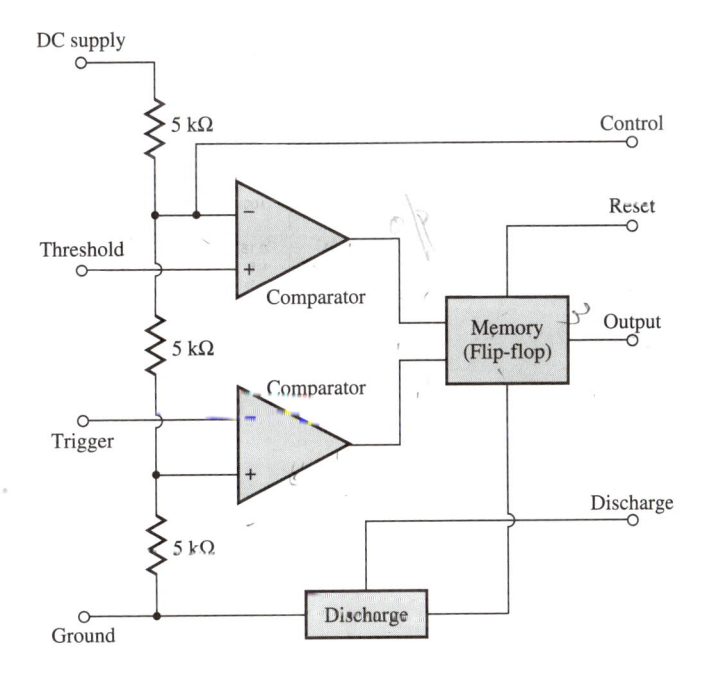

Figure 11-14 Simplified functional diagram of the LM555C timer.

11 Integrated Circuit Oscillators

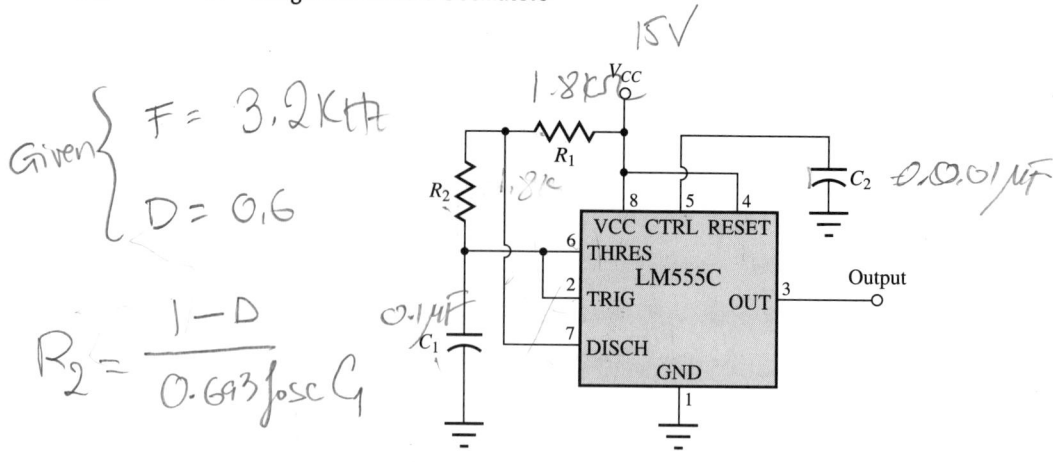

[Handwritten annotations:]
Given: $F = 3.2\,\text{KHz}$, $D = 0.6$

$R_2 = \dfrac{1-D}{0.693\, f_{osc}\, C_1}$

[Circuit annotations: $V_{CC} = 15\text{V}$, $R_1 = 1.8\,\text{k}\Omega$, $R_2 = 1.2\,\text{k}$, $C_1 = 0.1\,\mu\text{F}$, $C_2 = 0.001\,\mu\text{F}$]

Figure 11-15 Simple oscillator circuit using the LM555C.

The capacitor will discharge through resistor R_2 and the discharge switch connected to pin 7 until its voltage reaches $\tfrac{1}{3}V_{CC}$. The equation for the discharge of capacitor C_1 is

$$V(t) = \tfrac{2}{3}V_{CC}\,\epsilon^{-t/R_2 C_1} \qquad (11\text{-}17)$$

Substitute $\tfrac{1}{3}V_{CC}$ for $V(t)$ and solve for the discharge time:

$$t_{\text{discharge}} = 0.693\,R_2 C_1 \qquad (11\text{-}18)$$

The discharge time corresponds to the time when the output signal from the memory is low. When the capacitor voltage reaches the lower trigger point ($\tfrac{1}{3}V_{CC}$), the comparator triggers, the memory is set on, and the discharge switch is opened. The capacitor starts to charge for the next cycle.

The period of oscillation is simply the charge time plus the discharge time:

$$\text{period} = t_{\text{charge}} + t_{\text{discharge}} = 0.693 \times (R_1 + 2R_2) \times C_1 \qquad (11\text{-}19)$$

The frequency f_{osc} is 1/period:

$$f_{\text{osc}} = \frac{1}{0.693(R_1 + 2R_2)C_1} \qquad (11\text{-}20)$$

The duty cycle D is the on time divided by the period, and can be determined by dividing Equation 11-16 by Equation 11-19:

$$D = \frac{R_1 + R_2}{R_1 + 2R_2} \qquad (11\text{-}21)$$

To design component values for an LM555C-based oscillator, we normally select the oscillator frequency f_{osc}, the duty cycle D, and a suitable value for C_1. Solve Equations 11-20 and 11-21 simultaneously, first for R_2, then for R_1, to obtain the design equations

$$R_2 = \frac{1 - D}{0.693 f_{osc} C_1} \quad (11\text{-}22)$$

$$R_1 = \frac{2D - 1}{0.693 f_{osc} C_1} \quad (11\text{-}23)$$

(handwritten annotation: $= \frac{1 - 0.6}{0.693 (3.2k)(0.1\mu)} = 2.8k$)

Note that from Equations 11-21 and 11-23 the duty cycle must always be greater than 0.5.

This simple oscillator circuit is the one most often encountered for the LM555C. Unfortunately, it has a serious limitation—it can only be used to generate waves with duty cycles greater than 0.5. If we try to obtain a symmetrical square wave with a duty cycle of 0.5, we get $R_1 = 0$. This value would short the power supply during discharge.

EXAMPLE 11-5

Design a simple oscillator using an LM555C timer. The frequency should be 1200 Hz and the duty cycle 0.8. Use a 12 V supply.

Solution

Since only a simple 555 oscillator is required, we will use the circuit shown in Figure 11-15. This has been redrawn as Figure 11-16 with a parts list.

To design this circuit, given the frequency $f_{osc} = 1200$ Hz and the duty cycle $D = 0.80$, we must first select a suitable value for C_1 and then calculate R_1 and

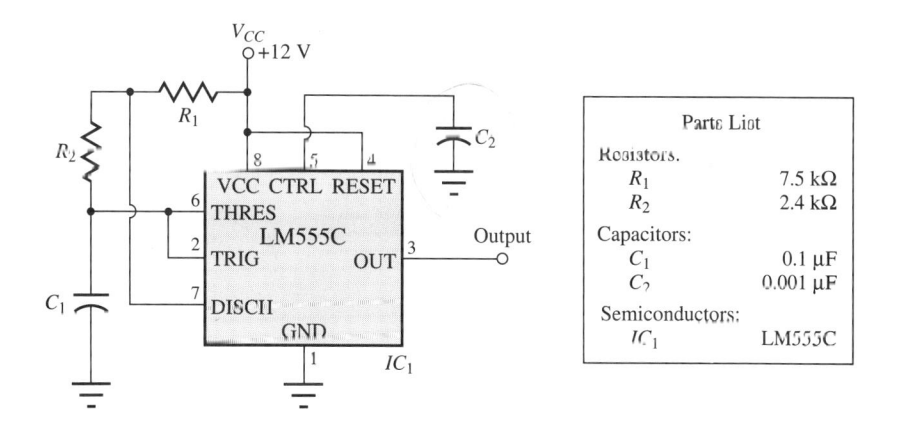

Figure 11-16

Parts List

Resistors:
- R_1 7.5 kΩ
- R_2 2.4 kΩ

Capacitors:
- C_1 0.1 μF
- C_2 0.001 μF

Semiconductors:
- IC_1 LM555C

R_2. Choose a value for C_1 of 0.1 µF. Solve first for R_1, using Equation 11-23:

$$R_1 = \frac{2D - 1}{0.693 f_{osc} C_1}$$

$$= \frac{2 \times 0.8 - 1}{0.693 \times 1200 \text{ Hz} \times 0.1 \text{ µF}} = 7.22 \text{ k}\Omega$$

Next, solve for R_2, using Equation 11-22:

$$R_2 = \frac{1 - D}{0.693 f_{osc} C_1}$$

$$= \frac{1 - 0.8}{0.693 \times 1200 \text{ Hz} \times 0.1 \text{ µF}} = 2.41 \text{ k}\Omega$$

Choose $R_1 = 7.5$ kΩ and $R_2 = 2.4$ kΩ as the closest standard values.

The oscillator will also require a bypass capacitor C_2 to be connected from pin 5 to ground. Use a value of 0.001 µF.

Our oscillator design is now complete.

The LM555C as a Square-Wave Oscillator

If we want to generate a true square wave using the LM555C, we must modify the circuit slightly. A diode is placed in series with R_2 and one in parallel with it. This circuit is illustrated in Figure 11-17. While the capacitor is charging, diode D_1 is forward-biased, thus shorting out R_2, and diode D_2 is reverse-biased, thereby preventing any current from flowing in R_2. The charge time is determined by R_1 only:

$$t_{charge} = 0.693 R_1 C_1 \tag{11-24}$$

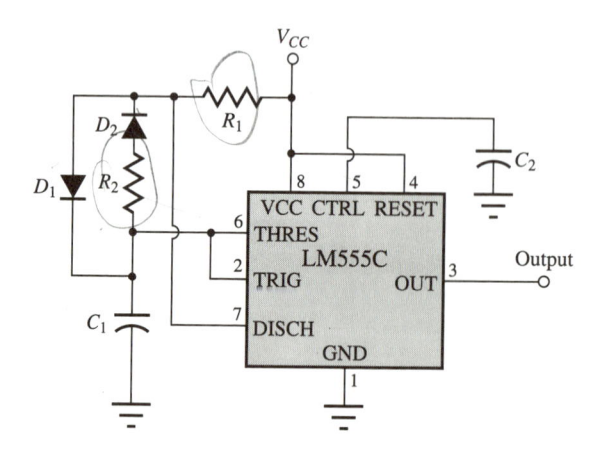

Figure 11-17 Square-wave oscillator circuit using the LM555C timer/oscillator.

When the capacitor is discharging, diode D_2 is forward-biased, and the capacitor discharges through R_2. The discharge time is

$$t_{\text{discharge}} = 0.693 R_2 C_1 \tag{11-18}$$

The frequency f_{osc} is

$$f_{\text{osc}} = \frac{1}{0.693(R_1 + R_2)C_1} \tag{11-25}$$

The duty cycle D is

$$D = \frac{R_1}{R_1 + R_2} \tag{11-26}$$

Note that this circuit has the advantage that the duty cycle can be any value from 0 to 1. For a square wave, the duty cycle is 0.5, the charge time equals the discharge time, and $R_1 = R_2$. In this case, the frequency is

$$f_{\text{osc}} = \frac{1}{1.386 R_1 C_1} \quad \text{or} \quad f_{\text{osc}} = \frac{1}{1.386 R_2 C_1} \tag{11-27}$$

To design the circuit for a square-wave output, we need only select a value for C_1 and calculate R_1 (or R_2) from Equation 11-27:

$$R_1 = \frac{1}{1.386 f_{\text{osc}} C_1} \tag{11-28}$$

To design the circuit for any other duty cycle, we solve Equations 11-25 and 11-26 simultaneously, first for R_1, then for R_2:

$$R_1 = \frac{D}{0.693 f_{\text{osc}} C_1} \tag{11-29}$$

$$R_2 = \frac{1 - D}{0.693 f_{\text{osc}} C_1} \tag{11-30}$$

This circuit is much more flexible than the simple circuit developed earlier.

At this point, it is interesting to compare the LM555C to the LM566C. The LM555C circuit in Figure 11-17 has an advantage that the duty cycle can be set to any desired value. Unfortunately, this is the only advantage. The circuit is more complex than the LM566, it cannot produce a triangle wave (the waveform across C_1 is at best a poor approximation to a triangle wave), and, to make the oscillator variable, it requires two potentiometers, both of which must be adjusted simultaneously.

There are other limitations for the LM555C oscillator circuit: the waveform becomes noticably distorted above 10 kHz, and the frequency varies somewhat if the supply voltage is changed. The signal output of the LM566 is good to at least 1 MHz, and the output frequency is determined solely by R_1 and C_1, not by the supply voltage. The poorer performance of the LM555C is simply because it was designed as a timer, not an oscillator.

EXAMPLE 11-6

Design an oscillator circuit using an LM555C timer to produce a square wave with a frequency of 1200 Hz. Use a 12 V power supply.

Solution

This is the same problem as Example 11-5 except that now we require a square-wave output. We cannot obtain a square wave from the circuit in Example 11-5. Instead, we must use the modified circuit in Figure 11-17. This circuit has been redrawn as Figure 11-18 with a parts list.

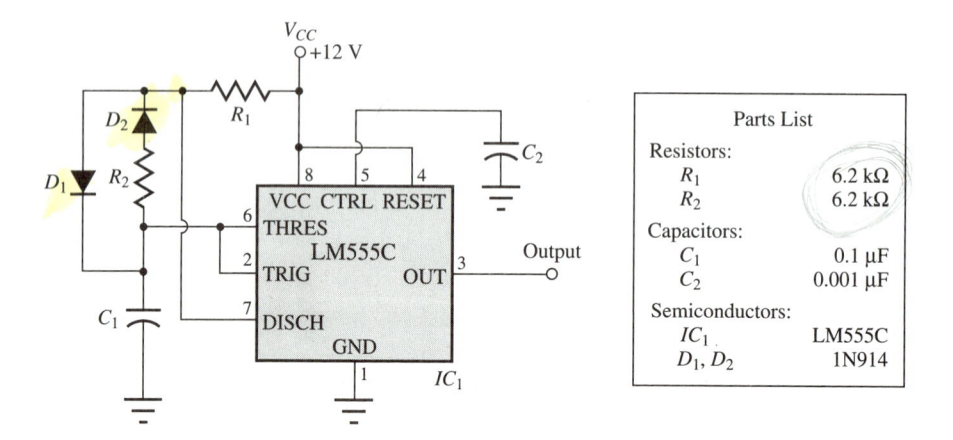

Figure 11-18

For a square wave, $R_1 = R_2$. Choose $C_1 = 0.1$ µF, and calculate R_1 from Equation 11-28:

$$R_1 = \frac{1}{1.386 f_{osc} C_1}$$

$$= \frac{1}{1.386 \times 1200 \text{ Hz} \times 0.1 \text{ µF}} = 6.01 \text{ k}\Omega$$

The closest standard value from Appendix A-1 is 6.2 kΩ. We will choose this value for both R_1 and R_2.

The oscillator will also require a bypass capacitor C_2 to be connected from pin 5 to ground. Use a value of 0.001 µF.

Our design is complete. Note that we could have used this circuit design as a solution for Example 11-5. It is left as an exercise for the reader to verify that the resistor values to give a 0.8 duty cycle are $R_1 = 10$ kΩ and $R_2 = 2.4$ kΩ.

The LM555C as a Timer

The preceding discussion showed that the LM555C can be used as an oscillator but does not perform as well as an integrated circuit designed specifically as an oscillator. We will now look at the application for which the LM555C was actually designed, namely an analog **timer.**

The circuit of the LM555C used as a timer is shown in Figure 11-19. The main elements in the timing circuit are resistor R_1 connected from V_{CC} to pin 6 and capacitor C_1 connected from pin 7 to ground. The time interval is simply the time for the capacitor to charge to $\frac{2}{3}V_{CC}$ through the resistor. The charging equation of C_1 is

$$V(t) = V_{CC}(1 - \epsilon^{-t/R_1C_1}) \qquad (11\text{-}31)$$

Substitute $\frac{2}{3}V_{CC}$ for $V(t)$ and solve for t:

$$t = 1.10 R_1 C_1 \qquad (11\text{-}32)$$

Normally, we choose C_1 and calculate R_1. Solve Equation 11-32 for R_1 to obtain our design equation:

$$R_1 = \frac{t}{1.10 C_1} \qquad (11\text{-}33)$$

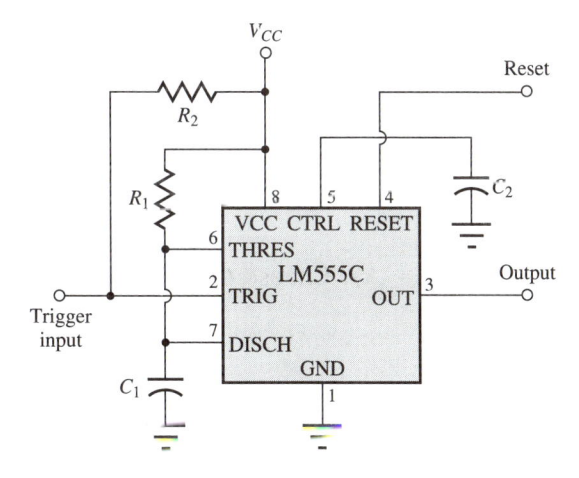

Figure 11-19 Timer circuit using the LM555C.

The rest of the circuit consists of capacitor C_2 connected from control (pin 5) to ground, and resistor R_2 connected as a pull-up resistor from V_{CC} to the trigger input (pin 2). Typical values are 0.001 μF for C_2 and 10 kΩ for R_2. The timer is started by grounding the trigger line (pin 2). The output from pin 3 immediately goes high for the duration of the time interval. The timer can be reset

514   11 Integrated Circuit Oscillators

at any time by connecting the reset line (pin 4) to ground. The output voltage during the timing interval is approximately 2 V less than the supply voltage.

This circuit is capable of generating timing pulses from a few microseconds to many minutes. The main limitation for very long time delays is the leakage in the large-value capacitor required for C_1.

EXAMPLE 11-7

Design a 10 s timer using an LM555C. Use a 12 V power supply.

Solution

This is the application for which the LM555C was designed. The circuit is shown in Figure 11-20 with a parts list.

Parts List

Resistors:
- R_1 91 kΩ
- R_2 10 kΩ

Capacitors:
- C_1 100 μF
- C_2 0.001 μF

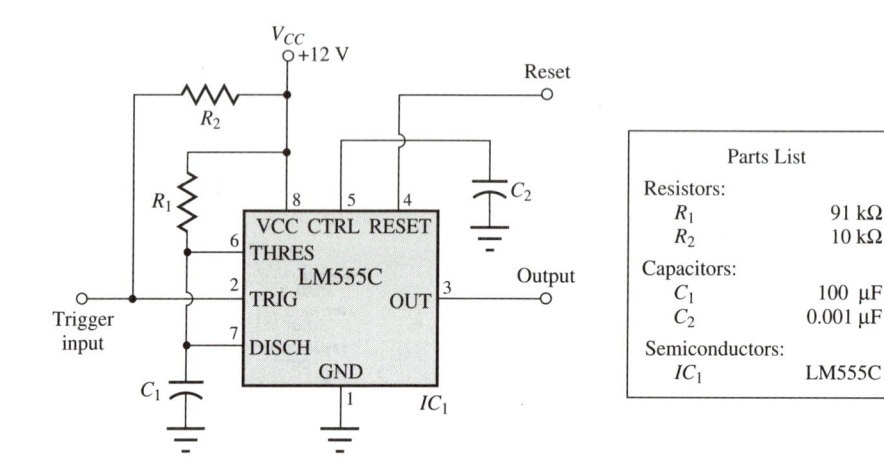

Semiconductors:
- IC_1 LM555C

Figure 11-20

We must design values for R_1 and C_1. A large value of C_1 will be required because 10 s is a relatively long timing interval. Choose C_1 to be 100 μF, and calculate R_1 from Equation 11-33:

$$R_1 = \frac{t}{1.10 C_1} = \frac{10 \text{ s}}{1.10 \times 100 \text{ μF}} = 90.9 \text{ kΩ}$$

The closest standard value of resistance is 91 kΩ. We will use this for R_1.

Use a value of 10 kΩ for the pull-up resistor R_2 and a value of 0.001 μF for capacitor C_2.

Our timer design is complete.

ICL8038
Precision Waveform Generator/Voltage Controlled Oscillator

GENERAL DESCRIPTION

The ICL8038 Waveform Generator is a monolithic integrated circuit capable of producing high accuracy sine, square, triangular, sawtooth and pulse waveforms with a minimum of external components. The frequency (or repetition rate) can be selected externally from .001Hz to more than 300kHz using either resistors or capacitors, and frequency modulation and sweeping can be accomplished with an external voltage. The ICL8038 is fabricated with advanced monolithic technology, using Schottky-barrier diodes and thin film resistors, and the output is stable over a wide range of temperature and supply variations. These devices may be interfaced with phase locked loop circuitry to reduce temperature drift to less than 250ppm/°C.

FEATURES

- Low Frequency Drift With Temperature — 250ppm/°C
- Simultaneous Sine, Square, and Triangle Wave Outputs
- Low Distortion — 1% (Sine Wave Output)
- High Linearity — 0.1% (Triangle Wave Output)
- Wide Operating Frequency Range — 0.001Hz to 300kHz
- Variable Duty Cycle — 2% to 98%
- High Level Outputs — TTL to 28V
- Easy to Use — Just A Handful of External Components Required

ORDERING INFORMATION

Part Number	Stability	Temp. Range	Package
ICL8038CCPD	250ppm/°C typ	0°C to +70°C	14 pin DIP
ICL8038CCJD	250ppm/°C typ	0°C to +70°C	14 pin CERDIP
ICL8038BCJD	180ppm/°C typ	0°C to +70°C	14 pin CERDIP
ICL8038ACJD	120ppm/°C typ	0°C to +70°C	14 pin CERDIP
ICL8038BMJD*	350ppm/°C max	−55°C to +125°C	14 pin CERDIP
ICL8038AMJD*	250ppm/°C max	−55°C to +125°C	14 pin CERDIP

*Add /883B to part number if 883 processing is required.

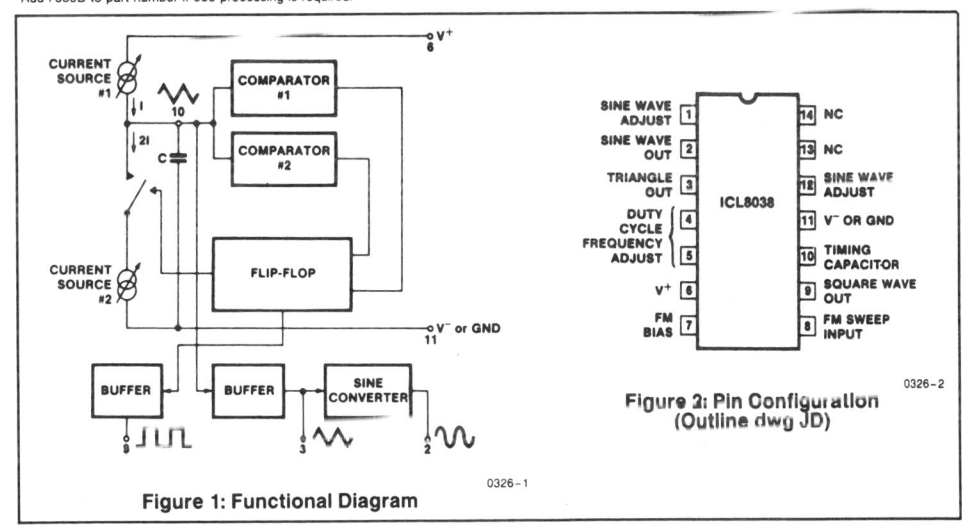

Figure 1: Functional Diagram

Figure 2: Pin Configuration (Outline dwg JD)

8-51

Figure 11-21 Data sheets for the ICL8038 function generator. *(Reprinted with permission of Harris Semiconductor)*

ICL8038

ABSOLUTE MAXIMUM RATINGS

Supply Voltage (V⁻ to V⁺) 36V
Power Dissipation(1) 750mW
Input Voltage (any pin) .... V⁻ to V⁺
Input Current (Pins 4 and 5) 25mA
Output Sink Current (Pins 3 and 9) 25mA

Storage Temperature Range .... −65°C to +150°C
Operating Temperature Range:
  8038AM, 8038BM .... −55°C to +125°C
  8038AC, 8038BC, 8038CC .... 0°C to +70°C
Lead Temperature (Soldering, 10sec) .... 300°C

NOTE: *Stresses above those listed under "Absolute Maximum Ratings" may cause permanent damage to the device. These are stress ratings only and functional operation of the device at these or any other conditions above those indicated in the operational sections of the specifications is not implied. Exposure to absolute maximum rating conditions for extended periods may affect device reliability.*

NOTE 1: *Derate ceramic package at 12.5mW/°C for ambient temperatures above 100°C.*

ELECTRICAL CHARACTERISTICS ($V_{SUPPLY} = \pm 10V$ or $+20V$, $T_A = 25°C$, $R_L = 10k\Omega$, Test Circuit Unless Otherwise Specified)

Symbol	General Characteristics	8038CC Min	8038CC Typ	8038CC Max	8038BC(BM) Min	8038BC(BM) Typ	8038BC(BM) Max	8038AC(AM) Min	8038AC(AM) Typ	8038AC(AM) Max	Units
V_{SUPPLY}	Supply Voltage Operating Range										
V⁺	Single Supply	+10		+30	+10		30	+10		30	V
V⁺, V⁻	Dual Supplies	±5		±15	±5		±15	±5		±15	V
I_{SUPPLY}	Supply Current ($V_{SUPPLY} = \pm 10V$)(2)										
	8038AM, 8038BM					12	15		12	15	mA
	8038AC, 8038BC, 8038CC		12	20		12	20		12	20	mA
Frequency Characteristics (all waveforms)											
f_{max}	Maximum Frequency of Oscillation	100			100			100			kHz
f_{sweep}	Sweep Frequency of FM Input		10			10			10		kHz
	Sweep FM Range(3)		35:1			35:1			35:1		
	FM Linearity 10:1 Ratio		0.5			0.2			0.2		%
$\Delta f/\Delta T$	Frequency Drift With Temperature(5) 8038 AC, BC, CC 0°C to 70°C		250			180			120		ppm/°C
	8038 AM, BM, −55°C to 125°C						350			250	
$\Delta f/\Delta V$	Frequency Drift With Supply Voltage (Over Supply Voltage Range)		0.05			0.05			0.05		%/V
Output Characteristics											
I_{OLK}	Square-Wave Leakage Current ($V_9 = 30V$)			1			1			1	μA
V_{SAT}	Saturation Voltage ($I_{SINK} = 2mA$)		0.2	0.5		0.2	0.4		0.2	0.4	V
t_r	Rise Time ($R_L = 4.7k\Omega$)		180			180			180		ns
t_f	Fall Time ($R_L = 4.7k\Omega$)		40			40			40		ns
ΔD	Typical Duty Cycle Adjust (Note 6)	2		98	2		98	2		98	%
$V_{TRIANGLE}$	Triangle/Sawtooth/Ramp Amplitude ($R_{TRI} = 100k\Omega$)	0.30	0.33		0.30	0.33		0.30	0.33		$\times V_{SUPPLY}$
	Linearity		0.1			0.05			0.05		%
Z_{OUT}	Output Impedance ($I_{OUT} = 5mA$)		200			200			200		Ω
V_{SINE}	Sine-Wave Amplitude ($R_{SINE} = 100k\Omega$)	0.2	0.22		0.2	0.22		0.2	0.22		$\times V_{SUPPLY}$
THD	THD ($R_S = 1M\Omega$)(4)		2.0	5		1.5	3		1.0	1.5	%
THD	THD Adjusted (Use Figure 6)		1.5			1.0			0.8		%

NOTES:
2. R_A and R_B currents not included.
3. $V_{SUPPLY} = 20V$; R_A and $R_B = 10k\Omega$, $f \cong 10kHz$ nominal; can be extended 1000 to 1. See Figures 7a and 7b.
4. 82kΩ connected between pins 11 and 12, Triangle Duty Cycle set at 50%. (Use R_A and R_B.)
5. Figure 3, pins 7 and 8 connected, $V_{SUPPLY} = \pm 10V$. See Typical Curves for T.C. vs V_{SUPPLY}.
6. Not tested, typical value for design purposes only.

NOTE: *All typical values have been characterized but are not tested.*

Figure 11-21 (*Continued*)

11-3 The ICL8038 Single-Chip Function Generator

Earlier we developed a function generator circuit using the LM566C voltage controlled oscillator. Most function generators can produce sine waves, triangle waves, and square waves over a range of frequencies from about 1 Hz to 100 kHz or 1 MHz. Our circuit using the LM566C was somewhat limited; there was no sine wave, and the lowest frequency was approximately 100 Hz because the outputs were capacitively coupled to eliminate DC offset. The generation of a sine wave was possible, but required the addition of a relatively complex wave-shaping circuit.

The simplest solution for designing a function generator is to use a function generator integrated circuit. There are a number on the market. The ICL8038 function generator manufactured by Intersil (now Harris Semiconductor) is probably the most widely used chip, and we will use it to design a function generator. The ICL8038 consists of a voltage controlled oscillator and a built-in wave-shaping circuit.

The data sheets for the ICL8038 are shown in Figure 11-21. There are a number of different versions of this chip, the main difference being in the quality of the output. The ICL8038CC, which has the most distortion (2%) and poorest linearity (0.1%) is the least expensive, costing only a few dollars. The ICL8038AC is the highest quality with minimum distortion (1%), highest linearity (0.05%), and maximum cost (typically five times as much as the least expensive version).

The ICL8038 produces sine-wave, triangle-wave, and square-wave outputs over a frequency range from 0.001 Hz to more than 200 kHz. It can operate from either a single supply between 10 and 30 V or from a dual supply between ±5 and ±15 V. The chip draws about 0.75 W of power and becomes quite warm during operation because of the power being dissipated in the resistive sine-wave-shaping circuit.

The pinout is shown on the data sheet, but has been reproduced to a larger scale in Figure 11-22. The schematic symbol is also shown in this figure. The hookup and operation of this function generator is quite simple, with typical application circuits shown in Figures 11-23 and 11-24.

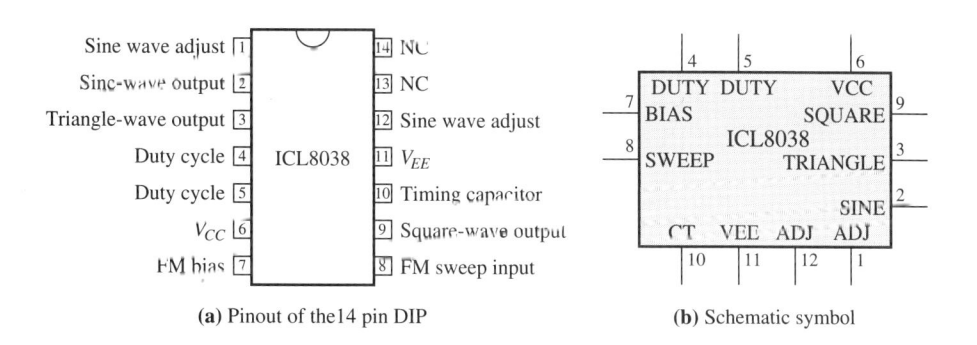

(a) Pinout of the 14 pin DIP

(b) Schematic symbol

Figure 11-22 Pinout and schematic symbol of the ICL8038 function generator.

518   11 Integrated Circuit Oscillators

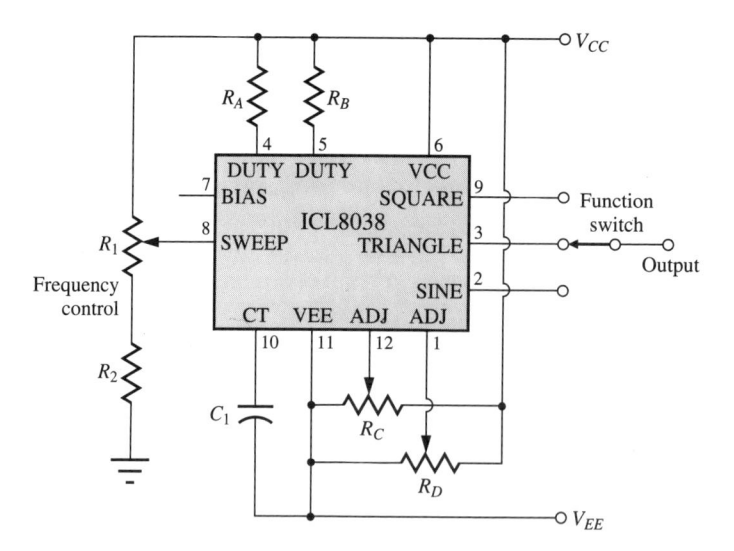

Figure 11-23 Simple function generator using the ICL8038.

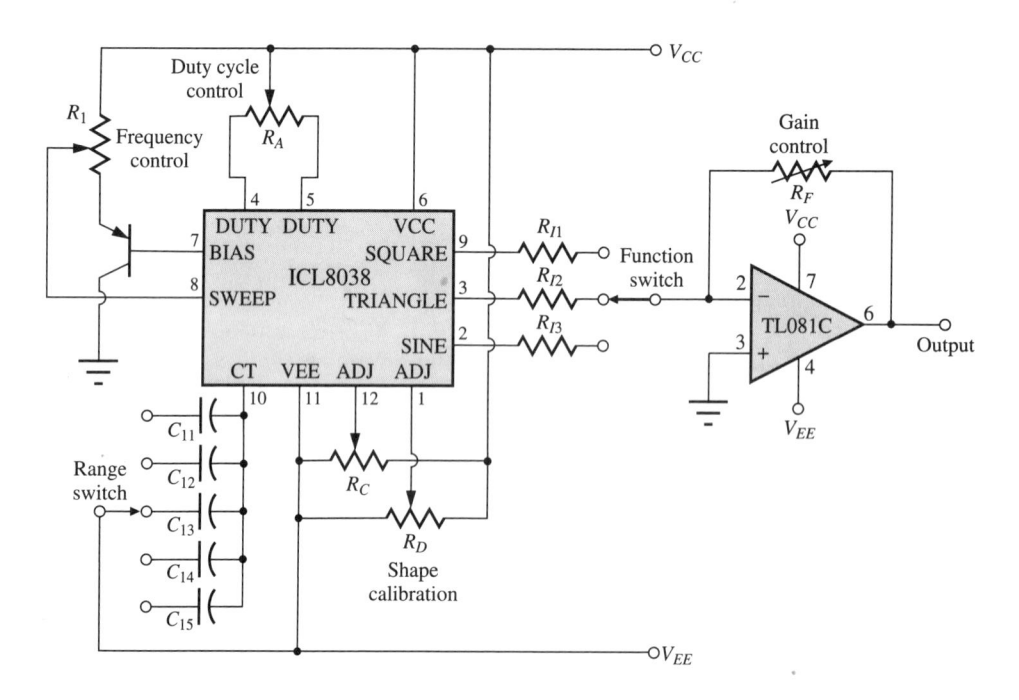

Figure 11-24 Enhanced function generator using the ICL8038. This function generator has decade frequency ranges from 1 Hz to 100 kHz, variable-gain output, bias frequency control, and variable duty cycle.

First we will consider the simple function generator in Figure 11-23. This function generator will produce a sine wave at pin 2, a triangle wave at pin 3, and a square wave at pin 9. The frequency of the function generator is determined by a timing capacitor C_1 connected between pin 10 and the negative supply, two timing resistors R_A and R_B connected from pins 4 and 5, respectively, to the positive supply, and a DC control signal at pin 8.

The frequency can be varied over a range of approximately 1000:1 by controlling the voltage applied to pin 8 (the "sweep" voltage input). The sweep voltage should be kept in the range

$$\tfrac{2}{3}V_{\text{supply}} + 2\text{ V} < V_{\text{sweep}} < V_{\text{supply}} \qquad (11\text{-}34)$$

The quantity V_{supply} is the total supply voltage across the ICL8038. For example, if the ICL8038 is operated from a ± 15 V supply, then $V_{\text{supply}} = 30$ V and V_{sweep} must be in the range from $+7$ to $+15$ V. The potentiometer R_1 and series resistor R_2 connected between the positive supply and ground provide a variable voltage to pin 8, allowing the frequency to be varied over the full operational range. The maximum frequency is obtained when the voltage is a minimum.

Timing resistors R_A and R_B connected from pins 4 and 5, respectively, to the positive supply are normally equal in value to give a duty cycle of 0.5. The manufacturer recommends a value of 10 kΩ for these resistors. If a variable duty cycle is required, then R_A and R_B can be replaced with a potentiometer, as we will discuss later. If the timing resistors are equal, then the frequency of the top of the range (with the voltage at pin 8 set to the minimum value) is

$$f_{\text{osc}} = \frac{0.3}{RC_1} \qquad (11\text{-}35)$$

where $R = R_A = R_B$, which is typically chosen as 10 kΩ.

Capacitor C_1 is selected to set the range. For example, if $R = 10$ kΩ, choosing $C_1 = 0.003$ µF sets the maximum frequency to 10 kHz. Since the ICL8038 has a 1000:1 frequency adjustment range, the minimum frequency would be approximately 10 Hz. If a number of ranges are required, several capacitors could be switched, as will be discussed later.

There are two adjustment terminals (pins 1 and 12) for optimizing the shape of the sine wave. If these pins are set to ground potential, the output sine-wave shape should be reasonable and adequate for most applications. If two 100 kΩ potentiometers are connected between the positive and negative supplies with the variable terminals connected to pins 1 and 12, the sine-wave shape can be optimized by adjusting these calibration controls.

Although the basic circuit in Figure 11-23 gives a usable function generator, it is quite easy to enhance its performance by a few simple changes. This has been done in the circuit in Figure 11-24.

One enhancement is to provide a variable duty cycle by placing a potentiometer between pins 4 and 5, with the positive supply connected to the variable terminal. When the variable terminal is at the center of the resistance range, the output waveforms will have a duty cycle of 0.5, and a frequency given by Equation 11-35 where R is half the potentiometer value. The frequency will change somewhat as the duty cycle is changed.

Another enhancement is to add a range switch to allow frequencies to be selected in decade ranges. This is done by using different values for C_1, differing by factors of 10, each of which can be selected by the range switch. For example, the selection of 3 µF, 0.3 µF, 0.03 µF, 0.003 µF, and 300 pF capacitors gives frequency decade ranges of 1 to 10 Hz, 10 to 100 Hz, 100 Hz to 1 kHz, 1 to 10 kHz, and 10 to 100 kHz, respectively.

Thus far, we have not mentioned pin 7, labelled "FM bias" on the pinout. This pin outputs a bias voltage to set the top end of the frequency range. For a single output frequency, connect pin 8 to pin 7 and calculate the frequency using Equation 11-35. For a range of frequencies, the output from pin 7 is used to bias the base of a PNP transistor. The sweep voltage is obtained from a potentiometer R_1 between the positive supply and the emitter of the transistor, as shown in Figure 11-24. This voltage eliminates the need for calculating potentiometer and series resistor values as was necessary for the simpler circuit in Figure 11-23. The resistance value of the potentiometer is not important. A value of 1.0 kΩ for R_1 is a typical choice.

One final enhancement is to add a buffer amplifier to the output of the function generator. Using a potentiometer for the feedback resistor R_F allows the output amplitude of the function generator to be controlled. This enhancement was used earlier in the simple function generator that we designed with the LM566C voltage controlled oscillator, and the design procedure here is the same. Notice that separate input resistors are used for each waveform in Figure 11-24. This allows the gain to be separately set for each waveform to equalize output amplitudes. The amplitude of the square wave produced by the ICL8038 is approximately 0.5 V_{supply}, the amplitude of the triangle wave is approximately 0.3 V_{supply}, and the amplitude of the sine wave is approximately 0.2 V_{supply}.

This enhanced function generator is typical of the circuits given in amateur magazines and project books for a "build-your-own" function generator project, and does produce a perfectly acceptable function generator. Other possible enhancements include substituting a multiturn potentiometer for R_1 to give more precise frequency adjustment, adding a push-pull power amplifier with an output impedance of 50 Ω as an output stage, and adding a summing amplifier to allow a DC offset to be added.

EXAMPLE 11-8

Design a function generator using the ICL8038 function generator integrated circuit to give output frequencies from 10 Hz to 100 kHz in 4 decade ranges. Use a ±15 V power supply.

Solution

Although a function generator is a relatively complex piece of electronic equipment, the ICL8038 function generator integrated circuit actually makes the design quite easy. The circuit is shown in Figure 11-25.

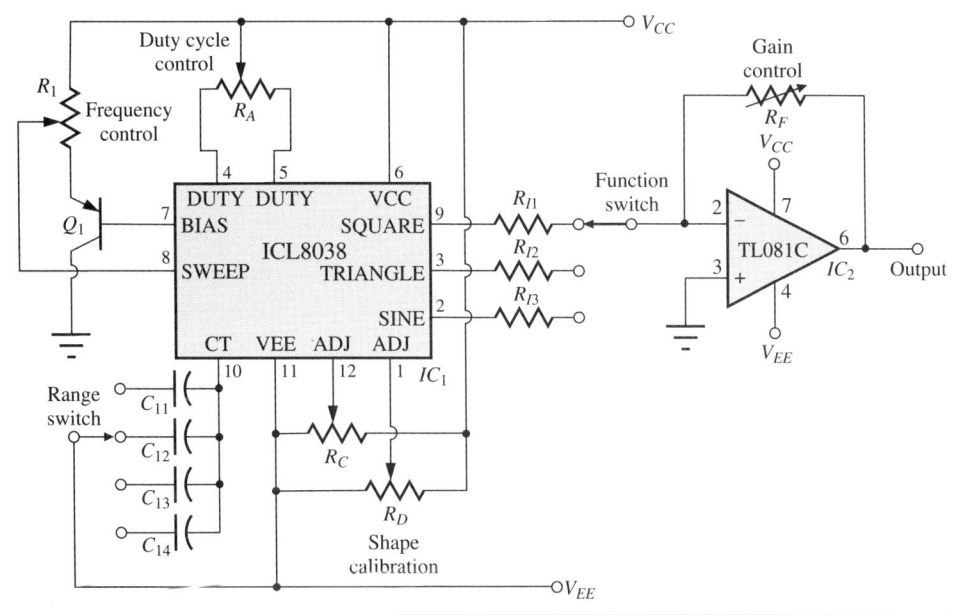

11-3 The ICL8038 Single-Chip Function Generator

Figure 11-25

Parts List

Resistors:
- R_1 — 1.0 kΩ pot
- R_A — 20 kΩ pot
- R_C — 100 kΩ pot
- R_D — 100 kΩ pot
- R_F — 10 kΩ pot
- R_{I1} — 5.1 kΩ
- R_{I2} — 3.6 kΩ
- R_{I3} — 2.2 kΩ

Capacitors:
- C_{11} — 270 pF
- C_{12} — 0.0027 μF
- C_{13} — 0.027 μF
- C_{14} — 0.27 μF

Semiconductors:
- IC_1 — ICL8038
- IC_2 — TL081C
- Q_1 — 2N4403

First, we must select the timing resistors R_A and R_B. We will use a 20 kΩ potentiometer to allow us to adjust the duty cycle. At a duty cycle of 0.5, the wiper arm is in the middle of its range; hence $R_A = R_B = R = 10$ kΩ, and the frequency is given by Equation 11-35:

$$f_{osc} = \frac{0.3}{RC_1}$$

We will design C_{11} for the highest decade range, 10 to 100 kHz. For this range, the maximum frequency is 100 kHz. Solve Equation 11-35 for C_1, which we will call C_{11}:

$$C_{11} = \frac{0.3}{Rf_{osc}} = \frac{0.3}{10 \text{ k}\Omega \times 100 \text{ kHz}} = 300 \text{ pF}$$

Choose the closest standard value of 270 pF. For the other decades, we will choose $C_{12} = 0.0027$ μF for the 1.0 to 10 kHz range, $C_{13} = 0.027$ μF for the 100 Hz to 1.0 kHz range, and $C_{14} = 0.27$ μF for the 10 to 100 Hz range.

Next, we will design the frequency control. Select a frequency control potentiometer R_1. Any potentiometer between 100 Ω and 10 kΩ is suitable. We will use a 1.0 kΩ potentiometer. One useful option here is to use a multiturn potentiometer, with counter dial, to give precise frequency control. For transistor Q_1, any PNP small-signal transistor is suitable. We will use a 2N4403.

Now select shape calibration potentiometers R_C and R_D. The manufacturer recommends 100 kΩ for these. These are calibration potentiometers, so trim pots would normally be used. Multiturn trim pots could be used, but they are not necessary because the waveform shape is relatively insensitive to exact resistance value.

Finally, we must design the buffer amplifier. First choose the gain control potentiometer R_F. Any value between 1.0 and 100 kΩ is suitable. We will use a 10 kΩ potentiometer. Since the square-wave output is approximately $0.55 V_{supply}$, or 15 V peak-to-peak, the maximum gain that we can use is slightly less than 2. The input resistor for the square wave, R_{I_1}, should be larger than 5 kΩ. We will use a value of 5.1 kΩ. The triangle-wave output is approximately $0.3 V_{supply}$, or 9 V peak-to-peak. The maximum gain must be slightly less than 3, so R_{I_2} should be greater than 3.3 kΩ. We will use a value of 3.6 kΩ. The sine-wave output is approximately $0.2 V_{supply}$, or 6 V peak-to-peak. The maximum gain must be slightly less than 5, hence R_{I_3} should be greater than 2.0 kΩ. We will use a value of 2.2 kΩ. With this choice of resistors, all three waveforms should have approximately the same peak-to-peak output amplitude, and the output amplitude should be adjustable from a maximum of approximately 28 V peak-to-peak to less than 0.1 V peak-to-peak.

Our function generator is designed. With the least expensive components, it should be possible to construct this function generator, including a suitable power supply (use the circuit in Figure 8-17), for less than $20.

11-4 The LM565C Phase-Locked Loop

The final oscillator we will introduce is the **phase-locked loop (PLL)**. In some respects, a phase-locked loop oscillator is similar to a voltage controlled oscillator, such as the LM566C. Unlike the LM566C, however, the frequency of oscillation is controlled not by an input voltage but by an input frequency. In its simplest mode of operation, the output frequency of a phase-locked loop is "locked" to the frequency of the input signal. It might seem that an oscillator circuit that simply echos the input frequency achieves very little. Wrong—the phase-locked loop is one of the most useful circuits in electronics. Whole books are written on applications of phase-locked loops.

To understand the importance of the phase-locked loop, we have to look at how the output frequency is controlled. Figure 11-26 shows a block diagram of

11-4 The LM565C Phase-Locked Loop

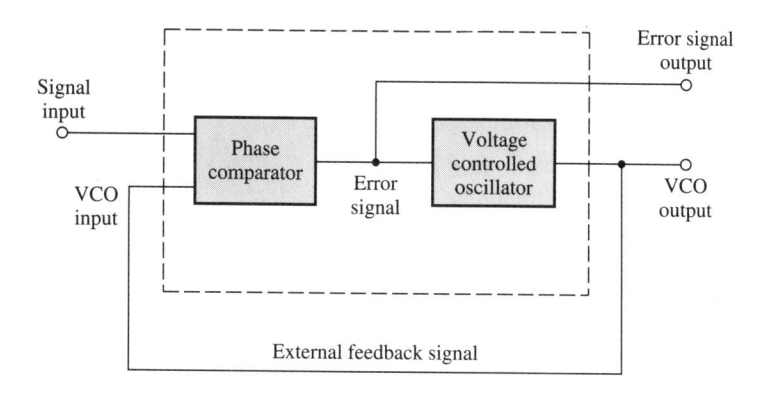

Figure 11-26 Block diagram of a phase-locked loop.

a simple phase-locked loop. The phase-locked loop contains a voltage controlled oscillator. The output signal from this oscillator is fed back to the input. At the input, a phase comparator compares the frequency of the input signal to the frequency of the voltage controlled oscillator. If the frequencies are different, the phase comparator produces a DC **error signal** proportional to the frequency difference. This error signal acts as a control voltage for the voltage controlled oscillator, forcing it to the same frequency as the input signal.

Two things make the phase-locked loop a useful circuit. First, the error signal is available as a separate output. If the input signal is changing in frequency, the error signal changes in amplitude with the frequency change. This allows us to use the phase-locked loop as an FM demodulator. Second, the feedback from the internal voltage controlled oscillator must be externally connected back to the phase comparator. A frequency divider circuit can be inserted into the feedback loop, reducing the frequency of the feedback signal by some integer amount. Since both signals at the input to the phase comparator must have the same frequency, the output frequency will be locked to some integer multiple of the input frequency. This allows us to use the phase-locked loop as a frequency converter or multiplier.

The Phase Comparator

The heart of a phase-locked loop is the **phase comparator** (or, more correctly, **phase detector**). An understanding of the operation of a phase comparator is vital to the understanding of the operation of a phase-locked loop. The phase comparator compares two input AC waves and produces a DC output voltage proportional to the phase difference. In fact, there is no actual comparison as in the case of the comparators discussed in Chapter 5, but rather a multiplication of the two inputs. The multiplication of two sine waves produces a sine wave with a frequency equal to the sum of the input frequencies, another sine wave with a frequency equal to the difference of the input frequencies, and a DC term proportional to the cosine of the phase difference. The DC component (cosine term) is zero for a phase difference of 90°; hence, for zero frequency difference the input waves must differ in phase by 90°. For a phase-locked loop where the frequencies of the two signals are the same, the difference frequency is not present. The sum frequency term is always present and must be eliminated by a low-pass filter. Additional frequency terms are present if other waveforms (triangle wave or square wave) are used. A more detailed discussion of phase detector circuits is beyond the scope of this text, and the reader is referred to a text on communication theory for additional information.

Phase-Locked Loop Operation

First, refer to Figure 11-26, showing the complete phase-locked loop with frequency feedback. The phase comparator output, suitably filtered to remove high frequency components, is used as a control voltage for the voltage controlled oscillator. The **free-running frequency** of the voltage controlled oscillator is the frequency determined by the external timing resistor and timing capacitor alone, in the absence of any error signal. Assume that the input frequency is exactly equal to the free-running frequency. With a phase difference of 90°, the error signal from the phase comparator will be zero, and the voltage controlled oscillator will continue to operate at its free-running frequency.

Now assume the input frequency is lower than the free-running frequency. The phase difference will increase from 90°, producing a positive increasing error signal from the phase comparator. Recall that for a voltage controlled oscillator such as the LM566C, the frequency decreases as the control voltage increases. The positive error signal reduces the frequency of the voltage controlled oscillator until it is the same as the input frequency. The phase difference, however, does not drop back to 90°. If the phase difference dropped back to 90°, the control voltage would drop to zero and the voltage controlled oscillator signal would resume its original free-running frequency. Instead, the phase difference remains greater than 90°, the error signal remains at a constant positive value, and the voltage controlled oscillator signal remains locked to the input frequency. Exactly the same analysis can be applied to a free-running frequency less than the input frequency.

There is a maximum error signal level, corresponding to a maximum phase difference of 180° (or 0°), and consequently a maximum frequency range over which the voltage controlled oscillator can be driven. This maximum frequency range is called the **lock range** and corresponds to the maximum range of frequencies over which the voltage controlled oscillator in the phase-locked loop can lock onto the input frequency. The lock range is symbolized by $\pm f_L$.

There is another range of frequencies, called the **capture range**, and symbolized by $\pm f_C$, over which the voltage controlled oscillator of the phase-locked loop can *initially* lock onto the input frequency. The capture range is always *less than or equal to* the lock range. The phase-locked loop may not be able to automatically lock onto an input frequency outside of the capture range, but, once locked, it will be able to stay locked as the input frequency varies over the entire lock range.

LM565/LM565C Phase Locked Loop

Industrial Blocks

General Description

The LM565 and LM565C are general purpose phase locked loops containing a stable, highly linear voltage controlled oscillator for low distortion FM demodulation, and a double balanced phase detector with good carrier suppression. The VCO frequency is set with an external resistor and capacitor, and a tuning range of 10:1 can be obtained with the same capacitor. The characteristics of the closed loop system—bandwidth, response speed, capture and pull in range—may be adjusted over a wide range with an external resistor and capacitor. The loop may be broken between the VCO and the phase detector for insertion of a digital frequency divider to obtain frequency multiplication.

The LM565H is specified for operation over the $-55°C$ to $+125°C$ military temperature range. The LM565CH and LM565CN are specified for operation over the $0°C$ to $+70°C$ temperature range.

Features

- 200 ppm/°C frequency stability of the VCO
- Power supply range of ±5 to ±12 volts with 100 ppm/% typical
- 0.2% linearity of demodulated output
- Linear triangle wave with in phase zero crossings available
- TTL and DTL compatible phase detector input and square wave output
- Adjustable hold in range from ±1% to > ±60%.

Applications

- Data and tape synchronization
- Modems
- FSK demodulation
- FM demodulation
- Frequency synthesizer
- Tone decoding
- Frequency multiplication and division
- SCA demodulators
- Telemetry receivers
- Signal regeneration
- Coherent demodulators.

Schematic and Connection Diagrams

Metal Can Package

Order Number LM565H or LM565CH
See NS Package H10C

Dual-In-Line Package

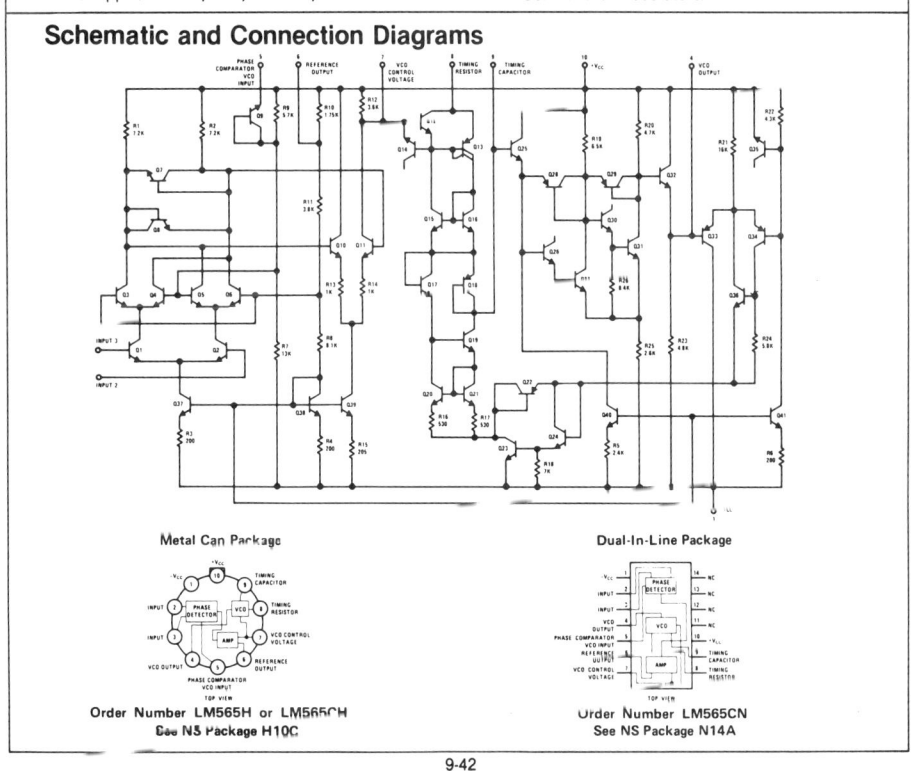

Order Number LM565CN
See NS Package N14A

Figure 11-27 Data sheets for the LM565 phase-locked loop. *(Reprinted with permission of National Semiconductor Corporation)*

Absolute Maximum Ratings

Supply Voltage	±12V
Power Dissipation (Note 1)	300 mW
Differential Input Voltage	±1V
Operating Temperature Range LM565H	−55°C to +125°C
LM565CH, LM565CN	0°C to 70°C
Storage Temperature Range	−65°C to +150°C
Lead Temperature (Soldering, 10 sec)	300°C

Electrical Characteristics (AC Test Circuit, $T_A = 25°C$, $V_C = ±6V$)

PARAMETER	CONDITIONS	LM565 MIN	LM565 TYP	LM565 MAX	LM565C MIN	LM565C TYP	LM565C MAX	UNITS
Power Supply Current			8.0	12.5		8.0	12.5	mA
Input Impedance (Pins 2, 3)	−4V < V_2, V_3 < 0V	7	10			5		kΩ
VCO Maximum Operating Frequency	C_o = 2.7 pF	300	500		250	500		kHz
Operating Frequency Temperature Coefficient			−100	300		−200	500	ppm/°C
Frequency Drift with Supply Voltage			0.01	0.1		0.05	0.2	%/V
Triangle Wave Output Voltage		2	2.4	3	2	2.4	3	V_{p-p}
Triangle Wave Output Linearity			0.2	0.75		0.5	1	%
Square Wave Output Level		4.7	5.4		4.7	5.4		V_{p-p}
Output Impedance (Pin 4)			5			5		kΩ
Square Wave Duty Cycle		45	50	55	40	50	60	%
Square Wave Rise Time			20	100		20		ns
Square Wave Fall Time			50	200		50		ns
Output Current Sink (Pin 4)		0.6	1		0.6	1		mA
VCO Sensitivity	f_o = 10 kHz	6400	6600	6800	6000	6600	7200	Hz/V
Demodulated Output Voltage (Pin 7)	±10% Frequency Deviation	250	300	350	200	300	400	mV_{pp}
Total Harmonic Distortion	±10% Frequency Deviation		0.2	0.75		0.2	1.5	%
Output Impedance (Pin 7)			3.5			3.5		kΩ
DC Level (Pin 7)		4.25	4.5	4.75	4.0	4.5	5.0	V
Output Offset Voltage $|V_7 - V_6|$			30	100		50	200	mV
Temperature Drift of $|V_7 - V_6|$			500			500		μV/°C
AM Rejection		30	40			40		dB
Phase Detector Sensitivity K_D		0.6	.68	0.9	0.55	.68	0.95	V/radian

Note 1: The maximum junction temperature of the LM565 is 150°C, while that of the LM565C and LM565CN is 100°C. For operation at elevated temperatures, devices in the TO-5 package must be derated based on a thermal resistance of 150°C/W junction to ambient or 45°C/W junction to case. Thermal resistance of the dual-in-line package is 100°C/W.

Figure 11-27 *(Continued)*

11-4 The LM565C Phase-Locked Loop

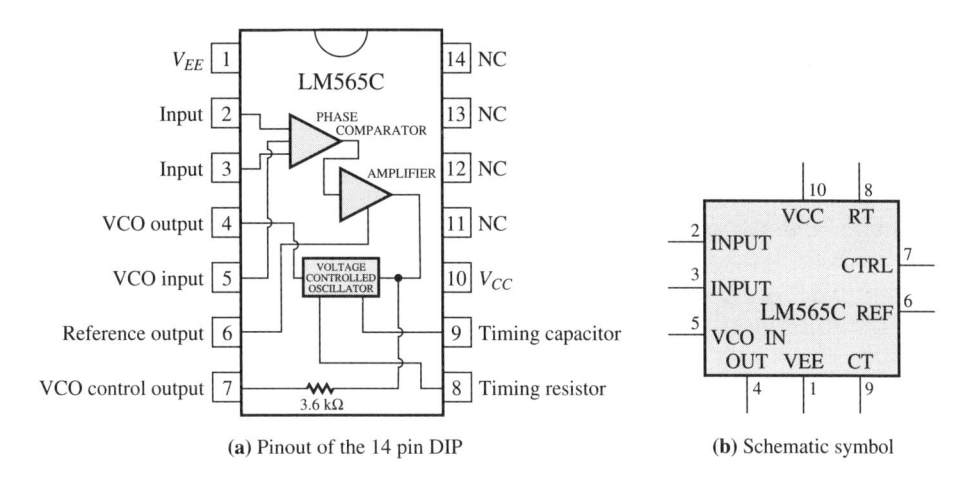

Figure 11-28 Pinout and schematic symbol for the LM565C phase-locked loop.

The LM565C Phase-Locked Loop

There are a number of phase-locked loop integrated circuits available. The LM565 phase-locked loop that we will be studying in this section is one of the simplest and easiest to use. Data sheets for the LM565 phase-locked loop are shown in Figure 11-27, the pinout and schematic symbol in Figure 11-28.

As with most of the other devices that we have studied, the LM565 is available in two versions: the military version and the commercial version. We will use the commercial version, the LM565C, for all of the following discussion. The LM565C is designed for dual-supply operation and can operate at any voltage from ±5 to ±12 V (pins 10 and 1). It can also be operated from a single supply if input pins 2 and 3 are suitably biased to $\frac{1}{2}V_{CC}$.

The basic circuit for operating the LM565C is shown in Figure 11-29. Pins 2 and 3 are differential inputs. A DC bias current must flow into each input (similar to bipolar op-amps such as the LM741C), so equal resistors R_B with typical values between 1 and 10 kΩ should be connected from pin 2 to ground and from pin 3 to ground. These biasing resistors set the input impedance for each input to R_B. The input signal is capacitively coupled to either pin 2 or pin 3 through C_{in}. This capacitor should be chosen so that the capacitive reactance at the lowest operating frequency is less than the input impedance. Make $X_C \leq R_B$ at the lowest operating frequency f_{min}; hence, the equation for determining C_{in} is

$$C_{in} = \frac{1}{2\pi f_{min} R_B} \qquad (11\text{-}36)$$

We have used this equation (or a similar equation) before.

The free-running frequency f_0 (sometimes called the idle frequency) of the internal voltage controlled oscillator is determined by timing resistor R_1, connected between pin 8 and the positive supply (V_{CC}), and timing capacitor C_1,

528   **11 Integrated Circuit Oscillators**

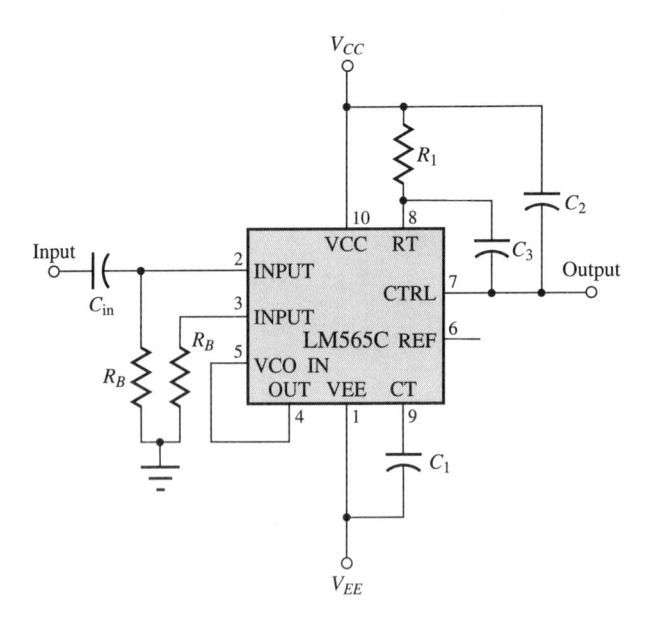

Figure 11-29   Basic phase-locked loop circuit using the LM565C.

connected between pin 9 and the negative supply (V_{EE}). The free-running frequency is

$$f_0 = \frac{1}{3.3 R_1 C_1} \tag{11-37}$$

To design the circuit for the desired free-running frequency, we normally choose C_1 between 100 pF and 0.1 μF and solve Equation 11-37 for R_1:

$$R_1 = \frac{1}{3.3 f_0 C_1} \tag{11-38}$$

The voltage controlled oscillator produces a square-wave output at pin 4 (VCO output) and a triangle-wave output at pin 9 (the timing capacitor pin). The square-wave voltage is always greater than zero, varying between approximately 0 V and V_{CC}. The triangle wave is symmetric about 0 V with a peak-to-peak amplitude approximately one-third that of the square wave.

The VCO output from pin 4 is normally connected to the VCO input of the phase comparator (pin 5). The phase comparator compares the phase of this feedback signal to the input signal applied to either pin 2 or pin 3. The external connection between pins 4 and 5 allows a frequency-dividing circuit to be inserted between these pins. The possibilities of this will be discussed later.

The error signal from the phase comparator (used as the voltage controlled oscillator control voltage) is available at pin 7. As discussed earlier, the phase comparator produces high frequency AC components which must be removed from the error signal by a low-pass filter. Filtering is accomplished by connecting capacitor C_2 from pin 7 to V_{CC}, paralleling an internal 3.6 kΩ resistor R_{int}.

11-4 The LM565C Phase-Locked Loop

The capture range f_C, that is, the range of frequencies over which the voltage controlled oscillator frequency can initially be locked to or "captured by" the input frequency, is determined by the choice of C_2, and is

$$f_C = \pm \sqrt{\frac{f_L}{2\pi R_{int} C_2}} \qquad (11\text{-}39)$$

The quantity f_L is the lock range, that is, the range of frequencies over which the LM565 can lock onto the input frequency. The lock range is

$$f_L = \pm \frac{8 f_0}{V_{supply}} \qquad (11\text{-}40)$$

where V_{supply} is the total supply voltage. For example, if the LM565C is operated from a ± 12 V supply, V_{supply} would be 24 V. The supply voltage affects the lock range (and hence the capture range): the higher the supply voltage, the lower the lock range. The input voltage should have an amplitude greater than 0.1 V peak to achieve the full lock range.

Capacitor C_2 must be designed to set the capture range. Substitute f_L from Equation 11-40 into Equation 11-39 and solve for C_2:

$$C_2 = \frac{4 f_0}{\pi f_C^2 R_{int} V_{supply}} \qquad (11\text{-}41)$$

The phase-locked loop can lock over a broader range of frequencies than can be captured. To initially capture the voltage controlled oscillator signal, the frequency must be relatively close to the free-running frequency. Once captured, the phase-locked loop can track and stay locked onto the input frequency as it varies over a larger frequency range. If the input frequency exceeds the lock frequency range, however, the lock will be broken, and the voltage controlled oscillator frequency will drop back to the free-running frequency.

The nominal DC level of the output signal from pin 7 is $\frac{3}{4} V_{CC}$. This level occurs when the input signal frequency is equal to the voltage controlled oscillator free-running frequency. In some applications, such as FM demodulation, which we will discuss shortly, it is desirable to eliminate this DC level. Pin 6 (the reference output) provides a constant DC voltage equal to $\frac{3}{4} V_{CC}$. If the outputs from pin 7 and pin 6 are connected to a subtracting or differencing amplifier, then the $\frac{3}{4} V_{CC}$ voltage will appear as a common-mode voltage and will cancel, leaving only the frequency difference component of the error signal.

Finally, note capacitor C_3, connected between the error signal output (pin 7) and the timing resistor input (pin 8). This is necessary to suppress parasitic oscillations.

The LM565C Phase-Locked Loop as an FM Demodulator

Although several applications of the phase-locked loop have been suggested, we have not yet considered them in detail. The first application that we will investigate is the use of the LM565C phase-locked loop as a **demodulator** or detector of frequency-modulated signals.

11 Integrated Circuit Oscillators

Since the maximum operating frequency of the voltage controlled oscillator in the LM565C is only 500 kHz, it obviously cannot be used for demodulating commercial broadcast FM signals, which have a frequency of 10.7 MHz. The LM565C, however, can be used for demodulating frequency-modulated industrial control and measurement signals, for tone detection, and for various subcarrier applications.

The circuit for using the LM565C as an FM demodulator is shown in Figure 11-30. This is essentially the same circuit that we used to describe the operation of the LM565C, with a differencing amplifier added to the output to eliminate the $\frac{3}{4}V_{CC}$ component of the demodulated output.

To design the FM demodulator, we must know the carrier (or center) frequency and the frequency deviation. The frequency deviation is proportional to the amplitude of the modulating signal, and we need to know this to ensure that the FM signal stays within the lock range of the phase-locked loop.

First design the timing capacitor C_1 and timing resistor R_1 to set the center frequency of the voltage controlled oscillator f_0 to the center frequency of the FM signal. Choose a suitable value for C_1 and calculate R_1 from Equation 11-38. Next calculate the value of the input coupling capacitor C_{in} so that the capacitive reactance at the lowest input frequency is much less than the input impedance of the phase-locked loop. Use Equation 11-36, where the input impedance of the phase-locked loop is the biasing resistance R_B (typically chosen as 10 kΩ), and the minimum input frequency is the center frequency f_0 minus the FM deviation. Lastly, the filter capacitor C_2 must be designed so that the capture range is greater than or equal to the FM deviation. Use Equation 11-41, setting the capture frequency f_C equal to the frequency deviation.

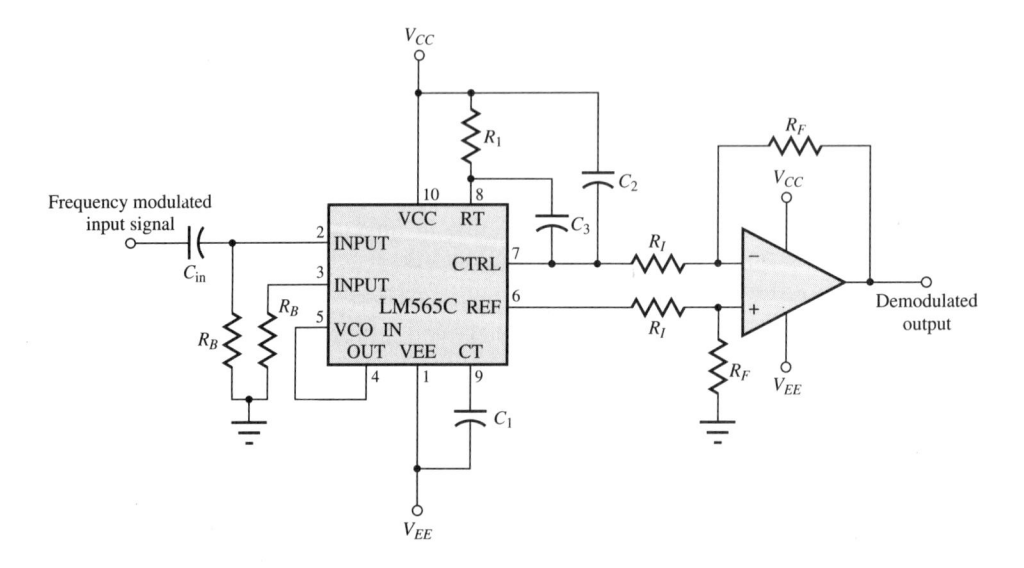

Figure 11-30 The LM565C phase-locked loop used as an FM detector.

11-4 The LM565C Phase-Locked Loop

The output amplifier, used to eliminate the $\frac{3}{4}V_{CC}$ offset voltage from pin 7 by using the $\frac{3}{4}V_{CC}$ reference voltage from pin 6 as a common-mode signal, is a differencing amplifier as discussed in Section 4-2. Both input resistors R_I are equal in value, as are the feedback resistor R_F and the corresponding resistor on the noninverting input. The input impedance, equal to $R_I + R_F$, must be much larger than the filter output impedance to ensure that the amplifier does not load the filter. Typically choose R_I as 10 kΩ or larger. The gain is determined as for a normal inverting amplifier. From the data sheet for the LM565C, the phase-locked loop demodulated output from pin 7 is only 300 mV peak-to-peak for a ±10% frequency deviation and will generally need to be amplified.

EXAMPLE 11-9

Design an FM demodulating circuit using an LM565C phase-locked loop to demodulate the signal produced by the FM modulator designed in Example 11-4. The signal from the output amplifier should have a peak-to-peak amplitude of 10 V. Use a ±12 V power supply for both the op-amp and the LM565C phase-locked loop.

Solution

The FM signal produced by the LM566C modulator circuit in Example 11-4 has a center frequency of 20 kHz and a frequency deviation of ±2222 Hz, or ±11.1% of the carrier frequency.

The circuit of the FM demodulator is shown in Figure 11-31.

First design the timing capacitor C_1 and timing resistor R_1 to set the free-running frequency of the phase-locked loop equal to the carrier frequency of the input FM signal. Make $f_0 = 20$ kHz. Choose C_1 as 0.001 μF and use Equation 11-38 to calculate R_1:

$$R_1 = \frac{1}{3.3 f_0 C_1} = \frac{1}{3.3 \times 20 \text{ kHz} \times 0.001 \text{ μF}} = 15.2 \text{ kΩ}$$

Choose R_1 as 15 kΩ, the closest standard value.

Next choose the input biasing resistors R_B to be 10 kΩ. This sets the input impedance of the phase-locked loop to 10 kΩ.

Calculate the value of the input coupling capacitor C_{in} so that the capacitive reactance is much less than this input impedance. Select f_{min} as the center frequency f_0 minus the FM deviation: $f_{min} = f_0 -$ deviation $= 20$ kHz $- 2.222$ kHz $= 17.78$ kHz. Now calculate C_{in} from Equation 11-36:

$$C_{in} = \frac{1}{2\pi f_{min} R_B} = \frac{1}{2\pi \times 17.78 \text{ kHz} \times 10 \text{ kΩ}} = 895 \text{ pF}$$

To ensure that the capacitive reactance is always less than the value assumed for this equation, choose C_{in} at least twice as large as this calculated value. We will choose $C_{in} = 0.0022$ μF.

532   11 Integrated Circuit Oscillators

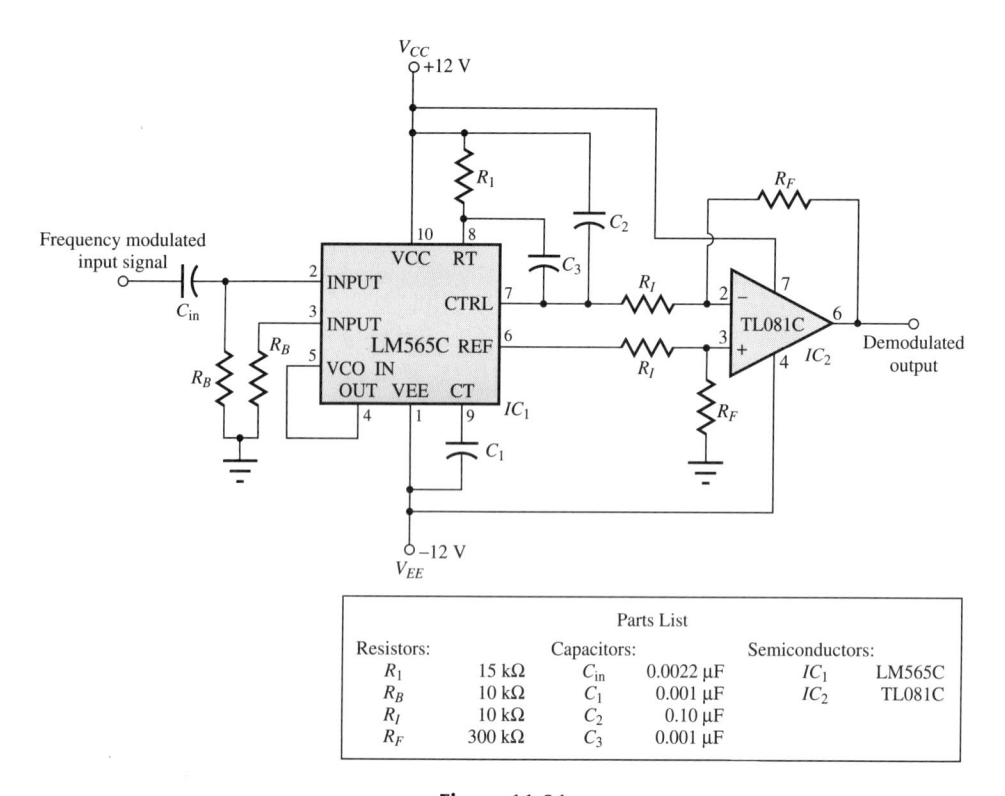

Figure 11-31

Parts List

Resistors:
R_1 15 kΩ
R_B 10 kΩ
R_I 10 kΩ
R_F 300 kΩ

Capacitors:
C_{in} 0.0022 µF
C_1 0.001 µF
C_2 0.10 µF
C_3 0.001 µF

Semiconductors:
IC_1 LM565C
IC_2 TL081C

Next, calculate the filter capacitor C_2 so that the capture range is at least as large as the frequency deviation. Use Equation 11-41 with f_C equal to the frequency deviation of 2222 Hz, and the total supply voltage $V_{supply} = 24$ V:

$$C_2 = \frac{4f_0}{\pi f_C^2 R_{int} V_{supply}}$$

$$= \frac{4 \times 20 \text{ kHz}}{\pi \times (2222 \text{ Hz})^2 \times 3.6 \text{ k}\Omega \times 24 \text{ V}} = 0.0597 \text{ µF}$$

To ensure that the capture range is larger than the deviation, choose a smaller value for C_2. We will use a value of 0.047 µF.

Lastly, we must design the output amplifier. From the data sheet for the LM565C, the PLL-demodulated output from pin 7 is 300 mV peak-to-peak for a ±10% frequency deviation. Since we have an 11.1% frequency deviation, the output will be 333 mV. We require an output of 10 V peak-to-peak, which means that the amplifier must have a gain of 30. Choose the input resistors R_I as 10 kΩ and calculate R_F from Equation 3-1:

$$R_F = A_V R_I = 30 \times 10 \text{ k}\Omega = 300 \text{ k}\Omega$$

This is a standard resistance value.

The design of the demodulator is complete. Before we leave this example, however, we should verify that the lock range and capture range are sufficient for the given frequency deviation. The lock range, given by Equation 11-40, is

$$f_L = \pm \frac{8 f_0}{V_{\text{supply}}} = \pm \frac{8 \times 20 \text{ kHz}}{24} = \pm 6.67 \text{ kHz}$$

The capture range, given by Equation 11-39, is

$$f_C = \pm \sqrt{\frac{f_L}{2\pi R_{\text{int}} C_2}} = \pm \sqrt{\frac{6.67 \text{ kHz}}{2\pi \times 3.6 \text{ k}\Omega \times 0.047 \text{ }\mu\text{F}}} = \pm 2.50 \text{ kHz}$$

In other words, the signal will be captured if the frequency is between 17.5 and 22.5 kHz, and it will be locked if the frequency falls between 13.33 and 26.67 kHz. Since the deviation of the input signal is ±2222 Hz, the maximum signal range is 17.78 to 22.22 kHz, well within both the capture and lock range. Note that the capture range is much less than the lock range.

The LM565C Phase-Locked Loop as a Frequency Converter

Another application of the phase-locked loop is as a **frequency converter.** It is sometimes necessary to convert one signal frequency to a higher frequency. For example, in frequency counters, at low frequencies (say less than 100 Hz) the standard frequency counter either has a very poor resolution or requires a very long count time to achieve a suitable resolution. If the input frequency is converted to a frequency 10, 100, or 1000 times higher, the resolution is correspondingly increased.

The method of using a phase-locked loop as a frequency converter is quite simple. The phase comparator produces an error signal which drives the voltage controlled oscillator to make the frequency *at the VCO input to the phase comparator* equal to the input frequency. Instead of feeding the voltage controlled oscillator output directly back to the phase comparator, however, the voltage controlled oscillator signal is first passed through a frequency divider. The signal seen by the phase comparator is the divided frequency. For example, if we pass the voltage controlled oscillator output through a 100:1 frequency divider, the voltage controlled oscillator output frequency must be 100 times as large as the input frequency to the phase comparator so that the feedback signal is the same frequency as the input signal.

To design a frequency converter, we need simply to design the basic phase-locked loop circuit and select a suitable frequency divider. Frequency dividers are available as decade scaling counters, such as the DS8629 divide-by-100 prescaler, or as programmable dividers, such as the DM8520 modulo-N divider. These are digital circuits, and a discussion of their operation is beyond the scope of this text. We will use a DS8629 in our examples, because all we need to do with this divider is to connect the frequency from the LM565C voltage controlled

11 Integrated Circuit Oscillators

oscillator to the input (pin 6 or pin 7) of the DS8629 and take the divided frequency from the output (pin 2). As with most digital circuits, the DS8629 requires a 5 V supply.

Figure 11-32 shows the circuit of a 100 times frequency converter. There is actually very little to design in this circuit. Choose input resistors R_B equal to 10 kΩ and calculate C_{in} from Equation 11-36. Since the output frequency of the voltage controlled oscillator is to be 100 times the input frequency, the timing resistor and timing capacitor for the voltage controlled oscillator must be selected to set the free-running frequency f_0 to 100 times the input frequency. The timing circuit is designed by first selecting a suitable value for C_1, and then calculating R_1 from Equation 11-38.

The output of the voltage controlled oscillator is the desired frequency, equal to 100 times the input frequency. The VCO control and reference outputs (pins 7 and 6, respectively) are not used, although capacitor C_3 (0.001 μF) should be connected from pin 7 to pin 8 to suppress parasitic oscillations, and filter capacitor C_2 should be connected from pin 7 to the positive supply. Normally, we want as large a capture range as possible. This requires that $f_C = f_L$. Substitute f_L from Equation 11-40 for f_C into Equation 11-41. This gives the minimum value of C_2, necessary for the maximum capture range, as

$$C_2 = \frac{1}{2\pi f_L R_{int}} \tag{11-42}$$

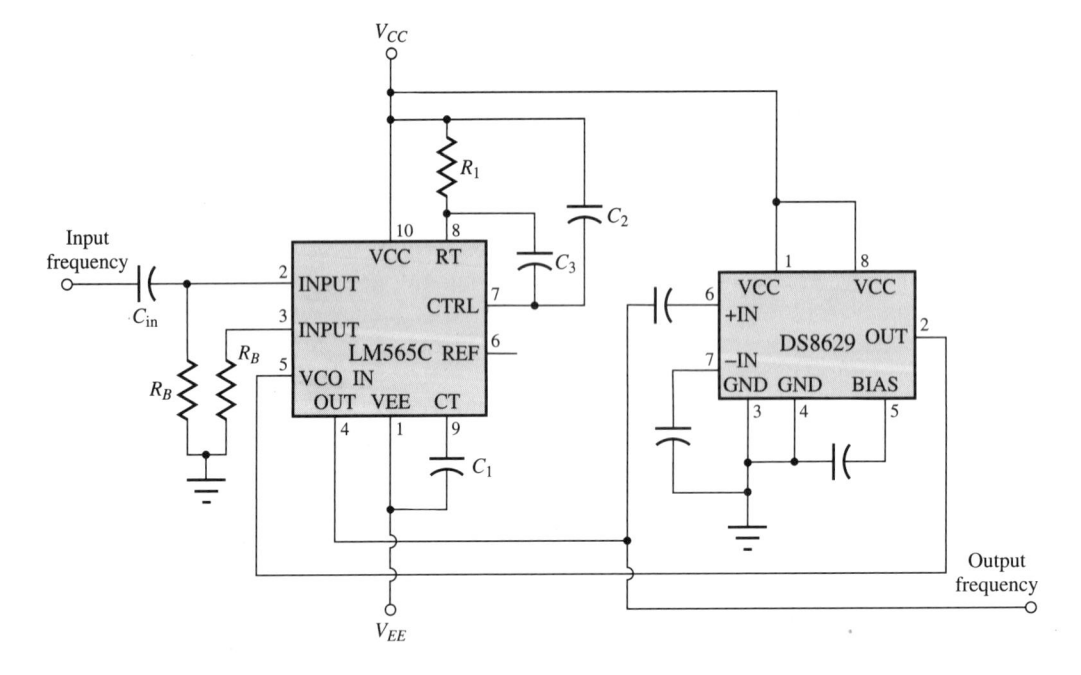

Figure 11-32 Frequency converter circuit using the LM565C phase-locked loop.

where

$$f_L = \pm \frac{8f_0}{V_{supply}} \tag{11-40}$$

by Equation 11-40. The feedback path from the VCO output (pin 4) to the VCO input (pin 5) must include the DS8629 frequency divider. All capacitors used with the DS8629 can be 0.01 μF.

EXAMPLE 11-10

Design a frequency converter circuit to take a 2.4 kHz input signal and convert it to a 240 kHz output. Use an LM565C phase-locked loop and a DS8629 divide-by-100 frequency divider. Power the circuit from a ±5 V supply.

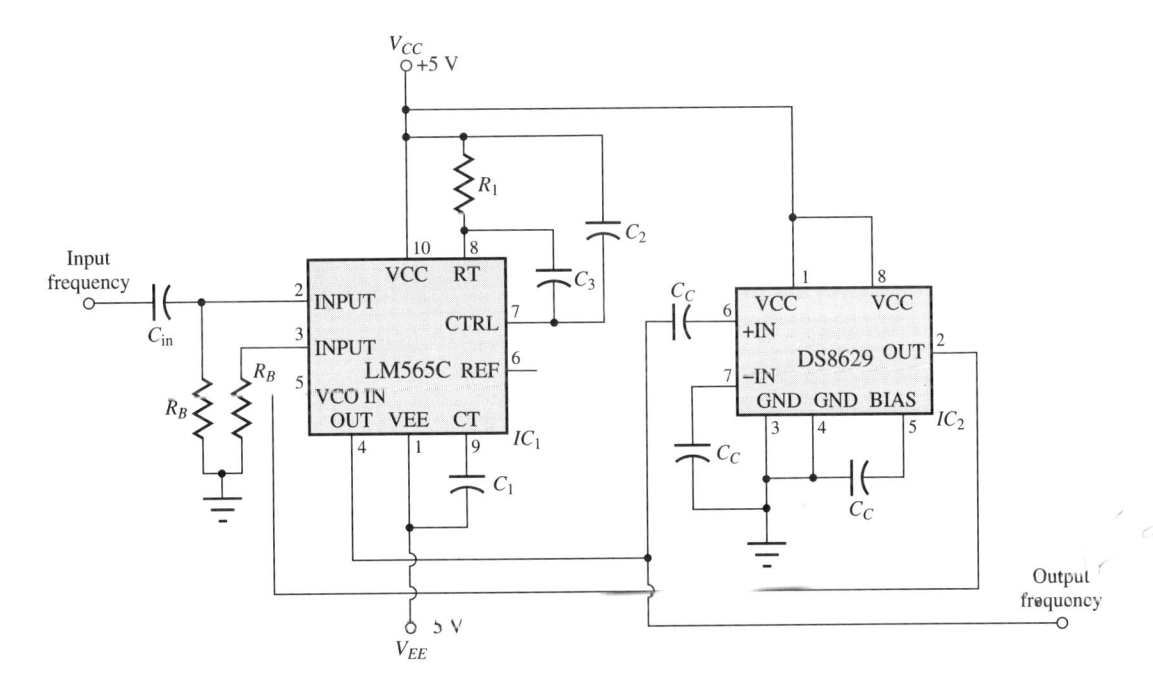

Parts List

Resistors:
- R_1 13 kΩ
- R_B 10 kΩ

Capacitors:
- C_{in} 0.015 μF
- C_C 0.01 μF
- C_1 100 pF
- C_2 270 pF
- C_3 0.001 μF

Semiconductors:
- IC_1 LM565C
- IC_2 DS8629

Figure 11-33

Solution

A frequency converter consists of a basic phase-locked loop circuit and a frequency divider as shown in Figure 11-33.

The basic phase-locked loop circuit requires the design of input resistors R_B, a timing resistor R_1, a timing capacitor C_1, and an input capacitor C_{in}.

Choose the input resistors R_B to be 10 kΩ, and calculate C_{in} from Equation 11-36, where f_{min} = 2.4 kHz:

$$C_{in} = \frac{1}{2\pi f_{min} R_B} = \frac{1}{2\pi \times 2.4 \text{ kHz} \times 10 \text{ k}\Omega} = 0.00663 \text{ μF}$$

Choose a value at least twice as large. We will use 0.015 μF.

The output frequency of the voltage-controlled oscillator must be approximately 240 kHz. The timing resistor and timing capacitor must be selected to make f_0 equal to this frequency. Choose C_1 as 100 pF and calculate R_1 from Equation 11-38:

$$R_1 = \frac{1}{3.3 f_0 C_1} = \frac{1}{3.3 \times 240 \text{ kHz} \times 100 \text{ pF}} = 12.6 \text{ k}\Omega$$

Choose R_1 = 13 kΩ.

The VCO control and reference outputs (pins 7 and 6, respectively) are not used, but we must connect a 0.001 μF capacitor C_3 from pin 7 to pin 8, to eliminate parasitic oscillations, and a filter capacitor C_2 from pin 7 to V_{CC}. Design the filter capacitor for maximum capture range. First, we will need to calculate the lock range f_L from Equation 11-40:

$$f_L = \pm \frac{8 f_0}{V_{supply}} = \pm \frac{8 \times 240 \text{ kHz}}{10 \text{ V}} = \pm 192 \text{ kHz}$$

Now calculate C_2 from Equation 11-42.

$$C_2 = \frac{1}{2\pi f_L R_{int}} = \frac{1}{2\pi \times 192 \text{ kHz} \times 3.6 \text{ k}\Omega} = 230 \text{ pF}$$

Since this is the minimum value of C_2 that can be used, we will select the closest larger value of 270 pF from Appendix A-2.

The feedback path from the voltage controlled oscillator output (pin 4) to the VCO input (pin 5) must include the DS8629 frequency divider as shown in Figure 11-33.

Our design is complete.

Summary

This chapter introduced integrated circuit oscillators. The simplest was the LM566C voltage controlled oscillator, which can be used to produce square-wave or triangle-wave outputs over a wide range of frequencies. The oscillator circuit requires simply a resistor, a capacitor, and a control voltage source. Voltage controlled oscillators can be used as simple tone generators, FM modulators, and voltage-to-frequency converters.

The LM555C timer/oscillator was also introduced. This flexible, although overused, integrated circuit can be used to perform pulse generation and timing applications.

The most complex oscillator circuit introduced was the ICL8038 function generator. This versatile device is capable of producing a sine wave, triangle wave, or square wave with any frequency between 0.01 Hz and 200 kHz, and, if necessary, the duty cycle can be varied to give a wide range of sawtooth and pulse waves. The design of a complete function generator using the ICL8038 was demonstrated.

Finally, the phase-locked loop was introduced. The phase-locked loop is a circuit that superficially simply produces an output frequency identical to the input frequency, but in reality it is a circuit with tremendous application. It can be used for FM demodulation and frequency conversion in analog, digital, and communications applications. The LM565 was introduced as a typical integrated circuit phase-locked loop.

Review Questions

11-1 The LM566C voltage controlled oscillator

1. How can the frequency of oscillation of the LM566C be varied? List three methods.
2. What does "voltage controlled oscillator" mean?
3. What forms of output are produced by the LM566C?
4. Over what range of input voltages is the output of the LM566C controllable?
5. What is the purpose of resistors R_2 and R_3 in the LM566C circuit?
6. What is a function generator?
7. Describe how the LM566C can be used to make a simple function generator.
8. What controls are available with this simple function generator? What waveforms are available?
9. What are the limitations of this function generator?

11-2 The LM555C timer/oscillator

10. For what purpose was the LM555C designed?
11. What is the internal circuit of the LM555C?
12. What are the trigger points of the internal comparators?
13. Describe the operation of the LM555C as a simple oscillator.
14. What is the limitation of this simple oscillator?
15. How can this simple oscillator be improved to produce a perfect square wave?
16. Describe the operation of the improved square-wave oscillator circuit using the LM555C timer.
17. How can a LM555C oscillator be made continuously variable in frequency? What is the limitation of this procedure?

18. Compare the performance of the LM555C as an oscillator to the performance of the LM566C voltage controlled oscillator.
19. Describe the circuit of the LM555C used as a timer.

11-3 The ICL8038 single-chip function generator

20. The ICL8038 is a single-chip function generator. Who manufactures it?
21. What versions of the ICL8038 are available? What are the differences between versions?
22. How is the frequency varied using the ICL8038? How is the range varied?
23. Describe how the duty cycle can be varied using the ICL8038.
24. How can the shape of the sine-wave output be optimized?
25. What is the purpose of the bias output (pin 7)?
26. Why is an additional amplifier usually included in the function generator circuit?

11-4 The LM565C phase-locked loop

27. What is a phase-locked loop?
28. Draw the block diagram of a phase-locked loop and explain how it operates.
29. What is a phase comparator?
30. Explain how the phase comparator in a phase-locked loop controls the output frequency of the voltage controlled oscillator.
31. Define lock range and capture range. Explain the difference. Which is larger?
32. Why must the LM565C phase-locked loop have identical resistors connecting each input to ground?
33. What is the free-running frequency of a phase-locked loop? By what other names is it known?
34. What determines the free-running frequency of the LM565C phase-locked loop?
35. What is the purpose of the reference output of the LM565C phase-locked loop?
36. On what two quantities does the lock range depend?
37. Name three applications of a phase-locked loop.
38. What is the purpose of the low-pass filter on the output of a phase-locked loop?
39. What is the purpose of the output amplifier in the FM demodulator using the LM565C phase-locked loop?
40. Explain the operation of a frequency converter.

Design and Analysis Problems

11-1 The LM566C voltage controlled oscillator

1. Design a tone generator using an LM566C voltage controlled oscillator to produce a square wave with a frequency of 5000 Hz. Use a 12 V power supply.
2. Design a frequency modulator using an LM566C voltage controlled oscillator. Use a carrier frequency of 200 kHz. The power supply is 12.5 V.
3. Analyze the circuit in Figure 11-34 to determine the
    (a) carrier frequency
    (b) modulating voltage if the frequency deviation is 2000 Hz.
    (c) frequency deviation if the modulating voltage is 0.15 V.

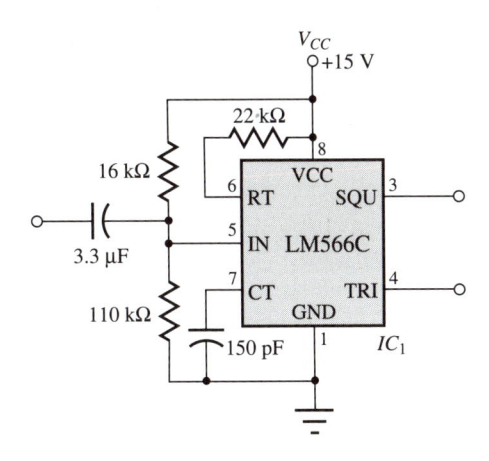

Figure 11-34

11-2 The LM555C timer/oscillator

4. Design a simple oscillator using an LM555C timer. Use the circuit in Figure 11-15. The frequency should be 3200 Hz and the duty cycle 0.60. Use a 15 V power supply.
5. Repeat Problem 4, but use the circuit in Figure 11-17.
6. Design a square-wave oscillator using an LM555C timer with a frequency of 600 Hz. Use a 12 V power supply.
7. Design a 1 s timer using an LM555C. Use a 12 V power supply. Modify the circuit so that it can also be used as a 10 s timer and as a 100 s timer. Calculate additional component values.
8. Analyze the circuit in Figure 11-35 to determine the
   (a) frequency of oscillation
   (b) duty cycle
   (c) amplitude of the output wave.

Figure 11-35

11-3 The ICL8038 single-chip function generator

9. Design a function generator using an ICL8038 function generator chip operating from ±15 V. The function generator should have
   (a) frequency ranges of 1 to 10 Hz, 10 to 100 Hz, 100 Hz to 1 kHz, 1 to 10 kHz, and 10 to 100 kHz
   (b) output amplitude for all ranges of 0.2 to 20 V peak-to-peak.

11-4 The LM565C phase-locked loop

10. Design an FM demodulating circuit using an LM565C phase-locked loop to demodulate an FM signal with a carrier frequency of 350 kHz and a frequency deviation of ±10 kHz. Use a ±9 V supply, and make the amplitude of the output amplifier 10 V peak-to-peak for the maximum input deviation.
11. Design a frequency converter to take a 1.5 kHz input signal and convert it to a 150 kHz output frequency. Use an LM565C phase-locked loop and a DS8926 frequency divider both operated from a ±5 V supply.
12. The circuit for a frequency converter using an LM565C phase-locked loop is shown in Figure 11-36. The supply voltage to the phase-locked loop is ±9 V and to the frequency divider is +5 V.
    (a) What is the free-running frequency of the phase-locked loop?
    (b) What is the output frequency if the input frequency is 150 Hz?
    (c) What is the lock range of the phase-locked loop?
    (d) What is the capture range of the phase-locked loop?
13. Design a frequency converter circuit to increase the resolution of the 10 to 100 Hz range of a frequency counter by a factor of 100—that is, to increase the input frequency by a factor of 100. Use an LM565C phase-locked loop and a DS8926 divide-by-100 counter.

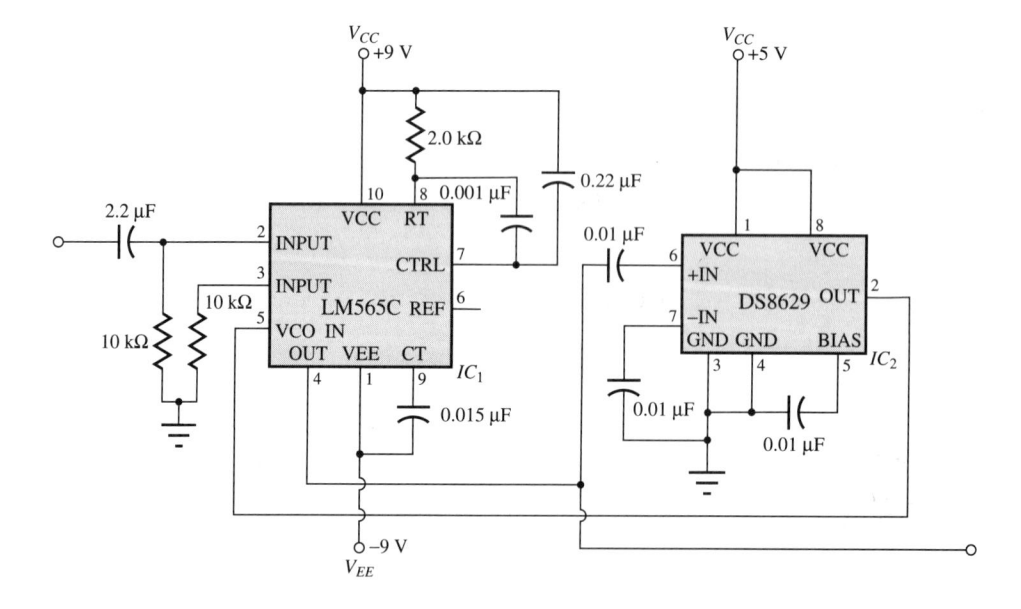

Figure 11-36

Audio Integrated Circuits 12

OUTLINE

12-1 The LM387 Audio Preamplifier
12-2 The LM1877 Audio Power Amplifier
12-3 Volume and Tone Control Circuits
12-4 AM and FM Radio Circuits

KEY TERMS

Audio amplifier
Preamplifier
NAB equalization
Audio mixer
Audio power amplifier
Volume control
Tone control
Balance control
Bass control
Treble control
Audio taper (logarithmic) potentiometer
Muting (or squelching)
Automatic frequency control (AFC)
Automatic gain control (AGC)
Preemphasis
Deemphasis

OBJECTIVES

On completion of this chapter, the reader should be able to:

- explain the purpose of a preamplifier and describe why a preamplifier circuit constructed using op-amps is inferior to a specially designed preamplifier integrated circuit.
- design amplifier circuits using the LM387 dual preamplifier, including a microphone amplifier, a tape record amplifier, a tape playback amplifier, and an audio mixer.
- design power amplifier circuits with the LM1877 dual-power amplifier, including a simple 2 W audio power amplifier, a 2 W per channel stereo amplifier, a 4 W bridge amplifier, and a high-power amplifier.
- describe the operation of tone and volume controls in audio systems, and use the LM1035 tone/volume/balance control integrated circuit as a stereo control.
- describe the construction of complete AM, FM, and FM stereo radio receivers using integrated circuit components.

Introduction

Many students enter the field of electronics because of an interest in audio systems. This final chapter is included especially for those students. It introduces a variety of special-purpose integrated circuits used in audio systems such as stereo amplifiers, PA systems, tape decks, and AM and FM radio receivers. Collectively these devices are referred to as "consumer integrated circuits" and represent a major subgroup of linear integrated circuits.

Many of the circuits discussed can be constructed from simple op-amps and similar building blocks, but their performance may not be satisfactory for the optimal reproduction of sound. Audio systems have special requirements which are often difficult to satisfy with ordinary op-amps. In particular, audio systems may require higher gain, lower distortion, lower noise levels, and larger power outputs than can be provided by simple op-amps. On the other hand, audio systems generally do not need the very high input impedance, the differential inputs, the high common-mode rejection, or the direct coupling provided by op-amps.

Most manufacturers of audio equipment either produce their own audio integrated circuits or purchase specially designed integrated circuits from major manufacturers. These devices are usually identified only by the audio manufacturer's part number. Replacements must be obtained from the audio manufacturer, and data sheets are usually not available. Even special-purpose devices produced by the major integrated circuit manufacturers, and listed in their data books, are often hard to obtain. We will discuss only integrated circuits that are readily available, so that any audio enthusiast reading this book can actually build the circuits.

In this chapter, we will develop circuits for simple audio preamplifiers, audio power amplifiers, and volume and tone controls, introducing integrated circuits designed specifically for these applications. Unfortunately, the field of consumer electronics is large, and it is not possible to cover the entire audio field in one chapter. Interesting topics and applications such as graphic equalizers, noise reduction systems, and video systems simply could not be included.

The final section briefly introduces communication integrated circuits and develops circuits for complete AM, FM, and FM stereo radio receivers. These radio circuits illustrate the construction of complete electronic systems using integrated circuits.

12-1 The LM387 Audio Preamplifier

Audio systems use many types of circuits, the most basic being the **audio amplifier.** Audio amplifiers are designed to operate in the normal range of audible sound, typically from 50 Hz to 20 kHz. The main requirement of an audio amplifier is that it accurately reproduce the input signal without adding appreciable noise or distortion. We will see that noise and distortion requirements place quite severe demands on the performance of audio amplifiers. The term "audio amplifier" actually refers to two types of amplifiers: the audio preamplifier and the audio

National Semiconductor

Audio/Radio Circuits

LM387/LM387A Low Noise Dual Preamplifier

General Description

The LM387 is a dual preamplifier for the amplification of low level signals in applications requiring optimum noise performance. Each of the two amplifiers is completely independent, with an internal power supply decoupler-regulator, providing 110 dB supply rejection and 60 dB channel separation. Other outstanding features include high gain (104 dB), large output voltage swing ($V_{CC}-2V$)p-p, and wide power bandwidth (75 kHz, 20 Vp-p). The LM387A is a selected version of the LM387 that has lower noise in a NAB tape circuit, and can operate on a larger supply voltage. The LM387 operates from a single supply across the wide range of 9V to 30V, the LM387A operates on a supply of 9V to 40V.

The amplifiers are internally compensated for gains greater than 10. The LM387, LM387A is available in an 8-lead dual-in-line package. The LM387, LM387A is biased like the LM381. See AN-64 and AN-104.

Features

- Low noise — 1.0 μV total input noise
- High gain — 104 dB open loop
- Single supply operation
- Wide supply range LM387 9 to 30V
  LM387A 9 to 40V
- Power supply rejection — 110 dB
- Large output voltage swing ($V_{CC} - 2V$)p-p
- Wide bandwidth 15 MHz unity gain
- Power bandwidth 75 kHz, 20 Vp-p
- Internally compensated
- Short circuit protected
- Performance similar to LM381

Schematic and Connection Diagrams

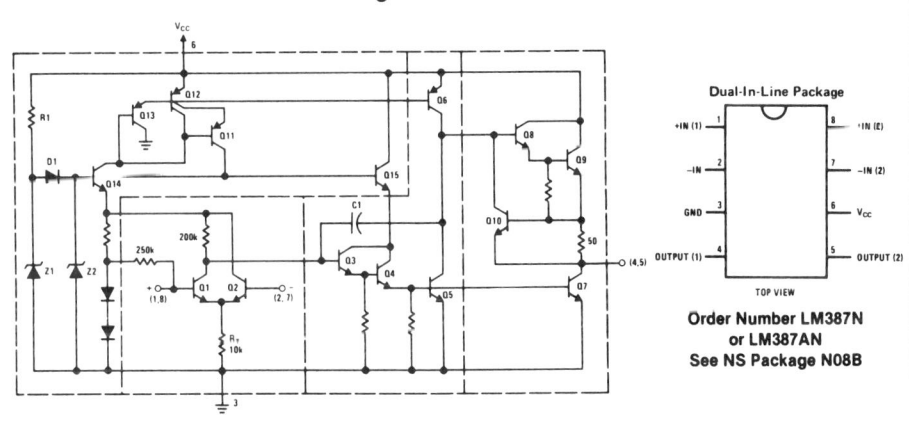

Dual-In-Line Package

TOP VIEW

Order Number LM387N or LM387AN
See NS Package N08B

Typical Applications

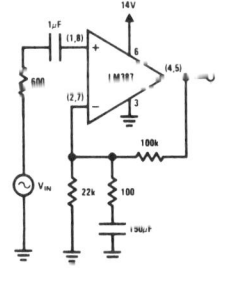

FIGURE 1. Flat Gain Circuit ($A_V = 1000$)

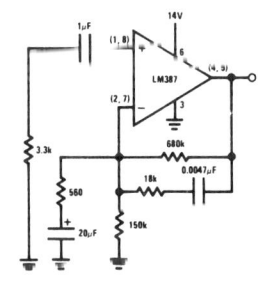

FIGURE 2. NAB Tape Circuit

10-44

Figure 12-1 Data sheets for the LM387 dual preamplifier. *(Reprinted with permission of National Semiconductor Corporation)*

Absolute Maximum Ratings

Supply Voltage
  LM387   +30V
  LM387A   +40V
Power Dissipation (Note 1)   660 mW

Operating Temperature Range   0°C to +70°C
Storage Temperature Range   −65°C to +150°C
Lead Temperature (Soldering, 10 seconds)   300°C

Electrical Characteristics

$T_A = 25°C$, $V_{CC} = 14V$, unless otherwise stated.

PARAMETER	CONDITIONS	MIN	TYP	MAX	UNITS
Voltage Gain	Open Loop, f = 100 Hz		160,000		V/V
Supply Current	LM387, V_{CC} 9–30V, $R_L = \infty$		10		mA
	LM387A, V_{CC} 9–40V, $R_L = \infty$		10		mA
Input Resistance					
Positive Input		50	100		kΩ
Negative Input			200		kΩ
Input Current					
Negative Input			0.5	3.1	µA
Output Resistance	Open Loop		150		Ω
Output Current	Source		8		mA
	Sink		2		mA
Output Voltage Swing	Peak-to-Peak		$V_{CC}-2$		V
Unity Gain Bandwidth			15		MHz
Large Signal Frequency Response	20 Vp-p ($V_{CC} > 24V$), THD $\leq$ 1%		75		kHz
Maximum Input Voltage	Linear Operation			300	mVrms
Supply Rejection Ratio Input Referred	f = 1 kHz		110		dB
Channel Separation	f = 1 kHz	40	60		dB
Total Harmonic Distortion	60 dB Gain, f = 1 kHz		0.1	0.5	%
Total Equivalent Input Noise (Flat Gain Circuit)	10–10,000 Hz LM387 Figure 1		1.0	1.2	µVrms
Output Noise NAB Tape Playback Circuit Gain of 37 dB	Unweighted LM387A Figure 2		400	700	µVrms

Note 1: For operation in ambient temperatures above 25°C, the device must be derated based on a 150°C maximum junction temperature and a thermal resistance of 187°C/W junction to ambient.

Typical Applications (Continued)

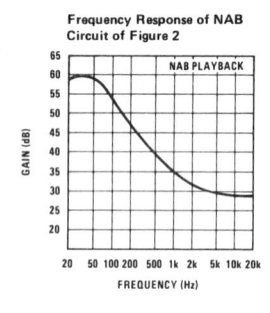

Two-Pole Fast Turn-ON NAB Tape Preamplifier

Frequency Response of NAB Circuit of Figure 2

10-45

Figure 12-1 (*Continued*)

power amplifier. This section will discuss audio preamplifiers. The next section will discuss audio power amplifiers.

The term **preamplifier** is used to describe low-power audio amplifiers used to boost the audio signal level as it comes from a microphone, tape deck, or phonograph pickup to a level suitable for driving a power amplifier, typically to the standard 1 V peak-to-peak (0.35 V RMS) audio line level. In function, however, they are simple amplifiers.

An inverting or noninverting amplifier using an op-amp such as the TL081C with the circuits described in Chapter 3 (see, for example, Figures 3-5 and 3-10) could be used for such an application. An op-amp, however, is not entirely satisfactory. First, the op-amp requires a dual supply often not available in audio equipment. Second, standard op-amps generally introduce considerable noise. Third, audio systems may require gains as large as 1000 at frequencies in excess of 20 kHz. (The maximum gain for a TL081 circuit at 20 kHz is approximately 150.) Finally, although the distortion introduced by an op-amp is very low for *small signals at low gains*, it can be objectional at the gain levels encountered in audio systems.

Admittedly, these various limitations can be minimized. As discussed in Chapter 3, a single supply circuit for a standard op-amp can be used to remove the dependency of an op-amp on a dual supply (see, for example, Figure 3-14). A low-noise op-amp such as the TL071 can be used to reduce noise levels. An uncompensated op-amp with external feedforward compensation can be used to provide a greater gain. Several stages of amplification can be used to minimize distortion. These solutions all, however, increase the circuit complexity. The simplest and best solution is to use a preamplifier integrated circuit designed specifically for audio applications.

The LM387 is typical of such audio preamplifiers. The data sheets for this integrated circuit are shown in Figure 12-1. The LM387 is a low-noise dual (for stereo applications) preamplifier packaged in an 8 pin DIP. It requires a single supply from 9 to 30 V, has a unity-gain frequency of 15 MHz (about five times higher than a standard op-amp), and a distortion of approximately 0.1% for a gain of 1000. The LM387 is internally compensated, but must be used with a minimum gain of 10.

In many ways, the LM387 resembles a standard bipolar op-amp and is represented schematically by the op-amp symbol. Although it can be operated as a noninverting preamplifier or as an inverting preamplifier, for most applications, as we will see later, it is used in the noninverting configuration. Therefore, the schematic symbol is normally drawn inverted to that of a standard op-amp. There are some subtle differences between the LM387 and an op-amp. The noninverting input is internally biased and, hence, can be AC coupled to the source without an external biasing circuit. The internal biasing network contains two diodes connected in series, which holds the DC voltage at the noninverting input 1.4 V above ground. The inverting input is not internally biased, and an external DC biasing path must be provided. The DC voltage at this input will be the same as the voltage at the noninverting input, as for any standard op-amp. This DC bias voltage is amplified by the LM387 and appears as a DC offset voltage at the output. Because of this 1.4 V offset, neither input can be connected directly to ground.

The LM387 as a Noninverting Preamplifier

Figure 12-2 shows the basic circuit for connecting the LM387 as a noninverting amplifier. The circuit resembles an op-amp noninverting amplifier, but there are several important differences. The feedback resistor R_F is the same as for an op-amp, but there is no R_I resistor. Instead there are two resistors, R_A and R_B, connected to the inverting input (pin 2 or pin 7). Resistor R_A is capacitively coupled to ground and is used to set the AC gain. Resistor R_B is connected directly to ground and is used to establish the DC bias current. This resistor must be designed before the feedback resistor R_F can be selected. The minimum bias current required by the LM387 is 0.5 µA, and R_B should be chosen so that at least ten times this current can flow through it. Because the voltage at the inverting input is 1.4 V, the maximum value for R_B is

$$R_B = \frac{1.4 \text{ V}}{10 \, I_{in}} = \frac{1.4 \text{ V}}{5 \, \mu A} = 280 \text{ k}\Omega \quad (12\text{-}1)$$

Any value of R_{B_2} can be chosen as long as it does not exceed 280 kΩ. Typically we will use 100 kΩ.

Once resistor R_B has been selected, resistor R_F must be designed to set the DC gain. The DC gain must be chosen so that the bias voltage of 1.4 V is amplified to give a DC output voltage of $V_{CC}/2$ to allow for the maximum swing in the amplified AC signal. Resistors R_F and R_B act like the feedback and input resistors in a noninverting amplifier. From Equation 3-5, the gain of a noninverting amplifier

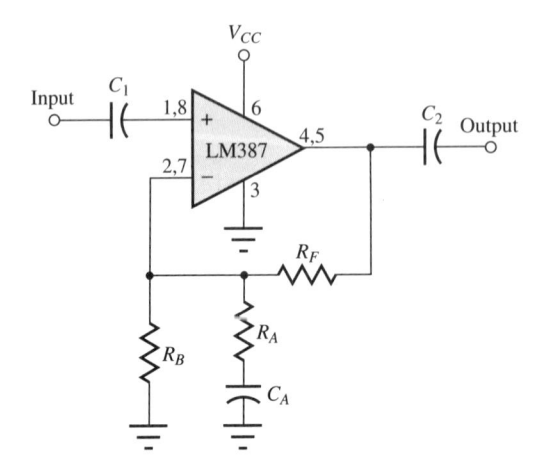

Figure 12-2 Basic noninverting amplifier circuit using the LM387 dual audio preamplifier.

in terms of R_F and R_B is

$$A_V = \frac{R_F}{R_B} + 1 \qquad (12\text{-}2)$$

The required voltage gain is output voltage $V_{CC}/2$ divided by the voltage at the inverting input of 1.4 V. Substitute this for the gain in Equation 12-2 to get

$$\frac{V_{CC}}{2.8} = \frac{R_F}{R_B} + 1 \qquad (12\text{-}3)$$

Solve Equation 12-3 for R_F:

$$R_F = R_B \times \left(\frac{V_{CC}}{2.8} - 1\right) \qquad (12\text{-}4)$$

This is the design equation for R_F.

At audio frequencies above the lower cutoff frequency established by C_A, the AC gain is determined by the parallel combination of R_A and R_B. In all practical preamplifier circuits using the LM387, R_B will be much greater than R_A and, thus, can be ignored in the design of R_A. In a noninverting amplifier, where the gain is described by Equation 3-5, the AC gain is

$$A_V = \frac{R_F}{R_A} + 1 \qquad (12\text{-}5)$$

To design R_A to set the AC gain to a specified value, solve Equation 12-5 for R_A to obtain

$$R_A = \frac{R_F}{A_V - 1} \qquad (12\text{-}6)$$

Since resistor R_A is AC-coupled to ground through C_A, the -3 dB low frequency cutoff f_{CL} of the amplifier is determined by C_A and occurs when the capacitive reactance X_{C_A} is equal to R_A. Equating these quantities and solving for C_A yields

$$C_A = \frac{1}{2\pi f_{CL} R_A} \qquad (12\text{-}7)$$

This is the design equation for C_A. The lower cutoff frequency f_{CL} is usually taken as 50 Hz for audio applications.

Because the LM387 operates from a single supply, it must be capacitively coupled to the source and to the load. The source capacitor C_1 is chosen such that the capacitive reactance X_{C_1} is equal to the sum of the source impedance R_S and the input impedance Z_{in} of the amplifier at the lowest operating frequency f_{CL} (generally 50 Hz). Setting X_{C_1} equal to $R_S + Z_{in}$ at f_{CL} and solving for C_1, we obtain

$$C_1 = \frac{1}{2\pi f_{CL}(R_S + Z_{in})} \qquad (12\text{-}8)$$

From the data sheets, the typical input impedance Z_{in} for the noninverting input is 100 kΩ.

Similarly, the output capacitor C_2 must be designed so that its capacitive reactance X_{C_2} is equal to the sum of the amplifier output impedance Z_{out} and the load impedance R_L at the lower cutoff frequency f_{CL}. The open-loop output impedance of the LM387 is listed as 150 Ω. With feedback, this becomes negligible compared with the input impedance of an audio power amplifier (typically 100 kΩ), so we omit it. Setting X_{C_2} equal to R_L at f_{CL} and solving for C_2, we obtain

$$C_2 = \frac{1}{2\pi f_{CL} R_L} \qquad (12\text{-}9)$$

The LM387 as an Inverting Preamplifier

Although the LM387 is normally used as a noninverting amplifier, it can be used as an inverting amplifier, as shown in Figure 12-3. The circuit in this case is somewhat confusing. Resistors R_F and R_B are in a different position in the circuit, but are the same as for the noninverting amplifier. These resistors still establish the DC gain and are calculated from the same equations as for the noninverting amplifier. However, resistor R_A and capacitor C_A are not used. Instead, we have resistor R_I connected to the source and resistor R_{F1} connected as a second feedback resistor.

The AC gain of the LM387 in this configuration is determined by the ratio of the feedback resistance to the input resistance as in a standard op-amp inverting amplifier, but in this case the AC feedback resistance is the combination of R_F, R_{F1}, and R_B. From the Thevenin equivalent circuit for these resistors, the effective feedback resistance of the amplifier is calculated to be $(R_{F1}R_F + R_{F1}R_B + R_F R_B)/R_B$, and the gain is

$$A_V = \frac{R_{F1}R_F + R_{F1}R_B + R_F R_B}{R_B(R_I + R_S)} \qquad (12\text{-}10)$$

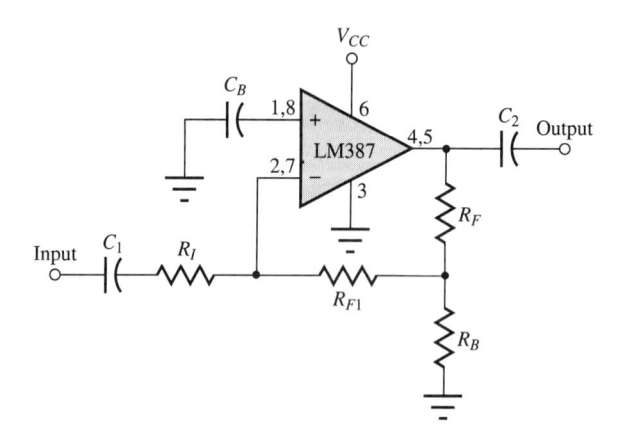

Figure 12-3 The LM387 dual audio preamplifier used as an inverting amplifier.

Notice the inclusion of R_S, the source resistance, in this equation. When we discussed amplifiers using op-amps in Chapter 3, we ignored the source resistance. In audio preamplifiers, the source resistance may be appreciable and must be included in all calculations.

Because the DC bias current must also flow through R_{F1}, there is a maximum value of R_{F1} that should be used to avoid any offset problems. The voltage drop across R_{F1} due to the bias current must be much less than the 1.4 V bias voltage. This will always be the case if we choose R_{F1} subject to the same condition by which we chose R_B, or simply

$$R_{F1\text{max}} < R_{B\text{max}} \tag{12-11}$$

Normally, we will make $R_{F1} = R_B$, and typically we will use a value of 100 kΩ.

Capacitor C_B is a bypass capacitor connecting the noninverting input to ground, and is chosen such that its capacitive reactance X_{C_B} is less than the minimum noninverting input impedance Z_{in} at the lowest operating frequency of the preamplifier f_{CL}. Solving the equation $X_{C_B} > Z_{\text{in}}$ for C_B, we obtain

$$C_B > \frac{1}{2\pi f_{CL} Z_{\text{in}}} \tag{12-12}$$

Substituting 50 kΩ for Z_{in} (obtained from the data sheets in Figure 12-1 as the minimum input impedance of the LM387) and 50 Hz for f_{CL} (a typical lower cutoff frequency for an audio preamplifier), we find that C_B should be greater than 0.064 μF. A value of 0.1 μF will be adequate for most applications.

Finally, the coupling capacitors C_1 and C_2 are calculated in exactly the same manner as for a noninverting amplifier by using Equations 12-8 and 12-9.

Microphone Preamplifier

We will now look at several typical applications of the LM387 audio preamplifier. The first, and simplest, application is a microphone preamplifier. There are two types of microphones in use: high impedance and low impedance. High-impedance microphones have a typical output impedance of 20 kΩ and signal levels approaching 1 V. Often no preamplifier is necessary, but if one is used then a simple op-amp buffer or low-gain amplifier, such as designed in Chapter 3, is adequate. Because of the large amplitude signal from the microphone, distortion and noise introduced by using an op-amp are negligible. High-impedance microphones are susceptible to hum pickup because of their high impedance and, consequently, are seldom used.

Low-impedance microphones have an impedance of typically 200 Ω and output signal levels around 1 mV. Usually they have a balanced three-wire output with two signal leads and a ground return to minimize noise pickup. The microphone can be connected to the preamplifier circuit in Figure 12-2 by using a matching transformer as in Figure 12-4(a) or by direct coupling with the preamplifier configured for a differential input, as in Figure 12-4(b). When the differential amplifier in Figure 12-4(b) is used, care must be taken to ensure that the microphone sees both inputs as identical for maximum common-mode noise rejection.

550 12 Audio Integrated Circuits

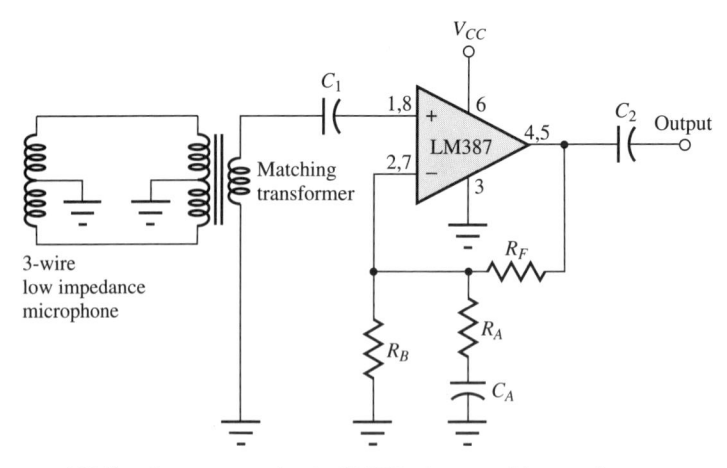

(a) Microphone connected to the LM387 using a matching transformer

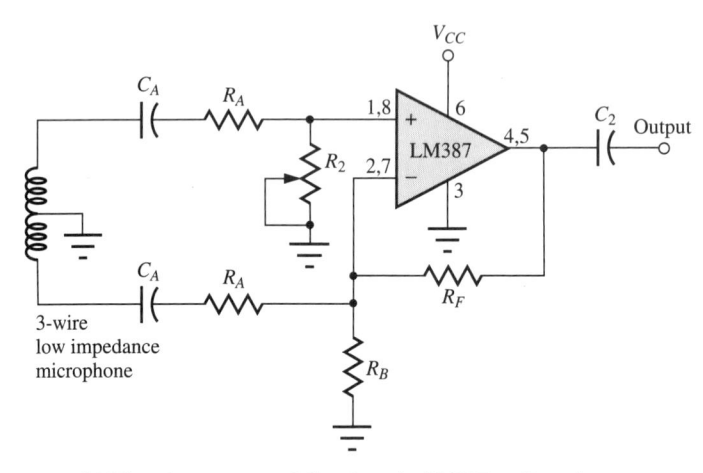

(b) Microphone connected directly to the LM387 configured as a differential amplifier

Figure 12-4 The LM387 used as a low-impedance microphone preamplifier.

Resistors R_A and capacitors C_A are used for each input. A variable resistor R_2 is connected from the noninverting input to ground. This resistor should have a value nominally equal to R_{B_2}, but because the two inputs have different input impedances it should be variable to allow exact matching of the inputs to optimize the common-mode rejection.

EXAMPLE 12-1

Design a microphone preamplifier to amplify the output of a microphone having an impedance of 200 Ω and a level of 3.0 mV RMS to drive a power amplifier requiring an input level of 0.35 V RMS into 50 kΩ. Use an LM387 dual preamplifier with a 12 V power supply. Assume a lower cutoff frequency of 50 Hz.

Solution

We will use the LM387 configured as a noninverting preamplifier and couple the microphone through a matching transformer. The circuit is shown in Figure 12-5 with a parts list.

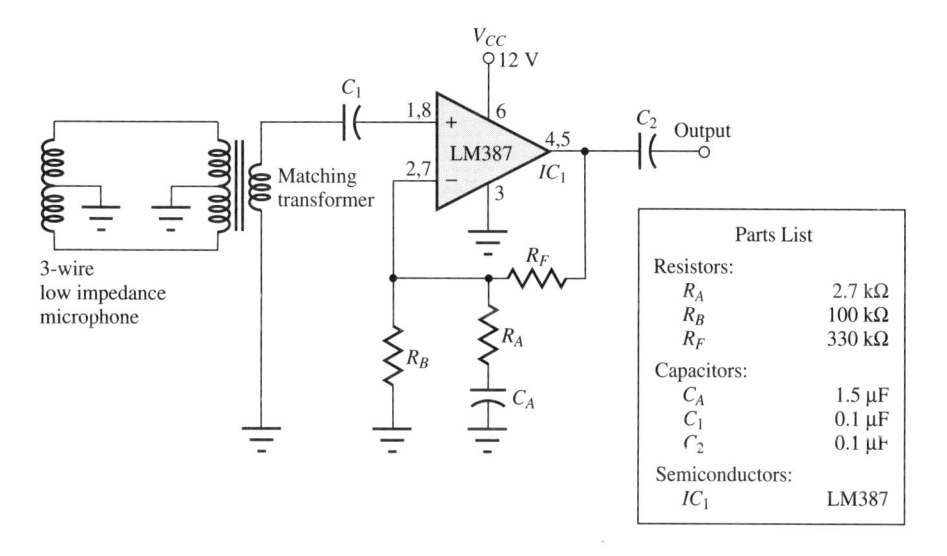

Figure 12-5

First we must design R_B and R_F to set the DC gain of the amplifier so that the output offset will be $V_{CC}/2$ for maximum output voltage swing. We calculated from Equation 12-1 that the maximum value for R_B is 280 kΩ. We will arbitrarily choose a value of 100 kΩ for R_B. Now we can design resistor R_F from Equation 12-4:

$$R_F = R_B \times \left(\frac{V_{CC}}{2.8} - 1\right) = 100 \text{ k}\Omega \times \left(\frac{12 \text{ V}}{2.8 \text{ V}} - 1\right) = 329 \text{ k}\Omega$$

Use a value of 330 kΩ for R_F.

The AC gain is determined by R_A, which is calculated from Equation 12-6. The AC gain is $V_{out}/V_{in} = 0.35 \text{ V}/3.0 \text{ mV} = 117$

$$R_A = \frac{R_F}{A_V - 1} = \frac{330 \text{ k}\Omega}{117 - 1} = 2.84 \text{ k}\Omega$$

We will use the standard value of 2.7 kΩ.

12 Audio Integrated Circuits

To design C_A to give a lower cutoff frequency $f_{CL} = 50$ Hz, use Equation 12-7:

$$C_A = \frac{1}{2\pi f_{CL} R_A} = \frac{1}{2\pi \times 50 \text{ Hz} \times 2.7 \text{ k}\Omega} = 1.18 \text{ μF}$$

Choose C_A as 1.5 μF.

Finally, we must calculate the input capacitor C_1 and the output capacitor C_2 from Equations 12-8 and 12-9, respectively. From the data sheets shown in Figure 12-1, the input impedance to the noninverting input Z_{in} is 100 kΩ and the load impedance is specified as 50 kΩ.

$$C_1 = \frac{1}{2\pi f_{CL}(R_S + Z_{in})} = \frac{1}{2\pi \times 50 \text{ Hz} \times 100 \text{ k}\Omega} = 0.0318 \text{ μF}$$

$$C_2 = \frac{1}{2\pi f_{CL} R_L} = \frac{1}{2\pi \times 50 \text{ Hz} \times 50 \text{ k}\Omega} = 0.0637 \text{ μF}$$

Choose both C_1 and C_2 as 0.1 μF.

Our preamplifier is designed.

Tape Record Preamplifier

The tape head used for magnetic recording is an inductive device. The inductive reactance of the head, and hence the amplitude of the signal recorded on the tape, increases with frequency at the rate of 20 dB per decade up to a frequency slightly above 1 kHz. At this point stray capacitance and other losses cause the reactance and signal level to decrease. The typical response of a record head is shown in Figure 12-6. In magnetic tape recording, the high frequency response of the input signal is usually boosted by using a preamplifier to compensate for the high frequency drop in gain of the head. The resultant response curve is shown in Figure 12-7(a). On playback, it is necessary to attenuate the high frequency signal level back to the original level by using an amplifier with the complementary response

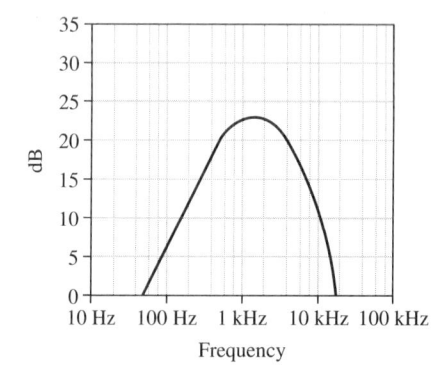

Figure 12-6 Typical response of a record head.

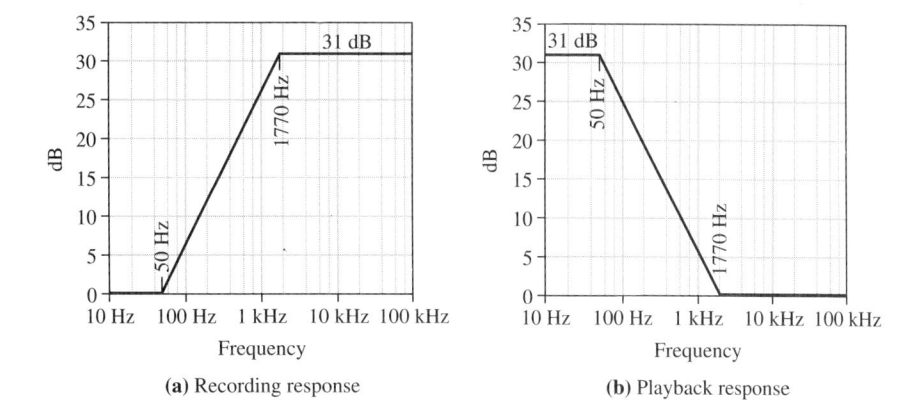

Figure 12-7 Standard NAB equalization response curve.

(a) Recording response

(b) Playback response

curve, the curve shown in Figure 12-7(b). This process of boosting the high frequency record signals and attenuating the high frequency response on playback is called **NAB (National Association of Broadcasters) equalization.** The curves shown correspond to open-reel tape speeds of $1\frac{7}{8}$ and $3\frac{3}{4}$ in./s and to cassette recorders. Slightly different curves, used for open-reel recording at $7\frac{1}{2}$ and 15 in./s, are not shown.

The curves in Figure 12-7(a), (b) are standardized curves used throughout the tape recording industry. For recording, starting at 50 Hz, the recorded signal amplitude increases at a rate of 20 dB per decade up to a frequency of 1770 Hz, by which point signals see a total boost of 31 dB (a gain of 35.48). Higher frequencies are recorded at a constant boost of 31 dB. For playback, the reverse is done. Frequencies lower than 1770 Hz are boosted by 20 dB per decade down to 50 Hz, by which point signals see a total boost of 31 dB (a gain of 35.48).

We will design a recording preamplifier using the LM387 to boost the level of the signal to the recording head to produce the standard NAB equalization response. At low frequencies (less than the frequency where the recording head gain starts to fall, approximately 1 kHz), the preamplifier requires a constant gain. Starting at a frequency of 1000 Hz, the preamplifier must boost the incoming signal at a rate of 20 dB per decade to generate the NAB response. This response is achieved by the circuit shown in Figure 12-8, where a series R-C circuit consisting of R_3 and C_3 has been connected to the noninverting input of a standard noninverting preamplifier.

The design of this tape record preamplifier is similar to the basic preamplifier described earlier. First set the DC gain by selecting a value for R_D less than the maximum value of 280 kΩ given by Equation 12-1. Then use this value to calculate R_F from Equation 12-4. Resistor R_A is designed by Equation 12-6 to set the low frequency gain, and capacitor C_A is designed by Equation 12-7 to set the lower cutoff frequency f_{CL}. Input capacitor C_1 and output capacitor C_2 are calculated from Equations 12-8 and 12-9, respectively.

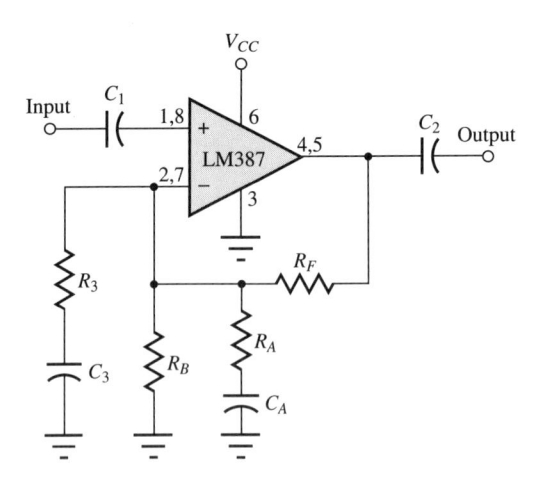

Figure 12-8 The LM387 used as a tape record preamplifier.

The difference in this preamplifier, however, is that the gain must start to increase at a frequency of approximately 1000 Hz to overcome the drop in gain of the recording head. In the circuit in Figure 12-8, this increasing gain is produced by C_3 connected to the noninverting input. This capacitor is effectively in parallel with R_A, and at frequencies above the critical frequency f_3, where $X_{C_3} = R_A$, causes the gain to increase by 20 dB per decade. Solving for C_3 gives

$$C_3 = \frac{1}{2\pi f_3 R_A} \qquad (12\text{-}13)$$

The circuit in Figure 12-8 also includes resistor R_3 in series with capacitor C_3. If this resistor is not included in the circuit, the gain will increase at 20 dB per decade for all frequencies above f_3. This is undesirable because at very high frequencies, the gain becomes very large and can cause the amplifier to oscillate. Resistor R_3 is included to prevent this. For high frequencies, where the capacitive reactance of C_3 is much less than R_3, the gain is determined by R_F and the parallel combination of R_A and R_3. The critical frequency, where R_3 begins to limit the gain, which we will call f_4, is determined by $X_{C_3} = R_3$. Solving this equation for R_3, we obtain

$$R_3 = \frac{1}{2\pi f_4 C_3} \qquad (12\text{-}14)$$

Normally, f_4 is set to approximately 20 kHz, the top end of the audio range.

Figure 12-9 shows the actual circuit that would be used for recording. Note the inclusion of two additional elements. Resistor R_4 is connected in series from the output of the amplifier to limit the current to the recording head. The parallel L-C filter is included to prevent the tape bias signal from feeding back to the amplifier. It is designed with a resonant frequency equal to the bias frequency.

12-1 The LM387 Audio Preamplifier

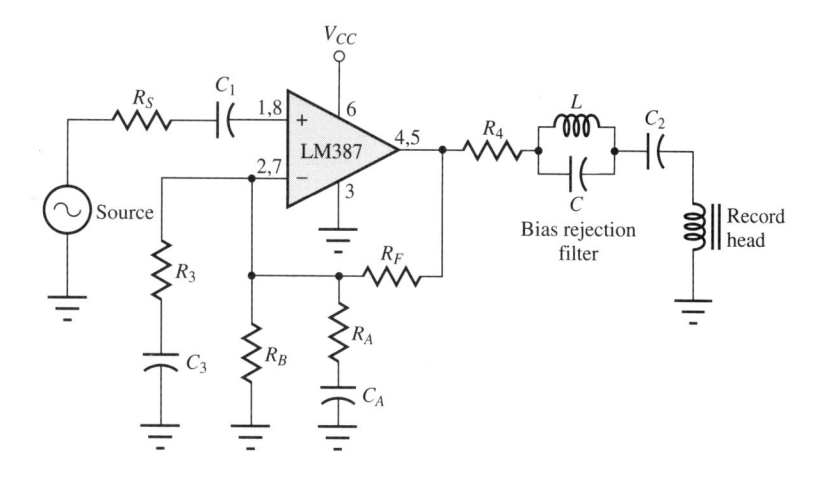

Figure 12-9 LM387 tape record preamplifier connected to source and recording head.

EXAMPLE 12-2

Design a tape record preamplifier using an LM387 dual audio preamplifier. Assume the standard recording head response shown in Figure 12-6 and design the preamplifier to give a standard NAB response. The input signal has a maximum amplitude of 20 mV RMS from a 400 Ω source, and the tape head requires a drive current of 25 µA RMS. The recorder response should be from 50 Hz to 20,000 Hz. The recorder has a 9 V power supply.

Solution

The circuit is shown in Figure 12-10.

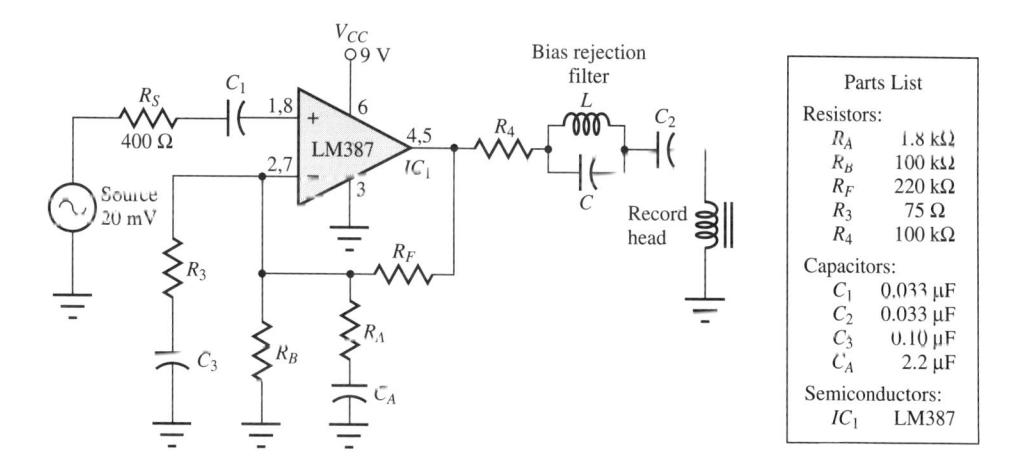

Parts List

Resistors:
R_A 1.8 kΩ
R_B 100 kΩ
R_F 220 kΩ
R_3 75 Ω
R_4 100 kΩ

Capacitors:
C_1 0.033 µF
C_2 0.033 µF
C_3 0.10 µF
C_A 2.2 µF

Semiconductors:
IC_1 LM387

Figure 12-10

First design the DC gain to set the output offset to $\frac{1}{2}V_{CC}$. The procedure is the same as in Example 12-1. Choose a value of 100 kΩ for R_B and calculate R_F from Equation 12-4:

$$R_F = R_B \times \left(\frac{V_{CC}}{2.8} - 1\right) = 100 \text{ k}\Omega \times \left(\frac{9 \text{ V}}{2.8 \text{ V}} - 1\right) = 221 \text{ k}\Omega$$

We will use a 220 kΩ resistor for R_F.

The low frequency (less than 1000 Hz) AC gain is determined by R_A, which is calculated from Equation 12-6. We must first determine the desired AC gain. The input signal has a maximum amplitude of 20 mV RMS. This is $2 \times \sqrt{2} \times 20$ mV = 56.6 mV peak-to-peak. The maximum output peak-to-peak signal is approximately 2 V less than the supply voltage, or 9 V − 2 V = 7 V. The required AC gain is V_{out}/V_{in} = 7.0 V/56.6 mV = 123.7:

$$R_A = \frac{R_F}{A_V - 1} = \frac{220 \text{ k}\Omega}{123.7 - 1} = 1.79 \text{ k}\Omega$$

We will choose the standard value of 1.8 kΩ.

Design C_A for a lower cutoff frequency f_{CL} of 50 Hz by using Equation 12-7:

$$C_A = \frac{1}{2\pi f_{CL} R_A} = \frac{1}{2\pi \times 50 \text{ Hz} \times 1.8 \text{ k}\Omega} = 1.77 \text{ }\mu\text{F}$$

Choose C_A = 2.2 μF as the closest larger standard value.

The frequency f_3 where the gain of the preamplifier must start to increase to compensate for recording head losses is 1000 Hz. This gain increase is achieved by the addition of C_3. Capacitor C_3 is given by Equation 12-13:

$$C_3 = \frac{1}{2\pi f_3 R_A} = \frac{1}{2\pi \times 1000 \text{ Hz} \times 1.8 \text{ k}\Omega} = 0.0884 \text{ }\mu\text{F}$$

We will choose the closest standard value of 0.1 μF.

Resistor R_3, which limits the gain at frequencies above 20 kHz, is calculated from Equation 12-14:

$$R_3 = \frac{1}{2\pi f_4 C_3} = \frac{1}{2\pi \times 20 \text{ kHz} \times 0.1 \text{ }\mu\text{F}} = 79.6 \text{ }\Omega$$

The closest standard value is 75 Ω.

Resistor R_4 is used to limit the current from the amplifier to the recording head. Since the tape head requires an RMS drive current of 25 μA and the output voltage is 7 V peak-to-peak or $(7/2)/\sqrt{2}$ = 2.47 V RMS, R_4 is

$$R_4 = \frac{V_{out}}{I_{drive}} = \frac{2.47 \text{ V}}{25 \text{ }\mu\text{A}} = 98.8 \text{ k}\Omega$$

We will use a 100 kΩ resistor for R_4.

Finally, we must calculate C_1 and C_2. Input capacitor C_1 is calculated from Equation 12-8, using a source resistance R_S of 400 Ω and an input resistance Z_{in}

of 100 kΩ (from the data sheet). Output capacitor C_2 is calculated from Equation 12-9 for a load resistance equal to R_4:

$$C_1 = \frac{1}{2\pi f_{CL}(R_S + Z_{in})} = \frac{1}{2\pi \times 50\text{ Hz} \times 100.4\text{ k}\Omega} = 0.0317\ \mu\text{F}$$

$$C_2 = \frac{1}{2\pi f_{CL} R_L} = \frac{1}{2\pi \times 50\text{ Hz} \times 100\text{ k}\Omega} = 0.0318\ \mu\text{F}$$

Choose $C_1 = C_2 = 0.033\ \mu\text{F}$.

Our preamplifier is designed.

Tape Playback Preamplifier

We will now design a tape playback preamplifier. The signal recorded on the tape has the NAB response shown in Figure 12-7(a) due to the combined effects of the recording head response and the recording amplifier response. On playback, we want a level output response. The playback preamplifier must boost the amplitude at frequencies lower than 1770 Hz at a rate of 20 dB per decade down to 50 Hz to give the required level response.

The circuit of the LM387 used as a playback preamplifier is shown in Figure 12-11. This circuit is similar to the basic preamplifier circuit in Figure 12-2, but has a series filter circuit consisting of R_4 and C_4 in parallel with R_F. At very low frequencies, capacitor C_4 has a very large capacitive reactance, and, hence, the filter branch has a very high impedance compared with R_F. The audio gain is determined by R_F alone from Equation 12-5. At very high frequencies, capacitor C_4 has a very small capacitive reactance, and the audio gain is determined from

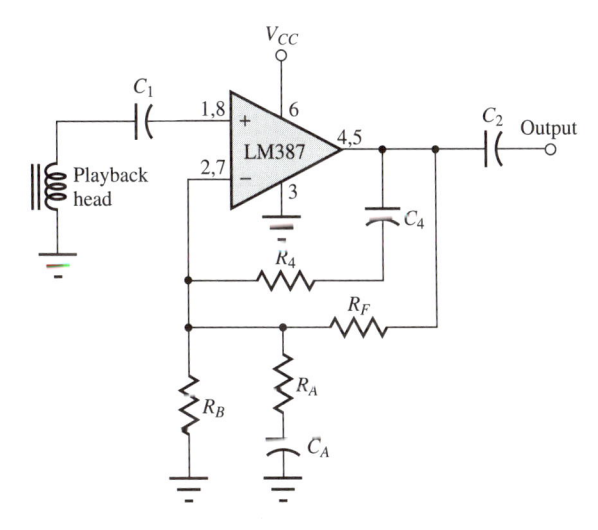

Figure 12-11 Circuit of the LM387 used as a NAB tape playback preamplifier.

the parallel value of R_F and R_4. Because R_4 will always be much less than R_F, the parallel resistance is approximately R_4. Replace R_F in Equation 12-5 with R_4 to get the equation for high frequency gain:

$$A_V = \frac{R_4}{R_A} + 1 \qquad (12\text{-}15)$$

The high frequency gain required is determined from the input level required by the power amplifier divided by the output level supplied by the tape deck (the head sensitivity).

To design the preamplifier circuit, first design R_B and R_F to set the DC gain. This is done in exactly the same manner as for the previous amplifiers. Select a value of R_B less than the maximum value of 280 kΩ given by Equation 12-1. Use this value of R_B to calculate R_F from Equation 12-4.

The lower critical frequency of the NAB equalization curve, $f_1 = 50$ Hz, is determined by setting $X_{C_4} = R_F$. Solve this equation to determine C_4:

$$C_4 = \frac{1}{2\pi f_1 R_F} \qquad (12\text{-}16)$$

The upper critical frequency of the NAB equalization curve, $f_2 = 1770$ Hz, is determined by setting $X_{C_4} = R_4$. Solve this equation to determine R_4:

$$R_4 = \frac{1}{2\pi f_2 C_4} \qquad (12\text{-}17)$$

Finally, use the value of R_4 to determine the value of R_A required to set the high frequency gain of the amplifier. Solving Equation 12-15 for R_A, we obtain

$$R_A = \frac{R_4}{A_V - 1} \qquad (12\text{-}18)$$

Capacitors C_A, C_1, and C_2 are determined as for a simple preamplifier using the LM387.

EXAMPLE 12-3

Design an NAB tape playback amplifier using an LM387 dual-audio preamplifier. The playback head response has a sensitivity of 1200 μV, and the preamplifier drives a power amplifier requiring 0.35 V input into 40 kΩ. The recorder response should be from 50 Hz to 20 kHz. The recorder has a 9 V power supply.

Solution

The circuit is shown in Figure 12-12.

As in the previous preamplifiers, we will first set the DC gain by designing resistors R_B and R_F. We will arbitrarily choose a value of 100 kΩ for R_B and

12-1 The LM387 Audio Preamplifier

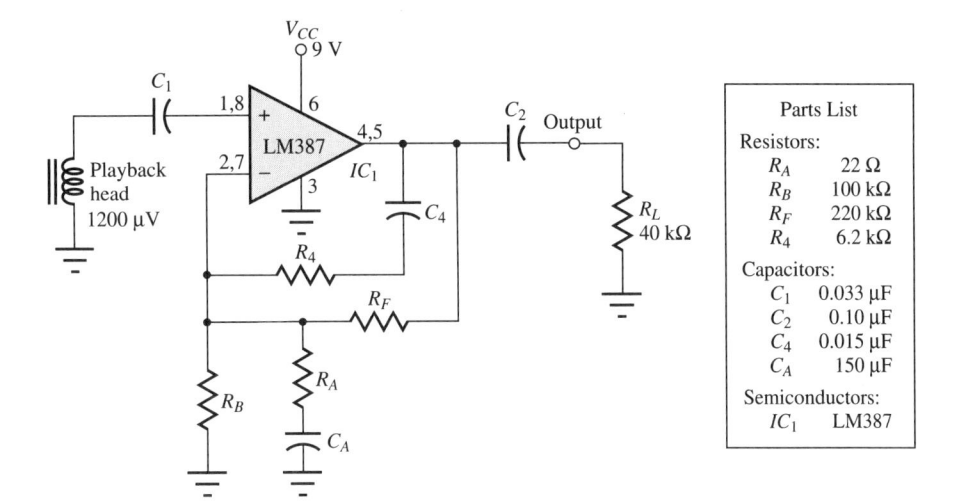

Figure 12-12

calculate R_F from Equation 12-4:

$$R_F = R_B \times \left(\frac{V_{CC}}{2.8} - 1\right) = 100 \text{ k}\Omega \times \left(\frac{9 \text{ V}}{2.8 \text{ V}} - 1\right) = 221 \text{ k}\Omega$$

We will choose $R_F = 220 \text{ k}\Omega$.

NAB equalization is determined by resistor R_4 and capacitor C_4. Calculate C_4 from Equation 12-16 to set the lower critical frequency of the NAB equalization curve, f_1, to 50 Hz:

$$C_4 = \frac{1}{2\pi f_1 R_F} = \frac{1}{2\pi \times 50 \text{ Hz} \times 220 \text{ k}\Omega} = 0.0145 \text{ }\mu\text{F}$$

Choose $C_4 = 0.015 \text{ }\mu\text{F}$.

Calculate R_4 from Equation 12-17 to set the upper critical frequency of the NAB equalization curve, f_2, to 1770 Hz:

$$R_4 = \frac{1}{2\pi f_2 C_4} = \frac{1}{2\pi \times 1770 \text{ Hz} \times 0.015 \text{ }\mu\text{F}} = 5.99 \text{ k}\Omega$$

Choose $R_4 = 6.2 \text{ k}\Omega$.

Finally, we will use this value of R_4 to determine the value of R_A to set the high frequency gain of the amplifier. We will first have to determine the required high frequency gain. Since the required output is 0.35 V for an input of 1200 μV, the gain is 0.35 V/1200 μV = 292. Calculate R_A from Equation 12-18:

$$R_A = \frac{R_4}{A_V - 1} = \frac{6.2 \text{ k}\Omega}{292 - 1} = 21.3 \text{ }\Omega$$

We will choose the standard value of 22 Ω for R_A.

Design C_A to give a lower cutoff frequency $f_{CL} = 50$ Hz, using Equation 12-7:

$$C_A = \frac{1}{2\pi f_{CL} R_A} = \frac{1}{2\pi \times 50 \text{ Hz} \times 22 \text{ }\Omega} = 145 \text{ }\mu\text{F}$$

Choose $C_A = 150$ µF as the closest larger standard value.

Finally, we must calculate the input capacitor C_1 and the output capacitor C_2. Calculate C_1 from Equation 12-8. We will assume the source resistance R_S of the playback head to be zero, and from the data sheets the input resistance Z_{in} is 100 kΩ.

$$C_1 = \frac{1}{2\pi f_{CL}(R_S + Z_{in})} = \frac{1}{2\pi \times 50 \text{ Hz} \times 100 \text{ k}\Omega} = 0.0318 \text{ }\mu\text{F}$$

Calculate C_2 from Equation 12-9 for a load resistance equal to the 40 kΩ input impedance of the power amplifier:

$$C_2 = \frac{1}{2\pi f_{CL} R_L} = \frac{1}{2\pi \times 50 \text{ Hz} \times 40 \text{ k}\Omega} = 0.0796 \text{ }\mu\text{F}$$

Choose $C_1 = 0.033$ µF and $C_2 = 0.1$ µF.

Our preamplifier is designed.

Audio Mixer

The final application circuit that we will introduce for the LM387 preamplifier is that of an **audio mixer.** Audio mixers are circuits for combining two or more inputs, and are used for applications such as public address systems, tape recording, and guitar amplifiers. We will only consider a simple two-channel mixer, although more channels can be added with little difficulty. The mixer circuit using the LM387 audio preamplifier is shown in Figure 12-13. Note that we are using the

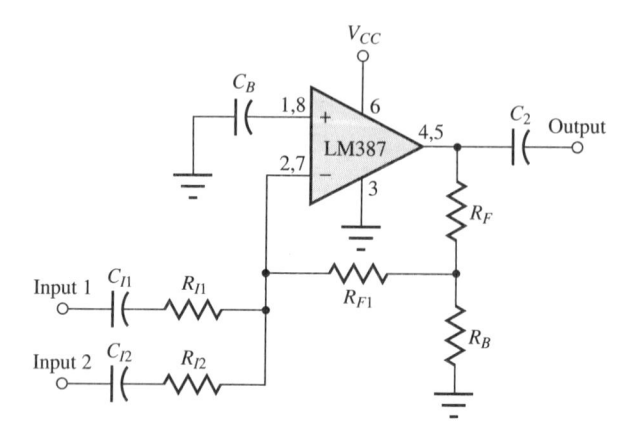

Figure 12-13 Two-channel audio mixing circuit using the LM387 preamplifier.

inverting input of the preamplifier, taking advantage of the low input impedance of this input to minimize crosstalk between channels (this configuration is the same as that for summing circuits in Chapter 4).

Resistors R_B and R_F are designed to set the DC gain as in previous preamplifier circuits. Select a value of R_B less than the maximum value of 280 kΩ given by Equation 12-1. Use this value to calculate R_F using Equation 12-4. Inverting preamplifiers have a second feedback resistor R_{F1} that must be chosen with a value less than the maximum value of R_B. Normally, set this resistor equal to R_B.

The AC gain of an inverting preamplifier using the LM387 was described by Equation 12-10, and depends on R_F, R_{F1}, R_B, R_I, and R_S. Equation 12-10 is normally solved for R_I to give the design equation for this resistor:

$$R_I = \frac{R_{F1}R_F + R_{F1}R_B + R_FR_B}{R_BA_V} - R_S \qquad (12\text{-}19)$$

The gain A_V is determined by dividing the desired output voltage level by the individual input source voltage levels:

$$A_V = \frac{\text{desired output voltage level}}{\text{source voltage level}} \qquad (12\text{-}20)$$

The gain must be calculated for each input to the mixer and, in general, is different for each input. As a consequence, separate resistors R_I must be designed for each input.

Input coupling capacitors C_I (called C_1 earlier) must be designed for each input from Equation 12-8 with the R_I and R_S values for each input. The output coupling capacitor C_2 is calculated from Equation 12-9. Capacitor C_B connected to the noninverting input is simply a bypass capacitor to ensure that this input is at effective AC ground and is calculated from Equation 12-12. As mentioned earlier, any value greater than 0.1 μF is adequate for this capacitor.

If a variable gain is required to allow for different input devices or for producing different mixing levels, R_I on each input can be replaced by a potentiometer. In this case, the capacitors C_I should be chosen much larger than the design value to allow for small values of R_I when the potentiometer is at the low end of its range.

EXAMPLE 12-4

Design a two-channel mixer using an LM387 dual-audio preamplifier. One input is from a microphone with an impedance of 600 Ω and an output level of 12 mV. The other input is the signal from a tape deck preamplifier having negligible output impedance and a signal level of 350 mV. The mixer must deliver a signal level of 10 V to a 10 kΩ load. Use an 18 V power supply.

Solution

The circuit is shown in Figure 12-14.

12 Audio Integrated Circuits

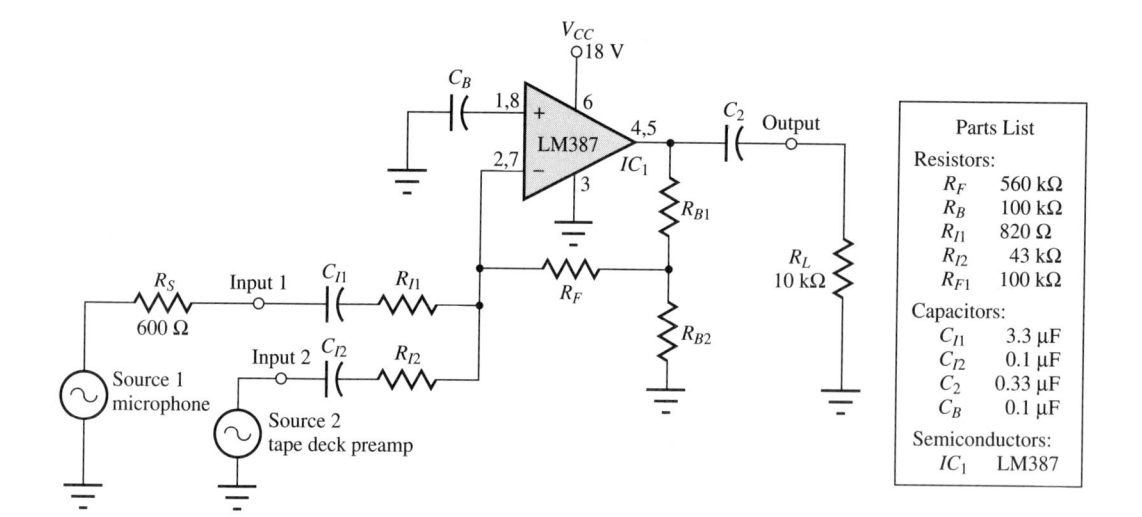

Figure 12-14

First design R_B and R_F. This will be the same as in previous examples. We will arbitrarily select R_B as 100 kΩ, less than the maximum value of 280 kΩ required by Equation 12-1, and calculate R_F from Equation 12-4:

$$R_F = R_B \times \left(\frac{V_{CC}}{2.8} - 1\right) = 100\text{ k}\Omega \times \left(\frac{18\text{ V}}{2.8\text{ V}} - 1\right) = 543\text{ k}\Omega$$

Choose the nearest standard value of 560 kΩ.

The voltage gain must be calculated for each input to the mixer by using Equation 12-20:

$$A_V = \frac{\text{desired output voltage level}}{\text{source voltage level}}$$

For the microphone input, the voltage gain required is $A_{V_1} = 10/0.012 = 833$, whereas for the tape deck input the voltage gain required is $A_{V_2} = 10/0.35 = 28.6$.

We will select a value of $R_{F1} = 100$ kΩ (equal to R_B). Calculate R_I for each input by Equation 12-19:

$$R_I = \frac{R_{F1}R_F + R_{F1}R_B + R_F R_B}{R_B A_V} - R_S$$

For the microphone input, the required gain is 833 and the source resistance is 600 Ω:

$$R_{I_1} = \frac{100\text{ k}\Omega \times 560\text{ k}\Omega + 100\text{ k}\Omega \times 100\text{ k}\Omega + 560\text{ k}\Omega \times 100\text{ k}\Omega}{100\text{ k}\Omega \times 833} - 600\text{ }\Omega$$

$$= 1465\text{ }\Omega - 600\text{ }\Omega = 864\text{ }\Omega$$

Choose the closest standard value of 820 Ω.

For the tape deck input, the required gain is 28.6 and the source resistance is negligible:

$$R_{I_2} = \frac{100 \text{ k}\Omega \times 560 \text{ k}\Omega + 100 \text{ k}\Omega \times 100 \text{ k}\Omega + 560 \text{ k}\Omega \times 100 \text{ k}\Omega}{100 \text{ k}\Omega \times 28.6} - 0 \text{ }\Omega$$

$$= 42.7 \text{ k}\Omega$$

Choose the closest standard value of 43 kΩ.

Next, input coupling capacitors C_I must be designed for each input by using Equation 12-8 and assuming that the lower cutoff frequency is 50 Hz:

$$C_I = \frac{1}{2\pi f_{CL}(R_S + R_I)}$$

First for the microphone input, where $R_S = 600 \text{ }\Omega$ and $R_I = 1.6 \text{ k}\Omega$,

$$C_{I_1} = \frac{1}{2\pi \times 50 \text{ Hz} \times (600 \text{ }\Omega + 820 \text{ }\Omega)} = 2.24 \text{ }\mu\text{F}$$

Choose the standard larger value of 3.3 μF.

Next for the tape deck input, where $R_S = 0 \text{ }\Omega$ and $R_I = 43 \text{ k}\Omega$,

$$C_{I_2} = \frac{1}{2\pi \times 50 \text{ Hz} \times 43 \text{ k}\Omega} = 0.0740 \text{ }\mu\text{F}$$

Choose the larger standard value of 0.1 μF.

We will arbitrarily select bypass capacitor C_B as 0.1 μF. This will be more than adequate according to Equation 12-12.

Finally, the output coupling capacitor C_2 is calculated from Equation 12-9, where the load resistance is 10 kΩ.

$$C_2 = \frac{1}{2\pi f_{CL} R_L} = \frac{1}{2\pi \times 50 \text{ Hz} \times 10 \text{ k}\Omega} = 0.318 \text{ }\mu\text{F}$$

Choose the larger standard value of 0.33 μF.

The design of the mixer circuit is complete.

12-2 The LM1877 Audio Power Amplifier

The second audio integrated circuit that we will study is the **audio power amplifier.** The audio power amplifier is an amplifier designed for the audio range of frequencies that has a relatively low voltage gain but a very high current gain and, consequently, a large power gain. Audio power amplifiers are designed to drive loads of a few ohms, typically speakers with impedances of 4 or 8 Ω.

There are a wide variety of audio power amplifiers, ranging in power output from less than 1 W to more than 100 W. Most of these are special-application devices, custom-manufactured for a specific piece of audio equipment, and thus are not readily available. Some major manufacturers of integrated circuits, however, such as National Semiconductor, Texas Instruments, and Motorola, produce lines of general-purpose audio power amplifiers. These are relatively low power (typically less than 10 W), but are readily available and usually quite inexpensive. We will use one of these, the LM1877 produced by National Semiconductor, as a typical example. High-power audio amplifiers (with power ratings of more than 10 W) are usually difficult to obtain and are expensive. As we will see, however, this restriction in high-power devices generally does not limit our options in designing audio power amplifiers.

The LM1877 is a dual 2 W power amplifier, ideal for stereo applications. The data sheets for the LM1877 are shown in Figure 12-15, and the pinout and schematic symbol are shown in Figure 12-16. The LM1877 can operate from any single supply from 6 to 24 V or from a dual supply from ± 3 to ± 12 V. Although it is intended to be used as a stereo amplifier delivering up to 2 W per channel to an 8 Ω load (speaker), each amplifier can be operated independently, or the two amplifiers can be operated in a bridge configuration to provide 4 W of power to an 8 Ω speaker.

As in all power devices, the rated power is for a case temperature of 25°C, and the actual output power must be derated for higher case temperatures. The LM1877 would normally be operated with a heatsink. In fact, any power level above approximately 0.75 W/channel requires some form of heatsink. The distortion of the LM1877 depends on the power output. For power levels less than 1 W, the total harmonic distortion is less than 0.1%, whereas for 2 W output the distortion can be as large as 10%.

The maximum input signal level is ± 0.7 V, or 1.4 V peak-to-peak. The maximum peak-to-peak output voltage depends on the power supply V_{CC} and is

$$\text{maximum output voltage} = V_{CC} - 6 \text{ V} \qquad (12\text{-}21)$$

The output power that the amplifier delivers to a load is the square of the RMS output voltage divided by the load resistance. The maximum power for any value of V_{CC} is

$$\text{maximum output power} = \frac{(V_{CC} - 6 \text{ V})^2}{8R_L} \qquad (12\text{-}22)$$

For a supply voltage of, say, 20 V, the maximum output voltage is 14 V peak-to-peak, and the corresponding output power is approximately 3 W, more than the 2 W maximum for the LM1877. In normal operation, this is not a problem because most audio signals have fluctuating levels, and the average value is considerably less than the maximum. This limit is important to recognize, however, if the LM1877 is used for applications such as an amplifier for an alarm or a function generator, where the output would be at a sustained constant level.

The circuit for the LM1877 is similar to a standard op-amp, but it is internally compensated only for gains greater than 10 and requires compensating circuitry on the input and on the output. It is designed to be used as a noninverting amplifier, with an open-loop input impedance of 4 MΩ and an open-loop gain of 70 dB (approximately 3000). As in an op-amp, feedback is used to reduce and stabilize the gain, and in so doing it increases the input impedance. In all practical circuits, we can assume that the input impedance is infinite.

National Semiconductor

Audio/Radio Circuits

LM1877 Dual Power Audio Amplifier

General Description

The LM1877 is a monolithic dual power amplifier designed to deliver 2W/channel continuous into 8Ω loads. The LM1877 is designed to operate with a low number of external components, and still provide flexibility for use in stereo phonographs, tape recorders and AM–FM stereo receivers, etc. Each power amplifier is biased from a common internal regulator to provide high power supply rejection, and output Q point centering. The LM1877 is internally compensated for all gains greater than 10.

Features

- 2W/channel
- −65 dB ripple rejection, output referred
- −65 dB channel separation, output referred
- Wide supply range, 6–24V
- Very low cross-over distortion
- Low audio band noise
- AC short circuit protected
- Internal thermal shutdown

Applications

- Multi-channel audio systems
- Stereo phonographs
- Tape recorders and players
- AM–FM radio receivers
- Servo amplifiers
- Intercom systems
- Automotive products

Connection Diagram

Dual-In-Line Package

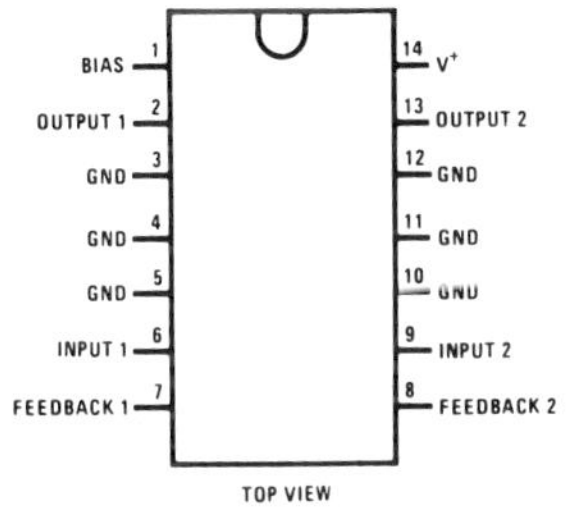

Pin	Name	Pin	Name
1	BIAS	14	V⁺
2	OUTPUT 1	13	OUTPUT 2
3	GND	12	GND
4	GND	11	GND
5	GND	10	GND
6	INPUT 1	9	INPUT 2
7	FEEDBACK 1	8	FEEDBACK 2

TOP VIEW

Order Number LM1877N
See NS Package N14A

Equivalent Schematic Diagram

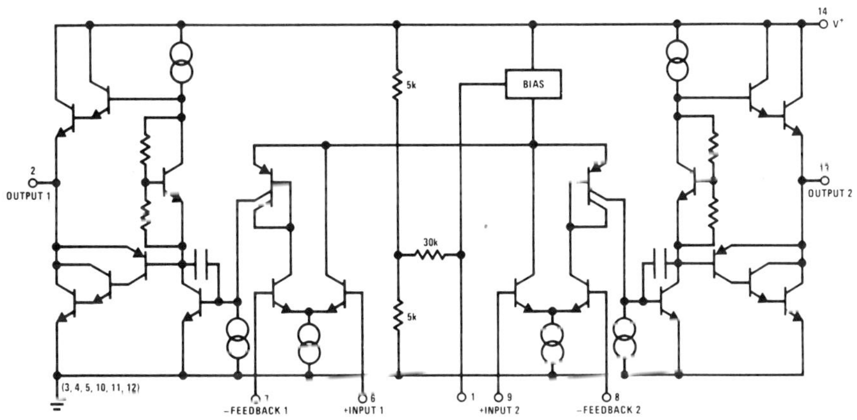

Figure 12-15 Data sheets for the LM1877 dual power audio amplifier. *(Reprinted with permission of National Semiconductor Corporation)*

Absolute Maximum Ratings

Supply Voltage	26V
Input Voltage	±0.7V
Operating Temperature	0°C to +70°C
Storage Temperature	−65°C to +150°C
Junction Temperature	150°C
Lead Temperature (Soldering, 10 seconds)	300°C

Electrical Characteristics

V_S = 20V, T_A = 25°C, R_L = 8Ω, A_V = 50 (34 dB) unless otherwise specified

PARAMETER	CONDITIONS	MIN	TYP	MAX	UNITS
Total Supply Current	P_O = 0W		25	50	mA
Output Power LM1877N	THD = 10% V_S = 20V, R_L = 8Ω	2.0			W
Total Harmonic Distortion LM1877	f = 1 kHz, V_S = 14V				
	P_O = 50 mW/Channel		0.075		%
	P_O = 500 mW/Channel		0.045		%
	P_O = 1W/Channel		0.055		%
Output Swing	R_L = 8Ω		V_S−6		Vp-p
Channel Separation	C_F = 50 μF, C_{IN} = 0.1 μF, f = 1 kHz, Output Referred				
	V_S = 20V, V_O = 4 Vrms	−50	−70		dB
	V_S = 7V, V_O = 0.5 Vrms		−60		dB
PSRR Power Supply Rejection Ratio	C_F = 50 μF, C_{IN} = 0.1 μF, f = 120 Hz, Output Referred				
	V_S = 20V, V_{RIPPLE} = 1 Vrms	−50	−65		dB
	V_S = 7V, V_{RIPPLE} = 0.5 Vrms		−40		dB
Noise	Equivalent Input Noise R_S = 0, C_{IN} = 0.1 μF, BW = 20 Hz–20 kHz		2.5		μV
	Output Noise Wideband R_S = 0, C_{IN} = 0.1 μF, A_V = 200		0.80		mV
Open Loop Gain	R_S = 0, f = 100 kHz, R_L = 8Ω		70		dB
Input Offset Voltage			15		mV
Input Bias Current			50		nA
Input Impedance	Open Loop		4		MΩ
DC Output Level	V_S = 20V	9	10	11	V
Slew Rate			2.0		V/μs
Power Bandwidth			65		kHz
Current Limit			1.0		A

Note 1: For operation at ambient temperature greater than 25°C, the LM1877 must be derated based on a maximum 150°C junction temperature using a thermal resistance which depends upon device mounting techniques.

10-168

Figure 12-15 (*Continued*)

12-2 The LM1877 Audio Power Amplifier

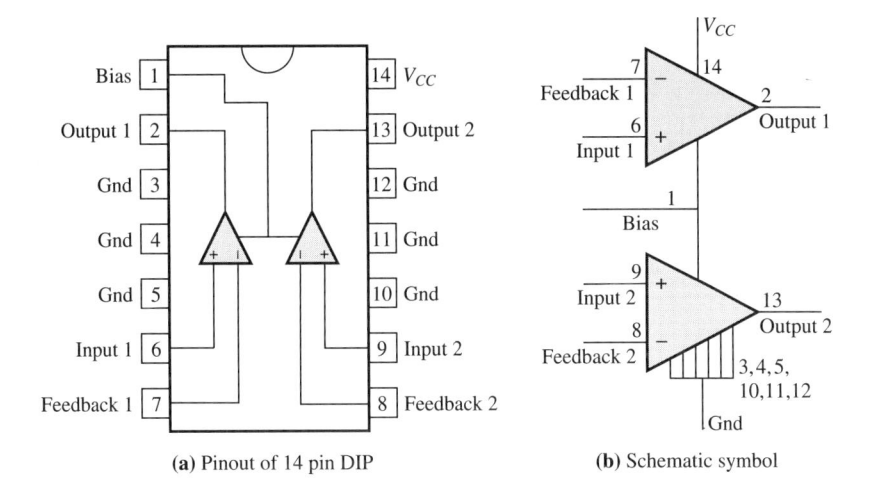

Figure 12-16 Pinout and schematic symbol for the LM1877 dual power amplifier.

(a) Pinout of 14 pin DIP

(b) Schematic symbol

The LM1877 as a Simple Power Amplifier

The circuit of a simple noninverting power amplifier using half of the LM1877 is shown in Figure 12-17. The voltage gain is determined by feedback resistor R_F from the output (pin 2 or 13) to the feedback (inverting) input (pin 7 or 8) and by input resistor R_I from the feedback input through capacitor C_I to ground. Equation 3-5 is used, which is the same equation we used for a noninverting amplifier for a standard op-amp. Normally a value is chosen for R_F, and Equation 3-5 is solved for R_I to give the design equation for this resistor:

$$R_I = \frac{R_F}{A_V - 1} \qquad (12\text{-}23)$$

The gain A_V required by the amplifier is determined by the ratio of output signal level to input signal level:

$$A_V = \frac{\text{output voltage level}}{\text{source voltage level}} \qquad (12\text{-}24)$$

If the output voltage is calculated from the power supply voltage by using Equation 12-21, the value is in peak-to-peak volts, and it is important to express the input in the same units when using Equation 12-24.

Because the LM1877 is designed to operate from a single supply, resistor R_I must be connected to AC (or signal) ground through capacitor C_I. This capacitor determines the lower cutoff frequency f_{CL} of the amplifier. This is the frequency where the capacitive reactance X_{C_I} equals R_I. Solving for C_I gives

$$C_I = \frac{1}{2\pi f_{CL} R_I} \qquad (12\text{-}25)$$

568 12 Audio Integrated Circuits

(a) Amplifier using a single supply

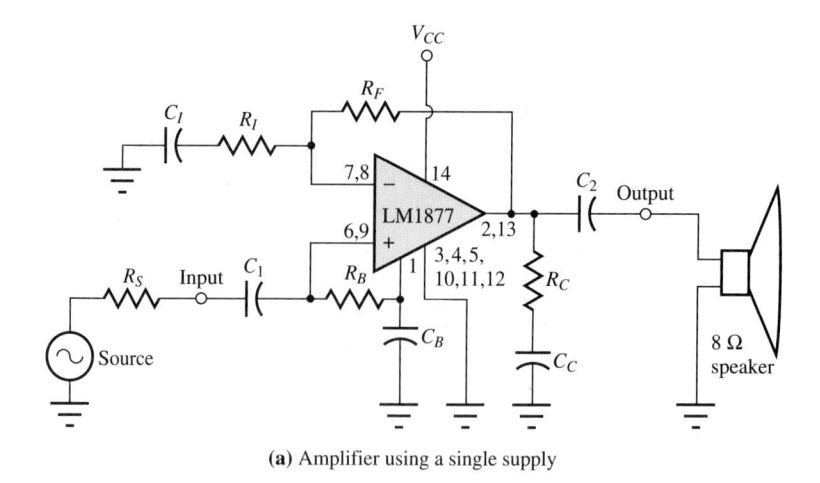

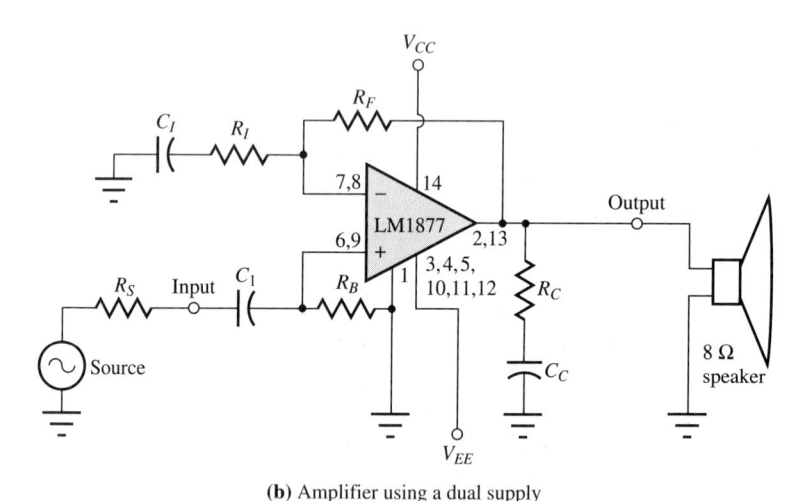

(b) Amplifier using a dual supply

Figure 12-17 Simple noninverting power amplifier using the LM1877.

Capacitor C_I should always be chosen larger than this value to ensure that the capacitive reactance is negligible at all operating frequencies.

The manufacturer recommends using a compensating network consisting of a 2.7 Ω resistor R_C in series with a 0.1 μF capacitor C_C connected from the output (pin 2 or 13) to ground. The output is also capacitively coupled to the load (speaker) with output capacitor C_2. This capacitor is designed such that its capacitive reactance X_{C_2} equals the load impedance R_L at the lowest operating frequency f_{CL}. This gives the following design equation for C_2:

$$C_2 = \frac{1}{2\pi f_{CL} R_L} \tag{12-26}$$

Notice that we did not include Z_{out}, the output impedance of the amplifier, in this equation. Although the output impedance is not given on the data sheet, it can be assumed to be a few ohms. With feedback, it becomes even smaller and, hence, can be neglected in calculating C_2. For an 8 Ω speaker as load and a cutoff frequency of 50 Hz, C_2 is typically 500 μF.

The signal input is connected to the noninverting input of the amplifier (pin 6 or 9) through coupling capacitor C_1. Since the input impedance of the amplifier is essentially infinite, we can use any value of capacitor. Choose a 0.1 μF value for C_1.

The noninverting input must be connected to pin 1 through a resistor. Pin 1 is labelled BIAS on the pinout in Figure 12-16. Because the LM1877 uses bipolar technology, a bias current must flow into each input. For the inverting or feedback input, the bias current flows from the output through R_F. For the noninverting input, however, the source is capacitively coupled and, without pin 1, there is no path for DC bias current to flow. The bias pin must be decoupled to ground by a large-value capacitor C_B, for which the manufacturer recommends a value of 250 μF. To prevent the input signal from being shorted to ground by this capacitor, insert a bias resistor R_B between the input and the bias pin. The value of R_B is not critical, and any large value will be satisfactory. We will use a value of 1.0 MΩ for all of the following designs. If both amplifiers are used, separate bias paths must be established for each noninverting input (pins 7 and 8).

The LM1877 has a number of pins labelled as ground (3, 4, 5, 10, 11, and 12). For operation from a single supply, these are all connected to supply ground. For operation from a dual-polarity supply, they are all connected to the negative supply. All other ground connections still go to ground. Figure 12-17(a) shows the LM1877 connected to a single supply, and Figure 12-17(b) shows the LM1877 connected to a dual supply. When the LM1877 is operated from a dual supply, the bias output (pin 1) is connected directly to ground. It does not need to be capacitively coupled through C_B because the AC ground and system ground are the same in this mode of operation.

EXAMPLE 12-5

Design a simple audio power amplifier using half of an LM1877 dual-power audio amplifier to drive an 8 Ω speaker. Assume that the RMS input voltage from the preamplifier is 0.35 V, and design the power amplifier for maximum output voltage level. Use a single 18 V power supply.

Solution

The circuit is shown in Figure 12-18.

First we must determine the amplifier gain required. The input from the preamplifier has an RMS value of 0.35 V, or a peak-to-peak value of 1.0 V. The maximum output voltage is given by Equation 12-21:

maximum output voltage = V_{CC} − 6 V = 18 V − 6 V = 12 V

12 Audio Integrated Circuits

Parts List

Resistors:
- R_I 9.1 kΩ
- R_F 100 kΩ
- R_B 1.0 MΩ
- R_C 2.7 Ω

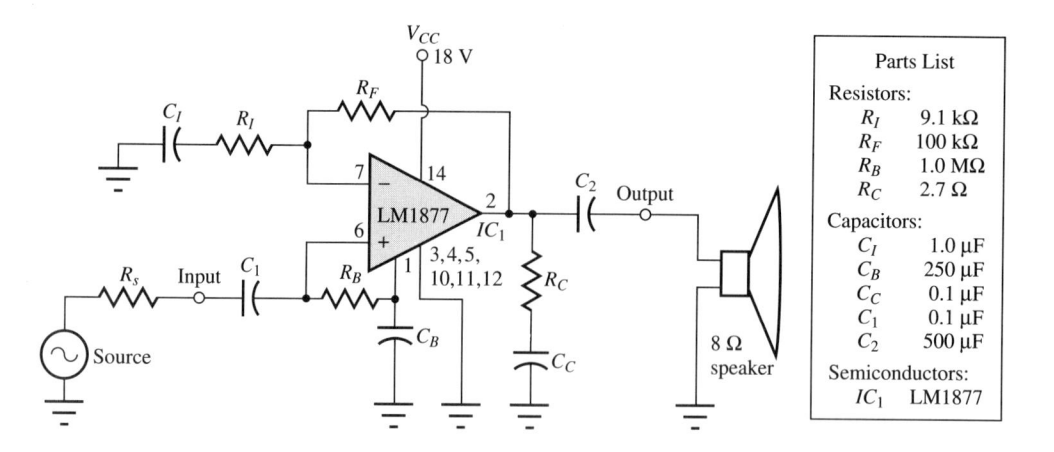

Capacitors:
- C_I 1.0 µF
- C_B 250 µF
- C_C 0.1 µF
- C_1 0.1 µF
- C_2 500 µF

Semiconductors:
- IC_1 LM1877

The required voltage gain, given by Equation 12-24, is

$$A_V = \frac{\text{output voltage level}}{\text{source voltage level}} = \frac{12 \text{ V}}{1 \text{ V}} = 12$$

This value will be used to determine R_I. Choose R_F as 100 kΩ and calculate R_I from Equation 12-23:

$$R_I = \frac{R_F}{A_V - 1} = \frac{100 \text{ k}\Omega}{12 - 1} = 9.09 \text{ k}\Omega$$

Choose the closest standard value of 9.1 kΩ.

Now calculate capacitor C_I from Equation 12-25, assuming a lower critical frequency of 50 Hz:

$$C_I = \frac{1}{2\pi f_{CL} R_I} = \frac{1}{2\pi \times 50 \text{ Hz} \times 9.1 \text{ k}\Omega} = 0.350 \text{ µF}$$

Capacitor C_I should always be chosen larger than this value to ensure the capacitive reactance is negligible at all operating frequencies, so we will choose a value of 1.0 µF.

Calculate the output coupling capacitor C_2 from Equation 12-26, where the load impedance R_L is 8 Ω and the lowest operating frequency f_{CL} is 50 Hz:

$$C_2 = \frac{1}{2\pi f_{CL} R_L} = \frac{1}{2\pi \times 50 \text{ Hz} \times 8 \text{ }\Omega} = 398 \text{ µF}$$

Choose the larger standard value of 500 µF.

We do not need to calculate the input coupling capacitor C_1 because the input impedance of the amplifier is essentially infinite. We will simply select a 0.1 µF value for C_1.

Use the recommended values of 1.0 MΩ for the bias resistor R_B, 250 μF for the bias capacitor C_B, 2.7 Ω for the compensating resistor R_C, and 0.1 μF for the compensating capacitor C_C.

Finally, connect all the ground pins (pins 3, 4, 5, 10, 11, and 12) to ground. The power amplifier design is complete.

The LM1877 as a Stereo Amplifier

The LM1877 consists of two identical amplifiers and is designed basically to be used as a stereo amplifier. The circuit for this application is shown in Figure 12-19. Each amplifier is connected by using the same circuit as described for a single amplifier. Each amplifier circuit has its own biasing, its own compensation network, and its own feedback network. For biasing, both bias resistors are connected to pin 1 and decoupled through a single decoupling capacitor C_B. The gains of the two amplifiers can be made different if necessary.

Figure 12-19(a) shows the stereo amplifier using a single supply, and Figure 12-19(b) shows the same circuit with a dual supply. When using a dual supply, the ground pins (pins 3, 4, 5, etc.) are connected to the negative supply, and the bias output (pin 1) is connected directly to ground (it does not need to be capacitively coupled through C_B).

(a) Amplifier using a single supply

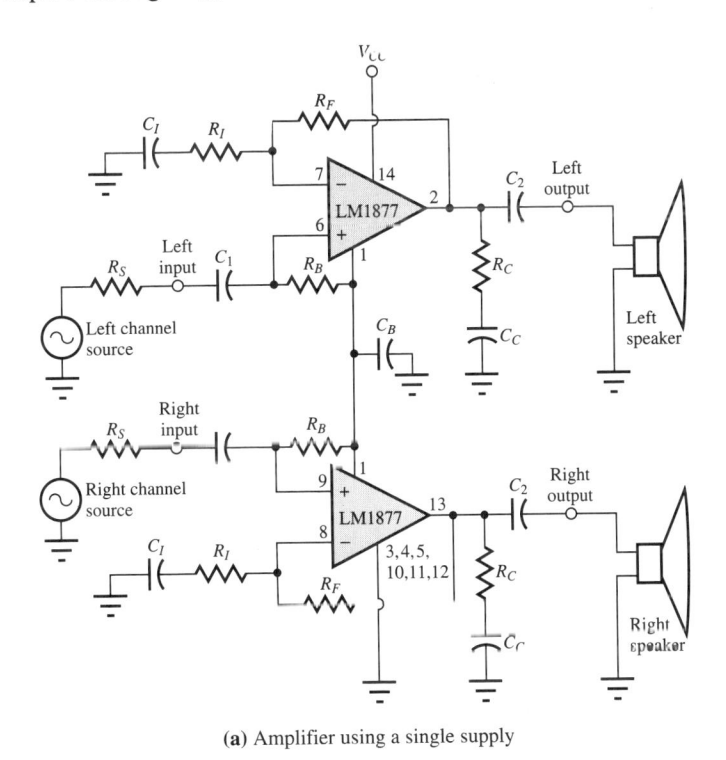

Figure 12-19 The LM1877 used as a stereo amplifier.

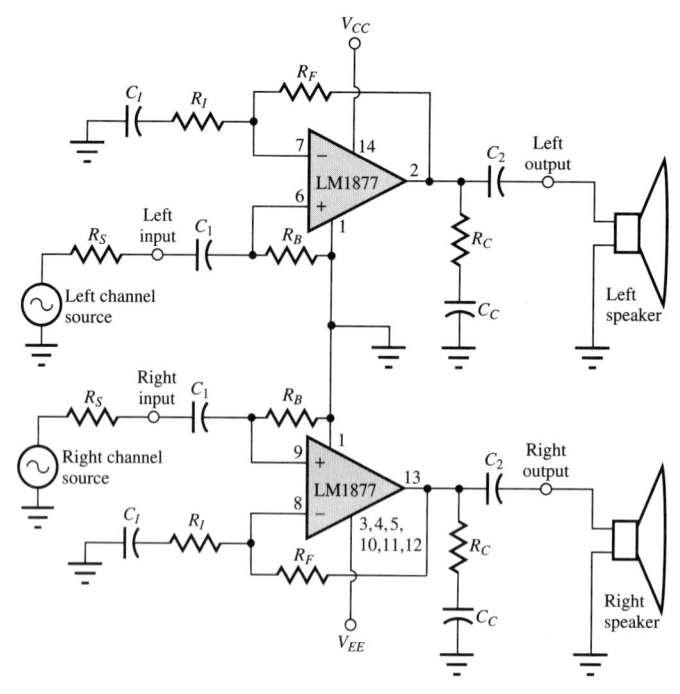

(b) Amplifier using a dual supply

Figure 12-19 (*Continued*)

EXAMPLE 12-6

Design a stereo audio power amplifier using an LM1877 dual-power audio amplifier. Each output will drive an 8 Ω speaker. Assume that the input voltage from the preamplifier is 0.35 V RMS and design the power amplifiers for maximum output voltage level. Use a dual ±12 V power supply.

Solution

The circuit is shown in Figure 12-20.

The overall design is similar to the simple power amplifier designed in the previous example. First determine the gain required from the amplifier. The input from the preamplifier has an RMS value of 0.35 V, hence a peak-to-peak value of 1.0 V. The maximum output voltage from Equation 12-21 is

maximum output voltage = $V_{CC} - 6\,V = 24\,V - 6\,V = 18\,V$

The required voltage gain, given by Equation 12-25, is

$$A_V = \frac{\text{output voltage level}}{\text{source voltage level}} = \frac{18\,V}{1.0\,V} = 18$$

12-2 The LM1877 Audio Power Amplifier

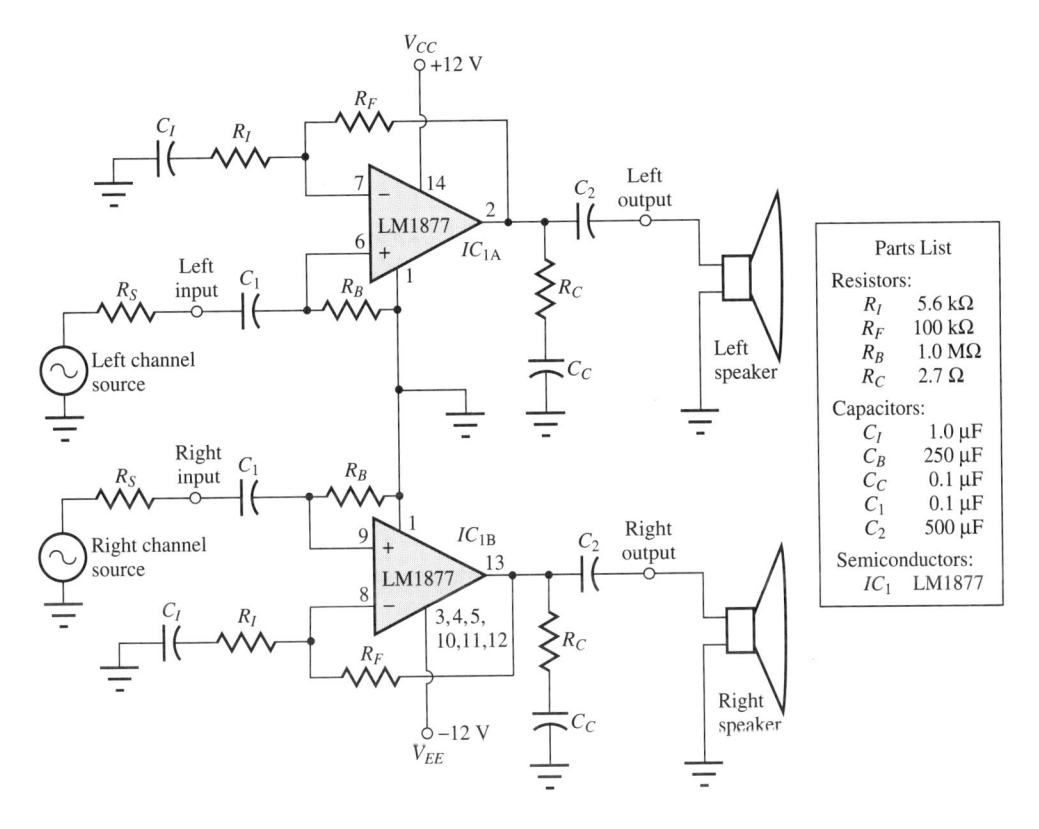

Figure 12-20

Parts List
Resistors:
 R_I 5.6 kΩ
 R_F 100 kΩ
 R_B 1.0 MΩ
 R_C 2.7 Ω

Capacitors:
 C_I 1.0 µF
 C_B 250 µF
 C_C 0.1 µF
 C_1 0.1 µF
 C_2 500 µF

Semiconductors:
 IC_1 LM1877

Select R_F as 100 kΩ and calculate R_I from Equation 12-23:

$$R_I = \frac{R_F}{A_V - 1} = \frac{100 \text{ k}\Omega}{18 - 1} = 5.88 \text{ k}\Omega$$

Choose the closest standard value of 5.6 kΩ from Appendix A-1.

Now calculate capacitor C_I from Equation 12-25, assuming a lower critical frequency of 50 Hz:

$$C_I = \frac{1}{2\pi f_{CL} R_I} = \frac{1}{2\pi \times 50 \text{ Hz} \times 5.6 \text{ k}\Omega} = 0.568 \text{ µF}$$

Capacitor C_I should always be chosen larger than this value, so we will choose a value of 1.0 µF.

Calculate output capacitor C_2 from Equation 12-26 for a load impedance R_L of 8 Ω and a cutoff frequency f_{CL} of 50 Hz:

$$C_2 = \frac{1}{2\pi f_{CL} R_L} = \frac{1}{2\pi \times 50 \text{ Hz} \times 8 \text{ }\Omega} = 398 \text{ µF}$$

Choose the larger standard value of 500 µF.

Select a value of 0.1 µF for coupling capacitor C_1. Use the recommended values of 1.0 MΩ for the bias resistor R_B, 250 µF for the bias capacitor C_B, 2.7 Ω for the compensating resistor R_C, and 0.1 µF for the compensating capacitor C_C.

Finally, connect all the ground pins (pins 3, 4, 5, 10, 11, and 12) to the negative supply.

The stereo power amplifier is now designed.

The LM1877 Used as a Bridge Amplifier

The two amplifiers in the LM1877 can be configured as a single bridge amplifier to supply 4 W to an 8 Ω load. This application is shown in Figure 12-21. The outputs of the two amplifiers are connected to either side of the 8 Ω speaker with the output compensating network connected in parallel across the speaker.

For this configuration to work, the output signal from one amplifier must be a maximum when the output signal from the other amplifier is a minimum; that is, the outputs must be 180° out of phase. This is accomplished by supplying the input signal to only one amplifier, the left amplifier as shown in Figure 12-21. The right amplifier input at pin 9 is effectively ground because it is connected to the bias pin and thus to ground through the bias capacitor C_B. The input network consisting of R_I and C_I goes from the feedback input of the left amplifier (pin 7) to the feedback of the right amplifier (pin 8). This allows the feedback signal from the left amplifier to serve as input for the right amplifier, introducing a phase shift of 180° and making the amplifiers operate in a push-pull mode as they drive the speaker. The component design is the same as for the simple power amplifier described earlier.

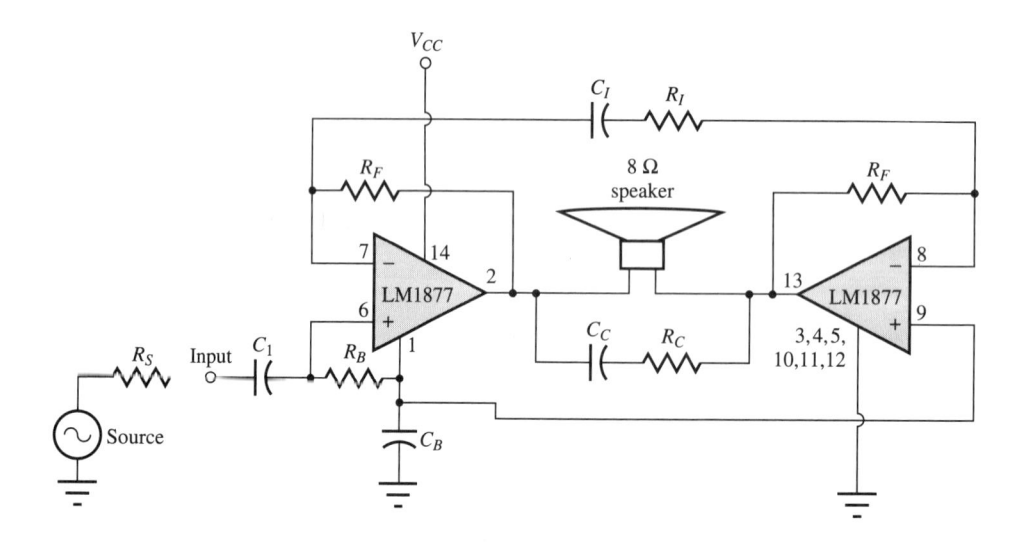

Figure 12-21 The LM1877 used as a 4 W bridge amplifier.

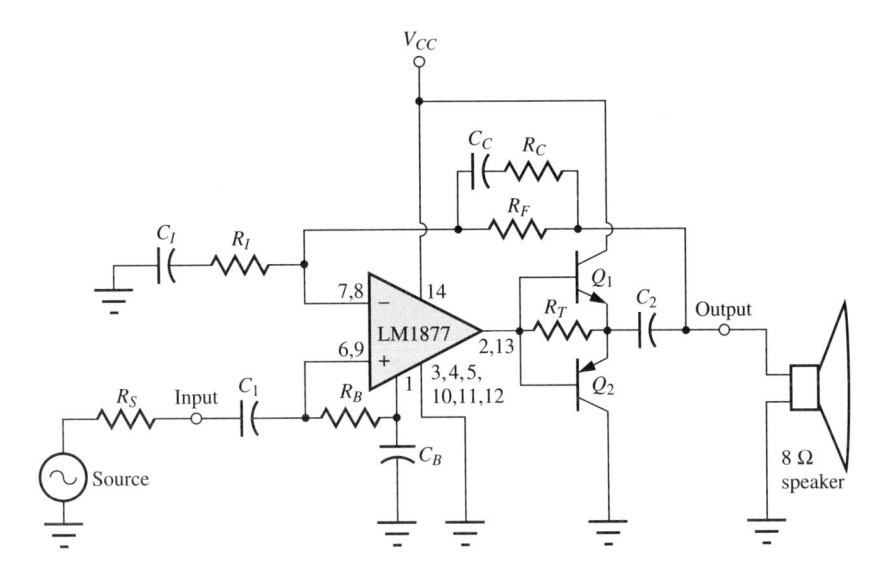

Figure 12-22 The LM1877 used as a high-power amplifier.

High-Power Amplifier

The maximum output power available from the LM1877 is 2 W/channel (or 4 W if the amplifiers are connected in bridge configuration), but with the addition of external power transistors, it is possible to increase the output power well beyond this limit. As shown in Figure 12-22, the LM1877 can be used to drive a complementary pair of power transistors, with the output power limited by the current gains of the transistors.

The resistor R_T connected between the bases and emitters of the power transistors is used to bias the bases at 0.7 V to minimize crossover distortion. It has a typical value of 5 Ω. Each transistor conducts for alternate halves of the input cycle and is nonconducting for the other half-cycle. This is known as class B operation, and it allows high operating efficiency for the power transistors. The circuit in Figure 12-22 is a minimal circuit, and for optimal performance a more elaborate discrete component power amplifier should be used. The design of discrete component power amplifiers is beyond the scope of this book, and the reader is referred to any standard text on discrete circuit design.

The design of the circuit for the LM1877 power amplifier portion of the circuit is similar to the basic power amplifier designed earlier. There are a few minor differences. There is no compensating network on the output. The input impedance of the external discrete component power amplifier is approximately equal to the transistor β times the load impedance. This is much larger than the 8 Ω of a speaker, so the basic compensating network would overcompensate the integrated circuit amplifier. Instead, the manufacturer recommends a compensating network consisting of an 82 pF compensating capacitor C_C in series with a 27 kΩ compensating resistor R_C in parallel with the feedback resistor. The feedback

576 12 Audio Integrated Circuits

signal is taken from the output of the discrete component amplifier, not from the output of the integrated circuit amplifier.

12-3 Volume and Tone Control Circuits

With most audio amplifier systems, circuitry is provided for varying the loudness of the output sound level (**volume control**), for varying the low frequency gain relative to the high frequency gain (**tone control**), and for varying the relative gain between the two channels in stereo systems (**balance control**). Of these, the tone control is the most complex, requiring adjustment of the low frequency gain (**bass control**) and of the high frequency gain (**treble control**). All of these controls are easy to implement, usually as part of the preamplifier circuit.

Simple Tone and Volume Controls

We will first look at simple controls for a single-channel amplifier (monaural system). Obviously, there is no need for a balance control. The volume control is usually a potentiometer across the input to the power amplifier, with the input signal picked off the wiper arm as shown in Figure 12-23(a). Since the human ear has a logarithmic response, an **audio taper (logarithmic) potentiometer** should be used so that the change in sound intensity as perceived by the ear will vary linearly with respect to the setting of the volume control. Bass and treble controls are shown in Figure 12-23(b) and 12-23(c), respectively. The potentiometers in these should also be audio taper.

Consider the bass control first. This control can attenuate or boost low frequency signals relative to midband frequencies. In the circuit in Figure 12-23(b), for midrange and high frequencies, capacitors C_{1B} and C_{2B} effectively short out potentiometer R_{PB}, and the attenuation is

$$\text{attenuation} = \frac{R_{2B}}{R_{1B} + R_{2B}} \tag{12-27}$$

(a) Volume control

(b) Bass tone control

(c) Treble tone control

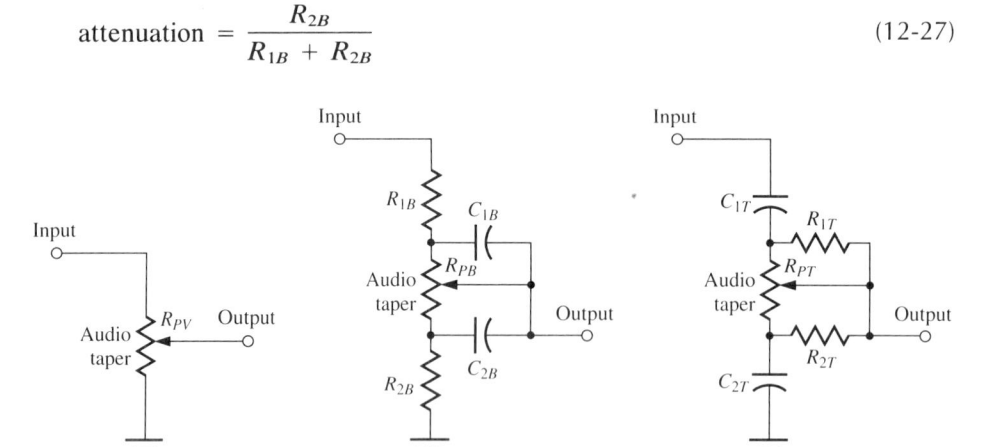

Figure 12-23 Volume and tone control circuits.

At low frequencies, the capacitors and potentiometer boost the signal or cut it relative to this attenuation.

To design this bass control, first determine values for R_{1B}, R_{2B}, and R_{PB}. The ratio of R_{1B}/R_{2B} and R_{PB}/R_{1B} sets the amount of "boost" and "cut." If both ratios are 10, then the boost and cut in decibels are each 20 log 10 = 20 dB, a typical value for a bass tone control. Choose the capacitors such that $X_{C_{1B}} = R_{1B}$ and $X_{C_{2B}} = R_{2B}$ at frequency f_B, where the bass attenuation is to effectively start (the −3 dB point). Solve for C_{1B} and C_{2B}:

$$C_{1B} = \frac{1}{2\pi f_B R_{1B}} \tag{12-28}$$

$$C_{2B} = \frac{1}{2\pi f_B R_{2B}} \tag{12-29}$$

The treble control in Figure 12-23(c) is similar. The resistor and capacitor dividers are reversed from the bass control, and the control attenuates or boosts *high frequency signals* relative to the midband frequencies. For midrange and low frequencies, capacitors C_{1T} and C_{2T} have high impedances and give an attenuation of

$$\text{attenuation} = \frac{C_{1T}}{C_{1T} + C_{2T}} \tag{12-30}$$

To design this circuit, first choose a value for potentiometer R_{PT}. Design capacitor C_{1T} such that $X_{C_{1T}} = R_{PT}$ at the frequency f_T where the treble attenuation or boost is to effectively start (the −3 dB point). Solving for C_{1T} gives

$$C_{1T} = \frac{1}{2\pi f_T R_{PT}} \tag{12-31}$$

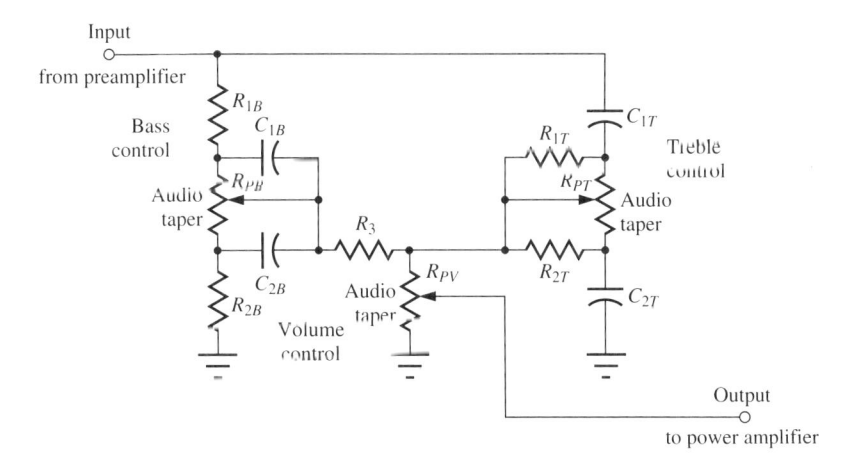

Figure 12-24 Combinations of tone controls and volume control in a typical circuit.

578   12   Audio Integrated Circuits

The ratio of C_{2T}/C_{1T} sets the amount of boost and cut. If the ratio is 10, then the boost and cut are each 20 dB. Choose resistors R_{1T} and R_{2T} such that $R_{1T} = X_{C_{1T}}$ and $R_{2T} = X_{C_{2T}}$ at frequency f_T. The following equations give R_{1T} and R_{2T}:

$$R_{1T} = \frac{1}{2\pi f_T C_{1T}} \tag{12-32}$$

$$R_{2T} = \frac{1}{2\pi f_T C_{2T}} \tag{12-33}$$

Figure 12-24 shows how the gain, bass, and treble controls can be incorporated into a typical audio amplifier circuit. The tone control circuit is usually placed between the preamplifier and the power amplifier, and the overall circuit gain is designed to compensate for the midband attenuation of the tone controls.

Tone controls are sometimes incorporated in the feedback loop of a separate preamplifier. There is little advantage in this for most applications, and cost is an additional amplifier stage and more complicated circuit design.

EXAMPLE 12-7

Design a single-channel tone/volume control network. The bass control should become effective at 600 Hz with a boost/cut of 20 dB, and the treble control should become effective at 1500 Hz with a boost/cut of 20 dB.

Solution

The circuit is shown in Figure 12-25. Use 100 kΩ audio taper potentiometers for all of the controls.

First design the bass circuit. The ratio of R_{1B}/R_{2B} and R_{PB}/R_{1B} sets the amount of boost or cut. Thus, for a boost/cut of 20 dB, the ratio is simply 10. Since R_{PB} has been chosen as 100 kΩ, $R_{1B} = R_{PB}/10 = 10$ kΩ, and $R_{2B} = R_{1B}/10 = 1.0$ kΩ. These are all standard values.

Now calculate capacitors C_{1B} and C_{2B} from Equations 12-28 and 12-29 for a frequency f_B of 600 Hz:

$$C_{1B} = \frac{1}{2\pi f_B R_{1B}} = \frac{1}{2\pi \times 600 \text{ Hz} \times 10 \text{ k}\Omega} = 0.0265 \text{ }\mu\text{F}$$

$$C_{2B} = \frac{1}{2\pi f_B R_{2B}} = \frac{1}{2\pi \times 600 \text{ Hz} \times 1.0 \text{ k}\Omega} = 0.265 \text{ }\mu\text{F}$$

Choose values of 0.027 μF for C_{1B} and 0.27 μF for C_{2B}.

Next design the treble circuit. Since we have chosen R_{PT} as 100 kΩ, we can design capacitor C_{1T} from Equation 12-31 for a frequency f_T of 1500 Hz:

$$C_{1T} = \frac{1}{2\pi f_T R_{PT}} = \frac{1}{2\pi \times 1500 \text{ Hz} \times 100 \text{ k}\Omega} = 0.00106 \text{ }\mu\text{F}$$

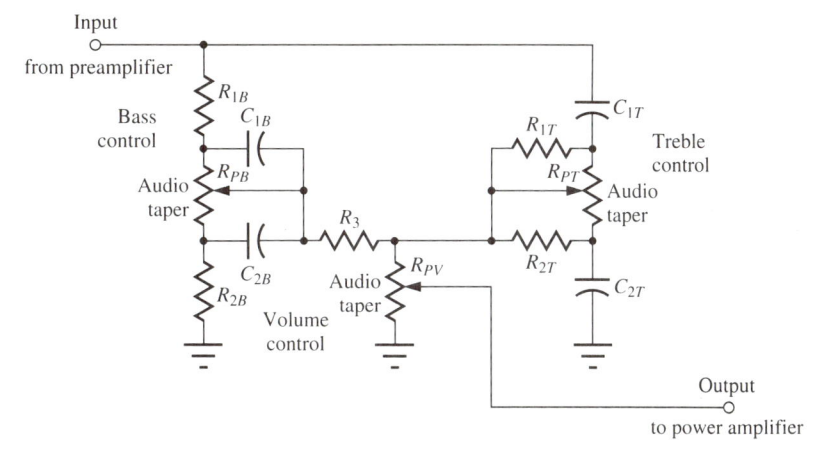

12-3 Volume and Tone Control Circuits

Figure 12-25

Parts List

Resistors:
- R_{1B} 10 kΩ
- R_{2B} 1.0 kΩ
- R_{1T} 110 kΩ
- R_{2T} 11 kΩ
- R_{PB} 100 kΩ pot
- R_{PT} 100 kΩ pot
- R_{PV} 100 kΩ pot
- R_3 10 kΩ

Capacitors:
- C_{1B} 0.027 µF
- C_{2B} 0.27 µF
- C_{1T} 0.001 µF
- C_{2T} 0.01 µF

Choose a value of 0.001 µF. The ratio of C_{2T}/C_{1T} sets the amount of boost/cut. For a boost/cut of 20 dB, the ratio is 10, and a value of 0.01 µF will be used for capacitor C_{2T}.

Now design resistors R_{1T} and R_{2T} from Equations 12-32 and 12-33 for a frequency $f_T = 1500$ Hz:

$$R_{1T} = \frac{1}{2\pi f_T C_{1T}} = \frac{1}{2\pi \times 1500 \text{ Hz} \times 0.001 \text{ µF}} = 106 \text{ k}\Omega$$

$$R_{2T} = \frac{1}{2\pi f_T C_{2T}} = \frac{1}{2\pi \times 1500 \text{ Hz} \times 0.01 \text{ µF}} = 10.6 \text{ k}\Omega$$

Choose values of 110 kΩ for R_{1T} and 11 kΩ for R_{2T}.

Finally, select a value for resistor R_3. This resistor isolates the bass and treble controls to prevent one control from loading the other. We will arbitrarily choose a value of 10 kΩ.

The volume/tone control circuit is designed.

The LM1035 Dual Tone/Volume/Balance Control

The basic controls we have discussed are widely used and, for a monaural system, are entirely satisfactory. In a stereo system, however, two complete sets of tone/volume controls are necessary, resulting in a relatively complicated circuit. Furthermore, in a stereo system, the tone/volume controls for both channels must be varied simultaneously. This requires the use of ganged potentiometers and matched components, making the tone control system relatively costly. Integrated circuit technology has produced an alternative: the LM1035 tone/volume/balance control.

The data sheets for the LM1035 tone/volume/balance control are shown in Figure 12-26. The pinout of the chip is shown on the first page of the data sheets. The LM1035 operates from any supply between 8 and 18 V, provides control for both channels simultaneously, and only requires the connection of several external capacitors. The volume, bass, and treble, for both channels, and the balance are all controlled by DC voltages applied to four control pins (pin 12 for volume, pin 14 for bass, pin 4 for treble, and pin 9 for balance). This DC voltage (typically from 0 to 5.4 V) can be supplied by an external control circuit or obtained from potentiometers connected to an internal 5.4 V zener regulated supply (pin 17). A linear variation in voltage produces the proper logarithmic audio response in the volume, bass, or treble, so standard linear potentiometers can be used.

A typical stereo tone control circuit using the LM1035 is shown in Figure 12-27. The bass and treble responses for each channel are defined by capacitors C_B and C_T, respectively. For channel 1, C_B is connected from pin 6 to ground, and C_T is connected from pin 3 to ground. For channel 2, C_B is connected from pin 15 to ground, and C_T is connected from pin 18 to ground. For standard operation with approximately 15 dB boost or cut, the manufacturer recommends $C_B = 0.39$ μF and $C_T = 0.01$ μF. Increased boost and cut can be obtained by reducing C_B and proportionally increasing C_T.

Capacitors C_1 and C_2 are the input and output coupling capacitors, respectively. The input capacitor C_1 is designed such that its impedance X_{C_1} at the lowest operating frequency f_{CL} is equal to the sum of the input impedance Z_{in} and the source impedance R_S. The resulting design equation for C_1 is

$$C_1 = \frac{1}{2\pi f_{CL}(Z_{in} + R_S)} \quad (12\text{-}34)$$

The input impedance Z_{in} for each channel is typically 30 kΩ, and the source resistance R_S depends on the preamplifier or other input circuit. The output capacitor C_2 is designed such that its impedance X_{C_2} at the lowest operating frequency f_{CL} is equal to the sum of the output impedance Z_{out} and the load impedance R_L. The resulting design equation for C_2 is

$$C_2 = \frac{1}{2\pi f_{CL}(Z_{out} + R_L)} \quad (12\text{-}35)$$

Audio/Radio Circuits

National Semiconductor

LM1035 Dual DC Operated Tone/Volume/Balance Circuit

General Description

The LM1035 is a DC controlled tone (bass/treble), volume and balance circuit for stereo applications in car radio, TV and audio systems. An additional control input allows loudness compensation to be simply effected.

Four control inputs provide control of the bass, treble, balance and volume functions through application of DC voltages from a remote control system or, alternatively, from four potentiometers which may be biased from a zener regulated supply provided on the circuit.

Each tone response is defined by a single capacitor chosen to give the desired characteristic.

Features

- Wide supply voltage range, 8V to 18V
- Large volume control range, 80 dB typical
- Tone control, ± 15 dB typical
- Channel separation, 75 dB typical
- Low distortion, 0.05% typical for an input level of 1 Vrms
- High signal to noise, 80 dB typical for an input level of 1 Vrms
- Few external components required

Block and Connection Diagram

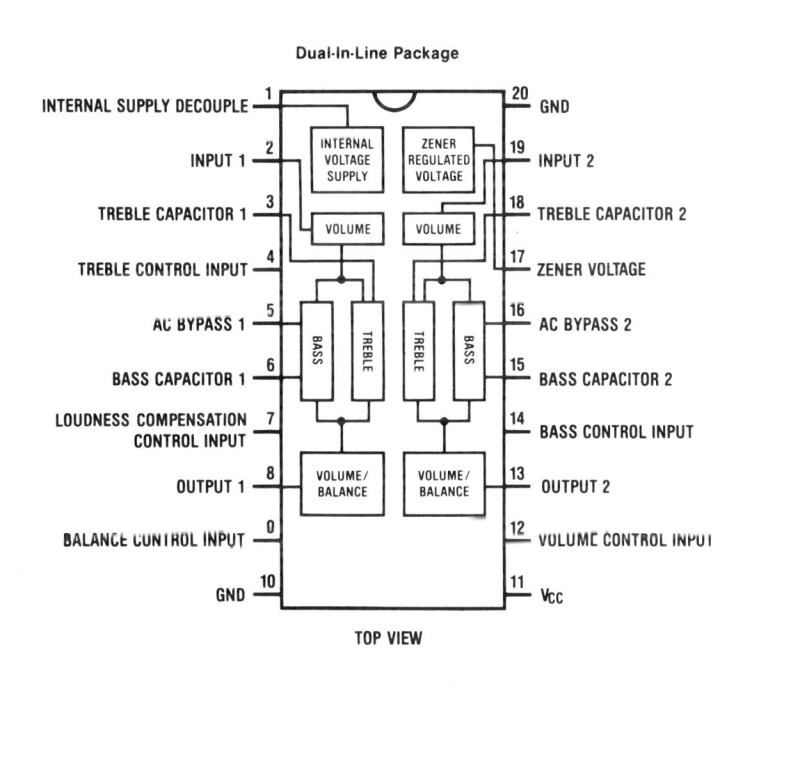

Dual-In-Line Package

Pin	Function	Pin	Function
1	INTERNAL SUPPLY DECOUPLE	20	GND
2	INPUT 1	19	INPUT 2
3	TREBLE CAPACITOR 1	18	TREBLE CAPACITOR 2
4	TREBLE CONTROL INPUT	17	ZENER VOLTAGE
5	AC BYPASS 1	16	AC BYPASS 2
6	BASS CAPACITOR 1	15	BASS CAPACITOR 2
7	LOUDNESS COMPENSATION CONTROL INPUT	14	BASS CONTROL INPUT
8	OUTPUT 1	13	OUTPUT 2
9	BALANCE CONTROL INPUT	12	VOLUME CONTROL INPUT
10	GND	11	V$_{CC}$

TOP VIEW

10-75

Figure 12-26 Data sheets for the LM1035 Tone/Volume/Balance Control. *(Reprinted with permission of National Semiconductor Corporation)*

Absolute Maximum Ratings

Supply Voltage	20V
Control Pin Voltage (Pins 4, 7, 9, 12, 14)	V_{CC}
Operating Temperature Range	0°C to +70°C
Storage Temperature Range	−65°C to +150°C
Lead Temperature (Soldering, 10 seconds)	300°C

Electrical Characteristics V_{CC} = 12V, T_A = 25°C (unless otherwise stated)

Parameter	Conditions	Min	Typ	Max	Units
Supply Voltage Range	Pin 11	8		18	V
Supply Current			35	45	mA
Zener Regulated Output Voltage	Pin 17		5.4		V
Current				5	mA
Maximum Output Voltage	Pins 8, 13; f = 1 kHz				
	V_{CC} = 8V		1.3		Vrms
	V_{CC} = 12V	2	2.5		Vrms
	V_{CC} = 18V		3.5		Vrms
Maximum Input Voltage (Note 1)	Pins 2, 19; f = 1 kHz Flat Response	2	2.5		Vrms
Input Resistance	Pins 2, 19; f = 1 kHz	20	30		kΩ
Output Resistance	Pins 8, 13; f = 1 kHz		20		Ω
Maximum Gain	V(Pin 12) = V(Pin 17); f = 1 kHz	−2	0	2	dB
Volume Control Range	f = 1 kHz	70	80		dB
Gain Tracking	f = 1 kHz				
	0 dB through −40 dB		1	3	dB
	−40 dB through −60 dB		2		dB
Balance Control Range	Pins 8, 13; f = 1 kHz		+1		dB
			−26	−20	dB
Bass Control Range (Note 2)	f = 40 Hz, C_b = 0.39 μF				
	V(Pin 14) = V(Pin 17)	12	15	18	dB
	V(Pin 14) = 0V	−12	−15	−18	dB
Treble Control Range (Note 2)	f = 16 kHz, C_t = 10 nF				
	V(Pin 4) = V(Pin 17)	12	15	18	dB
	V(Pin 4) = 0V	−12	−15	−18	dB
Total Harmonic Distortion	f = 1 kHz, V_i = 1 Vrms, Maximum Gain		0.05	0.2	%
Channel Separation	f = 1 kHz, Maximum Gain		75		dB
Signal/Noise Ratio	Unweighted 100 Hz−20 kHz				
	Maximum Gain, 0 dB = 1 Vrms		80		dB
	CCIR/ARM (Note 3)				
	Gain = 0 dB	76	80		dB
	Gain = −20 dB		64		dB
Output Noise Voltage at Minimum Gain	CCIR/ARM (Note 3)		25	35	μV
Supply Ripple Rejection	200 mVrms, 1 kHz Ripple		40		dB
Control Input Currents	Pins 4, 7, 9, 12, 14 (V = 0V)		−0.6	−2.5	μA
Frequency Response	−1 dB (Flat Response 20 Hz−16 kHz)		250		kHz

Note 1: The maximum permissible input level is dependent on tone and volume settings. See Application Notes.
Note 2: The tone control range is defined by capacitors C_b and C_t. See Application Notes.
Note 3: Measured with a CCIR filter with a 0 dB level at 2 kHz and an average responding meter.

Figure 12-26 (*Continued*)

12-3 Volume and Tone Control Circuits

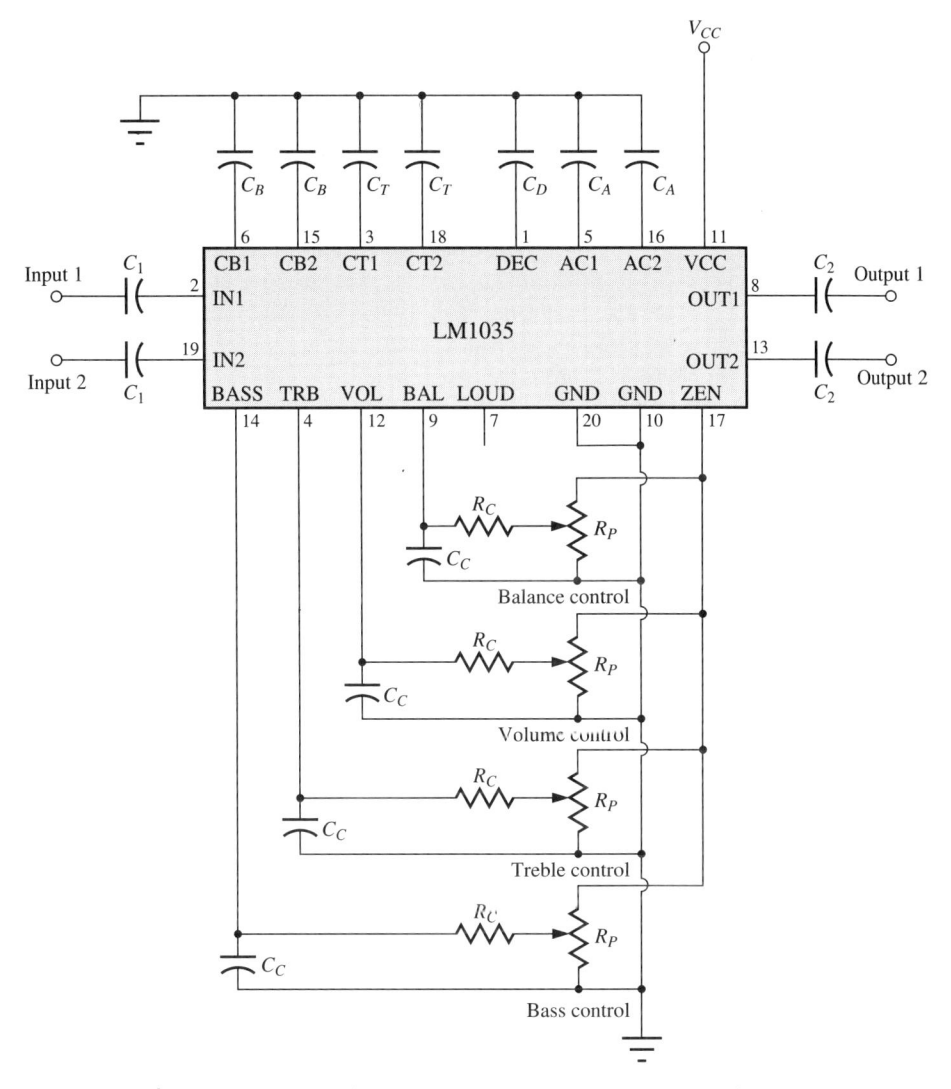

Figure 12-27 Tone/volume/balance control circuit using the LM1035.

The output impedance Z_{out} for each channel is typically 20 Ω, whereas the load resistance R_L depends on the input impedance of the power amplifier or other loading circuit.

If the internal zener-regulated voltage source is used with potentiometers to generate the control voltages for the volume, bass, treble, and balance, as shown in Figure 12-27, then the potentiometers must be chosen so as not to load the voltage source. The voltage source is capable of supplying a maximum current of 5 mA. If we allow a maximum current through each potentiometer of 1 mA, with 5.4 V across the potentiometer, the minimum allowable resistance of the potentiometer is 5.4 V/1.0 mA = 5.4 kΩ. We should choose a larger resistance

584   **12   Audio Integrated Circuits**

for all the potentiometers. There is also a restriction on the maximum resistance of the potentiometers. If the potentiometer resistance is too large, it will be loaded by the control input resistance. The input impedance of each control input is approximately 1 MΩ; hence, the control potentiometer must have a much smaller value. Typically, the control potentiometers should have values between 20 and 100 kΩ.

Figure 12-27 shows the control potentiometers decoupled from the control inputs with series resistor R_C and capacitor C_C to ground. These ensure that there is no AC signal present at the control inputs. The manufacturer recommends R_C = 47 kΩ and C_C = 0.22 μF. Three other decoupling or bypass capacitors are required: a supply decoupling capacitor C_D connected to pin 1, with a typical value of 50 μF, and two AC bypass capacitors C_A connected to pins 5 and 16, with typical values of 10 μF.

Although there are a number of external components with this circuit, there is very little to design. Most of the values are suggested by the manufacturer. The circuit is very easy to use. The four linear potentiometers provide the four controls for volume, bass, treble, and balance, respectively. Either monaural or stereo operation is possible.

EXAMPLE 12-8

Design a stereo tone/volume/balance control system using an LM1035 integrated circuit tone control placed after a preamplifier which has zero output impedance and before a power amplifier which has an input impedance of 1 MΩ. Both the bass control and the treble control should have a boost/cut of 15 dB.

Solution

The circuit is shown in Figure 12-28.

This is a simple circuit to design because most of the component values are suggested by the manufacturer. First select suitable potentiometers for all of the controls. We will select 50 kΩ linear potentiometers. Note that we use linear potentiometers instead of the audio taper potentiometers employed in the previous example because the logarithmic response is built into the LM1035.

Select control resistors R_C = 47 kΩ and control capacitors C_C = 0.22 μF according to the manufacturer's recommendation.

Select the bass capacitors C_B = 0.39 μF and the treble capacitors C_T = 0.01 μF to give 15 dB boost/cut.

Select the supply bypass capacitor C_D = 50 μF and the two AC bypass capacitors C_A = 10 μF, again according to manufacturer's recommendations.

The only components to design in the circuit are the input capacitors C_1 and the output capacitors C_2. To design input capacitor C_1, we use Equation 12-34 with Z_{in} = 30 kΩ, R_S = 0, and f_{CL} = 50 Hz:

$$C_1 = \frac{1}{2\pi f_{CL}(Z_{in} + R_S)} = \frac{1}{2\pi \times 50 \text{ Hz} \times 30 \text{ k}\Omega} = 0.106 \text{ μF}$$

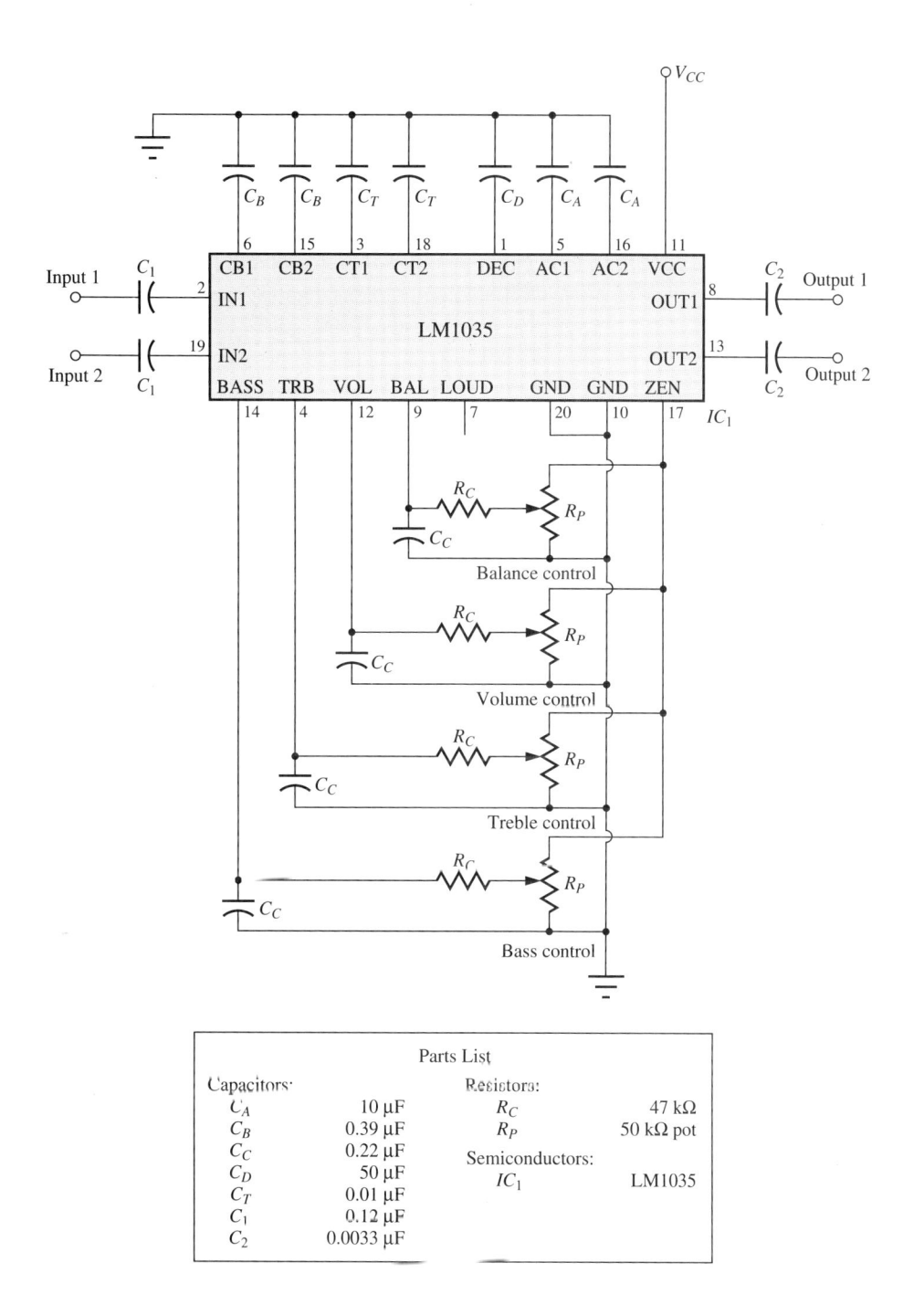

Figure 12-28

Parts List			
Capacitors:		Resistors:	
C_A	10 µF	R_C	47 kΩ
C_B	0.39 µF	R_P	50 kΩ pot
C_C	0.22 µF	Semiconductors:	
C_D	50 µF	IC_1	LM1035
C_T	0.01 µF		
C_1	0.12 µF		
C_2	0.0033 µF		

To design output capacitor C_2, we use Equation 12-35 with $Z_{out} = 20 \, \Omega$, $R_L = 1.0 \, M\Omega$, and $f_{CL} = 50 \, Hz$:

$$C_2 = \frac{1}{2\pi f_{CL}(Z_{out} + R_L)} = \frac{1}{2\pi \times 50 \text{ Hz} \times 1.0 \text{ M}\Omega} = 0.00318 \, \mu F$$

We will choose values of 0.12 μF for C_1 and 0.0033 μF for C_2.

Our tone/volume/balance control circuit design is complete.

12-4 AM and FM Radio Circuits

To conclude this chapter, we will look at three complete systems using audio integrated circuits: an AM radio, a simple FM monaural radio, and an FM stereo radio. These systems illustrate the application of some of the audio circuits introduced earlier. They also illustrate the use of integrated circuits in complete, and relatively complex, electronic systems. Although the field of radio communications is outside the scope of this book, we will introduce three communication integrated circuits as a final illustration of the range of applications of linear integrated circuits.

The LM3820 Single-Chip AM Receiver

National Semiconductor produces a single chip containing the local oscillator, mixer, intermediate frequency amplifier, and automatic gain control sections of an AM radio. All that is necessary is to connect several external transformers and capacitors, a diode detector and filter, and an audio section to make a complete AM radio.

The data sheets for the LM3820 single-chip AM radio are shown in Figure 12-29. The LM3820 can be powered with any supply from 5 to 16 V, although it is actually designed to operate from a 12 V car battery. Internal regulation is provided if the supply voltage is greater than approximately 8 V. A basic radio receiver circuit using the LM3820 is shown in Figure 12-30. There is nothing to design for this circuit if the manufacturer's recommended transformers and capacitors are used. We will describe the various parts of the circuit and briefly discuss the operation of an AM receiver.

The *radio frequency* (RF) signal transmitted by the radio station is received by an antenna. For small AM receivers, a ferrite rod antenna is generally used. The secondary of the antenna transformer is connected across pins 11 and 12 (RF input), with pin 11 (the RF bypass) connected to ground by, at least, a 1.0 μF tantalum capacitor. The output of the RF amplifier (pin 13) is connected to the power supply through a 220 Ω resistor and to the mixer input (pin 1) by a 0.01 μF capacitor. The *mixer* is a circuit that combines the RF signal with an internally produced *local oscillator* (LO) signal to produce an *intermediate frequency* (IF) signal. The connection between the local oscillator and the mixer is internal to the chip. The oscillator tank coil input (pin 2), however, must be connected to

National Semiconductor

Audio/Radio Circuits

LM3820 AM Radio System

General Description

The LM3820 is a 3-stage AM radio IC consisting of an RF amplifier, oscillator, mixer, IF amplifier, AGC detector, and zener regulator.

The device was originally designed for use in slug-tuned auto radio applications, but is also suitable for capacitor-tuned portable radios.

The LM3820 is an improved replacement for the LM1820.

Features

- Input protection diodes
- Good control on sensitivity
- Improved S/N and tweet
- Versatile building-block approach
- Gain-controlled RF stage
- Cascode IF amplifier
- Regulated supply
- Pin compatible with LM1820

Connection Diagram

Dual-In-Line Package

Pin	Signal	Pin	Signal
1	MIXER INPUT	14	MIXER OUTPUT
2	OSCILLATOR TANK	13	RF OUTPUT
3	V+	12	RF INPUT
4	MIXER BYPASS	11	RF BYPASS
5	AGC DRIVE	10	AGC CAPACITOR
6	IF OUTPUT	9	RF GROUND
7	IF INPUT	8	SUBSTRATE AND IF AMPL GROUND

TOP VIEW

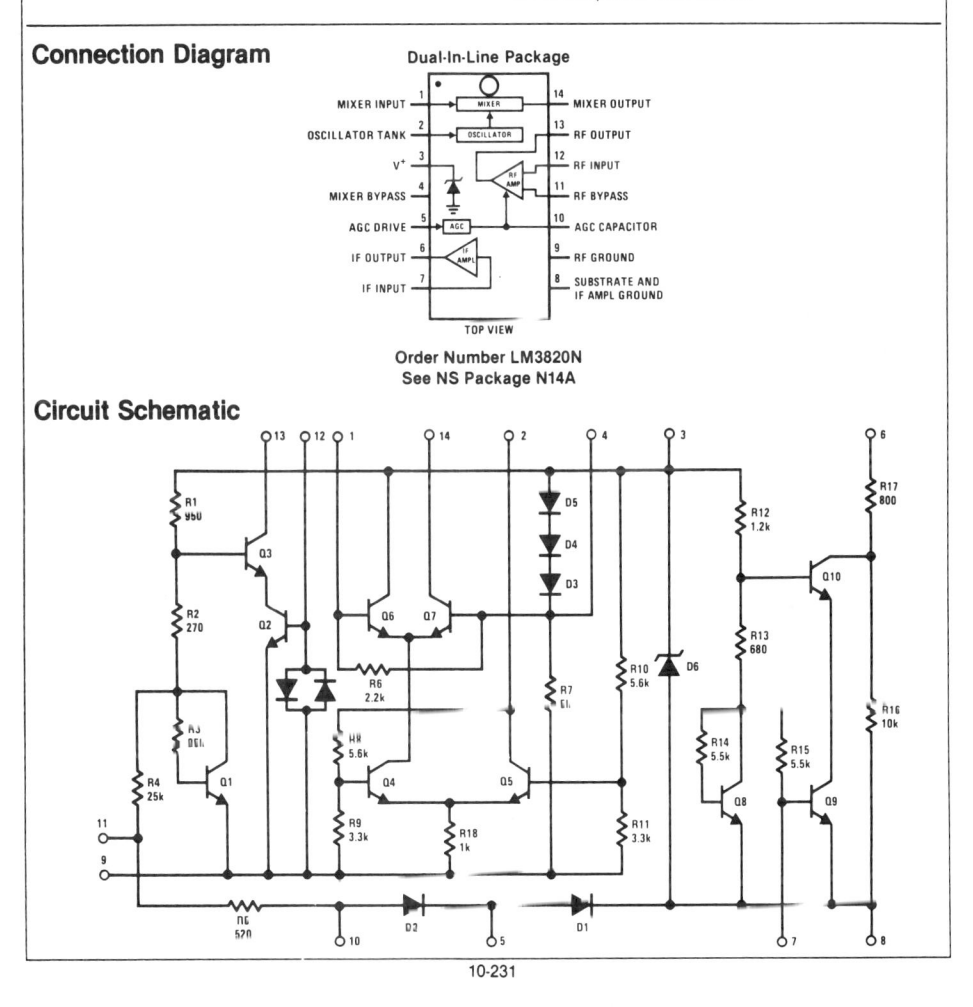

Order Number LM3820N
See NS Package N14A

Circuit Schematic

Figure 12-29 Data sheets for the LM3820 AM radio integrated circuit. *(Reprinted with permission of National Semiconductor Corporation)*

Absolute Maximum Ratings

Power Dissipation (Note 1)	700 mW	Operating Temperature Range	−25°C to 85°C
Supply Voltage	16V	Storage Temperature Range	−65°C to 150°C
Current into Supply Terminal (Pin 3)	35 mA	Lead Temperature (Soldering, 10 seconds)	300°C

Electrical Characteristics (*Figure 1*, $T_A = 25°C$, $V_S = 6V$ unless noted)

Parameter	Conditions	Min	Typ	Max	Units
Supply Current (I_S)	No RF Input	12	18	24	mA
Internal Zener Voltage (V_Z)		7.0	7.5	8.0	V
Input Sensitivity	f = 1 MHz, 30% Mod 400 Hz Measure RF Input Level for 10 mV Audio Output with Tuning Peaked	15	35	70	µV
Signal to Noise Ratio	f = 1 MHz, 30% Mod 1 kHz (S + N)/N at Audio Output with 100 µV RF Input	22	28	—	dB
Overload Distortion	f = 1 MHz, 90% Mod 1 kHz THD at Audio Output with 30 mV RF Input	—	6	10	%

Note 1: Above $T_A = 25°C$, derate based on $T_{J(MAX)} = 150°C$ and $\theta_{JA} = 180°C/W$.

Typical Applications

FIGURE 1. Capacitor-Tuned Test Fixture

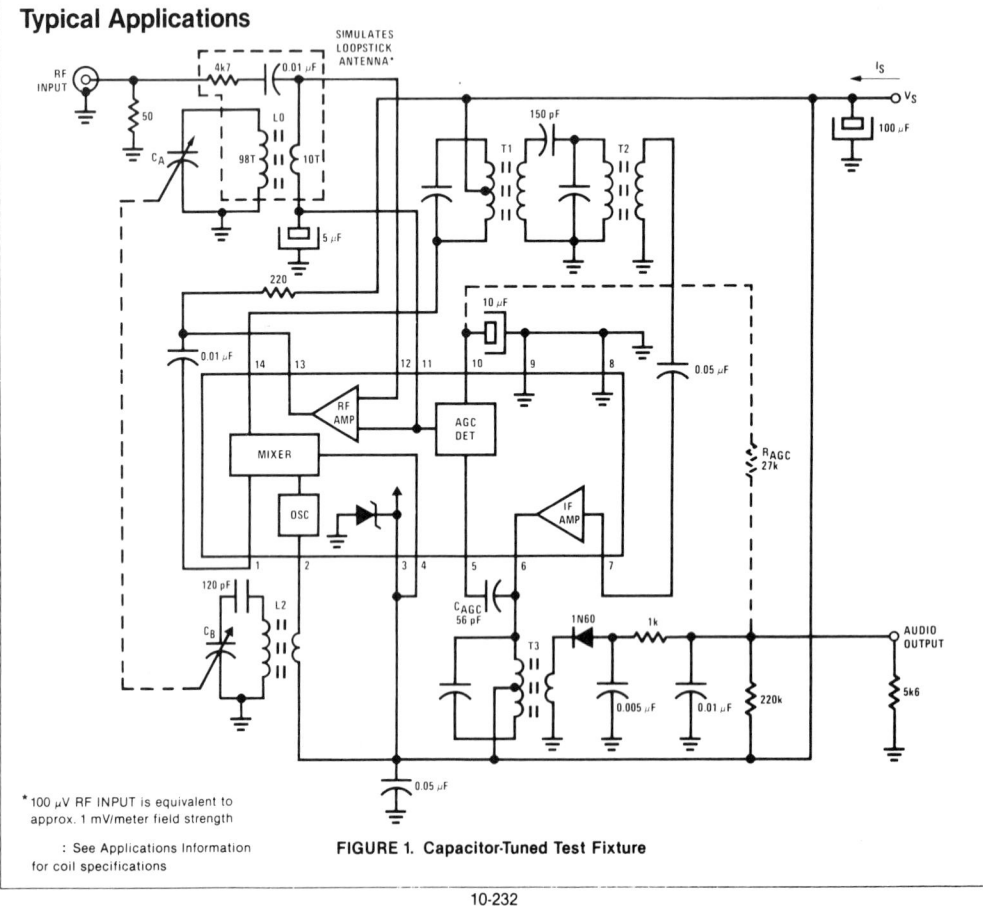

* 100 µV RF INPUT is equivalent to approx. 1 mV/meter field strength

† : See Applications Information for coil specifications

Figure 12-29 (*Continued*)

588

12-4 AM and FM Radio Circuits

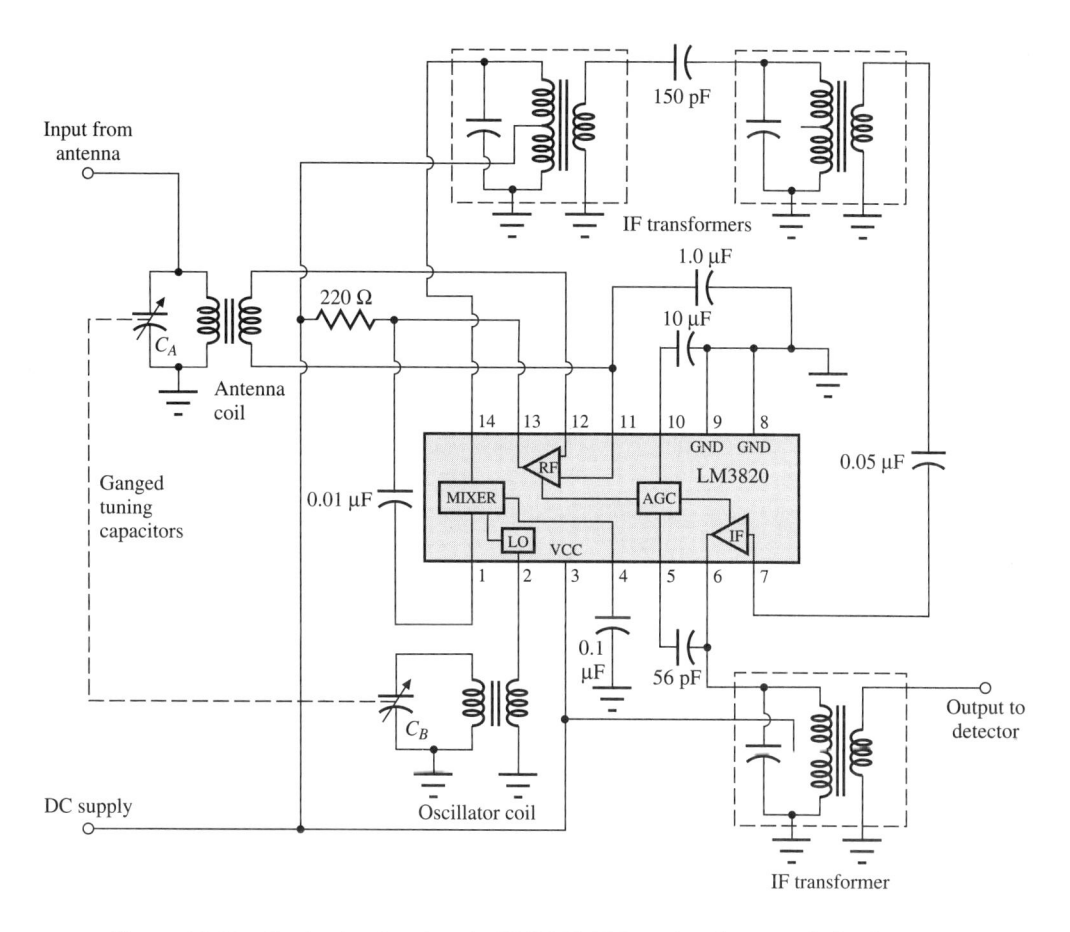

Figure 12-30 Basic circuit using the LM3820 AM receiver integrated circuit.

the secondary of the oscillator coil. The mixer bypass (pin 4) is connected to ground through a 0.1 μF capacitor.

Both the primary of the antenna coil and the primary of the oscillator coil are tuned by ganged capacitors C_A and C_B. These two capacitors are varied together to set the tuned frequency of the antenna and the local oscillator frequency, respectively. The two tuning capacitors share a common grounded frame, and each has parallel trimmer capacitors (not shown in the diagram) for fine alignment of the tuned frequency. Although the exact values of the tuning capacitors will depend on the chosen antenna and oscillator coil, typically they will have values of approximately 20 to 100 pF.

The mixer output consists of signals at two different frequencies: a frequency equal to the sum of the RF and local oscillator frequencies, and a frequency equal to the difference between the same two frequencies. The intermediate frequency is the difference frequency and must be separated from the sum frequency. This is done by interstage IF transformers tuned to the desired intermediate frequency. The mixer output from pin 14 goes to the primary of the first interstage IF transformer. The center tap of this transformer is connected to the DC supply. The secondary of this transformer is connected by a 150 pF capacitor to a second identical IF transformer, and thence to the IF amplifier input (pin 7) through a 0.05 µF capacitor.

The intermediate frequency is constant (generally 455 kHz), and most of the amplification of the radio signal is provided at this frequency. The output of the IF amplifier is connected to the primary of the third IF transformer. Like the first IF transformer, the center tap is connected to the DC supply. The output of this transformer goes directly to an external detector circuit. All three transformers are standard AM-IF transformers and are each tuned to 455 kHz. The tuning capacitors shown within the dashed squares representing the metal RF shields of the IF transformers are built into each transformer module.

The output of the IF amplifier is also connected to the input of the automatic gain control (pin 5) through a 56 pF capacitor C_{AGC}. The **automatic gain control (AGC)** varies the amount of amplification to ensure that strong nearby stations and weak distant stations have approximately the same output amplitudes. The value of capacitor C_{AGC} is important because it determines the response time of the AGC circuit. The AGC circuit also requires a 10 µF tantalum (or 50 µF aluminum) capacitor connected from pin 10 to ground. Pins 8 and 9 should also be grounded. The DC supply voltage is connected to pin 3.

A designer experienced in AM radio circuits may vary the configuration of the IF transformers, oscillator coil, and antenna, and specify other device values for optimal performance. Such design is beyond the scope of this text, and the circuit in Figure 12-30 is adequate for most purposes.

A Complete AM Radio

The LM3820 AM RF amp/local oscillator/mixer/IF can be combined with a simple diode detector and half of an LM1877 2 W amplifier to form a complete AM radio. The circuit is shown in Figure 12-31. The radio operates from 12 V provided by a car battery or a line-operated supply. If a line-operated supply is used, it can be designed according to the procedures outlined in Chapter 8.

A detector circuit must be connected between the output of the third IF transformer and the input to the audio amplifier. The detector separates the audio signal from the IF carrier. The detector circuit in Figure 12-31 is a standard type used on most AM radios. Diode D_1 is any signal diode; a 1N914 is suitable. Filter capacitors C_{F_1} and C_{F_2} are 0.005 µF and 0.01 µF, respectively, and filter resistors R_{F1} and R_{F2} are 1.0 kΩ and 220 kΩ, respectively. The output of the filter is connected directly to the input of the audio amplifier, and the signal level is large enough to drive the LM1877 power amplifier directly.

The LM1877 power amplifier circuit is similar to that designed in Example 12-5, except that the supply voltage is 12 V instead of 18 V and the gain is set at

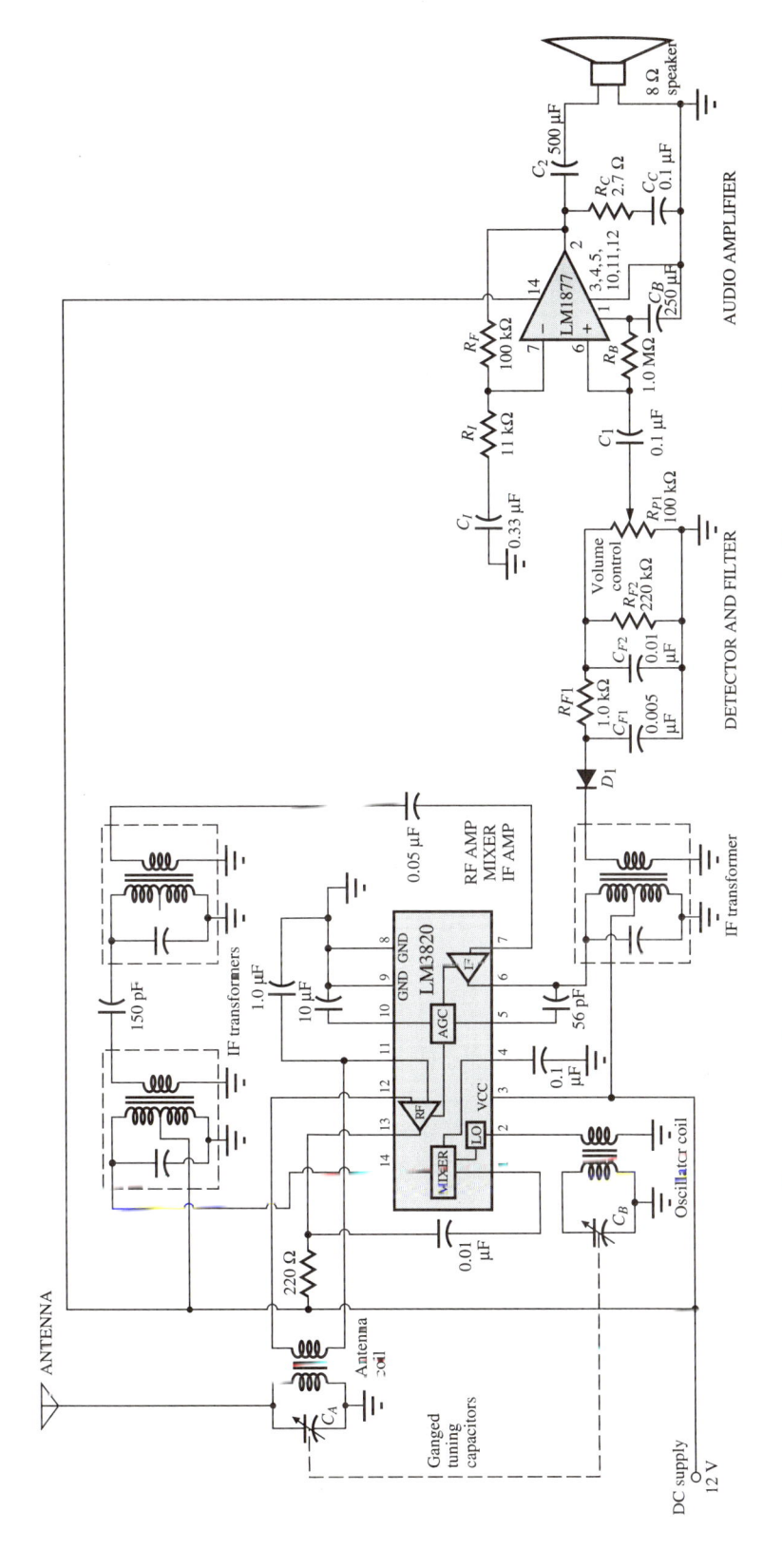

Figure 12-31 Complete AM radio using a LM3820 with a LM1877 power amplifier.

10. The design is not shown here, but the reader should be able to verify the circuit values by the procedures developed earlier. Note the volume control potentiometer between the detector and amplifier input. Although no tone control is provided in Figure 12-31, a control circuit such as developed in Example 12-7 could easily be added between the detector and power amplifier. Since the tone control would attenuate the signal, the gain of the audio amplifier would have to be increased.

Two adjustments must be made in the circuit for proper operation. First, the three IF transformers all have to be aligned to a frequency of 455 kHz. This is accomplished by injecting a 10 to 100 µV amplitude-modulated 455 kHz test signal at the input to the first IF transformer and then adjusting the tuning slug for each transformer in turn until the output from the detector is a maximum. The other adjustment is to align the main tuning dial. This is accomplished by selecting a strong station and adjusting the trimmer capacitors on the main tuning capacitors connected to the antenna coil and local oscillator coil until the station is received at the correct point on the tuning dial. For more information on the alignment of AM radio receivers, the reader is referred to any standard text on electronic communications.

The LM3189 FM IF/Limiter/Detector

As for the AM radio, integrated circuits are available to simplify the design of FM radios. One example is the LM3189 FM IF amplifier/limiter/detector integrated circuit produced by National Semiconductor. Data sheets for the LM3189 are shown in Figure 12-32. Unlike the LM3820 AM radio circuit, however, the LM3189 is not a complete FM receiver. It is simply the IF section of an FM receiver and requires an additional tuner section, as we will discuss shortly. As the IF section for an FM receiver, however, the LM3189 only requires two external coils, a muting control potentiometer, several resistors, and several capacitors. Although the circuitry is relatively complex, there is nothing to design— all of the circuit component values, and the circuit configuration is provided by the manufacturer.

The typical FM IF/limiter/detector circuit using the LM3189 is shown in Figure 12-33. The IF output of the tuner section is connected to the IF input of the LM3189 (pin 1). Bias current must be supplied to the IF amplifier. The bias current is supplied from pin 3 to pin 1 through resistor R_B. Pin 3 is bypassed to ground through C_B. Resistor R_B and capacitor C_B prevent any signal from reaching the bias supply, and the manufacturer suggests values of 51 Ω for R_B and 0.01 µF for C_B. These values are not critical. Pin 2 is used for decoupling the unused input of the differential amplifier stage, and the manufacturer recommends a 0.01 µF capacitor to pin 2. Again, the capacitor value is not critical.

Pins 5 and 12 form the basis for muting control. Although muting is optional (leave these pins disconnected), it is a useful feature for any FM receiver. **Muting** (or **squelching**) eliminates the audio signal when the receiver is tuned between channels or when it is tuned to a station that is too weak for the limiter to remove the amplitude-modulated noise. This produces a quieting effect as the receiver is

National Semiconductor

Audio/Radio Circuits

LM3189 FM IF System

General Description

The LM3189N is a monolithic integrated circuit that provides all the functions of a comprehensive FM IF system. The block diagram of the LM3189N includes a three stage FM IF amplifier/limiter configuration with level detectors for each stage, a doubly balanced quadrature FM detector and an audio amplifier that features the optional use of a muting (squelch) circuit.

The advanced circuit design of the IF system includes desirable deluxe features such as programmable delayed AGC for the RF tuner, an AFC drive circuit, and an output signal to drive a tuning meter and/or provide stereo switching logic. In addition, internal power supply regulators maintain a nearly constant current drain over the voltage supply range of +8.5V to +16V.

The LM3189N is ideal for high fidelity operation. Distortion in an LM3189N FM IF system is primarily a function of the phase linearity characteristic of the outboard detector coil.

The LM3189N has all the features of the LM3039N plus additions.

The LM3189N utilizes the 16-lead dual-in-line plastic package and can operate over the ambient temperature range of −40°C to +85°C.

Features

- Exceptional limiting sensitivity: 12 μV typ at −3 dB point
- Low distortion: 0.1% typ (with double-tuned coil)
- Single-coil tuning capability
- Improved (S + N)/N ratio
- Externally programmable recovered audio level
- Provides specific signal for control of inter-channel muting (squelch)
- Provides specific signal for direct drive of a tuning meter
- On channel step for search control
- Provides programmable AGC voltage for RF amplifier
- Provides a specific circuit for flexible audio output
- Internal supply voltage regulators
- Externally programmable ON channel step width, and deviation at which muting occurs

Block Diagram

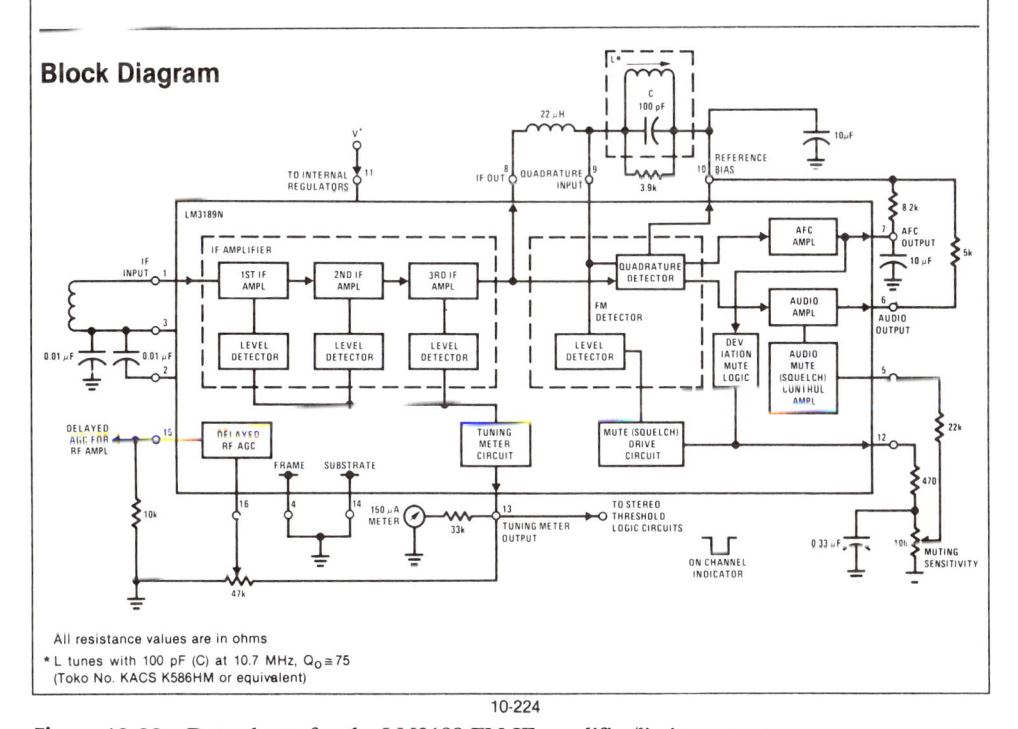

All resistance values are in ohms
* L tunes with 100 pF (C) at 10.7 MHz, $Q_O \cong 75$
  (Toko No. KACS K586HM or equivalent)

10-224

Figure 12-32 Data sheets for the LM3189 FM IF amplifier/limiter. *(Reprinted with permission of National Semiconductor Corporation)*

Absolute Maximum Ratings

Supply Voltage Between Pin 11 and Pins 4, 14	16V
DC Current Out of Pin 12	5 mA
DC Current Out of Pin 13	5 mA
DC Current Out of Pin 15	2 mA
Power Dissipation (Note 2)	1390 mW
Operating Temperature Range	−40°C to +85°C
Storage Temperature Range	−65°C to +150°C
Lead Temperature (Soldering, 10 seconds)	300°C

Electrical Characteristics $T_A = 25°C$, $V^+ = 12V$

Symbol	Parameter	Conditions (see single-tuned test circuit)	Min	Typ	Max	Units
STATIC (DC) CHARACTERISTICS						
I_{11}	Quiescent Circuit Current		20	31	44	mA
	DC Voltages:	No Signal Input, Non Muted				
V1	Terminal 1 (IF Input)		1.2	2.0	2.4	V
V2	Terminal 2 (AC Return to Input)		1.2	2.0	2.4	V
V3	Terminal 3 (DC Bias to Input)		1.2	2.0	2.4	V
V15	Terminal 15 (RF AGC)		7.5	9.5	11	V
V10	Terminal 10 (DC Reference)		5	5.75	6	V
DYNAMIC CHARACTERISTICS						
$V_i(lim)$	Input Limiting Voltage (−3 dB Point)			12	25	µV
AMR	AM Rejection (Term. 6)	$V_{IN} = 0.1V$	45	55		dB
$V_O(AF)$	Recovered AF Voltage (Term. 6)	AM Mod. = 30%, $f_o = 10.7$ MHz, $f_{mod} = 400$ Hz, Deviation ±75 kHz	325	500	650	mV
THD	Total Harmonic Distortion (Note 1) Single Tuned (Term. 6) Double Tuned (Term. 6)	$V_{IN} = 0.1V$		0.5 0.1	1	% %
S+N/N	Signal plus Noise to Noise Ratio (Term. 6)		65	80		dB
f_{DEV}	Deviation Mute Frequency	$f_{mod} = 0$		±40		kHz
V16	RF AGC Threshold			1.25		V
V12	On Channel Step	$V_{IN} = 0.1V$	$f_{DEV} < ±40$ kHz $f_{DEV} > ±40$ kHz	0 5.6		V

Note 1: THD characteristics are essentially a function of the phase characteristics of the network connected between terminals 8, 9, and 10.

Note 2: For operation in ambient temperatures above 25°C, the device must be derated based on a 150°C maximum junction temperature and a thermal resistance of 90°C/W junction to ambient.

Connection Diagram

Dual-In-Line Package

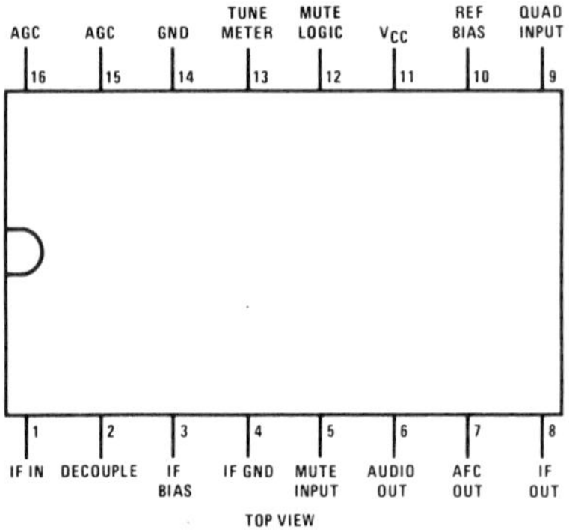

Pin 16: AGC, Pin 15: AGC, Pin 14: GND, Pin 13: TUNE METER, Pin 12: MUTE LOGIC, Pin 11: V_{CC}, Pin 10: REF BIAS, Pin 9: QUAD INPUT

Pin 1: IF IN, Pin 2: DECOUPLE, Pin 3: IF BIAS, Pin 4: IF GND, Pin 5: MUTE INPUT, Pin 6: AUDIO OUT, Pin 7: AFC OUT, Pin 8: IF OUT

TOP VIEW

Order Number LM3189N
See NS Package N16E

10-225

Figure 12-32 (*Continued*)

12-4 AM and FM Radio Circuits

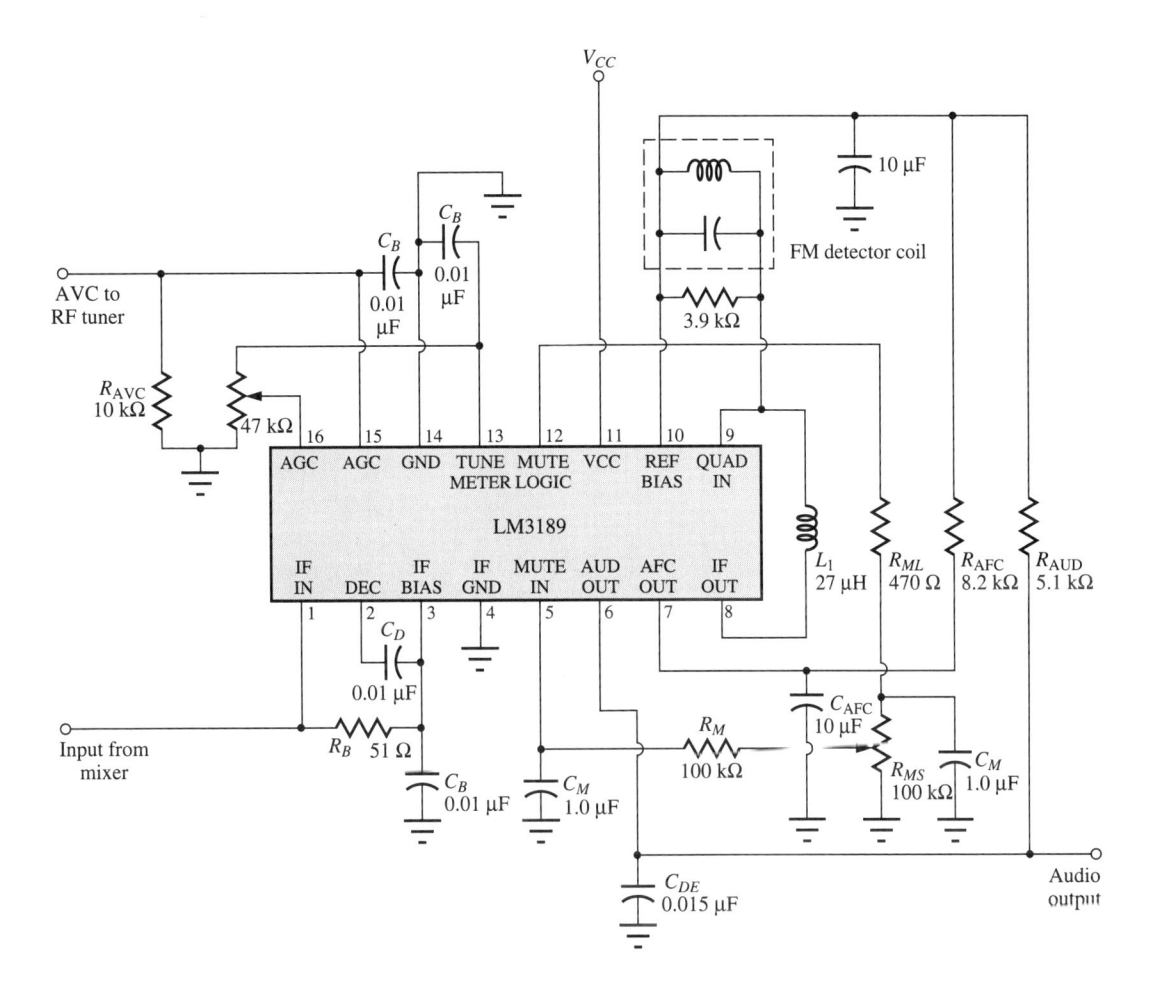

Figure 12-33 FM IF/limiter/detector circuit using the LM3189 integrated circuit.

tuned, with audio output occurring only when the receiver is tuned to a station within the detection range of the receiver. Without muting, there would be a background of noise whenever the tuner is not tuned to a specific station. At high output volume levels, this noise can be quite unpleasant. The muting level is set by feeding a portion of the mute logic output (pin 12) to the mute input (pin 5). The recommended muting circuit shown in Figure 12-33 uses potentiometer R_{MS} connected between pin 12 and ground as a voltage divider to send a portion of the output from pin 12 to pin 5. Pin 5 is isolated from the potentiometer by resistor R_M. Potentiometer R_{MS} and resistor R_M can have any value between 10 and 100 kΩ. Both pin 5 and the potentiometer should be bypassed to ground by capacitors C_M having values of at least 1.0 μH. Finally, the potentiometer should be isolated from the mute logic output (pin 12) by a 470 Ω resistor.

A quadrature detector is included in the LM3189. The IF output (from pin 8) is connected to the detector (pin 9) through a 27 μH coil, and the detector is biased from pin 10 through the detector coil. A quadrature detector is a relatively complex circuit, and a detailed analysis is beyond the scope of this text. The reader is referred to a text on electronic communications for a more detailed discussion of the theory and operation of quadrature detectors. The detector portion of the circuit in Figure 12-33 is that recommended by the manufacturer.

Automatic frequency control (AFC) is used in FM receivers to prevent the local oscillator frequency from drifting. The control signal is derived from the quadrature detector and is connected to the AFC through pin 7.

The automatic gain control circuitry is connected between pins 13, 15, and 16 and is used to control the gain of the RF amplifier module directly from pin 15. The circuit shown allows for delayed AGC to the RF amplifier and for output to a tuning meter.

A Simple FM Radio

The LM3189 FM IF/detector can be combined with an LM1877 2 W amplifier, a tone/volume control, and a modular FM tuner to form a complete FM radio. The circuit is shown in Figure 12-34. The radio operates from 12 V provided by a car battery or a line-operated supply. If a line-operated supply is used, it can be designed according to the procedures outlined in Chapter 8.

A Waller 32SN2F1-30 modular tuner, produced by the Waller Corporation, Crystal Lake, Illinois, is used as an RF amplifier, local oscillator, and mixer. This module is *not an integrated circuit*, but is constructed of discrete components and used to provide a very low noise level. This module receives the FM signals from the antenna, tunes the desired station, provides a stage of RF amplification, and converts the signal to the 10.7 MHz IF frequency used in FM receivers. Other modular tuner units would also be satisfactory. To further simplify the circuit, two 10.7 MHz ceramic filters (Toko CFS-30AE-10 or equivalent) are used to filter the IF signal between the tuner module and the IF amplifier input.

The IF section is exactly as described in the previous section. The audio output is taken from pin 6 and goes to the LM1877 power amplifier through a tone/volume control network. Capacitor C_{DE} is used to supply deemphasis to the audio output and must have a value of 0.015 μF. High frequencies in the audio signal are boosted before transmission by a process called **preemphasis** in an effort to reduce noise pickup. When the audio signal is demodulated, this preemphasis must be removed by a low-pass filter in a process called **deemphasis.**

The tone/volume control is the same network designed in Example 12-7. The LM1877 power amplifier circuit is the same as that used for the AM radio, except that the gain is set at 20 instead of 10 to compensate for the attenuation introduced by the tone control.

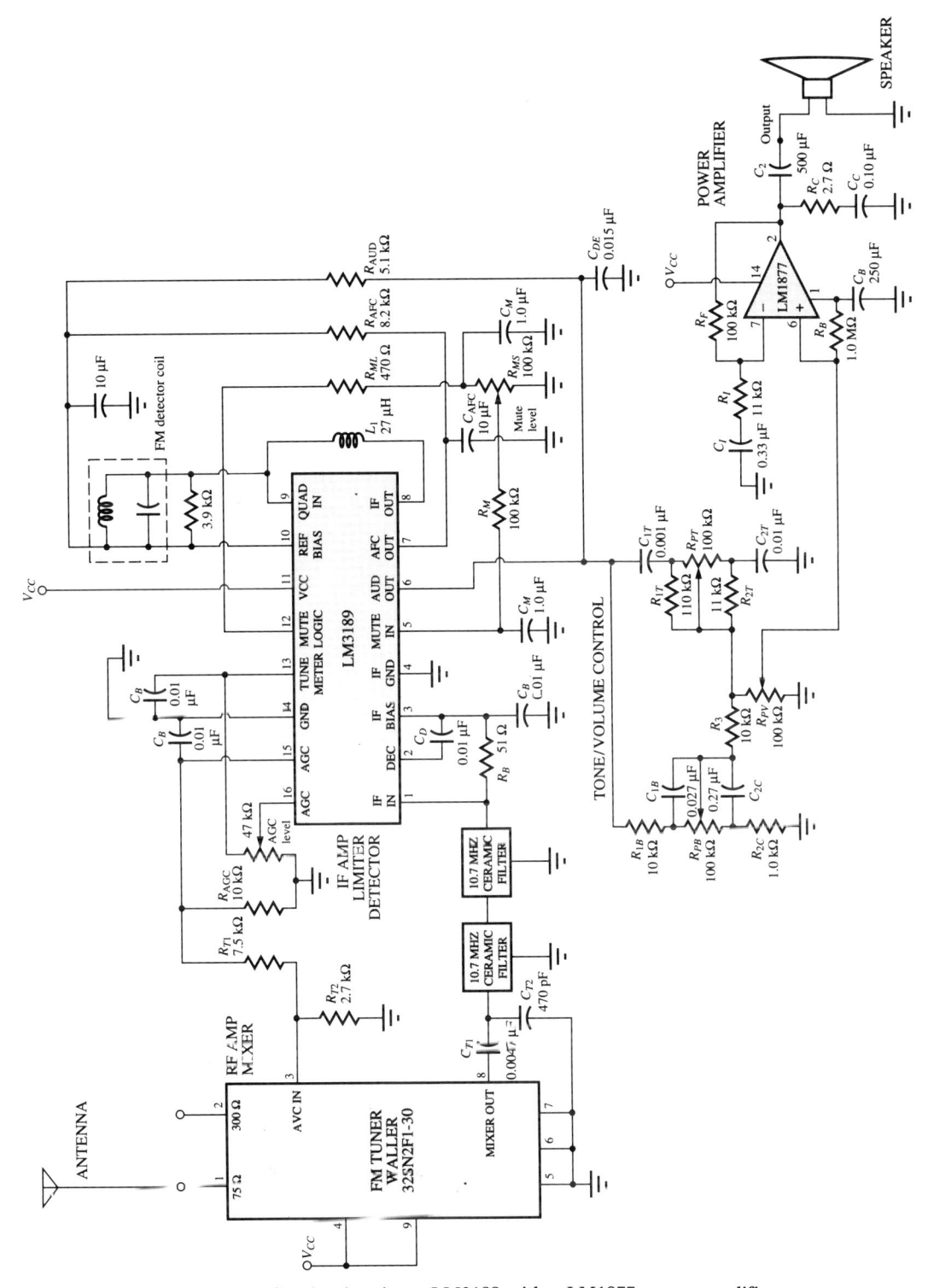

Figure 12-34 FM radio circuit using a LM3189 with a LM1877 power amplifier.

The LM1800 Stereo Demodulator

Because of the superior noise qualities of FM and the wide audio bandwidth available for FM transmission, FM stations frequently transmit stereo audio signals and many FM receivers include the provision for stereo reception.

The process of decoding stereo information from the received FM signal is relatively complex, and the reader is referred to any text on communications for a description of stereo encoding/decoding. Fortunately, integrated circuits are available which will perform the complete decoding and will provide fully decoded left and right outputs. One of these is the LM1800 phase-locked loop FM stereo decoder produced by National Semiconductor.

The data sheets for this integrated circuit are shown in Figure 12-35. As with the LM3189 IF chip, the circuit required by this integrated circuit is predesigned by the manufacturer, and it is only necessary to add a few external components. No transformers or tuning coils are required. All of the tuning is accomplished by a single potentiometer. The LM1800 can operate from any power supply between 10 and 18 V.

The LM1800 is essentially a decoding circuit that takes the signal from the FM detector (such as the audio output signal from the LM3189) and translates it to left and right channel signals. The LM1800 includes a phase-locked loop, similar to those discussed in Chapter 11, for generating the 38 kHz stereo carrier signal and for extracting the $L - R$ signal. The LM1800 contains an electronic switch controlled by the 19 kHz pilot signal contained in a stereo transmission. This switch connects the output of the phase-locked loop $L - R$ signal to the decoder only if the 19 kHz signal is sufficiently strong to indicate good stereo reception. Otherwise, the $L + R$ signal (or for monaural transmissions, the single received signal) goes to both outputs.

The typical circuit using the LM1800 as a stereo demodulator is shown in Figure 12-36. The composite input from the FM detector is applied to pin 1 through a 10 μF coupling capacitor. Deemphasis networks consisting of a 3.9 kΩ resistor in parallel with a 0.022 μF capacitor are connected between pin 3 (left load and deemphasis) and ground and between pin 6 (right load and deemphasis) and ground. The listed values must be used to ensure proper deemphasis. The decoded left and right channel outputs are taken from pins 3 and 4, respectively.

The audio amp output (pin 2) must be connected to the phase detector input (pin 12) by a 0.033 μF capacitor and to ground through a 0.0025 μF capacitor. A 0.33 μF capacitor is connected across the threshold filter pins (pins 9 and 10). A 0.22 μF capacitor is connected across the loop filter pins (pins 13 and 14) in parallel with a 0.47 μF capacitor in series with a 3.3 kΩ resistor. A 5.0 kΩ potentiometer in series with a 22 kΩ resistor is connected between the VCO control (pin 15) and ground and is used for adjusting the oscillator frequency. A 390 pF bypass capacitor is also connected from pin 15 to ground. Pin 11, the pilot monitor, is used for calibration; the VCO control is used to adjust the frequency measured at this pin to 19 kHz.

Provision is made for a stereo indicator lamp, which lights when a stereo signal is received. This lamp is connected from the power supply to pin 7, the lamp driver switch, which is capable of sinking up to 100 mA.

National Semiconductor

Audio/Radio Circuits

LM1800 Phase-Locked Loop FM Stereo Demodulator

General Description

The LM1800 is a second generation integrated FM stereo demodulator using phase locked loop techniques to regenerate the 38 kHz subcarrier. The numerous features integrated on the die make possible a system delivering high fidelity sound while still meeting the cost requirements of inexpensive stereo receivers. More information available in AN-81.

Features

- Automatic stereo/monaural switching
- 45 dB power supply rejection
- No coils, all tuning performed with single potentiometer
- Wide operating supply voltage range
- Excellent channel separation
- Emitter follower output buffers

Connection Diagram

Order Number LM1800N
See NS Package N16A

Typical Application

Typical Performance Characteristics

10-111

Figure 12-35 Data sheets for the LM1800 stereo demodulator. *(Reprinted with permission of National Semiconductor Corporation)*

Absolute Maximum Ratings

Supply Voltage	18V
Power Dissipation (Note 3)	715 mW
Operating Temperature Range	0°C to +70°C
Operating Supply Voltage Range	+10V to +18V
Storage Temperature Range	−65°C to +150°C
Lead Temperature (Soldering, 10 seconds)	300°C

Electrical Characteristics (Note 1)

PARAMETER	CONDITIONS	MIN	TYP	MAX	UNITS
Supply Current	Lamp "off"		21	30	mA
Lamp Driver Saturation	100 mA Lamp Current		1.3	1.8	V
Lamp Driver Leakage			1.0		nA
Pilot Level for Lamp "ON"	Pin 11 Adjusted to 19.00 kHz		15	20	mVrms
Pilot Level for Lamp "OFF"	Pin 11 Adjusted to 19.00 kHz	3.0	7.0		mVrms
Stereo Lamp Hysteresis		3.0	6.0		dB
Stereo Channel Separation	100 Hz (Note 2)		40		dB
	1000 Hz (Note 2)	30	45		dB
	10000 Hz (Note 2)		45		dB
Monaural Channel Unbalance	200 mVrms, 1000 Hz Input		0.3	1.5	dB
Monaural Voltage Gain	200 mVrms, 400 Hz Input	140	200	260	mVrms
Total Harmonic Distortion	500 mVrms, 1000 Hz Input		0.4	1.0	%
Total Harmonic Distortion	500 mVrms, 1000 Hz Input, 1800A Only		0.1	0.3	%
Capture Range	25 mVrms of Pilot	±2.0		±6.0	% of f_o
Supply Ripple Rejection	200 mVrms of 200 Hz Ripple	35	45		dB
Dynamic Input Resistance		20	45		kΩ
Dynamic Output Resistance		900	1300	2000	Ω
SCA Rejection	(Note 4)		70		dB
Ultrasonic Freq. Rejection	Combined 19 and 38 kHz, Ref. to Output		33		dB

Note 1: T_A = 25°C and V^+ = 12V unless otherwise specified.
Note 2: The stereo input signal is made by summing 123 mVrms LEFT or RIGHT modulated signal with 25 mVrms of 19 kHz pilot tone, measuring all voltages with an average responding meter calibrated in rms. The resulting waveform is about 800 mVp-p.
Note 3: For operation in ambient temperatures above 25°C, the device must be derated based on a 150°C maximum junction temperature and a thermal resistance of 175°C/W junction to ambient.
Note 4: Measured with a stereo composite signal consistency of 80% stereo, 10% pilot and 10% SCA as defined in the FCC Rules on Broadcasting.
Note 5: VCO "OFF" curve represents the distortion attainable using good 19 kHz and 38 kHz filters.

Typical Performance Characteristics (Continued)

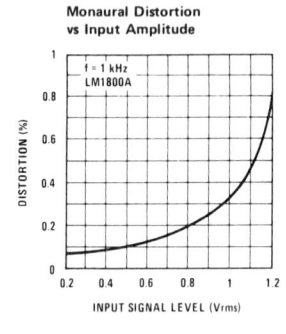

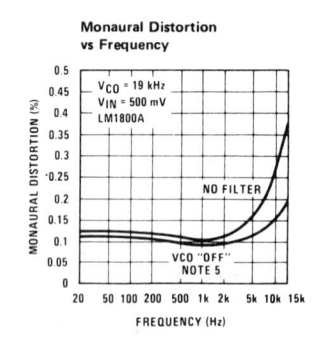

Figure 12-35 (*Continued*)

12-4 AM and FM Radio Circuits 601

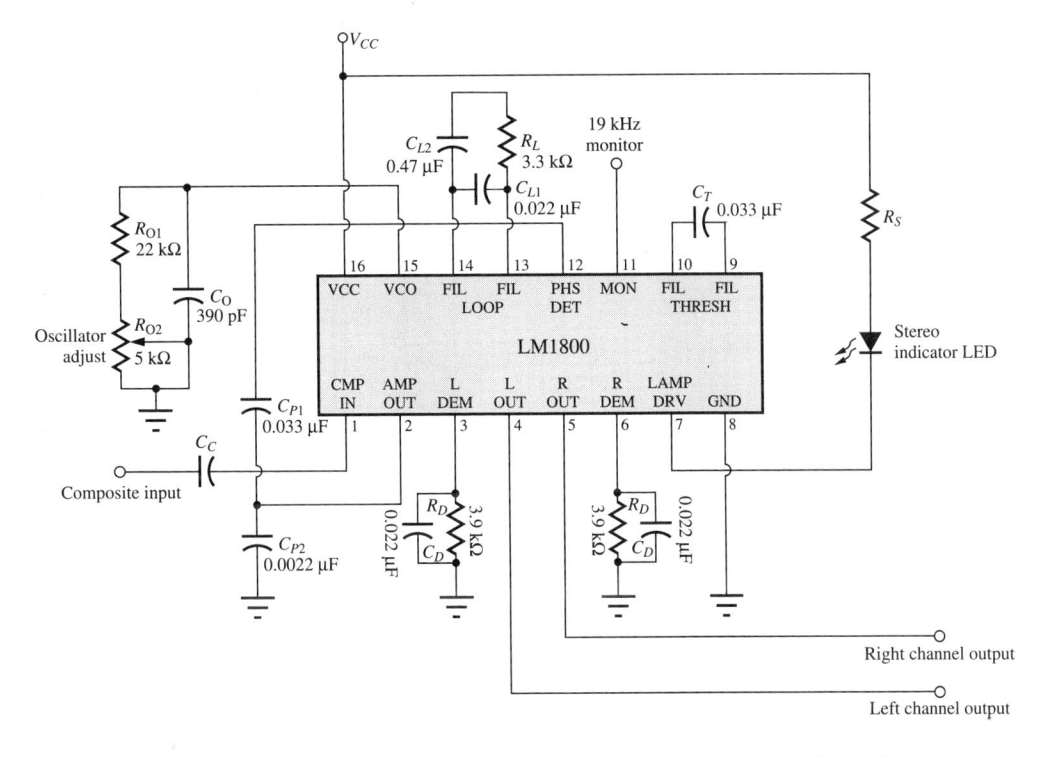

Figure 12-36 Stereo demodulator circuit using the LM1800 integrated circuit.

A Stereo FM Radio

The LM3189 FM IF/detector and the LM1800 stereo demodulator integrated circuits can be combined with an LM1035 tone/volume/balance control, an LM1877 dual 2 W amplifier, and a modular FM tuner unit to form a complete FM radio. The circuit for this complete radio is shown in Figure 12-37. The radio operates from a 12 V battery or DC power supply. As with the monaural FM receiver, a Waller 32SN2F1-30 modular tuner/RF amp/mixer/local oscillator is used for the receiver front end, coupled to the LM3189 IF section through two 10.7 MHz ceramic filters.

Note that it is necessary to include a phase adjustment circuit consisting of a 0.0022 μF capacitor in parallel with a 25 kΩ potentiometer between the detector and the stereo demodulator input. This control corrects for excess phase shift through the IF amplifier to yield maximum channel separation. This is one of only two adjustments that must be made in the circuit. The other is the adjustment of the VCO control on the LM1800 to give a pilot frequency of exactly 19 kHz. All other critical controls are built into the FM tuner and are adjusted at the factory or eliminated by the use of fixed-frequency ceramic filters.

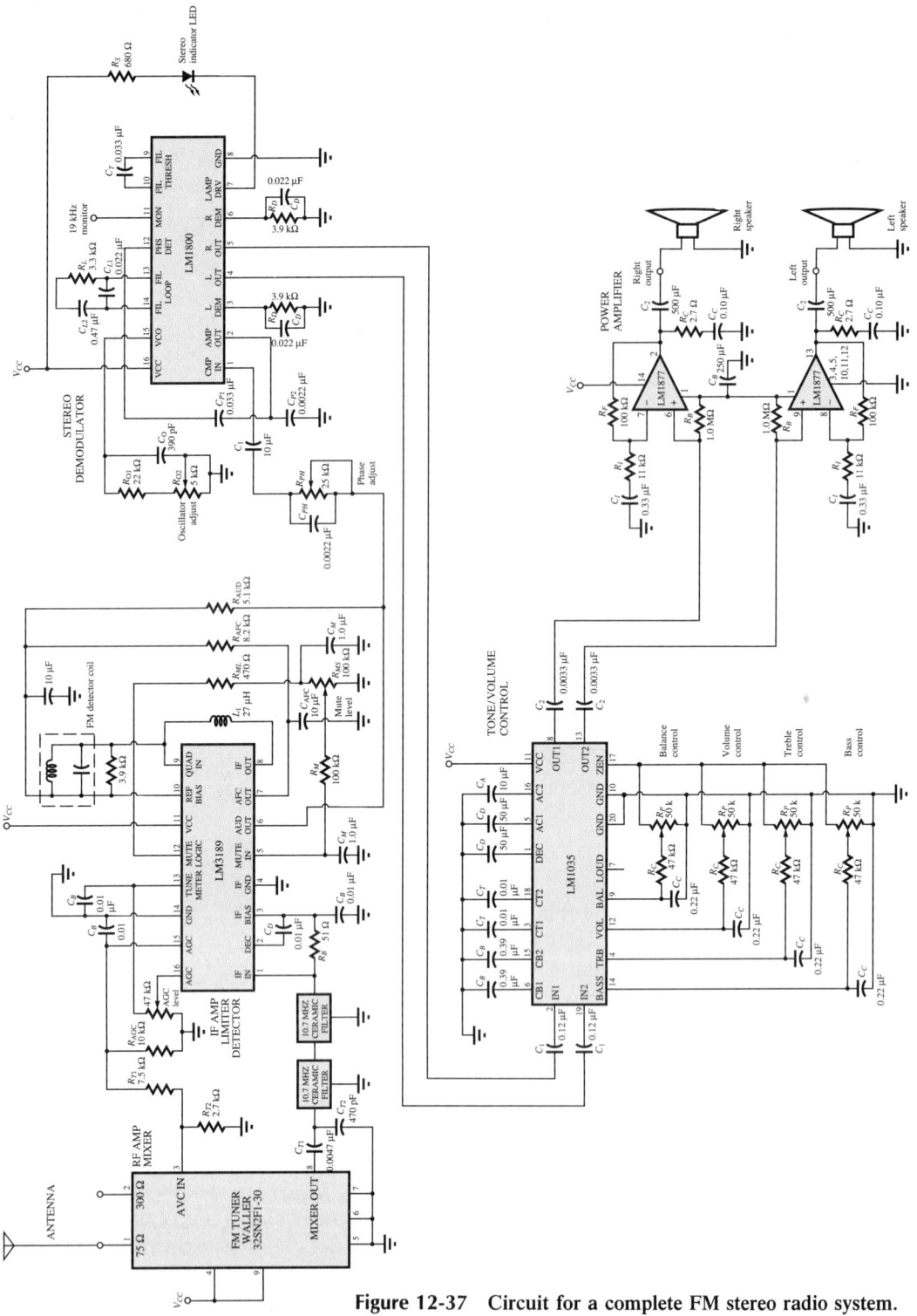

Figure 12-37 Circuit for a complete FM stereo radio system.

Summary

This chapter on audio circuits presented some integrated circuits used in the home entertainment field. These are generally classed as consumer integrated circuits. First we introduced the LM387 dual preamplifier and showed how it could be used as a microphone preamplifier, as a tape record amplifier to give an NAB response characteristic, as a tape playback amplifier to convert the NAB response back to a flat response, and as a mixer. All of these functions could have been performed with standard op-amps, but the LM387 is designed specifically for this type of task. It has superior noise characteristics, bandwidth, and can operate from a single supply.

Next, the LM1877 dual 2 W power amplifier was introduced. Although 2 W is a relatively low power amplifier, it is adequate for many applications. If necessary, the power can be increased to 4 W by connecting the two amplifiers in a bridge configuration or to 20 W or more by using external driver transistors.

Volume and tone control circuits were introduced, including the LM1035 tone and volume control. Although simple R-C tone controls are satisfactory for monaural systems, integrated circuit controls such as the LM1035 are definitely superior for stereo systems. Not only is the circuit simpler, but also the control is far simpler. With only three linear potentiometers the LM1035 can control the volume, bass response, and treble response of both channels simultaneously.

We concluded with a discussion of AM and FM radio circuits. These served to illustrate practical applications of the audio circuits and demonstrated additional applications of integrated circuits in the communication field. Most importantly, however, we saw how entire systems, including an AM radio receiver, a monaural FM radio receiver, and a stereo FM radio receiver could be constructed by using integrated circuit technology.

Review Questions

12-1 The LM387 audio preamplifier

1. What is the frequency range of a typical audio amplifier?
2. What are the two types of audio amplifiers?
3. List three applications for preamplifiers.
4. Why is it preferable to use a special preamplifier integrated circuit for audio circuits rather than a preamplifier constructed from op-amps?
5. How does the LM387 audio preamplifier differ from a standard op-amp such as the TL081?
6. Explain the function of resistor R_B in LM387 preamplifier circuits.
7. What are the circuit differences between the LM387 used as a noninverting amplifier and the LM387 used as an inverting amplifier?
8. Differentiate between the two types of microphones. Which is most widely used?
9. What is the disadvantage of high-impedance microphones?

10. What sort of hookup is usually used for a low-impedance microphone? Why?
11. Sketch the typical frequency response of a tape recording head.
12. How is the nonlinear response of the recording head corrected?
13. What do the letters NAB stand for?
14. What is NAB equalization?
15. Sketch an NAB equalization curve indicating the critical frequencies.
16. What additions must be made to a simple recording preamplifier circuit to produce NAB equalization?
17. What additions must be made to a simple playback preamplifier to produce a linear response from an NAB equalized tape recording?
18. What is an audio mixer?
19. Give two specific applications of audio mixing.
20. Why should mixers always use the inverting input of an integrated circuit amplifier?
21. How is the gain for each input determined in an audio mixer?

12-2 The LM1877 audio power amplifier

22. What characteristics are required in an audio power amplifier?
23. What is the typical load for an audio power amplifier?
24. In what power ranges are integrated circuit power amplifiers found? Which are most readily available?
25. What are the power supply requirements of an LM1877 audio power amplifier?
26. Although the maximum rated power of the LM1877 is listed as 2 W/channel, what must be done to actually achieve this power level?
27. How is the output distortion of an LM1877 affected by power level?
28. Although the LM1877 can be operated from supply voltages as large as 26 V, supply voltages greater than about 18 V produce power levels greater than 2 W. Why is this usually not a problem?
29. What is the minimum gain of the LM1877?
30. What is the usual configuration of amplifier circuits using the LM1877 (inverting or noninverting)?
31. What is the difference between the LM1877 circuit and a standard op-amp circuit for the same configuration?
32. What compensating network is necessary with the LM1877?
33. What is the purpose of the bias pin in the LM1877? Why is it necessary?
34. The LM1877 can be operated from a single supply or from a dual supply. What is the difference in the circuit?
35. Why does the bias pin not need to be AC-coupled to ground when the LM1877 is used with a dual supply?
36. How can the LM1877 be configured to provide an output of 4 W?
37. Describe how the two amplifiers are connected so that there is a 180° phase shift between their outputs.
38. How can the output power level in a 2 W amplifier such as the LM1877 be increased to 20 W or more?
39. What limits the maximum power output in this case?
40. Why is a different compensating network required when the LM1877 is used to drive an external amplifier?
41. From where must the feedback be taken when the LM1877 is used to drive an external amplifier?

12-3 Volume and tone control circuits

42. What controls are generally found on a stereo amplifier?
43. What is the low frequency portion of an audio signal called?
44. What is the high frequency portion called?
45. Describe the response of the human ear to sound.
46. What special types of potentiometers are necessary in volume and tone controls to enable the ear to hear a linear response?
47. Describe a simple resistor/capacitor network for producing a bass tone control. Explain how the circuit works.
48. Describe a simple resistor/capacitor network for producing a treble tone control. Explain how the circuit works.
49. What is the midband attenuation produced by a bass tone control? by a treble tone control?
50. Where, in an audio amplifier circuit, are the tone and volume controls usually placed?
51. What change in amplifier design is necessary when tone controls are incorporated in an amplifier circuit?
52. What is the major limitation of simple resistor/capacitor tone controls?
53. How is tone/volume/balance controlled by the LM1035 integrated circuit tone control?
54. How many separate controls are required for stereo operation of the LM1035?
55. What is the purpose of the zener output on the LM1035?
56. What is the advantage of linear control such as is provided by the LM1035?
57. What is the bass and treble boost/cut provided using the recommended components with the LM1035?
58. How can the bass and treble boost/cut be increased?
59. What are the only components that must be designed with the LM1035?
60. What restriction is placed on the potentiometers used to control the tone/volume/balance with the LM1035?

12-4 AM and FM radio circuits

61. What circuits are required to convert the incoming RF signal to an IF signal?
62. What is the standard AM IF frequency?
63. What is automatic gain control? Why is it necessary?
64. What is the process of separating the audio signal from the IF signal called?
65. What is the most common method of AM detection in AM radio receivers?
66. What functions in an AM radio can be performed with the LM3820 integrated circuit?
67. What specific components need to be designed when using an LM3820 AM radio chip?
68. What circuits must be added to the LM3820 to make a complete AM radio?
69. What adjustments must be made to an AM radio so that it will perform correctly?
70. What functions are performed by the FM tuner?
71. Why is the FM tuner section usually made of discrete components?
72. What is the IF frequency of an FM receiver?
73. What FM radio functions are performed by an LM3189 integrated circuit?
74. What additional components or devices are required to make a complete FM radio?
75. What additional components or devices are required to make a stereo radio receiver?
76. What is the purpose of muting or squelching in an FM receiver?
77. What type of detector is used in an FM radio?
78. What is preemphasis? Why is it used?
79. What circuitry is required in an FM receiver to remove the preemphasis of the received signal?

80. What is the function of an LM1800 integrated circuit?
81. What components must be designed in a circuit using the LM1800?
82. List the integrated circuits that can be used to make a complete FM stereo radio.
83. Why is a phase adjustment circuit necessary?
84. What two adjustments must be made for an FM stereo receiver to function properly?

Design and Analysis Problems

12-1 The LM387 audio preamplifier

1. Design a simple noninverting preamplifier to give an output level of 5.0 V to a 40 kΩ load for a 12 mV source with an impedance of 600 Ω. Use an LM387 dual preamplifier with a 20 V power supply.
2. Design a differential microphone preamplifier to amplify the output of a microphone with an impedance of 200 Ω and a level of 5.0 mV to drive a power amplifier requiring an input level of 3.0 V into 500 kΩ. Use an LM387 dual preamplifier with an 18 V power supply.
3. Design a tape record preamplifier using an LM387 dual audio preamplifier. Assume the standard recording head response shown in Figure 12-6, and design the preamplifier to give a standard NAB response. The input signal has a maximum amplitude of 10 mV RMS from a source with an impedance of 600 Ω, and the tape head requires a drive current of 20 μA RMS. The recorder response should be from 100 to 12,000 Hz. The recorder has a 12 V power supply.
4. Design an NAB tape playback amplifier using an LM387 dual audio preamplifier. The playback head response has a sensitivity of 800 μV, and the preamplifier drives a power amplifier requiring 0.35 V RMS input into 150 kΩ. The recorder response should be from 60 to 12,000 Hz. The recorder has a 12 V power supply.
5. Design a two-channel mixer using an LM387 dual audio preamplifier. One input has an impedance of 200 Ω and a signal level of 8.0 mV. The other input has an impedance of 500 Ω and a signal level of 35 mV. The mixer must deliver a signal level of 12 V to a 100 kΩ load. Use a 24 V power supply.
6. Design a three-channel mixer using an LM387 dual audio preamplifier. The first input has an impedance of 3300 Ω and a signal level of 75 mV. The second input has an impedance of 200 Ω and a signal level of 15 mV. The third input has an impedance of 1200 Ω and a signal level of 30 mV. The mixer must deliver a signal level of 8 V to a 30 kΩ load. Use a 15 V power supply.

12-2 The LM1877 audio power amplifier

7. Design a simple audio power amplifier using half of an LM1877 dual-power audio amplifier to drive an 8 Ω speaker. Assume that the peak input voltage from the preamplifier is 0.5 V, and design the power amplifier for maximum output voltage level. Use a 24 V power supply.
8. Design a stereo audio power amplifier using an LM1877 dual-power amplifier. Each output will drive an 8 Ω speaker. Assume that the RMS input voltage from the preamplifier is 0.35 V, and design for maximum output voltage level. Use a dual ±12 V power supply.

9. Design a 4 W amplifier using both halves of an LM1877 dual-power audio amplifier in a bridge configuration to drive an 8 Ω speaker. The input signal from the preamplifier has a maximum amplitude of 0.50 V peak. Use an 18 V supply.

12-3 Volume and tone control circuits

10. Design a single channel tone/volume control system using standard control circuits to be placed after a preamplifier with an effectively zero output impedance and before a power amplifier with an input impedance of 10 MΩ. The bass control should become effective starting at 450 Hz and have a boost/cut of 25 dB, the treble control should become effective starting at 1800 Hz with a boost/cut of 15 dB.
11. Design a stereo tone/volume/balance control system using an LM1035 integrated circuit tone control. The source impedance is 1200 Ω, and the load impedance is 300 kΩ. Both the bass control and the treble control should have a boost/cut of 15 dB.

Reference Data A

NOTE: This appendix presents reference data for components and devices such as resistors, capacitors, and transistors used in the various circuits introduced in the text.

OUTLINE

A-1 Standard Resistor Values
A-2 Typical Capacitor Values
A-3 Power Transformer Data
A-4 Zener Diode Data
A-5 Transistor Data

A-1 Standard Resistor Values

Resistors are produced by a number of manufacturers, including Ohmite, IRC, Allen-Bradley, Phillips, Dale, and TWR, and are readily available from any electronics distributor.

Carbon Composition Resistors

Carbon composition resistors are the most widely used general purpose resistors, manufactured in the standard values shown in Table A-1 with $\pm 5\%$ tolerance and available in power ratings of $\frac{1}{8}$ W, $\frac{1}{4}$ W, $\frac{1}{2}$ W, 1 W, and 2 W. Carbon film resistors are available in the same standard values.

Precision Resistors

Precision resistors are generally wirewound with a standard tolerance of 1% and are available in the values shown in Table A-2. Only one decade is shown. For other decade values, multiply by 10, 100, 1 k, 10 k, 100 k, or 1 MΩ.

Power Resistors

Power resistors refer to resistors designed to handle in excess of 2 W of power. They are usually wirewound and do not adhere to the sequence for low-power resistors. Typical power ratings are 3 W, 5 W, 10 W, 12 W, 20 W, 25 W, and up,

Table A-1 Standard low-power resistor values

1.0 Ω	10 Ω	100 Ω	1.0 kΩ	10 kΩ	100 kΩ	1.0 MΩ	10 MΩ
	11 Ω	110 Ω	1.1 kΩ	11 kΩ	110 kΩ		
1.2 Ω	12 Ω	120 Ω	1.2 kΩ	12 kΩ	120 kΩ	1.2 MΩ	12 MΩ
	13 Ω	130 Ω	1.3 kΩ	13 kΩ	130 kΩ		
1.5 Ω	15 Ω	150 Ω	1.5 kΩ	15 kΩ	150 kΩ	1.5 MΩ	15 MΩ
1.6 Ω	16 Ω	160 Ω	1.6 kΩ	16 kΩ	160 kΩ		
1.8 Ω	18 Ω	180 Ω	1.8 kΩ	18 kΩ	180 kΩ	1.8 MΩ	18 MΩ
2.0 Ω	20 Ω	200 Ω	2.0 kΩ	20 kΩ	200 kΩ		
2.2 Ω	22 Ω	220 Ω	2.2 kΩ	22 kΩ	220 kΩ	2.2 MΩ	22 MΩ
2.4 Ω	24 Ω	240 Ω	2.4 kΩ	24 kΩ	240 kΩ		
2.7 Ω	27 Ω	270 Ω	2.7 kΩ	27 kΩ	270 kΩ	2.7 MΩ	
3.0 Ω	30 Ω	300 Ω	3.0 kΩ	30 kΩ	300 kΩ		
3.3 Ω	33 Ω	330 Ω	3.3 kΩ	33 kΩ	330 kΩ	3.3 MΩ	
3.6 Ω	36 Ω	360 Ω	3.6 kΩ	36 kΩ	360 kΩ		
3.9 Ω	39 Ω	390 Ω	3.9 kΩ	39 kΩ	390 kΩ	3.9 MΩ	
4.3 Ω	43 Ω	430 Ω	4.3 kΩ	43 kΩ	430 kΩ		
4.7 Ω	47 Ω	470 Ω	4.7 kΩ	47 kΩ	470 kΩ	4.7 MΩ	
5.1 Ω	51 Ω	510 Ω	5.1 kΩ	51 kΩ	510 kΩ		
5.6 Ω	56 Ω	560 Ω	5.6 kΩ	56 kΩ	560 kΩ	5.6 MΩ	
6.2 Ω	62 Ω	620 Ω	6.2 kΩ	62 kΩ	620 kΩ		
6.8 Ω	68 Ω	680 Ω	6.8 kΩ	68 kΩ	680 kΩ	6.8 MΩ	
7.5 Ω	75 Ω	750 Ω	7.5 kΩ	75 kΩ	750 kΩ		
8.2 Ω	82 Ω	820 Ω	8.2 kΩ	82 kΩ	820 kΩ	8.2 MΩ	
9.1 Ω	91 Ω	910 Ω	9.1 kΩ	91 kΩ	910 kΩ		

Table A-2 Precision resistor values for one decade

1.00	1.47	2.15	3.16	4.64	6.81
1.02	1.50	2.21	3.24	4.75	6.98
1.05	1.54	2.26	3.32	4.87	7.15
1.07	1.58	2.32	3.40	4.99	7.32
1.10	1.62	2.37	3.48	5.11	7.50
1.13	1.65	2.43	3.57	5.23	7.68
1.15	1.69	2.49	3.65	5.36	7.87
1.18	1.74	2.55	3.74	5.49	8.06
1.21	1.78	2.61	3.83	5.62	8.25
1.24	1.82	2.67	3.92	5.76	8.45
1.27	1.87	2.74	4.02	5.90	8.66
1.30	1.96	2.87	4.22	6.19	9.09
1.37	2.00	2.94	4.32	6.34	9.31
1.40	2.05	3.01	4.42	6.49	9.53
1.43	2.10	3.09	4.53	6.65	9.76

but vary with manufacturer. Available resistor values also vary with manufacturer. Table A-3 lists a typical sequence of available values.

Table A-3 Power resistor values

0.10 Ω	1.0 Ω	10 Ω	100 Ω	1.0 kΩ	10 kΩ	100 kΩ
	1.2 Ω	12 Ω	120 Ω	1.2 kΩ	12 kΩ	
0.15 Ω	1.5 Ω	15 Ω	150 Ω	1.5 kΩ	15 kΩ	150 kΩ
0.20 Ω	2.0 Ω	20 Ω	200 Ω	2.0 kΩ	20 kΩ	200 kΩ
0.25 Ω	2.5 Ω	25 Ω	250 Ω	2.5 kΩ	25 kΩ	
0.30 Ω	3.0 Ω	30 Ω	300 Ω	3.0 kΩ	30 kΩ	
	4.0 Ω	40 Ω	400 Ω	4.0 kΩ		
0.50 Ω	5.0 Ω	50 Ω	500 Ω	5.0 kΩ	50 kΩ	
	7.5 Ω	75 Ω	750 Ω	7.5 kΩ	75 kΩ	

Potentiometers

Potentiometers or variable resistors are available in a variety of sizes and power ratings. There are standard control potentiometers, multiturn potentiometers, and trim pots. The values available differ between manufacturers. Table A-4 lists a typical sequence of available values.

Table A-4 Potentiometer values

10 Ω	100 Ω	1.0 kΩ	10 kΩ	100 kΩ	1.0 MΩ
20 Ω	200 Ω	2.0 kΩ	20 kΩ	200 kΩ	2.0 MΩ
		2.5 kΩ	25 kΩ	250 kΩ	
50 Ω	500 Ω	5.0 kΩ	50 kΩ	500 kΩ	5.0 MΩ

A-2 Typical Capacitor Values

Capacitors are produced by a number of manufacturers, including Centralab, ITT, Mallory, Matsuo, Micronics, Phillips, and United Chemi-Con, and are readily available from any electronics distributor. There are a number of different types of capacitors, each type having different characteristics and recommended for different applications. The listing in this appendix is not intended to be comprehensive, but only to present a working set of representative types.

Mica and Ceramic Capacitors

Mica and ceramic capacitors are generally small-value capacitors intended for high-frequency applications. Most manufacturers produce them in 10% tolerance with typical voltage ratings of 63 V, 100 V, 300 V, 500 V, 600 V, and 1000 V. Capacitor values are not standardized like resistor values and vary between manufacturers. The range of ceramic capacitor values listed in Table A-5 is typical. Mica capacitors are sometimes found with values up to 1 μF.

Table A-5 Standard mica and ceramic capacitor values (pF)

	10	100	1000	10,000
	12	120	1200	
	15	150	1500	
1.8	18	180	1800	
2.2	22	220	2200	22,000
2.7	27	270	2700	
3.3	33	330	3300	
3.9	39	390	3900	
4.7	47	470	4700	
5.6	56	560		
6.8	68	680		
8.2	82	820		

Mylar and PVC Film Capacitors

Mylar and PVC film capacitors are medium-value capacitors ranging from 0.001 µF to over 1 µF. They are constructed from thin plastic sheets coated with aluminum, wound into a compact cylinder, and encased in epoxy. They possess some intrinsic inductance, which limits their high-frequency performance, but are useful for filtering, coupling, and timing applications. Typical voltage ratings are 50 V, 100 V, 200 V, 400 V, 600 V, and 1000 V. Table A-6 lists values of PVC capacitors produced by several manufacturers.

Table A-6 Standard mylar and PVC capacitor values (µF)

0.001	0.01	0.1	1.0
0.0012	0.012	0.12	
0.0015	0.015	0.15	1.5
0.0018	0.018	0.18	
0.0022	0.022	0.22	
0.0027	0.027	0.27	
0.0033	0.033	0.33	
0.0039	0.039	0.39	
0.0047	0.047	0.47	
0.0056	0.056	0.56	
0.0068	0.068	0.68	
0.0082	0.082	0.82	

Tantalum Capacitors

Tantalum capacitors are very low-inductance, medium- to high-value, electrolytic capacitors constructed from a slug of powdered tantalum immersed in an electrolyte. They are used for moderately high-frequency applications, particularly as bypass capacitors in voltage regulator circuits. They are available in values from a fraction of a microfarad to several hundred microfarads and in voltage ratings of 6.3 V, 10 V, 16 V, 20 V, 25 V, 35 V, and 50 V. Table A-7 lists a typical range of values produced by several manufacturers.

Table A-7 Standard tantalum electrolytic capacitor values (μF)

0.10	1.0	10	100
0.15	1.5	15	150
0.22	2.2	22	220
0.33	3.3	33	330
0.47	4.7	47	
0.68	6.8	68	

Aluminum Electrolytic Capacitors

Most large-value capacitors are aluminum electrolytics, constructed from thin sheets of aluminum wound in tight coils and immersed in an electrolyte. Values range from a few microfarads to almost a farad. Because of their construction, they have considerable inductance, and high-frequency performance is limited. If both large capacitance and high-frequency performance are required, a tantalum or mica disk capacitor is connected in parallel with the aluminum electrolytic. Aluminum electrolytic capacitors are available in voltage ratings of 10 V, 16 V, 25 V, 40 V, 63 V, and 100 V, with typical values as listed in Table A-8.

Table A-8 Standard aluminum electrolytic capacitor values (μF)

1.0	10	100	1000	10,000
1.5	15	150	1500	15,000
2.2	22	220	2200	22,000
3.3	33	330	3300	33,000
4.7	47	470	4700	47,000
6.8	68	680	6800	68,000

A-3 Power Transformer Data

Power transformers are manufactured by a number of companies, including Hammond, Signal, SPC Technology, Stancor, and Triad. They are readily available from most electronics distributors. For the power supplies described in this text, low-voltage rectifier transformers are used, with current ratings from 0.25 A to

5 A and center-tapped output voltages from 6 to 30 V. A selection of transformers, taken from the Hammond catalog, covering current ranges from 0.5 A to 5.0 A and voltage ranges from 5.0 V to 30.0 V, are listed in Table A-9. Hammond catalog numbers have been included.

Table A-9 Data on selected center-tapped power transformers

Full Load Voltage (V)	No Load Voltage (V)	Current (A)	Catalog Number	Full Load Voltage (V)	No Load Voltage (V)	Current (A)	Catalog Number
5.0	5.83	0.50	166G5	16.0	17.6	1.0	166J16
5.0	6.17	1.0	166J5	16.0	17.0	2.2	166L16
5.0	5.40	3.0	167M5	16.0	17.4	3.0	166M16
5.0	5.43	6.0	167Q5	16.0	17.4	5.0	167P16
6.3	7.80	0.6	166G6	18.0	19.6	1.5	166K18
6.3	7.25	1.2	166K6	18.0	19.4	3.0	165M18
6.3	7.14	2.0	166L6	18.0	18.7	5.0	165P18
6.3	6.90	4.0	166N6	20.0	22.0	0.50	166G20
6.3	6.70	6.0	165Q6	20.0	22.0	1.0	166J20
10.0	11.0	0.50	166G10	20.0	21.4	2.0	166L20
10.0	11.0	1.0	166J10	20.0	20.8	3.0	167M20
10.0	10.8	2.0	166L10	20.0	21.0	5.0	167P20
10.0	10.5	3.0	166M10	25.0	28.0	0.50	166G25
10.0	10.7	4.0	165N10	25.0	27.5	1.0	166J25
10.0	10.7	5.0	165P10	25.0	26.6	3.0	165M25
12.6	14.5	0.50	166G12	25.0	26.6	4.0	167N25
12.6	14.3	1.0	166J12	25.0	27.2	5.0	165P25
12.6	13.8	2.0	166L12	28.0	32.0	0.50	166G28
12.6	13.4	4.0	166N12	28.0	30.8	1.0	166J28
14.0	16.1	0.50	166G14	28.0	29.8	2.0	165L28
14.0	16.0	1.0	166J14	28.0	29.6	3.0	167M28
14.0	15.3	2.0	166L14	30.0	34.2	0.50	166G30
14.0	15.1	4.0	167N14	30.0	31.6	1.5	167K30
16.0	18.0	0.50	166G16	30.0	32.0	3.0	166M30

A-4 Zener Diode Data

Zener diodes are used for voltage reference applications in a number of examples in this book. There are a number of different families of zener diodes. The most popular of these is the 1N4728–1N4764 series, produced by most major manufacturers. Data sheets are shown in Figure A-1 (reproduced with the permission of Motorola, Inc.).

MOTOROLA SEMICONDUCTORS

P.O. BOX 20912 • PHOENIX, ARIZONA 85036

Designers Data Sheet

1N4728, A thru 1N4764, A

1.0 WATT ZENER REGULATOR DIODES
3.3–100 VOLTS

ONE WATT HERMETICALLY SEALED GLASS SILICON ZENER DIODES

- Complete Voltage Range — 2.4 to 100 Volts
- DO-41 Package — Smaller than Conventional DO-7 Package
- Double Slug Type Construction
- Metallurgically Bonded Construction
- Nitride Passivated Die

Designer's Data for "Worst Case" Conditions

The Designers Data sheets permit the design of most circuits entirely from the information presented. Limit curves — representing boundaries on device characteristics — are given to facilitate "worst case" design.

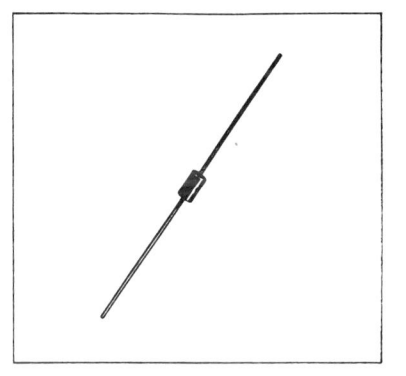

*MAXIMUM RATINGS

Rating	Symbol	Value	Unit
DC Power Dissipation @ T_A = 50°C	P_D	1.0	Watt
Derate above 50°C		6.67	mW/°C
Operating and Storage Junction Temperature Range	T_J, T_{stg}	−65 to +200	°C

MECHANICAL CHARACTERISTICS

CASE: Double slug type, hermetically sealed glass

MAXIMUM LEAD TEMPERATURE FOR SOLDERING PURPOSES: 230°C, 1/16" from case for 10 seconds

FINISH: All external surfaces are corrosion resistant with readily solderable leads.

POLARITY: Cathode indicated by color band. When operated in zener mode, cathode will be positive with respect to anode.

MOUNTING POSITION: Any

FIGURE 1 — POWER TEMPERATURE DERATING CURVE

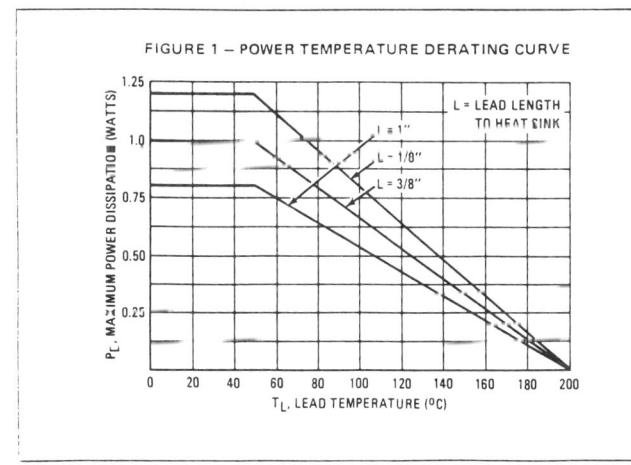

NOTE:
1. POLARITY DENOTED BY CATHODE BAND
2. LEAD DIAMETER NOT CONTROLLED WITHIN "F" DIMENSION.

DIM	MILLIMETERS MIN	MILLIMETERS MAX	INCHES MIN	INCHES MAX
A	4.07	5.20	0.160	0.205
B	2.04	2.71	0.080	0.107
D	0.71	0.86	0.028	0.034
F	—	1.27	—	0.050
K	27.94	—	1.100	—

All JEDEC dimensions and notes apply.

CASE 59-03 (DO-41)

*Indicates JEDEC Registered Data
▲Trademark of Motorola Inc.

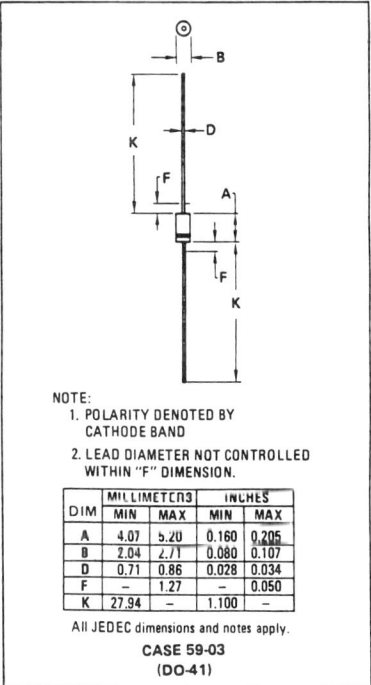

©MOTOROLA INC., 1978   DS 7039 R1

Figure A-1

*ELECTRICAL CHARACTERISTICS (T_A = 25°C unless otherwise noted) V_F = 1.2 V max, I_F = 200 mA for all types.

JEDEC Type No. (Note 1)	Nominal Zener Voltage V_Z @ I_{ZT} Volts (Notes 2 and 3)	Test Current I_{ZT} mA	Maximum Zener Impedance (Note 4) Z_{ZT} @ I_{ZT} Ohms	Z_{ZK} @ I_{ZK} Ohms	I_{ZK} mA	Leakage Current I_R µA Max	V_R Volts	Surge Current @ T_A = 25°C i_r — mA (Note 5)
1N4728	3.3	76	10	400	1.0	100	1.0	1380
1N4729	3.6	69	10	400	1.0	100	1.0	1260
1N4730	3.9	64	9.0	400	1.0	50	1.0	1190
1N4731	4.3	58	9.0	400	1.0	10	1.0	1070
1N4732	4.7	53	8.0	500	1.0	10	1.0	970
1N4733	5.1	49	7.0	550	1.0	10	1.0	890
1N4734	5.6	45	5.0	600	1.0	10	2.0	810
1N4735	6.2	41	2.0	700	1.0	10	3.0	730
1N4736	6.8	37	3.5	700	1.0	10	4.0	660
1N4737	7.5	34	4.0	700	0.5	10	5.0	605
1N4738	8.2	31	4.5	700	0.5	10	6.0	550
1N4739	9.1	28	5.0	700	0.5	10	7.0	500
1N4740	10	25	7.0	700	0.25	10	7.6	454
1N4741	11	23	8.0	700	0.25	5.0	8.4	414
1N4742	12	21	9.0	700	0.25	5.0	9.1	380
1N4743	13	19	10	700	0.25	5.0	9.9	344
1N4744	15	17	14	700	0.25	5.0	11.4	304
1N4745	16	15.5	16	700	0.25	5.0	12.2	285
1N4746	18	14	20	750	0.25	5.0	13.7	250
1N4747	20	12.5	22	750	0.25	5.0	15.2	225
1N4748	22	11.5	23	750	0.25	5.0	16.7	205
1N4749	24	10.5	25	750	0.25	5.0	18.2	190
1N4750	27	9.5	35	750	0.25	5.0	20.6	170
1N4751	30	8.5	40	1000	0.25	5.0	22.8	150
1N4752	33	7.5	45	1000	0.25	5.0	25.1	135
1N4753	36	7.0	50	1000	0.25	5.0	27.4	125
1N4754	39	6.5	60	1000	0.25	5.0	29.7	115
1N4755	43	6.0	70	1500	0.25	5.0	32.7	110
1N4756	47	5.5	80	1500	0.25	5.0	35.8	95
1N4757	51	5.0	95	1500	0.25	5.0	38.8	90
1N4758	56	4.5	110	2000	0.25	5.0	42.6	80
1N4759	62	4.0	125	2000	0.25	5.0	47.1	70
1N4760	68	3.7	150	2000	0.25	5.0	51.7	65
1N4761	75	3.3	175	2000	0.25	5.0	56.0	60
1N4762	82	3.0	200	3000	0.25	5.0	62.2	55
1N4763	91	2.8	250	3000	0.25	5.0	69.2	50
1N4764	100	2.5	350	3000	0.25	5.0	76.0	45

* Indicates JEDEC Registered Data.

NOTE 1 — Tolerance and Type Number Designation. The JEDEC type numbers listed have a standard tolerance on the nominal zener voltage of ±10%. A standard tolerance of ±5% on individual units is also available and is indicated by suffixing "A" to the standard type number.

NOTE 2 — Specials Available Include:

A. Nominal zener voltages between the voltages shown and tighter voltage tolerances,
B. Matched sets.

For detailed information on price, availability, and delivery, contact your nearest Motorola representative.

NOTE 3 — Zener Voltage (V_Z) Measurement. Motorola guarantees the zener voltage when measured at 90 seconds while maintaining the lead temperature (T_L) at 30°C ± 1°C, 3/8" from the diode body.

NOTE 4 — Zener Impedance (Z_Z) Derivation. The zener impedance is derived from the 60 cycle ac voltage, which results when an ac current having an rms value equal to 10% of the dc zener current (I_{ZT} or I_{ZK}) is superimposed on I_{ZT} or I_{ZK}.

NOTE 5 — Surge Current (i_r) Non-Repetitive. The rating listed in the electrical characteristics table is maximum peak, non-repetitive, reverse surge current of 1/2 square wave or equivalent sine wave pulse of 1/120 second duration superimposed on the test current, I_{ZT}, per JEDEC registration; however, actual device capability is as described in Figures 4 and 5.

APPLICATION NOTE

Since the actual voltage available from a given zener diode is temperature dependent, it is necessary to determine junction temperature under any set of operating conditions in order to calculate its value. The following procedure is recommended:

Lead Temperature, T_L, should be determined from

$$T_L = \theta_{LA} P_D + T_A$$

θ_{LA} is the lead-to-ambient thermal resistance (°C/W) and P_D is the power dissipation. The value for θ_{LA} will vary and depends on the device mounting method. θ_{LA} is generally 30 to 40°C/W for the various clips and tie points in common use and for printed circuit board wiring.

The temperature of the lead can also be measured using a thermocouple placed on the lead as close as possible to the tie point. The thermal mass connected to the tie point is normally large enough so that it will not significantly respond to heat surges generated in the diode as a result of pulsed operation once steady-state conditions are achieved. Using the measured value of T_L, the junction temperature may be determined by:

$$T_J = T_L + \Delta T_{JL}.$$

ΔT_{JL} is the increase in junction temperature above the lead temperature and may be found as follows:

$$\Delta T_{JL} = \theta_{JL} P_D$$

θ_{JL} may be determined from Figure 3 for dc power conditions. For worst-case design, using expected limits of I_Z, limits of P_D and the extremes of $T_J (\Delta T_J)$ may be estimated. Changes in voltage, V_Z, can then be found from:

$$\Delta V = \theta_{VZ} \Delta T_J$$

θ_{VZ}, the zener voltage temperature coefficient, is found from Figure 2.

Under high power-pulse operation, the zener voltage will vary with time and may also be affected significantly by the zener resistance. For best regulation, keep current excursions as low as possible.

Surge limitations are given in Figure 5. They are lower than would be expected by considering only junction temperature, as current crowding effects cause temperatures to be extremely high in small spots resulting in device degradation should the limits of Figure 5 be exceeded.

MOTOROLA Semiconductor Products Inc.

Figure A-1 (*Continued*)

A-5 Transistor Data

Transistors are used in a variety of applications throughout this book. The choice of transistors for these applications is not critical, and the transistors used were selected simply because they were inexpensive and readily available.

Small-Signal Transistors

General-purpose small-signal transistors are used for switching, current limiting in power supplies, and in logarithmic amplifiers. Figure A-2 shows the data sheets for the 2N4400/2N4401/2N4402/2N4403 series of general-purpose, small-signal transistors (reproduced with the permission of Motorola, Ltd.).

Power Transistors

Power transistors are used as pass transistors in linear power supplies and as switching transistors in switch-mode regulators. Figure A-3 shows the data sheets for the TIP31/TIP32 medium-power power transistors (reproduced with the permission of Texas Instruments). Figure A-4 shows the data sheets for the TIP33/TIP34 high-power power transistors (reproduced with the permission of Texas Instruments).

2N4400 / 2N4401

CASE 29-04, STYLE 1
TO-92 (TO-226AA)

GENERAL PURPOSE TRANSISTORS

NPN SILICON

MAXIMUM RATINGS

Rating	Symbol	Value	Unit
Collector-Emitter Voltage	V_{CEO}	40	Vdc
Collector-Base Voltage	V_{CBO}	60	Vdc
Emitter-Base Voltage	V_{EBO}	6.0	Vdc
Collector Current — Continuous	I_C	600	mAdc
Total Device Dissipation @ T_A = 25°C Derate above 25°C	P_D	625 5.0	mW mW/°C
Total Device Dissipation @ T_C = 25°C Derate above 25°C	P_D	1.5 12	Watt mW/°C
Operating and Storage Junction Temperature Range	T_J, T_{stg}	−55 to −150	°C

THERMAL CHARACTERISTICS

Characteristic	Symbol	Max	Unit
Thermal Resistance, Junction to Case	$R_{\theta JC}$	83.3	°C/W
Thermal Resistance, Junction to Ambient	$R_{\theta JA}$	200	°C/W

ELECTRICAL CHARACTERISTICS (T_A = 25°C unless otherwise noted.)

Characteristic	Symbol	Min	Max	Unit
OFF CHARACTERISTICS				
Collector-Emitter Breakdown Voltage(1) (I_C = 1.0 mAdc, I_B = 0)	$V_{(BR)CEO}$	40	—	Vdc
Collector-Base Breakdown Voltage (I_C = 0.1 mAdc, I_E = 0)	$V_{(BR)CBO}$	60	—	Vdc
Emitter-Base Breakdown Voltage (I_E = 0.1 mAdc, I_C = 0)	$V_{(BR)EBO}$	6.0	—	Vdc
Base Cutoff Current (V_{CE} = 35 Vdc, V_{EB} = 0.4 Vdc)	I_{BEV}	—	0.1	μAdc
Collector Cutoff Current (V_{CE} = 35 Vdc, V_{EB} = 0.4 Vdc)	I_{CEX}	—	0.1	μAdc
ON CHARACTERISTICS(1)				
DC Current Gain (I_C = 0.1 mAdc, V_{CE} = 1.0 Vdc) 2N4401	h_{FE}	20	—	—
(I_C = 1.0 mAdc, V_{CE} = 1.0 Vdc) 2N4400 2N4401		20 40	— —	
(I_C = 10 mAdc, V_{CE} = 1.0 Vdc) 2N4400 2N4401		40 80	— —	
(I_C = 150 mAdc, V_{CE} = 1.0 Vdc) 2N4400 2N4401		50 100	150 300	
(I_C = 500 mAdc, V_{CE} = 2.0 Vdc) 2N4400 2N4401		20 40	— —	
Collector-Emitter Saturation Voltage (I_C = 150 mAdc, I_B = 15 mAdc) (I_C = 500 mAdc, I_B = 50 mAdc)	$V_{CE(sat)}$	— —	0.4 0.75	Vdc
Base-Emitter Saturation Voltage (I_C = 150 mAdc, I_B = 15 mAdc) (I_C = 500 mAdc, I_B = 50 mAdc)	$V_{BE(sat)}$	0.75 —	0.95 1.2	Vdc
SMALL-SIGNAL CHARACTERISTICS				
Current-Gain — Bandwidth Product (I_C = 20 mAdc, V_{CE} = 10 Vdc, f = 100 MHz) 2N4400 2N4401	f_T	200 250	— —	MHz
Collector-Base Capacitance (V_{CB} = 5.0 Vdc, I_E = 0, f = 100 kHz)	C_{cb}	—	6.5	pF

MOTOROLA SMALL-SIGNAL TRANSISTORS, FETs AND DIODES

Figure A-2(a)

2N4400, 2N4401

ELECTRICAL CHARACTERISTICS (continued) ($T_A = 25°C$ unless otherwise noted.)

Characteristic		Symbol	Min	Max	Unit
Emitter-Base Capacitance ($V_{BE} = 0.5$ Vdc, $I_C = 0$, $f = 100$ kHz)		C_{eb}	—	30	pF
Input Impedance ($I_C = 1.0$ mAdc, $V_{CE} = 10$ Vdc, $f = 1.0$ kHz)	2N4400 2N4401	h_{ie}	0.5 1.0	7.5 15	k ohms
Voltage Feedback Ratio ($I_C = 1.0$ mAdc, $V_{CE} = 10$ Vdc, $f = 1.0$ kHz)		h_{re}	0.1	8.0	$\times 10^{-4}$
Small-Signal Current Gain ($I_C = 1.0$ mAdc, $V_{CE} = 10$ Vdc, $f = 1.0$ kHz)	2N4400 2N4401	h_{fe}	20 40	250 500	—
Output Admittance ($I_C = 1.0$ mAdc, $V_{CE} = 10$ Vdc, $f = 1.0$ kHz)		h_{oe}	1.0	30	μmhos
SWITCHING CHARACTERISTICS					
Delay Time	$V_{CC} = 30$ Vdc, $V_{EB} = 2.0$ Vdc, $I_C = 150$ mAdc, $I_{B1} = 15$ mAdc	t_d	—	15	ns
Rise Time		t_r	—	20	ns
Storage Time	$V_{CC} = 30$ Vdc, $I_C = 150$ mAdc, $I_{B1} = I_{B2} = 15$ mAdc	t_s	—	225	ns
Fall Time		t_f	—	30	ns

(1) Pulse Test: Pulse Width ≤ 300 μs, Duty Cycle ≤ 2.0%.

SWITCHING TIME EQUIVALENT TEST CIRCUITS

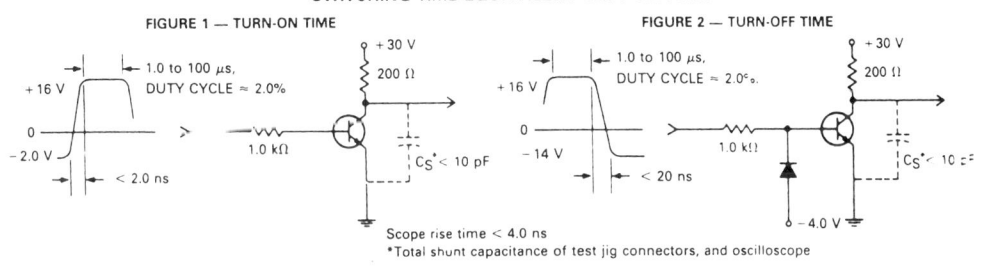

FIGURE 1 — TURN-ON TIME

FIGURE 2 — TURN-OFF TIME

Scope rise time < 4.0 ns
*Total shunt capacitance of test jig connectors, and oscilloscope

TRANSIENT CHARACTERISTICS
——— 25°C   – – – 100°C

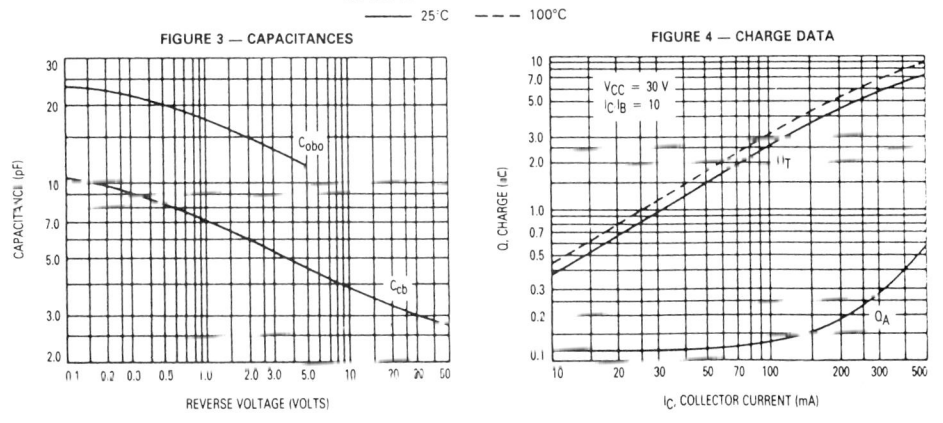

FIGURE 3 — CAPACITANCES

FIGURE 4 — CHARGE DATA

MOTOROLA SMALL-SIGNAL TRANSISTORS, FETs AND DIODES

Figure A-2(a)   *(Continued)*

2N4402 2N4403

CASE 29-04, STYLE 1
TO-92 (TO-226AA)

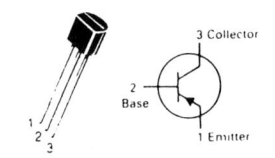

GENERAL PURPOSE TRANSISTORS

PNP SILICON

MAXIMUM RATINGS

Rating	Symbol	Value	Unit
Collector-Emitter Voltage	V_{CEO}	40	Vdc
Collector-Base Voltage	V_{CBO}	40	Vdc
Emitter-Base Voltage	V_{EBO}	5.0	Vdc
Collector Current — Continuous	I_C	600	mAdc
Total Device Dissipation @ T_A = 25°C Derate above 25°C	P_D	625 5.0	mW mW/°C
Total Device Dissipation @ T_C = 25°C Derate above 25°C	P_D	1.5 12	Watt mW/°C
Operating and Storage Junction Temperature Range	T_J, T_{stg}	−55 to −150	°C

THERMAL CHARACTERISTICS

Characteristic	Symbol	Max	Unit
Thermal Resistance, Junction to Case	$R_{\theta JC}$	83.3	°C/W
Thermal Resistance, Junction to Ambient	$R_{\theta JA}$	200	°C/W

ELECTRICAL CHARACTERISTICS (T_A = 25°C unless otherwise noted.)

Characteristic		Symbol	Min	Max	Unit
OFF CHARACTERISTICS					
Collector-Emitter Breakdown Voltage(1) (I_C = 1.0 mAdc, I_B = 0)		$V_{(BR)CEO}$	40	—	Vdc
Collector-Base Breakdown Voltage (I_C = 0.1 mAdc, I_E = 0)		$V_{(BR)CBO}$	40	—	Vdc
Emitter-Base Breakdown Voltage (I_E = 0.1 mAdc, I_C = 0)		$V_{(BR)EBO}$	5.0	—	Vdc
Base Cutoff Current (V_{CE} = 35 Vdc, V_{BE} = 0.4 Vdc)		I_{BEV}	—	0.1	μAdc
Collector Cutoff Current (V_{CE} = 35 Vdc, V_{BE} = 0.4 Vdc)		I_{CEX}	—	0.1	μAdc
ON CHARACTERISTICS					
DC Current Gain (I_C = 0.1 mAdc, V_{CE} = 1.0 Vdc)	2N4403	h_{FE}	30	—	—
(I_C = 1.0 mAdc, V_{CE} = 1.0 Vdc)	2N4402 2N4403		30 60	— —	
(I_C = 10 mAdc, V_{CE} = 1.0 Vdc)	2N4402 2N4403		50 100	— —	
(I_C = 150 mAdc, V_{CE} = 2.0 Vdc)(1)	2N4402 2N4403		50 100	150 300	
(I_C = 500 mAdc, V_{CE} = 2.0 Vdc)(1)	Both		20	—	
Collector-Emitter Saturation Voltage(1) (I_C = 150 mAdc, I_B = 15 mAdc) (I_C = 500 mAdc, I_B = 50 mAdc)		$V_{CE(sat)}$	— —	0.4 0.75	Vdc
Base-Emitter Saturation Voltage(1) (I_C = 150 mAdc, I_B = 15 mAdc) (I_C = 500 mAdc, I_B = 50 mAdc)		$V_{BE(sat)}$	0.75 —	0.95 1.3	Vdc
SMALL-SIGNAL CHARACTERISTICS					
Current-Gain — Bandwidth Product (I_C = 20 mAdc, V_{CE} = 10 Vdc, f = 100 MHz)	2N4402 2N4403	f_T	150 200	— —	MHz
Collector-Base Capacitance (V_{CB} = 10 Vdc, I_E = 0, f = 140 kHz)		C_{cb}	—	8.5	pF
Emitter-Base Capacitance (V_{BE} = 0.5 Vdc, I_C = 0, f = 140 kHz)		C_{eb}	—	30	pF
Input Impedance (I_C = 1.0 mAdc, V_{CE} = 10 Vdc, f = 1.0 kHz)	2N4402 2N4403	h_{ie}	750 1.5k	7.5k 15k	ohms

MOTOROLA SMALL-SIGNAL TRANSISTORS, FETs AND DIODES

Figure A-2(b)

2N4402, 2N4403

ELECTRICAL CHARACTERISTICS (continued) ($T_A = 25°C$ unless otherwise noted.)

Characteristic		Symbol	Min	Max	Unit
Voltage Feedback Ratio ($I_C = 1.0$ mAdc, $V_{CE} = 10$ Vdc, f = 1.0 kHz)		h_{re}	0.1	8.0	$\times 10^{-4}$
Small-Signal Current Gain ($I_C = 1.0$ mAdc, $V_{CE} = 10$ Vdc, f = 1.0 kHz)	2N4402 2N4403	h_{fe}	30 60	250 500	—
Output Admittance ($I_C = 1.0$ mAdc, $V_{CE} = 10$ Vdc, f = 1.0 kHz)		h_{oe}	1.0	100	μmhos

SWITCHING CHARACTERISTICS

Delay Time	($V_{CC} = 30$ Vdc, $V_{BE} = 2.0$ Vdc, $I_C = 150$ mAdc, $I_{B1} = 15$ mAdc)	t_d	—	15	ns
Rise Time		t_r	—	20	ns
Storage Time	($V_{CC} = 30$ Vdc, $I_C = 150$ mAdc, $I_{B1} = I_{B2} = 15$ mAdc)	t_s	—	225	ns
Fall Time		t_f	—	30	ns

(1) Pulse Test: Pulse Width ≤ 300 μs, Duty Cycle ≤ 2.0%.

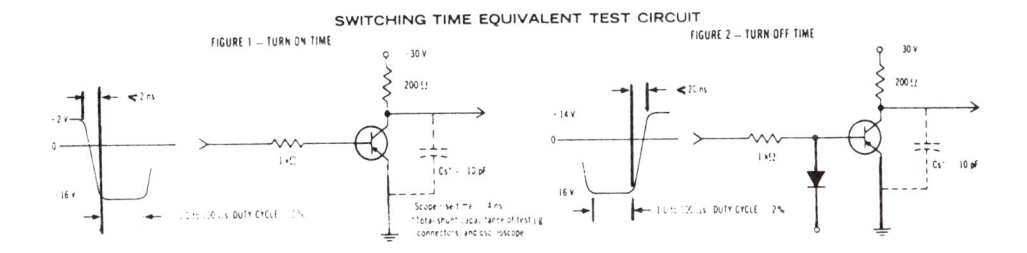

SWITCHING TIME EQUIVALENT TEST CIRCUIT

FIGURE 1 — TURN ON TIME

FIGURE 2 — TURN OFF TIME

TRANSIENT CHARACTERISTICS
——— 25°C – – – 100°C

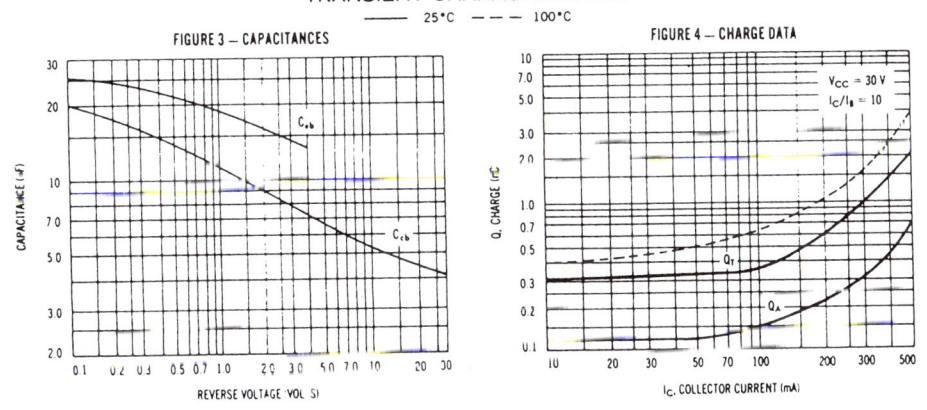

FIGURE 3 — CAPACITANCES

FIGURE 4 — CHARGE DATA

MOTOROLA SMALL-SIGNAL TRANSISTORS, FETs AND DIODES

Figure A-2(b) *(Continued)*

A-14 Reference Data

TYPES TIP31, TIP31A, TIP31B, TIP31C
N-P-N SINGLE-DIFFUSED MESA SILICON POWER TRANSISTORS

electrical characteristics at 25°C case temperature

PARAMETER		TEST CONDITIONS	TIP31 MIN	TIP31 MAX	TIP31A MIN	TIP31A MAX	TIP31B MIN	TIP31B MAX	TIP31C MIN	TIP31C MAX	UNIT
$V_{(BR)CEO}$	Collector-Emitter Breakdown Voltage	$I_C = 30$ mA, $I_B = 0$, See Note 6	40		60		80		100		V
I_{CEO}	Collector Cutoff Current	$V_{CE} = 30$ V, $I_B = 0$		0.3		0.3					mA
		$V_{CE} = 60$ V, $I_B = 0$						0.3		0.3	
I_{CES}	Collector Cutoff Current	$V_{CE} = 40$ V, $V_{BE} = 0$		0.2							mA
		$V_{CE} = 60$ V, $V_{BE} = 0$				0.2					
		$V_{CE} = 80$ V, $V_{BE} = 0$						0.2			
		$V_{CE} = 100$ V, $V_{BE} = 0$								0.2	
I_{EBO}	Emitter Cutoff Current	$V_{EB} = 5$ V, $I_C = 0$		1		1		1		1	mA
h_{FE}	Static Forward Current Transfer Ratio	$V_{CE} = 4$ V, $I_C = 1$ A, See Notes 6 and 7	25		25		25		25		
		$V_{CE} = 4$ V, $I_C = 3$ A, See Notes 6 and 7	10	50	10	50	10	50	10	50	
V_{BE}	Base-Emitter Voltage	$V_{CE} = 4$ V, $I_C = 3$ A, See Notes 6 and 7		1.8		1.8		1.8		1.8	V
$V_{CE(sat)}$	Collector-Emitter Saturation Voltage	$I_B = 375$ mA, $I_C = 3$ A, See Notes 6 and 7		1.2		1.2		1.2		1.2	V
h_{fe}	Small-Signal Common-Emitter Forward Current Transfer Ratio	$V_{CE} = 10$ V, $I_C = 0.5$ A, $f = 1$ kHz	20		20		20		20		
$|h_{fe}|$	Small-Signal Common-Emitter Forward Current Transfer Ratio	$V_{CE} = 10$ V, $I_C = 0.5$ A, $f = 1$ MHz	3		3		3		3		

NOTES: 6. These parameters must be measured using pulse techniques. $t_w = 300$ μs, duty cycle ≤ 2%.
7. These parameters are measured with voltage-sensing contacts separate from the current-carrying contacts and located within 0.125 inch from the device body.

thermal characteristics

PARAMETER		MAX	UNIT
$R_{\theta JC}$	Junction-to-Case Thermal Resistance	3.125	°C/W
$R_{\theta JA}$	Junction-to-Free-Air Thermal Resistance	62.5	

switching characteristics at 25°C case temperature

PARAMETER		TEST CONDITIONS†	TYP	UNIT
t_{on}	Turn-On Time	$I_C = 1$ A, $I_{B(1)} = 100$ mA, $I_{B(2)} = -100$ mA,	0.5	μs
t_{off}	Turn-Off Time	$V_{BE(off)} = -4.3$ V, $R_L = 30$ Ω, See Figure 1	2	

†Voltage and current values shown are nominal; exact values vary slightly with transistor parameters.

Figure A-3(a)

Reference Data   A-15

TYPES TIP31, TIP31A, TIP31B, TIP31C
N-P-N SINGLE-DIFFUSED MESA SILICON POWER TRANSISTORS

BULLETIN NO. DL-S 7511402, DECEMBER 1970–REVISED FEBRUARY 1975

FOR POWER-AMPLIFIER AND HIGH-SPEED-SWITCHING APPLICATIONS
DESIGNED FOR COMPLEMENTARY USE WITH TIP32, TIP32A, TIP32B, TIP32C

- 40 W at 25°C Case Temperature
- 3 A Rated Collector Current
- Min f_T of 3 MHz at 10 V, 500 mA

mechanical data

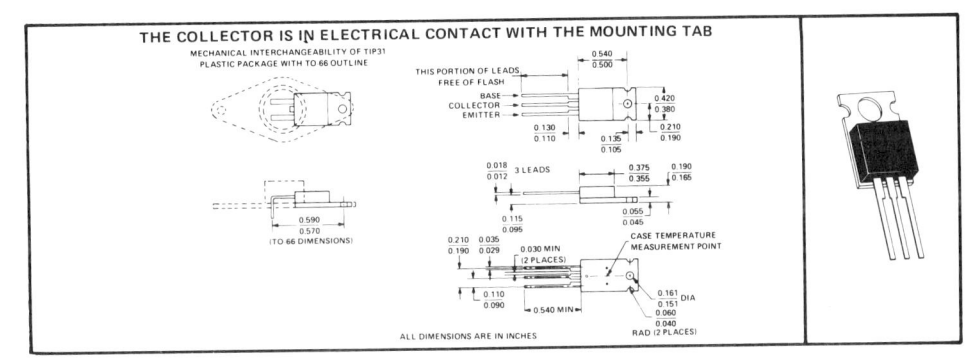

THE COLLECTOR IS IN ELECTRICAL CONTACT WITH THE MOUNTING TAB

ALL DIMENSIONS ARE IN INCHES

absolute maximum ratings at 25°C case temperature (unless otherwise noted)

	TIP31	TIP31A	TIP31B	TIP31C
Collector-Base Voltage	40 V	60 V	80 V	100 V
Collector-Emitter Voltage (See Note 1)	40 V	60 V	80 V	100 V
Emitter-Base Voltage	5 V	5 V	5 V	5 V
Continuous Collector Current	← 3 A →
Peak Collector Current (See Note 2)	← 5 A →
Continuous Base Current	← 1 A →
Safe Operating Region at (or below) 25°C Case Temperature	← See Figure 12 →
Continuous Device Dissipation at (or below) 25°C Case Temperature (See Note 3)	← 40 W →
Continuous Device Dissipation at (or below) 25°C Free-Air Temperature (See Note 4)	← 2 W →
Unclamped Inductive Load Energy (See Note 5)	← 32 mJ →
Operating Collector Junction Temperature Range	← −65°C to 150°C →
Storage Temperature Range	← −65°C to 150°C →
Lead Temperature 1/8 Inch from Case for 10 Seconds	← 260°C →

NOTES: 1. This value applies when the base-emitter diode is open-circuited.
2. This value applies for $t_W \leq 0.2$ ms, duty cycle $\leq 10\%$.
3. Derate linearly to 150°C case temperature at the rate of 0.32 W/°C or refer to Dissipation Derating Curve, Figure 10.
4. Derate linearly to 150°C free-air temperature at the rate of 16 mW/°C or refer to Dissipation Derating Curve, Figure 11.
5. This rating is based on the capability of the transistor to operate safely in the circuit of Figure 2. L = 20 mH, $R_{BB2} = 100\ \Omega$, $V_{BB2} = 0$ V, $R_3 = 0.1\ \Omega$, $V_{CC} = 10$ V, Energy $\approx I_C^2 L/2$.

Figure A-3(a) *(Continued)*

Reference Data

TYPES TIP32, TIP32A, TIP32B, TIP32C
P-N-P SINGLE-DIFFUSED MESA SILICON POWER TRANSISTORS
BULLETIN NO. DL-S 7511403, DECEMBER 1970—REVISED FEBRUARY 1975

**FOR POWER-AMPLIFIER AND HIGH-SPEED-SWITCHING APPLICATIONS
DESIGNED FOR COMPLEMENTARY USE WITH TIP31, TIP31A, TIP31B, TIP31C**

- 40 W at 25°C Case Temperature
- 3 A Rated Collector Current
- Min f_T of 3 MHz at 10 V, 500 mA

mechanical data

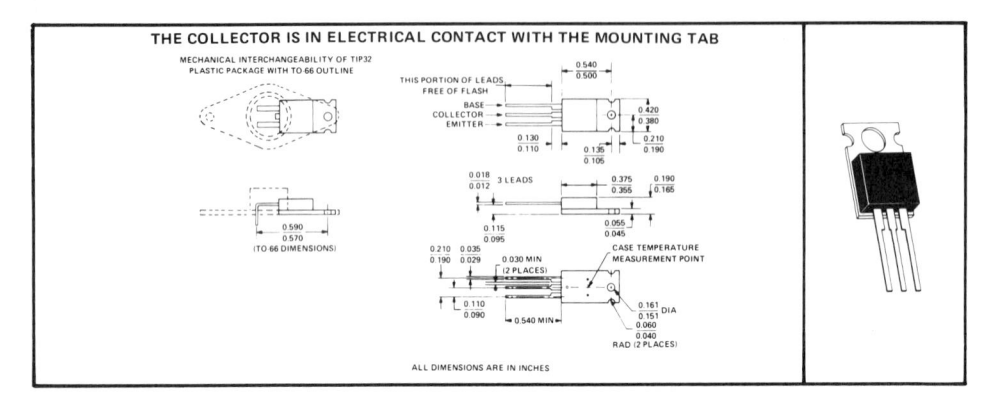

THE COLLECTOR IS IN ELECTRICAL CONTACT WITH THE MOUNTING TAB

ALL DIMENSIONS ARE IN INCHES

absolute maximum ratings at 25°C case temperature (unless otherwise noted)

	TIP32	TIP32A	TIP32B	TIP32C
Collector-Base Voltage	−40 V	−60 V	−80 V	−100 V
Collector-Emitter Voltage (See Note 1)	−40 V	−60 V	−80 V	−100 V
Emitter-Base Voltage	−5 V	−5 V	−5 V	−5 V
Continuous Collector Current	←——— −3 A ———→
Peak Collector Current (See Note 2)	←——— −5 A ———→
Continuous Base Current	←——— −1 A ———→
Safe Operating Region at (or below) 25°C Case Temperature	←——— See Figure 12 ———→
Continuous Device Dissipation at (or below) 25°C Case Temperature (See Note 3)	←——— 40 W ———→
Continuous Device Dissipation at (or below) 25°C Free-Air Temperature (See Note 4)	←——— 2 W ———→
Unclamped Inductive Load Energy (See Note 5)	←——— 32 mJ ———→
Operating Collector Junction Temperature Range	←——— −65°C to 150°C ———→
Storage Temperature Range	←——— −65°C to 150°C ———→
Lead Temperature 1/8 Inch from Case for 10 Seconds	←——— 260°C ———→

NOTES: 1. This value applies when the base-emitter diode is open-circuited.
2. This value applies for $t_w \leq 0.3$ ms, duty cycle $\leq 10\%$.
3. Derate linearly to 150°C case temperature at the rate of 0.32 W/°C or refer to Dissipation Derating Curve, Figure 10.
4. Derate linearly to 150°C free-air temperature at the rate of 16 mW/°C or refer to Dissipation Derating Curve, Figure 11.
5. This rating is based on the capability of the transistor to operate safely in the circuit of Figure 2. L = 20 mH, R_{BB2} = 100 Ω, V_{BB2} = 0 V, R_S = 0.1 Ω, V_{CC} = 10 V. Energy ≈ $I_C^2 L/2$.

Figure A-3(b)

Reference Data   A-17

TYPES TIP32, TIP32A, TIP32B, TIP32C
P-N-P SINGLE-DIFFUSED MESA SILICON POWER TRANSISTORS

electrical characteristics at 25°C case temperature

PARAMETER		TEST CONDITIONS	TIP32 MIN	TIP32 MAX	TIP32A MIN	TIP32A MAX	TIP32B MIN	TIP32B MAX	TIP32C MIN	TIP32C MAX	UNIT
$V_{(BR)CEO}$	Collector-Emitter Breakdown Voltage	$I_C = -30$ mA, $I_B = 0$, See Note 6	−40		−60		−80		−100		V
I_{CEO}	Collector Cutoff Current	$V_{CE} = -30$ V, $I_B = 0$		−0.3		−0.3					mA
		$V_{CE} = -60$ V, $I_B = 0$						−0.3		−0.3	
I_{CES}	Collector Cutoff Current	$V_{CE} = -40$ V, $V_{BE} = 0$		−0.2							mA
		$V_{CE} = -60$ V, $V_{BE} = 0$				−0.2					
		$V_{CE} = -80$ V, $V_{BE} = 0$						−0.2			
		$V_{CE} = -100$ V, $V_{BE} = 0$								−0.2	
I_{EBO}	Emitter Cutoff Current	$V_{EB} = -5$ V, $I_C = 0$		−1		−1		−1		−1	mA
h_{FE}	Static Forward Current Transfer Ratio	$V_{CE} = -4$ V, $I_C = -1$ A, See Notes 6 and 7	25		25		25		25		
		$V_{CE} = -4$ V, $I_C = -3$ A, See Notes 6 and 7	10	50	10	50	10	50	10	50	
V_{BE}	Base-Emitter Voltage	$V_{CE} = -4$ V, $I_C = -3$ A, See Notes 6 and 7		−1.8		−1.8		−1.8		−1.8	V
$V_{CE(sat)}$	Collector-Emitter Saturation Voltage	$I_B = -375$ mA, $I_C = -3$ A, See Notes 6 and 7		−1.2		−1.2		−1.2		−1.2	V
h_{fe}	Small-Signal Common-Emitter Forward Current Transfer Ratio	$V_{CE} = -10$ V, $I_C = -0.5$ A, $f = 1$ kHz	20		20		20		20		
$\|h_{fe}\|$	Small-Signal Common-Emitter Forward Current Transfer Ratio	$V_{CE} = -10$ V, $I_C = -0.5$ A, $f = 1$ MHz	3		3		3		3		

NOTES: 6. These parameters must be measured using pulse techniques. $t_w = 300$ μs, duty cycle ≤ 2%.
7. These parameters are measured with voltage-sensing contacts separate from the current-carrying contacts and located within 0.125 inch from the device body.

thermal characteristics

PARAMETER		MAX	UNIT
$R_{\theta JC}$	Junction-to-Case Thermal Resistance	3.125	°C/W
$R_{\theta JA}$	Junction-to-Free-Air Thermal Resistance	62.5	

switching characteristics at 25°C case temperature

PARAMETER		TEST CONDITIONS†	TYP	UNIT
t_{on}	Turn-On Time	$I_C = -1$ A, $I_{B(1)} = -100$ mA, $I_{B(2)} = 100$ mA,	0.3	μs
t_{off}	Turn-Off Time	$V_{BE(off)} = 4.3$ V, $R_L = 30$ Ω, See Figure 1	1.0	

†Voltages and current values shown are nominal; exact values vary slightly with transistor parameters.

Figure A-3(b)   (*Continued*)

TYPES TIP33, TIP33A, TIP33B, TIP33C
N-P-N SINGLE-DIFFUSED MESA SILICON POWER TRANSISTORS

BULLETIN NO. DL-S 7611404, DECEMBER 1970–REVISED JULY 1976

FOR POWER-AMPLIFIER AND HIGH-SPEED-SWITCHING APPLICATIONS
DESIGNED FOR COMPLEMENTARY USE WITH TIP34, TIP34A, TIP34B, TIP34C

- 80 W at 25°C Case Temperature
- 10 A Rated Collector Current
- Min f_T of 3 MHz at 10 V, 500 mA

mechanical data

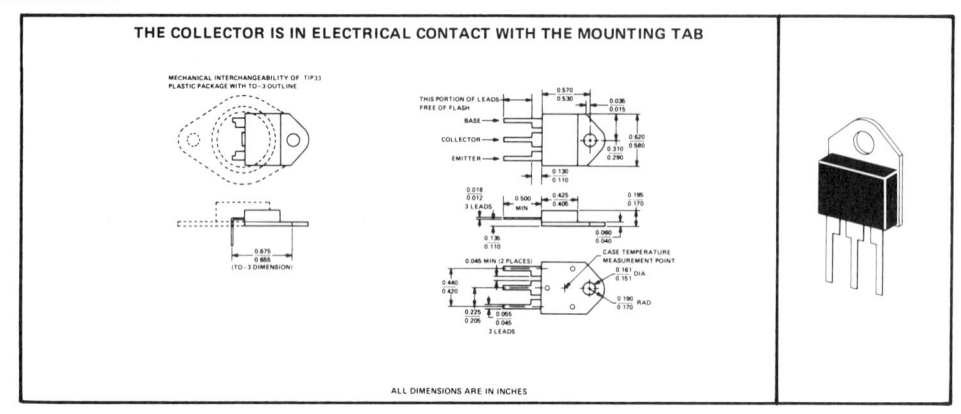

THE COLLECTOR IS IN ELECTRICAL CONTACT WITH THE MOUNTING TAB

ALL DIMENSIONS ARE IN INCHES

absolute maximum ratings at 25°C case temperature (unless otherwise noted)

	TIP33	TIP33A	TIP33B	TIP33C
Collector-Base Voltage	40 V	60 V	80 V	100 V
Collector-Emitter Voltage (See Note 1)	40 V	60 V	80 V	100 V
Emitter-Base Voltage	5 V	5 V	5 V	5 V
Continuous Collector Current	←―――― 10 A ――――→
Peak Collector Current (See Note 2)	←―――― 15 A ――――→
Continuous Base Current	←―――― 3 A ――――→
Safe Operating Region at (or below) 25°C Case Temperature	←―― See Figure 12 ――→
Continuous Device Dissipation at (or below) 25°C Case Temperature (See Note 3)	←―――― 80 W ――――→
Continuous Device Dissipation at (or below) 25°C Free-Air Temperature (See Note 4)	←―――― 3.5 W ――――→
Unclamped Inductive Load Energy (See Note 5)	←―――― 62.5 mJ ――――→
Operating Collector Junction Temperature Range	←―― −65°C to 150°C ――→
Storage Temperature Range	←―― −65°C to 150°C ――→
Lead Temperature 1/8 Inch from Case for 10 Seconds	←―――― 260°C ――――→

NOTES: 1. This value applies when the base-emitter diode is open-circuited.
2. This value applies for $t_w \leqslant 0.3$ ms, duty cycle $\leqslant 10\%$.
3. Derate linearly to 150°C case temperature at the rate of 0.64 W/°C or refer to Dissipation Derating Curve, Figure 10.
4. Derate linearly to 150°C free-air temperature at the rate of 28 mW/°C or refer to Dissipation Derating Curve, Figure 11.
5. This rating is based on the capability of the transistor to operate safely in the circuit of Figure 2. L = 20 mH, R_{BB2} = 100 Ω, V_{BB2} = 0 V, R_S = 0.1 Ω, V_{CC} = 10 V. Energy ≈ $I_C^2 L/2$.

Figure A-4(a)

TYPES TIP33, TIP33A, TIP33B, TIP33C
N-P-N SINGLE-DIFFUSED MESA SILICON POWER TRANSISTORS

electrical characteristics at 25°C case temperature

PARAMETER		TEST CONDITIONS		TIP33 MIN	TIP33 MAX	TIP33A MIN	TIP33A MAX	TIP33B MIN	TIP33B MAX	TIP33C MIN	TIP33C MAX	UNIT
$V_{(BR)CEO}$	Collector-Emitter Breakdown Voltage	I_C = 30 mA, See Note 6	I_B = 0,	40		60		80		100		V
I_{CEO}	Collector Cutoff Current	V_{CE} = 30 V,	I_B = 0		0.7		0.7					mA
		V_{CE} = 60 V,	I_B = 0						0.7		0.7	
I_{CES}	Collector Cutoff Current	V_{CE} = 40 V,	V_{BE} = 0		0.4							mA
		V_{CE} = 60 V,	V_{BE} = 0				0.4					
		V_{CE} = 80 V,	V_{BE} = 0						0.4			
		V_{CE} = 100 V,	V_{BE} = 0								0.4	
I_{EBO}	Emitter Cutoff Current	V_{EB} = 5 V,	I_C = 0		1		1		1		1	mA
h_{FE}	Static Forward Current Transfer Ratio	V_{CE} = 4 V, See Notes 6 and 7	I_C = 1 A,	40		40		40		40		
		V_{CE} = 4 V, See Notes 6 and 7	I_C = 3 A,	20	100	20	100	20	100	20	100	
V_{BE}	Base-Emitter Voltage	V_{CE} = 4 V, See Notes 6 and 7	I_C = 3 A,		1.6		1.6		1.6		1.6	V
		V_{CE} = 4 V, See Notes 6 and 7	I_C = 10 A,		3		3		3		3	
$V_{CE(sat)}$	Collector-Emitter Saturation Voltage	I_B = 0.3 A, See Notes 6 and 7	I_C = 3 A,		1		1		1		1	V
		I_B = 2.5 A, See Notes 6 and 7	I_C = 10 A,		4		4		4		4	
h_{fe}	Small-Signal Common-Emitter Forward Current Transfer Ratio	V_{CE} = 10 V, f = 1 kHz	I_C = 0.5 A,	20		20		20		20		
$|h_{fe}|$	Small-Signal Common-Emitter Forward Current Transfer Ratio	V_{CE} = 10 V, f = 1 MHz	I_C = 0.5 A,	3		3		3		3		

NOTES: 6. These parameters must be measured using pulse techniques. t_w = 300 µs, duty cycle ⩽ 2%.
7. These parameters are measured with voltage-sensing contacts separate from the current-carrying contacts and located within 0.125 inch from the device body.

thermal characteristics

	PARAMETER	MAX	UNIT
$R_{\theta JC}$	Junction-to-Case Thermal Resistance	1.56	°C/W
$R_{\theta JA}$	Junction-to-Free Air Thermal Resistance	35.7	

switching characteristics at 25°C case temperature

	PARAMETER	TEST CONDITIONS[†]			TYP	UNIT
t_{on}	Turn-On Time	I_C = 6 A,	$I_{B(1)}$ = 0.6 A,	$I_{B(2)}$ = −0.6 A,	0.6	µs
t_{off}	Turn-Off Time	$V_{BE(off)}$ = −4 V,	R_L = 5 Ω,	See Figure 1	1	

[†]Voltage and current values shown are nominal; exact values vary slightly with transistor parameters.

Figure A-4(a) *(Continued)*

TYPES TIP34, TIP34A, TIP34B, TIP34C
P-N-P SINGLE-DIFFUSED MESA SILICON POWER TRANSISTORS

BULLETIN NO. DL-S 7611405, DECEMBER 1970—REVISED JULY 1976

FOR POWER-AMPLIFIER AND HIGH-SPEED-SWITCHING APPLICATIONS
DESIGNED FOR COMPLEMENTARY USE WITH TIP33, TIP33A, TIP33B, TIP33C

- 80 W at 25°C Case Temperature
- 10 A Rated Collector Current
- Min f_T of 3 MHz at 10 V, 500 mA

mechanical data

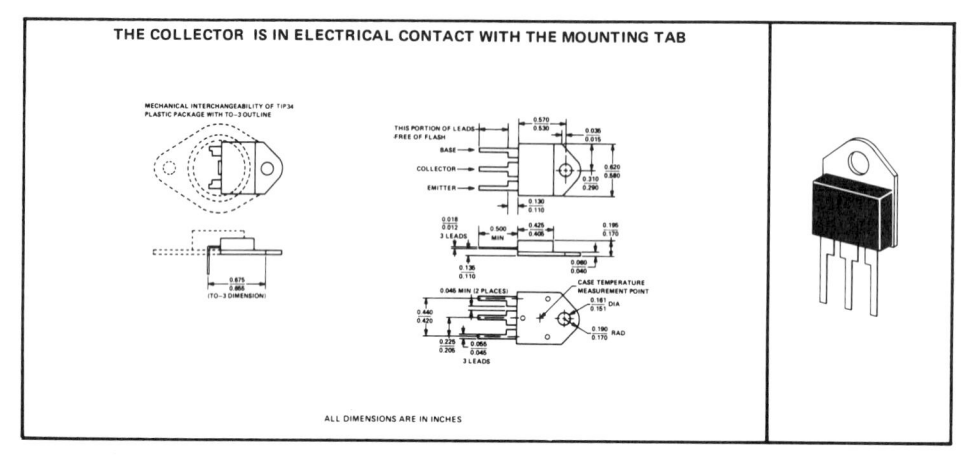

THE COLLECTOR IS IN ELECTRICAL CONTACT WITH THE MOUNTING TAB

ALL DIMENSIONS ARE IN INCHES

absolute maximum ratings at 25°C case temperature (unless otherwise noted)

	TIP34	TIP34A	TIP34B	TIP34C
Collector-Base Voltage	−40 V	−60 V	−80 V	−100 V
Collector-Emitter Voltage (See Note 1)	−40 V	−60 V	−80 V	−100 V
Emitter-Base Voltage	−5 V	−5 V	−5 V	−5 V
Continuous Collector Current	←	−10 A	→	
Peak Collector Current (See Note 2)	←	−15 A	→	
Continuous Base Current	←	−3 A	→	
Safe Operating Region at (or below) 25°C Case Temperature	←	See Figure 12	→	
Continuous Device Dissipation at (or below) 25°C Case Temperature (See Note 3)	←	80 W	→	
Continuous Device Dissipation at (or below) 25°C Free-Air Temperature (See Note 4)	←	3.5 W	→	
Unclamped Inductive Load Energy (See Note 5)	←	62.5 mJ	→	
Operating Collector Junction Temperature Range	←	−65°C to 150°C	→	
Storage Temperature Range	←	−65°C to 150°C	→	
Lead Temperature 1/8 Inch from Case for 10 Seconds	←	260°C	→	

NOTES:
1. This value applies when the base-emitter diode is open-circuited.
2. This value applies for $t_w \leq 0.3$ ms, duty cycle $\leq 10\%$.
3. Derate linearly to 150°C case temperature at the rate of 0.64 W/°C or refer to Dissipation Derating Curve, Figure 10.
4. Derate linearly to 150°C free-air temperature at the rate of 28 mW/°C or refer to Dissipation Derating Curve, Figure 11.
5. This rating is based on the capability of the transistor to operate safely in the circuit of Figure 2. L = 20 mH, R_{BB2} = 100 Ω, V_{BB2} = 0 V, R_S = 0.1 Ω, V_{CC} = 10 V. Energy ≈ 1/2 I_C^2L/2.

Figure A-4(b)

Reference Data A-21

TYPES TIP34, TIP34A, TIP34B, TIP34C
P-N-P SINGLE-DIFFUSED MESA SILICON POWER TRANSISTORS

electrical characteristics at 25°C case temperature

PARAMETER		TEST CONDITIONS	TIP34 MIN	TIP34 MAX	TIP34A MIN	TIP34A MAX	TIP34B MIN	TIP34B MAX	TIP34C MIN	TIP34C MAX	UNIT
$V_{(BR)CEO}$	Collector-Emitter Breakdown Voltage	$I_C = -30$ mA, $I_B = 0$, See Note 6	−40		−60		−80		−100		V
I_{CEO}	Collector Cutoff Current	$V_{CE} = -30$ V, $I_B = 0$		−0.7		−0.7					mA
		$V_{CE} = -60$ V, $I_B = 0$						−0.7		−0.7	
I_{CES}	Collector Cutoff Current	$V_{CE} = -40$ V, $V_{BE} = 0$		−0.4							mA
		$V_{CE} = -60$ V, $V_{BE} = 0$				−0.4					
		$V_{CE} = -80$ V, $V_{BE} = 0$						−0.4			
		$V_{CE} = -100$ V, $V_{BE} = 0$								−0.4	
I_{EBO}	Emitter Cutoff Current	$V_{EB} = -5$ V, $I_C = 0$		−1		−1		−1		−1	mA
h_{FE}	Static Forward Current Transfer Ratio	$V_{CE} = -4$ V, $I_C = -1$ A, See Notes 6 and 7	40		40		40		40		
		$V_{CE} = -4$ V, $I_C = -3$ A, See Notes 6 and 7	20	100	20	100	20	100	20	100	
V_{BE}	Base-Emitter Voltage	$V_{CE} = -4$ V, $I_C = -3$ A, See Notes 6 and 7		−1.6		−1.6		−1.6		−1.6	V
		$V_{CE} = -4$ V, $I_C = -10$ A, See Notes 6 and 7		−3		−3		−3		−3	
$V_{CE(sat)}$	Collector-Emitter Saturation Voltage	$I_B = -0.3$ A, $I_C = -3$ A, See Notes 6 and 7		−1		−1		−1		−1	V
		$I_B = -2.5$ A, $I_C = -10$ A, See Notes 6 and 7		−4		−4		−4		−4	
h_{fe}	Small-Signal Common-Emitter Forward Current Transfer Ratio	$V_{CE} = -10$ V, $I_C = -0.5$ A, f = 1 kHz	20		20		20		20		
$|h_{fe}|$	Small-Signal Common-Emitter Forward Current Transfer Ratio	$V_{CE} = -10$ V, $I_C = -0.5$ A, f = 1 MHz	3		3		3		3		

NOTES: 6. These parameters must be measured using pulse techniques. $t_w = 300$ μs, duty cycle ≤ 2%.
7. These parameters are measured with voltage-sensing contacts separate from the current-carrying contacts and located within 0.125 inch from the device body.

thermal characteristics

PARAMETER		MAX	UNIT
$R_{\theta JC}$	Junction-to-Case Thermal Resistance	1.66	°C/W
$R_{\theta JA}$	Junction-to-Free-Air Thermal Resistance	35.7	

switching characteristics at 25°C case temperature

PARAMETER		TEST CONDITIONS†			TYP	UNIT
t_{on}	Turn-On Time	$I_C = -6$ A,	$I_{B(1)} = -0.6$ A,	$I_{B(2)} = 0.6$ A,	0.4	μs
t_{off}	Turn-Off Time	$V_{BE(off)} = 4$ V,	$R_L = 5$ Ω,	See Figure 1	0.7	

†Voltage and current values shown are nominal; exact values vary slightly with transistor parameters.

Figure A-4(b) (*Continued*)

B | Design of Toroidal Inductors and Transformers

The switch-mode regulators introduced in Chapter 9 generally require special inductors and transformers. These inductors and transformers must handle relatively large amounts of current (typically several amperes) at moderately high frequencies (20 kHz to 100 kHz) with high efficiency. Although it is possible to purchase suitable prewound inductors and transformers for large-scale production purposes, inductors and transformers suitable for prototyping and experimentation may be difficult to obtain. Fortunately, these inductors and transformers are not difficult to design and construct by the procedures described in this appendix.

Inductors and transformers for switch-mode regulators are generally wound on either ferrite pot cores or powdered-iron toroidal cores. The ferrite pot cores are easier to wind and are inherently self-shielding, but the powdered-iron toroidal cores are capable of higher flux densities and are generally preferred. Both ferrite cores and powdered-iron cores are available in several grades with different frequency responses, and it is important to select a core with a frequency range suitable for the design switching frequency. It is also important to use a core material that has soft saturation. Cores that saturate abruptly can produce excessive peak currents in the switching transistors.

Because toroidal cores are more popular than ferrite cores, our designs will use toroidal cores exclusively. Suitable toroidal cores may be difficult to find. One company that handles a full line of toroidal cores is Neosid, and Figure B-1 is adapted from their catalog, with permission. We will use the data from this catalog as the basis for our inductor and transformer designs.

The frequency range of the cores listed in Figure B-1 are indicated by color. For switch-mode regulator applications, the yellow cores, with a frequency range of 50 Hz to 250 kHz, are the most suitable.

Inductors and transformers are wound with standard enamelled magnet wire, available from most electronics distributors. The sizes and resistances of the standard wire gauges of bare copper wire are listed in Table B-1. The enamel insulation is typically 0.001 in. (0.025 mm) thick, hence the diameter of enamelled wire is approximately 0.002 in. (0.050 mm) larger than the values listed in the table.

Design of Toroidal Inductors and Transformers A-23

NEOSID

Catalogue Listings

NEOSID PART NO.	DIMENSIONS O.D. in./mm.	I.D. in./mm.	HT. in./mm.	AL (μh/100 turns)	ℓ cm	A cm^2	V cm^3
N 37T 17BL 175 N 37T 17YW 350 N 37T 17GR 490	.375/9.5	.175/4.4	.190/4.8	175 350 490	2.19	.123	.268
N 44T 23BL 125 N 44T 23YW 250 N 44T 23GR 340	.440/11.2	.235/6.0	.160/4.1	125 250 340	2.68	.106	.284
N 50T 30BL 125 N 50T 30YW 250 N 50T 30GR 340	.500/12.7	.300/7.6	.190/4.8	125 250 340	3.19	.123	.392
N 56T 32BL 125 N 56T 32YW 250 N 56T 32GR 340	.560/14.2	.320/8.1	.190/4.8	125 250 340	3.51	.147	.516
N 56T 32BL 175 N 56T 32YW 350 N 56T 32GR 490	.560/14.2	.320/8.1	.250/6.4	175 350 490	3.51	.193	.677
N 68T 38BL 140 N 68T 38YW 280 N 68T 38GR 385	.680/17.3	.380/9.7	.190/4.8	140 280 385	4.23	.184	.778
N 68T 38BL 180 N 68T 38YW 360 N 68T 38GR 500	.680/17.3	.380/9.7	.250/6.4	180 360 500	4.23	.242	1.02
N 82T 51BL 150 N 82T 51YW 300 N 82T 51GR 410	.820/20.8	.510/13.0	.250/6.4	150 300 410	5.31	.249	1.32
N 82T 51BL 300 N 82T 51YW 600 N 82T 51GR 820	.820/20.8	.510/13.0	.500/12.7	300 600 820	5.31	.499	2.65
N 97T 50BL 310 N 97T 50YW 620 N 97T 50GR 875	.970/24.6	.500/12.7	.380/9.7	310 620 875	5.86	.576	3.37
N106T 57BL 235 N106T 57YW 470 N106T 57GR 665	1.060/26.9	.570/14.5	.310/7.9	235 470 665	6.50	.489	3.18
N106T 57BL 335 N106T 57YW 670 N106T 57GR 900	1.060/26.9	.570/14.5	.440/11.2	335 670 900	6.50	.694	4.51
N106T 57BL 440 N106T 57YW 880 N106T 57GR1200	1.060/26.9	.570/14.5	.580/14.7	440 880 1200	6.50	.911	5.92
N130T 79BL 270 N130T 79YW 540 N130T 79GR 765	1.300/33.0	.790/20.1	.440/11.2	270 540 765	8.34	.724	6.34
N157T 95BL 350 N157T 95YW 700 N157T 95GR 975	1.570/40.0	.950/24.1	.560/14.2	350 700 975	10.05	1.13	11.3
N185T 93BL 615 N185T 93YW1230 N185T 93GR1700	1.850/47.0	.930/23.6	.720/18.3	615 1230 1700	11.09	2.14	23.7
N200T125BL 300 N200T125YW 600 N200T125GR 820	2.000/50.8	1.250/31.8	.500/12.7	300 600 820	12.97	1.21	15.7
N200T125BL 600 N200T125YW1200 N200T125GR1640	2.000/50.8	1.250/31.8	1.000/25.4	600 1200 1640	12.97	2.42	31.4

Outside Diameter (O.D.)
Inside Diameter (I.D.)
Height (HT.)
Mean Magnetic Path Length (ℓ)
Cross-Sectional Area (A)

Core Coating Colour	Permeability Range (μ)	Nominal Permeability (μ)	Suggested Frequency Range
Blue (BL)	22 - 28	25	250kHz to 2MHz
Yellow (YW)	45 - 55	50	50Hz to 250kHz
Green (GR)	65 - 75	70	50Hz to 50kHz

Figure B-1 Toroidal core data.

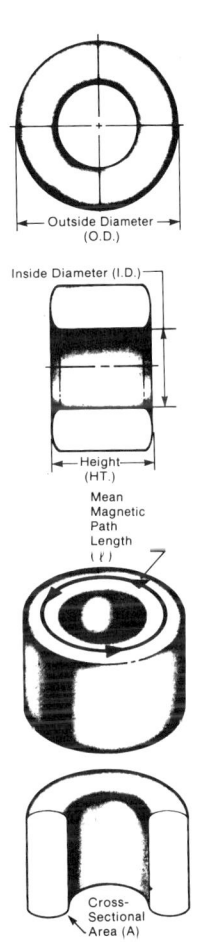

Table B-1 Standard wire data

Gauge	Diameter (in.)	Diameter (mm)	Resistance (Ω/ft at 25°C)	Resistance (Ω/m at 25°C)
1	0.2893	7.348	0.0001264	0.0004146
2	0.2576	6.543	0.0001593	0.0005225
3	0.2294	5.827	0.0002009	0.0006590
4	0.2043	5.189	0.0002533	0.0008308
5	0.1819	4.620	0.0003195	0.001048
6	0.1620	4.115	0.0004028	0.001321
7	0.1443	3.665	0.0005080	0.001666
8	0.1285	3.264	0.0006405	0.002101
9	0.1144	2.906	0.0008077	0.002649
10	0.1019	2.588	0.001018	0.003339
11	0.09074	2.305	0.001284	0.004212
12	0.08081	2.053	0.001619	0.005310
13	0.07196	1.828	0.002042	0.006698
14	0.06408	1.628	0.002575	0.008446
15	0.05707	1.450	0.003247	0.01065
16	0.05082	1.291	0.004094	0.01343
17	0.04526	1.150	0.005163	0.01693
18	0.04030	1.024	0.006510	0.02135
19	0.03589	0.9116	0.008210	0.02693
20	0.03196	0.8118	0.01035	0.03395
21	0.02846	0.7229	0.01305	0.04280
22	0.02535	0.6439	0.01646	0.05399
23	0.02257	0.5733	0.02076	0.06809
24	0.02010	0.5105	0.02617	0.08584
25	0.01790	0.4547	0.03300	0.1082
26	0.01594	0.4049	0.04162	0.1365
27	0.01420	0.3607	0.05248	0.1721
28	0.01264	0.3211	0.06617	0.2170
29	0.01126	0.2860	0.08344	0.2737
30	0.01003	0.2548	0.1052	0.3451
31	0.008928	0.2268	0.1327	0.4353
32	0.007950	0.2019	0.1673	0.5487
33	0.007080	0.1798	0.2110	0.6921
34	0.006305	0.1601	0.2660	0.8725
35	0.005615	0.1426	0.3350	1.099
36	0.005000	0.1270	0.4230	1.387
37	0.004453	0.1131	0.5334	1.749
38	0.003965	0.1007	0.6726	2.206
39	0.003531	0.08969	0.8481	2.782
40	0.003145	0.07988	1.069	3.506

Design of Inductors

We will assume that the required inductance has been calculated, using either Equation 9-12 for a step-down supply, Equation 9-27 for a step-up supply, or Equation 9-40 for a voltage-inverting supply. Two design decisions must be made. First, we must select the wire gauge to be used. This depends on the current that will be flowing: As a general rule of thumb, if the current is less than 2 A, use 20 gauge wire; if the current is between 2 and 8 A, use 18 gauge wire; if the current is between 8 and 12 A, use 16 gauge wire; and if larger than 12 A, use 14 gauge wire. Any wire gauge heavier than 18 gauge is relatively stiff and difficult to wind.

The second design decision is to select the toroidal core. The data sheets list a quantity A_L, the inductance in µH per 100 turns, for each core. The number of turns N necessary to give any required inductance L in µH is given by:

$$N = 100 \sqrt{\left(\frac{L}{A_L}\right)} \tag{B-1}$$

For optimum performance, the windings should be single-layer; hence the core size is chosen such that the desired inductance is achieved with a single layer of the selected wire size.

Figure B-2 shows the typical configuration of windings on a toroid core. Normally, a core size is chosen by engineering judgment (an intelligent guess), the number of windings calculated, and the suitability of the core verified by calcu-

(a) Face view of the windings

(b) Cross-sectional view of the windings

(c) Definition of winding length

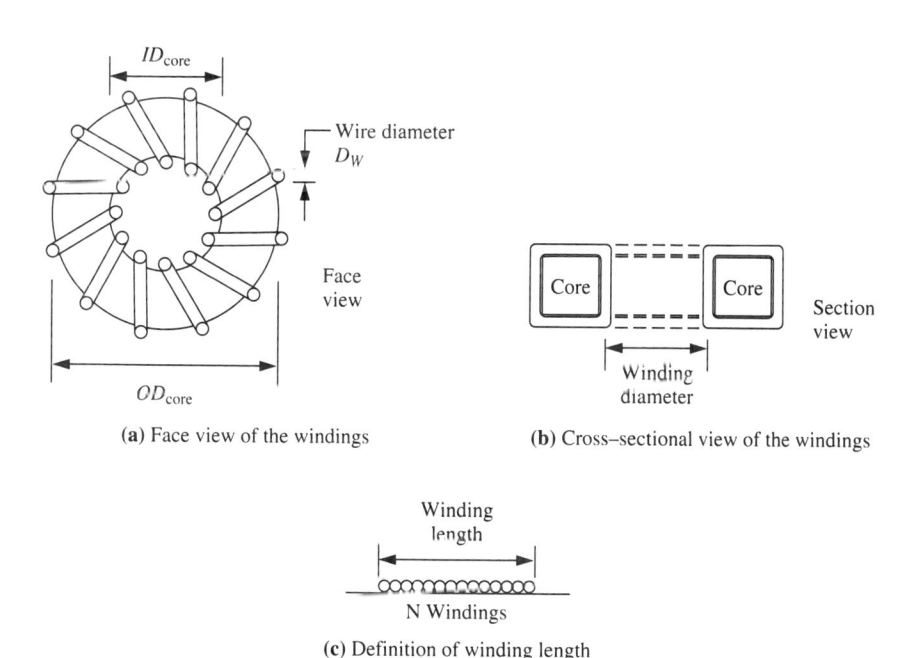

Figure B-2 Configuration of windings on a toroidal core.

lation. To verify that the windings will actually fit on the chosen core, we must verify that the winding circumference is larger than the winding length. The winding circumference is the space available on the inside of the core for windings and is illustrated in Figure B-2(a). First, determine the winding diameter as the distance measured from the center of the winding on one side to the center of the opposite winding, as illustrated in Figure B-2(b); the winding diameter is equal to the inside diameter of the core ID_{core} minus the diameter D_W of the wire:

$$\text{winding diameter} = ID_{core} - D_W \tag{B-2}$$

The winding circumference is π times this diameter:

$$\text{winding circumference} = \pi \times \text{winding diameter} \tag{B-3}$$

The winding length is the length of all the windings placed side by side and is equal to the number of windings N times the wire diameter D_W.

$$\text{winding length} = N \times D_W \tag{B-4}$$

The winding circumference must be larger than the winding length if a single-layer winding is to be wound on the core. If the winding circumference turns out to be much larger than the winding length, then generally too large a core has been chosen. The procedures for choosing core size and wire diameter are largely developed as a matter of practical experience.

We must also estimate the total length of wire required for the winding. The length is given approximately by the equation:

$$\text{wire length} = N \times (OD_{core} - ID_{core} + 2HT_{core} + 4D_W) \tag{B-5}$$

It is wise to add at least 10% to this value to account for leads and loose windings.

Finally, we should verify that the time constant of the inductor is larger than the period of the pulse. If the time constant of the inductor is less than the period of the pulse, the inductor current will become constant, and regulation will not be possible. The time constant of an inductor is L/R, where R is the internal resistance of the windings. The winding resistance is simply the wire resistance given in Table B-1, times the length. If the time constant is too short, either a higher switching frequency or a larger diameter wire will have to be selected. In actual operation, any external circuit resistance, such as resistance in the switching transistor, diode, printed circuit board traces, capacitor leads, or DC voltage source, will be added to the winding resistance to determine the time constant. It is very important to keep this external resistance to a minimum.

EXAMPLE B-1

Determine the size of a suitable core and the number of windings for an inductor in a switch-mode regulator if the inductor has an inductance of 17 μH and must pass a current of 10 A at a switching frequency of 80 kHz.

Solution

To design the inductor, we must choose both a core and a suitable diameter of wire. Since our current is 10 A, we should select at least 16 gauge wire.

For a frequency of 80 kHz, it is best to select the yellow-core-type material. We shall try core number N82T51YW300 from the data sheet shown in Figure B-1. This core has an A_L value of 300. Substitute the required inductance of 17 µH into Equation B-1 to calculate the number of turns:

$$N = 100 \sqrt{\left(\frac{L}{A_L}\right)} = 100 \sqrt{\frac{17}{300}} = 23.8$$

A value of 24 turns is reasonable.

Now we must verify that it is possible to wind the coil on the proposed core. From Table B-1 we find that 16 gauge wire has a diameter of 1.291 mm. Assume the enamel coating is 0.025 mm thick, hence the effective diameter is 1.291 mm + 0.05 mm = 1.34 mm. The winding diameter, given by Equation B-2, is:

winding diameter = $ID_{core} - D_W$
= 13.0 mm − 1.34 mm = 11.66 mm

The winding circumference from Equation B-3 is:

winding circumference = π × winding diameter
= 36.63 mm

The winding length from Equation B-4 is:

winding length = $N \times D_W$
= 24 × 1.34 mm = 32.16 mm

Since the winding length is less than the winding circumference, the winding will fit on the core, and our choice of core is satisfactory. If it had been larger, we would have had to select a larger core. If it had been much less, we would have had to select a smaller core.

Next, estimate the total length of wire by Equation B-5:

wire length = $N \times (OD_{core} - ID_{core} + 2 \times HT_{core} + 4D_W)$
= 24 × (20.8 mm − 13.0 mm + 2 × 6.4 mm + 4 × 1.34 mm)
= 623 mm

Since it is wise to add at least 10% to this value to account for leads and loose windings, we will require approximately 0.68 m of wire.

Finally, calculate the inductor time constant. The winding resistance is 0.01343 Ω/m × 0.68 m = 0.0091 Ω. The time constant is L/R = 17 µH/0.0091 Ω = 1.87 msec. This is much greater than the period of the pulse T = 1/80 kHz = 0.0125 msec; hence both the wire diameter and pulse frequency are adequate. Note that the actual time constant will be smaller when the external circuit resistance is included. This external resistance must be kept to a minimum if the regulator is to operate.

This example has illustrated how an inductor can be designed. It is largely a process of trial and error. Sometimes a different core will have to be chosen, sometimes a different wire diameter must be used.

Design of Transformers

Like inductors, transformers are designed by choosing a wire size on the basis of the current to be carried, selecting a core size by engineering judgment, and calculating the number of turns required. The number of turns in the primary winding must be as large as possible for efficient coupling and maximum inductance, yet must be kept small enough to prevent the magnetic field from causing core saturation. From Faraday's law, the voltage induced in a coil is proportional to the rate of change of flux:

$$V = \frac{\Delta \Phi}{\Delta t} \tag{B-6}$$

Rearrange this equation to solve for the flux in terms of the applied voltage $V = V_{DC} - V_{sat}$ and the pulse period $f_{osc} = 1/\Delta t$.

$$\Delta \Phi = \frac{(V_{DC} - V_{sat})}{f_{osc}} \tag{B-7}$$

The flux Φ is equal to the flux density B in the core times the cross-sectional area of the core A times the number of turns N: $\Phi = NAB$. Since the flux density B must not exceed the saturation flux density B_{sat}, $\Delta \Phi = NAB_{sat}$. Substitute this into Equation B-7 and solve for the number of turns N:

$$N = \frac{V_{DC} - V_{sat}}{B_{sat} f_{osc} A} \tag{B-8}$$

This is the design equation for toroidal transformers. Note that the cross-sectional area of the core A is in m² but is given on the data sheets shown in Figure B-1 in cm²—divide the given value by 10,000 to convert to m². The value of B_{sat} depends on the particular core material used. For the powdered iron toroidal cores shown in Figure B-1, the value for B_{max} is approximately 0.025 W/m².

The number of turns required for each secondary winding is quite easy to calculate. Because of the high frequencies, the transformer can be assumed to be operating almost ideally; hence the number of turns for each secondary is determined directly from the voltage ratio using the following equation:

$$N_{sec} = N_{pri} \times \frac{V_{sec}}{V_{pri}} \tag{B-9}$$

The secondary voltage V_{sec} is normally chosen several volts larger than the desired *regulated* output voltage. The primary voltage V_{pri} is simply $V_{DC} - V_{sat}$.

Use the same criteria for selecting wire size as were used for inductors; namely, if the current is less than 2 A, use 20 gauge wire; if the current is between

2 and 8 A, use 18 gauge wire; if the current is between 8 and 12 A, use 16 gauge wire; and if larger than 12 A, use 14 gauge wire. To determine the wire size for the primary, however, it is first necessary to determine the primary current. The component of primary current due to each secondary is given by the following equation:

$$I_{\text{pri}} = I_{\text{sec}} \times \frac{N_{\text{sec}}}{N_{\text{pri}}} \tag{B-10}$$

The secondary current I_{sec} is the desired output current I_{out} for the secondary. Add the primary currents for each secondary to find the total primary current. Use this current to determine the primary wire size.

The toroidal core is selected in the same way as for an inductor, by choosing a size such that the primary windings can be accommodated in a single layer. The primary should be wound first, covering the entire core. The secondary windings are wound on top of the primary winding. The order is not important.

The inductive time constant of the primary winding must be larger than the pulse length to ensure that the pulse is effectively coupled to the secondary. If the time constant is less than the pulse width, only a small fraction of the primary current will be transferred to the secondary, and there will be no regulation. To find the time constant of the primary, calculate the inductance of the primary by solving Equation B-1 for L:

$$L = A_L \times \left(\frac{N}{100}\right)^2 \tag{B-11}$$

Calculate the resistance from the wire length in the same way as for an inductor. The time constant is simply L/R.

EXAMPLE B-2

Design a transformer to allow a switch-mode regulator operating from a 12 V DC supply to give output voltages of 12 V at 1 A, 8 V at 2 A, and 5 V at 4 A. Use a switching frequency of 80 kHz.

Solution

Select a suitable toroidal core from the data sheet in Figure B-1. Because this transformer will handle a relatively large amount of current, and because four windings will be required, we should choose a relatively large core. We will try Neosid number N157T95YW700. Note that we choose a yellow core as having the most suitable frequency characteristics.

Calculate the number of primary windings from Equation B-8.

$$N = \frac{V_{\text{DC}} - V_{\text{sat}}}{B_{\text{sat}} f_{\text{osc}} A}$$

$$= \frac{12.0 \text{ V} - 1.0 \text{ V}}{0.025 \text{ W/m}^2 \times 80 \text{ kHz} \times 0.000113 \text{ m}^2} = 45 \text{ windings}$$

Design of Toroidal Inductors and Transformers

Note that we converted the area given on the data sheet in square centimeters to square meters. The B_{max} value of 0.025 W/m² is suggested as a typical value.

Now calculate the secondary windings. Use Equation B-9 with the primary voltage $V_{pri} = V_{DC} - V_{sat} = 12.0 \text{ V} - 1.0 \text{ V} = 11 \text{ V}$.

$$N_{sec} = N_{pri} \times \frac{V_{sec}}{V_{pri}}$$

For the 12 V secondary, we will choose a transformer output voltage of 16 V, 4 V more than the final regulated voltage. The number of turns is:

$$N_{12} = 45 \times \frac{16 \text{ V}}{11 \text{ V}} = 65 \text{ turns}$$

For the 8 V secondary, we will choose a transformer output voltage of 12 V. The number of turns is:

$$N_8 = 45 \times \frac{12 \text{ V}}{11 \text{ V}} = 49 \text{ turns}$$

Finally, for the 5 V secondary, we will choose a transformer output voltage of 8 V. The number of turns is:

$$N_5 = 45 \times \frac{8 \text{ V}}{11 \text{ V}} = 33 \text{ turns}$$

Now determine the wire sizes. First calculate the current in the primary winding using Equation B-10 for each secondary.

$$I_{pri} = I_{sec} \times \frac{N_{sec}}{N_{pri}}$$

For the 12 V secondary, we have 65 turns and a current of 1 A.

$$I_{pri} = 1 \text{ A} \times \frac{65}{45} = 1.44 \text{ A}$$

For the 8 V secondary, we have 49 turns and a current of 2 A.

$$I_{pri} = 2 \text{A} \times \frac{49}{45} = 2.18 \text{ A}$$

For the 5 V secondary, we have 33 turns and a current of 4 A.

$$I_{pri} = 4 \text{ A} \times \frac{33}{45} = 2.93 \text{ A}$$

The total primary current is 6.55 A. Choose 18 gauge wire for the primary, the 5 V secondary, and the 8 V secondary. Choose 20 gauge wire for the 12 V secondary.

Now we should verify that it is possible to place the primary winding on the proposed core. From Table B-1 we find that 18 gauge wire has a diameter of 1.024 mm. Assume the enamel coating is 0.025 mm thick; hence the effective diameter is 1.024 mm + 0.05 mm = 1.074 mm. The winding diameter is given by Equation B-2:

winding diameter = $ID_{core} - D_W$
= 24.1 mm − 1.744 mm = 22.4 mm

Calculate the winding circumference from Equation B-3:

winding circumference = $\pi \times$ winding diameter
= 70.2 mm

Finally, calculate the winding length from Equation B-4:

winding length = $N \times D_W$
= 45 × 1.074 mm = 48.3 mm

Since the winding length is less than the winding circumference, the winding will fit on the core, and our choice of core is satisfactory. It is left as an exercise for the reader to verify that the next smaller size core, the N130T79YW540, is too small. (Note that it will be necessary to recalculate the number of primary windings for this core.)

Next, estimate the total length of wire using Equation B-5:

wire length = $N \times (OD_{core} - ID_{core} + 2 \times HT_{core} + 4D_W)$

For the primary:

wire length = 45 × (40.0 mm − 24.1 mm
+ 2 × 14.2 mm + 4 × 1.07 mm)
= 2186 mm

Adding 10% to this value gives 2.4 m of wire.
For the 12 V secondary using 20 gauge wire:

wire length = 65 × (42.2 mm − 22.0 mm
+ 2 × 18.0 mm + 4 × 0.87 mm)
= 3658 mm

Adding 10% to this value gives 4.2 m of wire. Note that we added the thickness of the first winding to the coil *OD* and *HT* and subtracted it from the *ID* to compensate for the thickness of the first winding.
For the 8 V secondary:

wire length = 49 × (43.9 mm − 20.3 mm
+ 2 × 16.3 mm + 4 × 1.07 mm)
= 2963 mm

Adding 10% to this value gives 3.3 m of wire.
For the 5 V secondary:

wire length = 33 × (46.0 mm − 18.2 mm
+ 2 × 18.4 mm + 4 × 1.07 mm)
= 2273 mm

Adding 10% to this value gives 2.5 m of wire.

Finally, calculate the time constant due to the primary inductance. We have 45 turns on core N157T95YW700. The A_L value for this core is 700 μH for 100 turns. The inductance L, calculated using Equation B-11, is:

$$L = A_L \times \left(\frac{N}{100}\right)^2 = 700 \ \mu H \times \left(\frac{45}{100}\right)^2 = 142 \ \mu H$$

The winding resistance is 0.02135 Ω/m × 2.4 m = 0.0512 Ω. The time constant is L/R = 142 μH/0.0512 Ω = 2.77 msec. This is much greater than the period of the pulse T = 1/80 kHz = 0.125 msec; hence both the wire diameter and pulse frequency are adequate.

Winding of Toroidal Inductors and Transformers

Finally, we will present a few hints as to the actual winding of the toroidal inductors and transformers. If there are only a few windings, simply cut the necessary length of wire, thread it halfway through the core, and wind each half by threading the wire through the hole in the core.

If there are more than several feet of wire, however, simple winding becomes quite difficult. In this case, the use of a wrench-shaped bobbin is almost mandatory. In fact, a small open-end wrench that will fit through the hole in the toroidal core is ideal. Wind all the wire lengthwise over the wrench through the two open ends. Start winding at one end of the coil, passing the wrench through the hole in the toroidal core, unwinding lengths from the wrench as necessary.

The windings should be as tight as possible. Tape them in place using transformer tape. Black electrician's tape is satisfactory. If a number of layers of windings are necessary, as in a transformer, cover each successive layer with tape. Leave sufficient length at each end for leads, and make sure to strip the enamel insulation off the end of the lead. Inductors and transformers can be mounted vertically or horizontally. Vertical mounting is more common, since it requires less space.

Sine-Wave Shaping Circuit   C

We have already introduced several circuits that can be used to change the shape of a waveform. For example, a comparator can be used to convert a sine wave or triangle wave into a square wave, and an integrating circuit can be used to convert a square wave into a triangle wave. This appendix describes a simple passive circuit that can be used to convert a triangle wave to a sine wave, providing a frequency-independent method for producing a sine wave from a bootstrap oscillator or a voltage-controlled oscillator such as the LM566.

Figure C-1 shows a sine wave, a superimposed triangle wave, and a superimposed three-section straight-line approximation to the sine wave. In practice, the three-line approximation generates quite a good sine wave, with only a few percent distortion.

The circuit that we will use for generating the three-line approximation is shown in Figure C-2. This circuit uses switching diodes to vary the output of the triangle wave as a function of amplitude. The top two diodes shape the positive half of the wave, while the bottom two diodes shape the negative half of the wave. In designing this circuit, we will consider the top half only. The negative half is analyzed in exactly the same way, and the component values are the same for a symmetrical sine-wave output.

Resistors R_1, R_2, and R_3 form a voltage divider for biasing the diodes D_1 and D_2. Voltages V_1 and V_2 from the voltage divider determine where the triangle wave changes slope. Resistors R_{D1} and R_{D2}, in conjunction with R_{in}, determine the slope of the shaped triangle wave. The voltage divider resistors R_1, R_2, and R_3 must be smaller than the slope resistors R_{D1} and R_{D2} to prevent loading of the voltage divider.

If V_{in} is less than $V_1 + 0.7$, none of the diodes are conducting and $V_{out} = V_{in}$. If $(V_1 + 0.7) < V_{in} < (V_2 + 0.7)$, diode D_1 conducts, and V_{out} is given by:

$$V_{out} = V_{in} - (V_{in} - V_1 - 0.7) \times \frac{R_{in}}{R_{in} + R_{D1}}$$

$$= \frac{R_{D1}}{R_{in} + R_{D1}} \times V_{in} + \frac{R_{in}}{R_{in} + R_{D1}} \times (V_1 + 0.7)$$

(C-1)

The first term is the slope of the modified triangle wave. The second term is the break-point voltage and is a constant. For this portion of the curve, as V_{in} in-

Sine-Wave Shaping Circuit

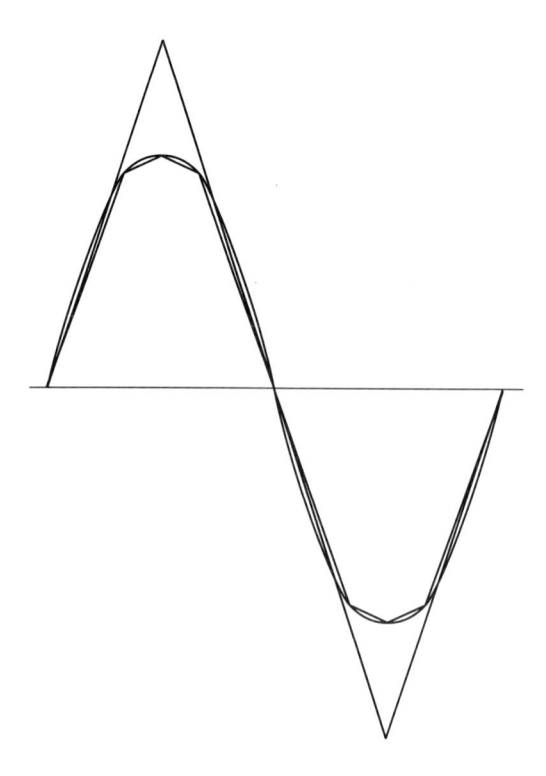

Figure C-1 Triangle wave, sine wave, and three-segment approximation to a sine wave.

creases, V_{out} also increases, but by the ratio $R_{D1}/(R_{in} + R_{D1})$. In other words, once D_1 starts to conduct, the slope of the output wave is less. Resistor R_{D1} must be designed to give the slope S_1 defined by:

$$S_1 = \frac{R_{D1}}{R_{in} + R_{D1}} \tag{C-2}$$

Solve Equation C-2 for R_{D1}:

$$R_{D1} = \frac{S_1 R_{in}}{1 - S_1} \tag{C-3}$$

This gives the design equation for R_{D1}. We will choose a value for R_{in} and estimate a suitable value for S_1.

If $V_{in} > (V_2 + 0.7)$, then D_2 conducts, and V_{out} is given by:

$$\begin{aligned} V_{out} &= V_{in} - (V_{in} - V_2 - 0.7) \times \frac{R_{in}}{R_{in} + R_{D2}} \\ &= \frac{R_{D2}}{R_{in} + R_{D2}} \times V_{in} + \frac{R_{in}}{R_{in} + R_{D2}} \times (V_2 + 0.7) \end{aligned} \tag{C-4}$$

As V_{in} continues to increase, V_{out} will also increase, but now by the ratio

Sine-Wave Shaping Circuit   A-35

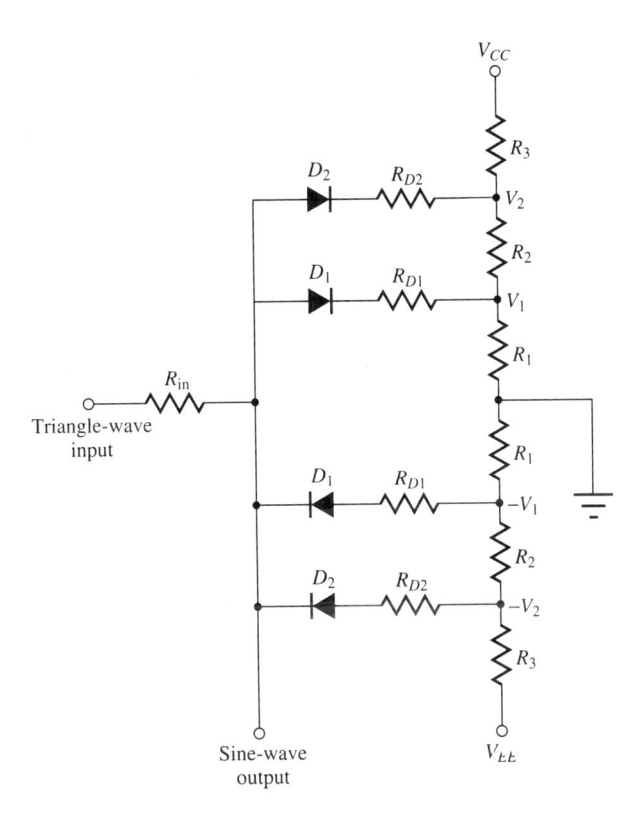

Figure C-2 Circuit for converting a triangle wave into a sine wave.

$R_{D2}/(R_{in} + R_{D2})$. If R_{D2} is smaller than R_{D1}, then the slope will be less. For the design of R_{D2}, we need the relative slope S_2 given as:

$$\text{relative slope} = \frac{R_{D2}}{R_{in} + R_{D2}} \tag{C-5}$$

Solve Equation C-5 for R_{D2}:

$$R_{D2} = \frac{S_2 R_{in}}{1 - S_2} \tag{C-6}$$

This gives the design equation for R_{D2}. As for R_{D1}, it is necessary to estimate a suitable value for S_2.

Finally, we must design the voltage divider consisting of R_1, R_2, and R_3. The standard equation for designing a voltage divider is:

$$R_X = R_T \times \frac{V_X}{V_T} \tag{C-7}$$

Resistance R_T is the total resistance $R_1 + R_2 + R_3$ and is chosen to be much smaller than R_{D2}. Resistor R_1, which determines V_1, is given by:

$$R_1 = R_T \times \frac{V_1}{V_T} \tag{C-8}$$

Resistor R_2, which determines V_2 is given by:

$$R_2 = R_T \times \frac{V_2 - V_1}{V_T} \tag{C-9}$$

Finally, R_3 is simply:

$$R_3 = R_T - R_1 - R_2 \tag{C-10}$$

Although this discussion considers a sine-wave approximation consisting of three line sections only, there is actually no limit to the number of line sections that can be used, and the sine wave can be approximated to any degree of precision. The formulas for calculating additional biasing voltages and diode resistors are similar to those derived here.

Although the design of the shaping circuit appears quite complicated, in fact, it is quite easy. The hardest part of the design is to determine where to place the break-points. This can be done graphically by trial and error to yield quite good results. Once the break-points are established, the slopes of the line sections can be determined.

This circuit is amplitude-dependent. The output wave shape depends very strongly on the amplitude of the input wave. If the amplitude is other than the design amplitude, then the break-points will occur in the wrong places, and a badly distorted sine wave will result. It is necessary to ensure that the triangle-wave oscillator gives a constant output amplitude.

EXAMPLE C-1

Design a wave-shaping circuit to convert a 20 V peak-to-peak triangle wave into a sine wave. The supply voltage is ±15 V.

Solution

The circuit that we will use is shown in Figure C-3. Since the bottom half of the circuit is identical to the top half with the exception of the polarity of the diodes, we need to design only the top half of the circuit.

The hardest part of the design procedure is to determine where to place the break-points. This is best done with a sketch. Figure C-4 shows a scaled sketch of the positive half of the triangle wave, of the sine wave that we want to shape, and of our approximate wave. The scale along the bottom is in degrees, since this circuit operates independently of frequency.

Choose an amplitude of 6 V for the sine wave. This is arbitrary, but a 6 V sine wave seems to fit the triangle wave quite well. Choose break-points at 30° and 60°. The break-voltage at 30° is approximately 3.4 V. Draw a line to represent

Sine-Wave Shaping Circuit A-37

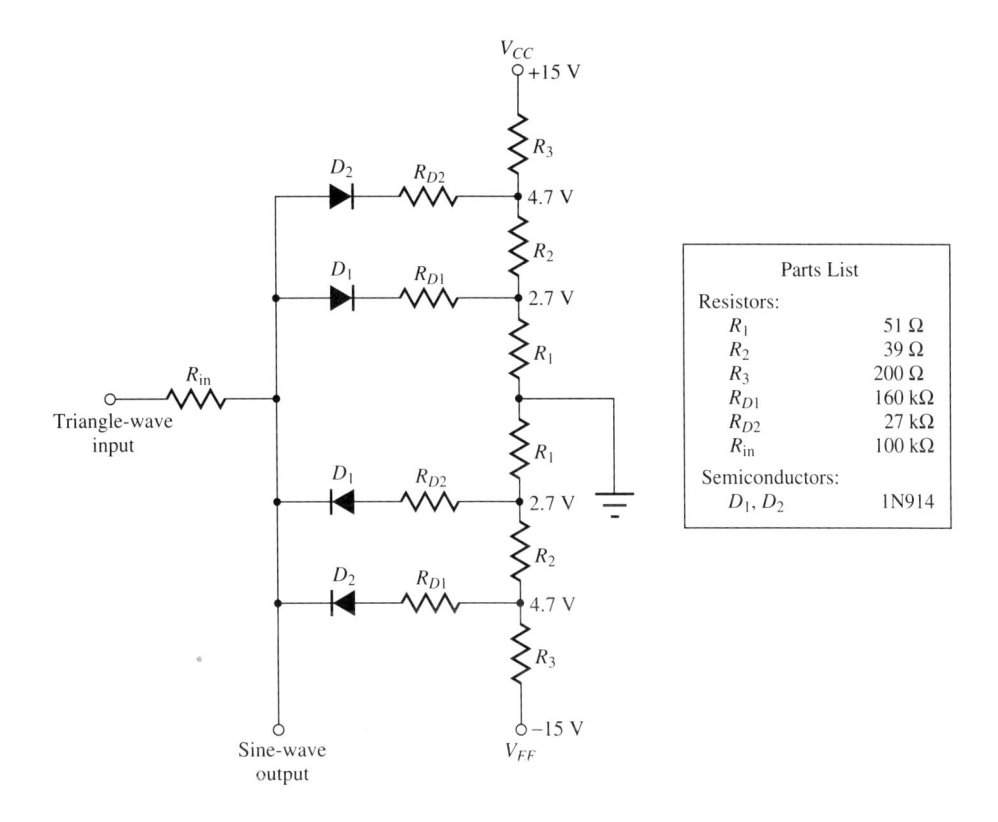

Figure C-3

Parts List
Resistors:
R_1 51 Ω
R_2 39 Ω
R_3 200 Ω
R_{D1} 160 kΩ
R_{D2} 27 kΩ
R_{in} 100 kΩ
Semiconductors:
D_1, D_2 1N914

the second portion of the curve through this point. This line has a slope S_1 of approximately 0.625 of the original triangle-wave slope. This number is calculated from the ratio of the rises of both lines from the first break-point to the second break-point. The second break-point voltage (at 60°) is approximately 5.4 V. Draw the third line segment to complete the approximation. This line has a slope S_2 of approximately 0.220 of the original triangle-wave slope.

Choose $R_{in} = 100$ kΩ. The slope of the second line section is $S_1 = 0.625$. Determine R_{D1} from Equation C 3:

$$R_{D1} = \frac{S_1 R_{in}}{1 - S_1} = \frac{0.625 \times 100 \text{ kΩ}}{1 - 0.625} = 167 \text{ kΩ}$$

Similarly, determine R_{D2} from Equation C-6:

$$R_{D2} = \frac{S_2 R_{in}}{1 - S_2} = \frac{0.22 \times 100 \text{ kΩ}}{1 - 0.22} = 28.2 \text{ kΩ}$$

We will choose values of 160 kΩ for R_{D1} and 27 kΩ for R_{D2}, respectively.

Now design the voltage divider. The supply voltage is 15 V with respect to the ground in the center of the divider. We will choose a value of 300 Ω for R_T.

Sine-Wave Shaping Circuit

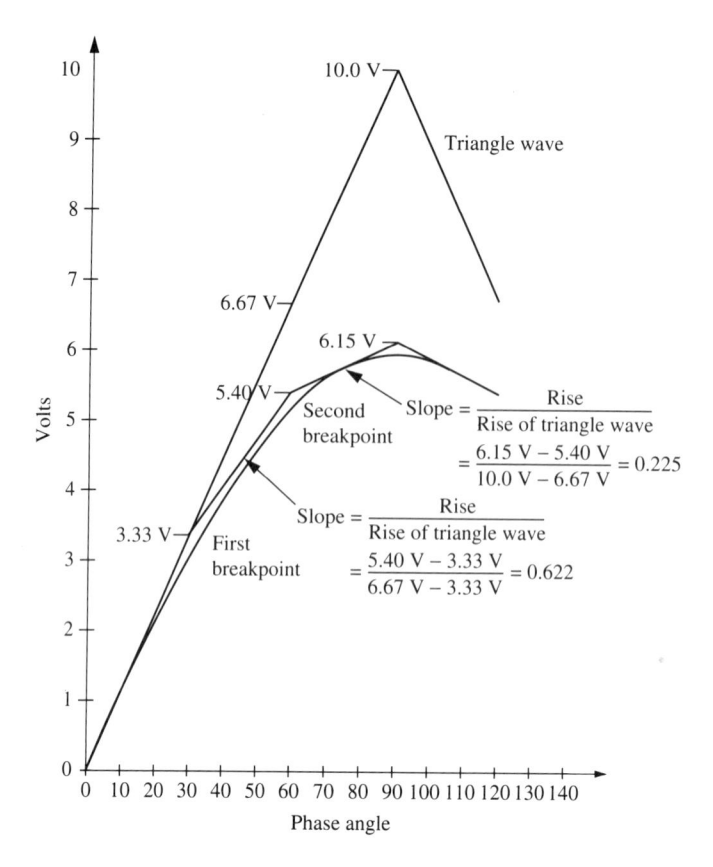

[Graph showing triangle wave peaking at 10.0 V and approximated sine wave with breakpoints]

- Triangle wave
- 10.0 V
- 6.67 V
- 6.15 V
- 5.40 V — Second breakpoint
- 3.33 V — First breakpoint

Slope = $\dfrac{\text{Rise}}{\text{Rise of triangle wave}}$ = $\dfrac{6.15\text{ V} - 5.40\text{ V}}{10.0\text{ V} - 6.67\text{ V}}$ = 0.225

Slope = $\dfrac{\text{Rise}}{\text{Rise of triangle wave}}$ = $\dfrac{5.40\text{ V} - 3.33\text{ V}}{6.67\text{ V} - 3.33\text{ V}}$ = 0.622

X-axis: Phase angle (0 to 140); Y-axis: Volts (0 to 10)

Figure C-4

The first break-point voltage is $(V_1 + 0.7) = 3.4$ V; hence $V_1 = 2.7$ V. Determine a value for R_1 from Equation C-8:

$$R_1 = R_T \times \frac{V_1}{V_T} = 300 \text{ }\Omega \times \frac{2.7 \text{ V}}{15.0 \text{ V}} = 54 \text{ }\Omega$$

Similarly we calculate $V_2 = 4.7$ V from the second break-point voltage of 5.4 V. Determine R_2 from Equation C-9:

$$R_2 = R_T \times \frac{V_2 - V_1}{V_T} = 300 \text{ }\Omega \times \frac{4.7 \text{ V} - 2.7 \text{ V}}{15.0 \text{ V}} = 40 \text{ }\Omega$$

Finally, calculate R_3 from Equation C-10:

$$R_3 = R_T - R_1 - R_2 = 300 \text{ }\Omega - 54 \text{ }\Omega - 40 \text{ }\Omega = 206 \text{ }\Omega$$

We will choose standard values of 51 Ω for R_1, 39 Ω for R_2, and 200 Ω for R_3.

Our design is complete. At this point, the circuit should be constructed to verify that a suitably shaped sine wave is produced. Although the linear approximation has some sharp corners, in practice these corners will be rounded due to the finite frequency response of the components, making the actual waveform an even closer approximation to a sine wave.

There is no best solution to this problem. The reader is urged to try different break-points and different sine-wave amplitudes to find equivalent or better solutions.

D | Glossary

This glossary contains definitions of all of the terms introduced as key terms in each chapter, as well as definitions of a number of other secondary terms.

AC coupling the process of connecting amplifiers using transformers (**transformer coupling**) or capacitors (**capacitor coupling**) to prevent DC bias signals from one stage from reaching the next stage.

Active filters filter circuits that use operational amplifiers, or other active devices, to increase the performance of simple R-C filters.

All-pass filter a filter circuit that passes all frequencies unattenuated but introduces a phase shift at some critical frequency.

Amplifier a circuit or device that takes some input signal and outputs the same signal, unchanged in shape, but increased in voltage, current, and power. See also **Gain**.

Amplitude the value of the waveform voltage (or current) at any time.

Amplitude modulation (AM) a process of transmitting radio communication signals in which the audio signal appears as a variation in the amplitude of the radio frequency carrier wave.

Analog computer an early type of computer that used linear electronic circuits to simulate a mathematical problem.

Analog integrated circuit see **Linear integrated circuit.**

Audio amplifier an amplifier specifically designed for reproducing sound.

Audio frequency (or AF) amplifier an amplifier specifically designed for use at frequencies less than 100 kHz.

Audio mixer a circuit used for combining two or more audio inputs, essentially a **summing amplifier.**

Audio power amplifier an audio amplifier with a low voltage gain, but a high current gain and high power gain, designed specifically to drive low impedance loads such as speakers.

Audio taper potentiometer a potentiometer in which the resistance varies logarithmically with the position of the wiper. These are used in audio systems to simulate the response of the ear to sound intensity.

Automatic volume control (AVC) a feedback system used in radio and television receivers to maintain an approximately constant output level for a wide range of input levels.

Automatic frequency control (AFC) a feedback system used in FM receivers to prevent the local oscillator frequency from drifting off the station being received.

Average amplitude the DC level at which the area of the wave above this level is equal to the area of the wave below this level.

Balance control the provision for varying the relative gain between the two channels of a stereo system.

Bandpass filter a filter circuit that passes only a range of frequencies, between a lower critical frequency and a higher critical frequency, rejecting all lower and higher frequencies.

Band-rejection filter a filter circuit that rejects a range of frequencies, between a lower critical frequency and an upper critical frequency, passing all lower and higher frequencies.

Bandwidth the range of frequencies over which an amplifier will have approximately full gain. For filter circuits, it is the range of frequencies passed by a bandpass filter or rejected by a band-rejection filter.

Barkhausen criteria a set of requirements that must be satisfied if an oscillator is to produce a perfect sine wave.

Bass control a tone control that boosts or attenuates the low frequency gain with respect to the midband gain.

Bessel response a filter response characterized by a linear variation of phase with frequency but relatively slow initial frequency roll-off per pole.

BiFET operational amplifier an operational amplifier constructed using field effect transistors in the input circuits and bipolar transistors for the rest of the circuit.

Bipolar operational amplifier an operational amplifier constructed using bipolar transistors only.

Bode plot a straight-line approximation to the frequency response of a filter or amplifier.

Bootstrap oscillator an oscillator circuit capable of simultaneously producing triangle waves and square waves.

Broadband amplifier a radio frequency amplifier specifically designed to receive a wide range of frequencies. Also called a **video amplifier**.

Broadband filter a bandpass or band-rejection filter that passes or rejects a very wide range of frequencies.

Butterworth response a filter response characterized by a flat frequency passband response and a roll-off of 20 dB per decade per pole. The phase response is nonlinear.

Capacitor coupling see **AC coupling**.

Capture range the range of frequencies over which the voltage-controlled oscillator of a phase-locked loop can initially lock onto the input frequency. The capture range is always *less than or equal to* the lock range.

Cauer response a filter response with an initial very rapid roll-off at the critical frequency but with ripple both in the passband and in the stopband and a very nonlinear phase response. These are also known as **elliptic filters**.

Center frequency the frequency of the middle of the passband or stopband of a filter, expressed as the geometric average of the upper critical frequency and the lower critical frequency. See also **Free-running frequency**.

Chebyshev response a filter response with a faster initial roll-off than a **Butterworth response** but a frequency response that is not flat.

Closed-loop gain the gain of an amplifier with a feedback loop connected between input and output.

Common-mode gain the amount of gain experienced by a signal present at both inputs to a differential amplifier.

Common-mode rejection ratio The ratio of the gain for a signal present at only one input of a differential amplifier to the gain for the same signal if it were present at both inputs.

Comparator an op-amp circuit or special integrated circuit that compares two input voltage levels.

Composite stereo signal the stereo signal as output by the FM detector stage, consisting of the $L + R$ channel signal, the upper and lower sidebands of the $L - R$ channel signal, and the pilot signal.

Critical frequency the frequency defining the passband of a filter.

Crystal oscillator a high-frequency oscillator circuit in which the frequency is determined by a quartz crystal.

Current gain see **Gain**.

Current-mode amplifier see **Norton amplifier**.

Cycle the repeating waveform shape.

Darlington configuration a circuit consisting of two transistors, with the emitter of the first connected directly to the base of the second to produce an effective transistor with a very high gain.

DC or direct coupling the process of connecting amplifiers directly without using transformers or capacitors to allow the amplifier to respond to very low-frequency signals, including DC.

Decade an increase or decrease in frequency by a factor of 10.

Decibel the gain of an amplifier expressed in logarithmic terms. A decibel is 20 times the logarithm (base 10) of the voltage or current gain or 10 times the logarithm (base 10) of the power gain.

De-emphasis the removal of the pre-emphasis boost from the audio signal after FM detection. See **pre-emphasis**.

Detector the circuit in communication receivers which separates the audio signal from the radio frequency carrier signal.

Differencing amplifier see **Subtracting amplifier**.

Differential amplifier an amplifier with two inputs that amplifies the difference between the two inputs.

Differentiating amplifier an amplifier that produces the derivative of the input waveform.

Differentiation an operation from calculus that determines the rate of change of one variable with respect to another.

Digital integrated circuit an integrated circuit designed to switch between discrete signal levels.

Dual in-line package (DIP) a type of package for integrated circuits consisting of a rectangular case with two rows of pins.

Duty cycle the ratio of the on time to the total period for a pulse waveform.

Electronic filter a circuit that passes a certain range of frequencies and rejects all other frequencies.

Elliptic filter see **Cauer response**.

Error amplifier see **Error detector**.

Error detector an amplifier circuit that provides an output signal that is proportional to the difference between two signal levels. Also called an **error amplifier**.

Error signal in a phase comparator, a DC signal proportional to the phase difference of the input signals.

Exponential amplifier an amplifier in which the output is proportional to the exponent of the input. This type of amplifier is used in conjunction with the **logarithmic amplifier** for square root extraction, squaring, multiplying, and dividing.

Feedback the process by which a portion of the output signal is routed back to the input of an amplifier. **Negative feedback** stabilizes and improves the performance of an amplifier. **Positive feedback** is necessary for oscillations to occur.

Filter see **Simple filter** and **Power supply filter**.

Floating regulator a fixed-voltage regulator in which the common voltage is not ground.

FM demodulator a circuit for detecting the audio or encoded information from a frequency-modulated signal.

Foldback current limiting a method of limiting the output current of a voltage regulator by reducing the output voltage.

Fourier analysis the procedure of describing any periodic waveform as a series of pure sinusoidal harmonics.

Free-running frequency the frequency of a phase-locked loop when it is not locked to the incoming signal. Also called the **idle frequency** or the **center frequency**.

Frequency the reciprocal of the period expressed in cycles per second or Hertz.

Frequency converter a circuit for converting one frequency to a different frequency, usually using a phase-locked loop.

Frequency modulation (FM) a process of radio communication in which the variations of the amplitude of the audio signal are encoded as variations of the carrier frequency.

Frequency modulator a circuit that generates an output signal whose frequency is proportional to the amplitude of the input signal. Also known as a **voltage-to-frequency converter**.

Frequency response a plot of decibels of attenuation against the logarithm of the frequency.

Function generator a test instrument capable of producing a variety of waveforms over a wide range of frequencies.

Gain the amount of amplification produced by an amplifier. **Voltage gain** is a measure of the voltage amplification. **Current gain** is a measure of the current amplification. **Power gain** is a measure of the power amplification.

Gallium arsenide a semiconducting material used for very high-speed integrated circuits.

Germanium a semiconductor material used for producing the first integrated circuits.

Half-power point the frequency at which the gain of the amplifier has dropped by 3 dB of the midband or design gain. Also known as the **−3 dB point**.

Harmonic components sinusoidal waves with frequencies that are integer multiples of the fundamental frequency of a periodic waveform.

High-pass filter a filter circuit that passes only frequencies greater than a specified minimum or critical frequency and that rejects all other frequencies.

Hysteresis an effect shown by a comparator when it switches at one voltage level when the voltage is increasing and at a different voltage level when the voltage is decreasing.

Idle frequency see **Free-running frequency**.

Input bias current the average of the currents into the two inputs of a differential or operational amplifier.

Input impedance the resistance or loading seen by a source connected to the input of an amplifier and defined as the input voltage divided by the input current.

Input offset current the difference in bias currents into the inputs of a differential or operational amplifier.

Input sensitivity the minimum input signal level that an amplifier is capable of amplifying.

Instrumentation amplifier a special-purpose amplifier with two precisely balanced inputs and high gain, designed for amplifying transducer signals.

Integrated circuit a complete circuit fabricated onto a single substrate material (usually **silicon**).

Integration an operation from calculus by which some variable quantity is summed over a period of time (or some other variable).

Integrating amplifier an amplifier circuit that produces the integral of the input waveform.

Intermediate frequency (IF) a fixed frequency used in radio receivers (455 kHz for AM, 10.7 MHz for FM) at which the incoming signal receives most of its amplification.

Inverting input one of the two inputs to a differential or operational amplifier. The output for signals applied to this input always are 180° out of phase with the input. The other input is the **noninverting input.**

Limiter a circuit in an FM receiver that clips the amplitude of an FM signal to reduce any amplitude-modulated noise.

Line regulation a measure of how well a power supply maintains a constant DC output for changes in the supply voltage.

Linear integrated circuit an integrated circuit designed to handle a continuous range of input and output signal levels. Also called an **Analog integrated circuit.**

Load regulation a measure of how well a power supply maintains a constant DC output for variations in load current.

Local oscillator (LO) a variable-frequency oscillator circuit in a standard AM or FM radio receiver that is used for tuning the radio frequency signal. In AM radios it is set to 455 kHz higher in frequency than the incoming signal. In FM radios, it is 10.7 MHz higher in frequency.

Lock range the maximum range of frequencies to which the voltage-controlled oscillator in a phase-locked loop can generate an identical frequency to the input signal.

Logarithmic amplifier an amplifier circuit where the output is proportional to the logarithm of the input voltage.

Lower trigger voltage the voltage at which a **Schmitt trigger** will switch as the input test voltage is decreasing.

Low-pass filter a filter circuit that passes only frequencies up to a specified maximum or **critical frequency** and rejects all higher frequencies.

Maximum peak-to-peak output voltage the maximum AC signal level that an amplifier is capable of outputting without appreciable distortion.

Maximum surge current the current that would flow through the rectifier diode in a power supply if the supply were switched on at the peak of the input voltage and with the filter capacitor across the output totally discharged.

Mesa process a process for producing transistors in which a single crystal of silicon is successively doped and etched to produce a mesa-like structure.

Metal-oxide-semiconductor field-effect transistor (MOSFET) a transistor using insulating layers instead of bipolar junctions.

Micromodule a device developed by the United States Army by stacking discrete functional devices, connecting them with fine wires, and encapsulating the entire circuit.

Midband gain the optimum or designed gain of an amplifier, usually achieved in the middle region of the passband.

Midrange frequency any frequency where a filter does not attenuate the input signal, or an amplifier has its full gain.

Mixer see **Audio mixer** or **Radio frequency mixer.**

Monolithic process a manufacturing process whereby all of the parts or components of an integrated circuit are fabricated onto a single piece of silicon.

Muting (squelching) a circuit for eliminating the audio signal when an FM radio receiver is tuned between channels or when it is tuned to a station that is too weak for the limiter to remove the amplitude-modulated noise.

NAB (National Association of Broadcasters) equalization the compensation provided in magnetic tape recording for nonlinearities in the recording/playback head.

Narrow-band filter a bandpass or band-rejection filter that passes or rejects a very small range of frequencies.

Negative feedback see **Feedback.**

Noise figure the amount of noise that an amplifier introduces to the signal that it is amplifying.

Noninverting input the second input to a differential or operational amplifier. The output for signals applied to this input is always in phase with the input. The other input is the **inverting input.**

Norton amplifier an amplifier circuit in which the input signal is current instead of voltage. Also known as a **current-mode amplifier.**

Nuvistor tube an extremely small form of vacuum tube, comparable in size to the first transistors.

Open-loop gain the gain of an operational amplifier without any feedback.

Operational amplifier (op-amp) an amplifier characterized by a differential input, very high gain, and extremely large input impedance.

Oscillation the production of a time-varying periodic output signal.

Output impedance the internal resistance of an amplifier as seen by a load connected to the output.

Output offset voltage the DC voltage at the output of an operational amplifier arising from unequal bias currents at the inputs.

Pass transistor a power transistor used in a voltage regulator to increase its current-handling capability.

Passive filters filter circuits that are constructed of networks of resistors and capacitors (*R-C* filters) or inductors and capacitors (*L-C* filters).

Peak amplitude the difference between the maximum value (either positive or negative) and the average value of a waveform.

Peak reverse voltage rating the maximum reverse voltage that a rectifier diode is capable of withstanding.

Peak-to-peak amplitude the difference between the positive maximum and the negative maximum of a waveform.

Period the time (measured in seconds, milliseconds, or microseconds) for a wave to complete a full cycle.

Periodic signal any time-varying signal that keeps repeating itself exactly over an indefinite period of time.

Phase the fraction of the period elapsed from the start of a cycle to some selected point in the cycle.

Phase comparator a circuit that compares two input AC waves, producing an output DC signal proportional to the phase difference.

Phase-locked loop an oscillator circuit where the output frequency is controlled by, or locked onto, the input frequency.

Phase response the variation of phase with frequency for a filter circuit.

Phase shift network a circuit that is designed for shifting the phase of the output signal with respect to the input signal, specifically used in sine-wave oscillator circuits.

Pilot signal a 19 kHz signal used for stereo decoding in FM transmission. This frequency is exactly half of the stereo carrier frequency. Also called a **subcarrier.**

Planar process a process for producing transistors that involves diffusing the different layers into a substrate of doped silicon.

Poles a measure of the number of reactive elements in a filter circuit. The more poles, the faster the roll-off.

Positive feedback see **Feedback.**

Power gain see **Gain.**

Power supply filter a circuit consisting of a capacitor or a capacitor and inductor used for reducing the variations in output voltage from a power supply.

Power supply impedance a measure of the internal resistance of a power supply.

Preamplifier a low-power audio amplifier used to boost the signal level from a microphone, tape deck, or phonograph pickup.

Pre-emphasis the process of boosting high audio frequencies before FM modulation. Used for noise reduction.

Process variable any physical characteristic of an industrial process such as temperature, pressure, level, or flow rate.

Programmable unijunction transistor (PUT) a device that acts like a diode, except that the voltage when conduction begins is determined by an external trigger level.

Pulse wave a waveform that switches between two DC levels, *with different times* for each level.

Pulse-width modulator a pulse-generating circuit that produces an output pulse width proportional to the input voltage.

Quadrature detector the type of detector used in FM radio receivers, requiring separate oscillator and detector coils.

Quality factor Q a measure of the selectivity of a bandpass or band-rejection filter, expressed as the ratio of the center frequency of the filter to the bandwidth of the filter.

Quiescent current the current into a voltage regulator used for powering the regulator circuit itself.

Radio frequency (RF) any frequency greater than 100 kHz. Specifically, the frequency transmitted by a radio station, carrying a modulated audio signal.

Radio frequency (or RF) amplifier an amplifier specifically designed for high-frequency applications such as radio or television systems.

Radio frequency (or RF) mixer a circuit for combining the local oscillator signal with the incoming radio frequency signal, producing sum and difference frequencies.

Rectifier a circuit designed for converting AC to DC.

Regulation a measure of how well a power supply maintains a constant output voltage for input voltage variations and load variations. See **Line regulation** and **Load regulation.**

Regulator a circuit for maintaining a constant output voltage for a power supply despite line and load variations.

Response time the time for a comparator to switch states.

Ripple rejection the amount that a regulator reduces the input ripple.

Ripple voltage the residual AC component of the rectified voltage in a power supply.

Roll-off the rate at which the output amplitude of a filter or amplifier decreases with frequency, usually expressed in dB per decade.

Sawtooth wave a waveform that increases linearly to its maximum voltage, then decreases linearly *at a different rate* to its minimum voltage.

Schmitt trigger a **comparator** circuit with **hysteresis.**

Sensing circuit a circuit for detecting the level of the output voltage (or current) of a power supply.

Silicon a semiconductor material widely used for producing integrated circuits.

Simple filter a resistor-capacitor network designed for passing a range of frequencies while rejecting all other frequencies. See also **Low-pass filter, High-pass filter, Band-pass filter,** and **Band-rejection filter.**

Sine wave a cyclically varying AC waveform having the shape of the trigonometric sine function.

Single in-line package (SIP) a type of package for power-handling integrated circuits where the pins are all in a single row.

Slew rate the rate at which an amplifier responds to a step change in the input signal.

Square wave a waveform that switches between two DC levels, *with equal time* for each level.

State-variable filter an active filter using integrating circuits as filters and offering the provision of adding or subtracting individual filter responses to generate any net filter response.

Subcarrier see **Pilot signal.**

Subtracting amplifier an amplifier that subtracts one input from the other. Also known as a **differencing amplifier.**

Summing amplifier an amplifier circuit in which the output is equal to the sum of a number of input voltages. Often used as an **audio mixer.**

Supply voltage rejection ratio the characteristic that describes the insensitivity of operational amplifiers to variations in the supply voltage.

Surface-mount package a type of package for integrated circuits consisting of a small rectangular case with contacts instead of pins, designed for automated assembly of circuit boards.

Switched-capacitor filter an active filter that uses switched capacitors to act as resistors.

Time varying any electronic signal that changes with time.

Timer a circuit that is triggered on for a preset time interval.

Tone control the provision for varying the low-frequency gain relative to the high-frequency gain in an audio amplifier.

Toroidal core a high-performance transformer core in the shape of a donut made from either powdered iron or ferrite.

Transducer a device for converting a process variable such as pressure or temperature into an electronic signal.

Transformer coupling see **AC coupling.**

Treble control a tone control that boosts or attenuates the high-frequency gain with respect to the midband gain.

Triangle wave a waveform that increases linearly to its maximum voltage, then decreases linearly *at the same rate* to its minimum voltage.

Tuned or narrow-band amplifier a radio frequency amplifier specifically designed to receive a single channel or narrow range of frequencies.

Tuner module a preassembled circuit consisting of an RF amplifier, a local oscillator, and a mixer, used as the input stage for FM radio and television receivers.

Unity-gain bandwidth the range of frequencies for which the amplifier gain is greater than one.

Universal filter an active filter that can be configured as any type of filter (low-pass, high-pass, bandpass, or band-rejection) and any class of characteristic response (Butterworth, Bessel, Chebyshev, or Cauer).

Upper trigger voltage the voltage at which a **Schmitt trigger** will switch as the input test voltage is increasing.

Video amplifier see **Broadband amplifier.**

Voltage controlled oscillator an oscillator in which the output frequency is proportional to an input control voltage.

Voltage converter a device or circuit that converts a positive input voltage to a negative output voltage. Also known as a **voltage inverter**.

Voltage gain see **Gain**.

Voltage inverter see **Voltage converter**.

Voltage reference a device that produces a constant output voltage used for reference and comparison.

Voltage-to-frequency converter see **Frequency modulator**.

Volume control the provision for varying the loudness of the output sound level in an audio amplifier.

Wave-shaping circuit a circuit used for converting one waveform to another, for example, for converting a square wave to a triangle wave or a triangle wave to a sine wave.

Waveform the shape of the voltage or current variation when plotted against time or as portrayed on the screen of an oscilloscope.

Wien bridge a bridge circuit in which one arm is a parallel R-C circuit and the other is a series R-C circuit, used for reactance measurement and for sine-wave oscillator circuits.

Window comparator a comparator that switches whenever the input voltage enters or leaves a specified voltage range.

Zener diode a device that acts as a normal diode in its forward-conducting mode but as a diode with a fixed nondestructive breakdown voltage in the reverse direction.

Answers to Design and Analysis Problems

E

Answers are provided in this section to all of the design and analysis problems. Many of the answers required specific assumptions to be made, and the answers will be different if different assumptions are made. Closest standard values are listed for all component values.

Chapter 2

1. (a) $A_V = 150$ (b) $Z_{in} = 6.84$ kΩ (c) $Z_{out} = 1.39$ kΩ (d) $A_I = 513$
   (e) $A_P = 76,950$
2. (a) table   (b) table   (c) table   (d) 3 dB $BW = 28.3$ kHz
   (e) Unity-gain $BW = 150$ kHz
3. $A_V = 33.3$
4. (a) $BW = 20$ Hz   (b) $BW = 250$ kHz   (c) $BW = 10$ kHz
5. (a) $Z_{in} = 12$ GΩ (b) $Z_{out} = 0.0125$ Ω

Chapter 3

1. (a) circuit   (b) $R_I = 2.0$ kΩ, $R_F = 39$ kΩ, $R_K = 2.0$ kΩ (c) $BW = 47.6$ kHz
2. (a) circuit   (b) $R_I = 1.0$ kΩ, $R_F = 100$ kΩ
3. (a) inverting   (b) $A_V = 6.06$ (c) $Z_{in} = 3.3$ kΩ (d) $BW = 495$ kHz
4. (a) circuit   (b) $R_I = 10$ kΩ, $R_F = 91$ kΩ, $R_K = 9.1$ kΩ
5. (a) circuit   (b) $R_I = 10$ kΩ, $R_F = 470$ kΩ
6. (a) noninverting   (b) $A_V = 6.0$ (c) $Z_{in} = \infty$ (d) $BW = 167$ kHz
7. (a) buffer   (b) $A_V = 1.0$ (c) $Z_{in} = 2.2$ MΩ (d) $BW = 3$ MHz
8. $R_I = 3.0$ kΩ, $R_F = 75$ kΩ, $R_{D1} = R_{D2} = 10$ kΩ, $C_{in} = 1.2$ µF, $C_{out} = 1.8$ µF, $C_D = 50$ µF
9. (a) $A_V = 6.67$ (b) $Z_{in} = 3.3$ kΩ (c) $V_{DC} = 7.5$ V
10. $R_I = 3.3$ kΩ, $R_F = 150$ kΩ, $R_K = 3.3$ kΩ
11. $R_I = 3.0$ kΩ, $R_F = 75$ kΩ, $R_{D1} = 3.9$ MΩ, $R_{D2} = 75$ kΩ, $C_{in} = 0.33$ µF, $C_{out} = 0.15$ µF
12. (a) $A_V = 14.7$ (b) $Z_{in} = 3.3$ kΩ (c) $V_{DC} = 9.57$ V
13. $R_I = 200$ kΩ, $R_F = 3.9$ MΩ, $R_B = 8.2$ MΩ
14. (a) $A_V = 8.18$ (b) $Z_{in} = 220$ kΩ (c) $V_{DC} = 7.48$ V
15. $R_I = 1.0$ kΩ, $R_F = 1.0$ MΩ, $R_G = 4.7$ MΩ
16. Connect pin 7 to pin 10.
17. $Z_{in} = 5.0$ MΩ, $A_V = 400$

Chapter 4

1. $R_A = R_B = R_C = R_D = R_F = 20$ kΩ
2. $A_{VA} = 0.667$, $A_{VB} = 1.0$, $A_{VC} = 2.0$, $V_{out} = 4.30$ V
3. $R_F = 15$ kΩ, $R_A = 7.5$ kΩ, $R_B = 4.3$ kΩ
4. $R_I = 24$ kΩ, $R_{IN} = 24$ kΩ
5. $R_I = 5.1$ kΩ, $R_{IN} = 75$ kΩ
6. $R_I = 5.6$ kΩ, $R_F = 150$ kΩ
7. $R_I = 1.1$ kΩ, $R_F = 27$ kΩ
8. $R_1 = 91$ kΩ, $R_2 = 1.0$ kΩ, $R_3 = 2.0$ MΩ, $R_4 = 2.0$ kΩ

Chapter 5

1. $R_{LED} = 470$ Ω
2. $R_{LED} = 470$ Ω, reverse LED
3. Use 1N4735 zener, $R_S = 150$ Ω, $R_1 = 8.2$ kΩ, $R_2 = 10$ kΩ, $R_{LED} = 470$ Ω
4. Use 1N4735 zener, $R_S = 47$ Ω, $R_1 = 3.3$ kΩ, $R_2 = 10$ kΩ, $R_{LED} = 270$ Ω
5. $V_{battery} = 20.85$ V
6. $R_1 = 120$ kΩ, $R_2 = 10$ kΩ, $R_3 = 15$ kΩ, $R_4 = 10$ kΩ, $R_{LED} = 470$ Ω, $V_{TL} = 4.69$ V
7. $R_1 = 7.5$ kΩ, $R_2 = 10$ kΩ, $R_3 = 15$ kΩ, $R_4 = 10$ kΩ, $R_{LED} = 470$ Ω, $R_B = 47$ kΩ
8. Use 1N4735 zener, $R_{SU} = 150$ Ω, use 1N4731 zener, $R_{SL} = 150$ Ω, $R_{LED} = 510$ Ω, $R_B = 47$ kΩ
9. $V_{TL} = 8.24$ V, $V_{TU} = 8.93$ V
10. Use 1N4735 zener, $R_S = 47$ Ω, $R_1 = 3.3$ kΩ, $R_2 = 10$ kΩ, $R_{LED} = 100$ Ω
11. $R_1 = 51$ kΩ, $R_2 = 10$ kΩ, $R_S = 120$ Ω, $R_{LED} = 100$ Ω, $V_{TL} = 2.68$ V
12. $R_1 = 15$ kΩ, $R_2 = 10$ kΩ, $R_3 = 39$ kΩ, $R_4 = 10$ kΩ, $R_{LED} = 510$ Ω, $R_B = 13$ kΩ

Chapter 6

1. Bode plot
2. Bode plot
3. Bode plot
4. Bode plot
5. Bode plot
6. Bode plot
7. $C_1 = 0.10$ μF, $C_2 = 0.039$ μF, $C_3 = 0.0056$ μF, $R = 4.7$ kΩ
8. $R_1 = 750$ Ω, $R_2 = 2.0$ kΩ, $R_3 = 13$ kΩ, $C = 0.10$ μF
9. $R = 13$ kΩ, $C = 0.10$ μF, $R_I = 10$ kΩ, $R_F = 10$ kΩ
10. $R = 11$ kΩ, $C = 0.001$ μF, $R_{I1} = 10$ kΩ, $R_{F1} = 1.5$ kΩ, $R_{I2} = 10$ kΩ, $R_{F2} = 12$ kΩ
11. $R = 2.2$ kΩ, $C = 1.0$ μF, $R_{I1} = 100$ kΩ, $R_{F1} = 6.2$ kΩ, $R_{I2} = 10$ kΩ, $R_{F2} = 5.6$ kΩ, $R_{I3} = 10$ kΩ, $R_{F3} = 15$ kΩ
12. $R = 9.1$ kΩ, $C_1 = 0.01$ μF, $C_2 = 0.0082$ μF, $C_3 = 0.0022$ μF, $C_4 = 0.018$ μF, $C_5 = 0.0018$ μF, $C = 0.01$ μF, $R_1 = 1.0$ kΩ, $R_2 = 1.3$ kΩ, $R_3 = 4.3$ kΩ, $R_4 = 560$ Ω, $R_5 = 5.6$ kΩ
13. $R_H = 6.2$ kΩ, $C_H = 0.10$ μF, $R_L = 2.2$ kΩ, $C_L = 0.01$ μF, $R_{I1} = 10$ kΩ, $R_{F1} = 1.5$ kΩ, $R_{I2} = 10$ kΩ, $R_{F2} = 12$ kΩ
14. $Q = 11.49$, $n = 4$, $C_1 = 0.01$ μF, $C_2 = 0.01$ μF, $R_1 = 300$ Ω, $R_2 = 27$ kΩ, $R_3 = 5.6$ kΩ
15. $Q = 4.22$, $n = 1$, $C_1 = 0.01$ μF, $C_2 = 0.01$ μF, $R_1 = 270$ Ω, $R_2 = 16$ kΩ, $R_3 = 1.3$ kΩ

Answers to Design and Analysis Problems   A-51

16. $Q = 5.48$, $n = 2$, $C_1 = 0.01$ μF, $C_2 = 0.01$ μF, $R_1 = 220$ Ω, $R_2 = 10$ kΩ, $R_3 = 5.1$ kΩ
17. (a) narrow-band bandpass   (b) $f_o = 605$ Hz   (c) $BW = 54$ Hz
18. $C_1 = 0.01$ μF, $R_1 = 3.6$ kΩ

Chapter 7

1. (a) circuit   (b) 10 Ω   (c) $V_P = 25.2$ V, $C_F = 2200$ μF
2. (a) $V_{DC} = 15.47$ V   (b) $V_{ripple} = 0.867$ V RMS   (c) $I_L = 1.547$ A
3. line regulation = 32 mV/V, load regulation = 478 mV/A
4. zener = 1N4744, $R_S = 510$ Ω
5. (a) circuit   (b) $R_F = 10$ kΩ, $R_I = 9.1$ kΩ, $R_S = 1.3$ kΩ
6. $V_{out} = 5.14$ V
7. (a) $V_{out} = 19.39$ V   (b) $I_{max} = 1.956$ A   (c) $I_{short} = 0.631$ A   (d) $V_{DC} = 33.56$ V
8. (a) circuit   (b) $C_F = 3000$ μF   (c) $V_{DC} = 16.89$ V, $V_{ripple} = 0.872$ V
   (d) $R_F = 10$ kΩ, $R_I = 1.1$ kΩ, $R_S = 11$ kΩ   (e) $V_{CE} = 4.19$ V, $I_C = 1.0$ A, $P = 4.19$ W   (f) $R_{sens} = 0.7$ Ω

Chapter 8

1. (a) circuit   (b) $R_1 = 2.2$ kΩ, $R_2 = 8.2$ kΩ, $R_3 = 1.8$ kΩ, $R_{sens} = 0.467$ Ω   (c) Power = 9.9 W
2. (a) $V_{out} = 4.05$ V   (b) $I_L = 1.4$ A   (c) $P = 7.35$ W   (d) $V_{ripple} = 0.569$ V   (e) line regulation = 0.405 mV/V   (f) load regulation = 1.22 mV/A
3. (a) circuit   (b) $C_F = 3000$ μF   (c) $R_1 = 5.1$ kΩ, $R_2 = 4.7$ kΩ, $R_3 = 2.4$ kΩ   (d) $R_{sens} = 3.3$ Ω, $R_4 = 300$ Ω, $R_5 = 680$ Ω   (e) $V_{CE} = 5.85$ V, $I_C = 2.0$ A, $P = 11.7$ W
4. (a) $V_{out} = 11.15$ V   (b) $I_L = 0.103$ A   (c) $V_{DC} = 26.79$ V   (d) $V_{ripple} = 10.5$ mV   (e) line regulation = 15 mV/V   (f) load regulation = 4.5 mV/A
5. (a) circuit   (b) $R_1 = 62$ Ω, $R_2 = 270$ Ω   (c) $P_{reg} = 5$ W
6. (a) $I_L = 2.8$ A   (b) $I_{reg} = 0.07$ A   (c) line reg = 7 mV/V   (d) load regulation = 11 mV/A   (e) $V_{ripple} = 0.237$ mV   (f) $V_{reg} = 5.04$ V   (g) $P_{reg} = 0.353$ W, $P_Q = 16.07$ W
7. (a) circuit   (b) $C_F = 10,000$ μF   (c) $R_P = 15$ Ω, $R_{sens} = 0.233$ Ω   (d) $P_{reg} = 0.351$ W, $P_Q = 23.2$ W
8. (a) $V_{out} = 13.8$ V   (b) $I_L = 2.33$ A   (c) line reg = 20 mV/V   (d) load regulation = 14 mV/A
9. (a) circuit   (b) $C_F = 10,000$ μF   (c) $V_{DC} = 22.74$ V, $V_{reg} = 6.34$ V, $V_Q = 7.04$ V   (d) $R_P = 15$ Ω, $R_{sens} = 0.233$ Ω   (e) $R_1 = 120$ Ω, $R_2 = 1.0$ kΩ pot   (f) $R_A = 1.1$ kΩ, $R_B = 330$ Ω   (g) $R_S = 3.9$ kΩ.

Chapter 9

1. (a) circuit   (b) $R_T = 10$ kΩ, $C_T = 0.0018$ μF   (c) $R_1 = 10$ kΩ, $R_2 = 10$ kΩ   (d) $R_{sens} = 0.040$ Ω   (e) $R_{B1} = 180$ Ω, $R_{B2} = 270$ Ω   (f) $L = 51.4$ μH, $C = 330$ μF
2. (a) circuit   (b) $L_5 = 179$ μH, $C_5 = 47$ μF, $L_8 = 233$ μH, $C_8 = 47$ μF, $L_{15} = 268$ μH, $C_{15} = 47$ μF   (c) $R_T = 10$ kΩ, $C_T = 0.0027$ μF   (d) $R_1 = 16$ kΩ, $R_2 = 7.5$ kΩ   (e) $R_{B1} = 220$ Ω, $R_{B2} = 1.0$ kΩ, $R_{B3} = 220$ Ω
3. same as 2 except $L_5 = 114$ μH, $L_{15} = 129$ μH

4. (a) $f_{osc} = 20.2$ kHz (b) $V_{out} = 30.0$ V

5. $L = 189$ µH, $C = 22$ µF, $R_T = 10$ kΩ, $C_T = 0.0033$ µF, $R_1 = 36$ kΩ, $R_2 = 5.6$ kΩ, $R_{B1} = 5.6$ kΩ, $R_{B2} = 1.0$ kΩ, $R_{B3} = 180$ Ω

6. (a) $f_{osc} = 11.4$ kHz (b) $V_{out} = 6.63$ V

7. (a) circuit (b) $R_1 = 2.0$ kΩ (c) $C_T = 560$ pF (d) $R_{sens} = 0.233$ Ω
 (e) $C = 2000$ µF, $L = 61.7$ µH

Chapter 10

1. $C_1 = 0.10$ µF, $R_1 = 3.0$ kΩ, $R_2 = 13$ kΩ, $R_3 = 11$ kΩ

2. $C_1 = 0.01$ µF, $R_1 = 9.1$ kΩ, $R_2 = 13$ kΩ, $R_3 = 11$ kΩ, $C_F = 0.01$ µF, $R_I = 9.1$ kΩ, $R_F = 200$ kΩ

3. (a) relaxation triangle-wave oscillator (b) $f_{osc} = 4456$ Hz (c) $V_{p\text{-}p\ square} = 22$ V, $V_{p\text{-}p\ triangle} = 3.74$ V

4. $C_1 = 0.10$ µF, $R_1 = 10$ kΩ, $R_2 = 20$ kΩ, $R_3 = 10$ kΩ

5. $C_1 = 0.10$ µF, $R_1 = 11$ kΩ, $R_2 = 10$ kΩ, $R_3 = 3.0$ kΩ

6. (a) bootstrap oscillator (b) $f_{osc} = 2.34$ kHz (c) $V_{p\text{-}p\ square} = 22$ V, $V_{p\text{-}p\ triangle} = 6.67$ V

7. $C_1 = 0.47$ µF, $R_1 = 10$ kΩ, $R_2 = 20$ kΩ, $R_3 = 10$ kΩ, $R_4 = 2.4$ kΩ

8. (a) bootstrap ramp oscillator (b) $D = 0.239$ (c) $f_{osc} = 2.31$ kHz
 (d) $V_{p\text{-}p\ pulse} = 28$ V, $V_{p\text{-}p\ ramp} = 14$ V

9. $C = 0.022$ µF, $R = 6.8$ kΩ, $R_{F1} = 2$ kΩ pot, $R_{F2} = 56$ Ω, $R_I = 33$ Ω

10. $C = 680$ pF, $R = 9.1$ kΩ, $R_{F1} = 2$ kΩ pot, $R_{F2} = 1.0$ kΩ, $R_I = 510$ Ω

11. (a) phase-shift oscillator (b) $A_{V\text{start}} = 48.8$ (c) $A_V = 46.5$ (d) $f_{osc} = 62.8$ Hz
 (e) $V_{p\text{-}p} = 8.0$ V

12. (a) Wien bridge oscillator (b) $A_{V\text{start}} = 4.35$ (c) $A_V = 3.43$ (d) $f_{osc} = 88.4$ Hz
 (e) $V_{p\text{-}p} = 10.8$ V

13. $C_1 = 0.0033$ µF, $C_2 = 0.0033$ µF, $R_1 = 2.0$ kΩ, $R_2 = 910$ kΩ, $R_3 = 16$ kΩ

14. $C_1 = 10$ µF, $R_1 = 390$ kΩ, $R_2 = 20$ kΩ, $R_3 = 10$ kΩ

15. (a) bootstrap sine-wave oscillator (b) $f_{osc} = 75.8$ Hz

16. $T = 0.0122$ sec, $V_{peak} = 7.13$ V

Chapter 11

1. $C_1 = 0.01$ µF, $R_1 = 5.1$ kΩ, $R_2 = 16$ kΩ, $R_3 = 110$ kΩ

2. $C_1 = 100$ pF, $R_1 = 12$ kΩ, $R_F = 13$ kΩ, $R_{IB} = 15$ kΩ, $R_{IS} = 13$ kΩ

3. (a) $f_{osc} = 75.8$ kHz (b) $\Delta V = 0.099$ V (c) $\Delta f = 6.06$ kHz

4. $C_1 = 0.01$ µF, $R_1 = 9.1$ kΩ, $R_2 = 18$ kΩ

5. $C_1 = 0.01$ µF, $R_1 = 27$ kΩ, $R_2 = 18$ kΩ

6. $C_1 = 0.1$ µF, $R_1 = 12$ kΩ, $R_2 = 12$ kΩ

7. $C_1 = 10$ µF, $R_1 = 91$ kΩ

8. (a) $f_{osc} = 2924$ Hz (b) $D = 0.714$ (c) $V_{out} = 10$ V

9. $R_A = R_B = 10$ kΩ, $R_C = R_D = 100$ kΩ pot, $R_1 = 1.0$ kΩ pot, $R_F = 10$ kΩ pot, $R_{I1} = 7.5$ kΩ, $R_{I2} = 4.7$ kΩ, $R_{I3} = 3.0$ kΩ, $C_{11} = 270$ pF, $C_{12} = 0.0027$ µF, $C_{13} = 0.027$ µF, $C_{14} = 0.27$ µF, $C_{15} = 2.7$ µF

10. $C_1 = 100$ pF, $R_1 = 8.2$ kΩ, $R_B = 10$ kΩ, $C_{in} = 56$ pF, $C_2 = 0.068$ µF, $R_I = 1.0$ kΩ, $R_F = 120$ kΩ

11. $C_1 = 100$ pF, $R_1 = 20$ kΩ, $R_B = 10$ kΩ, $C_{in} = 0.015$ µF, $C_2 = 390$ pF

12. (a) $f_o = 10.1$ kHz (b) $f_{out} = 15$ kHz (c) $f_L = 4.49$ kHz (d) $f_C = 950$ Hz

13. $C_1 = 0.01$ µF, $R_1 = 5.6$ kΩ, $R_B = 10$ kΩ, $C_{in} = 2.2$ µF, $C_2 = 0.01$ µF

Chapter 12

1. $R_B = 100\ \text{k}\Omega$, $R_F = 620\ \text{k}\Omega$, $R_A = 1.5\ \text{k}\Omega$, $C_A = 2.2\ \mu\text{F}$, $C_1 = 0.033\ \mu\text{F}$, $C_2 = 0.10\ \mu\text{F}$
2. $R_B = 100\ \text{k}\Omega$, $R_F = 560\ \text{k}\Omega$, $R_A = 910\ \Omega$, $C_A = 3.9\ \mu\text{F}$, $C_1 = 0.033\ \mu\text{F}$, $C_2 = 0.01\ \mu\text{F}$
3. $R_B = 100\ \text{k}\Omega$, $R_F = 330\ \text{k}\Omega$, $R_A = 910\ \Omega$, $C_A = 2.2\ \mu\text{F}$, $C_1 = 0.018\ \mu\text{F}$, $C_2 = 0.01\ \mu\text{F}$, $C_3 = 0.18\ \mu\text{F}$, $R_3 = 75\ \Omega$, $R_4 = 180\ \text{k}\Omega$
4. $R_B = 100\ \text{k}\Omega$, $R_F = 330\ \text{k}\Omega$, $R_A = 20\ \Omega$, $C_A = 150\ \mu\text{F}$, $C_1 = 0.033\ \mu\text{F}$, $C_2 = 0.022\ \mu\text{F}$, $C_4 = 0.01\ \mu\text{F}$, $R_4 = 9.1\ \text{k}\Omega$
5. $R_B = 100\ \text{k}\Omega$, $R_F = 750\ \text{k}\Omega$, $R_{F1} = 100\ \text{k}\Omega$, $C_{I1} = 3.3\ \mu\text{F}$, $C_{I2} = 1.0\ \mu\text{F}$, $C_2 = 0.033\ \mu\text{F}$, $R_{I1} = 820\ \Omega$, $R_{I2} = 4.3\ \text{k}\Omega$
6. $R_B = 100\ \text{k}\Omega$, $R_F = 430\ \text{k}\Omega$, $R_{F1} = 100\ \text{k}\Omega$, $C_{I1} = 0.39\ \mu\text{F}$, $C_{I2} = 2.2\ \mu\text{F}$, $C_{I3} = 1.0\ \mu\text{F}$, $C_2 = 0.12\ \mu\text{F}$, $R_{I1} = 5.6\ \text{k}\Omega$, $R_{I2} = 1.6\ \text{k}\Omega$, $R_{I3} = 2.4\ \text{k}\Omega$
7. $R_F = 100\ \text{k}\Omega$, $R_I = 5.6\ \text{k}\Omega$, $C_I = 1.0\ \mu\text{F}$, $C_1 = 0.1\ \mu\text{F}$, $C_2 = 500\ \mu\text{F}$
8. $R_F = 100\ \text{k}\Omega$, $R_I = 5.6\ \text{k}\Omega$, $C_I = 1.0\ \mu\text{F}$, $C_1 = 0.1\ \mu\text{F}$, $C_2 = 500\ \mu\text{F}$.
9. $R_F = 100\ \text{k}\Omega$, $R_I = 9.1\ \text{k}\Omega$, $C_I = 0.47\ \mu\text{F}$, $C_1 = 0.1\ \mu\text{F}$, $C_2 = 500\ \mu\text{F}$
10. $R_{PB} = R_{PT} = R_{PV} = 100\ \text{k}\Omega$, $R_{1B} = 5.6\ \text{k}\Omega$, $R_{2B} = 330\ \Omega$, $C_{1B} = 0.068\ \mu\text{F}$, $C_{2B} = 1.0\ \mu\text{F}$, $R_{1T} = 110\ \text{k}\Omega$, $R_{2T} = 18\ \text{k}\Omega$, $C_{1T} = 820\ \text{pF}$, $C_{2T} = 0.0047\ \mu\text{F}$
11. $C_1 = 0.12\ \mu\text{F}$, $C_2 = 0.012\ \mu\text{F}$

F Device Index

A number of integrated circuits were introduced in this text. Many are cited and used in applications throughout the text. This index lists all of the integrated circuits, indicating the pages on which reference is made to each device. The pages in **boldface** refer to data sheets.

MF6 Switched-Capacitor Filter, 261, **262–263**, 264–268

LH0038 Instrumentation Amplifier, 120, **121–124**, 125–131

TL081/TL080/TL084 Operational Amplifiers, 13, 15, 54, 66–74, **69–70**, 80, 85–87, 91–92, 93, 94, 106, 140–142, 143–144, 147–148, 152–153, 157, 161–162, 164, 167, 168, 175, 178, 181, 182–183, 188–189, 192–194, 233, 235, 240, 241, 243, 245, 248–249, 253, 254–255, 257–259, 305, 306, 309, 311, 313, 316, 454, 456, 463, 464, 466–467, 471–472, 475, 478, 480–481, 499, 503, 518, 521, 532, 545

LMF100 Universal Filter, 271, **272–273**, 274, 275

LM313 Voltage Reference, 297, **298–299**, 300–301, 304, 306–307, 309, 311, 313, 316, 372, 377–378

LM317 Variable Regulator, 364, **365–366**, 367–372, 377–378, 379

TL321 Single-Supply Operational Amplifier, 80, 98, **99–100**, 101–105, 131

LM387 Audio Preamplifier, 542, **543–544**, 563

LM555 Timer, 490, 503–504, **505–506**, 507–514

LM565 Phase Locked Loop, 22, **24–28**, 490, **525–526**, 527–537

LM566 Voltage Controlled Oscillator, 490, **491–492**, 493–503, 504, 511, 531

LM723 Variable Regulator, 325, **326–327**, 328–346, 352, 353, 370, 386, 432

LM741 Bipolar Operational Amplifier, 13, 16, 17, 18, 19, 53, 62–67, **63–64**, 68, 71, 72, 80, 82–85, 88–90, 93, 94, 98, 106, 175, 194, 198

TL810 True Comparator, 194, **195–197**, 198–206

LM1035 Tone and Volume Control, 580, **581–582**, 583–586, 601–603

LH1605 Switch-Mode Regulator, 432, **433–434**, 435–440

LM1800 Stereo Demodulator, 598, **599–600**, 601–603

LM1877 Power Amplifier, 563–564, **565–566**, 567–575, 590–591, 596–597, 601–603

LM3189 FM IF Amplifier, 592, **593–594**, 595–598, 601–603

LM3524 Pulse-Width Modulator, 402, **403–404**, 405–429, 440

LM3820 AM Radio, 586, **587–588**, 589–592

LM3900 Norton Operational Amplifier, 80, 106–107, **108–109**, 110–114, 131

ICL7660 Voltage Converter, **373–374**, 375–379

MC7800 Fixed Voltage Regulator, 346–347, **348–351**, 342–364, 370, 371, 372, 379, 402, 432

ICL8038 Function Generator, 490, **515–516**, 517–522, 537

LIFE SUPPORT POLICY

NATIONAL'S PRODUCTS ARE NOT AUTHORIZED FOR USE AS CRITICAL COMPONENTS IN LIFE SUPPORT DEVICES OR SYSTEMS WITHOUT THE EXPRESS WRITTEN APPROVAL OF THE PRESIDENT OF NATIONAL SEMICONDUCTOR CORPORATION. As used herein:

1. Life support devices or systems are devices or systems which, (a) are intended for surgical implant into the body, or (b) support or sustain life, and whose failure to perform, when properly used in accordance with instructions for use provided in the labeling, can be reasonably expected to result in a significant injury to the user.

2. A critical component is any component of a life support device or system whose failure to perform can be reasonably expected to cause the failure of the life support device or system, or to affect its safety or effectiveness.

| National Semiconductor Corporation 2900 Semiconductor Drive P.O. Box 58090 Santa Clara, CA 95052-0090 Tel: (408) 721-5000 TWX: (910) 339-9240 | National Semiconductor GmbH Industriestrasse 10 D-8080 Furstenfeldbruck West Germany Tel: (0-81-41) 103-0 Telex: 527-649 Fax: (08141) 103554 | National Semiconductor Japan Ltd. Sanseido Bldg. 5F 4-15 Nishi Shinjuku Shinjuku-Ku, Tokyo 160, Japan Tel: 3-299-7001 FAX: 3-299-7000 | National Semiconductor Hong Kong Ltd. Suite 513, 5th Floor Chinachem Golden Plaza, 77 Mody Road, Tsimshatsui East, Kowloon, Hong Kong Tel: 3-7231290 Telex: 52996 NSSEA HX Fax: 3-3112536 | National Semicondutores Do Brasil Ltda. Av. Brig. Faria Lima, 1383 6.0 Andor-Conj. 62 01451 Sao Paulo, SP, Brasil Tel: (55/11) 212-5066 Fax: (55/11) 211-1181 NSBR BR | National Semiconductor (Australia) PTY. Ltd. 1st Floor, 441 St. Kilda Rd. Melbourne, 3004 Victory, Australia Tel: (03) 267-5000 Fax: 61-3-267/458 |

National does not assume any responsibility for use of any circuitry described, no circuit patent licenses are implied and National reserves the right at any time without notice to change said circuitry and specifications.

Index

AC-coupled amplifier, 36, 44, 45, 48–49, A-40
Active filter, 212, 213, 220, 227, A-40
   Broadband bandpass, 243–246
   Broadband band-rejection, 246–250
   Frequency response, 228–229, 242–244, 247–248
   High-pass equal component, 237, 239, 240, 241–242
   High-pass unity gain, 231–233, 236–237
   Low-pass equal component, 237, 238, 240–241
   Low-pass unity gain, 229–231, 233–235
   Narrow-band bandpass, 250–256, 257, 477–479
   Narrow-band band-rejection, 256–259
   Phase response, 228–229
AM radio, 7, 211, 224, 542, 586–592
Amplification (*see* Gain)
Amplifier, 6, 35–48, 59, 105–106, 449, A-40
(*see also specific types*)
   Characteristics, 35–36
   Classification, 36–37
   Equivalent circuit, 41, 43
   Frequency response, 44–48, 59–61
   Gain, 38–39, 42–43, 45–48
   Input impedance, 39–40, 42–43
   Output impedance, 39–43
   Schematic symbol, 37–38
Amplitude, 447, 471, 473, 475–476, 480, A-40
Amplitude modulation (AM) (*see* Modulation)
Analog computer, 53, A-40
Analog Devices Inc., 12, 14
Analog integrated circuit (*see* Linear integrated circuit)
Analog-to-digital converter, 6
Audio amplifier, 6, 36, 220, 542, A-40
   Audio power amplifier, 542, 545, 563–576, 590–592, 596–597, 601–603
   Audio preamplifier, 542–563, 576, 578, 584, A-46
Audio frequency amplifier (AF amp), 36, 542, A-40
Audio line level, 545

Audio mixer, 139, 141–142, 560–563, A-40
Audio taper potentiometer, 576, 578–579, 580, 584, A-40
Automatic frequency control, 596, A-40
Automatic gain control, 590, 596, A-40
Automatic volume control (*see* Automatic gain control)

Balance control, 576, 580, 583–585, A-41
Bandpass filter, 211, 212, 216–217, 224–225, 243–246, 250–256, A-41 (*see also* Passive filter, Active filter)
Band rejection filter, 211, 212, 226–227, 246–250, 256–259, A-41 (*see also* Passive filter, Active filter)
Bandwidth, 45, 55, 59–61, A-41
   3 dB bandwidth, 44–48, 55, 59–60, 82, 85, 87–90, 92, 93, 112
   For filters, 214, 242–245, 247–248, 251–254, 256–258
   Unity-gain bandwidth, 44–48, 55, 59–60, 93–94, 128
Bardeen, John, 3
Barkhausen criteria, 449, 467–469, A-41
Bell Telephone Labs, 3
Bessel filter response, 228–229, 271, 274, A-41
Biasing, 36, 48, 49, 83–84, 89–90, 104, 110–111, 125, 127, 330, 407, 410, 414, 416–418, 420, 422, 423–425, 426–427, 430–431, 471, 474, 494, 501, 527, 546, 549, 569, 592
BiFET technology, 54, 67–74, 80, 85–87, 90–92, A-41
Bipolar technology, 5, 53, 62–67, 80, 82–84, 88–90, 330, 407, 545, A-41
Bleeder current, 377–378, 411–412, 424–425
Bode plot, 213–220, A-41
Bootstrap oscillator, 459–467, 476–479, 482, 490, A-41
Brattain, Walter, 3
Bridge amplifier, 574 (*see also* Audio power amplifier)
Bridge rectifier (*see* Rectifier)

I-1

Index

Broadband amplifier, 36, A-41
Broadband filter, 242, A-41 (*see also* Active filter)
Buffer amplifier, 80, 93–94, 113, 117, 267, 333, 470–472, 518, 520–522
Butterworth frequency response, 228–229, 233, 234, 236, 240, 241, 243, 247, 248, 261, 271, 274, A-41

Calibration, 519
Capacitive coupling, 36, 44, 48–49, 95–98, 103–105, 111–112, 114, 498–500, 501, 527, 530, 534, 547–549, 569, 580, A-40
Capacitor, 150–153, 155–157, 221–227, 230–242, 244–256, 259–261, 269, 469–470, 473, 477–478, 577–579, A-3–A-5
  Bypass capacitor, 96, 98, 127, 129, 265, 435, 510, 569, 571, 584, 592, 598
  Compensating capacitor, 66, 71, 72–74, 167–168, 330, 568, 574
  Coupling capacitor, 95–98, 103–105, 110, 112, 114, 498–500, 527, 547–548, 552, 560, 563, 568–570, 573, 580
  Filter capacitor, 284–289, 353, 356, 368, 376, 529, 532, 534, 536
  Switch-mode capacitor, 389–390, 394–395, 397–398, 400, 408, 410, 414–415, 420, 422, 425, 431–432, 436, 438–439
  Timing capacitor, 264, 268, 407, 409, 418, 423, 429, 436, 439, 451–452, 458, 461, 479–480, 493, 497–499, 504, 510, 513–514, 518–522, 528
  Tuning capacitor, 589, 592
Capture range, 524, 529–530, 533–535, 561, A-41
Carrier, 500–503, 530
Cauer filter response, 228–229, 271, 274, A-41
Center frequency, 242–243, 246, 248, 250–251, 253, 254, 256, 258, 490, A-41
Chebyshev filter response, 228–229, 271, 274, A-41
Common-mode gain, 48, 51, A-41
Common-mode rejection ratio, 48, 51, 65, 71, 115, 125, 128, A-41
Comparator, 6, 173–206, 388, 450, 459–460, 476, A-41
  Level-sensing comparator, 173, 179–184, 200–202, 302
  Op-amps used as, 173–177
  True comparator, 194–206
  Used as switches, 175–177
  Window comparator, 173, 190–194, 204–206, 504–508, A-48
  With hysteresis, 173, 184–190, 203–204
  Zero-crossing comparator, 173, 177–179, 199–200
Compensation, 54, 72–74, 320, 332, 408, 564, 568, 574, 575
Critical frequency, 214, 215, 217, 219–224, A-42 (*see also* Lower critical frequency, Upper critical frequency)
Cross-reference guide, 18

Crosstalk, 143, 144
Current limiting, 311–319, 328, 329, 337–344, 346, 358–361, 364, 371–372, 407–408, 423–424, 436–437, 438–440 (*see also* Foldback limiting)
Current-mode op-amp (*see* Norton op-amp)
Cycle, 447, A-42

Data Books, 16–17, 22, 23, 29
D.A.T.A.BOOKS, 18, 20, 29–31
Data sheet, 16 (*see also individual data sheets*)
Date stamp, 15
DC offset, 139
Decade, 212, A-42
Decibel, 38–39, 45–48, 212, 217–218, A-42
De-emphasis, 596, 598, A-42
Demodulator, 529, A-42
  AM demodulator, 590–592
  FM demodulator, 529–533, 592, 595–596
  Stereo, 598–601
Detector (*see* Demodulator)
Device number, 10–16 (*see also Device Index*)
Differencing amplifier, 116–117, 139, 530–532
Differential amplifier, 35, 48–52, 101, 549–550, A-42
Differential gain, 51
Differentiating amplifier, 139, 153–157, A-42
Differentiation, 153–156, A-42
Digital integrated circuit, 6, 185, A-42
Digital-to-analog converter, 6
Diode, 165–168, 191–194, 376, 418, 457–458, 465–467, 510–512 (*also see* Zener diode)
  Catch diode, 387, 390, 392, 396, 398–399, 401–402, 421, 425–429
  Detector, 590
  Light emitting diode (LED), 175–176, 182–183, 188–189, 191–194, 199–206
  Protective diode, 367–368, 369, 378
  Rectifier diode, 281–285, 286
  Wave shaping diode, A-33–A-37
DIP (dual in-line package), 7, 8, 10, 54, 62, 68, 107, 198, 261, 264, 325, 328, 375, 405, 545, A-42
Direct-coupled (DC) amplifier, 37, 44, 45, A-42 (*see also* Differential amplifier)
Distortion, 59, 152, 156, 467, 470, 477, 517, 542, 545, 564, A-33
Double regulation, 402
Dropout voltage, 304, 330, 347
Duty cycle, 392–394, 397, 399, 446, 456–459, 465–467, 508–510, 511–512, 519, A-42

ECG Semiconductor Master Replacement Guide, 18, 21, 29, 31, 32
Effective series resistance (ESR), 390
Electronic filter, 211, A-42
Elliptic filter response (*see* Cauer filter response)

Error amplifier, 301, 328, 329, 333, 388, 402, A-42
Exar Corporation, 12, 14
Exponential amplifier, 139, 163–165, A-42

Fairchild Semiconductor, 3, 5, 10, 11, 12, 13, 14, 62
Faraday's law, A-28
Feedback, 56–62, 127, A-42
   Effect on bandwidth, 59
   Effect on distortion, 59
   Effect on input impedance, 61–62
   Effect on output impedance, 61–62
   Negative feedback, 56–62, 72–74, 80, 87, 468, 574, 575
   Positive feedback, 56, 449, 468, 473
Feedback factor, 57, 58, 468, 470, 473
Feedforward compensation, 73–74, 167, 168
Filter (see Power supply filter, Active filter, Electronic filter, Switched-capacitor filter, Universal filter)
Floating ground, 96, 461
FM radio, 7, 211, 224, 226, 542, 596–603
Foldback current limiting, 313–319, 329, 338–339, 340–342, 343, A-43
Fourier analysis, 468–469, 477, A-43
Free-running frequency, 524, 527–529, 531, 534, 536, A-43
Frequency, 447, 449, 452, 458, 461, 469, 473, 477, 480, 494, 508, 511, 519, 528, A-43
Frequency converter, 533–537, A-43
Frequency counter, 6, 533
Frequency deviation, 502–503, 530–533
Frequency divider, 533–536
Frequency modulation (FM) (see Modulation)
Frequency response (see Active filter frequency response, Amplifier frequency response, Op-amp frequency response, Passive filter frequency response)
Function generator, 35, 158, 446, 448, 490, 497–500, 515–522, 537, A-43

Gain, 38, 42–43, 51, 56–59, 72, 449, 470–471, 473–474, A-43
   Closed-loop gain, 57–62, 81, 83, 86, 88, 90, 92, 97, 101–102, 110, 115, 118–119, 125, 143, 144–146, 152, 156, 468, A-41
   Open-loop gain, 57–59, 65, 68, 71, 166, 173–174, 184, 198, 468, A-45
   Variable, 118, 125–127, 561
Gallium-arsenide, 5, 6, A-43
Germanium, 3, 6, A-43
Ground sense, 127
Guard drive amplifier, 125, 129

Half power point, 45, A-43
Harris Semiconductor, 10, 11, 517
Heatsink, 10, 309, 337, 342, 346, 347, 357, 358, 360, 362, 386–387, 564
Heiman, Frederic P., 5
High-pass filter, 211, 212, 214, 216–217, 222–224, 231–233, 236–237, 239, 240, 241–242, A-43 (see also Passive filter, Active filter)
Hoerni, Jean, 3
Hofstein, Steven R., 5
Hysteresis, 173, 184–190, 198, 203–204, A-43

Idle frequency (see Free-running frequency)
Inductor, 387, 389, 390–402, 408–410, 414, 415, 420, 421, 425, 426, 431, 435–436, 438, 439, A-22, A-25–A-28, A-32
Input bias current, 48, 51, 65, 71, A-43
Input impedance, 39–40, 61–62, 65, 68, 82, 83, 84, 86, 88, 89, 93, 96, 101–102, 116–117, 127, 145–146, 548, 549, 564, 580, 584, A-43
Input offset current, 48, 52, 65, 66, 71, A-43
Input sensitivity, 48, A-43
Instrumentation amplifier, 36, 80, 114–130, 144, A-43
Integrated circuit, 2, 5, 37, 49–53, 325, 388, 490, 542, A-43
Integrating amplifier, 139, 149–153, 269–271, 454–456, 458, A-44
Integration, 149–150, 153, A-44
Intermediate frequency (IF) signal, 586, 589–590, 592, 596, 598, 601, A-44
Intersil, 10, 11, 12, 14, 517
Inverting amplifier, 56, 57, 80–87, 93, 95, 103–105, 110–114, 139–141, 144–145, 267, 498–502, 545, 548–549, 560–563
Inverting comparator, 177–178, 180, 181, 185, 190, 194

Kilby, Jack S., 3, 4

Limiter, 592, A-44
Line regulation (see Regulation)
Linear integrated circuit, 6–7, A-44
Linear regulator, 6, 7, 22, 23, 281, 289–290, 402
   Floating regulator, 361–364, A-42
   Multipin regulator, 325–346
   Op-amp regulator, 301–319, 325, 328
   Three-terminal regulator, 346–372
   Variable regulator, 318, 330, 364–372
   Zener diode regulator, 295–296, 301
Linear Technology Corp., 12, 14
Load regulation (see Regulation)
Local oscillator, 586, 589, 592, 601, A-44
Lock range, 524, 529–530, 533–536, A-44
Logarithmic amplifier, 139, 158–162, 165, A-44
Logic gates, 6
Logo, 13–15
Low noise op-amp, 117, 120
Lower critical frequency, 216, 225–227
Lower trigger voltage, 185–189, 204, 450–453, 455, 504, 507–508, A-44
Low-pass filter, 211, 212–222, 229–231, 233–235, 237, 238, 240–241, A-44 (see also Passive filter, Active filter)

Manufacturers, 10–11 (*also see specific manufacturers*)
Maximum steady-state current rating, 283
Maximum surge current (*see* Surge current)
Mesa process, 3, A-44
Micromodule, 3, A-44
Microphone, 549–551, 561–562
Microprocessor, 5
Mixer (*see* Audio mixer, Radio frequency mixer)
Modulation, 388, 537
   Amplitude modulation, 586, 590, A-40
   Frequency modulation, 490, 500–503, 531, 596, A-43
Monolithic, 5, A-44
MOSFET (metal-oxide-semiconductor field-effect transistor), 5, 39, A-44
MOSFET technology, 5
Motorola Semiconductor, 10, 11, 12, 13, 14, 18, 22, 67, 68, 120, 346, 563
Muting, 592, 595, A-44

Narrow-band amplifier, 36
Narrow-band filter, 242, A-45 (*see also* Active filter)
National Association of Broadcasters (NAB) equalization, 553, 555, 557–559, A-45
National Semiconductor, 10, 11, 12, 13, 14, 22, 62, 67, 68, 261, 297, 432, 563–564, 586, 592, 598
Noise, 115–116, 125, 184, 542, 545
Noise figure, 48, 125, 130
Noise reduction, 158, 169
Noninverting amplifier, 56, 57, 80, 87–92, 101–103, 112–113, 115, 142–144, 145–146, 303–304, 306, 329, 545–548, 551, 564, 567–569
Noninverting comparator, 177, 180, 181, 190, 193
Norton op-amp, 106–114, A-45
Noyce, Robert, 3
Nuvistor tube, 2, A-45

Offset adjustment, 66, 68, 83, 125, 128–129
Offset drift, 130
Op-amp regulator (*see* Linear regulator)
Operational amplifier (op-amp), 5, 6, 12, 36, 53–62, 139, 220, 264, 267, 468, 542, A-45
   Bandwidth, 55–56, 59–61, 66–67, 68, 71
   Characteristics, 53, 65, 68, 71
   Feedback, 56–62
   Frequency response, 55–56, 66–67, 71, 72–74, 96
   Power supply requirements, 54, 62, 68, 354
   Schematic symbol, 54
   Single-supply, 80, 139
   Types, 53
   Used as comparators, 173–177
Oscillation, 72, 211, 446, 449, 529, A-45
Oscillator, 7, 261, 264–265, 268, 375, 388, 395, 397, 400, 407, 435–436, 446–481, 490–537
(*see also specific types of oscillator*)

Pulse oscillator, 457–459, 465–466, 508–510
Sawtooth oscillator, 458–459, 465–467, 479–481
Sine-wave oscillator, 467–479, 482, 499, 500, 517–522
Square-wave oscillator, 375, 449–454, 459–463, 468, 478, 493–500, 510–512, 517–522, 536
Triangle-wave oscillator, 454–456, 459–465, 493–495, 497–500, 517–522, 536
Variable frequency, 452–453, 455, 462, 471, 475, 480, 490, 493–494, 497–500, 517–522
Output impedance, 40–43, 61–62, 82, 96, 548, 569, 580, A-45
Output offset voltage, 48, 49, 50, 51, 101, 103–105, 111, 129, A-45
Output sense, 127

Part number, 15
Passive filter, 212, A-45
   All-pass filter, 227, A-40
   Bandpass filter, 211, 212, 216–217, 224–225
   Band-rejection filter, 211, 212, 226–227
   Frequency response, 212, 215, 219–227
   High-pass filter, 211, 212, 214, 216–217, 222–224
   Low-pass filter, 211, 212–222, 524, 528, 590–591
   Phase response, 212
Peak reverse voltage rating, 282, A-45
Period, 447, 451, 461, 480, 508, A-45
Periodic waveform, 152, 156, 446–447, A-45
Phase, 447, 468–469, 473, 574, 601, A-45
Phase comparator, 523–524, 528, 598, A-45
Phase compensation, 74
Phase detector (*see* Phase comparator)
Phase-locked loop (PLL), 22, 220, 490, 522–537, 598, A-45
Phase-shift oscillator, 3, 469–473, 482
Phillips (*see* ECG)
Planar process, 3, A-45
PLCC (plastic leaded chip carrier), 9
Polarity switch, 113
Poles, 219–220, A-46
Power amplifier, 9
Power supply, 7, 281 (*see also* Rectifier, Linear regulator, Switch-mode regulator)
Power supply filter, 284–286, 344–345, 355–356, 387, A-46
Power supply impedance, 291, A-46
Power supply rejection ratio, 48, 54, 128, 130
Preamplifier (*see* Audio amplifier)
Precision Monolithics Inc., 12, 14
Precision rectifier, 139, 165–169
Pre-emphasis, 596, A-46
Prefixes, 12
Process variable, 114, A-46
Programmable unijunction transistor (PUT), 479–482, A-46
Pulse wave, 446, 448, A-46 (*see also* Oscillator)

Pulse-width modulator, 7, 386–388, 390–392, 395–396, 398–399, 400–432, 435, 457, A-46
Push-pull amplifier, 574, 575

Quadrature detector, 227, 595–596, A-46 (*see also* Demodulator)
Quality factor, 242–243, 245, 248, 250–254, 256, 258, 264, 266, A-46
Quiescent current, 361–364, A-46

Radio frequency amplifier (RF amp), 36, 586, A-46
Radio frequency mixer, 586, 589–590, A-46
Radio frequency (RF) signal, 446, 586, 589–590, A-46
RCA (Radio Corporation of America), 2, 5, 10, 11, 12, 14
Rectifier, 165, 281–289, A-46
  Bridge rectifier, 282, 284–285, 288, 290, 292, 344, 354, 355–356
  Half-wave rectifier, 166, 282, 285
  Full-wave rectifier, 282–283, 285, 286
  Precision rectifier, 165–168
Reference voltage, 173, 177, 179–183, 190
Regulation, 281, 289–291, A-46
  Line regulation, 290, 292–293, 295, 332, 342, 344, 347, 352–353, 370, A-44
  Load regulation, 290, 292–293, 295, 332, 342, 344, 352–353, 370, A-44
Regulator (*see* Linear regulator, Switch-mode regulator)
Regulator efficiency, 386–387
Relaxation oscillator, 449–459, 482
Replacement guide, 18
Resistor values, A-1–A-3
Response time, 198, A-46, (*see also* Slew rate)
Ripple rejection, 291, 292–293, 295, 332, 342, 344, 352–353, 370, A-46
Ripple voltage, 285–288, 292, 344–345, 389, 394–395, 398, 400, A-46
Roll-off, 45, 212, 217, 219, A-46

Saturation voltage, 55–56, 166, 174, 184–185, 304, 450–455, 459–465
Sawtooth wave, 446, 448, A-46, (*see also* Oscillator)
Schematic symbol, 37, 54, 110, 329, 493, 504, 507, 517, 527, 545
Schmitt trigger, 173, 185, A-46, (*see also* Hysteresis)
Selection guide, 18, 22
Sensing circuit, 301, 311–318, 327–342, 358–359, 371, 407–408, 436, 440, A-46
Shockley, William, 3
Shockley equation, 158
Shutdown, 408
Signal compression, 158
Signetics Corporation, 12, 14
Silicon, 3, 6, 7, A-46
Silicon General, Inc., 12, 14

Sine wave, 446, 448, 520, A-33–A-39, A-47 (*see also* Oscillator)
Single-supply amplifier, 94–98, 101–105, 110–114
Single-supply comparator, 188–189, 192–194, 201–206
SIP (single in-line package), 8–9, A-47
Slave regulator, 407
Slew rate, 48, 67, 68, 72, 129, 166, 174–175, 198, A-47
SOIC (small outline integrated circuit), 9
Source impedance, 81, 89, 91
Square root extraction, 158, 165, 169
Square wave, 149–150, 153–155, 446, 448, 467, 520, 524, A-33, A-47 (*see also* Oscillator)
Squelching (*see* Muting)
State-variable filter, 269–271, A-47 (*see also* Universal filter)
Stereo, 222, 226, 542, 545, 564, 571–574, 576, 580, 584–585, 598–603
Subtracting amplifier (*see* Differencing amplifier)
Summing amplifier, 138–144, 227, 247, 249–250, 257–260, 269–271, 501–502, 520, A-40, A-47
Superposition theorem, 142
Surface mount device, 8–9, A-47
Surge current, 283, 286, 287, 288, A-44
Switched-capacitor filter, 211, 259, 275, A-47
Switch-mode regulator, 7, 375, 377, 386–440, 446
  Multiple-output regulator, 401–402, 425–432
  Single-package regulator, 432–440
  Step-down regulator, 387, 389, 390–395, 401, 406–414, 425, 435–440
  Step-up regulator, 387, 389, 395–398, 401, 415–420, 425
  Voltage-inverting regulator, 387, 389, 398–401, 421–425

Tape recording, 552–560
Texas Instruments, 3, 10, 11, 12, 13, 14, 15, 16, 17, 18, 67, 68, 563
Thermocouple, 114, 115, 128, 129
Thevenin equivalent circuit, 548
Time-base generator, 495–497
Timer, 503–504, 513–514, A-47
TO-3 package, 8–9, 325, 346, 347, 364, 432
TO-5 package, 7–9
TO-39 package, 8–9, 347, 364
TO-92 package, 8–9, 347
TO-100 package, 328
TO-202 package, 8–9, 347
TO-220 package, 8–9, 325, 346, 364, 432
Tone control, 576–586, 596–597, 601–602, A-47, (*see also* Balance Control, Volume control)
  Bass control, 576–580, 583–585, A-41
  Treble control, 576–580, 583–585, A-47
Tone generator, 493–495

Toroidal core, 389, 401, 408, A-22–A-23, A-24–A-32, A-47
Transducer, 114, A-47
Transformer, 36, 281–284, 401
  Coupling transformer, 36, A-40
  IF transformer, 589–590, 592
  Matching transformer, 550–551
  Power transformer, 281–289, 344, 355–356, A-5–A-6
  Switch-mode transformer, 401–402, 425–432, A-28–A-32
Transformer winding resistance, 284, A-32
Transistor, 3, 5, 48, 49, 51, 105, 106, 120, 127, 158–160, 163–164, 297, 436–437, 439, 575, A-8–A-21
  Bypass transistor, 311–314, 316–317, 319, 359–361, 371
  Darlington transistor, 307–308, 310–313, 316–318, 412, A-42
  Inverting transistor, 191–192, 416–420, 428–432
  Pass transistor, 307–319, 328, 329, 335–337, 339–342, 345–346, 357–359, 371, A-45
  Switching transistor, 176–177, 194, 206, 387–389, 390, 392, 395–396, 398–399, 401–402, 405, 410–414, 416–420, 422–432
Triangle wave, 149–150, 153–155, 388, 446, 448, 467, 520, 524, A-33–A-38, A-47, (*see also* Oscillator)
Tuner, 596–597, 601–602, A-47

Uncompensated op-amps, 72–74, 167, 168
United States Army, 3

Unity-gain amplifier (*see* Buffer amplifier)
Universal filter, 211, 261, 269–275, A-47
Upper critical frequency, 216, 225–227
Upper trigger voltage, 185–189, 203, 450–453, 455, 504, 507–508, A-47

Vacuum tube, 2
Variable gain, 118
Video amplifier, 36, A-47
Voltage-controlled oscillator (VCO), 490–503, 522–524, 527–529, 533–537, A-47
Voltage converter, 325, 372–379, A-48
Voltage divider, 56–57, 87, 95, 145, 180, 182, 186–187, 201, 202, 302, 314–316, 319, 329–331, 334, 372, 375, 406, 460–462, 469–470, 494–495, 496, 498, 501, A-33–A-38
Voltage reference, 180, 294–307, 319, 328, 329, 388, 407, 435, A-48
Voltage-to-frequency converter (*see* Voltage controlled oscillator)
Volume control, 576, 578–579, 580, 581–586, 591–592, 596–597, 601–602, A-48

Waveform, 447–448, A-48
Waveform shaping, 151, 169, 467, 499, 500, 517, A-33–A-39, A-48
Wien bridge oscillator, 473–476, 482, A-48
Window voltage, 190–192
Wire data, A-23

Zener diode, 180–183, 188–189, 190, 294–297, 301–306, 470–476, A-6–A-7, A-48
Zener diode regulator (*see* Linear regulator)

Pulse-width modulator, 7, 386–388, 390–392, 395–396, 398–399, 400–432, 435, 457, A-46
Push-pull amplifier, 574, 575

Quadrature detector, 227, 595–596, A-46 (*see also* Demodulator)
Quality factor, 242–243, 245, 248, 250–254, 256, 258, 264, 266, A-46
Quiescent current, 361–364, A-46

Radio frequency amplifier (RF amp), 36, 586, A-46
Radio frequency mixer, 586, 589–590, A-46
Radio frequency (RF) signal, 446, 586, 589–590, A-46
RCA (Radio Corporation of America), 2, 5, 10, 11, 12, 14
Rectifier, 165, 281–289, A-46
  Bridge rectifier, 282, 284–285, 288, 290, 292, 344, 354, 355–356
  Half-wave rectifier, 166, 282, 285
  Full-wave rectifier, 282–283, 285, 286
  Precision rectifier, 165–168
Reference voltage, 173, 177, 179–183, 190
Regulation, 281, 289–291, A-46
  Line regulation, 290, 292–293, 295, 332, 342, 344, 347, 352–353, 370, A-44
  Load regulation, 290, 292–293, 295, 332, 342, 344, 352–353, 370, A-44
Regulator (*see* Linear regulator, Switch-mode regulator)
Regulator efficiency, 386–387
Relaxation oscillator, 449–459, 482
Replacement guide, 18
Resistor values, A-1–A-3
Response time, 198, A-46, (*see also* Slew rate)
Ripple rejection, 291, 292–293, 295, 332, 342, 344, 352–353, 370, A-46
Ripple voltage, 285–288, 292, 344–345, 389, 394–395, 398, 400, A-46
Roll-off, 45, 212, 217, 219, A-46

Saturation voltage, 55–56, 166, 174, 184–185, 304, 450–455, 459–465
Sawtooth wave, 446, 448, A-46, (*see also* Oscillator)
Schematic symbol, 37, 54, 110, 329, 493, 504, 507, 517, 527, 545
Schmitt trigger, 173, 185, A-46, (*see also* Hysteresis)
Selection guide, 18, 22
Sensing circuit, 301, 311–318, 327–342, 358–359, 371, 407–408, 436, 440, A-46
Shockley, William, 3
Shockley equation, 158
Shutdown, 408
Signal compression, 158
Signetics Corporation, 12, 14
Silicon, 3, 6, 7, A-46
Silicon General, Inc., 12, 14

Sine wave, 446, 448, 520, A-33–A-39, A-47 (*see also* Oscillator)
Single-supply amplifier, 94–98, 101–105, 110–114
Single-supply comparator, 188–189, 192–194, 201–206
SIP (single in-line package), 8–9, A-47
Slave regulator, 407
Slew rate, 48, 67, 68, 72, 129, 166, 174–175, 198, A-47
SOIC (small outline integrated circuit), 9
Source impedance, 81, 89, 91
Square root extraction, 158, 165, 169
Square wave, 149–150, 153–155, 446, 448, 467, 520, 524, A-33, A-47 (*see also* Oscillator)
Squelching (*see* Muting)
State-variable filter, 269–271, A-47 (*see also* Universal filter)
Stereo, 222, 226, 542, 545, 564, 571–574, 576, 580, 584–585, 598–603
Subtracting amplifier (*see* Differencing amplifier)
Summing amplifier, 138–144, 227, 247, 249–250, 257–260, 269–271, 501–502, 520, A-40, A-47
Superposition theorem, 142
Surface mount device, 8–9, A-47
Surge current, 283, 286, 287, 288, A-44
Switched-capacitor filter, 211, 259–275, A-47
Switch-mode regulator, 7, 375, 377, 386–440, 446
  Multiple-output regulator, 401–402, 425–432
  Single-package regulator, 432–440
  Step-down regulator, 387, 389, 390–395, 401, 406–414, 425, 435–440
  Step-up regulator, 387, 389, 395–398, 401, 415–420, 425
  Voltage-inverting regulator, 387, 389, 398–401, 421–425

Tape recording, 552–560
Texas Instruments, 3, 10, 11, 12, 13, 14, 15, 16, 17, 18, 67, 68, 563
Thermocouple, 114, 115, 128, 129
Thevenin equivalent circuit, 548
Time-base generator, 495–497
Timer, 503, 504, 513–514, A-47
TO-3 package, 8–9, 325, 346, 347, 364, 432
TO-5 package, 7–9
TO-39 package, 8–9, 347, 364
TO-92 package, 8–9, 347
TO-100 package, 328
TO-202 package, 8–9, 347
TO-220 package, 8–9, 325, 346, 364, 432
Tone control, 576, 586, 596–597, 601–602, A-47, (*see also* Balance Control, Volume control)
  Bass control, 576–580, 583–585, A-41
  Treble control, 576–580, 583–585, A-47
Tone generator, 493–495

Toroidal core, 389, 401, 408, A-22–A-23, A-24–A-32, A-47
Transducer, 114, A-47
Transformer, 36, 281–284, 401
   Coupling transformer, 36, A-40
   IF transformer, 589–590, 592
   Matching transformer, 550–551
   Power transformer, 281–289, 344, 355–356, A-5–A-6
   Switch-mode transformer, 401–402, 425–432, A-28–A-32
Transformer winding resistance, 284, A-32
Transistor, 3, 5, 48, 49, 51, 105, 106, 120, 127, 158–160, 163–164, 297, 436–437, 439, 575, A-8–A-21
   Bypass transistor, 311–314, 316–317, 319, 359–361, 371
   Darlington transistor, 307–308, 310–313, 316–318, 412, A-42
   Inverting transistor, 191–192, 416–420, 428–432
   Pass transistor, 307–319, 328, 329, 335–337, 339–342, 345–346, 357–359, 371, A-45
   Switching transistor, 176–177, 194, 206, 387–389, 390, 392, 395–396, 398–399, 401–402, 405, 410–414, 416–420, 422–432
Triangle wave, 149–150, 153–155, 388, 446, 448, 467, 520, 524, A-33-A-38, A-47, (*see also* Oscillator)
Tuner, 596–597, 601–602, A-47

Uncompensated op-amps, 72–74, 167, 168
United States Army, 3

Unity-gain amplifier (*see* Buffer amplifier)
Universal filter, 211, 261, 269–275, A-47
Upper critical frequency, 216, 225–227
Upper trigger voltage, 185–189, 203, 450–453, 455, 504, 507–508, A-47

Vacuum tube, 2
Variable gain, 118
Video amplifier, 36, A-47
Voltage-controlled oscillator (VCO), 490–503, 522–524, 527–529, 533–537, A-47
Voltage converter, 325, 372–379, A-48
Voltage divider, 56–57, 87, 95, 145, 180, 182, 186–187, 201, 202, 302, 314–316, 319, 329–331, 334, 372, 375, 406, 460–462, 469–470, 494–495, 496, 498, 501, A-33–A-38
Voltage reference, 180, 294–307, 319, 328, 329, 388, 407, 435, A-48
Voltage-to-frequency converter (*see* Voltage controlled oscillator)
Volume control, 576, 578–579, 580, 581–586, 591–592, 596–597, 601–602, A-48

Waveform, 447–448, A-48
Waveform shaping, 151, 169, 467, 499, 500, 517, A-33–A-39, A-48
Wien bridge oscillator, 473–476, 482, A-48
Window voltage, 190–192
Wire data, A-23

Zener diode, 180–183, 188–189, 190, 294–297, 301–306, 470–476, A-6–A-7, A-48
Zener diode regulator (*see* Linear regulator)